Long Term Evolution

IN BULLETS

Companion website: www.lte-bullets.com

Comments and requests regarding the content of this book can be sent to contact@lte-bullets.com
These will be considered for future editions.

Chris Johnson has worked on LTE for the past 5 years. He has worked in mobile telecommunications for the last 18 years, spending most of this time focused on UMTS. Chris has been employed by Nokia Siemens Networks for the past 12 years. He is currently a Principal Engineer at Nokia Siemens Networks and is based in the UK. Chris has previously authored the 'Radio Access Networks for UMTS – Principles and Practice' title and has contributed content within the 'Radio Network Planning and Optimisation for UMTS' and 'HSDPA/HSUPA for UMTS' titles. These titles have been published by John Wiley & Sons, Inc.

The content of this book represents the understanding of the author. It does not necessarily represent the view nor opinion of the author's employer. Descriptions are intended to be generic and do not represent the implementation of any individual vendor.

The LTE™ and LTE Advanced™ logos have been used with kind permission from ETSI. LTE is a trade mark of ETSI.

The author would like to acknowledge his employer, Nokia Siemens Networks UK Limited for providing the opportunities to gain valuable project experience. The author would also like to thank his managers from within Nokia Siemens Networks UK Limited for supporting participation within projects which have promoted continuous learning and development. These include Andy King and Stuart Davis. The author would also like to thank Jyri Lamminmaki for providing opportunities to work on global LTE activities within Nokia Siemens Networks.

The author would like to acknowledge colleagues from within Nokia Siemens Networks who have supported and encouraged the development of material for this book. These include Lorena Serna Gonzalez, Tomas Novosad, Harri Holma and Peter Muszynski. In addition, the author would like to thank colleagues from outside Nokia Siemens Networks who have also supported the development of this book. These include Michele Garelli, Pinaki Roychowdhury, Ling Soon Leh, Gianluca Formica and Balan Muthiah.

In addition, the author would like to thank those who have provided feedback regarding the first edition of the book, and those who have provided suggestions for improvement. These include Mouaffac Ambriss, Sandro Grech, Jose M. Hernando, Paul Piggin, Gagan Rath, Paul L. Thompson, Cyril Valadon, Mahendran Venkatasamy, Eyal Verbin, Jiang Wie and Wilson Zheng.

The author would also like to offer special thanks to his parents who provided a perfect working environment during the weeks spent in Scotland. He would also like to thank them for their continuous support and encouragement.

Edition 2, version 1

Copyright © 2012 Chris Johnson, Northampton, England

All rights reserved. Any unauthorised copying will constitute an infringement of copyright.

1 ABBREVIATIONS

16QAM	16 Quadrature Amplitude Modulation
3GPP	3rd Generation Partnership Project
64QAM	64 Quadrature Amplitude Modulation
AAS	Active Antenna System
ABS	Almost Blank Subframe
AC	Access Class
AC	Admission Control
AM	Acknowledged Mode
A-MPR	Additional Maximum Power Reduction
AMR	Adaptive Multi-Rate
ANR	Automatic Neighbour Relations
AoA	Angle of Arrival
APN	Access Point Name
APN-AMBR	APN Aggregate Maximum Bit Rate
ARFCN	Absolute Radio Frequency Channel Number
ARP	Allocation and Retention Priority
ARQ	Automatic Repeat Request
AS	Access Stratum
BCCH	Broadcast Control Channel
BCD	Binary Coded Decimal
BCH	Broadcast Channel
BER	Bit Error Rate
BI	Back-off Indicator
BLER	Block Error Rate
BM-SC	Broadcast Multicast Service Centre
BSC	Base Station Controller
BSIC	Base Station Identity Code
BTS	Base Transceiver Station
CA	Carrier Aggregation
CB	Coordinated Beamforming
CBC	Cell Broadcast Center
CC	Component Carrier
CCCH	Common Control Channel
CCE	Control Channel Element
CCO	Cell Change Order
CD	Collision Detection
CDD	Cyclic Delay Diversity
CDM	Code Division Multiplexing
CFI	Control Format Indicator
CGI	Cell Global Identity
CIF	Carrier Indicator Field
CLI	Calling Line Identification
CMAS	Commercial Mobile Alert Service
CMR	Codec Mode Request
CoMP	Co-ordinated Multi-Point transmission
CPC	Continuous Packet Connectivity
CQI	Channel Quality Indicator
CRC	Cyclic Redundancy Check
CRI	Contention Resolution Identity
C-RNTI	Cell Radio Network Temporary Identifier
CRS	Common Reference Signal
CS	Circuit Switched
CS	Coordinated Scheduling
CSFB	CS Fallback
CSG	Closed Subscriber Group
CSI	Channel State Information
CSI RS	CSI Reference Signal
CSMA	Carrier Sense Multiple Access
CW	Code Word
DAI	Downlink Assignment Index
DC	Direct Current
DCCH	Dedicated Control Channel
DCH	Dedicated Channel
DCI	Downlink Control Information
DFT	Discrete Fourier Transform
DiffServ	Differentiated Services
DL-SCH	Downlink Shared Channel
DM-RS	Demodulation Reference Signal
DPI	Deep Packet Inspection
DPS	Dynamic Point Selection
DRB	Data Radio Bearers
DRX	Discontinuous Receive
DSCP	Differentiated Services Code Point
DTCH	Dedicated Traffic Channel
DTM	Dual Transfer Mode
DTX	Discontinuous Transmit
DwPTS	Downlink Pilot Time Slot
EARFCN	E-UTRA Absolute RF Channel Number
EBB	Eigen Based Beamforming
ECGI	E-UTRAN Cell Global Identifier
ECI	E-UTRAN Cell Identifier
ECM	EPS Connection Management
ECN	Explicit Congestion Notification
EHPLMN	Equivalent HPLMN
EIR	Equipment Identity Register
EIRP	Effective Isotropic Radiated Power
EMM	EPS Mobility Management
EPC	Evolved Packet Core
EPRE	Energy Per Resource Element
EPS	Evolved Packet System
E-RAB	E-UTRAN Radio Access Bearer
ESM	EPS Session Management
E-SMLC	Enhanced Serving Mobile Location Centre
ETWS	Earthquake and Tsunami Warning System
E-UTRAN	Evolved UMTS Radio Access Network
E-UTRA	Evolved UMTS Radio Access
FCC	Federal Communications Commission
FDD	Frequency Division Duplex
FDM	Frequency Division Multiplexing
FFT	Fast Fourier Transform
FGI	Feature Group Indicator
FMS	First Missing SDU
FSTD	Frequency Switch Transmit Diversity
FT	Fourier Transform
FTP	File Transfer Protocol
GAN	Generic Access Network
GANC	Generic Access Network Controller
GBR	Guaranteed Bit Rate
GNSS	Global Navigation Satellite System
GP	Guard Period
GPS	Global Positioning System
GRE	Generic Routing Encapsulation
GSMA	GSM Association
GTP-U	GPRS Tunnelling Protocol User Plane
GTPv2-C	GPRS Tunnelling Protocol Control Plane
GUMMEI	Globally Unique MME Identity
GUTI	Globally Unique Temporary UE Identity
HARQ	Hybrid Automatic Repeat Request
HFN	Hyper Frame Number
HI	HARQ Indicator
HLR	Home Location Register
HPLMN	Home Public Land Mobile Network

HRPD High Rate Packet Data
HSPA High Speed Packet Access
HSS Home Subscriber Server
HTTP Hypertext Transfer Protocol

ICIC Inter Cell Interference Coordination
IDFT Inverse Discrete Fourier Transform
IFFT Inverse Fast Fourier Transform
IFT Inverse Fourier Transform
IMEI International Mobile Equipment Identities
IMS IP Multimedia Subsystem
IMSI International Mobile Subscriber Identity
IoT Interference over Thermal
IP Internet Protocol
ISR Idle mode Signalling Reduction
ITU International Telecommunication Union

JP Joint Processing
JR Joint Reception
JT Joint Transmission

KSI Key Set Identifier

LAI Location Area Identity
LCID Logical Channel Identity
L-GW Local Gateway
LI Length Indicators
LIPA Local IP Access
LNA Low Noise Amplifier
LPP LTE Positioning Protocol
LSB Least Significant Bit
LSM Limited Service Mode
LTE Long Term Evolution

M2-AP M2 Application Protocol
M3-AP M3 Application Protocol
MAC Medium Access Control
MAC-I Message Authentication Code for Integrity
MBMS Multimedia Broadcast Multicast Service
MBR Maximum Bit Rate
MBSFN MBMS Single Frequency Network
MCC Mobile Country Code
MCCH Multicast Control Channel
MCE Multi-cell / Multicast Coordination Entity
MCH Multicast Channel
MCL Minimum Coupling Loss
MCS Modulation and Coding Scheme
MDT Minimisation of Drive Tests
MGRP Measurement Gap Repetition Period
MHA Mast Head Amplifiers
MIB Master Information Block
MIMO Multiple Input Multiple Output
MISO Multiple Input Single Output
MME Mobility Management Entity
MMEC MME Code
MMEGI MME Group Identity
MMEI MME Identity
MMTEL Multimedia Telephony
MNC Mobile Network Code
MO Mobile Originating
MPR Maximum Power Reduction
MPS Multimedia Priority Services
M-RNTI MBMS RNTI
MSB Most Significant Bit
MSI MCH Scheduling Information
MSIN Mobile Subscriber Identification Number
MSP MCH Scheduling Period
MT Mobile Terminating
MTCH Multicast Traffic Channel
M-TMSI MME Temporary Mobile Subscriber Identity
MU-MIMO Multi-User MIMO

NACC Network Assisted Cell Change
NAS Non-Access Stratum
NAT Network Address Translation
NCC Network Colour Code
NDI New Data Indicator
NNSF NAS Node Selection Function

OCC Orthogonal Covering Code
OFDMA Orthogonal Frequency Division Multiple Access
OSS Operations Support System
OTDOA Observed Time Difference of Arrival
OTT Over The Top

PA Power Amplifier
PAPR Peak to Average Power Ratio
PBCH Physical Broadcast Channel
PCEF Policy and Charging Enforcement Function
PCFICH Physical Control Format Indicator Channel
PCCH Paging Control Channel
PCEF Policy and Charging Enforcement Function
PCH Paging Channel
PCI Physical Channel Identity
PCI Physical layer Cell Identity
PCRF Policy and Charging Rules Function
PDCCH Physical Downlink Control Channel
PDCP Packet Data Convergence Protocol
PDN Packet Data Network
PDSCH Physical Downlink Shared Channel
PDU Packet Data Unit
PF Paging Frame
P-GW Packet Gateway
PHB Per Hop Behaviour
PHR Power Headroom Reports
PHICH Physical Hybrid ARQ Indicator Channel
PL Path Loss
PLMN Public Land Mobile Network
PMCH Physical Multicast Channel
PMI Precoding Matrix Indicator
P-MPR Power Management Maximum Power Reduction
PO Paging Occasion
PRACH Physical Random Access Channel
PRB Physical Resource Block
PRG Precoding Resource block Group
P-RNTI Paging RNTI
PRS Positioning Reference Signal
PS Packet Switched
PSD Power Spectral Density
PSS Primary Synchronisation Signal
PTI Precoding Type Indicator
PUCCH Physical Uplink Control Channel
PUSCH Physical Uplink Shared Channel
PWS Public Warning System

QCI QoS Class Identifier
QoS Quality of Service
QPSK Quadrature Phase Shift Keying

RA Random Access
RAPID Random Access Preamble Identity
RA-RNTI Random Access RNTI

RAR	Random Access Response
RAT	Radio Access Technology
RB	Resource Block
RBG	Resource Block Group
RF	Radio Frequency
RI	Rank Indication
RIM	RAN Information Management
RIP	Received Interference Power
RIV	Resource Indication Value
RLC	Radio Link Control
RLF	Radio Link Failure
RLM	Radio Link Monitoring
RN	Relay Node
RNC	Radio Network Controller
RNTI	Radio Network Temporary Identifier
RNTP	Relative Narrowband Transmit Power
ROHC	Robust Header Compression
R-PDCCH	Relay Physical Downlink Control Channel
RRC	Radio Resource Control
RRH	Remote Radio Head
RRM	Radio Resource Management
RSRP	Reference Signal Received Power
RSRQ	Reference Signal Received Quality
RSCP	Received Signal Code Power
RSSI	Received Signal Strength Indicator
RSTD	Reference Signal Time Difference
RTCP	Real time Transport Control Protocol
RTP	Real-time Transport Protocol
RTT	Round Trip Time
RV	Redundancy Version
S1-AP	S1 Application Protocol
S1-CP	S1 Control Plane
S1-UP	S1 User Plane
SAE	System Architecture Evolution
SAW	Stop and Wait
SC-FDMA	Single Carrier Frequency Division Multiple Access
SCTP	Stream Control Transmission Protocol
SDMA	Space Division Multiple Access
SDU	Service Data Unit
SFBC	Space Frequency Block Code
SFBC-FSTD	SFBC - Frequency Shift Tx Diversity
SFN	System Frame Number
SGsAP	SGs Application Protocol
S-GW	Serving Gateway
SIB	System Information Block
SIMO	Single Input Multiple Output
SINR	Signal to Interference plus Noise Ratio
SIP	Session Initiation Protocol
SI-RNTI	System Information RNTI
SISO	Single Input Single Output
SMS	Short Message Service
SMTP	Simple Mail Transfer Protocol
SN	Sequence Number
SON	Self Organising Network
SORTD	Spatial Orthogonal Resource Transmit Diversity
SPS	Semi Persistent Scheduling
SPS-RNTI	Semi Persistent Scheduling RNTI
SR	Scheduling Request
SRB	Signalling Radio Bearer
SRS	Sounding Reference Signal
SRVCC	Single Radio Voice Call Continuity
SSAC	Service Specific Access Class
SSPS	Semi-Static Point Selection
SSS	Secondary Synchronisation Signal
STBC	Space-Time Block Coding
S-TMSI	SAE TMSI
SU-MIMO	Single User MIMO
TA	Tracking Area
TAI	Tracking Area Identity
TAC	Tracking Area Code
TAU	Tracking Area Update
TBS	Transport Block Size
TCE	Trace Collection Entity
TCP	Transmission Control Protocol
TDD	Time Division Duplex
TEID	Tunnel Endpoint Identifier
TFT	Traffic Flow Template
TIN	Temporary Identity used in Next update
TM	Transparent Mode
TMGI	Temporary Mobile Group Identifier
TNL	Transport Network Layer
ToC	Table of Contents
ToP	Timing over Packet
TPC	Transmit Power Control
TPMI	Transmitted Precoding Matrix Indicator
TR	Trace Reference
TRSR	Trace Recording Session Reference
TTI	Transmission Time Interval
TTL	Time To Live
UCI	Uplink Control Information
UCS	Universal Character Set
UDP	User Datagram Protocol
UE	User Equipment
UE-AMBR	UE Aggregate Maximum Bit Rate
ULA	Uniform Linear Array
UL-SCH	Uplink Shared Channel
UM	Unacknowledged Mode
UpPTS	Uplink Pilot Time Slot
VANC	VoLGA Access Network Controller
VCC	Voice Call Continuity
VoIP	Voice over IP
VoLGA	Voice over LTE via Generic Access
VoLTE	Voice over LTE
VRB	Virtual Resource Block
X2-AP	X2 Application Protocol
X2-CP	X2 Control Plane
X2-UP	X2 User Plane

2 FUNDAMENTALS

2.1 INTRODUCTION

★ Long Term Evolution (LTE) starts from 3GPP release 8

★ 3GPP Technical Report 25.913 defines the key objectives of LTE as:

  - support for a flexible transmission bandwidth up to 20 MHz
  - peak downlink data rate of 100 Mbps when using 2 receive antenna at the UE
  - peak uplink data rate of 50 Mbps when using 1 transmit antenna at the UE
  - round trip time of less than 10 ms
  - downlink average spectrum efficiency improved 3 to 4 times relative to release 6 HSDPA
  - uplink average spectrum efficiency improved 2 to 3 times relative to release 6 HSUPA

★ LTE has a flat architecture which minimises the number of network elements

★ LTE is optimised for Packet Switched (PS) services but includes functionality to handle Circuit Switched (CS) services, e.g. CS fallback to UMTS

★ LTE supports the speech service using Voice over IP. Otherwise, the speech service can be supported by allowing the UE to fallback to UMTS, GSM or CDMA2000

★ LTE supports Multimedia Broadcast Multicast Services (MBMS) for the transmission of mobile TV

★ Frequency Division Duplex (FDD) and Time Division Duplex (TDD) versions of LTE have been standardised. Both allow channel bandwidths of up to 20 MHz

★ LTE allows inter-working with existing GSM, UMTS and CDMA2000 technologies

★ LTE uses QPSK, 16QAM and 64QAM modulation schemes with OFDMA (downlink) and SC-FDMA (uplink) multiple access technologies

★ LTE supports Multiple Input Multiple Output (MIMO) antenna technology in the downlink direction. 3GPP releases 8 and 9 do not support MIMO in the uplink direction

★ Existing spectrum allocations can be re-farmed for the introduction of LTE

★ LTE simplifies network planning by minimising the requirement for manually planned neighbour lists

★ LTE includes Self Organising Network (SON) functionality to help automate network configuration, optimisation, fault finding and fault handling

★ LTE Advanced starts from 3GPP release 10

★ LTE Advanced introduces Carrier Aggregation to provide wider effective channel bandwidths. It also introduces MIMO in the uplink direction, as well as increasing the number of antenna elements which can be used for MIMO in the downlink direction

★ Other technologies continue to develop in parallel to LTE, e.g. UMTS introduces HSPA+ with MIMO, 64QAM and Multi-Carrier Transmission allowing effective channel bandwidths of 10, 20 and 40 MHz

2.2 ARCHITECTURE

- ★ LTE refers to the Evolved UMTS Radio Access Network (E-UTRAN), whereas System Architecture Evolution (SAE) refers to the Evolved Packet Core (EPC). Figure 1 illustrates this division between radio access and core networks
- ★ LTE uses a flat architecture without a Radio Network Controller (RNC), nor Base Station Controller (BSC)
- ★ The LTE equivalent of a UMTS Node B is an 'evolved' Node B or eNode B. An eNode B is the Base Transceiver Station (BTS) for LTE. Radio resource management is completed by the eNode B
- ★ eNode B are connected to the Evolved Packet Core (EPC) using a Mobility Management Entity (MME) for control plane signalling, and a Serving Gateway for user plane data
- ★ The Serving Gateway is connected to a Packet Data Network (PDN) Gateway for connectivity to external networks including the public internet

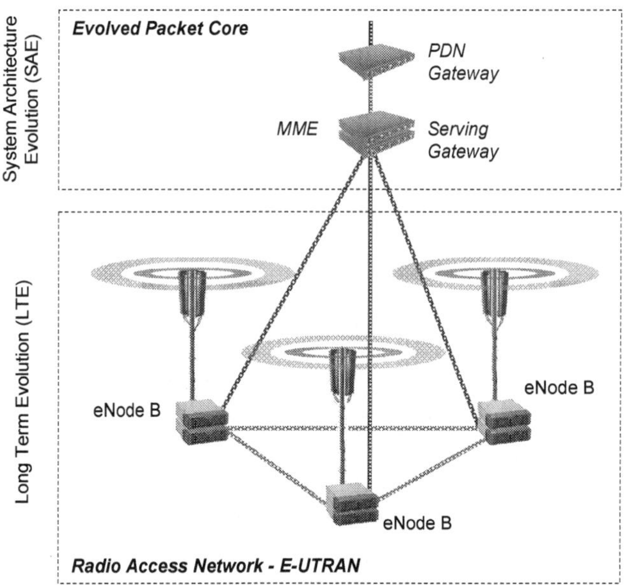

Figure 1 – Long Term Evolution (LTE) architecture

- ★ Figure 2 illustrates a radio access network which includes macro, micro, pico and femto BTS, as well as relays and repeaters. This type of radio access network is known as a heterogeneous network because it includes a range of different BTS types
- ★ The most common BTS type is the macro BTS, which 3GPP categorises as a 'wide area BTS'. 3GPP categorises pico BTS as 'local area BTS' and femto BTS as 'home BTS'. 3GPP does not specify a separate power class for micro BTS but a wide area BTS with reduced transmit power could be designed and used as a micro BTS. 3GPP also specifies separate requirements for repeaters and relays
- ★ Macro BTS are characterised by having their antenna above roof-top level so their coverage area is relatively large. Their transmit power is typically 20, 40 or 60 Watts, and they normally have more than a single sector
- ★ Micro BTS are characterised by having their antenna below roof-top level so their coverage tends to be limited by the neighbouring buildings. Their transmit power is typically 5 or 10 W, and they often have only a single sector
- ★ Pico BTS are designed to provide coverage and capacity across small areas. Their transmit power does not exceed 0.25 W so their antenna needs to be close to the source of traffic
- ★ Femto BTS are intended for use at home, or in small offices. Their transmit power does not exceed 0.1 W so they need to be used in areas where coverage from other BTS types is relatively weak, e.g. indoors. In contrast to other BTS types, the location of Femto BTS is not usually controlled by the network operator. End-users are free to place their Femto BTS wherever they like. The network architecture for Femto BTS also differs from other BTS types. Femto BTS are connected to the Evolved Packet Core using a Home eNode B Gateway (also known as a Femto Gateway). The connection between the Femto BTS and Femto Gateway typically uses a home broadband connection, e.g. ADSL
- ★ Repeaters can be used to extend the coverage of an existing BTS. They re-transmit the uplink and downlink signals without having to decode any of the content. Repeaters have one antenna directed towards the donor cell, and a second antenna directed towards the target coverage area. The target coverage area could be an indoor location so the second antenna could be indoors
- ★ Relays also rely upon an RF connection to a donor cell, but Relays differ from Repeaters because Relays have their own cells and their own protocol stack, i.e. a Relay is similar to a normal BTS but without a fixed transport connection. Relays decode signals and make radio resource management decisions

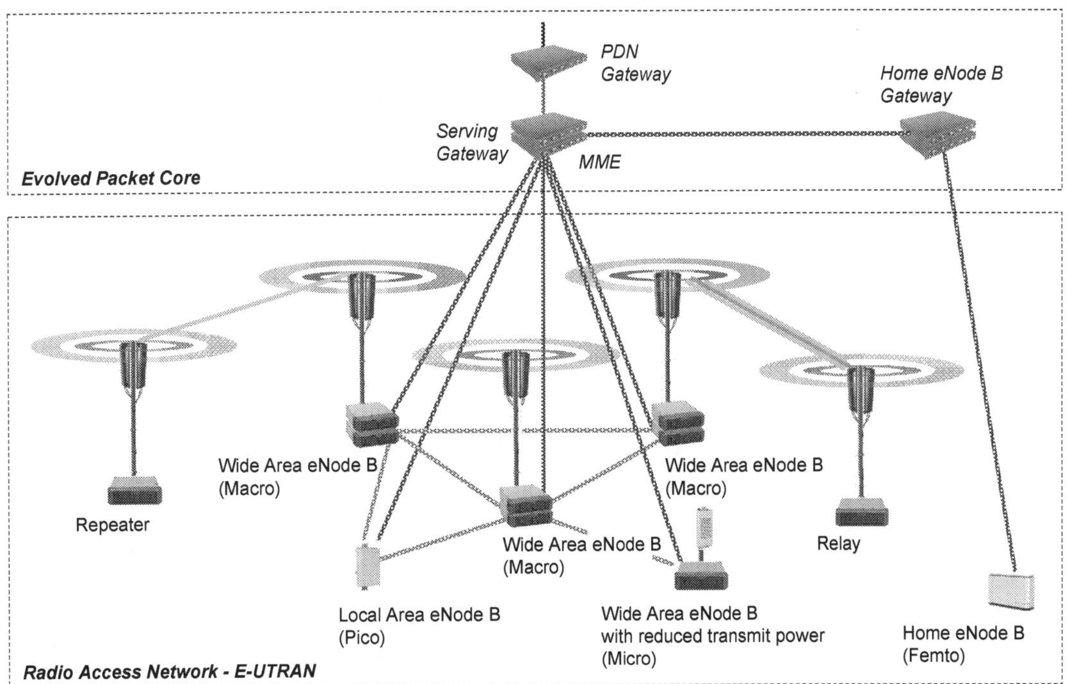

Figure 2 – Heterogeneous network for LTE

★ The various BTS types typically share the same channel bandwidth, so heterogeneous networks generate challenges in terms of co-channel interference and RF planning to achieve the intended coverage from each site. Traffic management can also be challenging when Closed Subscriber Groups (CSG) are used to ensure that only authorised subscribers can use certain BTS, e.g. only the family owning a Femto BTS can use that BTS. That BTS then appears as a source of interference to other subscribers

2.3 INTERFACES

- Figure 3 illustrates the most important interfaces for the radio access network
- The air-interface connection between the User Equipment (UE) and the eNode B is known as the Uu. The UE and eNode B make use of the Uu interface whenever they transmit or receive across the LTE air-interface
- The X2 interface connects one eNode B to another eNode B. This allows both signalling and data to be transferred between neighbouring eNode B
    - the control plane of the X2 (X2-CP) interface allows signalling between eNode B
    - the user plane of the X2 (X2-UP) interface allows the transfer of application data between eNode B
- The S1 interface connects an eNode B to the Evolved Packet Core (EPC). This allows both signalling and data to be transferred between the Evolved Packet Core (EPC) and Evolved UMTS Radio Access Network (E-UTRAN)
    - the control plane of the S1 (S1-MME) interface allows signalling with the MME
    - the user plane of the S1 (S1-UP) interface allows application data transfer through the Serving Gateway
- Application Protocols have been specified to define the signalling procedures and message types which can be sent across the X2 and S1 interfaces, i.e. X2-AP and S1-AP

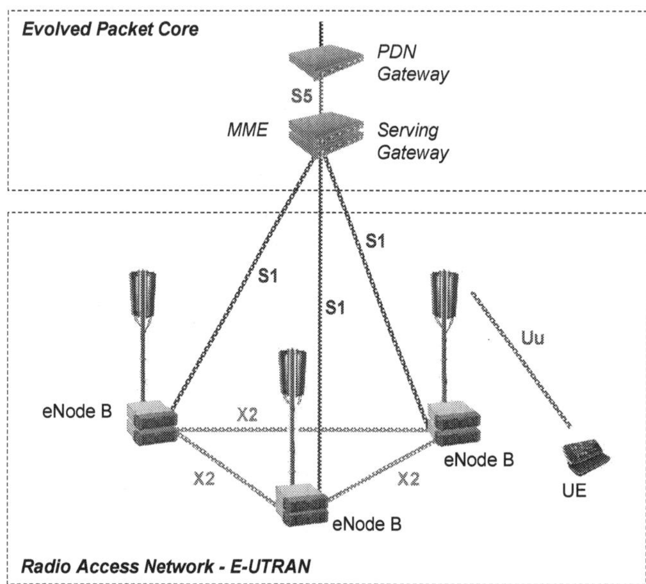

Figure 3 – Key interfaces for LTE (logical representation)

- Both the X2 and S1 interfaces are based upon IP
- Figure 3 illustrates a logical representation of the interfaces within E-UTRAN. In practice, the X2 and S1 are likely to use a single physical connection at the eNode B, i.e. a single Ethernet cable can be used for both the X2 and S1 interfaces
- Figure 4 illustrates an example physical representation of the X2 and S1 interfaces. The eNode B are connected to an IP backhaul transport network using a single Ethernet cable. This cable transfers information for both the X2 and S1 interfaces
- In the case of the X2 interface, the IP routers within the transport network receive data from one eNode B and direct it towards another eNode B. In the case of the S1 interface, the IP routers provide connectivity between the eNode B and the Evolved Packet Core
- The Ethernet connection between the eNode B and transport network could be based upon either an electrical or optical Gigabit Ethernet cable
- IP Quality of Service (QoS) can be used to differentiate and prioritise packets transferred across the IP backhaul
- Timing over Packet (ToP) can be used to provide the eNode B with synchronisation information. ToP is specified within IEEE 1588. Alternatively, Global Positioning System (GPS) satellites can be used, or a synchronisation signal can be provided by a co-sited BTS

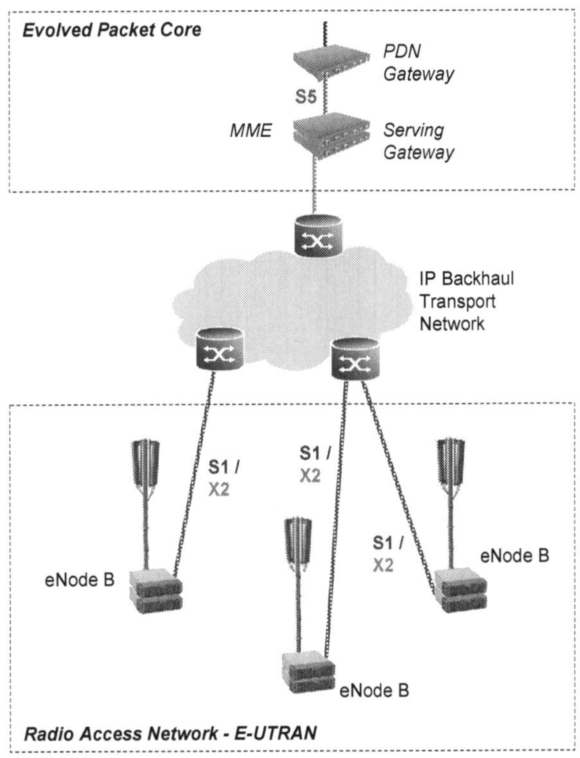

Figure 4 – Key interfaces for LTE (physical representation)

★ A more complete set of interfaces associated with LTE and the Evolved Packet Core is shown in Figure 5. Only a single eNode B is shown in this figure so the X2 interface does not appear. The control plane of the S1 interface is shown as the S1-MME, while the user plane is shown as the S1-UP

★ The S11 interface connects the MME to the Serving Gateway. This allows signalling information for mobility and bearer management to be transferred. Application data does not use the S11 interface

★ The S5 interface connects the Serving Gateway to the Packet Data Network (PDN) Gateway. Both control plane signalling and user plane data use the S5 interface. The PDN Gateway provides connectivity to the set of IP services so the S5 represents the main connection for application data across the Evolved Packet Core

★ The S8 interface is similar to the S5 interface but it terminates at a PDN Gateway belonging to a different PLMN. This interface is used by end-users who are roaming away from their home PLMN

★ The S6a interface connects the MME to the Home Subscriber Server (HSS). The HSS hosts a database containing subscription related information for the population of end-users. The HSS represents an evolution of the Home Location Register (HLR) used by earlier network architectures

★ The S13 interface connects the MME to the Equipment Identity Register (EIR). The EIR stores the International Mobile Equipment Identities (IMEI) of the end-user devices used within the network. These IMEI can be 'white listed', 'grey listed' or 'black listed' to control access to the network

★ The Gx interface connects the Policy and Charging Enforcement Function (PCEF) within the PDN Gateway to the Policy and Charging Rules Function (PCRF). The PCRF provides QoS and charging information to the PDN Gateway. The Gx interface is also known as the S7 interface in some references

★ The SGi interface provides connectivity between the PDN Gateway and a packet data network. The packet data network could be an external network (either public or private), or could belong to the operator. The SGi interface corresponds to the Gi interface in earlier network architectures

★ The S3 interface allows the transfer of control plane signalling between the MME and an SGSN. The SGSN could belong to either a UMTS or GPRS network. The main purpose of the signalling is to allow mobility between the various access technologies

★ The S4 interface allows the transfer of application data between the Serving Gateway and SGSN when a 'Direct Tunnel' is not established between the RNC and Serving Gateway. This interface may be used when a UE roams from the LTE network across to a UMTS network

★ The S2a interface provides connectivity between the PDN Gateway and a non-3GPP access technology. Figure 5 illustrates the non-3GPP technology as a wireless LAN. WiMax is a non-3GPP access technology which could be connected using the S2a interface

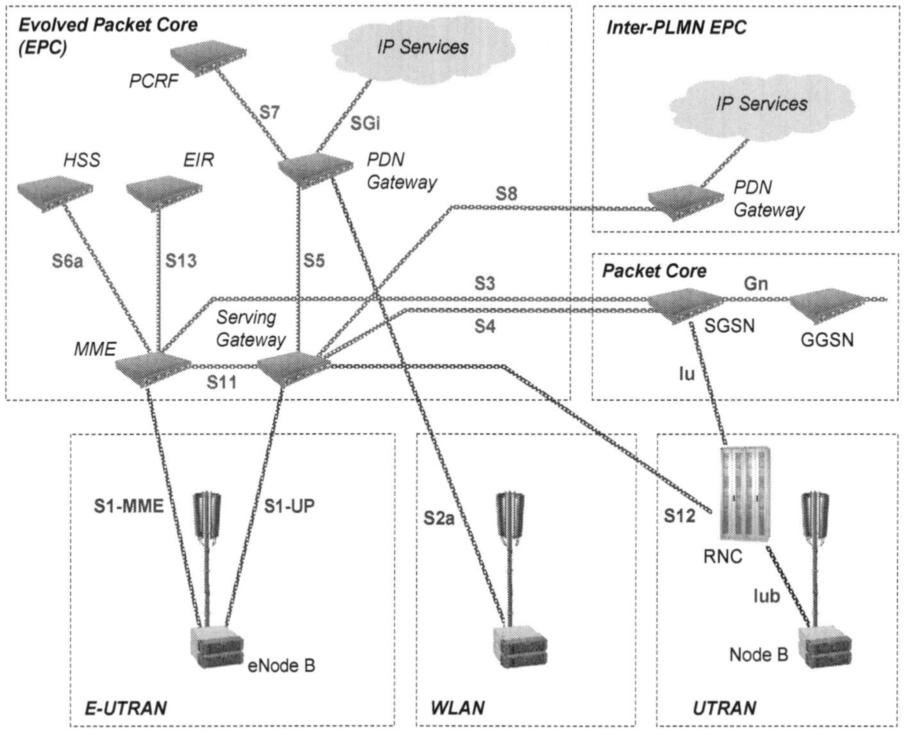

Figure 5 – Additional interfaces for LTE and the Evolved Packet Core

★ The S12 interface allows the transfer of application data between the Serving Gateway and RNC when a 'Direct Tunnel' is established. The S4 interface represents the alternative when a 'Direct Tunnel' is not established. Both the S12 and S4 interfaces are applicable when a UE roams from the LTE network across to a UMTS network

★ 3GPP References: TS 36.410, TS 36.420, TS 23.002, TS 23.402

2.4 CHANNEL BANDWIDTHS

★ 3GPP has specified a set of 6 channel bandwidths, ranging from 1.4 MHz to 20 MHz. These are presented in Table 1

	Channel Bandwidth					
	1.4 MHz	3 MHz	5 MHz	10 MHz	15 MHz	20 MHz
Number of Resource Blocks	6	15	25	50	75	100
Number of Subcarriers	72	180	300	600	900	1200
Uplink Subcarrier Bandwidth (MHz)	1.08	2.7	4.5	9.0	13.5	18.0
Downlink Subcarrier Bandwidth (MHz)	1.095	2.715	4.515	9.015	13.515	18.015

Table 1 – Channel bandwidths for LTE

★ A Resource Block represents the basic unit of resource for the LTE air-interface. The eNode B scheduler allocates Resource Blocks to UE when allowing data transfer

★ The subcarriers belong to the Orthogonal Frequency Division Multiple Access (OFDMA) technology in the downlink, and the Single Carrier Frequency Division Multiple Access (SC-FDMA) technology in the uplink

★ There are 12 subcarriers per Resource Block so the number of subcarriers equals 12 × number of Resource Blocks

★ Each subcarrier occupies 15 kHz so the total subcarrier bandwidth equals 15 kHz × number of subcarriers

★ The downlink subcarrier bandwidth includes an additional 15 kHz to accommodate a null subcarrier at the center of all other subcarriers. The null subcarrier provides 15 kHz of empty spectrum within which nothing is transmitted

★ The total subcarrier bandwidth is less than the channel bandwidth to allow for the roll-off of emissions and to provide some guard band

★ The larger channel bandwidths provide support for the higher throughputs. Smaller channel bandwidths provide support for lower throughputs but are easier to accommodate within existing spectrum allocations

★ 3GPP also specifies a subcarrier spacing of 7.5 kHz (in addition to the subcarrier spacing of 15 kHz). The subcarrier spacing of 7.5 kHz is only used in cells which are dedicated to Multimedia Broadcast Multicast Services (MBMS). There are 24 rather than 12 subcarriers per Resource Block when using the 7.5 kHz subcarrier spacing so the total bandwidth of a Resource Block remains the same

★ LTE Advanced provides support for Carrier Aggregation which allows multiple 'Component Carriers' to be used in parallel. This effectively increases the channel bandwidth to the sum of the individual Component Carriers

★ 3GPP References: TS 36.101, TS 36.104, TS 36.211

2.5 FREQUENCY AND TIME DIVISION DUPLEXING

- LTE has been specified to support both Frequency Division Duplexing (FDD) and Time Division Duplexing (TDD)
- The concepts of FDD and TDD are illustrated in Figure 6

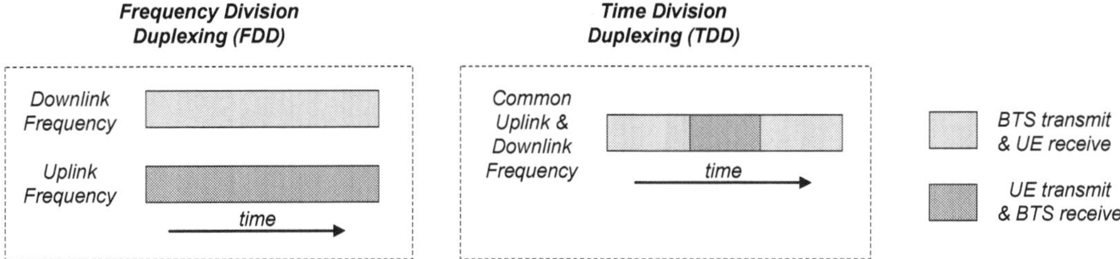

Figure 6 – FDD and TDD concepts

- FDD is based upon using two separate RF carriers for uplink and downlink transmission, i.e. the UE transmits using one RF carrier (uplink), while the BTS transmits using a different RF carrier (downlink)
- TDD is based upon using the same RF carrier for both the uplink and downlink transmissions. The UE and BTS cannot transmit simultaneously in the case of TDD because they share the same RF carrier
- FDD uses frame structure 'type 1', whereas TDD uses frame structure 'type 2'. These frame structures are presented in section 3.2
- TDD is attractive for systems where the data transfer is highly asymmetric because the ratio between the uplink and downlink transmissions can be adjusted appropriately and the RF carrier remains fully utilised. In the case of FDD, one of the RF carriers would be under utilised when the data transfer is highly asymmetric
- TDD devices benefit from not requiring a duplexer. This helps to reduce the cost of the device. A duplexer is required by FDD devices to extract the uplink signal from the antenna, while at the same time inserting the downlink signal into the antenna. Duplexers tend to increase the receiver noise figure in the receive direction and generate an additional loss in the transmit direction
- FDD is attractive for systems where the requirement for uplink and downlink capacity is relatively symmetric. FDD can offer higher throughputs because data transfer can be continuous in both directions. The capacity associated with a pair of FDD carriers is greater than the capacity associated with a single TDD carrier, but a greater quantity of spectrum is required
- FDD can be simpler to deploy in terms of synchronisation requirements. In general, it is not necessary for neighbouring FDD BTS to be time synchronised. Neighbouring TDD BTS require time synchronisation to limit levels of interference between uplink and downlink transmissions
- LTE also supports a third duplexing technology known as half duplex FDD. The concept of half duplex FDD is illustrated in Figure 7

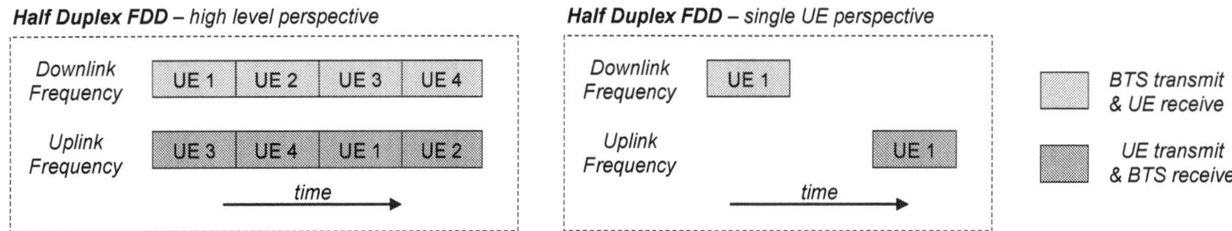

Figure 7 – Half duplex FDD concept

- In the case of half duplex FDD, the BTS is able to transmit and receive simultaneously, but the UE is not able to transmit and receive simultaneously. Both uplink and downlink RF carriers can be fully utilised by time multiplexing different UE
- Half duplex FDD uses frame structure 'type 1', i.e. the same frame structure as FDD. This frame structure is presented in section 3.2
- The BTS scheduler is responsible for providing half duplex operation by ensuring that UE do not need to transmit uplink data at the same time as receiving downlink data. This has to account for the requirements to send and receive acknowledgements after data has been transferred
- Similar to TDD, half duplex FDD can be an attractive solution because it avoids the requirement for a duplexer within the UE so helps to reduce the cost of devices. This argument is especially valid for operating bands which have small duplex separations (frequency separation between the uplink and downlink RF carriers). Duplexer design becomes more challenging and more expensive when the uplink and downlink operating bands are relatively close to each other

2.6 OPERATING BANDS

2.6.1 FDD

★ The LTE operating bands specified by 3GPP for Frequency Division Duplex (FDD) are shown in Table 2

★ The majority of these operating bands have also been specified for use by UMTS. This allows UMTS spectrum to be re-farmed for LTE, or for LTE and UMTS to share the same operating bands

★ LTE and UMTS share the same operating band numbering scheme, e.g. LTE operating band 1 is the same as UMTS operating band 1

★ Some LTE operating bands are the same as those used by GSM, e.g. operating band 8 corresponds to the E-GSM operating band. This allows GSM spectrum to be re-farmed for LTE, or for LTE and GSM to share the same operating bands

Operating Band	Uplink Band (MHz)	Downlink Band (MHz)	Duplex Spacing (MHz)	Bandwidth (MHz)	Other Uses
1	1920 – 1980	2110 – 2170	190	2 × 60	UMTS
2	1850 – 1910	1930 – 1990	80	2 × 60	UMTS, PCS 1900
3	1710 – 1785	1805 – 1880	95	2 × 75	UMTS, DCS 1800
4	1710 – 1755	2110 – 2155	400	2 × 45	UMTS
5	824 – 849	869 – 894	45	2 × 25	UMTS, GSM 850
6	830 – 840	875 – 885	45	2 × 10	UMTS
7	2500 – 2570	2620 – 2690	120	2 × 70	UMTS
8	880 – 915	925 – 960	45	2 × 35	UMTS, E-GSM 900
9	1749.9 – 1784.9	1844.9 – 1879.9	95	2 × 35	UMTS
10	1710 – 1770	2110 – 2170	400	2 × 60	UMTS
11	1427.9 – 1447.9	1475.9 – 1495.9	48	2 × 20	UMTS
12	699 – 716	729 – 746	30	2 × 17	UMTS, GSM 710
13	777 – 787	746 – 756	-31	2 × 10	UMTS
14	788 – 798	758 – 768	-30	2 × 10	UMTS
17	704 – 716	734 – 746	30	2 × 12	-
18	815 – 830	860 - 875	45	2 × 15	-
19	830 – 845	875 - 890	45	2 × 15	UMTS
20	832 – 862	791 - 821	-41	2 × 30	UMTS
21	1447.9 – 1462.9	1495.9 – 1510.9	48	2 × 15	UMTS
22	3410 – 3490	3510 – 3590	100	2 × 90	UMTS
23	2000 – 2020	2180 – 2200	180	2 × 20	-
24	1626.5 – 1660.5	1525 - 1559	-101.5	2 × 34	-
25	1850 – 1915	1930 - 1995	80	2 × 65	UMTS

Table 2 – FDD operating bands for LTE

★ The majority of operating bands have their uplink frequencies below their downlink frequencies. This approach helps to conserve UE battery power by allowing UE to transmit within the band which has the better radio propagation performance, i.e. radio propagation tends to be better at lower frequencies. Operating bands 13, 14, 20 and 24 have uplink bands which are higher than their downlink bands

★ The release 9 and 10 versions of the 3GPP specifications for LTE include a note to state that operating band 6 is not applicable. Operating band 6 remains applicable to UMTS but is superseded by operating bands 17 and 18 for LTE

★ Operating bands 15 and 16 are excluded from the 3GPP specifications for LTE. These operating bands are shown in Table 3, and have been defined by ETSI for use within Europe, Middle East and Africa. They are specified for UMTS within ETSI TS 102 735

Operating Band	Uplink Band (MHz)	Downlink Band (MHz)	Duplex Spacing (MHz)	Bandwidth (MHz)
15	1900 – 1920	2600 – 2620	700	2 × 20
16	2010 – 2025	2585 – 2600	575	2 × 15

Table 3 – FDD operating bands 15 and 16 specified by ETSI

2.6.2 TDD

★ The LTE operating bands specified by 3GPP for Time Division Duplex (TDD) are shown in Table 4

★ The majority of these operating bands have also been specified for use by UMTS. This allows UMTS spectrum to be re-farmed for LTE, or for LTE and UMTS to share the same operating bands

★ In the case of TDD, sharing the same operating band is more challenging due to both the BTS and UE transmitting and receiving on the same RF carrier. For example, if LTE and UMTS are allocated adjacent channels, a UMTS BTS could experience interference from an LTE BTS if the LTE BTS transmits while the UMTS BTS receives. In the case of FDD, the two BTS would be protected by the duplex spacing, i.e. there is a relatively large frequency separation between the transmit and receive channels

★ Coexistence of the two technologies is made simpler if they are synchronised so both sets of BTS transmit at the same time, and both populations of UE transmit at the same time. The transmit and receive patterns for LTE TDD have been specified to account for the coexistence requirements with other technologies

★ LTE and UMTS do not share the same operating band numbering scheme. LTE numbers its TDD operating bands from 33 upwards, whereas UMTS references its TDD operating bands as a, b, c, d, e and f

Operating Band	Uplink / Downlink Band (MHz)	Bandwidth (MHz)	Other Uses
33	1900 – 1920	1 × 20	UMTS
34	2010 – 2025	1 × 15	UMTS
35	1850 – 1910	1 × 60	UMTS
36	1930 – 1990	1 × 60	UMTS
37	1910 – 1930	1 × 20	UMTS
38	2570 – 2620	1 × 50	UMTS
39	1880 – 1920	1 × 40	UMTS
40	2300 - 2400	1 × 100	UMTS
41	2496 - 2690	1 × 194	-
42	3400 - 3600	1 × 200	-
43	3600 - 3800	1 × 200	-

Table 4 – TDD operating bands for LTE

★ Operating bands 33 to 40 were introduced within the release 8 version of the 3GPP specifications. Operating bands 41 to 43 were introduced within the release 10 version of the 3GPP specifications

★ 3GPP References: TS 36.101, TS 36.104

2.7 BEARER TYPES

- LTE provides an end-to-end service using the hierarchy of bearers shown in Figure 8
- An Evolved Packet System (EPS) bearer provides user plane connectivity between the UE and a Packet Data Network (PDN) Gateway
- An initial EPS bearer is established when the UE registers with the network using the Attach procedure. This EPS bearer is known as the 'default' EPS bearer and is used to provide *always-on* connectivity
- Other EPS bearers can be established to connect to other PDN Gateways, or to provide different Quality of Service (QoS) to the same PDN Gateway. These EPS bearers are known as 'dedicated' EPS bearers
- All user plane data transferred using the same EPS Bearer has the same QoS
- An EPS Bearer is generated from a combination of E-UTRAN Radio Access Bearer (E-RAB) and S5/S8 Bearer
- The S5 interface provides connectivity between a home Serving Gateway and a home PDN Gateway. The S8 interface provides roaming connectivity between a visited Serving Gateway and a home PDN Gateway

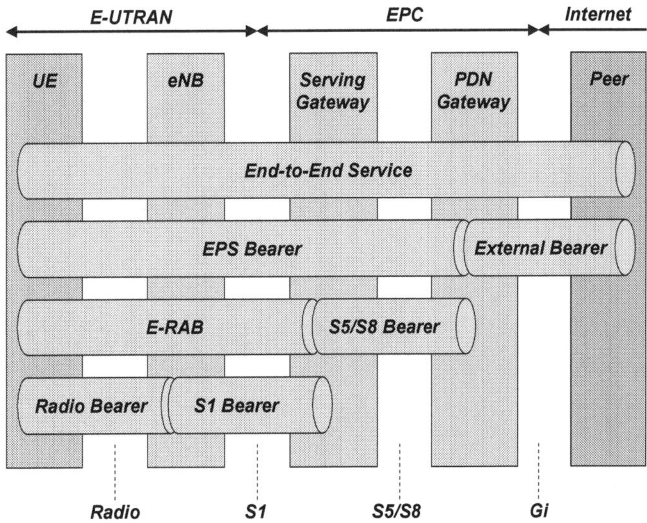

Figure 8 – Bearers for LTE

- An E-RAB is generated from a combination of Radio Bearer and S1 Bearer. Radio bearers provide the connection across the air-interface, whereas S1 bearers provide the connection across the transport network
- 3GPP References: TS 36.300

2.8 RADIO RESOURCE CONTROL STATES

- ★ The Radio Resource Control (RRC) state machine for LTE includes only 2 states: RRC Idle and RRC Connected. These 2 states are illustrated in Figure 9
- ★ RRC Idle is characterised by:
  - acquiring system information from the Broadcast Control Channel (BCCH)
  - UE controlled mobility based upon cell reselection
  - using a DRX cycle to monitor for paging messages which can be used to indicate:
    - incoming calls
    - system information changes
    - Earthquake and Tsunami Warning System (ETWS) notifications (if supported by the UE)
    - Commercial Mobile Alert Service (CMAS) notifications (if supported by the UE)
  - logging of measurements, location and time for UE configured to do so
- ★ RRC Connected is characterised by:
  - the ability to transfer data to and from the UE
  - using control channels to signal resource allocations
  - the UE reporting channel quality and other feedback information to the eNode B
  - network controlled mobility based upon handovers and cell change orders
  - using a UE specific DRX cycle at the lower layers if configured to do so
  - monitoring paging messages and/or System Information Block 1 (SIB1) to detect
    - system information changes
    - Earthquake and Tsunami Warning System (ETWS) notifications (if supported by the UE)
    - Commercial Mobile Alert Service (CMAS) notifications (if supported by the UE)
  - acquiring system information from the Broadcast Control Channel (BCCH)
- ★ The RRC connection establishment procedure allows the UE to make the transition from RRC Idle mode to RRC Connected mode
- ★ Figure 9 also illustrates mobility between LTE and UMTS, and mobility between LTE and GSM/GPRS
  - mobility between LTE and UMTS is based upon cell reselection and handovers
  - mobility between LTE and GSM/GPRS is based upon cell reselection, handovers, Cell Change Orders (CCO) and CCO with Network Assisted Cell Change (NACC)

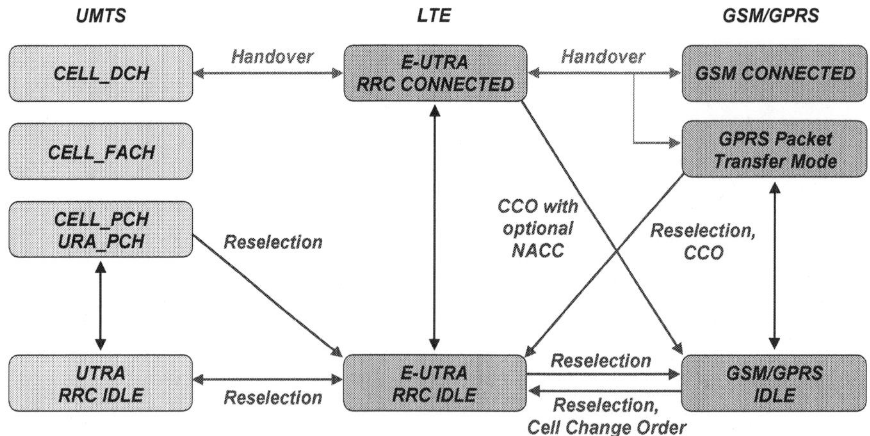

Figure 9 – RRC states for LTE and their interaction with UMTS and GSM/GPRS

- ★ Figure 10 illustrates mobility between LTE and CDMA2000 1×RTT (1 times Radio Transmission Technology), and mobility between LTE and CDMA2000 HRPD (High Rate Packet Data). Mobility is based upon handovers and cell reselections

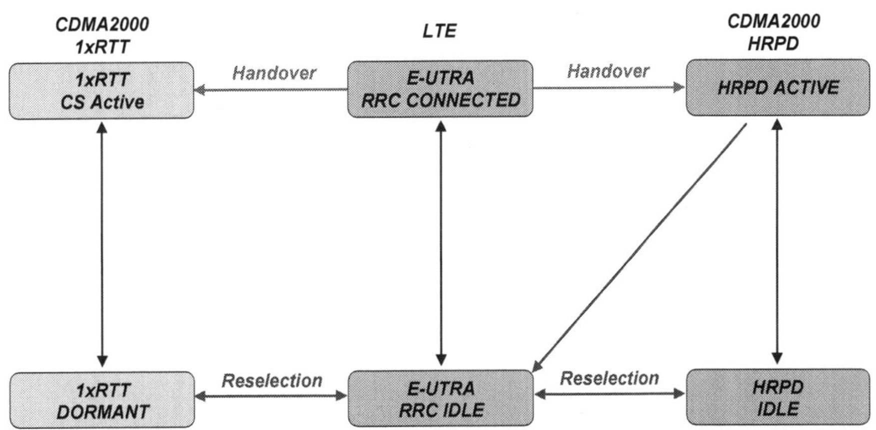

Figure 10 – RRC states for LTE and their interaction with CDMA2000

★ In addition to the state transitions illustrated in Figure 9 and Figure 10, LTE supports connection release with re-direction from E-UTRA RRC Connected mode to UMTS, GSM/GPRS and CDMA2000. Re-direction information is signalled to the UE within the RRC Connection Release message

★ 3GPP References: TS 36.331

2.9 SIGNALLING RADIO BEARERS

- Signalling Radio Bearers (SRB) are used to transfer RRC and Non-Access Stratum (NAS) signalling messages
  - RRC messages are used for signalling between the UE and eNode B
  - NAS messages are used for signalling between the UE and MME
- RRC messages can be used to encapsulate NAS messages for their transfer between UE and eNode B. The S1 Application Protocol (S1-AP) is used to transfer NAS messages between the eNode and MME
- 3 types of SRB have been defined for LTE:
  - SRB 0 transfers RRC messages which use the Common Control Channel (CCCH) logical channel
  - SRB 1 transfers RRC messages which use the Dedicated Control Channel (DCCH) logical channel
    - primarily allows RRC signalling between the UE and eNode B
    - messages may include a piggybacked NAS message
    - messages may be dedicated to transferring NAS messages if SRB 2 has not yet been configured
  - SRB 2 transfers RRC messages which use the DCCH logical channel
    - primarily allows NAS signalling between the UE and eNode B
    - messages may be dedicated to transferring NAS messages
    - messages may be used to transfer logged measurement information, e.g. recorded signal strength measurements
- The messages transferred by SRB 1 may include a piggybacked NAS message. This means that the message includes some RRC signalling content but also encapsulates a NAS message, e.g. the RRC Connection Setup Complete message includes a NAS message for the MME
- The Uplink Information Transfer and Downlink Information Transfer messages are dedicated to sending NAS messages and do not include any RRC signalling content. These messages are transferred using SRB 2 unless SRB2 has not yet been configured
- SRB 2 has lower priority than SRB 1 and is always configured after security activation
- After security activation, all messages transferred by SRB 1 and SRB 2 are integrity protected and ciphered by the Packet Data Convergence Protocol (PDCP) layer
- The set of SRB and their associated RRC messages are presented in Table 5
- SRB 0 uses transparent mode RLC while SRB 1 and 2 use acknowledged mode RLC
- SRB 0 transfers messages associated with RRC connection establishment and re-establishment. The UE is allocated a DCCH logical channel once an RRC connection has been established so SRB 1 and 2 are able to transfer subsequent messages
- Table 6 presents 2 additional RRC messages transferred by SRB 1. These messages have been separated because they are transferred between an eNode B and Relay Node (RN) rather than between an eNode B and UE. Relays are described in greater detail in section 30.7
- 3GPP References: TS 36.331

SRB	Direction	RRC Message	RLC Mode
SRB 0 (CCCH)	Downlink	RRC Connection Setup RRC Connection Reject RRC Connection Re-establishment RRC Connection Re-establishment Reject	Transparent
	Uplink	RRC Connection Request RRC Connection Re-establishment Request	
SRB 1 (DCCH)	Downlink	RRC Connection Reconfiguration RRC Connection Release Security Mode Command UE Capability Enquiry UE Information Request *DL Information Transfer (if no SRB 2)* Mobility From EUTRA Command Handover From EUTRA Preparation Request CS Fallback Parameter Response CDMA2000 Counter Check Logged Measurement Configuration	Acknowledged
	Uplink	RRC Connection Setup Complete Security Mode Complete Security Mode Failure RRC Connection Reconfiguration Complete RRC Connection Re-establishment Complete Measurement Report Inter-Frequency RSTD Measurement Indication UE Capability Information UE Information Response *UL Information Transfer (if no SRB 2)* UL Handover Preparation Transfer CS Fallback Parameters Request CDMA2000 Proximity Indication Counter Check Response MBMS Counting Response	
SRB 2 (DCCH)	Downlink	DL Information Transfer	
	Uplink	UL Information Transfer *UE Information Response (when logged measurement information is included)*	

Table 5 – RRC messages transferred by each SRB (between eNode B and UE)

SRB	Direction	RRC Message	RLC Mode
SRB 1 (DCCH)	Downlink	RN Reconfiguration	Acknowledged
	Uplink	RN Reconfiguration Complete	

Table 6 – RRC messages transferred by SRB 1 (between eNode B and Relay Node)

2.10 QUALITY OF SERVICE

- ★ Quality of Service (QoS) allows both subscribers and services to be differentiated. Premium subscribers can be prioritised over basic subscribers, while real time services can be prioritised over non-real time services
- ★ The importance of QoS increases during periods of congestion. An unloaded network is able to satisfy all subscribers and all services. As the network load increases, prioritisation determines which subscribers and services are able to maintain their performance, and which experience a loss in performance
- ★ QoS impacts admission control decisions. Connections with a guaranteed QoS are likely to require larger resource reservations during admission control. These connections may be blocked if insufficient resources are available, i.e. to protect existing connections with a guaranteed QoS. Connections without a guaranteed QoS are more likely to be admitted without considering the availability of resources, i.e. allowing them to use resources on a best-effort basis
- ★ QoS is applied between the UE and the PDN Gateway within the LTE network, i.e. it is applicable to an EPS bearer generated from a combination of an E-RAB and S5/S8 bearer
- ★ EPS bearers and E-RAB can be categorised as Guaranteed Bit Rate (GBR) or non-Guaranteed Bit Rate (non-GBR). Table 7 presents the QoS parameters associated with each bearer type

	GBR	Non-GBR
QoS Class Identifier (QCI)	✓	✓
Allocation and Retention Priority	✓	✓
Guaranteed Bit Rate	✓	
Maximum Bit Rate	✓	
APN Aggregate Maximum Bit Rate		✓
UE Aggregate Maximum Bit Rate		✓

Table 7 – QoS parameters associated with GBR and non-GBR bearers

- ★ The QoS Class Identifier (QCI) represents a pointer to a set of standardised QoS characteristics. 3GPP has adopted this approach to reduce both the signalling requirement and the maximum number of possible parameter combinations. Table 8 presents the relationship between QCI and the associated set of QoS characteristics

QCI	Resource Type	Priority	Packet Delay Budget	Packet Error Loss Rate	Example Services
1	GBR	2	100 ms	10^{-2}	Conversational Voice
2		4	150 ms	10^{-3}	Conversational Video (live streaming)
3		3	50 ms		Real Time Gaming
4		5	300 ms	10^{-6}	Non-Conversational Video (buffered streaming)
5	Non-GBR	1	100 ms		IMS Signalling
6		6	300 ms		Video (buffered streaming), TCP based applications
7		7	100 ms	10^{-3}	Voice, Video (live streaming), Interactive Gaming
8		8	300 ms	10^{-6}	Video (buffered streaming), TCP based applications
9		9			

Table 8 – Standardised QoS characteristics associated with each QCI

- ★ The QCI determines which bearers are categorised as GBR and which are categorised as non-GBR
- ★ The priority associated with each QCI is applied when forwarding packets across the network. High priority packets are transferred before low priority packets. A priority of 1 corresponds to the highest priority
- ★ The packet delay budget associated with each QCI defines an upper bound for the packet delay between the UE and the Policy and Charging Enforcement Function (PCEF) within the PDN Gateway. The delay budget figure is applicable to both the uplink and downlink with a 98 % confidence level. It is assumed that the average delay between the eNode B and PCEF is 20 ms so the remaining delay budget is available to the radio access network
- ★ The packet error loss rate defines the percentage of higher layer packets, e.g. IP packets, which are lost during periods that the network is not congested. The loss rate between the eNode B and PCEF is assumed to be negligible when the network is not congested so the requirement applies to the radio access network. Retransmissions at the MAC and RLC layers help to achieve the packet error loss rate requirements

★ The Allocation and Retention Priority (ARP) presented in Table 7 defines:

  - Pre-emption Capability (shall not trigger pre-emption, may trigger pre-emption). This characteristic determines whether or not a new connection request is allowed to pre-empt an existing connection
  - Pre-emption Vulnerability (not pre-emptable, pre-emptable). This characteristic determines whether or not an existing connection is allowed to be pre-empted by a new connection request
  - Priority (1 to 15) where 15 corresponds to no priority, 14 corresponds to the lowest priority and 1 corresponds to the highest priority. This characteristic can be used to identify which of the existing pre-emptable connections should be targeted for pre-emption. If the connection request has relatively low priority then it may not be able to pre-empt any of the existing connections. This priority is independent of the priority defined in Table 8 which is associated with packet handling of existing connections

★ The Guaranteed Bit Rate (GBR) in Table 7 defines the minimum bit rate which can be expected to be made available to the bearer when required. It can be configured with values between 0 and 10 000 Mbps. The GBR can be specified independently for the uplink and downlink

★ The Maximum Bit Rate (MBR) in Table 7 defines the maximum bit rate which can be expected to be made available to the bearer when required. It can be configured with values between 0 and 10 000 Mbps. The MBR can be specified independently for the uplink and downlink

★ The APN Aggregate Maximum Bit Rate (APN-AMBR) in Table 7 defines the maximum allowed throughput for an individual UE based upon the sum of its non-GBR bearers to a specific APN, i.e. the total non-GBR throughput generated by a UE to a specific APN is not allowed to exceed this limit. It can be configured with values between 1 kbps and 65 280 Mbps, and can be specified independently for the uplink and downlink

★ The UE Aggregate Maximum Bit Rate (UE-AMBR) in Table 7 defines the maximum allowed throughput for a UE based upon the sum of all its non-GBR bearers. It can be viewed as a limit placed upon the sum of all APN-AMBR belonging to a specific UE. The MME sets the UE-AMBR to the sum of the APN-AMBR of all active APN up to the value of the subscribed UE-AMBR. It can be configured with values between 0 and 10 000 Mbps, and can be specified independently for the uplink and downlink

★ 3GPP References: TS 36.300, TS 23.401, TS 23.203

2.11 MOBILITY MANAGEMENT STATES

★ EPS Mobility Management (EMM) states are maintained by the Non-Access Stratum (NAS) layers within the UE and MME. These states determine whether or not a UE is reachable, and whether or not a UE can receive services

★ The main two EMM states are:

- EMM-DEREGISTERED
- EMM-REGISTERED

These states are introduced within 3GPP TS 23.401. Additional EMM states and sub-states are described within 3GPP TS 24.301. The main two EMM states are illustrated in Figure 11

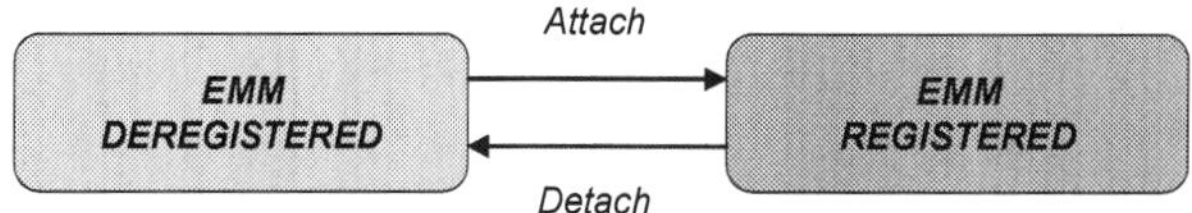

Figure 11 – Main EPS mobility management states

★ A UE is in the EMM-DEREGISTERED state when it is first switched on. The MME does not have knowledge of the UE location so the UE cannot be paged. In addition, the UE cannot have any user plane bearers while in the EMM-DEREGISTERED state

★ UE attempt to move into the EMM-REGISTERED state whenever possible. This is achieved by completing the attach procedure. This procedure registers the UE with the MME and establishes a default bearer for application data transfer. UE can also move into the EMM-REGISTERED state by completing a Tracking Area Update after an incoming inter-system transition

★ The location of UE in the EMM-REGISTERED state is known by the MME to at least an accuracy of the Tracking Area list allocated to the UE, i.e. the set of Tracking Areas with which the UE is registered. If the Tracking Area list includes only a single Tracking Area then the location of the UE is known to be within that Tracking Area. Depending upon the EPS Connection Management (ECM) state, the location of a UE can also be known by the MME to the accuracy of a single eNode B

★ The UE always has at least one active Packet Data Network (PDN) connection when in the EMM-REGISTERED state (the default bearer is established during the Attach procedure). De-activating all PDN connections causes the UE to move back to the EMM-DEREGISTERED state

★ UE can also make the transition from the EMM-REGISTERED state to the EMM-DEREGISTERED state by completing the Detach procedure, or by having a Tracking Area update rejected

★ 3GPP References: TS 23.401, TS 24.301

2.12 CONNECTION MANAGEMENT STATES

★ EPS Connection Management (ECM) states are maintained by the Non-Access Stratum (NAS) layers within the UE and MME. These states determine the signalling connectivity between the UE and Evolved Packet Core (EPC)

★ The two ECM states are:

- ECM-IDLE
- ECM-CONNECTED

These states are introduced within 3GPP TS 23.401. Procedures associated with these states are described in 3GPP TS 24.301. This specification refers to the same two states as EMM-IDLE and EMM-CONNECTED, i.e. different terminology is used in different specifications. The two ECM states are illustrated in Figure 12

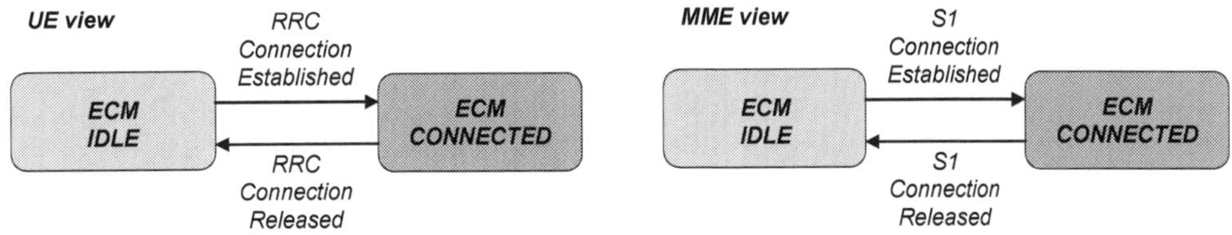

Figure 12 – EPS connection management states

★ A UE in ECM-IDLE state does not have a Non-Access Stratum (NAS) signalling connection to the MME

★ UE complete cell selection and cell reselection when in the ECM-IDLE state

★ The location of UE in the ECM-IDLE state is known by the MME to an accuracy of the Tracking Area list allocated to the UE, i.e. the set of Tracking Areas with which the UE is registered

★ When a UE is in both the EMM-REGISTERED and ECM-IDLE states then typical procedures include:

- Tracking Area Updates (TAU) triggered by mobility and the periodic TAU timer (T3412)
- TAU triggered for MME load balancing, i.e. relocating the UE's signalling connection to a different MME
- TAU triggered for the UE to signal a change of its Core Network Capability or its UE specific DRX cycle
- responding to paging messages by performing a Service Request procedure
- performing the Service Request procedure when uplink user plane data is to be sent

★ The UE and MME enter the ECM-CONNECTED state after a NAS signalling connection has been established. From the UE perspective, this corresponds to establishing an RRC connection between the UE and eNode B, while from the MME perspective, this corresponds to establishing an S1 connection between the eNode B and MME

★ Initial NAS messages which can be used to initiate a transition from ECM-IDLE to ECM-CONNECTED are Attach Request, Tracking Area Update Request, Service Request and Detach Request, i.e. the Attach, Tracking Area Update, Service Request and Detach procedures all require the UE to be in RRC connected mode

★ The location of UE in the ECM-CONNECTED state is known by the MME to an accuracy of the serving eNode B

★ UE mobility is handled using handovers rather than cell reselection when UE are in the ECM-CONNECTED state

★ UE continue to complete Tracking Area Updates (TAU) due to mobility when in the ECM-CONNECTED state, i.e. when the UE moves into a cell which does not belong to a Tracking Area within the UE's list of registered Tracking Areas

★ UE make the transition from ECM-CONNECTED to ECM-IDLE when their signalling connection is released, or when their signalling connection fails

★ 3GPP References: TS 23.401, TS 24.301

2.13 EVOLUTION OF 3GPP SPECIFICATIONS

★ The capabilities of Long Term Evolution (LTE) are enhanced with each version of the 3GPP specifications

★ The main capabilities associated with releases 8, 9 and 10 are presented in Figure 13

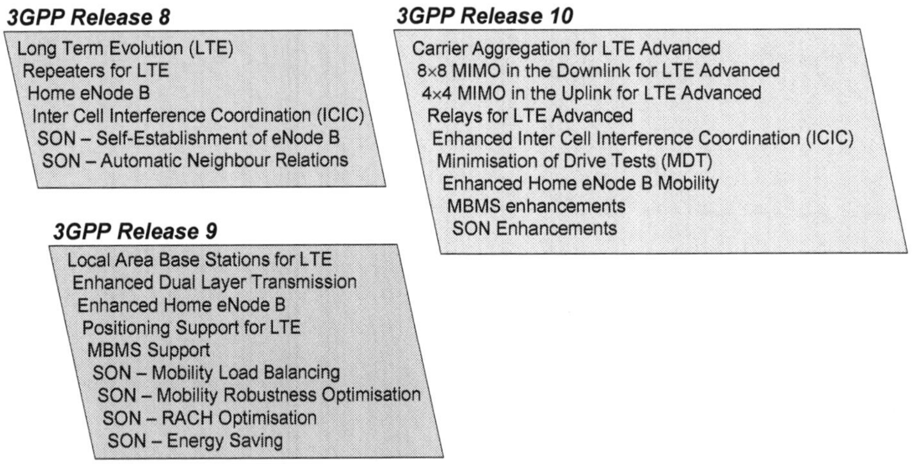

3GPP Release 8	3GPP Release 10
Long Term Evolution (LTE) Repeaters for LTE Home eNode B Inter Cell Interference Coordination (ICIC) SON – Self-Establishment of eNode B SON – Automatic Neighbour Relations	Carrier Aggregation for LTE Advanced 8×8 MIMO in the Downlink for LTE Advanced 4×4 MIMO in the Uplink for LTE Advanced Relays for LTE Advanced Enhanced Inter Cell Interference Coordination (ICIC) Minimisation of Drive Tests (MDT) Enhanced Home eNode B Mobility MBMS enhancements SON Enhancements

3GPP Release 9
Local Area Base Stations for LTE Enhanced Dual Layer Transmission Enhanced Home eNode B Positioning Support for LTE MBMS Support SON – Mobility Load Balancing SON – Mobility Robustness Optimisation SON – RACH Optimisation SON – Energy Saving

Figure 13 – Main capabilities of LTE associated with 3GPP releases 8, 9 and 10

★ The following capabilities were introduced within the release 8 version of the 3GPP specifications:

- LTE itself was introduced, including the specification of its physical layer, layers 2 and 3 and its various interfaces. Both RF and conformance testing requirements were specified. LTE was specified with a maximum downlink capability based upon a 20 MHz channel bandwidth with 4×4 MIMO and 64QAM. The uplink did not support multi antenna transmission but supported 64QAM
- LTE Repeaters were introduced for FDD using the same set of operating bands and channel bandwidths as an eNode B. Repeaters were specified within 3GPP TS 36.106 while their conformance testing was specified within TS 36.143
- Home eNode B (also known as Femto cells) were first introduced. Home eNode B are small cells typically used in domestic or small office environments. They connect to the Evolved Packet Core (EPC) via a Home eNode B gateway. The connection between the Home eNode B and gateway is typically a fixed broadband connection, e.g. DSL or cable. Closed Subscriber Groups (CSG) were specified to control access to Home eNode B
- support for Inter Cell Interference Coordination (ICIC) was introduced. ICIC allows neighbouring eNode B to exchange load information to help co-ordinate the use of both uplink and downlink resources, e.g. one eNode B uses resources at the top of the channel bandwidth while a second eNode B uses resources at the bottom of the channel bandwidth. This creates a trade-off between improving the signal to noise ratio and reducing the quantity of resources available to each eNode B
- the Self-Establishment of eNode B component of Self Organising Networks (SON) was introduced. This capability allows the eNode B to have plug-and-play functionality. After physically connecting the eNode B and switching on, it is able to automatically connect to the element manager and download software, as well as radio and transport configuration data. It may also be capable of establishing X2 and S1 interfaces
- the Automatic Neighbour Relations (ANR) component of SON was introduced. This capability allows the eNode B to automatically update its neighbour database based upon the mobility of UE within its cells, i.e. neighbour relations are added as they are used for the first time. Support is included for intra-frequency, inter-frequency and inter-system neighbour relations

★ The following capabilities were introduced within the release 9 version of the 3GPP specifications:

- Local Area Base Stations (also known as pico cells) were introduced. These eNode B have lower transmit power capabilities than the standard Wide Area Base Stations (also known as macro cells). Local Area Base Stations provide coverage across relatively small areas but can be used to increase network capacity at traffic hotspots
- Enhanced Dual Layer Transmission refers to user specific beamforming combined with 2×2 MIMO (allowing the transmission of 2 parallel streams of data). The release 8 version of the specifications introduced user specific beamforming but it was limited to the transmission of a single stream of data
- the specification of Home eNode B continued within the release 9 version of the 3GPP specifications. The concept of a 'whitelist' was introduced to help ensure that UE belonging to a Closed Subscriber Group (CSG), i.e. registered to use a Home eNode B, always camp on their Home eNode B rather than remaining camped on another RF carrier. The release 9 version of the specifications also added enhanced connected mode mobility for inbound handovers onto Home eNode B
- Positioning Reference Signals (PRS) were introduced to allow the UE to complete Observed Time Difference of Arrival (OTDOA) measurements. These measurements are completed from multiple eNode B to improve the accuracy of the location estimate. The release 9 version of the specifications also introduced Enhanced Cell Identity positioning. This approach combines

the location of the serving cell with measurements such as timing advance and receive–transmit time differences at the UE and eNode B

- Multimedia Broadcast Multicast Services (MBMS) were introduced. Some aspects of MBMS were already defined within the release 8 version of the specifications, e.g. the PMCH physical channel was defined. However, complete support for MBMS was provided by the release 9 version of the specifications. MBMS can be used to transmit downlink video services to groups of UE
- the Mobility Load Balancing component of SON was introduced. Mobility load balancing is used to trigger handovers from loaded cells towards less loaded cells with the objective of maximising the overall system capacity. It can also be used to optimise the handover parameter set
- the Mobility Robustness Optimisation component of SON was introduced. This allows the UE to report information regarding radio link failures. This information can be sent to the cell at which the radio link failure occurred and used to optimise the relevant parameter set
- the RACH Optimisation component of SON was introduced. This provides support for tuning the configuration and resources used by the random access procedure. The UE can be requested to report the number of preambles used to gain access, and whether or not contention was detected
- the Energy Saving component of SON was introduced. This allows capacity layers to be placed in a dormant state to conserve power consumption. Neighbouring eNode B are informed of any actions to allow handovers and neighbour list management to be handled appropriately

★ The release 10 version of the specifications introduces components of LTE Advanced. These are described in greater detail in section 2.14. Overall, the following capabilities were introduced within the release 10 version of the 3GPP specifications:

- Carrier Aggregation allows a single connection to use multiple RF carriers, known as Component Carriers. Signalling is defined to support the aggregation of up to 5 Component Carriers. RF capabilities are initially defined to support the aggregation of up to 2 Component Carriers providing an effective channel bandwidth of up to 40 MHz. This capability helps LTE Advanced to achieve both its peak and average throughput requirements
- 8×8 MIMO in the downlink allows the transmission of 8 parallel streams of data towards a single UE. This capability helps LTE Advanced to achieve its downlink peak spectral efficiency target, i.e. it increases the bits per second per Hz performance. This complements the existing 4×4 MIMO, 2×2 MIMO, transmit diversity and single antenna transmission schemes
- 4×4 MIMO in the uplink allows the transmission of 4 parallel streams of data towards an eNode B. This capability helps LTE Advanced to achieve its uplink peak spectral efficiency target, i.e. it increases the bits per second per Hz performance. The release 10 version of the specifications also includes 2×2 MIMO in the uplink. Prior to the release 10 version of the specifications, only single antenna transmission was possible in the uplink
- Relays are intended to provide extended LTE coverage and capacity with relatively low cost. They differ from Repeaters because a Repeater simply re-transmits the uplink and downlink RF signals to extend the coverage of a donor cell. A Relay has its own cells and is able to decode messages before forwarding them, i.e. a relay has its own MAC, RLC, PDCP and RRC layers. The cells provided by a Relay have their own Physical layer Cell Identities, Synchronisation Signals and Reference Signals
- Enhanced Inter Cell Interference Coordination (ICIC) builds upon the capabilities of ICIC introduced within the release 8 version of the specifications. The concept of Almost Blank Subframes (ABS) is introduced to allow eNode to reduce their transmissions during certain time intervals. This helps to improve the signal to noise ratio conditions for neighbouring eNode B, but reduces the total resources available for transmission
- Minimisation of Drive Tests (MDT) provides solutions for recording measurements from the UE perspective without the requirement for relatively expensive and time consuming drive testing. Measurements are collected from the population of subscribers which have provided their consent. Measurements can target individual UE, or groups of UE within a specific area. Measurements can be requested in RRC Idle mode or RRC Connected mode
- Enhanced Home eNode B mobility within the release 10 version of the specifications focuses upon mobility from one Home eNode B to another Home eNode B. This scenario is viewed as being particularly important for office environments where multiple Home eNode B may be used to provide coverage and capacity
- MBMS Enhancements provide additional capabilities such as UE counting to help optimise the transmission of downlink services, e.g. services can be disabled if there are no UE, sent using point-to-point transmissions if there is a small number of UE, and sent as broadcast transmissions if there is an increased number of UE. Allocation and Retention Priority (ARP) for MBMS was also introduced within the release 10 version of the specifications
- SON Enhancements define improvements for Mobility Robustness Optimisation and Mobility Load Balancing. Inter-system support was added for Mobility Robustness Optimisation, while inter-system support was enhanced for Mobility Load Balancing

2.14 LTE ADVANCED

★ LTE Advanced is an evolved version of LTE with increased capabilities and improved performance. It is introduced within the release 10 version of the 3GPP specifications, in contrast to LTE which was introduced within the release 8 version of the specifications

★ The requirements for LTE-Advanced are specified within 3GPP TR 36.913, whereas the original requirements for LTE are specified within 3GPP TR 25.913. The requirements for LTE-Advanced have been defined to satisfy the requirements of IMT-Advanced specified by the Radiocommunications division of the International Telecommunication Union (ITU-R)

★ Table 9 compares some of the key requirements for LTE-Advanced with those for LTE

		LTE Advanced Requirements	LTE Requirements
Peak Throughput	Uplink	500 Mbps	50 Mbps
	Downlink	1 Gbps	100 Mbps
Peak Spectrum Efficiency	Uplink	15 bps/Hz	2.5 bps/Hz
	Downlink	30 bps/Hz	5.0 bps/Hz
Average Spectrum Efficiency	Uplink	2.0 bps/Hz/cell	0.7 bps/Hz/cell
	Downlink	3.7 bps/Hz/cell	1.7 bps/Hz/cell
Control Plane Latency	From Idle Mode	50 ms	100 ms
	From Connected Mode DRX	10 ms	50 ms
User Plane Latency		< 5 ms	< 5 ms

Table 9 – Comparison of requirements for LTE and LTE-Advanced

★ Peak throughput requirements for LTE Advanced are 10 times greater than those for LTE. These improvements are fundamentally achieved using a combination of increased bandwidth and increased multiple antenna transmission capability

- o the maximum bandwidth of 20 MHz in LTE, evolves to a maximum bandwidth of up to 100 MHz in LTE Advanced
- o in the downlink direction, 4×4 MIMO in LTE evolves to 8×8 MIMO in LTE Advanced (although the 100 Mbps peak throughput requirement for LTE was based upon the assumption of 2 rather than 4 receive antenna at the UE)
- o in the uplink direction, single antenna transmission in LTE evolves to 4×4 MIMO in LTE Advanced

★ Peak spectrum efficiency requirements for LTE Advanced are 6 times greater than those for LTE. Spectrum efficiency is a measure of throughput per unit of bandwidth (measured in terms of bps/Hz). Increasing throughput by simply increasing the quantity of spectrum does not improve the spectrum efficiency. The spectrum efficiency requirements for LTE Advanced are primarily achieved using the increased multiple antenna transmission capability

★ Average spectrum efficiency represents the total throughput of all users, divided by the total bandwidth, divided by the number of cells. It is then expressed in units of bps/Hz/cell. 3GPP TR 36.913 specifies a range of average spectrum efficiency requirements for a range of antenna configurations. The figures in Table 9 are applicable to 2×4 transmission in the uplink (2 transmit antenna at the UE and 4 receive antenna at the eNode B), and to 4×4 transmission in the downlink. These figures are also specified to be applicable when using a 10 MHz channel bandwidth and a 500 m inter-site distance

★ In the case of LTE, the average spectrum efficiency requirements were not specified using absolute values, but were specified in terms of a relative improvement when compared to the release 6 version of UMTS

- o in the uplink direction, it was assumed that both UMTS and LTE used a single transmit antenna at the UE and 2 receive antenna at the BTS. The uplink LTE average spectrum efficiency was then specified to be 2 to 3 times greater than the UMTS average spectrum efficiency
- o in the downlink direction, it was assumed that UMTS used a single transmit antenna at the Node B with 2 receive antenna at the UE, while LTE used 2 transmit antenna at the eNode B with 2 receive antenna at the UE. The downlink LTE average spectrum efficiency was then specified to be 3 to 4 times greater than the UMTS average spectrum efficiency

The LTE average spectrum efficiency figures presented in Table 9 represent typical results which fit the requirements defined by 3GPP TR 25.913

★ LTE Advanced is specified to have reduced control plane latencies relative to LTE. Control plane latencies represent the delay in moving the UE into a state where it is ready to transfer data with a user plane connection. Control plane latencies are defined for initial conditions of RRC Idle mode, and the Discontinuous Reception (DRX) substate of RRC Connected mode

★ The user plane latency represents the one-way delay between the IP layer in the UE and the IP layer in the eNode B. The latency requirement is applicable to both the uplink and downlink directions so the effective round trip time requirement is < 10 ms. The user plane latency for LTE Advanced was specified to be better than LTE but without defining a specific value

- ★ The main solutions for LTE-Advanced are:
  - Carrier Aggregation
  - 8×8 MIMO in the downlink
  - 4×4 MIMO in the uplink
  - Enhanced uplink transmission
  - Relays
  - Heterogeneous networks
  - Co-ordinated Multi-Point transmission (CoMP)
- ★ Carrier Aggregation increases the channel bandwidth by combining multiple RF carriers. Each individual RF carrier is known as a Component Carrier. The release 10 version of the 3GPP specifications defines signalling to support up to 5 Component Carriers, i.e. a maximum combined channel bandwidth of 100 MHz. From the RF perspective, a maximum of 2 Component Carriers have been defined initially. Component Carriers do not need to be adjacent and can be located in different operating bands. The release 10 version of the 3GPP specifications defines individual Component Carriers to be backwards compatible so they can be used by release 8 and release 9 devices. Carrier Aggregation is described in section 18
- ★ 8×8 MIMO in the downlink requires 8 transmit antenna at the eNode B and 8 receive antenna at the UE. It provides support for the simultaneous transmission of 8 parallel streams of data. These streams of data are generated from 2 transport blocks which are processed by the physical layer before a serial to parallel conversion generates 4 streams from each block. 2 rather than 8 transport blocks are used to help keep the signalling overhead to a minimum. 8×8 MIMO for the downlink is described in section 4.5.1
- ★ 4×4 MIMO in the uplink requires 4 transmit antenna at the UE and 4 receive antenna at the eNode B. It provides support for the simultaneous transmission of 4 parallel streams of data. These streams of data are generated from 2 transport blocks which are processed by the physical layer before a serial to parallel conversion generates 2 streams from each block. 4×4 MIMO for the uplink is described in section 12.4.1
- ★ Enhanced uplink transmission refers to the support for non-contiguous Resource Block allocations. The release 8 and 9 versions of the 3GPP specifications only support contiguous Resource Block allocations to help minimise the requirement for power amplifier back-off. Minimising the requirement for back-off improves amplifier efficiency and allows the UE to transmit with greater power
  - LTE Advanced provides the option to allocate non-contiguous Resource Blocks. Allocating non-contiguous Resource Blocks provides the eNode B scheduler with greater flexibility and increases the potential for a frequency selective scheduling gain. These benefits provide potential for improved throughputs and spectrum efficiency
  - in addition, the release 8 and 9 versions of the 3GPP specifications do not support simultaneous transmission of the PUCCH and PUSCH physical channel because it would result in the transmission of non-contiguous Resource Blocks. LTE Advanced provides the option for simultaneous PUCCH and PUSCH transmission

  The allocation of non-contiguous Resource Blocks in the uplink is described in section 9.5
- ★ Relays use a donor cell belonging to an eNode B to provide connectivity towards the core network. They provide coverage and capacity at locations which do not have transport connections. A Relay has its own cells, Physical layer Cell Identities, Synchronisation Signals and Reference Signals. From the UE perspective, a relay appears to be the same as an eNode B. Relays are described in section 30.7
- ★ A network composed of multiple site types (macro, micro, pico, femto, relays and repeaters) is known as a heterogeneous network. Heterogeneous networks provide increased deployment flexibility. Microcells and picocells can be used when site acquisition for macrocells is difficult. The capacity gain from increasing the density of macrocells tends to saturate as levels of intercell interference increase. Microcells and picocells can be deployed with greater densities to further increase network capacity. Repeaters and relays can provide coverage and capacity at locations without a transport network connection. Femto can provide coverage at locations outside the reach of the main network. Heterogeneous networks are described in section 30
- ★ Coordinated Multi-Point (CoMP) transmission in the downlink and reception in the uplink are LTE-Advanced solutions to help improve the cell edge throughput and spectrum efficiency performance. 3GPP progressed CoMP as a study item during the timescales of release 10 development but the resultant work item aims to include CoMP within the release 11 version of the specifications. CoMP is described in section 29
- ★ The release 10 version of the 3GPP specifications introduces additional UE categories to provide support for LTE Advanced. UE category 8 is able to support up to 5 Component Carriers, 8×8 MIMO in the downlink direction and 4×4 MIMO in the uplink direction. UE categories 1 to 5 were introduced within the release 8 version of the specifications. Nevertheless, 3GPP release 10 UE which are category 1 to 5 may still support Carrier Aggregation. UE categories are described in sections 18.3 and 19.1

2.15 3GPP SPECIFICATIONS LIST

★ The '36' series of 3GPP technical specifications focuses upon LTE radio technology

★ Many of the specifications have similar titles to the equivalent '25' series for UMTS

★ Other series of specifications also include relevant information, e.g. TS 24.301 specifies the Non-Access Stratum (NAS) protocol

★ A sample of the 3GPP specifications applicable to LTE is presented in Table 10

Number	Title	Number	Title
36.101	UE radio transmission and reception	36.422	X2 signalling transport
36.104	BS radio transmission and reception	36.423	X2 application protocol
36.106	FDD repeater radio transmission	36.424	X2 data transport
36.116	Relay radio transmission and reception	36.440	Aspects and principles for interfaces for MBMS
36.133	Requirements for support of RRM	36.441	Layer 1 for interfaces supporting MBMS
36.201	LTE physical layer, general description	36.442	Signalling transport for interfaces for MBMS
36.211	Physical channels and modulation	36.443	M2 application protocol
36.212	Multiplexing and channel coding	36.444	M3 application protocol
36.213	Physical layer procedures	36.445	M1 data transport
36.214	Physical layer measurements	36.446	M1 user plane protocol
36.216	Physical layer for relaying	23.002	Network architecture
36.300	E-UTRAN overall description	23.003	Numbering, addressing and identification
36.302	Services provided by the physical layer	23.041	Technical realisation of Cell Broadcast Service
36.304	UE procedures in idle mode	23.122	NAS functions for MS in idle mode
36.306	UE radio access capabilities	23.203	Policy and charging control architecture
36.321	MAC protocol specification	23.206	Voice call continuity between CS and IMS
36.322	RLC protocol specification	23.216	Single radio voice call continuity
36.323	Packet data convergence protocol	23.272	CS fallback in EPS
36.331	RRC protocol specification	23.401	GPRS enhancements for E-UTRAN access
36.401	Architecture description	24.301	NAS protocol for evolved packet system
36.410	S1 layer 1 general aspects and principles	29.118	MME to VLR SGs interface specification
36.411	S1 layer 1	29.272	MME and SGSN Diameter protocol
36.412	S1 signalling transport	29.274	Evolved GPRS tunnelling protocol
36.413	S1 application protocol	29.280	Sv interface for SRVCC
36.414	S1 data transport	29.281	GPRS tunnelling protocol user plane
36.420	X2 general aspects and principles	32.422	Trace control and configuration management
36.421	X2 layer 1	37.320	Radio measurement collection for MDT

Table 10 – Sample of 3GPP specifications applicable to LTE

★ These specifications are available to download from the 3GPP internet site: www.3gpp.org

★ Versions are available for each release of the 3GPP specifications, e.g. releases 8, 9 and 10

3 DOWNLINK AIR-INTERFACE

3.1 MULTIPLE ACCESS

- Multiple access techniques allow resources to be shared between a group of users
- The downlink of LTE uses Orthogonal Frequency Division Multiple Access (OFDMA). This is in contrast to Frequency, Time or Code Division Multiple Access. These multiple access techniques are presented in Table 11

	FDMA	TDMA	CDMA	OFDMA
Resource Allocation	RF Carriers	Time Slots	Codes	Subcarriers
Example System	AMPS	GSM	UMTS	LTE

Table 11 – Multiple access techniques

- An OFDMA signal is based upon a set of orthogonal subcarriers. Resources are shared by allocating a subset of the subcarriers to each user
- The subcarriers are orthogonal because when sampling one subcarrier at its peak, all other subcarriers have zero amplitude
- The subcarriers are tightly packed to help make efficient use of the spectrum

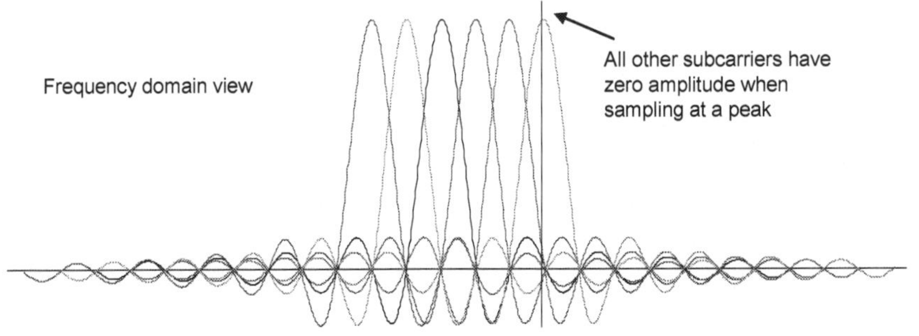

Figure 14 – Orthogonal subcarriers for OFDMA

- Information is transferred by modulating each subcarrier. For example, the modulation applied to each subcarrier could be QPSK, 16QAM or 64QAM
- The Inverse Fast Fourier Transform (IFFT) is used to generate a time domain OFDMA symbol from a combination of the modulated subcarriers
- A significant benefit of OFDMA is its tolerance to propagation channel delay spread
- OFDMA is tolerant to delay spread because its time domain symbols have relatively long durations. This is possible because data is transferred in parallel (across subcarriers) rather than in serial
  - a typical OFDMA symbol for LTE has a duration of 71.35 μs (defined as 1 / subcarrier spacing of 15 kHz + an overhead for the Cyclic Prefix)
  - a 10 MHz OFDMA channel could have 600 subcarriers spaced at 15 kHz generating a modulation symbol rate of 600 / 71.35 μs = 8.4 Msps
  - if transmitting data in serial, 8.4 Msps would require a symbol period of $1 / 8.4 \times 10^6 = 0.12$ μs
- The reduced impact of delay spread means that fading is flat in the frequency domain and receiver equalization becomes simpler
- A drawback of OFDMA is its relatively high Peak to Average Power Ratio (PAPR) generated by summing large numbers of subcarriers. High PAPR means that power amplifiers have to operate with increased back-off which leads to reduced efficiency
- Reduced power efficiency has an impact upon battery powered handheld devices. This has resulted in a different multiple access scheme being selected for the uplink of LTE
- A second drawback of OFDMA is its sensitivity to frequency offsets which could be introduced by Doppler, or by frequency inaccuracies within local oscillators at the transmitter and receiver

3.2 FRAME STRUCTURE

★ The frame structure considers only the time domain

★ 3GPP TS 36.211 specifies frame structures 'Type 1' and 'Type 2'

★ Frame structure 'Type 1' is applicable to FDD (both full and half duplex), whereas frame structure 'Type 2' is applicable to TDD

★ In both cases, radio frames are numbered using their System Frame Number (SFN)

★ The SFN is defined using a string of 10 bits which provides a range from 0 to 1023, i.e. the SFN cycles once every 10.24 seconds

★ The 8 most significant bits of the SFN are broadcast within the Master Information Block (MIB) of the system information on the Broadcast Channel (BCH)

★ The 2 least significant bits are deduced from the 4 radio frame cycle used to transmit the BCH, i.e. the first radio frame within the cycle corresponds to 00, the second to 01, the third to 10 and the forth to 11

3.2.1 FDD

★ Frame structure 'Type 1' is illustrated in Figure 15

★ A single 10 ms radio frame includes 10 subframes of 1 ms

★ Each 1 ms subframe includes 2 slots of 0.5 ms

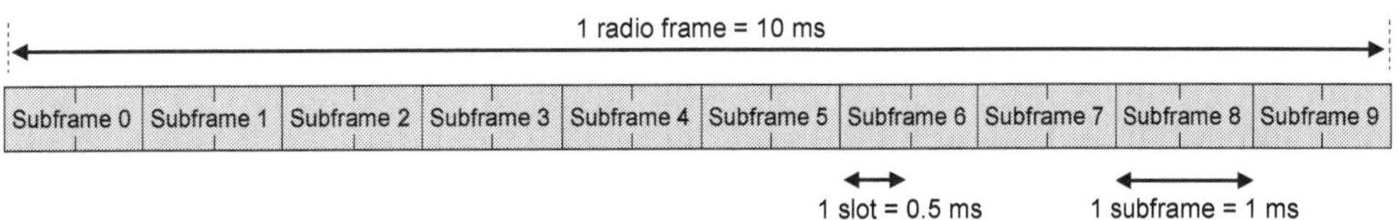

Figure 15 – Frame structure 'Type 1'

★ This frame structure is applicable to both the uplink and downlink. When measured at the UE, the timing of the uplink radio frame precedes the timing of the downlink radio frame to account for the propagation delay. The precise timing is determined by the Timing Advance instructions signalled by the eNode B (described in section 24.2)

★ In the case of FDD, neighbouring eNode B do not have to be synchronised with one another

★ 3GPP References: TS 36.211

3.2.2 TDD

★ Frame structure 'Type 2' is illustrated in Figure 16

★ A single 10 ms radio frame includes 10 subframes of 1 ms

★ Each 1 ms subframe includes 2 slots of 0.5 ms

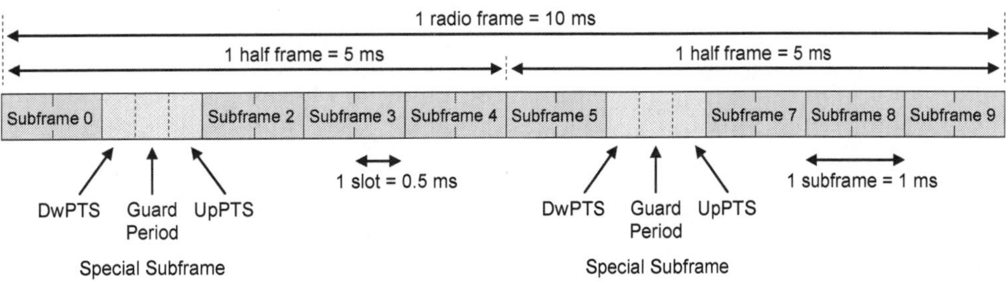

Figure 16 – Frame structure 'Type 2'

★ In the case of TDD each radio frame consists of 2 half-frames of 5 subframes each

★ Subframes can be either uplink subframes, downlink subframes or special subframes. Special subframes include the fields: Downlink Pilot Time Slot (DwPTS), Guard Period (GP) and Uplink Pilot Time Slot (UpPTS)

★ Special subframes are used when switching from downlink to uplink transmission. They are not used when switching from uplink to downlink transmission

★ Figure 16 illustrates 2 special subframes, but there can be either 1 or 2 special subframes within a radio frame

★ 3GPP has specified 7 allowed combinations of uplink, downlink and special subframes. These are presented in Table 12. The uplink-downlink configuration used by a cell is broadcast in System Information Block (SIB) 1 on the BCCH

Uplink - Downlink Configuration	DL-to-UL Switch Periodicity	Subframe Number									
		0	1	2	3	4	5	6	7	8	9
0	5 ms	D	S	U	U	U	D	S	U	U	U
1	5 ms	D	S	U	U	D	D	S	U	U	D
2	5 ms	D	S	U	D	D	D	S	U	D	D
3	10 ms	D	S	U	U	U	D	D	D	D	D
4	10 ms	D	S	U	U	D	D	D	D	D	D
5	10 ms	D	S	U	D	D	D	D	D	D	D
6	5 ms	D	S	U	U	U	D	S	U	U	D

Table 12 – Combinations of uplink, downlink and special subframes for TDD

★ Subframes 0 and 5 are always downlink subframes. These subframes include synchronisation signals and broadcast information

★ Configurations 0, 1, 2 and 6 have a 5 ms switching point periodicity. These configurations allow improved coexistence with TD-SCDMA (also known as 1.28 Mcps TDD or Low Chip Rate TDD) which uses a 5 ms subframe period

★ Configurations 3, 4 and 5 have a 10 ms switching point periodicity. These configurations allow improved coexistence with TD-CDMA which uses a 10 ms radio frame and can be configured to have a single switching point per radio frame

★ The configurations with 10 ms switching point periodicity have an increased number of subframes allocated to the downlink so are more appropriate for downlink dominated traffic profiles

★ Within the special subframes, the duration of the DwPTS, GP and UpPTS fields can be selected to suite the cell range and any coexistence requirements. The set of allowed values is presented in Table 13. The special subframe configuration used by a cell is broadcast in System Information Block (SIB) 1 on the BCCH

★ The DwPTS durations are defined in terms of downlink OFDMA symbols (see section 3.6) whereas the UpPTS durations are defined in terms of uplink SC-FDMA symbols (see section 11.6)

★ The headings in Table 13 use the terms: normal cyclic prefix and extended cyclic prefix. The choice of cyclic prefix has an impact upon the uplink and downlink symbol durations. Section 3.7 describes the use of cyclic prefixes in greater detail

Special Subframe Configuration	Normal Cyclic Prefix in Downlink				Extended Cyclic Prefix in Downlink			
	DwPTS	GP		UpPTS	DwPTS	GP		UpPTS
		Normal Cyclic Prefix in Uplink	Extended Cyclic Prefix in Uplink			Normal Cyclic Prefix in Uplink	Extended Cyclic Prefix in Uplink	
0	3 sym	714 μsec	702 μsec	1 sym	3 sym	679 μsec	667 μsec	1 sym
1	9 sym	285 μsec	273 μsec		8 sym	262 μsec	250 μsec	
2	10 sym	214 μsec	202 μsec		9 sym	179 μsec	167 μsec	
3	11 sym	143 μsec	131 μsec		10 sym	95 μsec	83 μsec	
4	12 sym	71 μsec	59 μsec		3 sym	607 μsec	583 μsec	2 sym
5	3 sym	643 μsec	619 μsec	2 sym	8 sym	191 μsec	167 μsec	
6	9 sym	214 μsec	190 μsec		9 sym	107 μsec	83 μsec	
7	10 sym	143 μsec	119 μsec					
8	11 sym	71 μsec	47 μsec					

Table 13 – Special subframe configurations

★ The Guard Period (GP) is necessary to accommodate:

- the round trip time propagation delay between the UE and eNode B
- the UE switching delay when changing from reception to transmission
- the BTS switching delay when changing from reception to transmission

★ The uplink and downlink subframe timing at the UE and eNode B is illustrated in Figure 17. This example assumes a UE at cell edge, i.e. at the maximum allowed distance from the eNode B

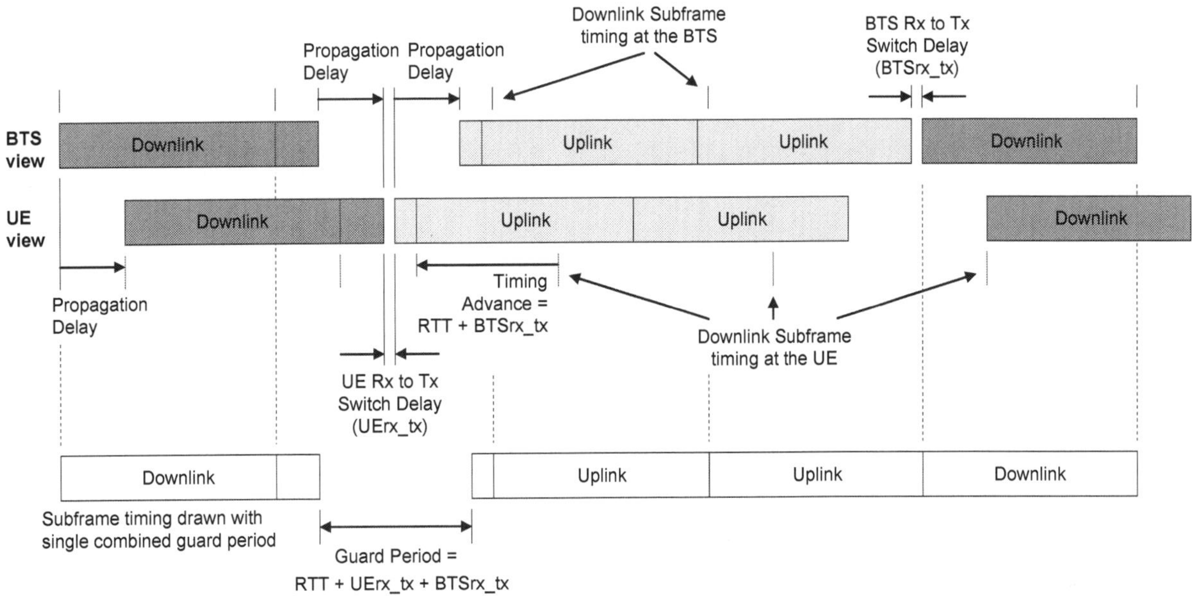

Figure 17 – Uplink and downlink subframe timing for TDD at the UE and eNode B

★ Downlink radio frames are received by the UE after experiencing the propagation delay. Timing Advance determines the point in time that the UE starts to transmit in the uplink direction. Timing Advance is relative to the downlink frame timing received by the UE. Timing Advance is described in section 24.2

★ Timing Advance cannot instruct the UE to start transmitting immediately after the downlink transmission has finished. It must allow some time for the UE to switch from receiving to transmitting. This typically requires 10 to 40 microseconds

★ Uplink radio frames are received by the eNode B after experiencing the propagation delay. The Timing Advance instruction must be large enough to ensure that there is time for the eNode B to switch from receiving to transmitting after the uplink transmission has finished. This typically requires 10 to 20 microseconds

★ The maximum cell range is determined by the duration of the Guard Period after subtracting the UE and eNode B switching times

★ Table 14 presents the maximum cell range associated with each special subframe configuration assuming a total of 40 microseconds for both the UE and eNode B switching times

Special Subframe Configuration	Normal Cyclic Prefix in Downlink		Extended Cyclic Prefix in Downlink	
	Normal Cyclic Prefix in Uplink	Extended Cyclic Prefix in Uplink	Normal Cyclic Prefix in Uplink	Extended Cyclic Prefix in Uplink
0	101 km	99 km	96 km	94 km
1	37 km	35 km	33 km	32 km
2	26 km	24 km	21 km	19 km
3	15 km	14 km	8 km	7 km
4	5 km	3 km	85 km	82 km
5	90 km	87 km	23 km	19 km
6	26 km	23 km	10 km	7 km
7	15 km	12 km		
8	5 km	1 km		

Table 14 – Maximum cell range associated with each special subframe configuration

★ The maximum cell range supported by the set of special subframe configurations is 101 km when using the normal cyclic prefix in both the uplink and downlink, and 94 km when using the extended cyclic prefix in both the uplink and downlink

★ In addition to accounting for the cell range, the special subframe configuration should be selected according to the requirements for coexistence with any neighbouring TDD technologies. Figure 18 illustrates an example for coexistence with TD-SCDMA

★ TD-SCDMA subframes include 7 time slots of 675 μs, a DwPTS of 75 μs, a guard period of 75 μs and a UpPTS of 125 μs. This generates the total subframe period of 5 ms. The example illustrated in Figure 18 assumes a 4:3 configuration where 4 time slots are used for downlink transmission and 3 time slots are used for uplink transmission

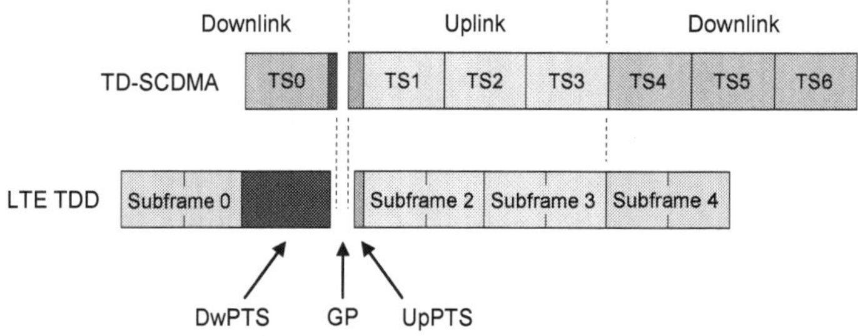

Figure 18 – Example coexistence of the TD-SCDMA and LTE TDD frame structures

★ The LTE TDD half frame includes 5 subframes of 1ms with the DwPTS, guard period and UpPTS accommodated within subframe 1. Selecting uplink-downlink configuration 1 provides support for 2 uplink subframes. Then assuming the normal cyclic prefix in both the uplink and downlink directions, special subframe configuration 2 provides support for:

- a DwPTS of 10 symbols (715 μs)
- a guard period of 214 μs
- a UpPTS of 1 symbol (71 μs)

★ These TD-SCDMA and LTE TDD figures mean that the two technologies can be synchronised to ensure that their uplink and downlink transmissions do not coincide. Interference levels are likely to increase if one technology transmits while the other receives

★ In the case of TDD, neighbouring eNode B have to be frame synchronised to ensure that their uplink and downlink transmissions do not coincide

★ 3GPP References: TS 36.211

3.3 RESOURCE BLOCKS

- ★ A Resource Block represents the basic unit of resource for the LTE air-interface. The eNode B scheduler allocates Resource Blocks to UE when allowing data transfer. Resource Blocks are specified within 3GPP TS 36.211
- ★ A Resource Block is defined in both the time and frequency domains. Its structure is illustrated in Figure 19
- ★ A Resource Block occupies a single 0.5 ms slot in the time domain. The 0.5 ms slot is divided into 7 OFDMA symbols when using the normal cyclic prefix, or 6 OFDMA symbols when using the extended cyclic prefix. Fewer symbols can be accommodated when using the extended cyclic prefix because the cyclic prefix occupies a greater percentage of the 0.5 ms slot
- ★ A Resource Block has 12 subcarriers which occupy 12 × 15 kHz = 180 kHz in the frequency domain
- ★ There is an additional Resource Block configuration which uses the extended cyclic prefix and has 3 symbols in the time domain. A 7.5 kHz subcarrier spacing is used so there are 24 subcarriers within the 180 kHz bandwidth. This configuration is used for Multimedia Broadcast Multicast Services (MBMS)

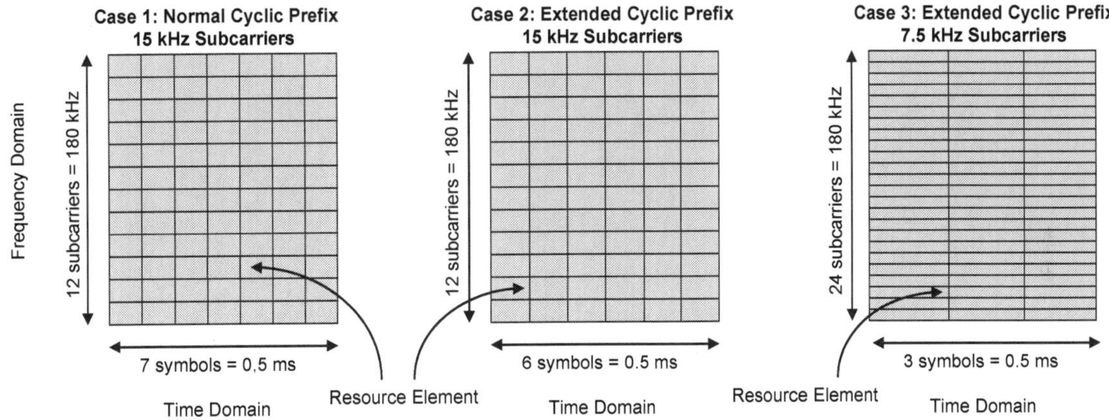

Figure 19 – Resource Blocks using normal and extended cyclic prefix

- ★ The grid generated by subcarriers in the frequency domain and symbols in the time domain defines a set of Resource Elements
- ★ A single Resource Block has 84 Resource Elements when using the normal cyclic prefix, and 72 Resource Elements when using the extended cyclic prefix
- ★ A single Resource Element can accommodate a single modulation symbol
- ★ A single Resource Block provides a modulation symbol rate of 168 ksps when using the normal cyclic prefix and 144 ksps when using the extended cyclic prefix
- ★ The number of Resource Blocks in the frequency domain depends upon the channel bandwidth. This dependency is presented in Table 15 (subcarrier figures assume the 15 kHz subcarrier spacing)

Channel Bandwidth	1.4 MHz	3 MHz	5 MHz	10 MHz	15 MHz	20 MHz
Resource Blocks in the frequency domain	6	15	25	50	75	100
Subcarriers in the frequency domain	72	180	300	600	900	1200
Total Subcarrier Bandwidth (MHz)	1.095	2.715	4.515	9.015	13.515	18.015
Normal cyclic Prefix Modulation symbol rate (Msps)	1.008	2.52	4.2	8.4	12.6	16.8
Extended cyclic Prefix Modulation symbol rate (Msps)	0.864	2.16	3.6	7.2	10.8	14.4

Table 15 – Resource Block characteristics as a function of the channel bandwidth

- ★ The total subcarrier bandwidth includes a 'null' subcarrier at the center of the channel bandwidth. This subcarrier is also known as a 'Direct Current' (DC) subcarrier because it is at 0 Hz when the OFDMA signal is generated at baseband
- ★ These modulation symbol rates allow an initial estimate of the LTE maximum throughput capability, e.g. 16.8 Msps corresponds to 100.8 Mbps when using 64QAM. This increases to 403.2 Mbps when 64QAM is combined with 4×4 MIMO
- ★ However, these bit rate figures assume that all Resource Elements are available for transferring data. In practice, not all Resource Elements are available for user plane data, i.e. there are control plane overheads
- ★ 3GPP References: TS 36.211

3.4 MODULATION

★ A different modulation scheme can be applied to each Resource Element, e.g. one Resource Element can be a QPSK modulation symbol while the adjacent Resource Element can be a 64QAM modulation symbol

★ Resource Elements can be allocated to either a Physical Channel or a Physical Signal

★ In the case of Physical Channels, the modulation scheme applied to a specific Resource Element depends upon the type of Physical Channel to which the Resource Element has been allocated. Table 16 presents the relationship between Physical Channel and modulation scheme

Physical Channel	Modulation Scheme	Bits per Modulation Symbol
PBCH	QPSK	2
PCFICH	QPSK	2
PDCCH	QPSK	2
PHICH	BPSK	1
PDSCH	QPSK, 16QAM, 64QAM	2, 4, 6
PMCH	QPSK, 16QAM, 64QAM	2, 4, 6

Table 16 – Modulation schemes applied to each downlink Physical Channel

★ The PDSCH modulation scheme depends upon the RF channel conditions and the quantity of data to be transferred. The modulation scheme is selected by the eNode B link adaptation

★ The PMCH modulation scheme depends upon the throughput requirements of the broadcast service

★ 64QAM provides the greatest throughput by generating a single modulation symbol from a group of 6 bits. 64QAM requires good signal to noise ratio conditions at the receiver to avoid misinterpreting one 64QAM symbol for another

★ Bits are mapped onto the modulation symbols using Gray coding. This approach helps to minimise bit errors by mapping the bits such that neighbouring modulation symbols differ by only a single bit. If the receiver misinterprets a modulation symbol for its neighbour then only a single bit error is introduced. The concept of Gray coding for 16QAM is illustrated in Figure 20. The first two bits identify the quadrant while the second two bits identify the location within the quadrant

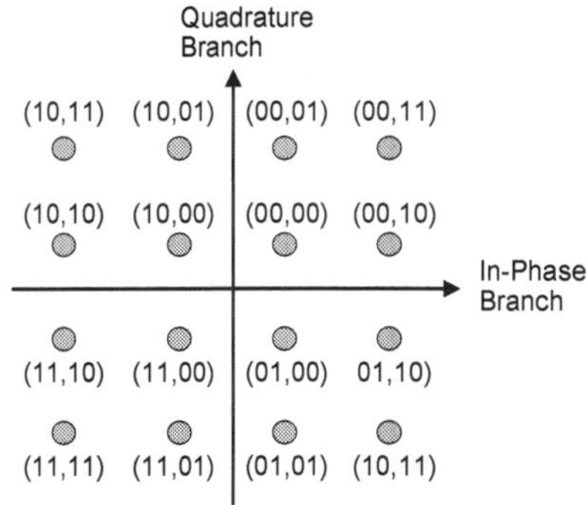

Figure 20 – Modulation constellation for 16QAM

★ In the case of Physical Signals, modulation is not necessary because the signal itself is a series of complex numbers which can be mapped directly onto the appropriate set of Resource Elements

★ 3GPP References: TS 36.211

3.5 OFDMA SIGNAL GENERATION

★ The set of modulated Physical Channels and Physical Signals are mapped onto the grid of Resource Elements

★ As shown in Figure 21, the transmitter processes one column of Resource Elements at a time

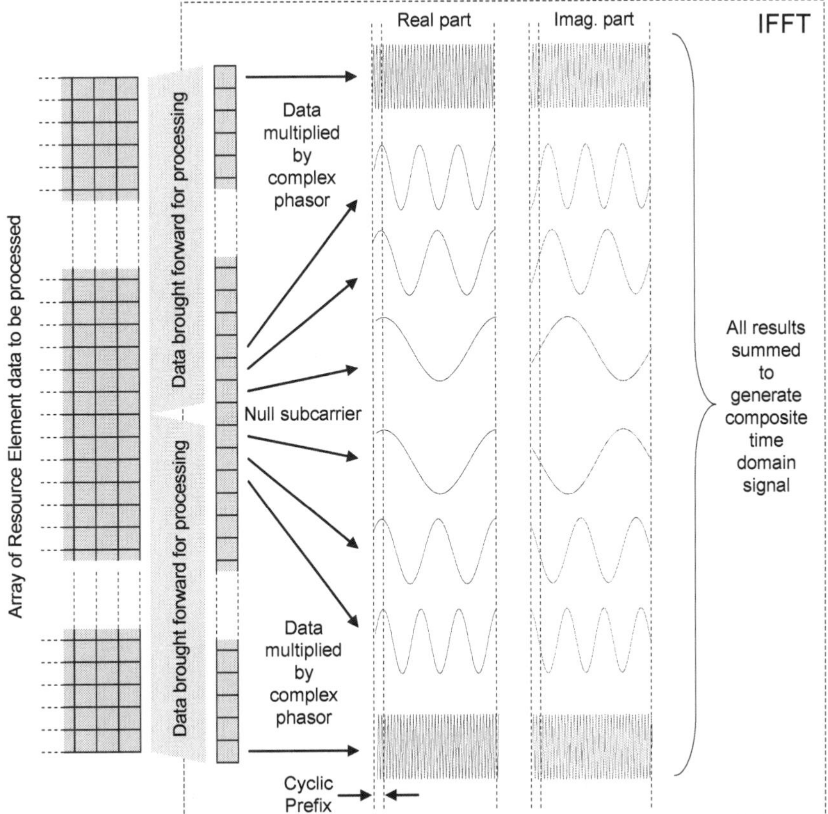

Figure 21 – Generation of the downlink OFDMA signal

★ The downlink signal is generated at baseband (centered around 0 Hz) rather than at RF

★ The column of Resource Elements is split into two halves. The top half is multiplied by a series of positive frequencies (complex phasors rotating in an anti-clockwise direction) while the bottom half is multiplied by a series of negative frequencies (complex phasors rotating in a clockwise direction). The concept of complex phasors with positive and negative frequencies is shown in Figure 22

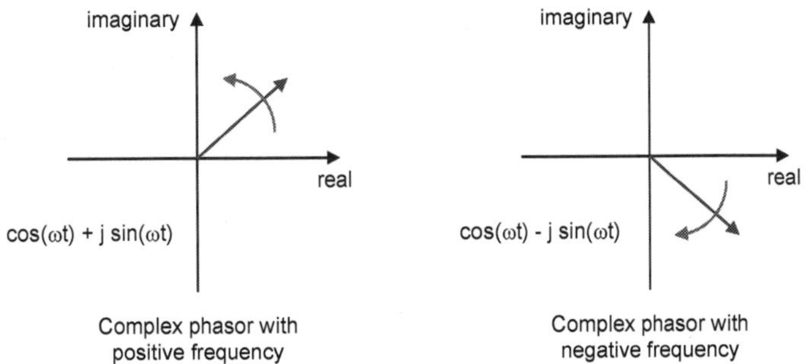

Figure 22 – Complex phasors with positive and negative frequencies

★ The first Resource Element from the top half (counting upwards) is multiplied by a complex phasor rotating at +15 kHz (1 × subcarrier spacing). The second Resource Element of the top half is multiplied by a complex phasor rotating at +30 kHz (2 × subcarrier spacing). Figure 23 illustrates the concept of multiplying a Resource Element by a complex phasor

★ The first Resource Element of the bottom half (counting downwards) is multiplied by a complex phasor rotating at -15 kHz (-1 × subcarrier spacing). The second Resource Element of the bottom half is multiplied by a complex phasor rotating at -30 kHz (-2 × subcarrier spacing)

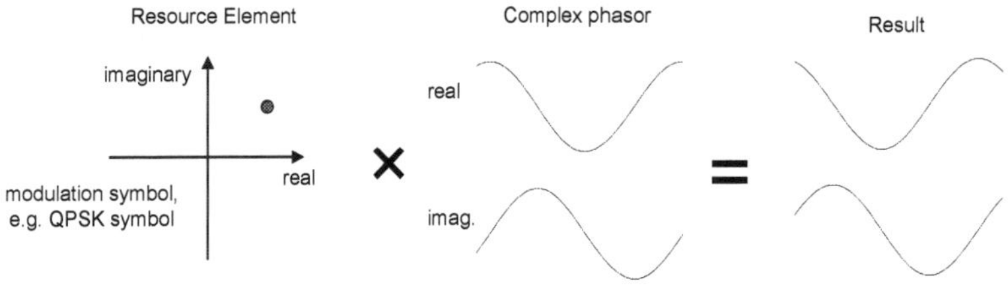

Resource Element | Complex phasor | Result

imaginary, real — modulation symbol, e.g. QPSK symbol × real / imag. =

Figure 23 – Multiplication between Resource Element and complex phasor

- ★ The null subcarrier is generated between the top and bottom halves by having a 30 kHz separation between the first subcarrier in the top half and the first subcarrier in the bottom half
- ★ When using the 7.5 kHz subcarrier spacing for Multimedia Broadcast Multicast Services (MBMS) the complex phasors rotate at +7.5 kHz for the first Resource Element in the upper half, + 15 kHz for the second Resource Element in the upper half, etc. The null subcarrier is generated between the upper and lower halves by having a 15 kHz separation
- ★ The duration of the multiplication corresponds to the duration of an OFDMA symbol, and can be divided into 2 parts:
  - a relatively short first part which corresponds to the cyclic prefix:

    Normal Cyclic Prefix
    - $160 / 3.072 \times 10^7 = 5.2$ µs in the first OFDMA symbol of a 0.5 ms slot
    - $144 / 3.072 \times 10^7 = 4.7$ µs in the remaining OFDMA symbols of a 0.5 ms slot

    Extended Cyclic Prefix (15 kHz subcarrier spacing)
    - $512 / 3.072 \times 10^7 = 16.7$ µs in all OFDMA symbols of a 0.5 ms slot

    Extended Cyclic Prefix (7.5 kHz subcarrier spacing)
    - $1024 / 3.072 \times 10^7 = 33.3$ µs in all OFDMA symbols of a 0.5 ms slot
  - a longer second part which corresponds to the main body of the OFDMA symbol

    Normal and Extended Cyclic Prefix (15 kHz subcarrier spacing)
    - $2048 / 3.072 \times 10^7 = 66.67$ µs

    Extended Cyclic Prefix (7.5 kHz subcarrier spacing)
    - $4096 / 3.072 \times 10^7 = 133.33$ µs
- ★ The longer second part always includes an integer number of cycles of the complex phasor, i.e. 15 kHz × 66.67 µs = 1, 30 kHz × 66.67 µs = 2, etc (or in the case of the 7.5 kHz subcarrier spacing, 7.5 kHz × 133.33 µs = 1, 15 kHz × 133.33 µs = 2, etc)
- ★ Once the complete column of Resource Elements has been multiplied by the appropriate complex phasors, the results are summed to generate the OFDMA symbol, i.e. the OFDMA symbol is a sum of complex phasors
- ★ High Peak to Average Power Ratios (PAPR) can be generated if a relatively large number of the complex phasors peak simultaneously
- ★ 3GPP TS 36.211 expresses the above series of operations mathematically using the equation below. This corresponds to the Inverse Fast Fourier Transform (IFFT)

$$s_l(t) = \sum_{k=-\lfloor 12\times N_{RB}^{DL}/2 \rfloor}^{-1} a_{k,l} \times e^{j2\pi k\times 15000\times(t-N_{CP}T_s)} + \sum_{k=1}^{\lceil 12\times N_{RB}^{DL}/2 \rceil} a_{k,l} \times e^{j2\pi k\times 15000\times(t-N_{CP}T_s)}$$

$$0 \le t < (N_{CP} + 2048)\times T_s$$

- ★ 3GPP uses the variable T_s as a minimum unit of time.
  - the main body of an OFDMA symbol includes 2048 samples when sampling at a rate defined by T_s = 66.67 µs / 2048 = 32.55 ns
  - these figures are applicable to the 15 kHz subcarrier spacing. The 7.5 kHz subcarrier spacing allows 4096 samples from the main body of an OFDMA symbol when sampling at a rate defined by T_s
- ★ Sampling at a rate defined by T_s is applicable to the 20 MHz channel bandwidth, in which case the maximum subcarrier frequency is 15 kHz × 12 subcarriers per resource block × 50 resource blocks = 9 MHz. The 9 MHz subcarrier completes 600 cycles within the main body of an OFDMA symbol (number of cycles = the number of subcarriers, i.e. 12 subcarriers per resource block × 50 resource blocks). This means that when 2048 samples are taken from an OFDMA symbol there are at least 3 samples per cycle of the 9 MHz subcarrier
- ★ Lower sampling rates can be used for smaller channel bandwidths

3.6 OFDMA SYMBOL

- There are 7 OFDMA symbols per slot when using the normal Cyclic Prefix, and either 6 or 3 OFDMA symbols per slot when using the extended Cyclic Prefix. The case of 3 OFDMA symbols per slot is applicable when using the 7.5 kHz subcarrier spacing for Multimedia Broadcast Multicast Services (MBMS). These slot structures are illustrated in Figure 24
- Each OFDMA symbol has a Cyclic Prefix followed by the main body of the symbol
- When using the normal Cyclic Prefix, the duration of the first prefix within a slot is 160 × Ts, whereas the duration of subsequent prefixes within a slot is 144 × Ts

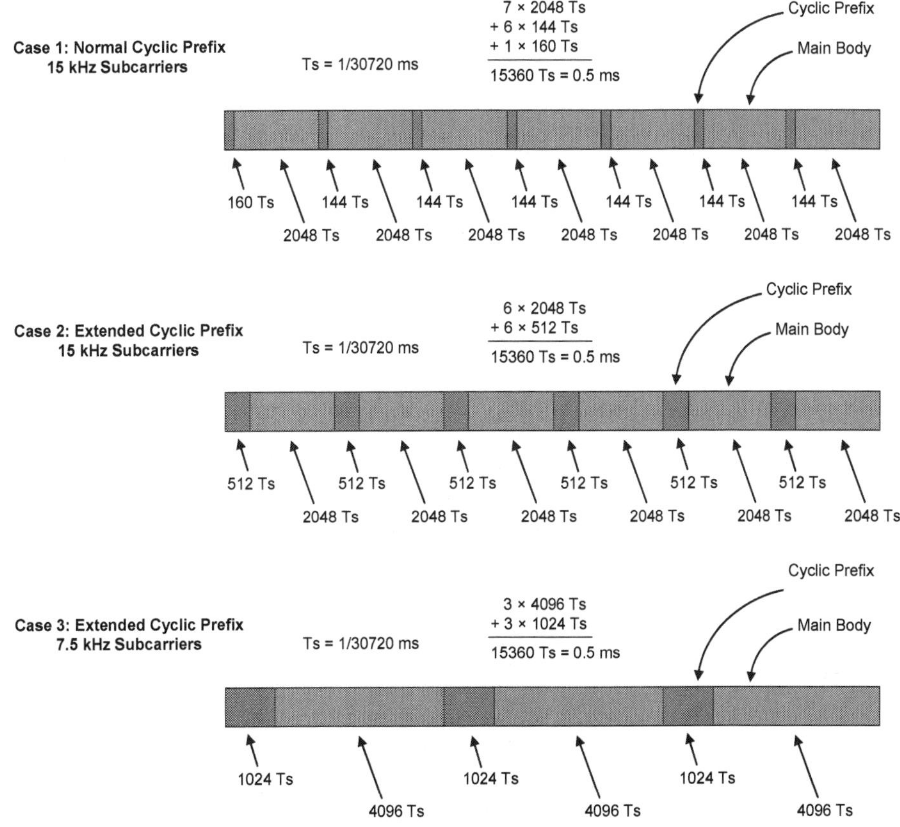

Figure 24 – Structure of the symbols within a 0.5 ms slot

- When using the extended Cyclic Prefix with the 15 kHz subcarrier spacing the duration of the prefix is always 512 × Ts
- When using the 15 kHz subcarrier spacing the duration of the main body of the symbol is always 2048 × Ts = 66.67 µs
- 66.67 µs is equal to the period of a 15 kHz cosine, i.e. 1/15000 = 66.67 µs
- The value of 2048 has been selected to ensure that the sampling rate is a multiple of 3.84 MHz. This allows re-use of the rates used for UMTS signal processing
- A sampling rate of 8 × 3.84 = 30.72 MHz is applicable to the 20 MHz channel bandwidth. A sampling rate of 30.72 MHz generates 30.72 × 66.67 = 2048 samples within the main body of the OFDMA symbol
- Lower sampling rates can be used for the smaller channel bandwidths, e.g. the 5 MHz channel bandwidth can use a sampling rate of 2 × 3.84 = 7.68 MHz to generate 512 samples within the main body of an OFDMA symbol
- The number of samples within the OFDMA symbol defines the size of the Fast Fourier Transform (FFT) at the receiver. The set of FFT sizes is presented in Table 17

Channel Bandwidth	1.4 MHz	3 MHz	5 MHz	10 MHz	15 MHz	20 MHz
Sampling Rate (Mega samples per second)	0.5 × 3.84 = 1.92	1 × 3.84 = 3.84	2 × 3.84 = 7.68	4 × 3.84 = 15.36	6 × 3.84 = 23.04	8 × 3.84 = 30.72
Samples within main body of symbol	128	256	512	1024	1536	2048
FFT Size	128	256	512	1024	1536	2048

Table 17 – Sampling rates and FFT sizes as a function of the channel bandwidth

3.7 CYCLIC PREFIX

★ Delay spread is generated by the set of different paths between the transmitter and receiver when those paths have different delays. For example, a signal following a direct line-of-sight path would arrive before a different version of the same signal which is reflected by a distant building

★ Time domain receivers typically synchronise with each delay spread component and adjust their individual timings before combining. When using a Rake receiver, each finger belonging to the Rake synchronises with a specific delay spread component. The number of delay spread components which can be combined is then limited by the number of Rake fingers. Any delay spread components which are not combined appear as interference

★ LTE receivers do not need to synchronise with individual delay spread components, i.e. it is not necessary to adjust the timing of delay spread components, nor is it necessary to do any combining of delay spread components. An LTE receiver can operate directly on the aggregate received signal without considering delay spread components

★ The Cyclic Prefix represents a guard period at the start of each OFDMA symbol which provides protection against multi-path delay spread. The cyclic prefix also represents an overhead which should be minimised

★ The duration of the cyclic prefix should be greater than the duration of the multi-path delay spread

★ LTE specifies both normal and extended cyclic prefix lengths. The normal cyclic prefix is intended to be sufficient for the majority of scenarios, while the extended cyclic prefix is intended for scenarios with particularly high delay spread. Durations for the normal and extended cyclic prefix are presented in Table 18

	Normal Cyclic Prefix		Extended Cyclic Prefix	
	15 kHz subcarriers		15 kHz subcarriers	7.5 kHz subcarriers
	160 Ts	144 Ts	512 Ts	1024 Ts
Duration	5.2 μs	4.7 μs	16.7 μs	33.3 μs
Equivalent Distance	1.6 km	1.4 km	5 km	10 km
Overhead	160 / 2048 = 7.8 %	144 / 2048 = 7.0 %	512 / 2048 = 25 %	1024 / 4096 = 25 %

Table 18 – Cyclic prefix lengths for the downlink of LTE

★ The cyclic prefix is generated by copying the end of the main body of the OFDMA symbol. This is shown in Figure 25

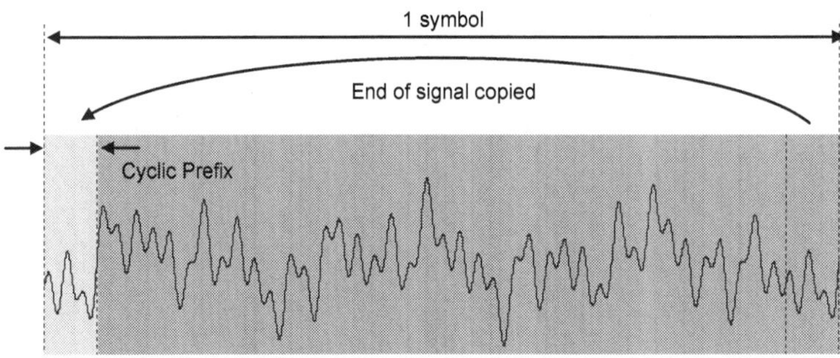

Figure 25 – Generating the cyclic prefix

★ The signal is always continuous at the interface between the cyclic prefix and the main body of the symbol. This results from the main body of the symbol always including an integer number of subcarrier cycles, i.e. 1 cycle at 15 kHz, 2 cycles at 30 kHz, etc

★ Figure 26 illustrates an example of 2 delay spread components. The second delay spread component is received later than the first delay spread component. A Fast Fourier Transform (FFT) processing window is defined at the receiver:

- o the processing window captures the main body of the OFDMA symbol belonging to the first delay spread component. The cyclic prefix belonging to the first delay spread component is discarded
- o the processing window captures part of the cyclic prefix and the majority of the main body of the OFDMA symbol belonging to the second delay spread component. Sections of the cyclic prefix and main body of the OFDMA symbol which fall outside the processing window are discarded

★ In the extreme case, where the delay spread is equal to the duration of the cyclic prefix then the FFT processing window fully captures the cyclic prefix belonging to the delay spread component and discards a section of the main body of the ODFMA symbol which has a duration equal to the cyclic prefix

★ The time domain representation of each delay spread component within the processing window is different (as shown in Figure 26). However, the frequency domain representation of each delay spread component within the processing window is identical

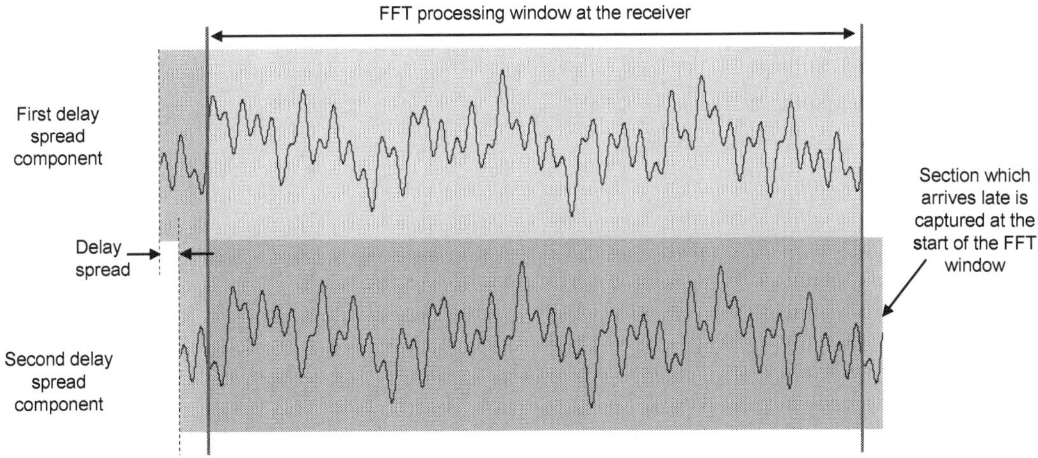

Figure 26 – Delay spread components captured by FFT processing window at the receiver

★ Moving a section of the time domain signal from the end, and adding it to the start does not change the frequency content of the signal, i.e. the signal includes the same set of frequency domain components and an FFT (which quantifies the set of frequency domain components included within a time domain signal) generates the same result

★ As long as the delay spread is less than the duration of the cyclic prefix, each delay spread component provides a complete representation of the signal within the FFT processing window, i.e. the same set of frequency components are generated by the FFT. This avoids the requirement to time synchronise with individual delay spread components prior to decoding

★ Figure 27 illustrates the aggregate signal captured by the receiver. The aggregate signal is the sum of all the delay spread components (only 2 are shown for simplicity). The receiver can operate directly on the aggregate signal without having to extract individual delay spread components because the frequency content of the sum of delay spread components, is the same as the frequency content of each individual delay spread component. The aggregate signal captures the energy from all delay spread components so generates a higher quality result from the FFT

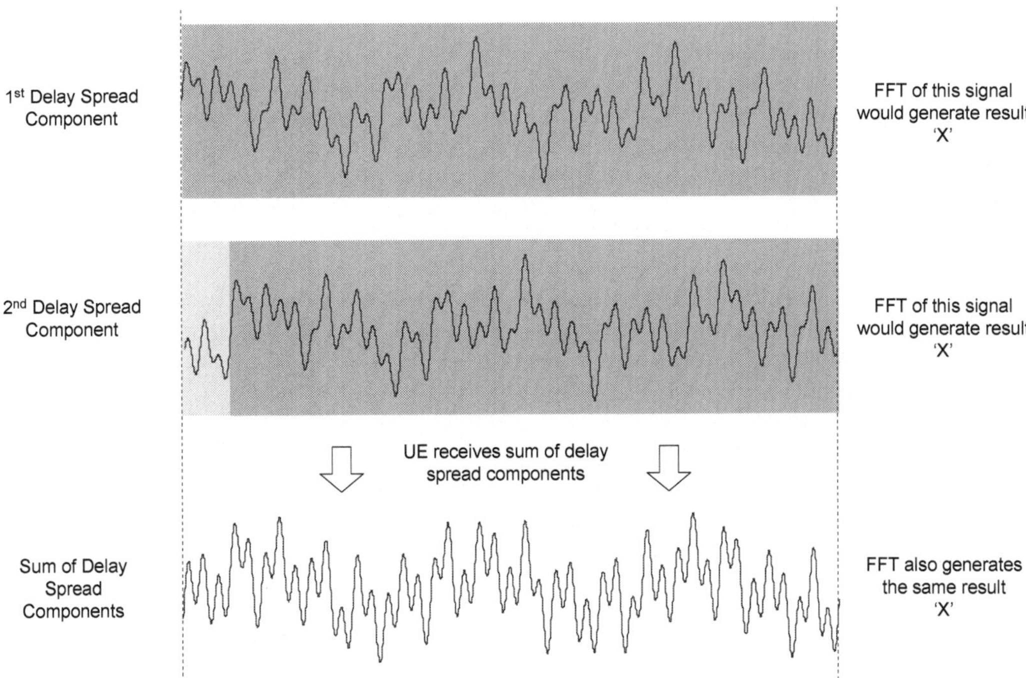

Figure 27 – Delay spread components captured by FFT processing window at the receiver

★ The equation shown in section 3.5 from 3GPP TS 36.211 includes the generation of the cyclic prefix. The time variable 't' accounts for N_{CP} samples which correspond to the cyclic prefix

3.8 WINDOWING

- ★ 3GPP has not explicitly specified the use of windowing for LTE
- ★ Equipment vendors may introduce windowing to help achieve the specified spectrum emission requirements
- ★ As shown in Figure 28, the transmit signal is likely to be discontinuous at the boundary between two OFDMA symbols

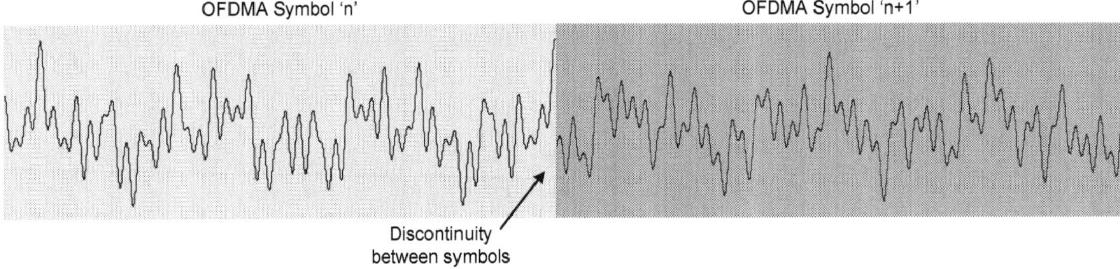

Figure 28 – Discontinuity between two adjacent OFDMA symbols

- ★ Discontinuities generate increased spectrum emissions outside the channel bandwidth
- ★ The impact of discontinuities can be reduced by using a windowing function to smooth the signal at inter-symbol boundaries. An example windowing function is illustrated in Figure 29

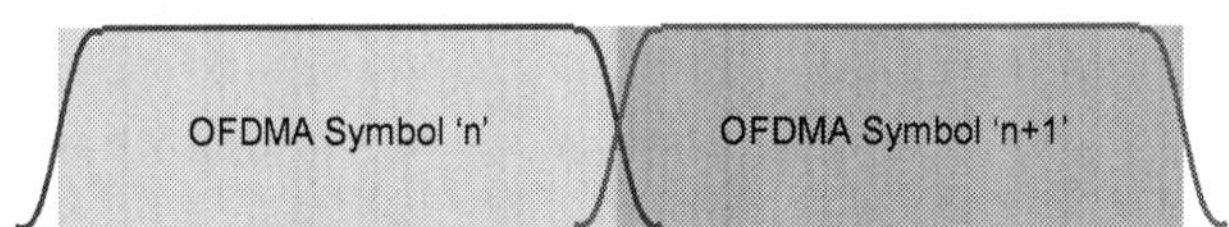

Figure 29 – Windowing used to smooth discontinuities at inter-symbol boundaries

- ★ Windowing represents a form of filtering which reduces the levels of unwanted spectrum emissions but also reduces the effectiveness of the cyclic prefix because the signal is distorted to some extent

3.9 INVERSE FOURIER TRANSFORM

★ In practice, OFDMA symbols are generated at the transmitter using the Inverse Fast Fourier Transform (IFFT)

★ The Inverse Fourier Transform (IFT) generates a continuous time domain signal from a continuous frequency domain signal:

$$f(t) = \frac{1}{2\pi} \int_{-\infty}^{+\infty} F(\omega) \times e^{j\omega t} d\omega$$

★ The IFT sums a series of complex phasors to generate an output signal

★ Digital signal processing is based upon sampled signals rather than continuous signals so a non-continuous version of the IFT is required

★ The Inverse Discrete Fourier Transform (IDFT) generates a sampled time domain signal from a sampled frequency domain signal:

$$f_n = \frac{1}{N} \sum_{k=0}^{N-1} F_k e^{j\frac{2\pi}{N}kn}$$

★ The Inverse Fast Fourier Transform (IFFT) provides the same translation as the IDFT but in a more computationally efficient manner

★ Use of the IFFT helps to reduce the baseband processing requirements and the associated power consumption

3.10 FOURIER TRANSFORM

★ OFDMA symbols are processed at the receiver using the Fast Fourier Transform (FFT)

★ The Fourier Transform (FT) generates a continuous frequency domain signal from a continuous time domain signal:

$$F(\omega) = \int_{-\infty}^{+\infty} f(t) \times e^{-j\omega t} dt$$

★ The FT is a correlator which compares a series of complex phasors with the input signal to determine whether or not those phasors are present within the signal

- o if the signal contains the frequency of the phasor then the FT generates a result at that frequency
- o otherwise the FT generates a zero at that frequency

★ Digital signal processing is based upon sampled signals rather than continuous signals so a non-continuous version of the FT is required

★ The Discrete Fourier Transform (DFT) generates a sampled frequency domain signal from a sampled time domain signal:

$$F_k = \sum_{n=0}^{N-1} f_n e^{-j\frac{2\pi}{N}kn}$$

★ The Fast Fourier Transform (FFT) provides the same translation as the DFT but in a more computationally efficient manner

★ Use of the FFT helps to reduce the baseband processing requirements and the associated power consumption

3.11 TRANSMITTER AND RECEIVER CHAIN

- The transmitter and receiver chain is illustrated in Figure 30
- The functions provided by the receiver are the inverse of those provided by the transmitter

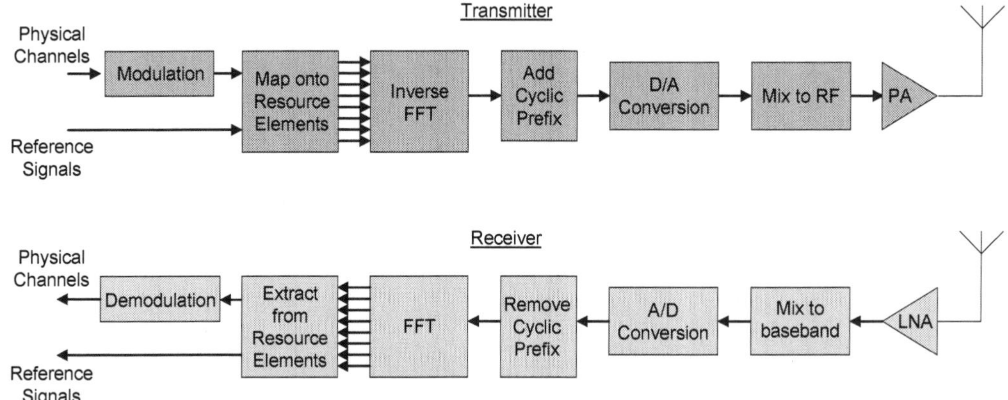

Figure 30 – Transmitter and receiver chain for OFDMA

- The transmitter includes the Inverse Fourier transform to translate each column of Resource Elements into an OFDMA symbol
- A cyclic prefix is added at the transmitter and removed at the receiver
- The receiver includes the Fourier Transform to translate each OFDMA symbol into a column of Resource Elements. This is illustrated in Figure 31

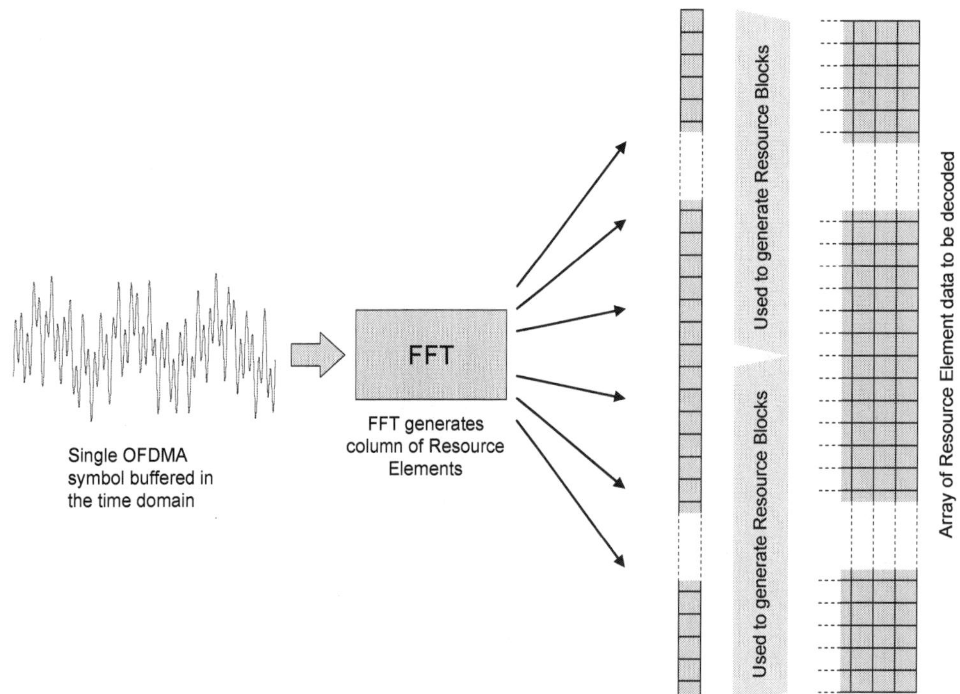

Figure 31 – Translation of received signal into Resource Elements using an FFT

- The transmitter includes a Power Amplifier (PA) to increase the signal strength to a level which can be transmitted across the target coverage area
- The receiver includes a Low Noise Amplifier (LNA) to help improve the received signal to noise ratio

4 DOWNLINK MULTIPLE ANTENNA TECHNOLOGIES

4.1 ANTENNA PORTS

★ 3GPP uses the concept of 'antenna ports'. It is important to differentiate between the concept of 'antenna ports' and physical antenna elements. Antenna ports map onto physical antenna elements

★ A downlink antenna port is defined by its associated Reference Signal. For example, antenna port 0 is associated with a cell specific Reference Signal, whereas antenna port 6 is associated with a positioning Reference Signal

★ The complete set of downlink antenna ports and their associated Reference Signals are presented in Table 19

Antenna Port	3GPP Release	Reference Signals	Application
0 to 3	8	Cell specific Reference Signals	Single stream transmission, transmit diversity, MIMO
4	8	MBSFN Reference Signal	Multimedia Broadcast Multicast Services (MBMS)
5	8	UE specific Reference Signal	Beamforming without MIMO
6	9	Positioning Reference Signal	Location based services
7 to 8	9	UE specific Reference Signals	Beamforming with MIMO; multi-user MIMO
9 to 14	10	UE specific Reference Signals	Beamforming with MIMO; multi-user MIMO
15 to 22	10	CSI Reference Signals	Channel State Information (CSI) reporting

Table 19 – Antenna ports and their associated Reference Signals

★ In some cases, there is a one-to-one mapping between antenna port and physical antenna element. This is the case when a single cross-polar antenna is used for downlink 2×2 MIMO, or downlink transmit diversity. In this case, antenna port 0 is mapped onto one physical antenna element, while antenna port 1 is mapped onto the other physical antenna element. This scenario is illustrated in Figure 32

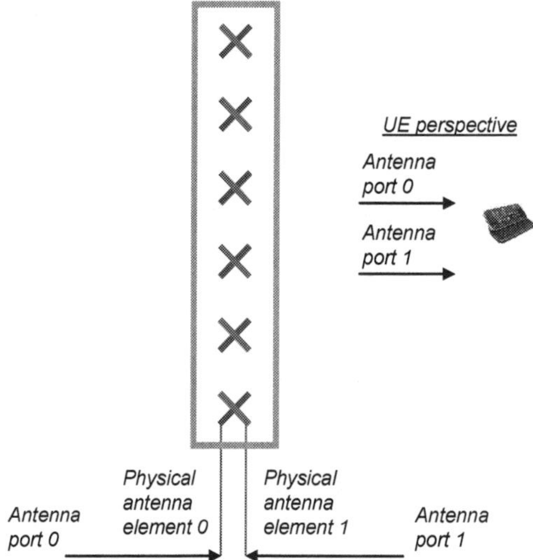

Figure 32 – Example of one-to-one mapping between antenna port and physical antenna elements

★ From the UE perspective, there are two downlink transmissions – one associated with the cell specific Reference Signal for antenna port 0, and another associated with the cell specific Reference Signal for antenna port 1

★ In other cases, a single antenna port is mapped onto multiple physical antenna elements. This approach is used for beamforming. Support for beamforming was introduced within the release 8 version of the specifications using antenna port 5. Beamforming uses multiple physical antenna elements to direct the downlink transmission towards a specific UE. This is typically achieved using an antenna array consisting of multiple columns of cross-polar antenna elements. This scenario is illustrated in Figure 33. Beamforming principles are described in section 33.6

★ The example shown in Figure 33 illustrates an antenna array with 8 physical antenna elements (4 columns of cross-polar pairs). Antenna port 5 is mapped onto all 8 of the physical antenna elements.

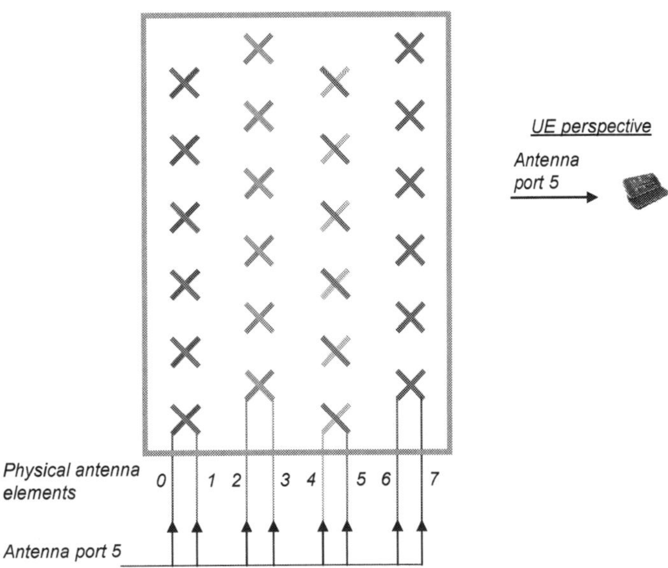

Figure 33 – Example of mapping 1 antenna port onto multiple physical antenna elements

★ From the UE perspective, there is a single downlink transmission originating from antenna port 5, and associated with the UE specific Reference Signal for antenna port 5

★ Antenna ports can be viewed as being virtual because they represent the downlink transmission from the UE perspective rather than the actual downlink transmission from physical antenna elements at the eNode B

★ Another example of the mapping between antenna ports and physical antenna elements is illustrated in Figure 34. This example is based upon an antenna array being used for downlink 2×2 MIMO, or downlink transmit diversity, i.e. the same transmission schemes as assumed for the first example in Figure 32

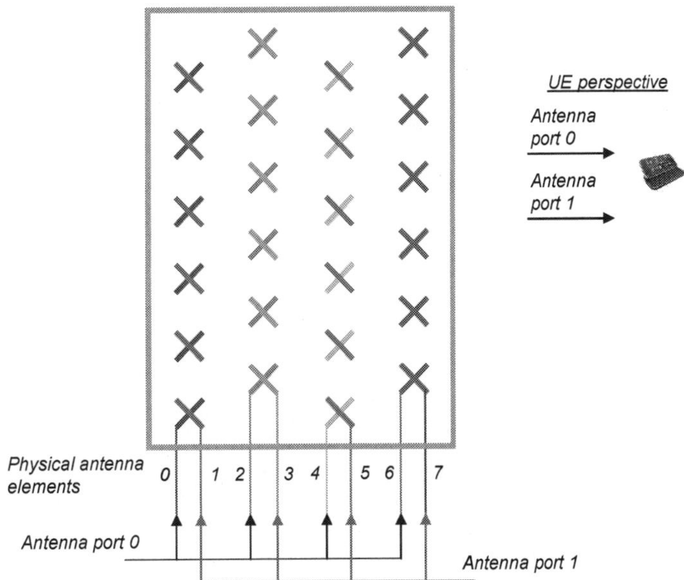

Figure 34 – Example of mapping two antenna ports onto multiple physical antenna elements

★ In this case, antenna ports 0 and 1 are each mapped onto multiple physical antenna elements. This demonstrates that the mapping between antenna ports and physical antenna elements depends upon the type of antenna being used. From the UE perspective, there are the same two downlink transmissions and no knowledge of the antenna type is required

4.2 TRANSMISSION MODES

- ★ The set of downlink transmission modes is specified within 3GPP TS 36.213:
  - o release 8 version introduced transmission modes 1 to 7
  - o release 9 version introduced transmission mode 8
  - o release 10 version introduced transmission mode 9
- ★ The eNode B signals the downlink transmission mode to the UE within an RRC Connection Setup, RRC Connection Reconfiguration or RRC Connection Re-establishment message
- ★ The set of downlink transmission modes is summarised in Table 20. These are applicable when the UE is addressed by its C-RNTI. A different set of transmission modes is defined for when the UE is addressed by its SPS-RNTI (see section 27.6.2)
- ★ The majority of transmission modes allow the eNode B to switch between transmission techniques without RRC signalling, e.g. transmission mode 3 allows the eNode B to dynamically switch between transmit diversity and open loop spatial multiplexing
- ★ Downlink Control Information (DCI) transferred on the PDCCH is used to signal resource allocations to the UE. Different DCI formats are used for different transmission techniques, e.g. a UE configured to use transmission mode 3 can expect to receive DCI format 1A when being instructed to use transmit diversity, and can expect to receive DCI format 2A when being instructed to use open loop spatial multiplexing. The content of each DCI format is described in section 9
- ★ The Search Space defines the set of PDCCH within which the UE checks for a resource allocation. This avoids the UE having to decode all PDCCH. Search Spaces are described in section 9.2

Mode	PDSCH Transmission Scheme	DCI Format	Search Space	Channel State Information Feedback from UE
1	Single Antenna Port, port 0	1A	Common and UE Specific	CQI
		1	UE Specific	
2	Transmit Diversity	1A	Common and UE Specific	CQI
		1	UE Specific	
3	Transmit Diversity	1A	Common and UE Specific	CQI, RI
	Open Loop Spatial Multiplexing, or Transmit Diversity	2A	UE Specific	
4	Transmit Diversity	1A	Common and UE Specific	CQI, RI, PMI
	Closed Loop Spatial Multiplexing, or Transmit Diversity	2	UE Specific	
5	Transmit Diversity	1A	Common and UE Specific	CQI, PMI
	Multi-User MIMO	1D	UE Specific	
6	Transmit Diversity	1A	Common and UE Specific	CQI, PMI
	Closed Loop Spatial Multiplexing using a Single Transmission Layer	1B	UE Specific	
7	Single Antenna Port, port 0, or Transmit diversity	1A	Common and UE Specific	CQI
	Single Antenna Port, port 5	1	UE Specific	
8	Single Antenna Port, port 0, or Transmit diversity	1A	Common and UE Specific	CQI (PMI & RI if instructed by eNode B)
	Dual Layer Transmission, port 7 and 8, or Single Antenna Port, port 7 or 8	2B	UE Specific	
9	Single Antenna Port, port 0, or Transmit diversity	1A	Common and UE Specific	CQI (PTI, PMI & RI if instructed by eNode B)
	Up to 8 Layer Transmission, ports 7-14	2C	UE Specific	

Table 20 – PDSCH transmission modes when using C-RNTI to address the UE

- ★ Transmission mode 1 provides support for single antenna port transmission. DCI formats 1 and 1A offer the standard and compact approaches to scheduling downlink Resource Blocks. Precoding is not applied to the downlink data and the UE is only required to provide feedback in terms of Channel Quality Indicators (CQI). CQI reports help the eNode B to schedule an appropriate bit rate for the current channel conditions
- ★ Transmission mode 2 provides support for transmit diversity. 3GPP has only specified open loop transmit diversity for LTE so the UE is only required to provide feedback in terms of CQI. Transmit diversity can be configured to use either 2 or 4 transmit antenna ports. In both cases, a single transport block is transferred during each subframe

★ Transmission mode 3 provides support for open loop spatial multiplexing. Transmission mode 3 allows the eNode B to dynamically switch between transmit diversity and open loop spatial multiplexing without having to use RRC signalling to complete a reconfiguration. The eNode B switches between these two transmission schemes according to the channel conditions. DCI format 2A is used to allocate resources for open loop spatial multiplexing. This DCI allows information to be signalled for either 1 or 2 transport blocks. It also allows precoding information to be included for 4×4 open loop spatial multiplexing. This precoding information signals the number of layers being used rather than the set of precoding weights. The UE is required to provide feedback in terms of both CQI and Rank Indication (RI). The RI provides the eNode B with a recommended number of layers. In this case, spatial multiplexing is categorised as open loop because the UE does not need to provide feedback in terms of a Precoding Matrix Indicator (PMI)

★ Transmission mode 4 provides support for closed loop spatial multiplexing. Transmission mode 4 allows the eNode B to dynamically switch between transmit diversity and closed loop spatial multiplexing without having to use RRC signalling to complete a reconfiguration. The eNode B switches between these two transmission schemes according to the channel conditions. DCI format 2 is used to allocate resources for closed loop spatial multiplexing. This DCI allows information to be signalled for either 1 or 2 transport blocks. It also allows precoding information to be included. The UE is required to provide feedback in terms of CQI, RI and a Precoding Matrix Indicator (PMI). The PMI provides the eNode B with a recommended set of precoding weights

★ Transmission mode 5 provides support for multi-user MIMO but limits the transmission to a single layer per UE (only a single transport block can be sent to each UE during each subframe). Transmission mode 5 allows the eNode B to dynamically switch between transmit diversity and multi-user MIMO without having to use RRC signalling to complete a reconfiguration. DCI format 1D is used to allocate resources for multi-user MIMO. Multi-user MIMO can transfer 2 transport blocks during a subframe and addresses those 2 transport blocks to 2 different UE. The UE is required to provide feedback in terms of CQI and PMI. The PMI can be used by the eNode B when selecting the precoding weights for multi-user MIMO. The UE is not required to report a Rank Indication (RI) because the multi-user MIMO transmission is always single layer

★ Transmission mode 6 provides support for closed loop spatial multiplexing using a single layer. This corresponds to a simplified version of transmission mode 4. Limiting transmission to a single layer removes the requirement for the UE to report a Rank Indication (RI). The UE is still able to signal its preferred precoding weights using the PMI. DCI format 1B is used to allocate the resources

★ Transmission mode 7 provides support for single layer beamforming. This transmission mode uses antenna port 5 to transmit a UE specific Reference Signal. Beamforming can be applied to the UE specific Reference Signal and to the set of allocated Resource Blocks. This directs them towards the appropriate UE. Beamforming cannot be applied to cell specific Reference Signals because they are used by all UE within the cell. UE specific and cell specific Reference Signals use different Resource Elements so both can be transferred within the same Resource Block. Transmission mode 7 can also be used for multi-user MIMO. The same set of Resource Blocks can be re-used for multiple UE by generating multiple antenna beams to isolate their transmissions. The UE only reports CQI for transmission mode 7

★ Transmission mode 8 allows 2 UE to simultaneously benefit from dual layer beamforming, i.e. a total of 4 layers can be transferred. Alternatively, those 4 layers can be used for single layer beamforming towards 4 UE. Dual layer beamforming allows throughputs to be doubled relative to single layer beamforming. Thus, this transmission mode supports a combination of beamforming and multi-user MIMO. Antenna ports 7 and 8, in combination with 2 scrambling identities are used to generate a set of 4 UE specific Reference Signals. The UE reports CQI by default, and the eNode B can request the UE to also report Precoding Matrix Indicators (PMI) and Rank Indicators (RI). PMI and RI are useful when supporting dual layer transmission. DCI format 2B was introduced within the release 9 version of the 3GPP specifications for the purposes of transmission mode 8

★ Transmission mode 9 provides support for single user beamforming with up to 8×8 MIMO. In terms of multi-user MIMO, transmission mode 9 is similar to transmission mode 8, i.e. 2 UE can simultaneously benefit from dual layer beamforming, or 4 UE can simultaneously benefit from single layer beamforming. When configured for single user beamforming antenna ports 7 to 14 are used. When configured for multi-user MIMO antenna ports 7 and 8 are used in combination with 2 scrambling identities. The UE reports CQI by default, and the eNode B can request the UE to also report Precoding Matrix Indicators (PMI) and Rank Indicators (RI). Transmission mode 9 also uses Precoder Type Indications (PTI) to signal the type of content within subsequent PMI reports. DCI format 2C was introduced within the release 10 version of the 3GPP specifications for the purposes of transmission mode 9

★ 3GPP References: TS 36.211, TS 36.212, TS 36.213

4.3 MULTIPLE INPUT MULTIPLE OUTPUT (MIMO)

- Multiple Input Multiple Output (MIMO) configurations benefit from multiple antenna elements at the transmitter and multiple antenna elements at the receiver. MIMO contrasts with receive diversity which only requires multiple antenna at the receiver, and transmit diversity which only requires multiple antenna at the transmitter
- The general concept of MIMO is illustrated in Figure 35. The Multiple Input component refers to multiple transmissions into the propagation channel, whereas the Multiple Output component refers to multiple signals being received out of the propagation channel
- Similar terminology is used for Single Input Multiple Output (SIMO), Multiple Input Single Output (MISO) and Single Input Single output (SISO)

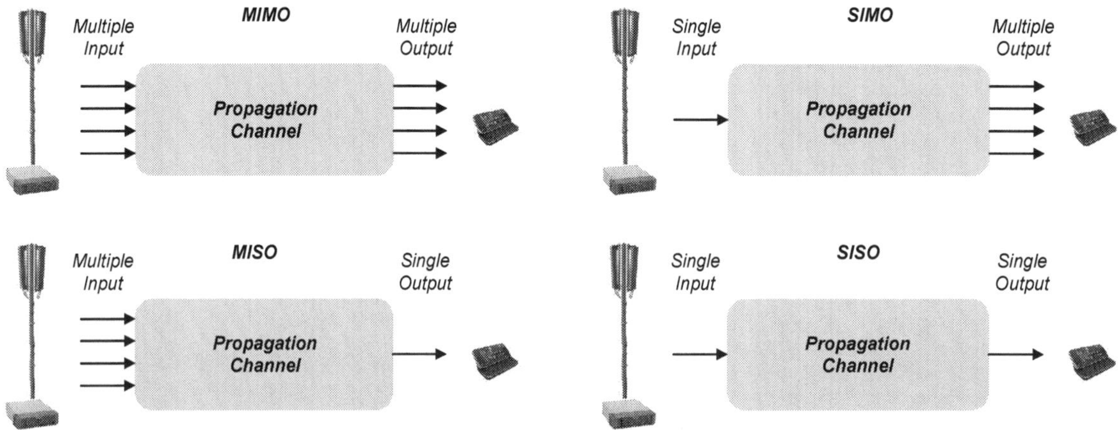

Figure 35 – General concept of MIMO

- The example of MIMO illustrated in Figure 35 assumes 4 antenna ports used for transmission and 4 antenna ports used for reception. This represents 4×4 MIMO. It is also possible to have an unequal number of antenna ports at the transmitter and receiver. Two examples of this are shown in Figure 36

Figure 36 – MIMO with unequal numbers of transmit and receive antenna ports

- The benefits of MIMO are summarised in Figure 37
  - diversity gain reduces the impact of fading when the fades on each propagation path are uncorrelated, i.e. one path may experience a fade while another path may not experience a fade. The receiver takes advantage of the paths which are not experiencing fades
  - array gain is achieved from the beamforming effect which is generated when transmitting from multiple antenna elements. Beamforming directs the transmitted signal towards the UE and improves the received signal to noise ratio
  - spatial multiplexing gain increases throughput by transferring multiple streams of data in parallel using the same set of time and frequency resources. Uncorrelated transmission paths allow the receiver to differentiate between the data streams

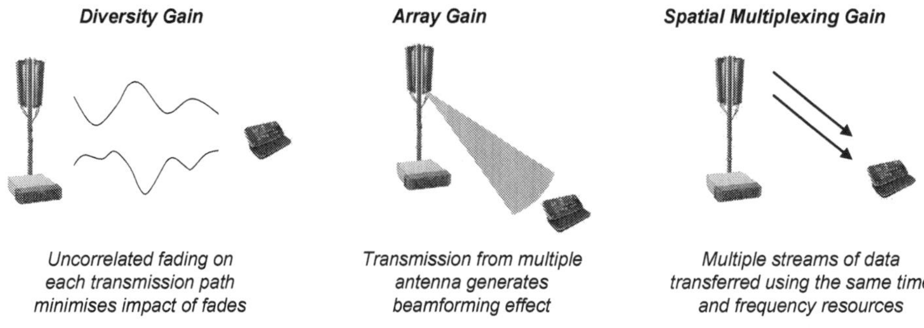

Figure 37 – Benefits of MIMO

★ UE experiencing good coverage (with high signal to noise ratios) can take advantage of the spatial multiplexing gain and can receive multiple parallel streams of data. The maximum number of parallel streams is given by the minimum of the number of transmit and receive antenna. For example, 2×2 MIMO, 4×2 MIMO and 2×4 MIMO are all capable of transferring a maximum of 2 parallel streams of data. Maximising throughput also relies upon having uncorrelated propagation paths between the transmit and receive antenna

★ UE in poor coverage (with a low signal to noise ratios) can take advantage of the diversity gain to help improve their signal to noise ratio. The magnitude of the diversity gain is dependant upon the number of receive antenna and the level of correlation between each of the propagation paths, i.e. the gain is maximised for a large number of receive antenna and uncorrelated propagation paths

★ This dependency upon channel conditions means that MIMO is used to transfer multiple parallel streams of data in good coverage conditions to maximise throughput, and is used to transfer a single stream of data in poor coverage conditions to maximise the diversity gain. These scenarios are illustrated in Figure 38. Both scenarios perform best when the propagation paths between transmit and receive antenna are uncorrelated

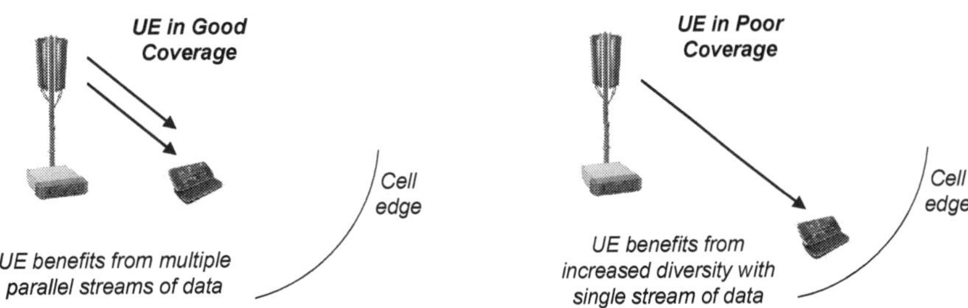

Figure 38 – MIMO scenarios for UE in good and poor coverage

★ The drawbacks of MIMO are its increased implementation complexity and increased hardware requirement. MIMO requires additional processing at both the transmitter and receiver. It also requires additional signalling in terms of feedback from the receiver and resource allocation information from the transmitter. MIMO requires additional power amplifiers at the transmitter and additional receive paths at the receiver. It also requires additional antenna elements at both the transmitter and receiver. These antenna elements may already be available if receive diversity is already in use

★ MIMO is referred to as spatial multiplexing within 3GPP TS 36.211. This specification has separate sections for spatial multiplexing, transmission on a single antenna port and transmit diversity

★ The release 8 and 9 versions of the 3GPP specifications support:

- 2×2 MIMO: 2 transmit antenna ports + 2 receive antenna ports
- 4×4 MIMO: 4 transmit antenna ports + 4 receive antenna ports

★ The release 10 version of the 3GPP specifications supports:

- 8×8 MIMO: 8 transmit antenna ports + 8 receive antenna ports

★ 3GPP has specified the following variants of MIMO:

- open loop spatial multiplexing
- closed loop spatial multiplexing
- multi-user MIMO

★ Open loop spatial multiplexing involves feedback from the UE in terms of Rank Indication (RI) and Channel Quality Indicator (CQI). It is categorised as 'open loop' because the UE is not required to provide feedback in terms of a Precoding Matrix Indicator (PMI)

★ Closed loop spatial multiplexing involves feedback from the UE in terms of RI, CQI and PMI. The UE selects a PMI to maximise the signal to noise ratio at its receiver. Applying the set of precoding weights at the eNode B represents a form of maximum ratio combining at the transmitter

★ MIMO can also be combined with beamforming to direct Resource Block allocations towards specific UE. Increasing the number of antenna elements horizontally allows increased directivity in the horizontal plane

★ MIMO can transfer either 1 or 2 codewords during each 1 ms subframe. A codeword is a transport block which has been processed by the Physical layer in terms of CRC addition, channel coding and rate matching

★ When 2 codewords are transferred, they do not need to be of equal size. CQI reporting, link adaptation and HARQ run independently for each codeword. In some cases, a single HARQ acknowledgement is used for multiple codewords due to the constraints in signalling capacity, e.g. ACK/NACK multiplexing for TDD

★ 4×4 MIMO cannot transfer 4 codewords during a 1 ms subframe, but 2 large codewords can be scheduled and subsequently divided into 4 sections. 8×8 MIMO can transfer 2 even larger codewords which can be divided into 8 sections. A maximum of 2 codewords has been standardised to optimise the trade-off between receiver processing and performance, and to reduce the signalling requirement in terms of CQI and HARQ acknowledgement reporting

- Generating a MIMO signal at the transmitter involves:
  - Layer Mapping – maps the modulated symbols belonging to either 1 or 2 codewords onto a number of 'layers' where the number of layers is less than or equal to the number of antenna ports
  - Precoding – applies coding to the 'layers' of modulated symbols prior to mapping onto Resource Elements and subsequent OFDMA signal generation
- The general structure of layer mapping and precoding is illustrated in Figure 39. The layer mapping and precoding operations are described in greater detail in subsequent sections

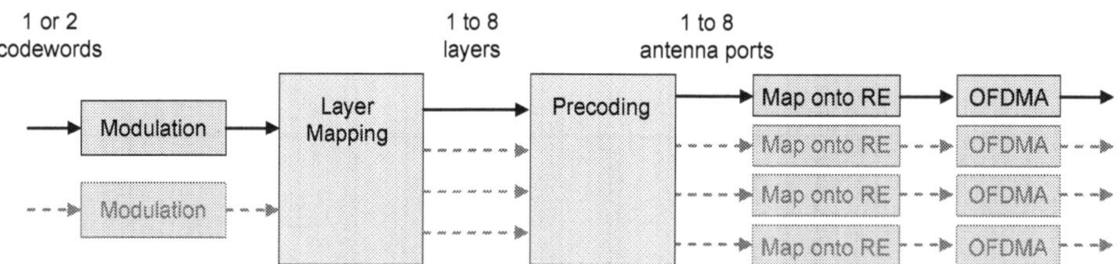

Figure 39 – Concepts of layer mapping and precoding

- 2×2 MIMO is limited to transmitting either 1 or 2 layers from 2 antenna ports
  - using 1 layer allows 1 codeword to be transferred during each 1 ms subframe and is applicable to when RF channel conditions are relatively poor, or when the eNode B does not have much data to transfer
  - using 2 layers allows 2 codewords to be transferred during each 1 ms subframe and is applicable to when RF channel conditions are good. Transferring 2 codewords across 2 layers allows peak connection throughputs to be doubled
- 4×4 MIMO can transmit 1 to 4 layers from 4 antenna ports
  - using 1 layer allows 1 codeword to be transferred during each 1 ms subframe and is applicable to when RF channel conditions are relatively poor, or when the eNode B does not have much data to transfer
  - using 4 layers allows 2 large codewords to be transferred during each 1 ms subframe and is applicable to when RF channel conditions are good. Transferring 2 large codewords across 4 layers allows peak connection throughputs to be increased by a factor of 4
  - an intermediate number of layers can be used when the channel conditions do not allow the maximum number of layers, i.e. the correlation between some of the propagation paths between the transmit antenna and receive antenna is too high
- 8×8 MIMO can transmit 1 to 8 layers from 8 antenna ports
  - using 1 layer allows 1 codeword to be transferred during each 1 ms subframe and is applicable to when RF channel conditions are relatively poor, or when the eNode B does not have much data to transfer
  - using 8 layers allows 2 very large codewords to be transferred during each 1 ms subframe and is applicable to when RF channel conditions are good. Transferring 2 very large codewords across 8 layers allows peak connection throughputs to be increased by a factor of 8
  - an intermediate number of layers can be used when the channel conditions do not allow the maximum number of layers, i.e. the correlation between some of the propagation paths between the transmit antenna and receive antenna is too high
- 3GPP References: TS 36.211, TS 36.213

4.4 LTE TECHNOLOGIES

4.4.1 TRANSMIT DIVERSITY

★ Transmit diversity increases resilience against the propagation channel

★ Transmit diversity requires multiple antenna elements at the transmitter and one or more antenna elements at the receiver

★ Transmit diversity was introduced within the release 8 version of the 3GPP specifications

★ The release 8 version of the 3GPP specifications introduced downlink transmit diversity based upon either 2 or 4 antenna ports at the eNode B

- o antenna ports 0 and 1 are used for 2-branch transmit diversity
- o antenna ports 0, 1, 2 and 3 are used for 4-branch transmit diversity

★ 3GPP has specified open loop transmit diversity so preferred precoding feedback from the UE is not required. The UE provides feedback in terms of Channel Quality Indicators (CQI) to help the eNode B select an appropriate transport block size

★ Transmit diversity transfers a single modulated codeword during each 1 ms subframe

★ After the codeword has been modulated, baseband processing for transmit diversity is completed in 2 phases

- o Layer Mapping – maps the modulated symbols onto a number of 'layers' where the number of layers equals the number of antenna ports (2 or 4)
- o Precoding – codes the modulated symbols belonging to each 'layer' to generate the symbol streams to be transmitted by each antenna port

★ Layer mapping allocates the modulated symbols to each layer in rotation, e.g. in the case of 2 antenna ports, even numbered symbols are mapped onto layer 1 while odd numbered symbols are mapped onto layer 2. Layer mapping is illustrated in Figure 40

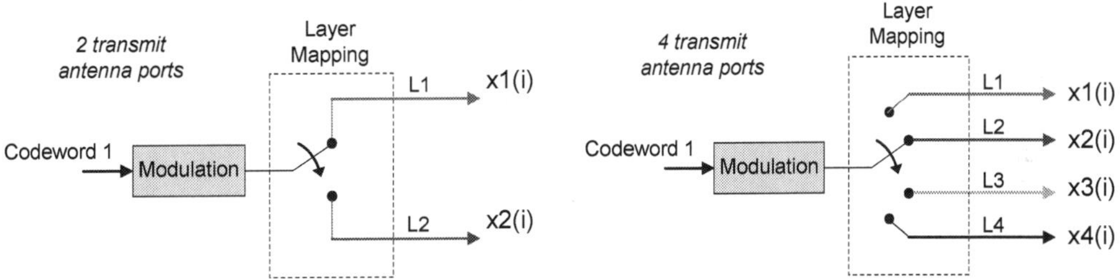

Figure 40 – Layer mapping for transmit diversity

★ When using 2 layers, the symbol rate per layer is one half of the original symbol rate. When using 4 layers, the symbol rate per layer is one quarter of the original symbol rate

★ Precoding for transmit diversity with 2 antenna ports is shown in Figure 41. Precoding generates 2 pairs of outputs for every pair of inputs. This results in a symbol rate per antenna which is equal to the original symbol rate prior to layer mapping

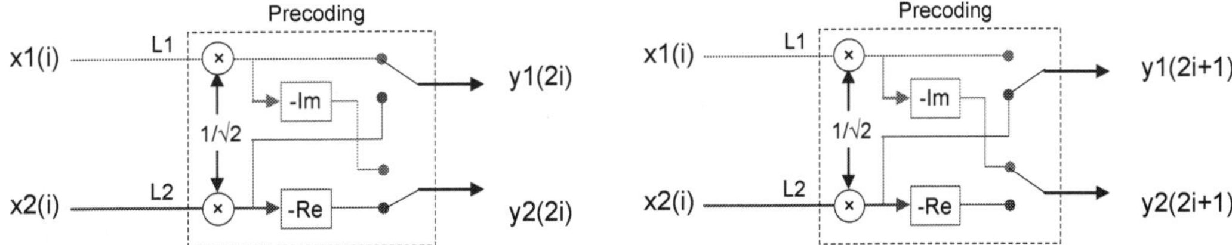

Figure 41 – Precoding for transmit diversity with 2 antenna ports

★ Precoding for 2 antenna ports scales the symbols belonging to each layer by $1/\sqrt{2}$, and

- o generates the symbol stream for antenna port 1 by time multiplexing symbols from each layer (alternating between layers)
- o generates the symbol stream for antenna port 2 by time multiplexing symbols from each layer after taking the complex conjugate of the symbols from layer 1 and the negative of the complex conjugate of the symbols from layer 2

★ The precoding process is fixed and non-adaptive as a result of transmit diversity for LTE being open loop

★ Precoding for transmit diversity with 4 antenna ports is shown in Figure 42. Precoding generates 4 sets of outputs for every set of inputs. This results in a symbol rate per antenna which is equal to the original symbol rate prior to layer mapping

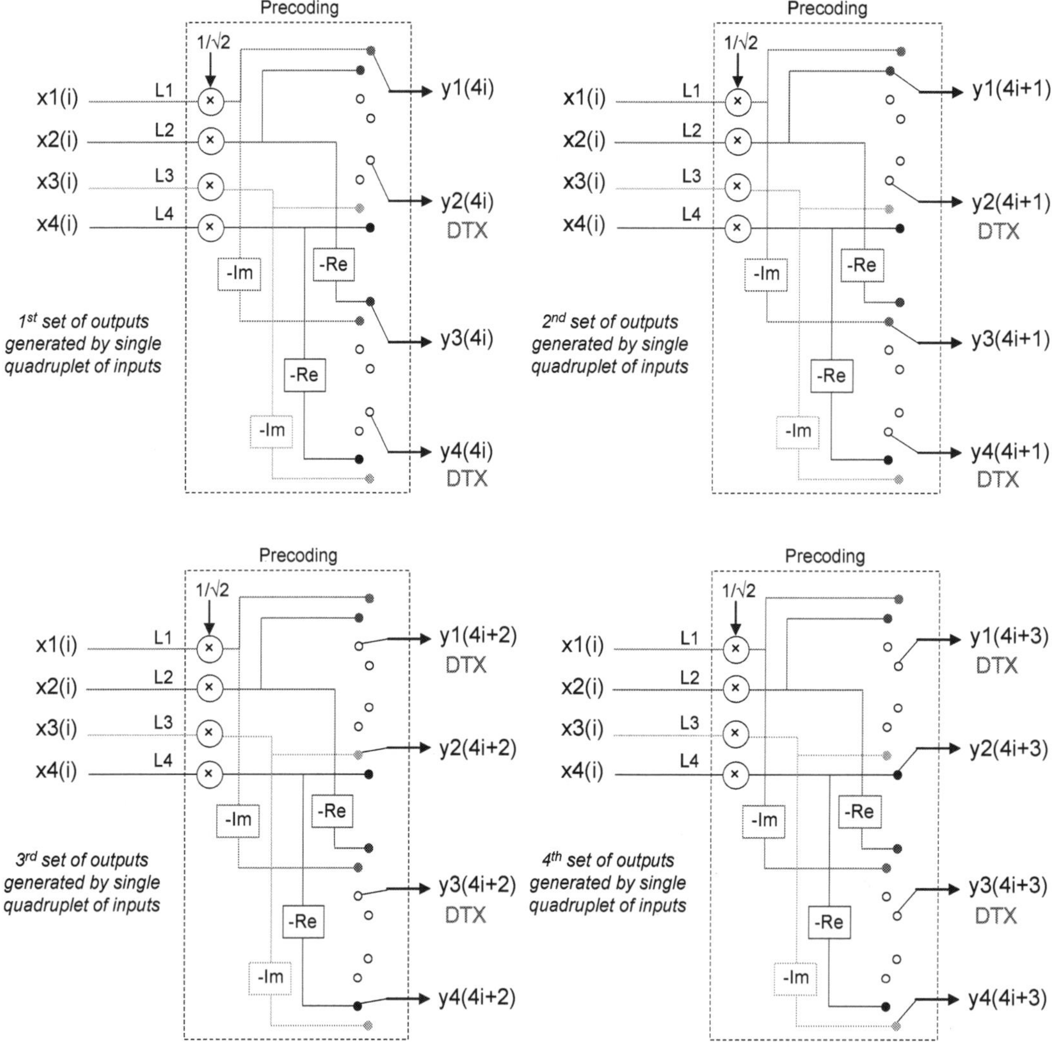

Figure 42 – Precoding for transmit diversity with 4 antenna ports

★ Precoding for 4 antenna elements scales the symbols belonging to each layer by 1/√2, and

- generates the symbol stream for antenna 1 by time multiplexing symbols from layers 1 and 2, with periods of DTX every 3rd and 4th symbol
- generates the symbol stream for antenna 2 by time multiplexing symbols from layers 3 and 4, with periods of DTX every 1st and 2nd symbol
- generates the symbol stream for antenna 3 by time multiplexing symbols from layer 2 after taking the negative of the complex conjugate, with symbols from layer 1 after taking the complex conjugate, with periods of DTX every 3rd and 4th symbol
- generates the symbol stream for antenna 4 by time multiplexing symbols from layer 4 after taking the negative of the complex conjugate, with symbols from layer 3 after taking the complex conjugate, with periods of DTX every 1st and 2nd symbol

★ The use of DTX for transmit diversity with 4 antenna means that antenna ports 1 and 3 use a different set of Resource Elements to antenna ports 2 and 4, i.e. Resource Elements which are occupied on antenna ports 1 and 3 are in DTX on antenna ports 2 and 4, and vice versa. This allows an increased transmit power to be applied to the Resource Elements which are occupied, i.e. the total downlink transmit power capability is shared across fewer Resource Elements

★ Figure 43 illustrates the output from precoding being mapped onto the set of Resource Elements before being used to generate the downlink OFDMA signal. Each antenna port has its own Resource Element grid

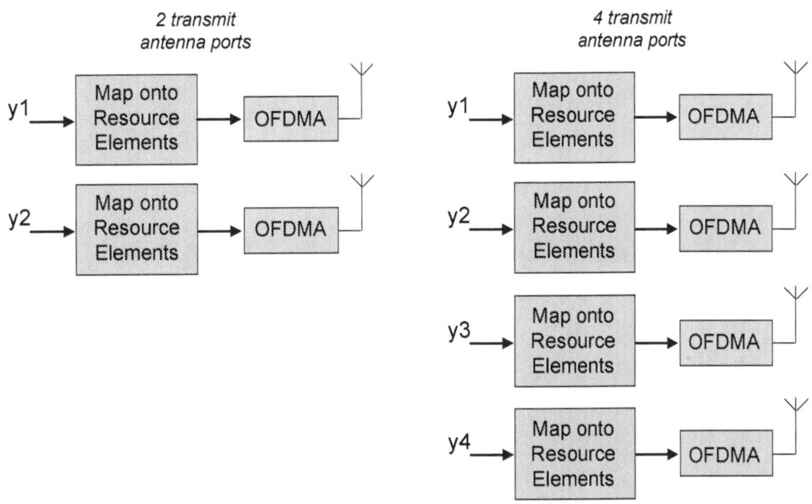

Figure 43 – Mapping onto Resource Elements and OFDMA signal generation after layer mapping and precoding

★ The LTE transmit diversity scheme for 2 antenna ports is known as a Space-Frequency Block Code (SFBC) approach

- the term 'space' is used because each modulated symbol is transmitted from multiple antenna elements, and those antenna elements have a physical separation
- the term 'frequency' is used because each modulated symbol is mapped onto multiple Resource Elements, and those Resource Elements have a frequency separation, i.e. they use different subcarriers

★ Figure 44 illustrates the flow of two example modulation symbols through the processing for transmit diversity with 2 antenna ports, i.e. using the SFBC approach

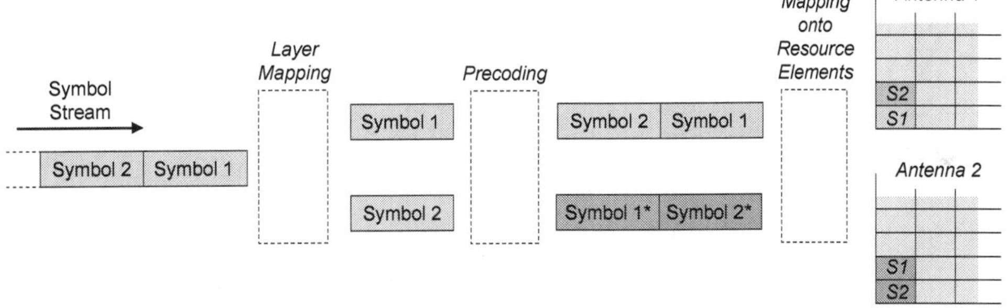

Figure 44 – Flow of two example modulation symbols through processing for transmit diversity

★ The LTE transmit diversity scheme for 4 antenna elements is known as a Space-Frequency Block Code - Frequency Shift Transmit Diversity (SFBC-FSTD) approach

★ Transmit diversity can be applied to the PDSCH, PBCH, PCFICH, PDCCH and PHICH physical channels

★ 3GPP References: TS 36.211, TS 36.213

4.4.2 OPEN LOOP SPATIAL MULTIPLEXING

★ Open loop spatial multiplexing was introduced within the release 8 version of the 3GPP specifications. It has not changed within the release 9 and 10 versions of the specifications

★ Open loop spatial multiplexing is applicable to transmission mode 3

★ Open loop spatial multiplexing relies upon the UE providing feedback in terms of:

- Rank Indication (RI) – suggested number of 'layers'
- Channel Quality Indicator (CQI) – suggested transport block size

★ This approach is categorised as 'open loop' because the UE is not required to provide feedback in terms of a Precoding Matrix Indicator (PMI)

★ 3GPP has specified Large Delay Cyclic Delay Diversity (CDD) as a solution for open loop spatial multiplexing

★ Precoding for Large Delay CDD is defined using the equation:

$$\begin{bmatrix} y^{(0)}(i) \\ \vdots \\ y^{(P-1)}(i) \end{bmatrix} = W(i) \times D(i) \times U \times \begin{bmatrix} x^{(0)}(i) \\ \vdots \\ x^{(v-1)}(i) \end{bmatrix}$$

where, P is the number of output antenna ports, i is the sample number and v is the number of input layers

★ The Cyclic Delay Diversity itself is generated using the D(i) component of this equation. In the case of 4 layers, 3GPP specifies D(i) as:

$$D(i) = \begin{bmatrix} 1 & 0 & 0 & 0 \\ 0 & e^{-j(\pi/2)i} & 0 & 0 \\ 0 & 0 & e^{-j(\pi)i} & 0 \\ 0 & 0 & 0 & e^{-j(3\pi/2)i} \end{bmatrix}$$

This matrix is in the same format as the identity matrix (identity matrix has a leading diagonal of all '1's and does not change the input after multiplication). However, the 2nd, 3rd and 4th leading diagonal elements are phasors with unity amplitude and a phase which increases for every input sample. Thus, the 2nd, 3rd and 4th rows of the input are multiplied by rotating phasors

★ The 1st leading diagonal element can be viewed as a phasor which does not rotate, so based upon the D(i) above:
- 1st phasor rotates by 0° for every input sample
- 2nd phasor rotates by 90° for every input sample
- 3rd phasor rotates by 180° for every input sample
- 4th phasor rotates by 270° for every input sample

★ If a time domain signal is multiplied by a rotating phasor then the frequency domain components of that signal are shifted

★ In the case of precoding for open loop spatial multiplexing, the rotating phasors are applied in the frequency domain (prior to the IFFT used to generate the time domain OFDMA signal). Applying the rotating phasors in the frequency domain causes the time domain components of the signal to be shifted, i.e. delay is introduced

★ The 4 phasors associated with D(i) shown above have different rates of rotation so each input signal experiences a different delay. The delay increases as the phasor's rate of rotation increases

★ An example of this increasing delay is illustrated in Figure 45 which shows a series of time histories after an input signal has been multiplied by a phasor in the frequency domain. Each time history corresponds to a phasor with a different rate of rotation

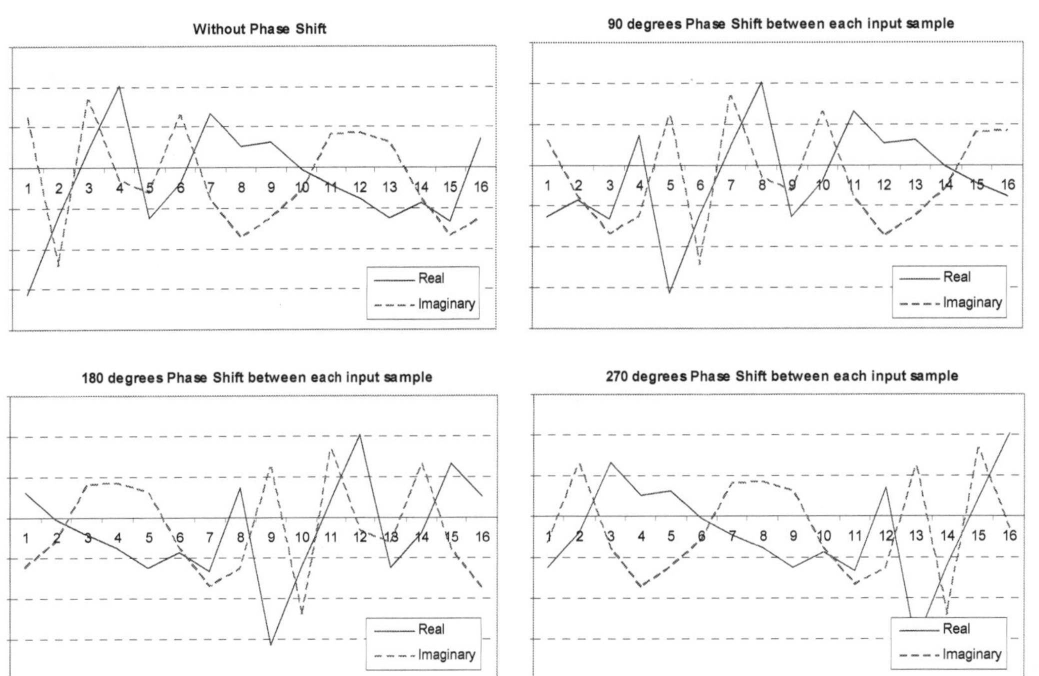

Figure 45 – Time domain result from IFFT after input signal has had a cyclic phase shift applied

- Precoding for open loop spatial multiplexing is referred to as 'Large Delay' CDD because the maximum possible delays are applied, i.e.
  - when there are 2 input layers, the phasors rotate 0° and 180° between input samples
  - when there are 3 input layers, the phasors rotate 0°, 120° and 240° between input samples
  - when there are 4 input layers, the phasors rotate 0°, 90°, 180° and 270° between input samples
- Introducing these delays generates diversity across the transmitted signals. The input samples belonging to each layer are multiplied by the 'U' matrix prior to having the CDD applied. The 'U' matrix ensures that each input provided to the CDD includes a combination of the original input layers. This means that each antenna port transmits a combination of all layers but with different time delays
- The 'U' matrix specified by 3GPP for 4 layers is:

$$U = \frac{1}{2}\begin{bmatrix} 1 & 1 & 1 & 1 \\ 1 & e^{-j(\pi/2)} & e^{-j(2\pi/2)} & e^{-j(3\pi/2)} \\ 1 & e^{-j(2\pi/2)} & e^{-j(4\pi/2)} & e^{-j(6\pi/2)} \\ 1 & e^{-j(3\pi/2)} & e^{-j(6\pi/2)} & e^{-j(9\pi/2)} \end{bmatrix}$$

The elements within this matrix are not a function of the sample number 'i'. They apply a fixed rotation to each input layer and do not generate a time delay in the same way as the D(i) matrix

- The product of the 'U' matrix with the input samples belonging to each layer is shown below. All elements within the 'U' matrix are non-zero so every row of the result includes a combination of input samples from all layers

$$U \times \begin{bmatrix} x^0(i) \\ x^1(i) \\ x^2(i) \\ x3(i) \end{bmatrix} = \frac{1}{2}\begin{bmatrix} x^0(i) + x^1(i) + x^2(i) + x^3(i) \\ x^0(i) + x^1(i)e^{-j(\pi/2)} + x^2(i)e^{-j(2\pi/2)} + x^3(i)e^{-j(3\pi/2)} \\ x^0(i) + x^1(i)e^{-j(2\pi/2)} + x^2(i)e^{-j(4\pi/2)} + x^3(i)e^{-j(6\pi/2)} \\ x^0(i) + x^1(i)e^{-j(3\pi/2)} + x^2(i)e^{-j(6\pi/2)} + x^3(i)e^{-j(9\pi/2)} \end{bmatrix}$$

- The W(i) matrix provides additional precoding dependent upon the number of antenna ports. This component of the precoding is described in the following sections

2×2 OPEN LOOP SPATIAL MULTIPLEXING

- 2×2 open loop spatial multiplexing always transfers 2 codewords during each subframe. Transmit diversity (section 4.4.1) provides the 2×2 open loop solution for transferring a single codeword
- UE can be switched between 2×2 transmit diversity and 2×2 open loop spatial multiplexing depending upon coverage conditions:
  - transmit diversity can be used to transfer 1 codeword when coverage is poor
  - open loop spatial multiplexing can be used to transfer 2 codewords when coverage is good
- 4×4 open loop spatial multiplexing can transfer either 1 or 2 codewords during each subframe
- Table 21 summarises the number of codewords and layers which are supported by 2×2 open loop spatial multiplexing

Number of Codewords	Number of Layers	Number of Antenna Ports
2	2	2

Table 21 – Codewords, layers and antenna ports for 2×2 open loop spatial multiplexing

- In the case of 2×2 open loop spatial multiplexing, 3GPP specifies the W(i) matrix as:

$$W(i) = \frac{1}{\sqrt{2}}\begin{bmatrix} 1 & 0 \\ 0 & 1 \end{bmatrix}$$

in this case, W(i) is fixed and is not actually a function of 'i'. The matrix section is the identity matrix so does not change the result when W(i) is multiplied with D(i), U and the input samples from each layer. The product of W(i), D(i) and U becomes:

$$W(i) \times D(i) \times U = \frac{1}{\sqrt{2}}\begin{bmatrix} 1 & 0 \\ 0 & 1 \end{bmatrix} \times \begin{bmatrix} 1 & 0 \\ 0 & e^{-j(\pi)i} \end{bmatrix} \times \frac{1}{\sqrt{2}}\begin{bmatrix} 1 & 1 \\ 1 & e^{-j(\pi)} \end{bmatrix} = \frac{1}{2}\begin{bmatrix} 1 & 1 \\ e^{-j(\pi)i} & e^{-j\pi(i+1)} \end{bmatrix}$$

This generates results of:

$$W(i) \times D(i) \times U = \frac{1}{2}\begin{bmatrix} 1 & 1 \\ 1 & -1 \end{bmatrix} \quad \text{when 'i' is even}$$

$$W(i) \times D(i) \times U = \frac{1}{2}\begin{bmatrix} 1 & 1 \\ -1 & 1 \end{bmatrix} \quad \text{when 'i' is odd}$$

- Figure 46 summarises both layer mapping and precoding for 2×2 open loop spatial multiplexing

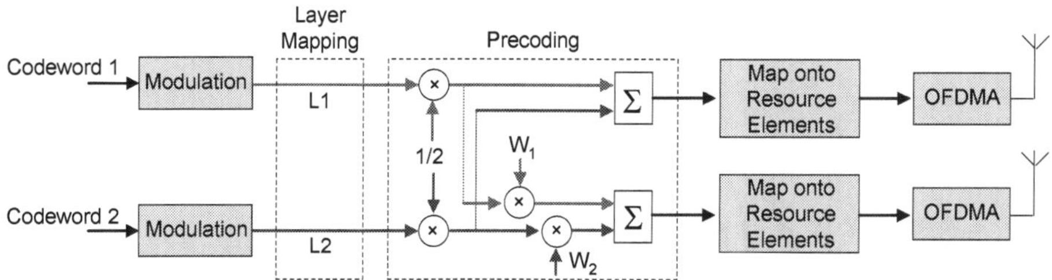

Figure 46 – 2×2 open loop spatial multiplexing

- Layer mapping is transparent and simply transfers the pair of codewords for precoding
- Precoding results in:
  - the 1st antenna transmitting (1st codeword × 1/2) + (2nd codeword × 1/2)
  - the 2nd antenna transmitting (1st codeword × 1/2 × precoding weight 1) + (2nd codeword × 1/2 × precoding weight 2)
- The precoding weights are selected from the 2 pairs of values presented in Table 22 (deduced from the previous matrix multiplication) depending upon whether an odd or even numbered modulation symbol is being transferred
- The deterministic selection of the precoding weights avoids the requirement to signal the weights being applied, i.e. the Downlink Control Information (DCI) does not need to provide the UE with any weight information

	Precoding Weight 1	Precoding Weight 2
Even numbered symbols	1	-1
Odd numbered symbols	-1	1

Table 22 – Codebook for 2×2 open loop spatial multiplexing

4×4 OPEN LOOP SPATIAL MULTIPLEXING

- Table 23 summarises the number of codewords and layers which are supported by 4×4 open loop spatial multiplexing. These are the same as those supported by 4×4 closed loop spatial multiplexing, with the exception that the number of layers is always > 1

Number of Codewords	Number of Layers	Number of Antenna Ports
1	2	4
2	2	
	3	
	4	

Table 23 – Codewords, layers and antenna ports for 4×4 open loop spatial multiplexing

- The options for layer mapping are illustrated in Figure 47

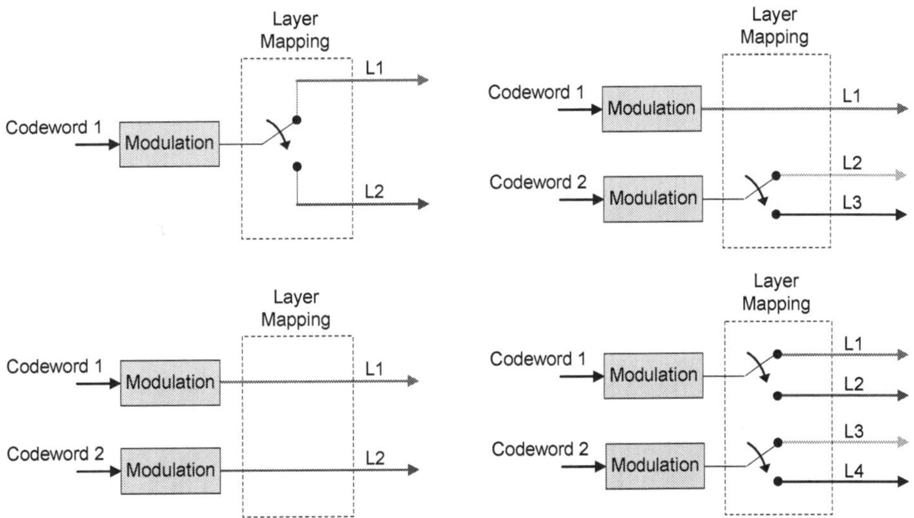

Figure 47 – Layer mapping for 4×4 open loop spatial multiplexing

★ 2 layers transferring a single codeword is most likely to be used when the UE is in poor coverage or has relatively little data to receive. 4 layers is most likely to be used when the UE is in good coverage and there is sufficient data to generate 2 large codewords

★ Figure 48 illustrates precoding for 2 layers. The total number of coding weights applied during precoding depends upon the number of layers. The number of coding weights is equal to 4 × number of layers

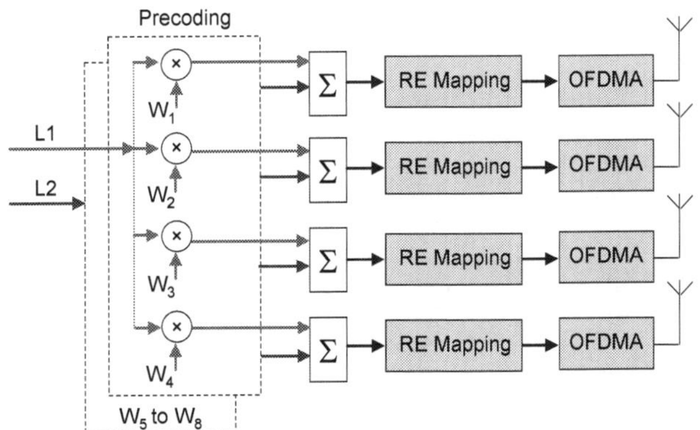

Figure 48 – 4×4 open loop spatial multiplexing with 1 codeword and 2 layers

★ Similar to 2×2 open loop spatial multiplexing, the selection of the coding weights is deterministic so does not need to be signalled to the UE within the DCI

★ The set of precoding weights is rotated for every 'n' sets of inputs from layer mapping, where 'n' is equal to the number of layers. Within Figure 48, when using 2 layers the precoding weights are changed after every 2 pairs of inputs. If the number of layers is 4 then the precoding weights are changed after every 4 quadruplets of inputs. The precoding weights themselves are extracted from the code book for 4×4 closed loop spatial multiplexing

4.4.3 CLOSED LOOP SPATIAL MULTIPLEXING

- ★ Closed loop spatial multiplexing relies upon the UE providing feedback in terms of:
  - o Rank Indication (RI) – suggested number of 'layers'
  - o Precoding Matrix Indicator (PMI) – suggested set of 'precoding weights'
  - o Channel Quality Indicator (CQI) – suggested transport block size
- ★ This information is signalled to the eNode B using either the PUCCH or PUSCH physical channels
- ★ The UE selects a PMI to maximise the signal to noise ratio at its receiver. Applying the set of precoding weights at the eNode B represents a form of maximum ratio combining at the transmitter
- ★ The eNode B may or may not follow the suggestions provided by the UE
  - o if the UE is in good coverage and suggests the use of 2 layers to increase the connection throughput, the eNode B may still schedule a single layer if there is not much data within its buffer
  - o if the UE is reporting a low CQI (corresponding to a small transport block size) but has been achieving a low Block Error Rate (BLER) then the eNode B may increase the reported CQI to schedule a larger transport block size. The eNode B may take the opposite action if the UE is experiencing a high BLER while reporting a high CQI
- ★ 3GPP has defined the following variants of closed loop spatial multiplexing within the release 8 version of the specifications
  - o 2×2 closed loop spatial multiplexing
  - o 4×4 closed loop spatial multiplexing

2×2 CLOSED LOOP SPATIAL MULTIPLEXING

- ★ Table 24 presents the number of codewords and layers which are supported by 2×2 closed loop spatial multiplexing.
  - o either 1 or 2 codewords can be transferred during each 1 ms subframe
  - o the number of layers is always equal to the number of codewords

Number of Codewords	Number of Layers	Number of Antenna Ports
1	1	2
2	2	

Table 24 – Codewords, layers and antenna ports for 2×2 closed loop spatial multiplexing

- ★ Layer mapping and precoding for the transfer of a single codeword is shown in Figure 49. Layer mapping simply transfers the codeword for precoding

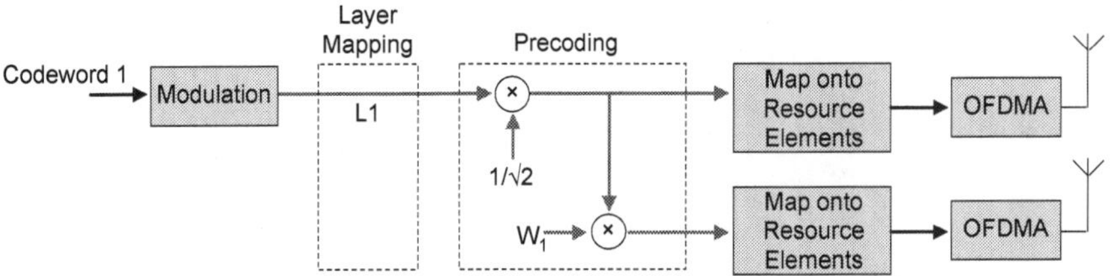

Figure 49 – 2×2 closed loop spatial multiplexing with 1 codeword

- ★ Precoding for the transfer of a single codeword results in:
  - o the 1st antenna transmitting, (codeword × 1/√2)
  - o the 2nd antenna transmitting (codeword × 1/√2 × precoding weight)
- ★ The precoding weight for the transmission of a single codeword is selected from the set of values presented in Table 25

Codebook Index	Precoding Weight	Codebook Index	Precoding Weight
0	1	2	j
1	-1	3	-j

Table 25 – Codebook for 2×2 closed loop spatial multiplexing with 1 codeword

- The precoding weight used within a specific subframe is signalled to the UE using Downlink Control Information (DCI) format 2 on the PDCCH physical channel
- Layer mapping and precoding for the transfer of two codewords is shown in Figure 50. Layer mapping simply transfers the two codewords for precoding

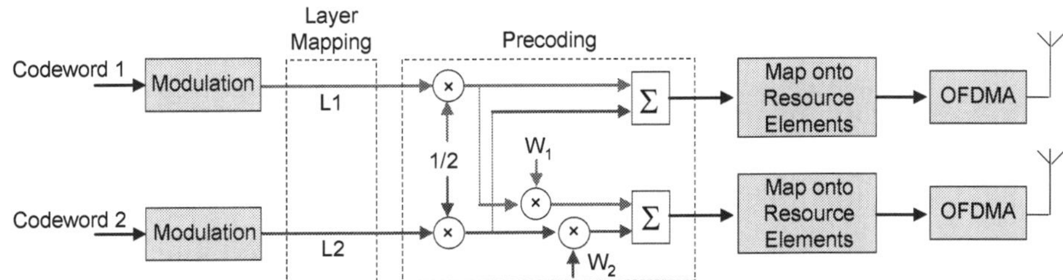

Figure 50 – 2×2 closed loop spatial multiplexing with 2 codewords

- Precoding for the transfer of two codewords results in:
    - the 1st antenna transmitting, (1st codeword × 1/2) + (2nd codeword × 1/2)
    - the 2nd antenna transmitting (1st codeword × 1/2 × precoding weight 1) + (2nd codeword × 1/2 × precoding weight 2)
- The precoding weights for the transmission of two codewords are selected from the pair of values presented in Table 26

Codebook Index	Precoding Weight 1	Precoding Weight 2
1	1	-1
2	j	-j

Table 26 – Codebook for 2×2 closed loop spatial multiplexing with 2 codewords

4×4 CLOSED LOOP SPATIAL MULTIPLEXING

- Table 27 presents the number of codewords and layers which are supported by 4×4 closed loop spatial multiplexing.
    - either 1 or 2 codewords can be transferred during each 1 ms subframe
    - the number of layers can range from 1 to 4

Number of Codewords	Number of Layers	Number of Antenna Elements
1	1	4
	2	
2	2	
	3	
	4	

Table 27 – Codewords, layers and antenna elements for 4×4 closed loop spatial multiplexing

- The options for layer mapping are presented in Figure 51

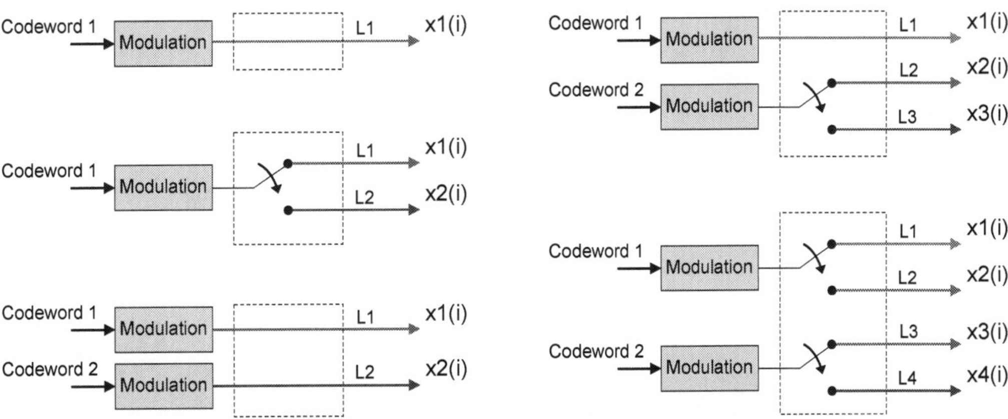

Figure 51 – Layer mapping for 4×4 closed loop spatial multiplexing

★ A single layer is most likely to be used when the UE is in poor coverage or has relatively little data to receive. 4 layers are most likely to be used when the UE is in good coverage and there is sufficient data to generate 2 large codewords

★ Figure 52 illustrates layer mapping and precoding for a single codeword and a single layer. The set of 4 weights is selected from 16 sets of 4 weights, i.e. the Downlink Control Information (DCI) requires 4 bits to signal the selected set of weights

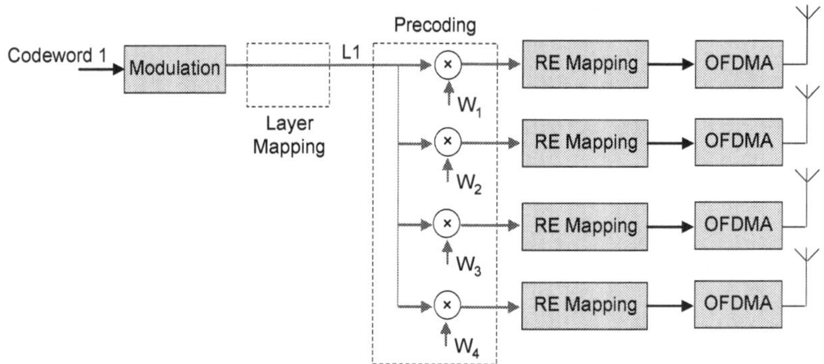

Figure 52 – 4×4 closed loop spatial multiplexing with 1 codeword and 1 layer

★ The number of coding weights applied during precoding depends upon the number of layers. The number of coding weights is equal to 4 × number of layers

★ Figure 53 illustrates layer mapping and precoding for 2 codewords and 4 layers. In this case, there is a total of 16 weights. The set of 16 weights is selected from 16 sets of 16 weights

★ Each of the layers contributes towards the signal transmitted from each antenna element

★ Similar to transmit diversity, the precoded data is mapped onto Resource Elements prior to OFDMA signal generation

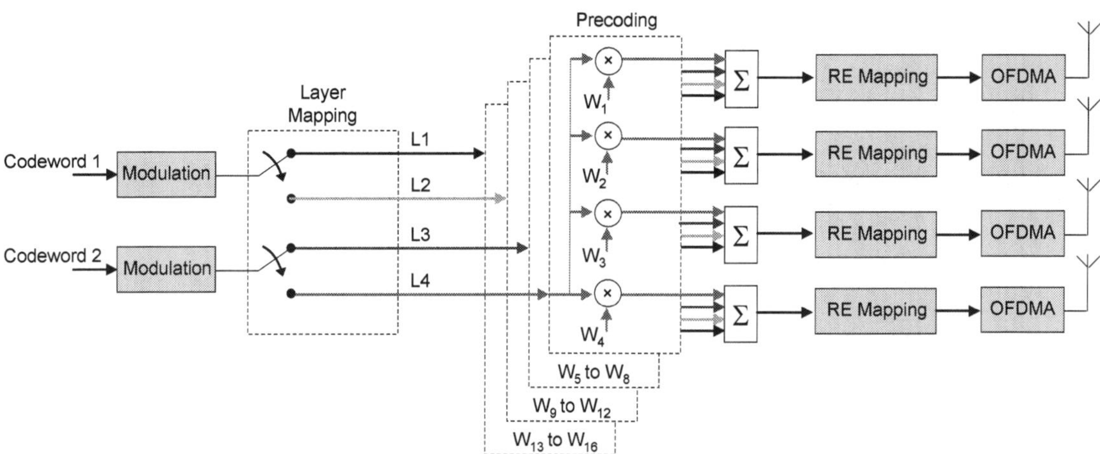

Figure 53 – 4×4 closed loop spatial multiplexing with 2 codewords and 4 layers

4.4.4 BEAMFORMING

★ The principles of beamforming are described in section 33.6, while the general concept is illustrated in Figure 54. An antenna array at the eNode B directs the downlink Resource Block transmissions towards the appropriate UE. This is achieved by adjusting the phase of the signal transmitted by each physical antenna element such that the transmissions combine constructively at the UE

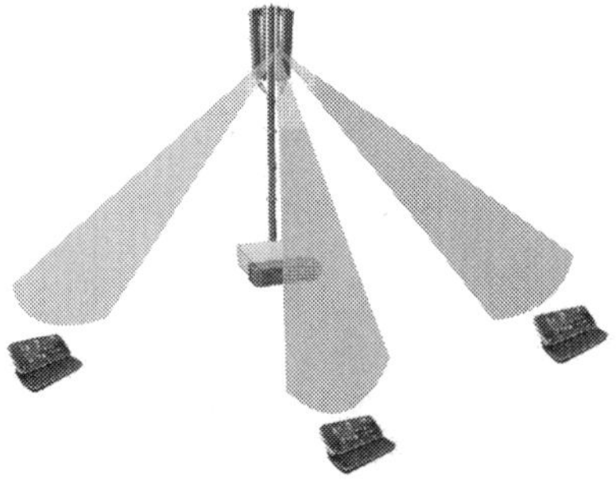

Figure 54 – General concept of beamforming

★ Beamforming increases the antenna gain in the direction of the UE and helps to decrease inter-cell interference. These characteristics help to increase both connection throughput and system capacity. Beamforming is especially beneficial to UE at cell edge where levels of inter-cell interference can be relatively high

★ Beamforming prefers the radio propagation channel from each physical antenna element to be correlated. This is in contrast to transmit diversity and MIMO which prefer uncorrelated transmissions

★ Figure 55 illustrates a typical antenna used for beamforming. The half-wavelength spacing between each column of antenna elements provides the correlated radio paths required by beamforming. The cross polarisation provides the uncorrelated radio paths required by transmit diversity and MIMO

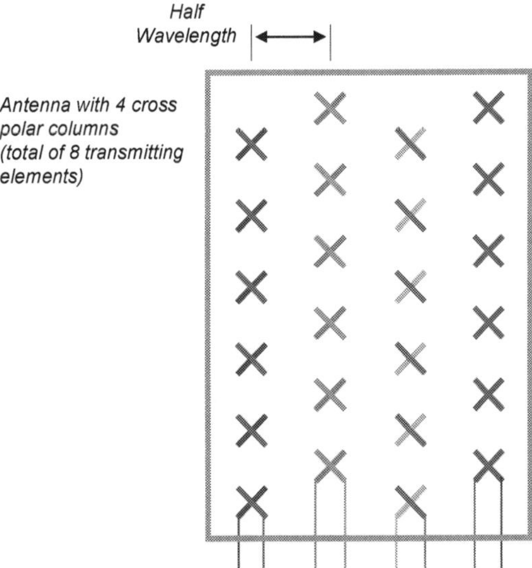

Figure 55 – 4-column cross polar antenna array

★ The physical size of this type of antenna depends upon the operating band, i.e. RF carriers within lower operating bands have longer wavelengths so require larger antenna

- Wavelengths are ~ 12 cm within the 2.6 GHz operating band and antenna arrays are typically 0.25 m × 1 m
- Wavelengths are ~ 35 cm within the 850 MHz operating band and antenna arrays are typically 0.7 m × 2.8 m

★ Increasing the number of antenna element columns would allow more directional beams to be generated but would also increase the physical size of the antenna. The use of antenna element columns allows beamforming in the horizontal direction rather than in the vertical direction

★ Support for beamforming has evolved throughout each release of the 3GPP specifications

★ Initial support for beamforming was introduced within the release 8 version of the specifications:

- supports single layer transmission to each UE

- uses transmission mode 7 with antenna port 5
- uses a UE specific Reference Signal, also known as a Demodulation Reference Signal (DM-RS)
- only CQI is reported by the UE (no Rank Indicator nor Precoding Matrix Indicator is reported)

★ A possible signal flow for beamforming based upon the release 8 version of the specifications is illustrated in Figure 56. A 4-column cross polar antenna is assumed so there are 8 parallel streams of signal processing. Each stream is mixed to RF and amplified before being input to the antenna array. This generates a requirement for 8 transceivers and 8 power amplifiers. Each power amplifier could have a 5 W transmit power capability providing a total of 40 W

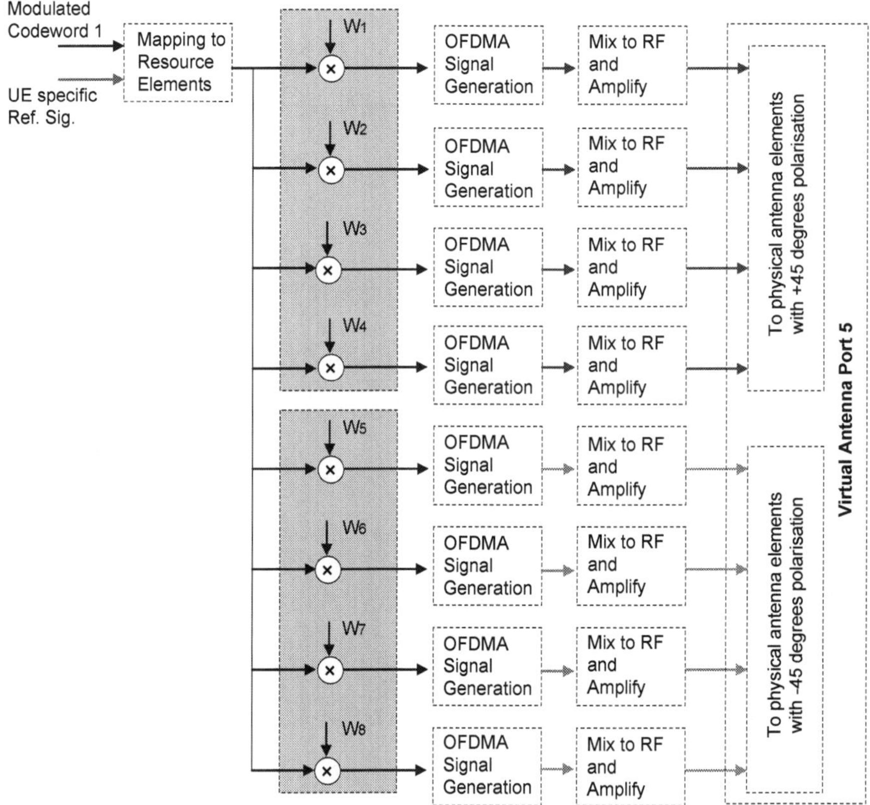

Figure 56 – Signal flow for single layer beamforming for single UE

★ The set of weights (w_1 to w_8) are not specified by 3GPP. These weights are implementation dependant and are generated by the eNode B. They adjust the phase of the transmitted signals to ensure that they combine constructively in the direction of the UE. 3GPP does not define any layer mapping nor precoding for beamforming within the release 8 version of the specifications

★ These weights at the transmitter are transparent to the UE and the UE requires no knowledge of them. As far as the UE is concerned, the eNode B is transmitting a single signal from virtual antenna port 5

★ Both the modulated data symbols and the UE specific Reference Signal are processed by the same set of weights, i.e. the UE specific Reference Signal is also directed towards the UE. This allows the UE specific Reference Signal to be used by the UE for channel estimation and subsequent demodulation

★ The weights shown in Figure 56 are divided into two groups – one group associated with each polarisation of the physical antenna elements. For a single polarisation, only a phase shift is required to achieve beamforming. When using both polarisations, both phase and amplitude shifts are required to achieve beamforming. Note that these weights and the choice of physical antenna are implementation dependent and that a different implementation could use a single polarisation

★ The set of weights can be generated by the eNode B using the Sounding Reference Signal (SRS) transmitted by the UE. In the case of TDD, the uplink and downlink radio propagation channels are the same (channel reciprocity) because the same RF carrier is shared between the uplink and downlink. This allows the eNode B to base its downlink transmission upon uplink propagation channel measurements

★ The details of the algorithms used to generate the beamforming weights are outside the scope of this book. A common approach is Eigen Based Beamforming (EBB). The set of weights do not necessarily have to be updated at a very high rate because the UE position does not change very rapidly. Updating the weights at a rate which is in the order of once per second, is known as long term beamforming. Updating the weights at a rate which is in the order of 100 times per second is known as short term beamforming. Short term beamforming can improve performance but requires increased quantities of computation and the resultant weights can be more susceptible to errors due to less averaging

★ Cell specific Reference Signals are not processed by the set of weights and are broadcast across the entire cell. Cell specific Reference Signals for single layer transmission are broadcast using antenna port 0. Antenna port 0 could be mapped onto the 4 physical antenna elements with +45 degrees polarisation. This leaves the 4 physical antenna elements with -45 degrees polarisation for antenna port 1 in case dual layer transmission is used in the cell. Transmission mode 7 allows fall-back to transmit diversity (without beamforming) so it is useful to have the ability to support both antenna ports 0 and 1

★ Beamforming can be simultaneously applied to multiple UE during the same subframe. For example, if the set of Resource Blocks are shared between 3 UE then the eNode can generate 3 beams and direct each Resource Block allocation to the appropriate UE. This is achieved by generating and applying the set of weights independently for each UE

★ Support for beamforming was enhanced within the release 9 version of the specifications:

- supports dual layer transmission to each UE
- uses transmission mode 8 with antenna ports 7 and 8
- uses new UE specific Reference Signals (new Demodulation Reference Signals (DM-RS))
- CQI, Rank Indicator and Precoding Matrix Indicator are reported by the UE

★ Dual layer beamforming is able to transfer 2 codewords per subframe rather than 1 codeword per subframe so is able to double the throughput relative to single layer beamforming

★ Transmission mode 8 is only specified for the normal cyclic prefix so dual layer beamforming is not possible when using the extended cyclic prefix

★ A possible signal flow for dual layer beamforming based upon the release 9 version of the specifications is illustrated in Figure 57. Similar to the single layer example already presented in this section, a 4-column cross polar antenna is assumed so there are 8 parallel streams of signal processing

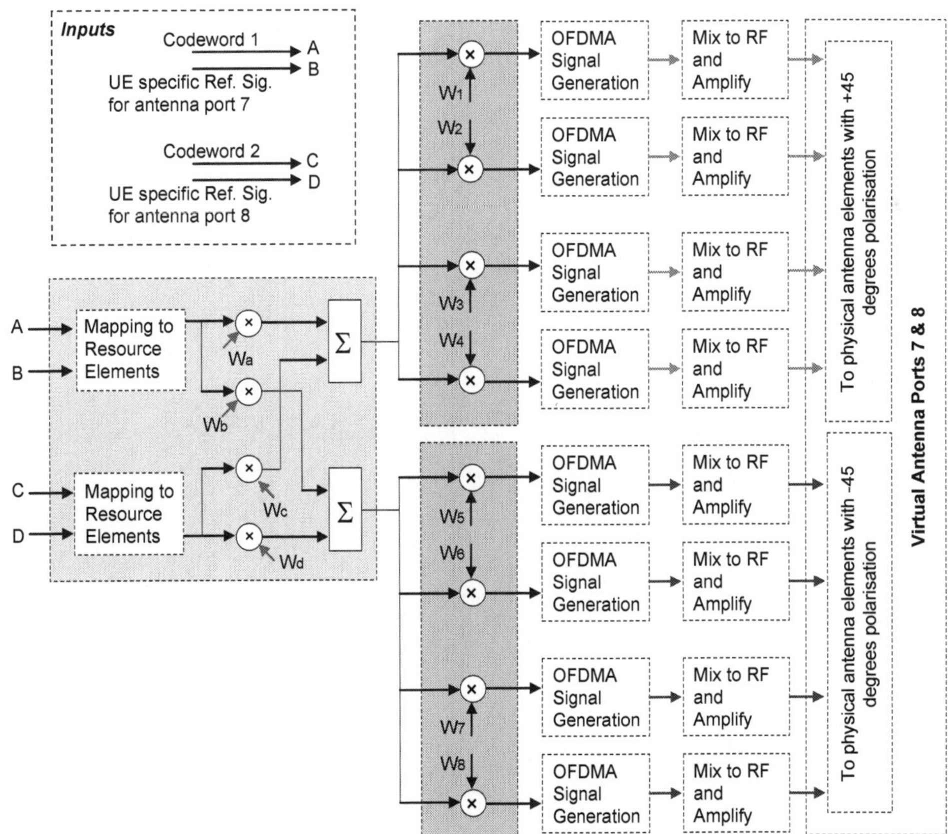

Figure 57 – Signal flow for dual layer beamforming for single UE

★ Codeword 1 and the UE specific Reference Signal for antenna port 7 are mapped onto one Resource Element grid, while codeword 2 and the UE specific Reference Signal for antenna port 8 are mapped onto another Resource Element grid. Codewords 1 and 2 have already been modulated at this stage

★ Precoding is then completed in two stages. This precoding is not specified by 3GPP so its implementation is dependent upon the network vendor. Figure 57 illustrates a first stage of precoding using complex weights wa, wb wc and wd. These complex weights could be based upon channel estimation from the uplink Sounding Reference Signal (SRS). Alternatively they could be based upon the Precoding Matrix Indicator (PMI) reported by the UE

★ The second stage of precoding uses weights w1 to w8. This stage of precoding is likely to be based upon channel estimation from the uplink Sounding Reference Signal (SRS). The first four weights generate the signals for the +45 degree antenna elements, while the second four weights generate the signals for the -45 degree antenna elements

★ Antenna ports 7 and 8 are mapped across both polarisations due to the mixing of signals within the first stage of precoding. From the UE perspective, there are two transmissions to receive on the downlink – a first layer of PDSCH transmissions associated with the UE specific Reference Signal for antenna port 7, and a second layer of PDSCH transmissions associated with the UE specific Reference Signal for antenna port 8

★ The pair of sequences used to define the UE specific Reference Signals for antenna ports 7 and 8 are mutually orthogonal. This means that they benefit from Code Division Multiplexing (CDM). The sequences are not dependent upon the UE identity, nor the Resource Block allocation

★ 3GPP defines the use of two scrambling identities which can be used to generate a second pair of sequences for the antenna port 7 and 8 UE specific Reference Signals. However, the sequences generated by one scrambling identity are not orthogonal to the sequences generated by the other scrambling identity. This concept is illustrated in Figure 58

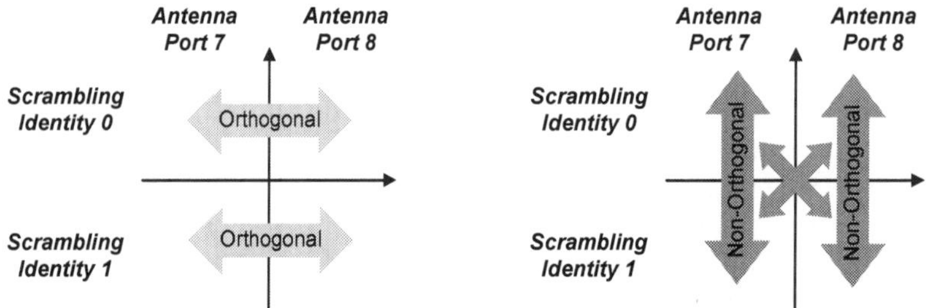

Figure 58 – Orthogonality between the sequences used to define the UE specific Reference Signals for antenna ports 7 and 8

★ The use of both scrambling identities allows 2 users with the same Resource Block allocation to simultaneously benefit from dual layer beamforming. Alternatively, 4 users with the same Resource Block allocation could simultaneously benefit from single layer beamforming. These multi-user MIMO scenarios are illustrated in Figure 59

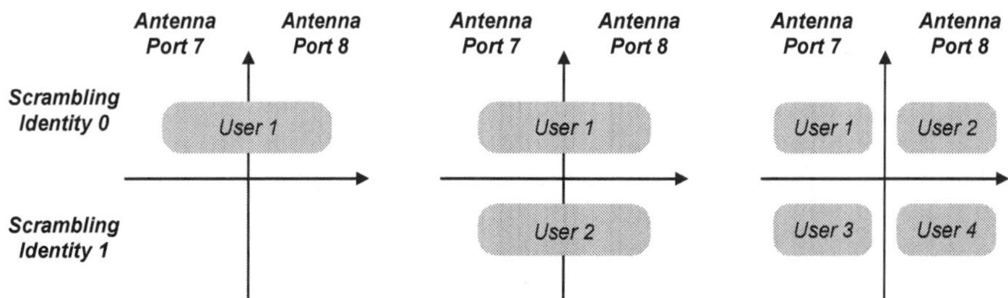

Figure 59 – Example scenarios using both scrambling identities associated antenna port 7 and 8 UE specific Reference Signals

★ Transmission mode 8 allows the eNode B to instruct the UE to generate Rank Indicator (RI) and Precoding Matrix Indicator (PMI) reports, in addition to CQI reports. The PMI reports may be used to define the weights applied within the first phase of precoding (although channel estimates from the uplink Sounding Reference Signal are likely to be used whenever available). The RI reports may be used to help the eNode B decide when to use dual layer transmission and when to use single layer transmission

★ Transmission mode 8 can be used in 3 ways:

- dual layer beamforming using both antenna ports 7 and 8
- single stream beamforming using either antenna port 7 or 8
- transmit diversity using antenna ports 0, 1, 2 and 3

The RI and CQI reported by the UE can be used to select between these 3 options. Dual layer beamforming is likely to be selected when the UE reports a high order rank and a high CQI. Single layer beamforming is likely to be selected when the UE reports a low order rank and a low CQI. Transmit diversity may be used when the UE reports a low order rank and a high CQI

★ Similar to single stream beamforming, dual stream beamforming can be simultaneously applied to multiple UE during the same subframe. For example, if the set of Resource Blocks are shared between 3 UE then the eNode can generate 3 dual layer beams and direct each Resource Block allocation to the appropriate UE. This is achieved by generating and applying the set of weights independently for each UE

4.4.5 MULTI-USER MIMO

★ Single User MIMO (SU-MIMO) addresses all layers to the same UE

★ Multi-User MIMO (MU-MIMO) addresses different layers to different UE. This allows multiple UE to use the same set of Resource Blocks during the same subframe, i.e. the same time and frequency resources are re-used by multiple UE

★ The general concept of multi-user MIMO is illustrated in Figure 60

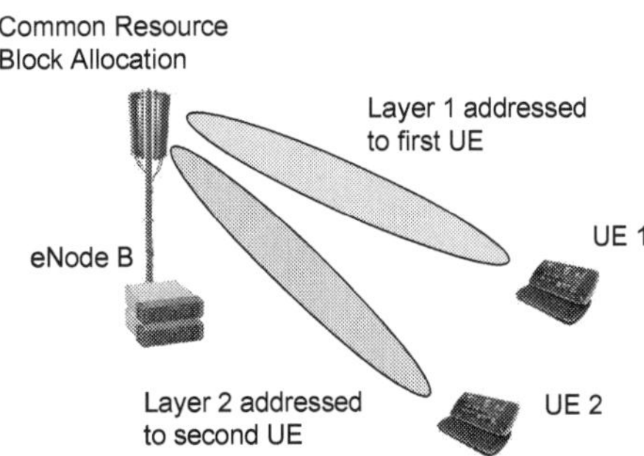

Figure 60 – Concept of multi-user MIMO

★ Multi-user MIMO aims to increase the overall system capacity rather than increasing the peak connection throughput

★ Multi-user MIMO was introduced within the release 8 version of the 3GPP specifications. The initial solution has been enhanced within the release 9 and 10 versions of the specifications. The release 10 version forms part of LTE Advanced

★ The release 8 version of the specifications provides 2 options for multi-user MIMO

- option 1:
  - allows simultaneous transmission to 2 users with a single layer per user. A single codeword is transferred to each user during each subframe
  - this option is based upon transmission mode 5 using antenna ports 0 and 1. At least 2 physical antenna elements must be available to support the pair of antenna ports. Transmission mode 5 was introduced specifically to support multi-user MIMO
  - the cell specific Reference Signal supports both CQI reporting and demodulation at the UE. UE specific Reference Signals are not used by this option
  - the UE reports both CQI and Precoding Matrix Indicators (PMI). The precoding weights applied by the eNode B are signalled to the UE within Downlink Control Information (DCI) format 1D
  - DCI format 1D also includes a power offset used to inform the UE that the downlink transmit power is being shared and that the downlink power relative to the Reference Signal has decreased. This information helps the UE receiver when demodulating
- option 2:
  - allows simultaneous transmission to 2 or more users (theoretically unlimited) with a single layer per user. A single codeword is transferred to each user during each subframe
  - this option is based upon transmission mode 7 using antenna port 5. Beamforming is used to direct the downlink data towards the appropriate UE. Beamforming requires multiple physical antenna elements. Transmission mode 7 was not introduced specifically to support multi-user MIMO
  - the cell specific Reference Signal supports CQI reporting, while a UE specific Reference Signal supports demodulation. The sequence used to generate the UE specific Reference Signal is a function of the UE identity (RNTI). This allows users to differentiate between their transmissions. However, the sequences are not orthogonal and are likely to interfere with each other even when the UE specific beams are directed in different directions
  - the UE reports only CQI and does not require any knowledge of the precoding applied by the eNode B to achieve the beamforming
  - DCI format 1 is used for transmission mode 7. This DCI does not include any information specifically for multi-user MIMO

★ The release 9 version of the 3GPP specifications defines a third solution for multi-user MIMO

- option 3:
  - supports dual layer transmission to 2 UE, or single layer transmission to 4 UE, i.e. up to 4 layers in total. Up to 2 codewords can be transferred to each user during each subframe

- this option is based upon transmission mode 8 using antenna ports 7 and 8. Beamforming is used to direct the downlink data towards the appropriate UE. Beamforming and multi-user MIMO for transmission mode 8 are described in greater detail in section 4.4.4
- the cell specific Reference Signal supports CQI reporting, while UE specific Reference Signals support demodulation. The sequences used to generate the UE specific Reference Signals are a function of a scrambling identity and an Orthogonal Covering Code (OCC). There are 2 scrambling identities and 2 OCC providing a total of 4 combinations to support the 4 layers
- the UE specific Reference Signals are a combination of orthogonal and quasi-orthogonal due to the pair of OCC being orthogonal while the pair of scrambling identities are not orthogonal
- the UE reports CQI by default, and the eNode B can request the UE to also report Precoding Matrix Indicators (PMI) and Rank Indicators (RI). PMI and RI are useful when supporting dual layer transmission
- DCI format 2B is used for transmission mode 8. This DCI signals the scrambling identity and antenna port(s) allocated to the UE. The UE is able to deduce its OCC from the allocated antenna port(s) because there is a one-to-one mapping between OCC and antenna port

★ The signalling overhead for Multi-user MIMO is small because UE are not explicitly informed about co-scheduled UE (with the exception of the power offset information within DCI format 1D)

★ The evolution of multi-user MIMO for LTE Advanced within the release 10 version of the specifications is described in section 4.5.2

4.5 LTE ADVANCED TECHNOLOGIES

★ LTE Advanced has an objective to increase the peak downlink spectrum efficiency to 30 bps/Hz. Carrier Aggregation has limited impact upon spectrum efficiency because increased throughputs are achieved by increasing the bandwidth. Multiple antenna transmission technologies allow the throughput to increase without increasing the bandwidth.

★ The release 10 version of the 3GPP specifications introduces transmission mode 9 to support the requirements of multiple antenna transmission for LTE Advanced. Transmission mode 9 provides support for:

  - spatial multiplexing with up to 8 layers of parallel data transfer which allows a single UE to benefit from 8×8 MIMO. Antenna ports 7 to 14 are used to transfer these 8 layers
  - a total of 4 layers of parallel data transfer for multi-user MIMO. These 4 layers can be shared between 2 UE so each UE receives 2 layers of data. Alternatively, they can be shared between 4 UE so each UE receives a single layer of data. The 4 layers associated with multi-user MIMO are based upon antenna ports 7 and 8 used in combination with a pair of scrambling identities
  - beamforming which can be applied to both the single user MIMO and multi-user MIMO scenarios

★ Transmission mode 9 also uses additional UE specific Reference Signals to support demodulation at the UE

  - the release 9 version of the 3GPP specifications defines UE specific Reference Signals for antenna ports 7 and 8
  - the release 10 version extends this to antenna ports 9 to 14, i.e. providing a total of 8 antenna ports for the transmission of UE specific Reference Signals (in addition to the original antenna port 5 specified within the release 8 version of the specifications)

★ The PDSCH is transferred using the same set of antenna ports as the UE specific Reference Signals so both experience the same propagation channel. This allows the PDSCH to be demodulated using measurements from the UE specific Reference Signal. Both the PDSCH and UE specific Reference Signals can have beamforming applied to direct them towards individual UE

★ UE specific Reference Signals are described in greater detail in section 5.2.3

★ Channel State Information (CSI) Reference Signals have been defined to improve the performance of link adaptation for UE using transmission mode 9

  - Resource Elements allocated to the cell specific Reference Signal often experience different levels of interference to those allocated to the PDSCH. This reduces the benefit of CQI reports based upon the cell specific Reference Signal
  - CSI Reference Signals occupy Resource Elements usually allocated to the PDSCH. This allows a more meaningful measure of channel quality but results in the PDSCH transmission being punctured

  CSI Reference Signals use antenna ports 15 to 22

★ CSI Reference Signals are broadcast across the entire cell area and do not have beamforming applied. This allows them to be used by all UE within the cell

★ CSI Reference Signals puncture the PDSCH belonging to release 8 and release 9 UE as well as the PDSCH belonging to release 10 UE. This has a negative impact upon the release 8 and release 9 UE which are not aware of the CSI Reference Signal

★ CSI Reference Signals are described in greater detail in section 5.2.5

★ The release 10 version of the 3GPP specifications also defines 'LTE Advanced' subframes during which UE specific and CSI Reference Signals are sent but cell specific Reference Signals are not sent. Cell specific Reference Signals are usually sent during all subframes supporting PDSCH transmission but this represents a special case. Avoiding transmission of the cell specific Reference Signal helps to reduce the signalling overhead

★ 'LTE Advanced' subframes are signalled as if they were MBSFN subframes. This prevents release 8 and release 9 UE from attempting to receive the PDSCH, because the PDSCH is not normally transmitted during MBSFN subframes. Release 10 UE using transmission mode 9 check for PDSCH transmissions during MBSFN subframes by scanning for their C-RNTI on the PDCCH

★ 3GPP References: TS 36.211, TS 36.213

4.5.1 OPEN AND CLOSED LOOP SPATIAL MULTIPLEXING

★ Transmission mode 9 provides support for LTE Advanced spatial multiplexing. Its capabilities include:
    - 2×2, 4×4 and 8×8 MIMO with beamforming
    - open and closed loop modes of operation
    - 2 stage codebook design for generating Precoding Matrix Indicators (PMI) at the UE
    - non-codebook based precoding at the eNode B
    - up to 8 Channel State Information (CSI) Reference Signals on antenna ports 15 to 22
    - up to 8 UE specific Reference Signals (also known as Demodulation Reference Signals (DM-RS)) on antenna ports 7 to 14
    - Physical Resource Block (PRB) Bundling to improve channel estimation from the UE specific Reference Signals

★ When using the open loop mode of operation, the UE provides feedback in terms of Channel Quality Indicator (CQI) reports. PMI and RI reports are not generated by the UE. This helps to reduce the uplink signalling overhead. Channel reciprocity between the uplink and downlink allows the eNode B to derive some channel state information from the uplink Sounding Reference Signal. CQI values are generated using measurements from the 3GPP release 8 Cell Specific Reference Signals when operating in open loop mode

★ The open loop mode of operation can be applied to high speed UE for whom the closed loop mode of operation may not be sufficiently responsive, i.e. channel conditions may have changed before the eNode B has been able to use the feedback from the UE. An open loop solution can use precoder cycling at the eNode B. This involves transmitting the downlink data using a range of different precoders. These precoders generate antenna beams in different directions to randomise the channel

★ When using the closed loop mode of operation, the UE provides feedback in terms of CQI, Rank Indication (RI) and Precoding Matrix Indicator (PMI) reports. CQI values are generated using measurements from the CSI Reference Signals

★ The eNode B instructs the UE whether or not to provide the RI and PMI reports using an RRC Connection Setup, RRC Connection Reconfiguration or RRC Connection Re-establishment message

★ Transmission mode 9 has been designed using the assumption that there are 2 main antenna solution alternatives at the eNode B:
    - an antenna array which includes 4 columns of cross polarised antenna elements, where columns are spaced by half a wavelength
    - an antenna array which includes 8 columns of co-polarised antenna elements, where columns are spaced by half a wavelength

★ Figure 61 illustrates an antenna array with 4 columns of cross polarised antenna elements.

★ This type of antenna design can be beneficial because it includes 2 distinct sets of co-polarised antenna elements
    - the propagation channel tends to be relatively well correlated between antenna elements belonging to the same set of co-polarised antenna elements. Beamforming performance improves when the propagation channel is correlated between antenna elements
    - the propagation channel is less correlated between the 2 sets of co-polarised antenna elements. MIMO performance improves when the propagation channel is uncorrelated between antenna elements, i.e. the rank of the propagation channel (equivalent to the number of MIMO layers which can be transmitted) increases as the number of uncorrelated propagation paths increases

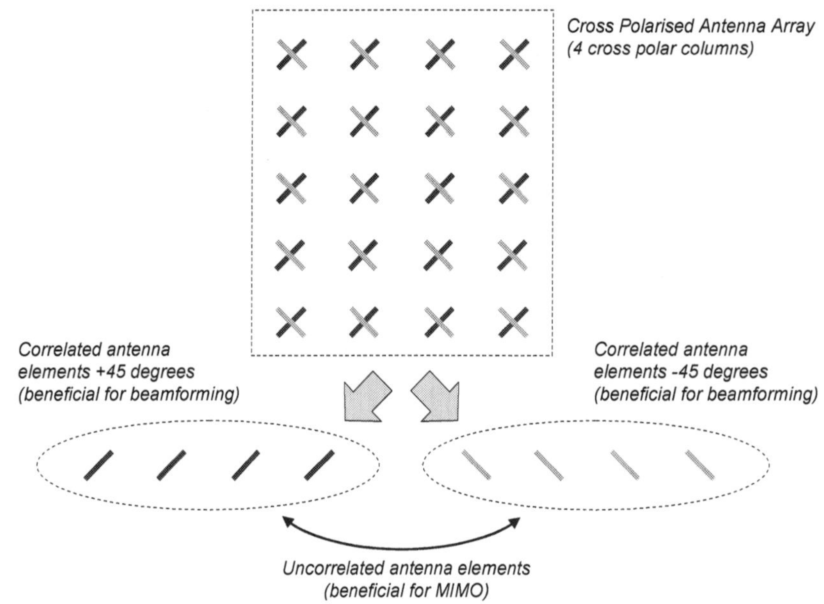

Figure 61 – Cross polarised antenna array with 4 columns of cross polarised antenna elements

★ Figure 62 illustrates an antenna array with 8 columns of co-polarised antenna elements. This type of antenna design is known as a Uniform Linear Array (ULA)

- increasing the number of antenna element columns from 4 to 8 allows beamforming to benefit from a narrower horizontal beamwidth, i.e. beamforming can be more selective
- the drawback of having all antenna elements with the same polarisation is that the propagation paths between the eNode B and UE are more likely to be correlated resulting in less scope for high rank MIMO transmissions, i.e. there will be less chance of being able to use 8×8 MIMO
- another drawback of the ULA is its physical size which is greater than the equivalent cross polar antenna because there are 8 columns of antenna elements rather than 4 (the columns are spaced by half a wavelength in both cases)

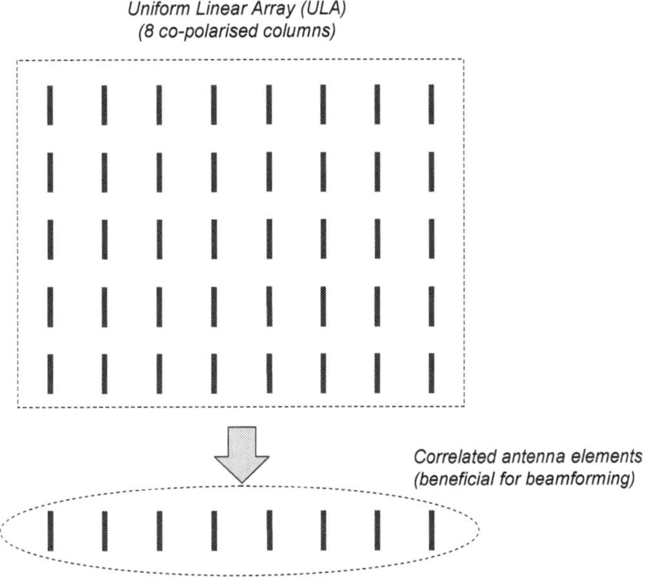

Figure 62 – Uniform Linear Array (ULA) antenna array with 8 columns of co-polarised antenna elements

★ Transmission mode 9 supports the transmission of 1 or 2 modulated codewords. Layer mapping processes the modulated codeword(s) to generate a specific number of layers. The primary layer mapping options for transmission mode 9 are illustrated in Figure 63

★ The general approach is that 1 codeword (CW) is used to generate 1 layer, while 2 codewords are used to generate 2 to 8 layers. Serial to parallel conversion is used to generate multiple layers from a single modulated codeword

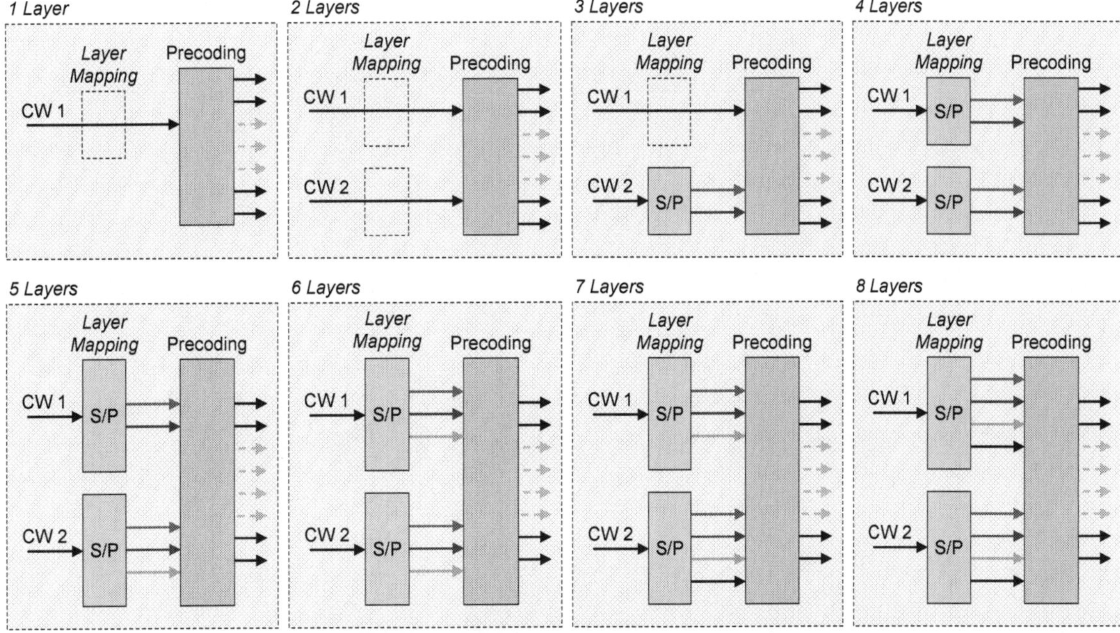

Figure 63 – Layer mapping options for transmission mode 9

★ The number of layers should match the current channel conditions between the set of transmit antenna and the set of receive antenna. Low correlation propagation paths between the transmit and receive antenna allows increased numbers of layers. This corresponds to a propagation channel with increased rank. When operating in closed loop mode, the UE suggests an appropriate number of layers using the Rank Indicators (RI)

★ The additional set of layer mapping options illustrated in Figure 64 has been defined to provide support for retransmissions. Retransmissions are completed using the same layer mapping as the original transmission. For example, if 2 codewords were originally used to generate 4 layers (2 layers per codeword), but only 1 codeword was received successfully then the other codeword is retransmitted after serial to parallel conversion to generate 2 layers

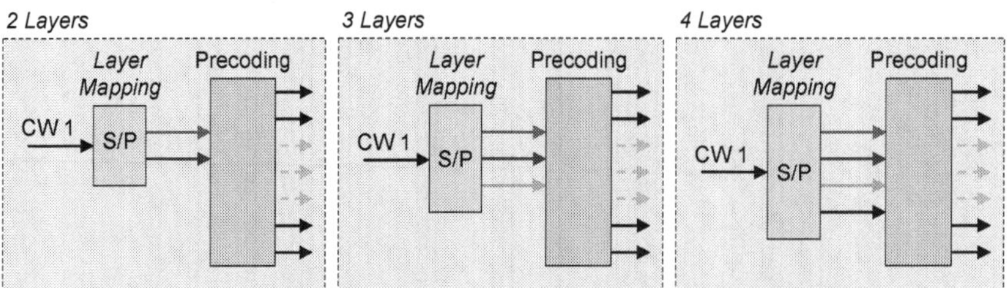

Figure 64 – Additional layer mapping options for transmission mode 9

★ Maximising spectrum efficiency and throughput requires the transmission of 2 large codewords using 8 layers. This corresponds to the maximum capability of 8×8 MIMO. Using a lower number of layers will result in lower spectrum efficiency and lower throughput

★ Transmission mode 9 is designed to use non-codebook based precoding. This means that 3GPP TS 36.211 does not specify any precoding matrices for transmission mode 9. The actual precoding applied to the downlink transmission within the eNode B is implementation dependent. In general, this precoding should provide both beamforming and MIMO capabilities

★ 3GPP TS 36.213 specifies a set of precoding matrices for the purposes of Precoding Matrix Indicator (PMI) feedback. The UE uses these precoding matrices to determine the PMI values to signal to the eNode B when operating in closed loop mode

★ The set of precoding matrices have been designed to support PMI reporting in 2 stages:

  - the first stage corresponds to reporting wideband longer term channel properties, e.g. the general direction of the UE relative to the eNode B. This stage is not expected to be frequency selective so a single wideband value is applicable to the complete channel

  - the second stage corresponds to reporting frequency selective short term channel properties, e.g. the precise direction of the UE relative to the eNode B, and the correlation between cross polar antenna elements. This stage is expected to be frequency selective so the specifications allow reporting of both wideband and sub-band PMI values

★ The precoding matrices are specified separately for each rank. The following description starts by considering the precoding matrices for rank 1, i.e. a single layer of transmission which corresponds to lower throughput but increased potential for beamforming and a reduced requirement for uncorrelated propagation paths

★ PMI reporting for single rank transmission defines the set of 32 beams illustrated in Figure 65. These 32 beams are organised into 16 overlapping groups, with 4 beams within each group.

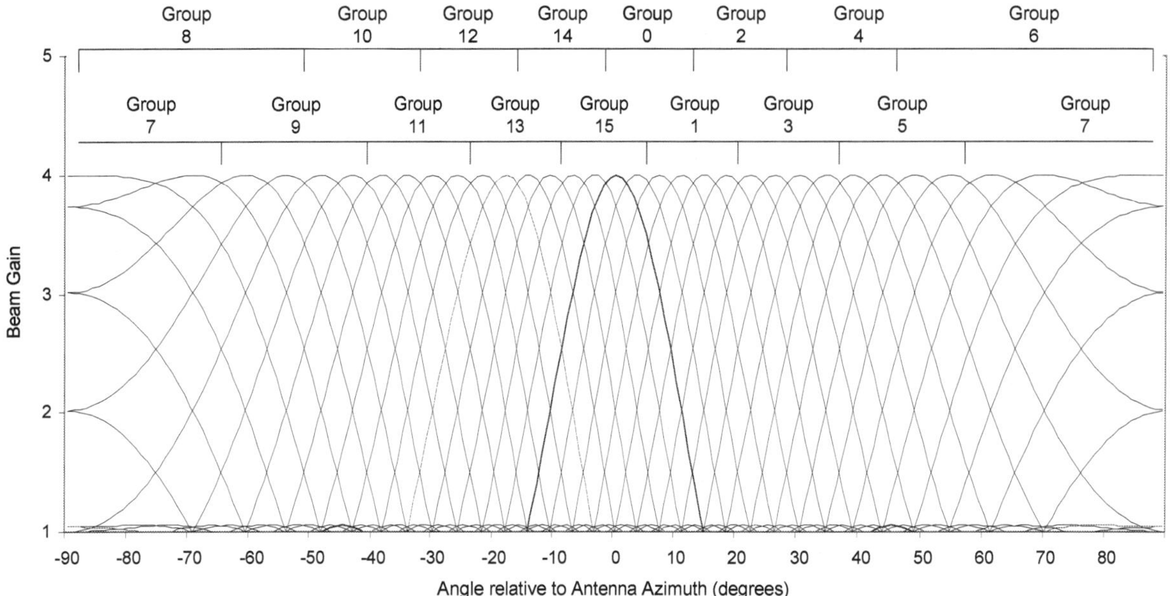

Figure 65 – Set of 32 beams used for transmission mode 9 PMI reporting for ranks 1 and 2

- ★ The set of 32 beams are distributed +/- 90 degrees relative to the direction of the antenna. The region of greatest interest is +/- 60 degrees when assuming a 3-sector BTS. Overlapping groups have been defined so the UE can always select a group with the preferred beam towards the center of the group
- ★ The first stage of PMI reporting corresponds to selecting 1 out of the 16 groups of beams. This provides the eNode B with information regarding the general direction of the UE, and this information is not likely to change rapidly over time. This means that the first stage of PMI reporting requires 4 bits of information to report the value of 'PMI_1' (represented by i_1 within 3GPP TS 36.213)
- ★ The second stage of PMI reporting corresponds to selecting 1 beam from the group of 4 beams, and also selecting 1 phase shift from a set of 4 phase shifts. The phase shift represents the phase difference between antenna elements 1 to 4, and antenna elements 5 to 8. Antenna element numbering is presented in Figure 66

Figure 66 – Numbering of antenna elements

- ★ The phase difference is more likely to be relatively small for the uniform linear array scenario because antenna elements 1 to 4 have the same polarisation as antenna elements 5 to 8
- ★ The second stage of PMI reporting for rank 1 transmission is illustrated in Figure 67. This stage effectively includes 2 sub-stages – selection of the beam and selection of the phase difference. A QPSK alphabet has been specified to represent the phase difference so both the beam selection and phase difference selection involve choosing 1 out of 4 options, i.e. a total of 16 overall combinations. This means that the second stage of PMI reporting requires 4 bits of information to report the value of 'PMI_2' (represented by i_2 within 3GPP TS 36.213)

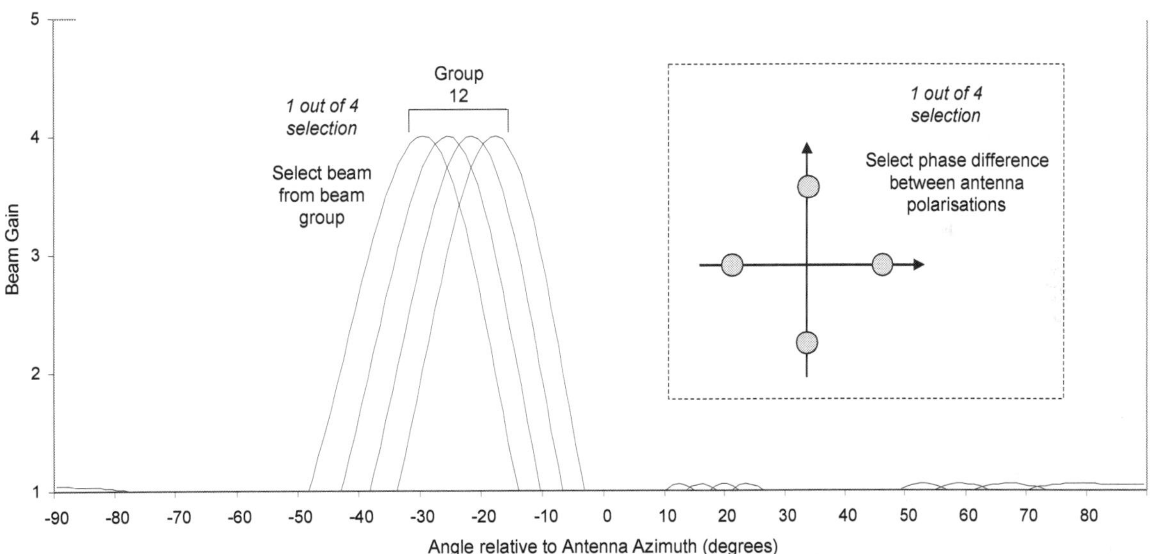

Figure 67 – Second stage of PMI reporting for transmission mode 9 rank 1

- ★ PMI reporting for rank 2 is based upon the same set of 32 beams divided into 16 overlapping groups of 4. The first stage of PMI reporting is the same as for rank 1. The second stage uses a BPSK alphabet rather than a QPSK alphabet to represent the phase difference between antenna elements 1 to 4 and antenna elements 5 to 8. This allows a separate beam to be selected for each layer
- ★ PMI reporting for ranks 3 and 4 is based upon 16 beams divided into 4 overlapping groups of 8. These 16 beams and their overlapping groups are illustrated in Figure 68
- ★ The number of beams is reduced as the rank increases because the beamforming horizontal beamwidth increases, and beamforming becomes less selective. This is caused by the total number of antenna elements being divided by the number of layers. As the number of layers increases, the antenna elements per layer decreases so the horizontal beamwidth increases
- ★ Using 4 groups of beams means that the first stage of PMI reporting for ranks 3 and 4 requires 2 bits of information
- ★ The second stage of PMI reporting for rank 3 involves selecting 1 out of 16 matrices to represent the short term channel conditions. This requires 4 bits of information when reporting
- ★ The second stage of PMI reporting for rank 4 involves selecting 1 out of 8 matrices to represent the short term channel conditions. This requires 3 bits of information when reporting

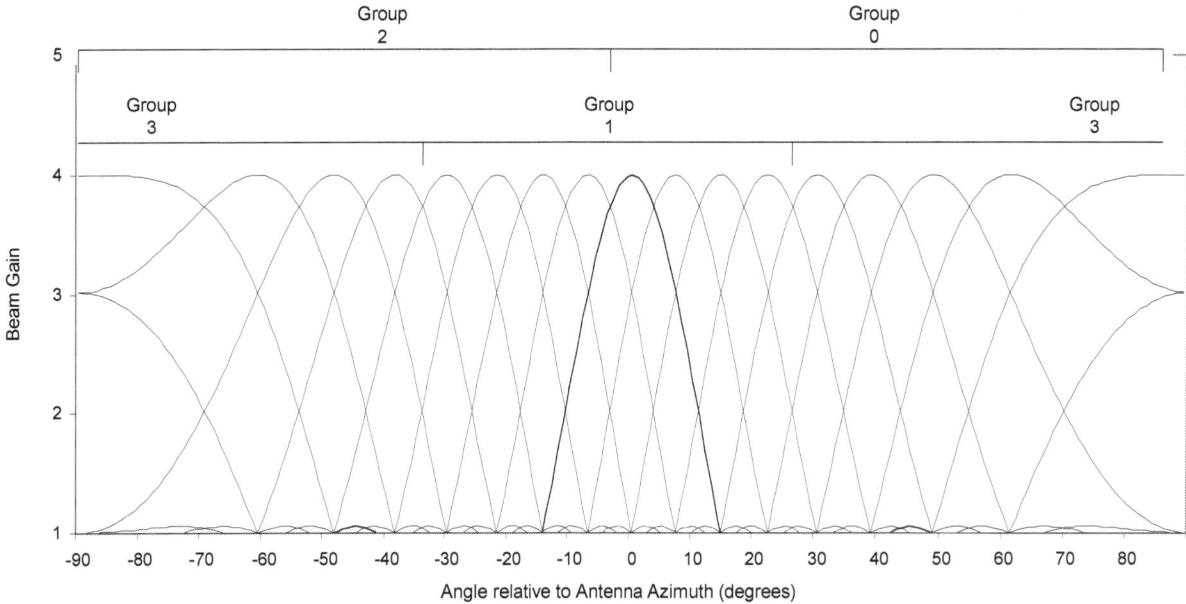

Figure 68 – Set of 16 beams used for transmission mode 9 PMI reporting for ranks 3 and 4

★ Precoding gain tends to decrease as the rank increases and the propagation channel between the transmit and receive antenna becomes less correlated. Consequently, fewer precoding matrices have been defined for the UE to select between when reporting PMI values for the higher ranks

★ PMI reporting for ranks 5 to 7 uses 4 matrices to represent the wideband, long term channel conditions and a single matrix to represent the frequency selective, short term channel conditions. This requires 2 bits of information when reporting the long term channel conditions. It is not necessary for the UE to report the short term channel conditions because only a single precoding matrix has been defined

★ PMI reporting for rank 8 uses a single matrix to represent the wideband, long term channel conditions and a single matrix to represent the frequency selective, short term channel conditions. In this case it is not necessary to report the long term nor short term channel conditions because only single precoding matrices have been defined. There is no scope for beamforming when using rank 8 with an antenna which has 8 physical antenna elements, i.e. there is only a single physical antenna element per rank

★ PMI values are reported to the eNode B during Channel State Information (CSI) reporting on the PUCCH and PUSCH. Both periodic and aperiodic reporting can be configured to report a combination of CQI, PMI and RI

★ Periodic CSI reporting on the PUCCH has been updated within the release 10 version of 3GPP TS 36.213 to accommodate the 2 stage PMI reporting for transmission mode 9. This allows a range of different approaches for reporting PMI_1 and PMI_2. Periodic CSI reporting on the PUCCH using modes 1-1 and 2-1 is applicable when the UE is configured to provide PMI reporting

★ PUCCH CSI reporting mode 1-1 corresponds to periodic wideband reporting. In this case, the UE can be configured to use submode 1 or submode 2. These submodes determine the content of each CSI report:

- Submode 1 uses 2 types of CSI report:
  - a 'type 5' report which signals a combination of RI and wideband PMI_1
  - a 'type 2b' report which signals a combination of wideband CQI and wideband PMI_2. This report also includes a differential wideband CQI if the current RI > 1 to provide CQI information for the second codeword
- Submode 2 also uses 2 types of CSI report:
  - a 'type 3' report which signals RI
  - a 'type 2c' report which signals a combination of wideband CQI, wideband PMI_1 and wideband PMI_2. This report also includes a differential wideband CQI if the current RI > 1 to provide CQI information for the second codeword

  The 'type 2c' report accommodates a relatively large quantity of information so it is necessary to subsample the PMI codebooks for PMI_1 and PMI_2. This means that not all PMI values are permitted and the UE must choose from a subset of the usual values. Subsampling reduces the resolution and accuracy of PMI reporting but avoids the size of the 'type 2c' report becoming too large

★ Submode 1 of reporting mode 1-1 has the benefit of not requiring PMI subsampling, whereas submode 2 has the benefit of using reports which are dedicated to the RI, i.e. an increased number of bits are used to signal the RI and its transmission reliability across the air-interface improves

★ The timing of the RI reports for both submodes 1 and 2 is defined by the equation:

$$\left(10 \times n_f + \lfloor n_s / 2 \rfloor - N_{OFFSET,CQI} - N_{OFFSET,RI}\right) \bmod \left(N_{pd} \cdot M_{RI}\right) = 0$$

where,

n_f is the radio frame number (0 to 1023), and n_s is the time slot number within a radio frame (0 to 19)

$N_{OFFSET,CQI}$ and N_{pd} are determined by the value of $I_{CQI/PMI}$ signalled to the UE by the RRC layer

$N_{OFFSET,RI}$ and M_{RI} are determined by the value of I_{RI} signalled to the UE by the RRC layer

★ The timing of the reports containing wideband CQI values is defined by the equation:

$$\left(10 \times n_f + \lfloor n_s / 2 \rfloor - N_{OFFSET,CQI}\right) \bmod \left(N_{pd}\right) = 0$$

★ The relative rate of reporting the RI and wideband CQI values is controlled using the M_{RI} parameter which can be allocated values of 1, 2, 4, 8, 16 and 32. Allocating a value of 1 means that the RI and wideband CQI reports are sent at the same rate

★ Figure 69 illustrates an example of the timing for reporting mode 1-1 when using submode 1. 'Type 5' reports containing the RI and wideband PMI_1 are sent less frequently than the 'type 2b' reports containing the wideband CQI and wideband PMI_2. This helps the eNode B to keep up-to-date with the CQI and PMI_2 values which are likely to change more frequently than the RI and PMI_1 values

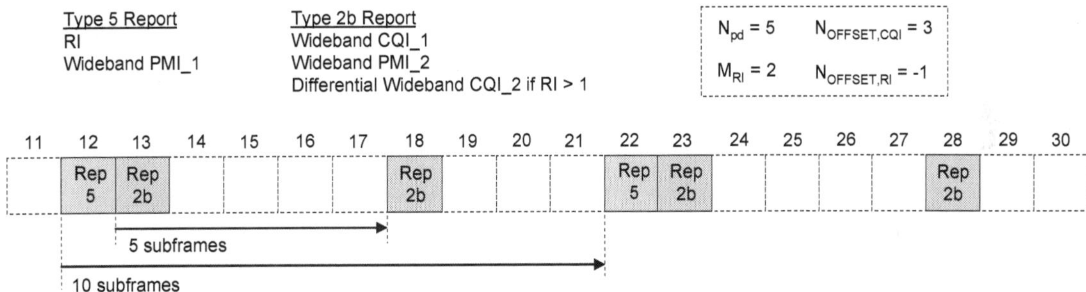

Figure 69 – Example timing for periodic PUCCH CSI reporting mode 1-1 using submode 1

★ Figure 70 illustrates an equivalent example of the timing for reporting mode 1-1 when using submode 2. In this case, it is 'type 3' reports containing the RI which are sent less frequently then the 'type 2c' reports containing the wideband CQI, wideband PMI_1 and wideband PMI_2

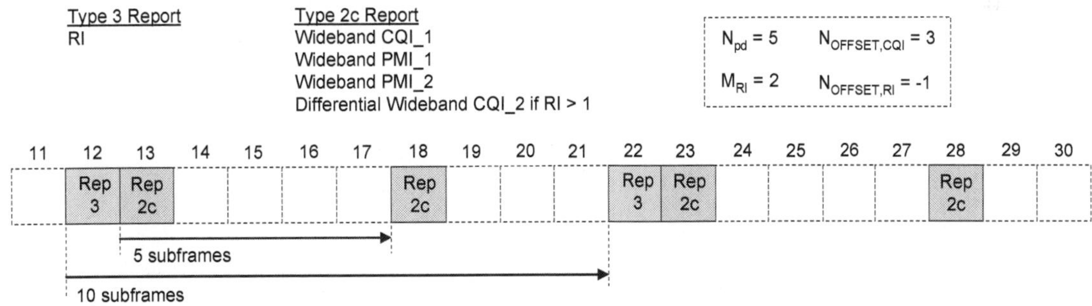

Figure 70 – Example timing for periodic PUCCH CSI reporting mode 1-1 using submode 2

★ PUCCH CSI reporting mode 2-1 corresponds to periodic sub-band reporting. The release 10 version of the 3GPP specifications introduces the Precoding Type Indicator (PTI) for this mode of reporting

- o the PTI is used to signal the type of PMI values within subsequent reports. It is signalled at the same time as the RI:
  - ▪ a 'type 6' report is used to signal the combination of RI and PTI during subframes which satisfy:

$$\left(10 \times n_f + \lfloor n_s / 2 \rfloor - N_{OFFSET,CQI} - N_{OFFSET,RI}\right) \bmod \left(H \cdot N_{pd} \cdot M_{RI}\right) = 0$$

where, H = J × K + 1. 'J' is the number of bandwidth parts, and 'K' is signalled to the UE by the RRC layer and can have values between 1 and 4. This equation indicates that the rate of reporting RI and PTI values can be relatively low compared to the rate of reporting CQI and PMI values which is defined by N_{pd}

★ A PTI value of '0' indicates that the UE has detected that PMI_1 is changing so it needs to be signalled to the eNode B. A PTI value of '1' indicates that the UE has detected that PMI_1 is not changing so reporting can focus on PMI_2. Focusing on PMI_2 provides scope for reporting a series of sub-band PMI_2 values rather than a single wideband value

- PTI = 0
  - a 'type 2a' report is used to signal a wideband PMI_1 during subframes which satisfy:

$$(10 \times n_f + \lfloor n_s / 2 \rfloor - N_{OFFSET,CQI}) \bmod (H' \cdot N_{pd}) = 0$$

  where, H' is signalled to the UE by the RRC layer and can have values of 2 or 4

  - a 'type 2b' report is used to signal a wideband CQI and a wideband PMI_2. This report also includes a differential wideband CQI if the current RI > 1 to provide CQI information for the second codeword. The 'type 2b' report is sent between 'type 2a' reports at a rate defined by N_{pd}
- PTI = 1
  - a 'type 2b' report is used to signal a wideband CQI and a wideband PMI_2. This report also includes a differential wideband CQI if the current RI > 1 to provide CQI information for the second codeword. The 'type 2b' report is sent during subframes which satisfy;

$$(10 \times n_f + \lfloor n_s / 2 \rfloor - N_{OFFSET,CQI}) \bmod (H \cdot N_{pd}) = 0$$

  - a 'type 1a' report is used to signal a sub-band CQI and a sub-band PMI_2. This report also includes a differential sub-band CQI if the current RI > 1 to provide CQI information for the second codeword. The PMI_2 value is generated after subsampling the codebook to reduce the number of possibilities and the corresponding number of bits required for signalling. The 'type 1a' report is sent between 'type 2b' reports at a rate defined by N_{pd}
  - For each 'type 1a' report, the UE selects the preferred sub-band from within a bandwidth part. The channel bandwidth is divided into a number of bandwidth parts (J), where the number of bandwidth parts is a function of the channel bandwidth. Table 28 presents the number of bandwidth parts for each channel bandwidth. The 'type 1a' reports cycle through each bandwidth part so the CQI and PMI values are reported for the preferred sub-band within each bandwidth part

Channel Bandwidth	Sub-band Size (Resource Blocks)	Bandwidth Parts, J
1.4 MHz	-	-
3 MHz	4	2
5 MHz	4	2
10 MHz	6	3
15 MHz	8	4
20 MHz	8	4

Table 28 – The sub-band size and the number of bandwidth parts for each channel bandwidth

★ Figure 71 illustrates an example of the timing for reporting mode 2-1 when the UE has selected a PTI value of 0. In this example, a 'Type 6' report containing the RI and PTI is sent once every 20 subframes. 'Type 2a' reports containing wideband PMI_1 values are sent once every 4 subframes, and 'type 2b' reports containing wideband CQI and wideband PMI_2 values are also sent once every 4 subframes

★ The key point for this example is that the UE has detected that the value of PMI_1 is changing so has selected a PTI value of 0 and is consequently reporting PMI_1, in addition to the CQI and PMI_2. All reported CQI and PMI values are wideband rather than sub-band

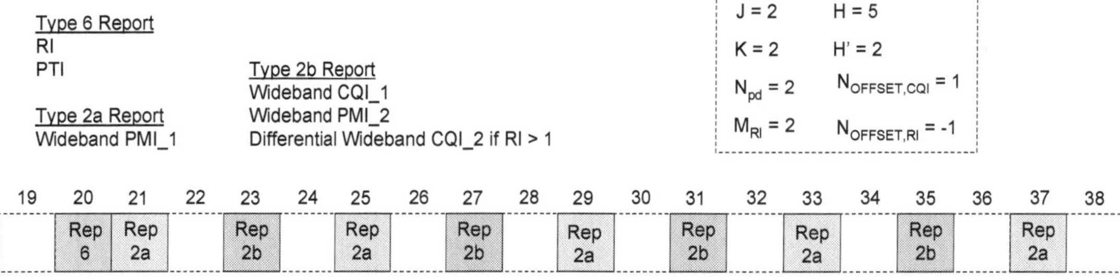

Figure 71 – Example timing for periodic PUCCH CSI reporting mode 2-1 with PTI = 0

★ Figure 72 illustrates an example of the timing for reporting mode 2-1 when the UE has selected a PTI value of 1. In this example, a 'Type 6' report containing the RI and PTI is sent once every 20 subframes. 'Type 2b' reports containing wideband CQI and wideband

PMI_2 values are sent once every 10 subframes. 'Type 1a' reports containing sub-band CQI and sub-band PMI_2 values are sent between the 'type 2b' reports. This example assumes 2 bandwidth parts so there are 2 cycles of the bandwidth parts between each 'type 2b' report

★ The key point for this example is that the UE has detected that the value of PMI_1 is not changing so has selected a PTI value of 1 and is consequently reporting CQI and PMI_2 in both wideband and sub-band formats

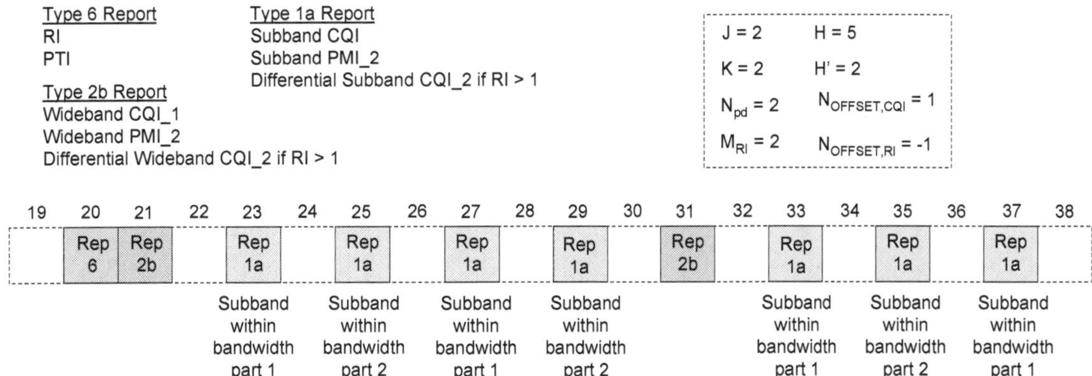

Figure 72 – Example timing for periodic PUCCH CSI reporting mode 2-1 with PTI = 1

★ Transmission mode 9 uses the concept of Physical Resource Block (PRB) Bundling when configured to provide RI/PMI feedback, i.e. when operating in closed loop mode

★ The objective of PRB Bundling is to improve downlink channel estimation at the UE. This is achieved by the eNode B applying the same precoding to 'bundles' of Resource Blocks. A bundle of Resource Blocks is a group of contiguous Resource Blocks

★ The UE can then assume that the precoding is the same within a bundle and channel estimation is improved because each channel estimate is based upon a group of Resource Blocks rather than a single Resource Block. When PRB bundling is not used, the eNode B is free to apply different precoding to each Resource Block and the UE has to estimate the channel response on a per Resource Block basis

★ The drawback of using PRB Bundling is a reduction in the flexibility of precoding at the eNode B, i.e. the eNode B is no longer free to apply different precoding to every Resource Block

★ PRB Bundling does not introduce any additional signalling. The UE can assume that it is applied when using transmission mode 9 in the closed loop mode of operation

★ The bundle size is known as the Precoding Resource block Group (PRG) size

★ The size of each PRG is a function of the channel bandwidth. Values have been chosen to suit the Resource Block Group size used during downlink resource allocation. Channel bandwidths less than 15 MHz use a PRG size equal to the Resource Block Group size. Channel bandwidths of 15 and 20 MHz have larger Resource Block Group sizes so use a PRG size which is half of the Resource Block Group size. Table 29 presents the PRG size for each channel bandwidth

Channel Bandwidth	Total Number of Resource Blocks	Resource Block Group Size (Resource Blocks)	PRG Size (Resource Blocks)	Number of PRG
1.4 MHz	6	1	1	6
3 MHz	15	2	2	8
5 MHz	25	2	2	13
10 MHz	50	3	3	17
15 MHz	75	4	2	38
20 MHz	100	4	2	50

Table 29 – The Precoding Resource block Group (PRG) size for each channel bandwidth

★ The channel bandwidth is divided into PRG starting from the lower end of the channel bandwidth. The final PRG may be smaller if the channel bandwidth is not an integer multiple of the PRG size

★ PRB bundling helps LTE Advanced to achieve its peak spectral efficiency requirements by helping to reduce the overhead generated by the UE specific Reference Signal. The UE uses the UE specific Reference Signal to complete channel estimation. PRB Bundling allows channel estimation to be completed across multiple Resource Blocks so the number of Resource Elements allocated to the UE specific Reference Signal per Resource Block can be reduced

4.5.2 MULTI-USER MIMO

★ The release 10 version of the 3GPP specifications provides a similar multi-user MIMO capability as release 9:
  - a total of 4 layers can be used for multi-user MIMO
    - 2 UE can simultaneously benefit from 2×2 MIMO
    - 4 UE can simultaneously benefit from single layer transmission
  - dynamic switching between single user MIMO and multi-user MIMO
  - support for beamforming and UE specific Reference Signals

★ Multi-user MIMO within release 10 benefits from enhanced Channel State Information (CSI) reporting. The UE is able to use the CSI Reference Signals rather than the cell specific Reference Signals when reporting CSI information

★ Transmission mode 9 provides support for both multi-user MIMO and single user MIMO

★ Similar to transmission mode 8 and multi-user MIMO within the release 9 version of the specifications, transmission mode 9 and multi-user MIMO within the release 10 version of the specifications generates its set of 4 layers using a combination of 2 orthogonal UE specific Reference Signal antenna ports and 2 UE specific Reference Signal scrambling sequences
  - antenna ports 7 and 8
  - scrambling identities 0 and 1

These antenna ports and scrambling identities are illustrated in Figure 73

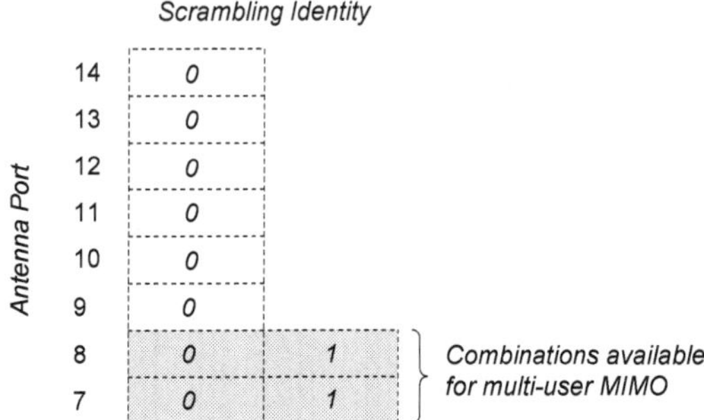

Figure 73 – Antenna ports and scrambling identities available for multi-user MIMO

★ The set of 4 transmission layers allow:
  - 2 users configured with dual layer transmission
  - 4 users configured with single layer transmission
  - 1 user configured with dual layer transmission, and 2 users configured with single layer transmission

These possibilities are illustrated in Figure 74

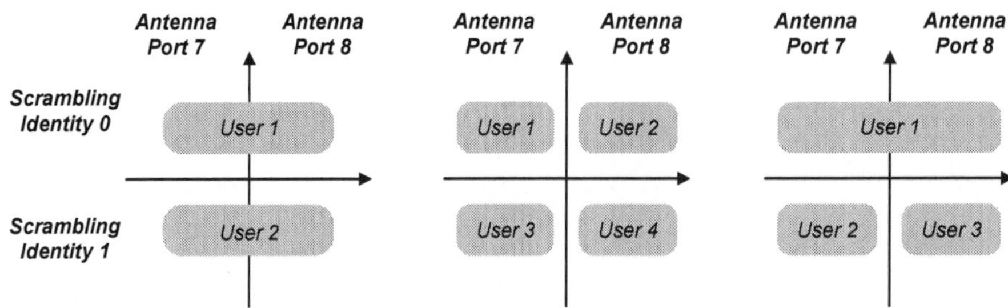

Figure 74 – Allocation of transmission layers for multi-user MIMO

★ 3GPP decided that there was not sufficient justification to increase the number of orthogonal antenna ports available for multi-user MIMO. The group of candidate UE for multi-user MIMO is expected to be relatively small because those UE need to have similar high geometries (ratio of wanted signal to the interference plus noise) and have data ready to be sent during the same subframe

★ DCI format 2C provides support for both multi-user MIMO and single user MIMO when using transmission mode 9. The content of DCI format 2C is presented in section 9.19

★ DCI format 2C includes 3 bits to specify the combination of allocated number of layers, antenna ports and scrambling identity. The mapping between the signalled value and the allocated resources is dependent upon the number of scheduled codewords. This mapping is presented in Table 30

1 Codeword			
Value	Number of Layers	Antenna Ports	Scrambling Identity
0	1	7	0
1			1
2		8	0
3			1
4	2	7 - 8	-
5	3	7 – 9	
6	4	7 - 10	
7	Reserved		

2 Codewords			
Value	Number of Layers	Antenna Ports	Scrambling Identity
0	2	7 – 8	0
1			1
2	3	7 – 9	-
3	4	7 – 10	
4	5	7 – 11	
5	6	7 – 12	
6	7	7 – 13	
7	8	7 - 14	

Table 30 – Antenna Ports, Scrambling Identity and Layers look-up table

★ Signalled values from 0 to 3 are applicable to multi-user MIMO when a singe codeword is scheduled, and signalled values of 0 and 1 are applicable when 2 codewords are scheduled

5 DOWNLINK SIGNALS

5.1 SYNCHRONISATION SIGNALS

- ★ Both the FDD and TDD versions of LTE broadcast Synchronisation Signals in the downlink direction:
  - o Primary Synchronisation Signal (PSS)
  - o Secondary Synchronisation Signal (SSS)
- ★ Synchronisation Signals are broadcast within every 10 ms radio frame
- ★ The UE uses the Synchronisation Signals to:
  - o achieve radio frame, subframe, slot and symbol synchronisation in the time domain
  - o identify the center of the channel bandwidth in the frequency domain
  - o deduce the Physical layer Cell Identity (PCI)
- ★ Detecting the Synchronisation Signals is a prerequisite to measuring the cell specific Reference Signals and decoding the Master Information Block (MIB) on the Physical Broadcast Channel (PBCH)

5.1.1 PRIMARY SYNCHRONISATION SIGNAL

- ★ The Primary Synchronisation Signal (PSS) is broadcast twice during every radio frame and both transmissions are identical
- ★ In the case of FDD:
  - o the PSS is broadcast using the central 62 subcarriers belonging to the last symbol of time slots 0 and 10
- ★ In the case of TDD:
  - o the PSS is broadcast using the central 62 subcarriers belonging to the third symbol of time slot 2 (subframe 1) and the third symbol of time slot 12 (subframe 6)
  - o subframe 1 is always a special subframe so the PSS is sent as part of the Downlink Pilot Time Slot (DwPTS)
  - o subframe 6, may or may not be a special subframe, depending upon the uplink-downlink subframe configuration. It is a special subframe for configurations 0, 1, 2 and 6. Otherwise it is a normal downlink subframe
- ★ The PSS is used to:
  - o achieve subframe, slot and symbol synchronisation in the time domain
  - o identify the center of the channel bandwidth in the frequency domain
  - o deduce a pointer towards 1 of 3 Physical layer Cell Identities (PCI)
    - ▪ PCI are organised into 168 groups of 3 so the Primary Synchronisation Signal identifies the position of the PCI within the group but does not identify the group itself
- ★ The PSS cannot be used to achieve radio frame synchronisation because both transmissions within the radio frame are identical and equally spaced in time

5.1.2 SECONDARY SYNCHRONISATION SIGNAL

- ★ The Secondary Synchronisation Signal (SSS) is broadcast twice within every radio frame. The two transmissions of the SSS are different so the UE can detect which is the first and which is the second
- ★ In the case of FDD:
  - o the SSS is broadcast using the central 62 subcarriers belonging to the second to last symbol of time slots 0 and 10
- ★ In the case of TDD:
  - o the SSS is broadcast using the central 62 subcarriers belonging to the last symbol of time slot 1 (subframe 0) and the last symbol of time slot 11 (subframe 5)
  - o both time slots 1 and 11 are always within normal downlink subframes
- ★ The SSS is used to:
  - o achieve radio frame synchronisation
  - o deduce a pointer towards 1 of 168 Physical layer Cell Identity (PCI) groups
    - ▪ allows the PCI to be deduced when combined with the pointer from the PSS

★ Figure 75 illustrates the position of the Primary and Secondary Synchronization Signals in both the time and frequency domains for the FDD 3 MHz channel bandwidth. Only the fist half of the radio frame is shown (time slots 0 to 9). The position in the frequency domain is the same for TDD

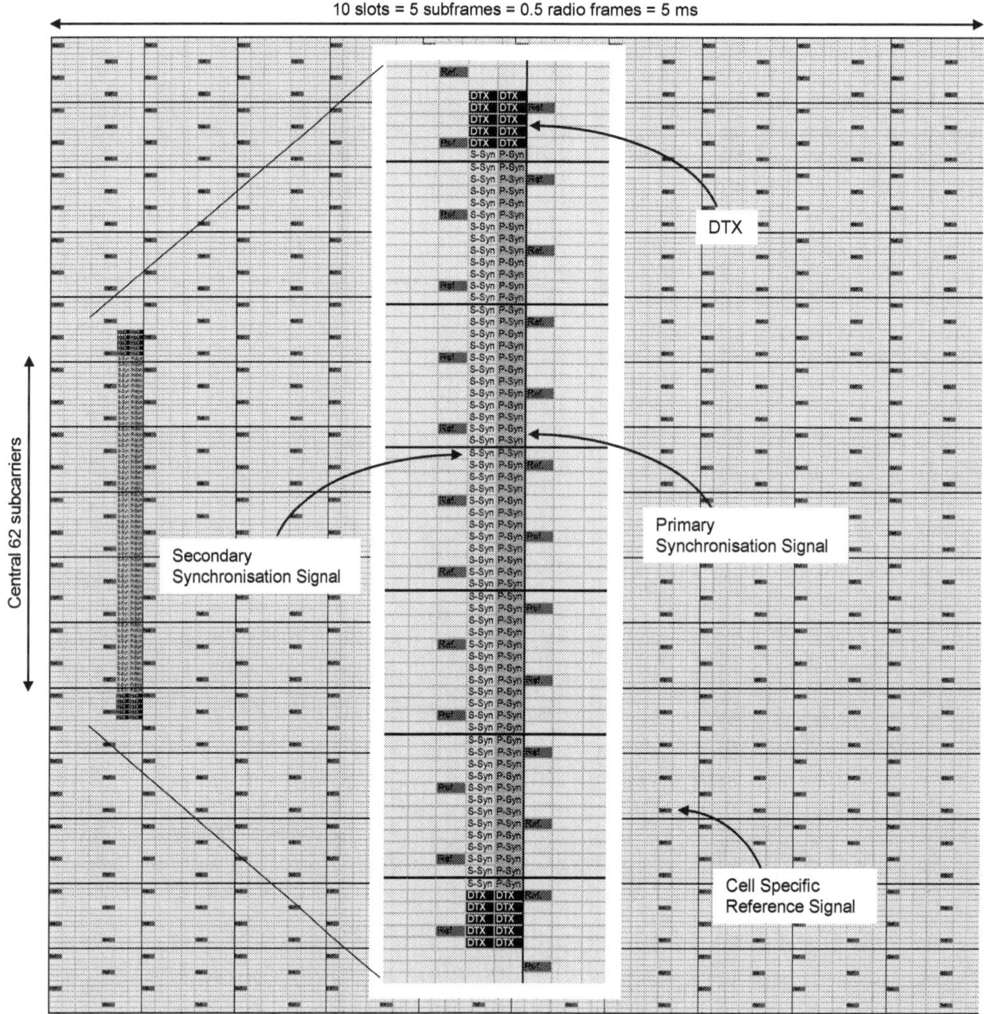

Figure 75 – FDD Synchronisation Signals within time slots 0 to 9 of a 3 MHz channel bandwidth

★ The 5 Resource Elements above and below the Synchronisation Signals are not used for transmission. They represent periods of Discontinuous Transmission (DTX). These are present for both FDD and TDD

★ The set of Resource Elements allocated to the Synchronisation Signals is independent of the channel bandwidth. The UE does not require any knowledge of the channel bandwidth prior to detecting the Synchronisation Signals. The downlink channel bandwidth is subsequently read from the Master Information Block (MIB) on the Physical Broadcast Channel (PBCH)

★ Figure 76 illustrates the timing of the PSS and SSS for FDD. This example assumes the normal cyclic prefix because there are 7 symbols within each time slot. The extended cyclic prefix follows a similar pattern except there are only 6 symbols within the time slot (the SSS and PSS remain within the last two symbols of the time slot)

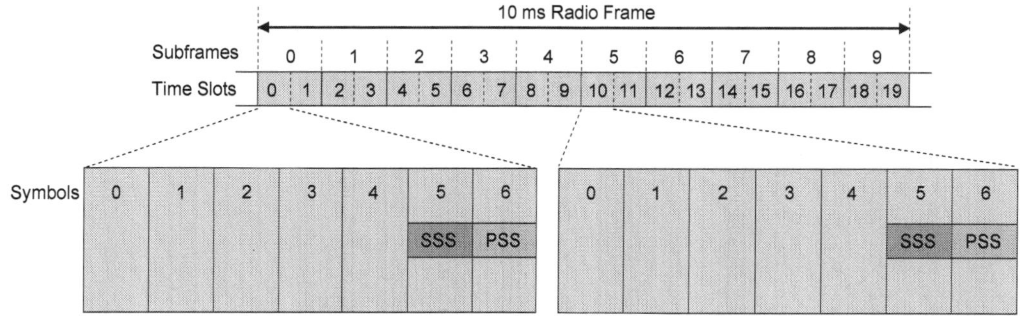

Figure 76 – Timing of Synchronisation Signals for FDD

★ Figure 77 illustrates the timing of the PSS and SSS for TDD. The example assumes the normal cyclic prefix, uplink-downlink subframe configuration 0 and special subframe configuration 0. The extended cyclic prefix follows a similar pattern except there are only 6 symbols within the time slot (the SSS remains within the last symbol of time slots 1 and 11, while the PSS remains within the third symbol of time slots 2 and 12)

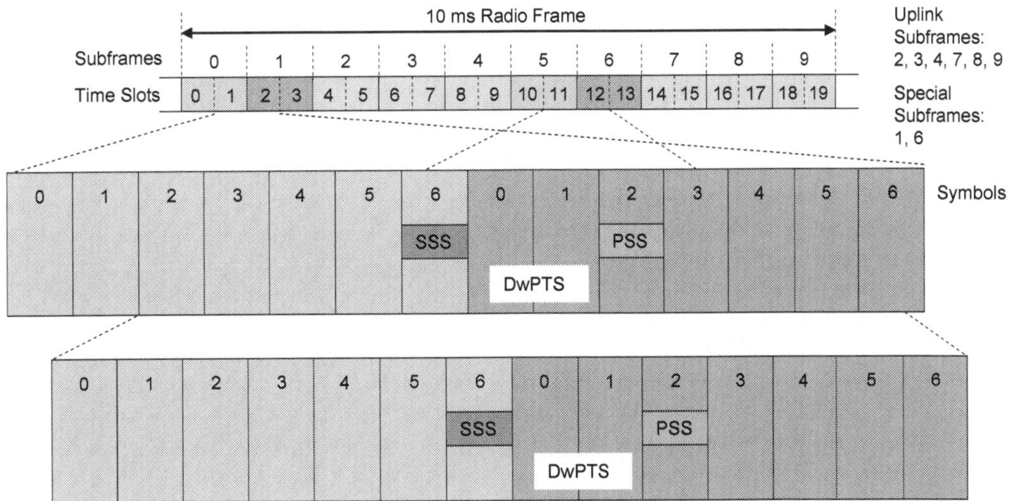

Figure 77 – Timing of Synchronisation Signals for TDD

★ In the case of TDD, the SSS and PSS are not in adjacent symbols. The first two symbols within time slots 2 and 12 are left available for the PCFICH, PHICH and PDCCH

★ Synchronisation Signals represent an overhead which reduces the number of Resource Elements available for user plane data

- overhead decreases for larger channel bandwidths
- overhead increases when using the extended cyclic prefix

★ In the case of FDD, the overhead generated by the Synchronisation Signals is presented in Table 31

	1.4 MHz	3 MHz	5 MHz	10 MHz	15 MHz	20 MHz
Sync. Resource Elements per 10 ms Radio Frame	2×144 = 288	2×144 = 288	2×144 = 288	2×144 = 288	2×144 = 288	2×144 = 288
Normal Cyclic Prefix						
Total Resource Elements per 10 ms Radio Frame	10080	25200	42000	84000	126000	168000
Overhead	2.9 %	1.1 %	0.7 %	0.3 %	0.2 %	0.2 %
Extended Cyclic Prefix						
Total Resource Elements per 10 ms Radio Frame	8640	21600	36000	72000	108000	144000
Overhead	3.3 %	1.3 %	0.8 %	0.4 %	0.3 %	0.2 %

Table 31 – Overhead generated by Synchronisation Signals (FDD)

★ In the case of TDD, the overhead generated by the Synchronisation Signals is the same as that for FDD when calculating the overhead relative to the total number of Resource Elements (both uplink and downlink)

★ However, the overhead for TDD is greater when calculating relative to the number of downlink Resource Elements. This calculation is multi-dimensional because the number of downlink Resource Elements depends upon the cyclic prefix length, the channel bandwidth, the uplink-downlink subframe configuration and the special subframe configuration

★ Table 32 presents the overheads when assuming uplink-downlink subframe configurations 0 and 5. These configurations represent those with the highest and lowest overheads, i.e. uplink-downlink subframe configuration 0 has the least downlink Resource Elements, while uplink-downlink subframe configuration 5 has the most downlink Resource Elements

★ In the worst case of uplink-downlink subframe configuration 0, using the extended cyclic prefix and special subframe configuration 0, the overhead for TDD reaches 13.3 %

★ 3GPP References: TS 36.211

		1.4 MHz	3 MHz	5 MHz	10 MHz	15 MHz	20 MHz
Sync. Sig. Resource Elements per 10 ms Radio Frame		2×144 = 288	2×144 = 288	2×144 = 288	2×144 = 288	2×144 = 288	2×144 = 288
Normal Cyclic Prefix							
Uplink-Downlink Subframe Config. 0	Special SF Config. 0	11.8 %	4.7 %	2.8 %	1.4 %	0.9 %	0.7 %
	Special SF Config. 1	8.7 %	3.5 %	2.1 %	1.0 %	0.7 %	0.5 %
	Special SF Config. 2	8.3 %	3.3 %	2.0 %	1.0 %	0.7 %	0.5 %
	Special SF Config. 3	8.0 %	3.2 %	1.9 %	1.0 %	0.6 %	0.5 %
	Special SF Config. 4	7.7 %	3.1 %	1.8 %	0.9 %	0.6 %	0.5 %
	Special SF Config. 5	11.8 %	4.7 %	2.8 %	1.4 %	0.9 %	0.7 %
	Special SF Config. 6	8.7 %	3.5 %	2.1 %	1.0 %	0.7 %	0.5 %
	Special SF Config. 7	8.3 %	3.3 %	2.0 %	1.0 %	0.7 %	0.5 %
	Special SF Config. 8	8.0 %	3.2 %	1.9 %	1.0 %	0.6 %	0.5 %
Uplink-Downlink Subframe Config. 5	Special SF Config. 0	3.5 %	1.4 %	0.8 %	0.4 %	0.3 %	0.2 %
	Special SF Config. 1	3.3 %	1.3 %	0.8 %	0.4 %	0.3 %	0.2 %
	Special SF Config. 2	3.3 %	1.3 %	0.8 %	0.4 %	0.3 %	0.2 %
	Special SF Config. 3	3.3 %	1.3 %	0.8 %	0.4 %	0.3 %	0.2 %
	Special SF Config. 4	3.2 %	1.3 %	0.8 %	0.4 %	0.3 %	0.2 %
	Special SF Config. 5	3.5 %	1.4 %	0.8 %	0.4 %	0.3 %	0.2 %
	Special SF Config. 6	3.3 %	1.3 %	0.8 %	0.4 %	0.3 %	0.2 %
	Special SF Config. 7	3.3 %	1.3 %	0.8 %	0.4 %	0.3 %	0.2 %
	Special SF Config. 8	3.3 %	1.3 %	0.8 %	0.4 %	0.3 %	0.2 %
Extended Cyclic Prefix							
Uplink-Downlink Subframe Config. 0	Special SF Config. 0	13.3 %	5.3 %	3.2 %	1.6 %	1.1 %	0.8 %
	Special SF Config. 1	10.0 %	4.0 %	2.4 %	1.2 %	0.8 %	0.6 %
	Special SF Config. 2	9.5 %	3.8 %	2.3 %	1.1 %	0.8 %	0.6 %
	Special SF Config. 3	9.1 %	3.6 %	2.2 %	1.1 %	0.7 %	0.5 %
	Special SF Config. 4	13.3 %	5.3 %	3.2 %	1.6 %	1.1 %	0.8 %
	Special SF Config. 5	10.0 %	4.0 %	2.4 %	1.2 %	0.8 %	0.6 %
	Special SF Config. 6	9.5 %	3.8 %	2.3 %	1.1 %	0.8 %	0.6 %
Uplink-Downlink Subframe Config. 5	Special SF Config. 0	4.0 %	1.6 %	1.0 %	0.5 %	0.3 %	0.2 %
	Special SF Config. 1	3.8 %	1.5 %	0.9 %	0.5 %	0.3 %	0.2 %
	Special SF Config. 2	3.8 %	1.5 %	0.9 %	0.5 %	0.3 %	0.2 %
	Special SF Config. 3	3.8 %	1.5 %	0.9 %	0.5 %	0.3 %	0.2 %
	Special SF Config. 4	4.0 %	1.6 %	1.0 %	0.5 %	0.3 %	0.2 %
	Special SF Config. 5	3.8 %	1.5 %	0.9 %	0.5 %	0.3 %	0.2 %
	Special SF Config. 6	3.8 %	1.5 %	0.9 %	0.5 %	0.3 %	0.2 %

Table 32 – Overhead generated by Synchronisation Signals (TDD uplink-downlink subframe configurations 0 and 5)

5.2 REFERENCE SIGNALS

- 3GPP TS 36.211 specifies 5 types of downlink Reference Signals:
  - Cell specific Reference Signals (also known as Common Reference Signals (CRS))
  - MBSFN Reference Signals
  - UE specific Reference Signals (also known as Demodulation Reference Signals (DM-RS))
  - Positioning Reference Signals (PRS)
  - Channel State Information Reference Signals (CSI-RS)
- Each Reference Signal uses its own subset of antenna ports. The antenna ports used by each Reference Signal are presented in Table 33

	Antenna Ports						
	0 to 3	4	5	6	7 to 8	9 to 14	15 to 22
Cell Specific	3GPP rel. 8	-	-	-	-	-	-
MBSFN	-	3GPP rel. 8	-	-	-	-	-
UE Specific	-	-	3GPP rel. 8	-	3GPP rel. 9	3GPP rel. 10	-
Positioning	-	-	-	3GPP rel. 9	-	-	-
CSI	-	-	-		-	-	3GPP rel. 10

Table 33 – Antenna ports used by each Reference Signal

- Cell specific Reference Signals are broadcast across the entire cell. They are used to support CQI reporting and demodulation. They can also be used to support Reference Signal Received Power (RSRP) and Reference Signal Received Quality (RSRQ) measurements for handover purposes
- MBSFN Reference Signals are used to support the reception of Multimedia Broadcast Multicast Services (MBMS). They are only defined for the extended cyclic prefix but can be applied to both the 15 kHz and 7.5 kHz subcarrier spacings
- UE specific Reference Signals can be directed towards individual UE using beamforming techniques. The same beamforming is applied to both the user plane data and the UE specific Reference Signal. This allows the UE to use the UE specific Reference Signal for channel estimation and subsequent demodulation. UE specific Reference Signals are only included within the Resource Blocks which have been allocated to the relevant UE. Initial support for UE specific Reference Signals was introduced within the release 8 version of the 3GPP specifications. Additional UE specific Reference Signals were introduced within the release 9 and 10 versions of the specifications
- Positioning Reference Signals are used to improve the 'hearability' of neighbouring cells when completing measurements for the downlink Observed Time Difference Of Arrival (OTDOA) positioning method. Hearability of co-channel neighbouring cells can be challenging when UE experience a high signal strength from the serving cell
- CSI Reference Signals are used to improve the reporting of Channel State Information (CSI) when using transmission mode 9. CSI Reference Signals occupy Resource Elements usually allocated to the PDSCH. This allows a more meaningful measure of channel quality but results in the PDSCH transmission being punctured. CSI Reference Signals are broadcast across the entire cell area and do not have beamforming applied. This allows them to be used by all UE within the cell

5.2.1 CELL SPECIFIC REFERENCE SIGNALS

- The cell specific Reference Signal (also known as the Common Reference Signal (CRS)) is:
  - equivalent to the CPICH in UMTS networks
  - used to support CQI reporting, demodulation, cell selection, cell reselection and handover
  - allocated Resource Elements which are distributed in both the time and frequency domains
  - broadcast from antenna ports 0 to 3 during subframes supporting PDSCH transmission (with the exception that the cell specific Reference Signal is not broadcast during the data section of MBSFN subframes used for PDSCH transmission to 3GPP release 10 devices. These are so called 'LTE Advanced' subframes which benefit from a reduced Reference Signal overhead)
- The cell specific Reference Signal is only defined for the 15 kHz subcarrier spacing, i.e. it is not supported for the 7.5 kHz subcarrier spacing used for Multimedia Broadcast Multicast Services (MBMS)
- The sequence used to generate each cell specific Reference Signal is a function of the Physical layer Cell Identity (PCI) and the cyclic prefix duration (normal or extended)
- The mapping of the cell specific Reference Signal onto the set of Resource Elements is the same for both FDD and TDD
- The Resource Elements allocated to the cell specific Reference Signal are dependent upon the Physical layer Cell Identity. This dependency is illustrated in Figure 78

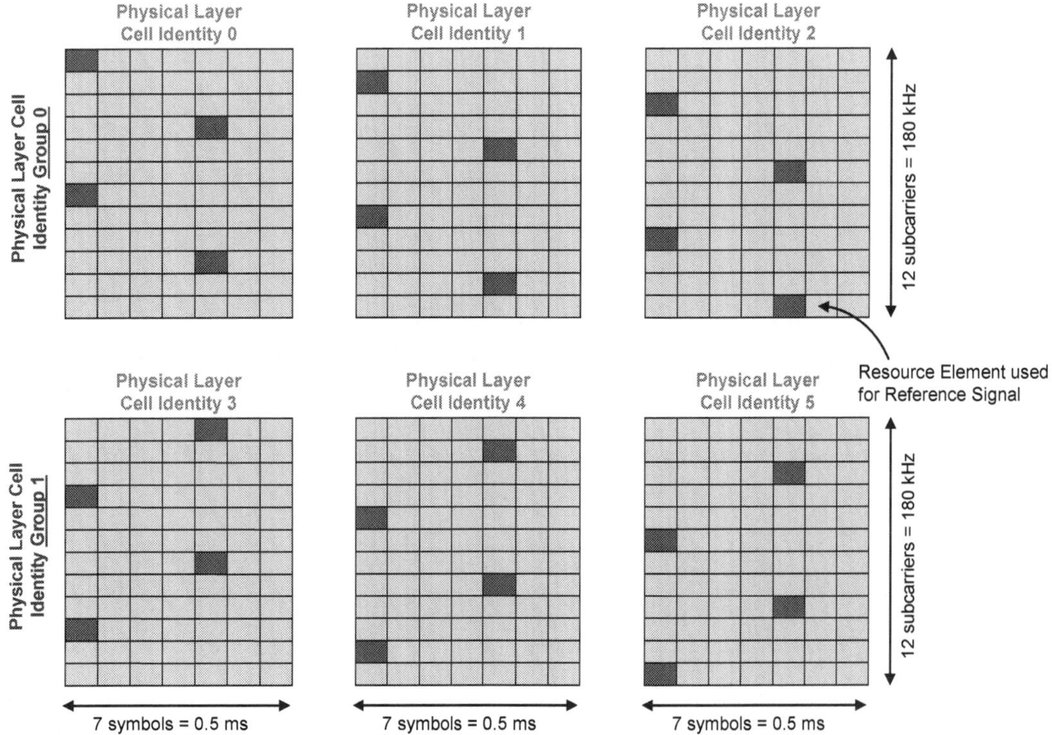

Figure 78 – Cell specific Reference Signal as a function of the Physical layer Cell Identity

- The Resource Element allocation cycles once every 6 Physical layer Cell Identities, e.g. identity 6 has the same Resource Element allocation as identity 0. This corresponds to 1 cycle for every 2 Physical layer Cell Identity Groups
- Once the UE has decoded the Primary and Secondary Synchronisation Signals and consequently identified the Physical layer Cell Identity, then it is able to deduce both the Resource Elements allocated to the Reference Signal, and the sequence used to generate the Reference Signal
- The Resource Element allocation for the cell specific Reference Signal also depends upon the number of transmit antenna ports being used, i.e. whether transmit diversity or spatial multiplexing is being used
- Cell specific Reference Signals are allocated Resource Elements from each antenna port when multiple antenna ports are used. The dependence upon the number of antenna ports is illustrated in Figure 79. This figure is applicable to the normal cyclic prefix because there are 7 symbols within each Resource Block
- When a Resource Element is allocated to the cell specific Reference Signal on one antenna port, the corresponding Resource Elements on the other antenna ports are left empty, i.e. DTX is applied to help improve the quality of the received Reference Signal

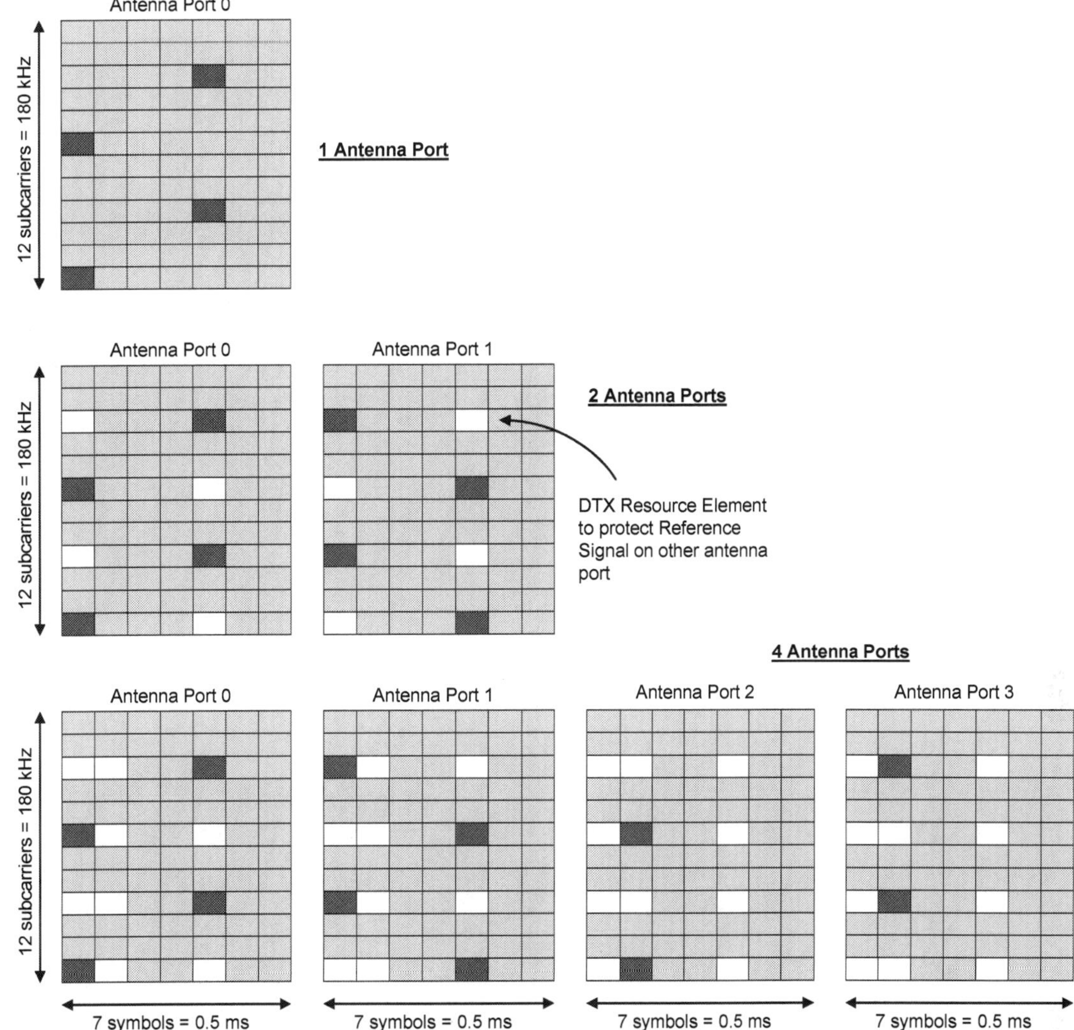

Figure 79 – Impact of the number of antenna ports upon the cell specific Reference Signal (normal cyclic prefix)

★ The equivalent Resource Element mapping for the extended cyclic prefix is illustrated in Figure 80. In this case there are only 6 symbols within each Resource Block. The number of Resource Elements allocated to the cell specific Reference Signal remains the same so the overhead for the extended cyclic prefix is greater

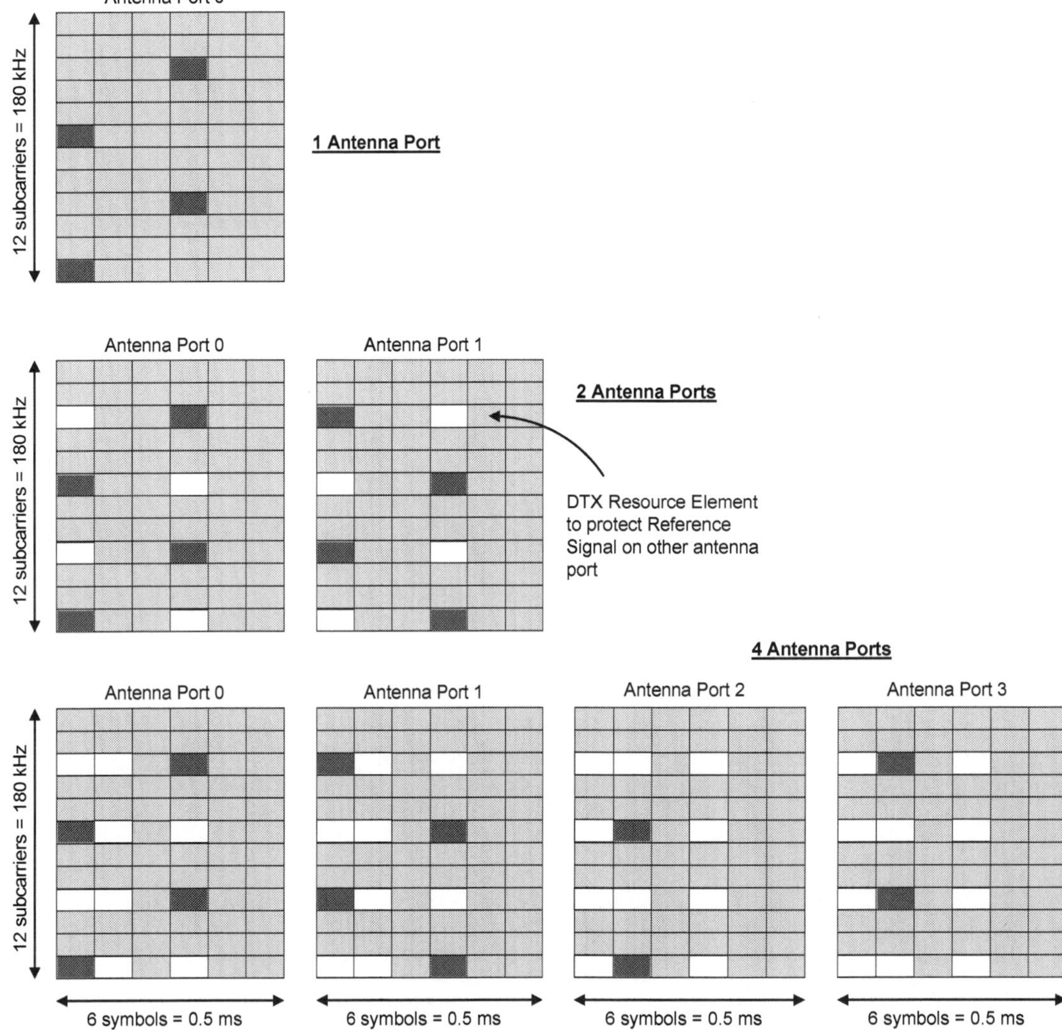

Figure 80 – Impact of the number of antenna ports upon the cell specific Reference Signal (extended cyclic prefix)

- ★ The cell specific Reference Signal represents an overhead which reduces the number of Resource Elements available for user plane data:
  - o overhead increases when multiple transmit antenna ports are used
  - o overhead increases when using the extended cyclic prefix
- ★ In the case of FDD, the overhead generated by the cell specific Reference Signal is presented in Table 34. Figures are presented for 1, 2 and 4 antenna ports

	1 Antenna Port	2 Antenna Ports	4 Antenna Ports
Reference Signal Resource Elements per Resource Block	4	8	12
Overhead (normal cyclic prefix)	4 / 84 = 4.8 %	8 / 84 = 9.5 %	12 / 84 = 14.3 %
Overhead (extended cyclic prefix)	4 / 72 = 5.6 %	8 / 72 = 11.1 %	12 / 72 = 16.7 %

Table 34 – Overhead generated by cell specific Reference Signals (FDD)

- ★ In general, the overheads generated by the cell specific Reference Signals are the same for both FDD and TDD. The overheads only change during the TDD special subframes because the eNode B transmits only a subset of the OFDMA symbols
- ★ Table 35 presents the overheads for the TDD special subframes when using single antenna port transmission. The calculations are completed over a 180 kHz bandwidth (12 subcarriers)

Special Subframe Configuration	Normal Cyclic Prefix				Extended Cyclic Prefix			
	DwPTS symbols	Total Resource Elements	Cell Specific Ref. Signal Res. Ele.	Overhead	DwPTS symbols	Total Resource Elements	Cell Specific Ref. Signal Res. Ele.	Overhead
0	3 sym	36	2	5.6 %	3 sym	36	2	5.6 %
1	9 sym	108	6	5.6 %	8 sym	96	6	6.3 %
2	10 sym	120	6	5.0 %	9 sym	108	6	5.6 %
3	11 sym	132	6	4.5 %	10 sym	120	8	6.7 %
4	12 sym	144	8	5.6 %	3 sym	36	2	5.6 %
5	3 sym	36	2	5.6 %	8 sym	96	6	6.3 %
6	9 sym	108	6	5.6 %	9 sym	108	6	5.6 %
7	10 sym	120	6	5.6 %				
8	11 sym	132	6	4.5 %				

Table 35 – Overhead generated by cell specific Reference Signals during TDD special subframes (1 antenna port)

★ Table 36 presents the overheads for the TDD special subframes when using 4 antenna port transmission. The calculations are completed over a 180 kHz bandwidth (12 subcarriers). The figures for the total Resource Elements and cell specific Reference Signal Resource Elements are summed across all 4 antenna ports

Special Subframe Configuration	Normal Cyclic Prefix				Extended Cyclic Prefix			
	DwPTS symbols	Total Resource Elements	Cell Spec. Ref. Sign. Res. Ele.	Overhead	DwPTS symbols	Total Resource Elements	Cell Spec. Ref. Sign. Res. Ele.	Overhead
0	3 sym	144	32	22.2 %	3 sym	144	32	22.2 %
1	9 sym	432	80	18.5 %	8 sym	384	80	20.8 %
2	10 sym	480	80	16.7 %	9 sym	432	80	18.5 %
3	11 sym	528	80	15.2 %	10 sym	480	96	20.0 %
4	12 sym	576	96	16.7 %	3 sym	144	32	22.2 %
5	3 sym	144	32	22.2 %	8 sym	384	80	20.8 %
6	9 sym	432	80	18.5 %	9 sym	432	80	18.5 %
7	10 sym	480	80	16.7 %				
8	11 sym	528	80	15.2 %				

Table 36 – Overhead generated by cell specific Reference Signals during TDD special subframes (4 antenna ports)

★ In this case, the overheads start to exceed 20%, but these figures are only applicable during TDD special subframes, i.e. the subframes used when switching from downlink transmission to uplink transmission

★ 3GPP References: TS 36.211

5.2.2 MBSFN REFERENCE SIGNALS

- ★ MBMS Single Frequency Network (MBSFN) Reference Signals were introduced within the release 8 version of the 3GPP specifications
- ★ MBSFN Reference Signals provide support to UE receiving MBMS services. They are only broadcast within Resource Blocks allocated to the PMCH physical channel
- ★ MBSFN Reference Signals are
  - o only defined for the extended cyclic prefix because the delay spread between transmissions received from multiple MBMS cells is expected to be relatively large (UE receive MBMS transmissions simultaneously from multiple cells)
  - o defined for both the 15 kHz and 7.5 kHz subcarrier spacings
  - o defined for both FDD and TDD
- ★ Antenna port 4 is used to transmit MBSFN Reference Signals (single antenna port transmission)
- ★ The sequence used to generate the MBSFN Reference Signals is a function of the MBSFN area identity
- ★ The Resource Elements allocated to the MBSFN Reference Signals are fixed, and are not a function of the Physical layer Cell Identity (PCI) nor MBSFN area identity
- ★ Figure 81 illustrates the MBSFN Reference Signals for the 15 kHz and 7.5 kHz subcarrier spacings

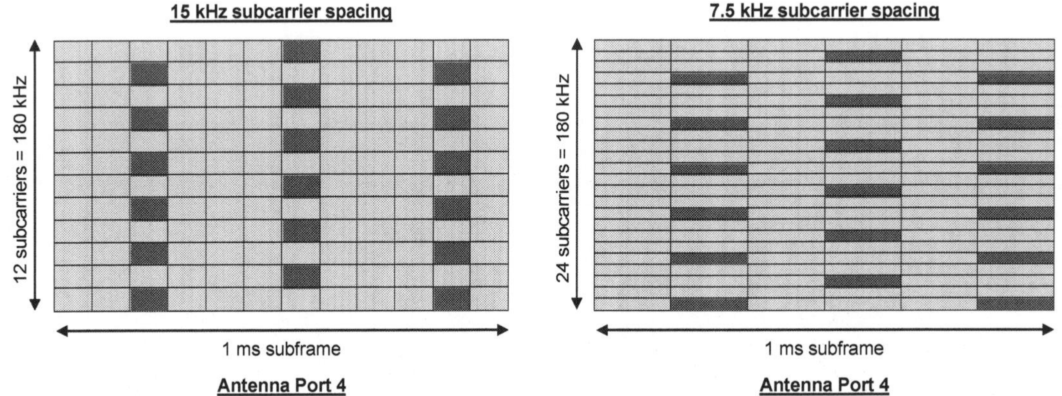

Figure 81 – MBSFN Reference Signals (antenna port 4)

- ★ The overhead generated by the MBSFN Reference Signals is presented in Table 37. The Resource Element figures in Table 37 are summed across a pair of Resource Blocks, i.e. across a 1 ms subframe as shown in Figure 81

	15 kHz subcarrier spacing	7.5 kHz subcarrier spacing
Reference Signal Resource Elements	18	18
Total Resource Elements	144	144
Overhead	18 / 144 = 12.5 %	18 / 144 = 12.5 %

Table 37 – Overhead generated by the MBSFN Reference Signals

- ★ 3GPP References: TS 36.211

5.2.3 UE SPECIFIC REFERENCE SIGNALS

★ UE specific Reference Signals are also known as Demodulation Reference Signals (DM-RS). They can be directed towards individual UE using beamforming techniques, and are used by the UE to support demodulation of the PDSCH which the eNode B transmits from the same antenna port, i.e. the same beamforming is applied to both the UE specific Reference Signal and the PDSCH

★ Initial support for UE specific Reference Signals was introduced within the release 8 version of the 3GPP specifications. Additional UE specific Reference Signals were introduced within the release 9 and 10 versions of the specifications:

- o release 8 supports UE specific Reference Signals on antenna port 5, allowing:
  - single user, single-layer beamforming
  - unlimited multi-user single-layer beamforming using quasi-orthogonal UE specific Reference Signals
- o release 9 adds support for UE specific Reference Signals on antenna ports 7 and 8, allowing:
  - single user, dual-layer beamforming for 2×2 MIMO
  - 2 user dual-layer beamforming using orthogonal and quasi-orthogonal UE specific Reference Signals
  - 4 user, single-layer beamforming using orthogonal and quasi-orthogonal UE specific Reference Signals
- o release 10 adds support for UE specific Reference Signals on antenna ports 9 to 14, allowing
  - single user, 8-layer beamforming for 8×8 MIMO

★ The eNode B transmits UE specific Reference Signals within the Resource Blocks allocated to the associated PDSCH. For example, if the eNode B allocates Resource Blocks 1 to 12 to a UE, then the eNode B transmits both the UE specific Reference Signal and the PDSCH within those 12 Resource Blocks. The same UE specific Reference Signal could be used for another UE during the same subframe in a different Resource Block allocation

★ UE specific Reference Signals are applicable to both FDD and TDD, using either the normal or extended cyclic prefix (although antenna ports 9 to 14 are not supported with the extended cyclic prefix)

★ UE specific Reference Signals coexist with cell specific Reference Signals, i.e. both types of Reference Signal are included within Resource Blocks used by UE specific Reference Signals

★ UE specific Reference Signals are only defined for the 15 kHz subcarrier spacing. They are not supported for the 7.5 kHz subcarrier spacing used by Multimedia Broadcast Multicast Services (MBMS)

★ The sequence used to generate each UE specific Reference Signal is a function of:

- o antenna port 5: the Physical layer Cell Identity (PCI) and RNTI of the UE to which the PDSCH is being sent
- o antenna ports 7 and 8: the PCI, scrambling identity and Orthogonal Covering Code (OCC)
- o antenna ports 9 to 14: the PCI and OCC

★ Orthogonal Covering Codes (OCC) are used to keep the UE specific Reference Signals orthogonal from each other at the transmitter. These codes are applied in the time domain to help maintain the orthogonality across the air-interface radio propagation channel

★ The PCI, RNTI and scrambling identity allow the UE to differentiate between UE specific Reference Signals but do not provide orthogonality between the sequences

★ An example Resource Element allocation for the UE specific Reference Signal on antenna port 5 is illustrated in Figure 82

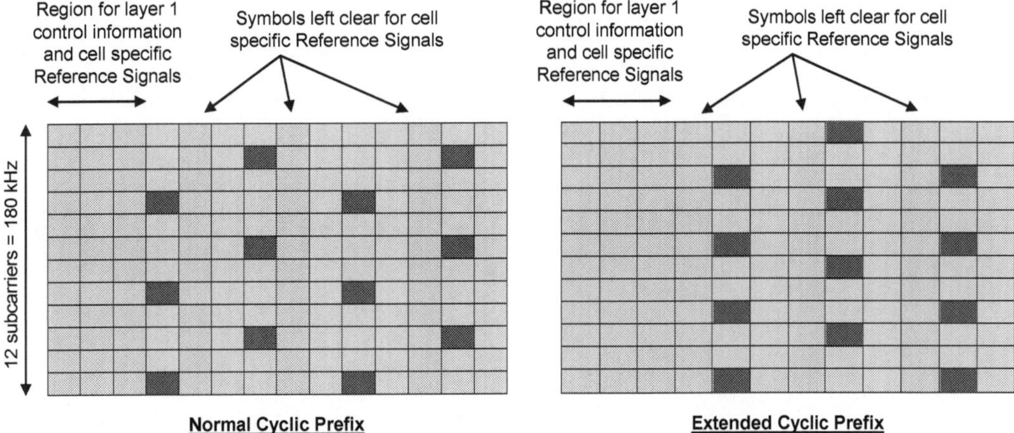

Figure 82 – UE specific Reference Signals (antenna port 5)

- This example assumes PCI mod 3 = 0. If PCI mod 3 = 1 then the vertical position of the UE specific Reference Signals would move up 1 Resource Element. Likewise, if PCI mod 3 = 2 then the vertical position of the UE specific Reference Signals would move up 2 Resource Elements
- Thus, the Resource Element allocation cycles once every 3 Physical layer Cell Identities, e.g. identity 3 has the same Resource Element allocation as identity 0. This corresponds to 1 cycle for every Physical layer Cell Identity Group
- The symbols selected for the UE specific Reference Signals do not coincide with the symbols used by the Cell specific Reference Signals. This helps the two Reference Signals to coexist. Symbols at the start of each subframe are also left clear for the PCFICH, PHICH and PDCCH physical channels
- The UE specific Reference Signals illustrated in Figure 82 allow single user, single-layer beamforming. In this context, the term 'single user' refers to a specific Resource Block allocation, i.e. a set of Resource Blocks is allocated to a single user. Another set of Resource Blocks could be allocated to a second user during the same subframe and re-use the same UE specific Reference Signal
- UE specific Reference Signals on antenna port 5 can also support unlimited multi-user single-layer beamforming. This is possible because the sequence used to generate the UE specific Reference Signal is a function of the UE identity (RNTI). This allows users to differentiate between their transmissions. However, the sequences are not orthogonal to one another and are likely to interfere with each other even when the UE specific beams are directed in different directions
- Resource Element allocations for the UE specific Reference Signals on antenna ports 7 and 8 are illustrated in Figure 83. In contrast to antenna port 5, the UE specific Reference Signals on antenna ports 7 and 8 have fixed positions which are not a function of the Physical layer Cell Identity (although the sequences mapped onto the Resource Elements are a function of the Physical layer Cell Identity)

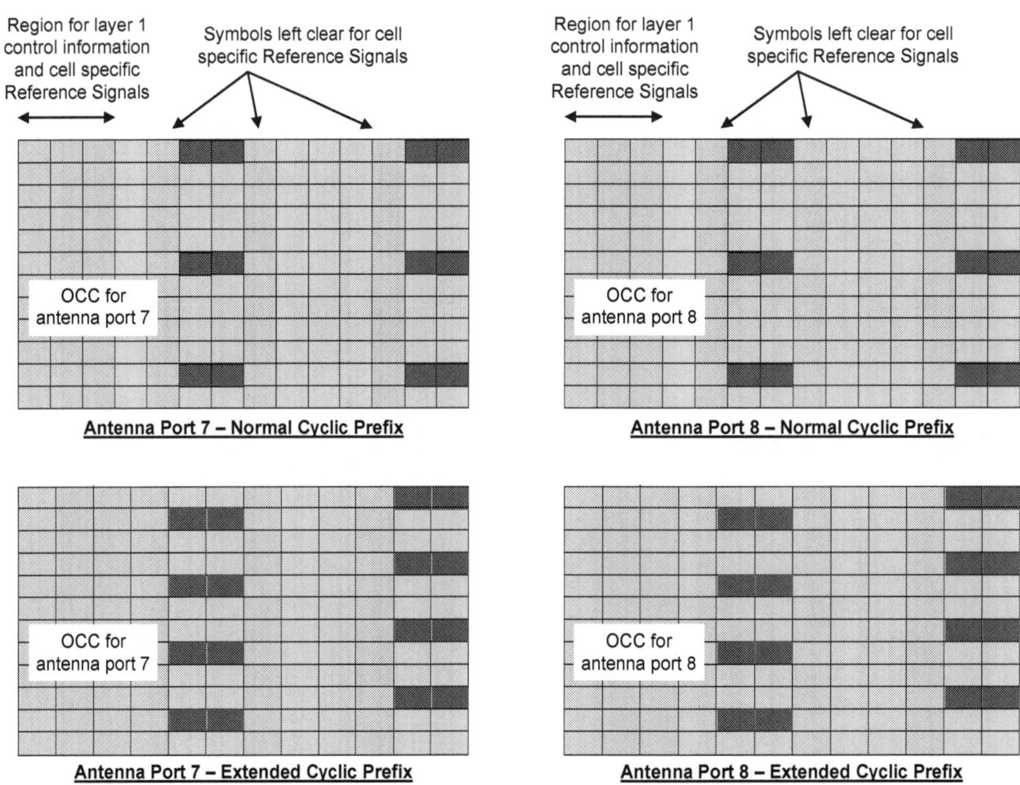

Figure 83 – UE specific Reference Signals (antenna ports 7 and 8)

- The Resource Element allocations for the UE specific Reference Signals are the same for both antenna ports 7 and 8. Orthogonality is achieved by applying Orthogonal Covering Codes (OCC) to the sequences mapped onto the UE specific Reference Signal Resource Elements, i.e. antenna ports 7 and 8 use different OCC which are orthogonal to each another. This provides support for single user dual-layer beamforming while maintaining orthogonality at the transmitter
- The concept of applying OCC is illustrated in Figure 84. 3GPP TS 36.211 specifies one 4-digit OCC for antenna port 7 and another 4-digit OCC for antenna port 8. These OCC are applied in the time domain to minimise the impact of the radio propagation channel, i.e. they are more robust when applied in the time domain rather than the frequency domain

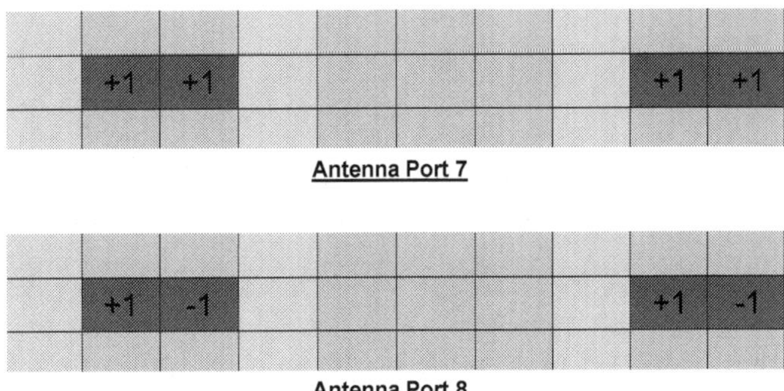

Figure 84 – Orthogonal Covering Codes (OCC) applied to antenna ports 7 and 8

★ The introduction of scrambling identities provides support for 2 user dual-layer transmission, and increases the total number of layers from 2 to 4. 3GPP TS 36.211 specifies scrambling identities of 0 and 1. Scrambling identities allow users to differentiate between UE specific Reference Signals but they do not provide orthogonality. Scrambling identities impact the selection of the sequence used for the UE specific Reference Signals prior to applying the OCC

★ The 4 layers provided by the combination of OCC and scrambling identities can also be used to support 4 user single-layer transmission. The UE specific Reference Signals for those 4 users would then use a mix of orthogonal and non-orthogonal sequences

★ UE specific Reference Signals for antenna ports 7 and 8 are specified separately for TDD special subframes. The mapping onto Resource Elements is illustrated in Figure 85. The mapping is the same for both antenna ports 7 and 8

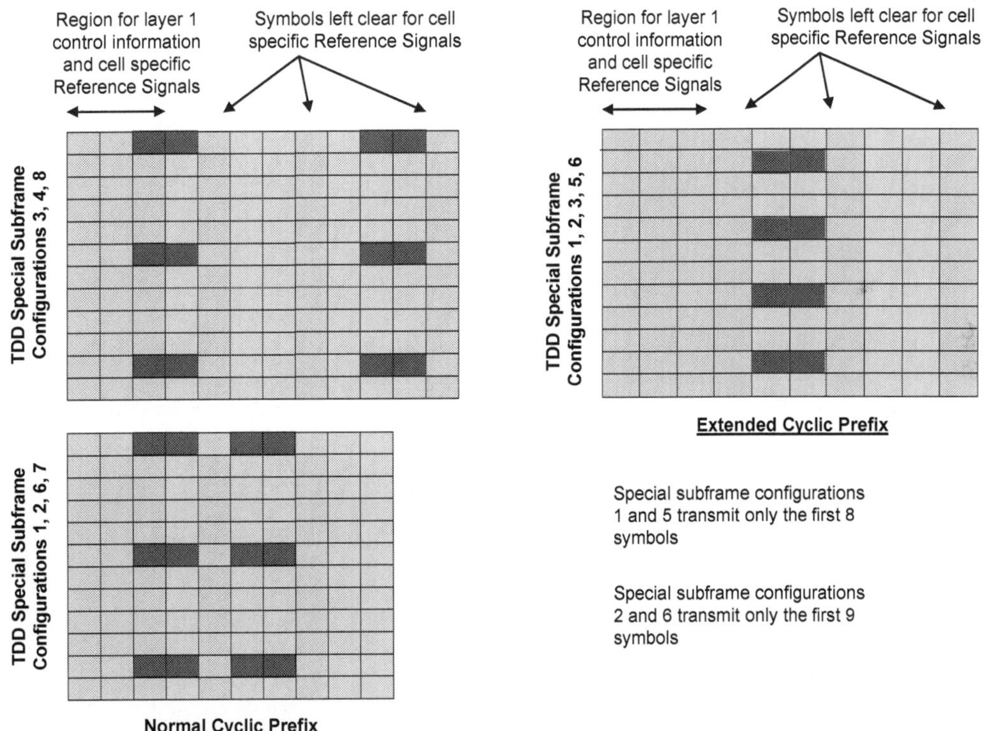

Figure 85 – UE specific Reference Signals for TDD special subframes (antenna ports 7 and 8)

★ Resource Element allocations for UE specific Reference Signals on antenna ports 11 and 13 are the same as for antenna ports 7 and 8. Resource Element allocations for UE specific Reference Signals on antenna ports 9, 10, 12 and 14 follow the same pattern but are shifted downwards by a single subcarrier. These Resource Element allocations are illustrated in Figure 86 for normal subframes. Antenna ports 9 to 14 are only defined for the normal cyclic prefix

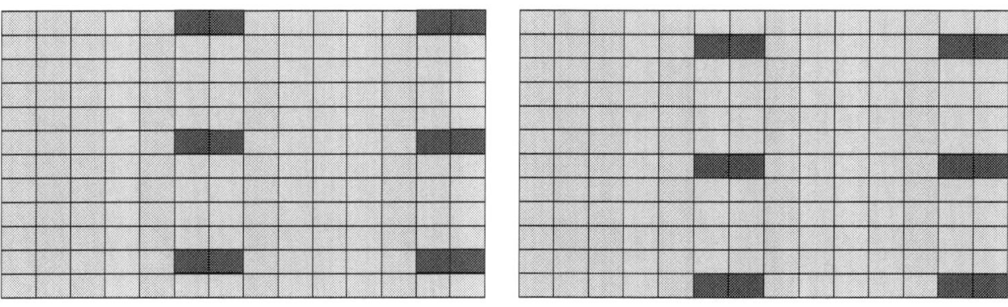

Figure 86 – UE specific Reference Signals for antenna ports 7 to 14 (normal cyclic prefix)

★ The frequency shift used for antenna ports 9, 10, 12 and 14 provides support for Frequency Division Multiplexing (FDM) which can be used in combination with the Code Division Multiplexing (CDM) provided by the Orthogonal Cover Code (OCC). The use of FDM allows some of the OCC to be re-used across multiple antenna ports. The OCC associated with each antenna port are presented in Table 38. Antenna ports 9 and 10 use the same OCC as antenna ports 7 and 8

Antenna Port	OCC
7	+1 +1 +1 +1
8	+1 -1 +1 -1
11	+1 +1 -1 -1
13	+1 -1 -1 +1

Antenna Port	OCC
9	+1 +1 +1 +1
10	+1 -1 +1 -1
12	-1 -1 +1 +1
14	-1 +1 +1 -1

Table 38 – OCC for antenna ports 7 to 14 (normal cyclic prefix)

★ Scrambling identities are not defined for antenna ports 9 to 14 so that additional dimension is not available when using those antenna ports

★ Antenna ports 7 to 14 provide support for single user 8-layer transmission, i.e. 8×8 MIMO for LTE Advanced

★ 3GPP References: TS 36.211

5.2.4 POSITIONING REFERENCE SIGNALS

- Positioning Reference Signals were introduced within the release 9 version of the 3GPP specifications. Their main objective is to improve the 'hearability' of neighbouring cells when completing measurements for the downlink Observed Time Difference Of Arrival (OTDOA) positioning method
- 3GPP recognised that the hearability of the existing cell specific Reference Signals was not sufficient to support the OTDOA positioning method. Hearability can be challenging as a result of neighbouring cells being co-channel with the serving cell, especially at locations where the serving cell signal strength is high
- The OTDOA positioning method makes use of Reference Signal Time Difference (RSTD) measurements from the UE. The RSTD quantifies the subframe timing difference between a reference cell and a neighbouring cell. The accuracy of the positioning calculation is improved if the UE can provide RSTD measurements from an increased number of cells
- RSTD is measured in units of Ts (1/30720 ms) and is reported to the Enhanced Serving Mobile Location Centre (E-SMLC) where the location calculation is completed. The E-SMLC is a network element within the Enhanced Packet Core
- Positioning Reference Signals are able to coexist with both the cell specific Reference Signals and the physical layer control information at the start of each subframe (PCFICH, PHICH, PDCCH). Positioning Reference Signals occupy an increased number of Resource Elements within a subframe relative to the cell specific Reference Signals to help improve RSTD measurement accuracy
- The sequence used to generate the positioning Reference Signal is a function of the Physical layer Cell Identity (PCI) and the cyclic prefix duration (normal or extended). Positioning Reference Signals are broadcast using antenna port 6. They are not mapped onto Resource Elements allocated to the PBCH, Primary Synchronisation Signal nor Secondary Synchronisation Signal
- In the case of TDD, positioning Reference Signals are not broadcast during special subframes, i.e. subframes including the DwPTS, guard period and UpPTS
- Positioning Reference Signals are only defined for the 15 kHz subcarrier spacing. They are not supported for the 7.5 kHz subcarrier spacing used by Multimedia Broadcast Multicast Services (MBMS)
- Figure 87 and Figure 88 illustrate examples of the positioning Reference Signal for normal and extended cyclic prefixes. In each case, there is a dependency upon the number of antenna ports used for the cell specific Reference Signal. Additional symbols are used by the cell specific Reference Signal when broadcast from 4 antenna ports

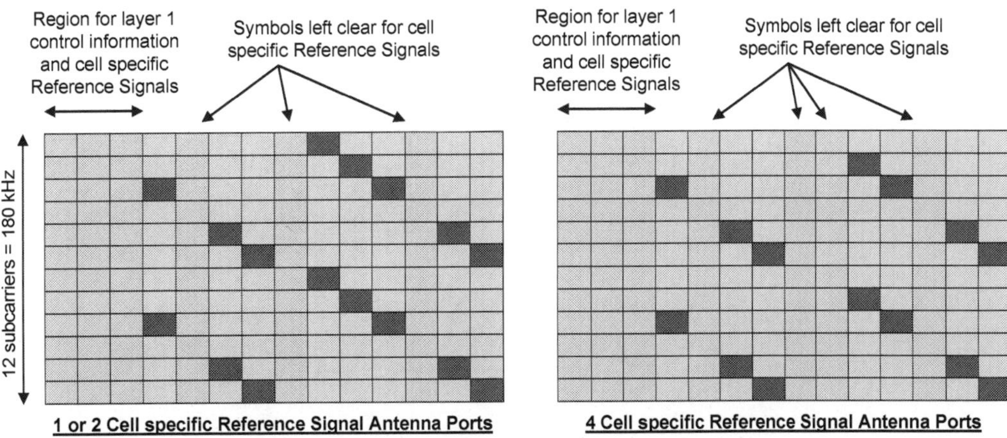

Figure 87 – Positioning Reference Signals (normal cyclic prefix)

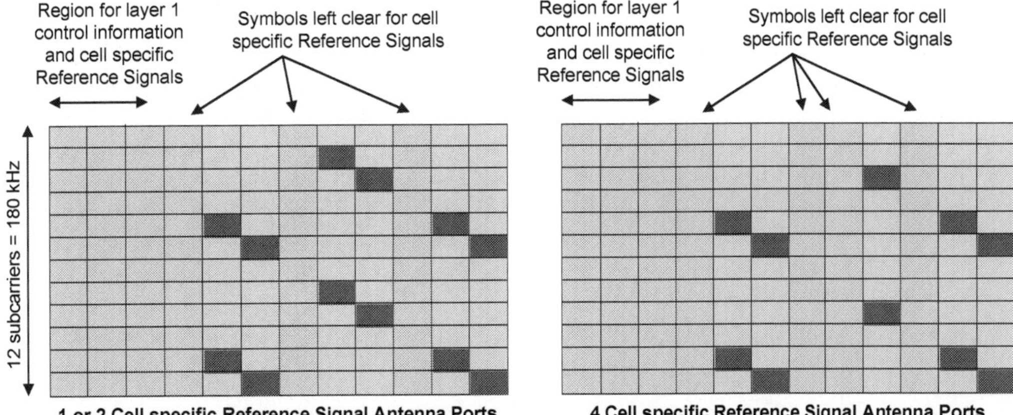

Figure 88 – Positioning Reference Signals (extended cyclic prefix)

- ★ Similar to the cell specific Reference Signal, the positioning Reference Signal shifts up and down the Resource Elements in the frequency domain according to the Physical layer Cell Identity. This vertical shift repeats for every 6 Physical layer Cell Identities, i.e. both the cell specific and positioning Reference Signals have a re-use pattern of 6 in terms of their Resource Element allocation. This shift is illustrated in Figure 78 for the cell specific Reference Signal
- ★ The positioning Reference Signal may not be broadcast across the complete channel bandwidth. The E-SMLC provides instructions to the eNode B regarding the bandwidth to be used. It also signals the same bandwidth information to the UE
- ★ The UE receives an LTE Positioning Protocol (LPP) Provide Assistance Data message from the E-SMLC. This message is packaged by the MME as a NAS message before being packaged by the eNode B as an RRC message. The Provide Assistance Data message includes information regarding both the reference and neighbouring cells. The reference cell does not have to be the current serving cell for the UE. The content of the reference cell information is presented in Table 39. Similar information is also provided for each of the neighbouring cells

Information Elements		
Physical layer Cell Identity		
Cell Global Identity		
EARFCN		
Antenna Port Configuration		
Cyclic Prefix Length		
Positioning Reference Signal (PRS) Information	PRS Bandwidth	
	PRS Configuration Index	
	Number of Consecutive Downlink Subframes	
	PRS Muting Information	CHOICE
		bit string of length 2 bits
		bit string of length 4 bits
		bit string of length 8 bits
		bit string of length 16 bits

Table 39 – Content of OTDOA Reference Cell Information

- ★ The positioning Reference Signal Bandwidth is signalled to the UE with a value of 6, 15, 25, 50, 75 or 100 Resource Blocks. The positioning Reference Signal bandwidth is always centered around the middle of the channel bandwidth
- ★ The positioning Reference Signal Configuration Index is used to define both a periodicity and subframe offset for the timing of the positioning Reference Signal. The look-up table presented in Table 40 is used to link the configuration index to the periodicity and subframe offset

PRS Configuration Index I_{PRS}	PRS Periodicity (subframes)	PRS Subframe Offset (subframes)
0 – 159	160	I_{PRS}
160 – 479	320	$I_{PRS} - 160$
480 – 1119	640	$I_{PRS} - 480$
1120 – 23399	1280	$I_{PRS} - 1120$

Table 40 – Positioning Reference Signal subframe configuration

- ★ The Number of Consecutive Downlink Subframes defines the number of subframes during which the positioning Reference Signal is broadcast within each positioning Reference Signal period. The Number of Consecutive Downlink Subframes can be configured with values of 1, 2, 4 or 6 subframes
- ★ The first subframe belonging to each group of consecutive positioning Reference Signal subframes is given by the condition:

$$(10 \times \text{SFN} + \lfloor \text{Slot Number} / 2 \rfloor - \text{PRS Subframe Offset}) \bmod (\text{PRS Periodicity}) = 0$$

For example, if the PRS Configuration Index is allocated a value of 200 then the equation becomes:

$$(10 \times \text{SFN} + \lfloor \text{Slot Number} / 2 \rfloor - 40) \bmod (320) = 0$$

And the PRS transmission starts from {SFN 4, subframe 0}, {SFN 36, subframe 0}, {SFN 68, subframe 0}, etc

- ★ Figure 89 illustrates a summary of the positioning Reference Signal timing. It also illustrates the location of the positioning Reference Signal in the frequency domain when the complete channel bandwidth is not used for positioning measurements. Each group of consecutive subframes containing the positioning Reference Signal is known as a PRS Positioning Occasion

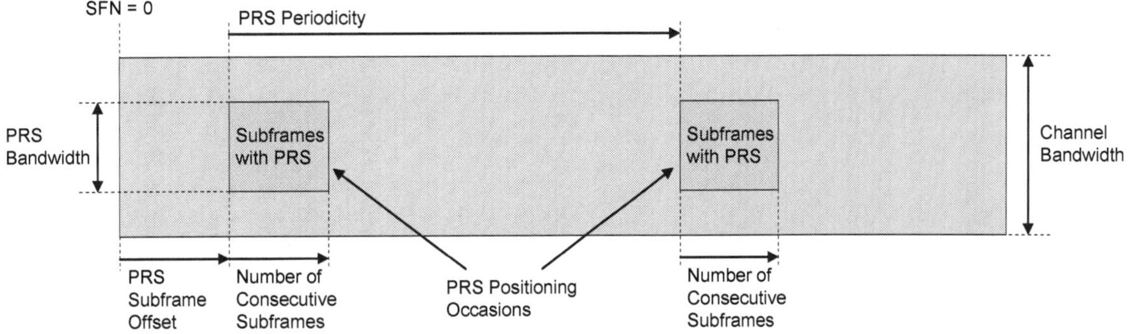

Figure 89 – Summary of positioning Reference Signal timing and bandwidth

★ Table 39 also shows that the OTDOA Reference Cell Information includes PRS Muting Information. This information is optional but can be used to indicate PRS Positioning Occasions where the positioning Reference Signal is not present. The PRS Muting information takes the form of a bit map which can have a length of either 2, 4, 8 or 16 bits. The length of the bit map defines the periodicity of the muting pattern. Each bit corresponds to a single PRS Positioning Occasion, and if a bit is set to 0 then the positioning Reference Signal is not present within that PRS Positioning Occasion

★ 3GPP References: TS 36.211, TS 36.305, TS 36.355, TS 36.133

5.2.5 CSI REFERENCE SIGNALS

- Channel State Information (CSI) Reference Signals were introduced within the release 10 version of the 3GPP specifications
- CSI Reference Signals are intended to improve the performance of link adaptation. Resource Elements allocated to the cell specific Reference Signal often experience different levels of interference to Resource Elements allocated to the PDSCH. This means that channel quality reporting based upon the cell specific Reference Signals does not always provide meaningful information when scheduling PDSCH resources
- The difference between the interference levels experienced by the cell specific Reference Signal and the PDSCH can be significant for unloaded networks with time synchronised cells. The cell specific Reference Signals at the start of the subframe experience inter-cell interference from the PCFICH, PDCCH and PHICH, while the remaining cell specific Reference Signals can experience inter-cell interference from other cell specific Reference Signals. Meanwhile, the PDSCH may experience only low levels of interference. This results in pessimistic CQI values when based upon the cell specific Reference Signals
- CSI Reference Signals occupy Resource Elements usually allocated to the PDSCH. This allows a more meaningful measure of channel quality but increases the overhead generated by Reference Signals. The increase in overhead is managed by limiting the number of Resource Elements allocated to the CSI Reference Signal
- CSI Reference Signals are only defined for the 15 kHz subcarrier spacing. They are not supported for the 7.5 kHz subcarrier spacing used by Multimedia Broadcast Multicast Services (MBMS)
- CSI Reference Signals are applicable to both FDD and TDD, using either the normal or extended cyclic prefix
- CSI Reference Signals have been designed to account for transmission through 1, 2, 4 or 8 antenna ports:
  - 1 antenna port: uses port 15
  - 2 antenna ports: uses ports 15 and 16
  - 4 antenna ports: uses ports 15, 16, 17 and 18
  - 8 antenna ports: uses ports 15, 16, 17, 18, 19, 20, 21 and 22
- The UE is signalled information regarding the CSI Reference Signal within either an RRC Connection Setup, RRC Connection Reconfiguration or RRC Connection Re-establishment message. Table 41 shows the information provided

<table>
<tr><th colspan="3">Information Elements</th></tr>
<tr><td rowspan="6">CSI Reference Signal</td><td colspan="2">CHOICE</td></tr>
<tr><td colspan="2">Release</td></tr>
<tr><td rowspan="4">Setup</td><td>Antenna Ports Count</td></tr>
<tr><td>CSI Reference Signal Configuration</td></tr>
<tr><td>Subframe Configuration</td></tr>
<tr><td>Pc</td></tr>
<tr><td rowspan="4">Zero Tx Power CSI Reference Signal</td><td colspan="2">CHOICE</td></tr>
<tr><td colspan="2">Release</td></tr>
<tr><td rowspan="2">Setup</td><td>Zero Tx Power Resource Configuration List</td></tr>
<tr><td>Zero Tx Power Subframe Configuration</td></tr>
</table>

Table 41 – Content of CSI Reference Signal Configuration information

- The Antenna Ports Count is signalled using a value of 1, 2, 4 or 8. This indicates the number of antenna ports from which the CSI Reference Signal will be sent
- The CSI Reference Signal Configuration has a value between 0 and 31. It points towards a row within a look-up table that specifies a 'reference' Resource Element. The actual Resource Elements used by the CSI Reference Signal are derived from this 'reference' Resource Element using antenna specific offsets. The look-up table for the normal cyclic prefix is presented as Table 42, whereas the equivalent look-up table for the extended cyclic prefix is presented as Table 43
- Within these look-up tables, (k', l') represents the 'reference' Resource Element, where k' defines the position in the frequency domain while l' defines the position in the time domain. The 'n_s mod 2' column indicates whether the 'reference' Resource Element is present within odd or even numbered time slots
- Note that when the 'reference' Resource Element is within the first time slot of a subframe (n_s mod 2 = 0) then the time domain position is always '5' for the normal cyclic prefix, and '4' for the extended cyclic prefix. This avoids the CSI Reference Signal from clashing with the PDCCH, PHICH and PCFICH symbols at the start of each subframe, or with a symbol used by the cell specific Reference Signal
- Likewise, when the 'reference' Resource Element is within the second time slot of a subframe (n_s mod 2 = 1) then the time domain position is chosen to avoid the CSI Reference Signal from clashing with a symbol used by the cell specific Reference Signal

	CSI Reference Signal Configuration	Number of CSI Reference Signals Configured					
		1 or 2		4		8	
		(k', l')	ns mod 2	(k', l')	ns mod 2	(k', l')	ns mod 2
Frame Structure Type 1 and 2	0	(9, 5)	0	(9, 5)	0	(9, 5)	0
	1	(11, 2)	1	(11, 2)	1	(11, 2)	1
	2	(9, 2)	1	(9, 2)	1	(9, 2)	1
	3	(7, 2)	1	(7, 2)	1	(7, 2)	1
	4	(9, 5)	1	(9, 5)	1	(9, 5)	1
	5	(8, 5)	0	(8, 5)	0		
	6	(10, 2)	1	(10, 2)	1		
	7	(8, 2)	1	(8, 2)	1		
	8	(6, 2)	1	(6, 2)	1		
	9	(8, 5)	1	(8, 5)	1		
	10	(3, 5)	0				
	11	(2, 5)	0				
	12	(5, 2)	1				
	13	(4, 2)	1				
	14	(3, 2)	1				
	15	(2, 2)	1				
	16	(1, 2)	1				
	17	(0, 2)	1				
	18	(3, 5)	1				
	19	(2, 5)	1				
Frame Structure Type 2 only	20	(11, 1)	1	(11, 1)	1	(11, 1)	1
	21	(9, 1)	1	(9, 1)	1	(9, 1)	1
	22	(7, 1)	1	(7, 1)	1	(7, 1)	1
	23	(10, 1)	1	(10, 1)	1		
	24	(8, 1)	1	(8, 1)	1		
	25	(6, 1)	1	(6, 1)	1		
	26	(5, 1)	1				
	27	(4, 1)	1				
	28	(3, 1)	1				
	29	(2, 1)	1				
	30	(1, 1)	1				
	31	(0, 1)	1				

Table 42 – CSI Reference Signal configuration (normal cyclic prefix)

	CSI Reference Signal Configuration	Number of CSI Reference Signals Configured					
		1 or 2		4		8	
		(k', l')	ns mod 2	(k', l')	ns mod 2	(k', l')	ns mod 2
Frame Structure Type 1 and 2	0	(11, 4)	0	(11, 4)	0	(11, 4)	0
	1	(9, 4)	0	(9, 4)	0	(9, 4)	0
	2	(10, 4)	1	(10, 4)	1	(10, 4)	1
	3	(9, 4)	1	(9, 4)	1	(9, 4)	1
	4	(5, 4)	0	(5, 4)	0		
	5	(3, 4)	0	(3, 4)	0		
	6	(4, 4)	1	(4, 4)	1		
	7	(3, 4)	1	(3, 4)	1		
	8	(8, 4)	0				
	9	(6, 4)	0				
	10	(2, 4)	0				
	11	(0, 4)	0				
	12	(7, 4)	1				
	13	(6, 4)	1				
	14	(1, 4)	1				
	15	(0, 4)	1				
Frame Structure Type 2 only	16	(11, 1)	1	(11, 1)	1	(11, 1)	1
	17	(10, 1)	1	(10, 1)	1	(10, 1)	1
	18	(9, 1)	1	(9, 1)	1	(9, 1)	1
	19	(5, 1)	1	(5, 1)	1		
	20	(4, 1)	1	(4, 1)	1		
	21	(3, 1)	1	(3, 1)	1		
	22	(8, 1)	1				
	23	(7, 1)	1				
	24	(6, 1)	1				
	25	(2, 1)	1				
	26	(1, 1)	1				
	27	(0, 1)	1				

Table 43 – CSI Reference Signal configuration (extended cyclic prefix)

★ As an example, consider the case of using 8 antenna ports with the normal cyclic prefix and assume that the UE is signalled a CSI Reference Signal configuration of 0. This results in a 'reference' Resource Element of (9, 5) within the first time slot of the subframe

★ The frequency domain position of the actual Resource Element used by the CSI Reference Signal is obtained by adding the offsets presented in Table 44. Based upon the example above, this generates Resource Elements of (9, 5), (9, 5), (3, 5), (3, 5), (8, 5), (8, 5), (2, 5), (2, 5) for the set of 8 antenna ports

Antenna Ports	Normal Cyclic Prefix	Extended Cyclic Prefix
15, 16	0	0
17, 18	-6	-3
19, 20	-1	-6
21, 22	-7	-9

Table 44 – Frequency domain offsets as a function of antenna port and cyclic prefix

★ The time domain position of the actual Resource Element used by the CSI Reference Signal is equal to the time domain position of the 'reference' Resource Element. However, a second Resource Element is allocated to the CSI Reference Signal. This second Resource Element has the same frequency domain position as the first Resource Element but a shifted time domain position. The shift in the time domain is +1 Resource Element for all cases except CSI Reference Signal configurations 20 to 31 when using the normal cyclic prefix. In this case, the shift is +2 Resource Elements

★ Figure 90 illustrates the CSI Reference Signal for the example already introduced in this section. This allows the channel quality to be measured from Resource Elements usually occupied by the PDSCH

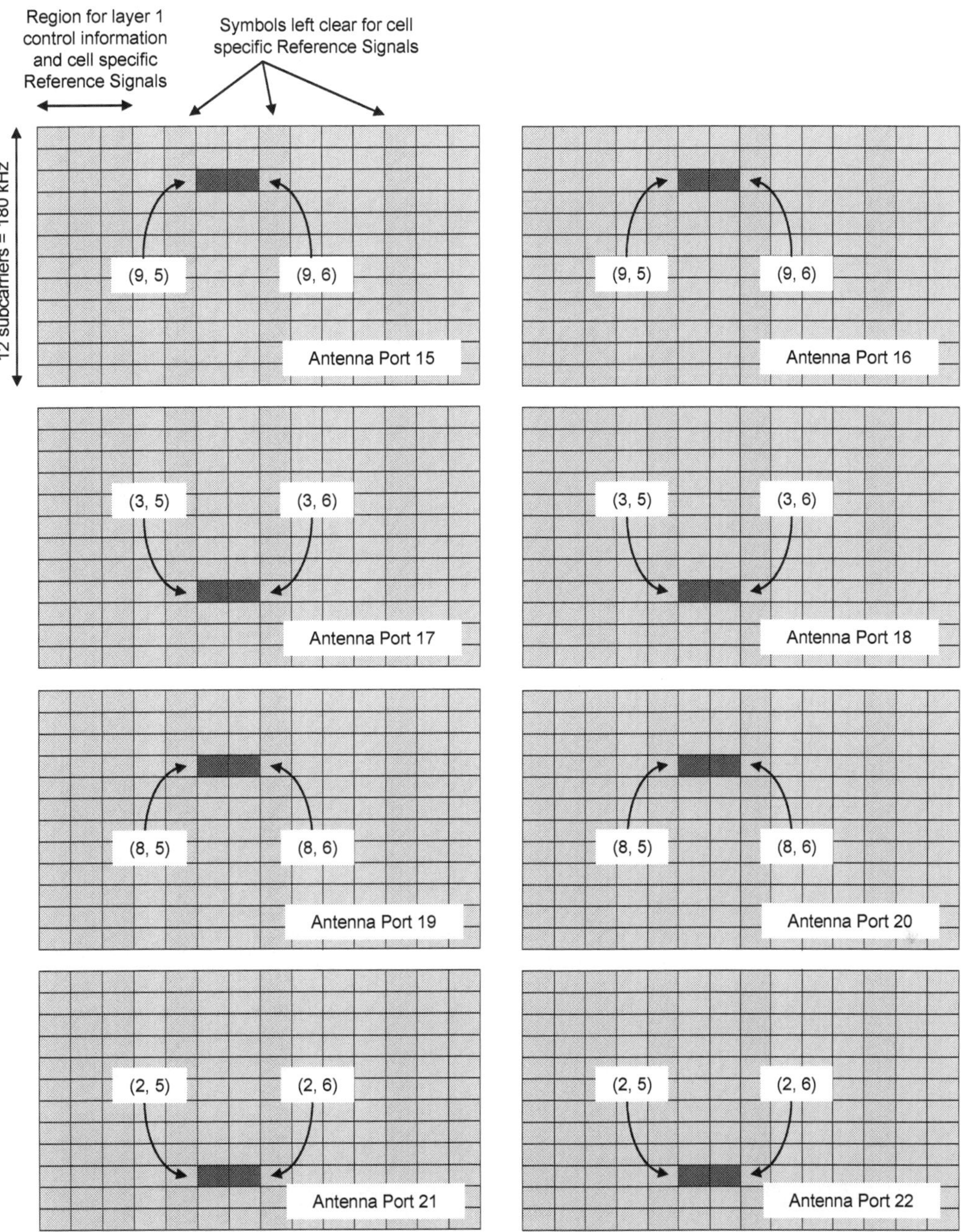

Figure 90 – CSI Reference Signal (normal cyclic prefix, CSI Reference Signal Configuration 0)

★ Figure 91 illustrates a second example of the CSI Reference Signal. This example is based upon the extended cyclic prefix and CSI Reference Signal Configuration 1

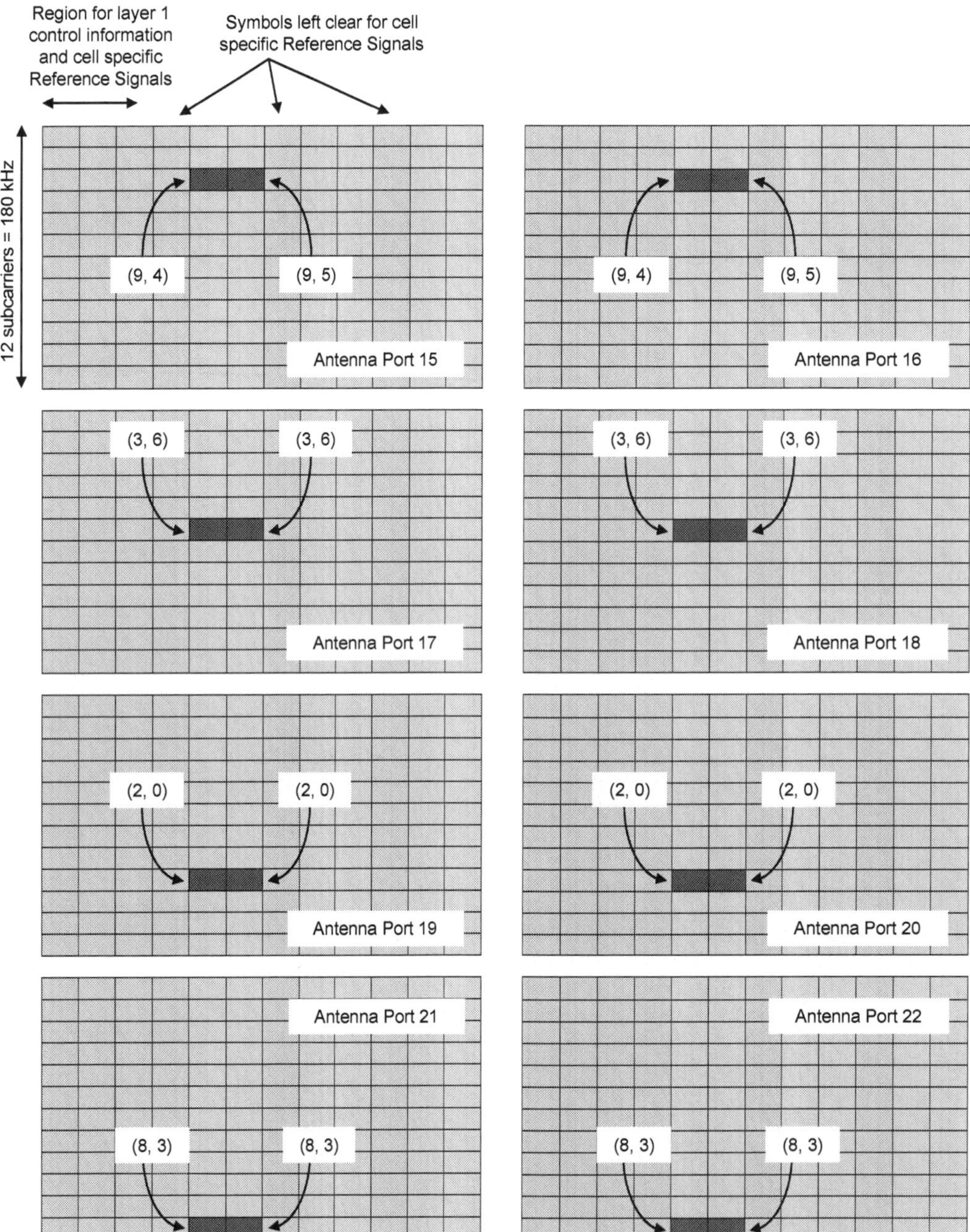

Figure 91 – CSI Reference Signal (extended cyclic prefix, CSI Reference Signal Configuration 1)

★ Returning to Table 41, the Subframe Configuration is used to define both a periodicity and subframe offset for the timing of the CSI Reference Signal. The look-up table presented in Table 45 is used to link the Subframe Configuration to the periodicity and subframe offset

CSI-RS Subframe Configuration I_{CSI-RS}	CSI-RS Periodicity (subframes)	CSI-RS Subframe Offset (subframes)
0 – 4	5	I_{CSI-RS}
5 – 14	10	$I_{CSI-RS} - 5$
15 – 34	20	$I_{CSI-RS} - 15$
35 – 74	40	$I_{CSI-RS} - 35$
75 – 154	80	$I_{CSI-RS} - 75$

Table 45 – CSI Reference Signal subframe configuration

★ The CSI Reference Signal occupies a single subframe per CSI-RS Period, i.e. the CSI Reference Signal can occupy one subframe per 5, 10, 20, 40 or 80 subframes. Within that subframe, the CSI Reference Signal uses all Resource Blocks in the frequency domain

★ The CSI Reference Signal is transmitted during subframes which satisfy the condition:

$$(10 \times \text{SFN} + \lfloor \text{Slot Number} / 2 \rfloor - \text{CSI-RS Subframe Offset}) \bmod (\text{CSI-RS Periodicity}) = 0$$

For example, if the CSI-RS Subframe Configuration is allocated a value of 7 then the equation becomes:

$$(10 \times \text{SFN} + \lfloor \text{Slot Number} / 2 \rfloor - 2) \bmod (10) = 0$$

And the CSI-RS transmission is included within {SFN 0, subframe 2}, {SFN 1, subframe 2}, {SFN 2, subframe 2}, etc

★ Figure 92 illustrates a summary of the CSI Reference Signal timing. It also illustrates the location of the CSI Reference Signal in the frequency domain, i.e. it occupies all Resource Blocks in the frequency domain

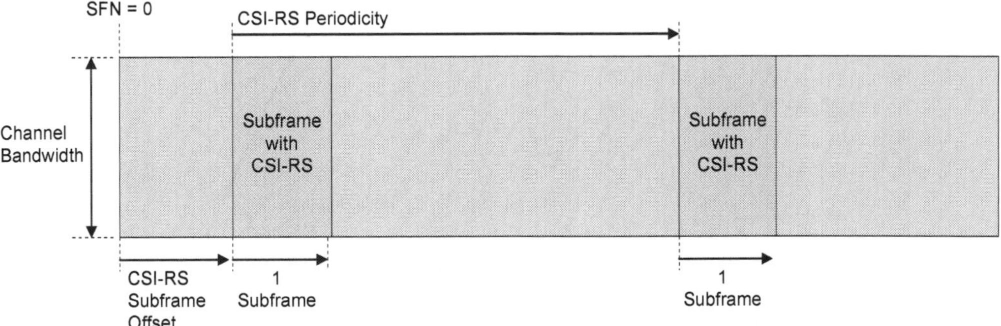

Figure 92 – Summary of CSI Reference Signal timing and bandwidth

★ Returning to Table 41, the variable Pc informs the UE of the ratio between the PDSCH Energy per Resource Element (EPRE) and the CSI Reference Signal EPRE. The value of Pc can be signalled with values between -8 and 15 dB

★ So far, this section has focused upon CSI Reference Signals with a finite transmit power, which allow the UE to measure a signal to noise ratio. CSI Reference Signals have also been specified to have zero transmit power, i.e. they represents gaps which are punctured out of the Resource Element grid. Zero transmit power CSI Reference Signals allow the UE to measure interference power without any wanted signal power

★ A single cell can be configured with:

- 0 or 1 CSI Reference Signal configurations using a finite transmit power
- 0 or more CSI Reference Signal configurations using zero transmit power

★ Table 41 illustrates that the UE can be provided with a Zero Tx Power Resource Configuration List. This information element is a bitmap with a length of 16 bits. Each bit corresponds to a row within the 4 CSI Reference Signal column of Table 42 for the normal cyclic prefix and Table 43 for the extended cyclic prefix. The first bit corresponds to configuration 0. If a bit is set to 1 then it indicates that the relevant CSI Reference Signal configuration is active with zero transmit power

★ If the zero transmit power, and non-zero transmit power CSI Reference Signal patterns overlap, then the non-zero transmit power pattern overrides the zero transmit power pattern

★ Table 41 also indicates that the UE is provided with a Zero Tx Power Subframe Configuration. This defines the periodicity and subframe offset for the timing of the zero transmit power CSI Reference Signal. The look-up table presented in Table 45 is used to link the Subframe Configuration to the periodicity and subframe offset (in the same way as for a non-zero transmit power CSI Reference Signal)

★ In the case of TDD, CSI Reference Signals are not transmitted within special subframes, i.e. subframes including the DwPTS, guard period and UpPTS

★ CSI Reference Signals are not transmitted within subframes where their transmission would collide with synchronisation signals, the PBCH or System Information Block type 1 messages. Nor are they transmitted during subframes configured for the transmission of paging messages

★ Resource Elements used for the transmission of CSI Reference Signals are not used for the transmission of the PDSCH on any of the other antenna ports

★ 3GPP References: TS 36.211

6 DOWNLINK PHYSICAL CHANNELS

6.1 PBCH

★ The Physical Broadcast Channel (PBCH) is used to broadcast the Master Information Block (MIB) using the BCH transport channel and BCCH logical channel

★ For both FDD and TDD, the PBCH is allocated the central 72 subcarriers belonging to the first 4 OFDMA symbols of the second time slot of every 10 ms radio frame (time slot 1 in subframe 0, with time slot numbering starting from 0)

★ Reference Signal Resource Elements (including those which would be allocated if antenna ports 0 to 3 were used, irrespective of the actual antenna ports used) are excluded from the PBCH allocation

★ In the case of TDD, subframe 0 is always a downlink subframe

★ The PBCH can be broadcast using only antenna port 0, or transmit diversity can be used to broadcast the PBCH using antenna ports {0, 1} or {0, 1, 2, 3}

★ Figure 93 illustrates the Resource Element allocation for the PBCH

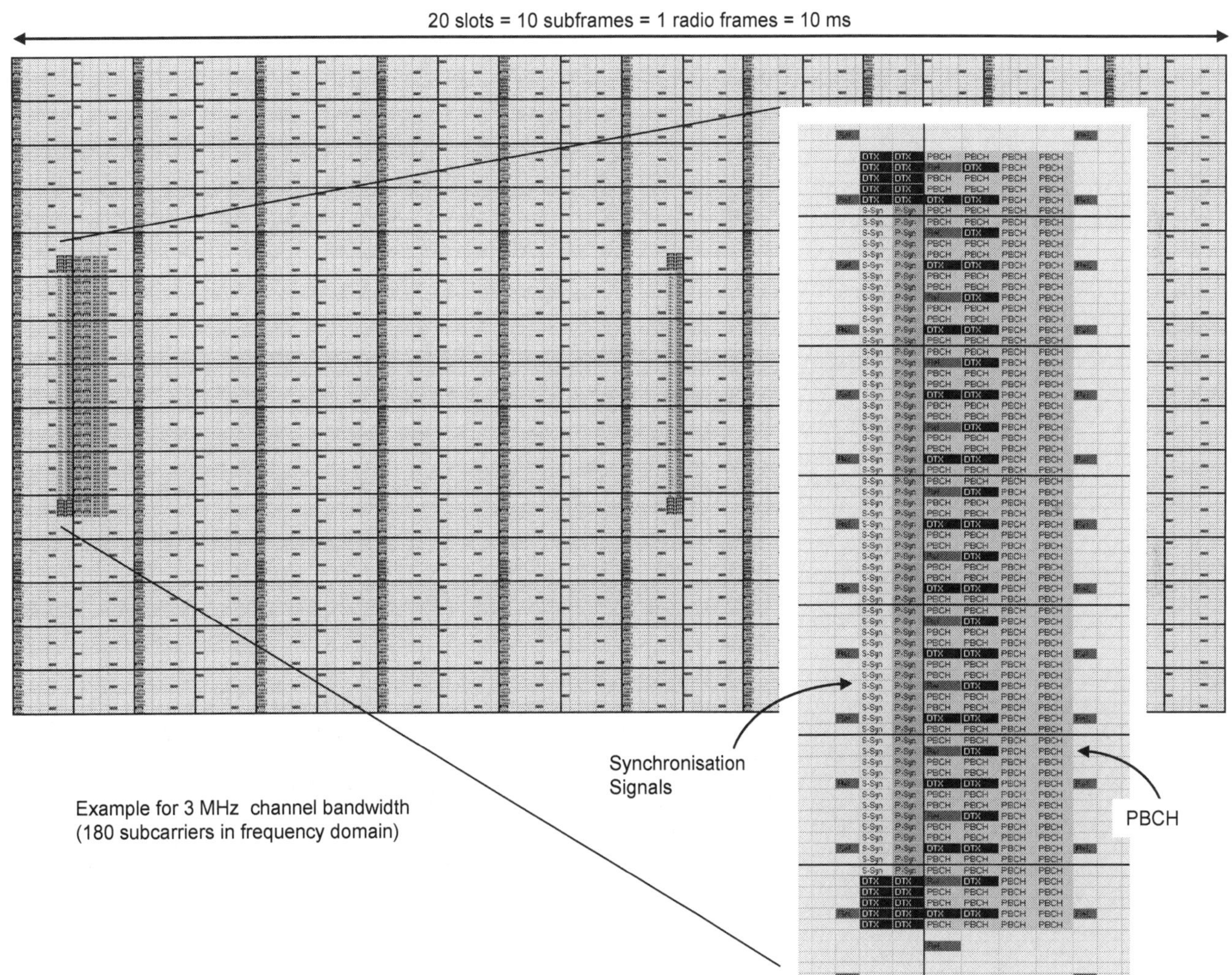

Figure 93 – PBCH Physical Channel within the first 4 OFDMA symbols of the second slot (normal cyclic prefix)

★ The PBCH occupies 240 Resource Elements when using the normal cyclic prefix, i.e. (72 × 4) – 48, where 48 is the number of Resource Elements allocated to the Reference Signal

★ The PBCH occupies 216 Resource Elements when using the extended cyclic prefix, i.e. (72 × 4) – 72, where 72 is the number of Resource Elements allocated to the Reference Signal (in this case, the third column of Reference Signals also overlaps with the set of PBCH Resource Elements)

★ The PBCH uses QPSK modulation so the 240 Resource Elements provide 480 bits when using the normal cyclic prefix, and the 216 Resource Elements provide 432 bits when using the extended cyclic prefix

★ The set of Resource Elements allocated to the PBCH is independent of the channel bandwidth. The UE does not require any knowledge of the channel bandwidth prior to decoding the PBCH. The downlink channel bandwidth is subsequently read from the Master Information Block (MIB) within the PBCH. This information is required to decode other physical channels

★ The PBCH represents an overhead which reduces the number of Resource Elements available for user plane data

- o overhead decreases for larger channel bandwidths
- o overhead is comparable for both the normal and extended cyclic prefix

★ The overhead generated by the PBCH for FDD is presented in Table 46. This overhead is relatively insignificant for the larger channel bandwidths but is significant for the smaller channel bandwidths

	1.4 MHz	3 MHz	5 MHz	10 MHz	15 MHz	20 MHz
Normal Cyclic Prefix						
PBCH Resource Elements per 10 ms Radio Frame	240	240	240	240	240	240
Overhead (normal cyclic prefix)	240 / 10080 = 2.4 %	240 / 25200 = 1.0 %	240 / 42000 = 0.6 %	240 / 84000 = 0.3 %	240 / 126000 = 0.2 %	240 / 168000 = 0.1 %
Extended Cyclic Prefix						
PBCH Resource Elements per 10 ms Radio Frame	216	216	216	216	216	216
Overhead (extended cyclic prefix)	216 / 8640 = 2.5 %	216 / 21600 = 1.0 %	216 / 36000 = 0.6 %	216 / 72000 = 0.3 %	216 / 108000 = 0.2 %	216 / 144000 = 0.2 %

Table 46 – Overhead generated by the PBCH Physical Channel (FDD)

★ In the case of TDD, the overhead generated by the PBCH is the same as that for FDD when calculating the overhead relative to the total number of Resource Elements (both uplink and downlink)

★ However, the overhead for TDD is greater when calculating relative to the number of downlink Resource Elements. This calculation is multi-dimensional because the number of downlink Resource Elements depends upon the cyclic prefix length, the channel bandwidth, the uplink-downlink subframe configuration and the special subframe configuration

★ Table 47 presents the overheads when assuming uplink-downlink subframe configurations 0 and 5. These configurations represent those with the highest and lowest overheads, i.e. uplink-downlink subframe configuration 0 has the least downlink Resource Elements, while uplink-downlink subframe configuration 5 has the most downlink Resource Elements

★ In the worst case of uplink-downlink subframe configuration 0, using the extended cyclic prefix and special subframe configuration 0, the overhead for TDD reaches 10.0 %

★ 3GPP References: TS 36.211

		1.4 MHz	3 MHz	5 MHz	10 MHz	15 MHz	20 MHz
Normal Cyclic Prefix							
PBCH Resource Elements		240	240	240	240	240	240
Uplink-Downlink Subframe Config. 0	Special SF Config. 0	9.8 %	3.9 %	2.4 %	1.2 %	0.8 %	0.6 %
	Special SF Config. 1	7.2 %	2.9 %	1.7 %	0.9 %	0.6 %	0.4 %
	Special SF Config. 2	6.9 %	2.8 %	1.7 %	0.8 %	0.6 %	0.4 %
	Special SF Config. 3	6.7 %	2.7 %	1.6 %	0.8 %	0.5 %	0.4 %
	Special SF Config. 4	6.4 %	2.6 %	1.5 %	0.8 %	0.5 %	0.4 %
	Special SF Config. 5	9.8 %	3.9 %	2.4 %	1.2 %	0.8 %	0.6 %
	Special SF Config. 6	7.2 %	2.9 %	1.7 %	0.9 %	0.6 %	0.4 %
	Special SF Config. 7	6.9 %	2.8 %	1.7 %	0.8 %	0.6 %	0.4 %
	Special SF Config. 8	6.7 %	2.7 %	1.6 %	0.8 %	0.5 %	0.4 %
Uplink-Downlink Subframe Config. 5	Special SF Config. 0	2.9 %	1.2 %	0.7 %	0.3 %	0.2 %	0.2 %
	Special SF Config. 1	2.8 %	1.1 %	0.7 %	0.3 %	0.2 %	0.2 %
	Special SF Config. 2	2.7 %	1.1 %	0.7 %	0.3 %	0.2 %	0.2 %
	Special SF Config. 3	2.7 %	1.1 %	0.7 %	0.3 %	0.2 %	0.2 %
	Special SF Config. 4	2.7 %	1.1 %	0.6 %	0.3 %	0.2 %	0.2 %
	Special SF Config. 5	2.9 %	1.2 %	0.7 %	0.3 %	0.2 %	0.2 %
	Special SF Config. 6	2.8 %	1.1 %	0.7 %	0.3 %	0.2 %	0.2 %
	Special SF Config. 7	2.7 %	1.1 %	0.7 %	0.3 %	0.2 %	0.2 %
	Special SF Config. 8	2.7 %	1.1 %	0.7 %	0.3 %	0.2 %	0.2 %
Extended Cyclic Prefix							
PBCH Resource Elements		216	216	216	216	216	216
Uplink-Downlink Subframe Config. 0	Special SF Config. 0	10.0 %	4.0 %	2.4 %	1.2 %	0.8 %	0.6 %
	Special SF Config. 1	7.5 %	3.0 %	1.8 %	0.9 %	0.6 %	0.5 %
	Special SF Config. 2	7.1 %	2.9 %	1.7 %	0.9 %	0.6 %	0.4 %
	Special SF Config. 3	6.8 %	2.7 %	1.6 %	0.8 %	0.5 %	0.4 %
	Special SF Config. 4	10.0 %	4.0 %	2.4 %	1.2 %	0.8 %	0.6 %
	Special SF Config. 5	7.5 %	3.0 %	1.8 %	0.9 %	0.6 %	0.5 %
	Special SF Config. 6	7.1 %	2.9 %	1.7 %	0.9 %	0.6 %	0.4 %
Uplink-Downlink Subframe Config. 5	Special SF Config. 0	3.0 %	1.2 %	0.7 %	0.4 %	0.2 %	0.2 %
	Special SF Config. 1	2.9 %	1.2 %	0.7 %	0.3 %	0.2 %	0.2 %
	Special SF Config. 2	2.9 %	1.1 %	0.7 %	0.3 %	0.2 %	0.2 %
	Special SF Config. 3	2.8 %	1.1 %	0.7 %	0.3 %	0.2 %	0.2 %
	Special SF Config. 4	3.0 %	1.2 %	0.7 %	0.4 %	0.2 %	0.2 %
	Special SF Config. 5	2.9 %	1.2 %	0.7 %	0.3 %	0.2 %	0.2 %
	Special SF Config. 6	2.9 %	1.1 %	0.7 %	0.3 %	0.2 %	0.2 %

Table 47 – Overhead generated by PBCH Physical Channel (TDD uplink-downlink subframe configurations 0 and 5)

6.2 PCFICH

★ The Physical Control Format Indicator Channel (PCFICH) is used at the start of each 1 ms downlink subframe to signal the number of symbols used for the PDCCH

★ The PCFICH is broadcast on the same set of antenna ports as the PBCH. This means that it can be broadcast using only antenna port 0, or transmit diversity can be used to broadcast the PCFICH using antenna ports {0, 1} or {0, 1, 2, 3}

★ The PCFICH occupies 16 Resource Elements within the first OFDMA symbol of each 1 ms downlink subframe. These 16 Resource Elements are divided into 4 quadruplets. The position of these 4 quadruplets within the first OFDMA symbol depends upon the downlink channel bandwidth and the Physical layer Cell Identity (PCI). An example PCFICH allocation is illustrated in Figure 94

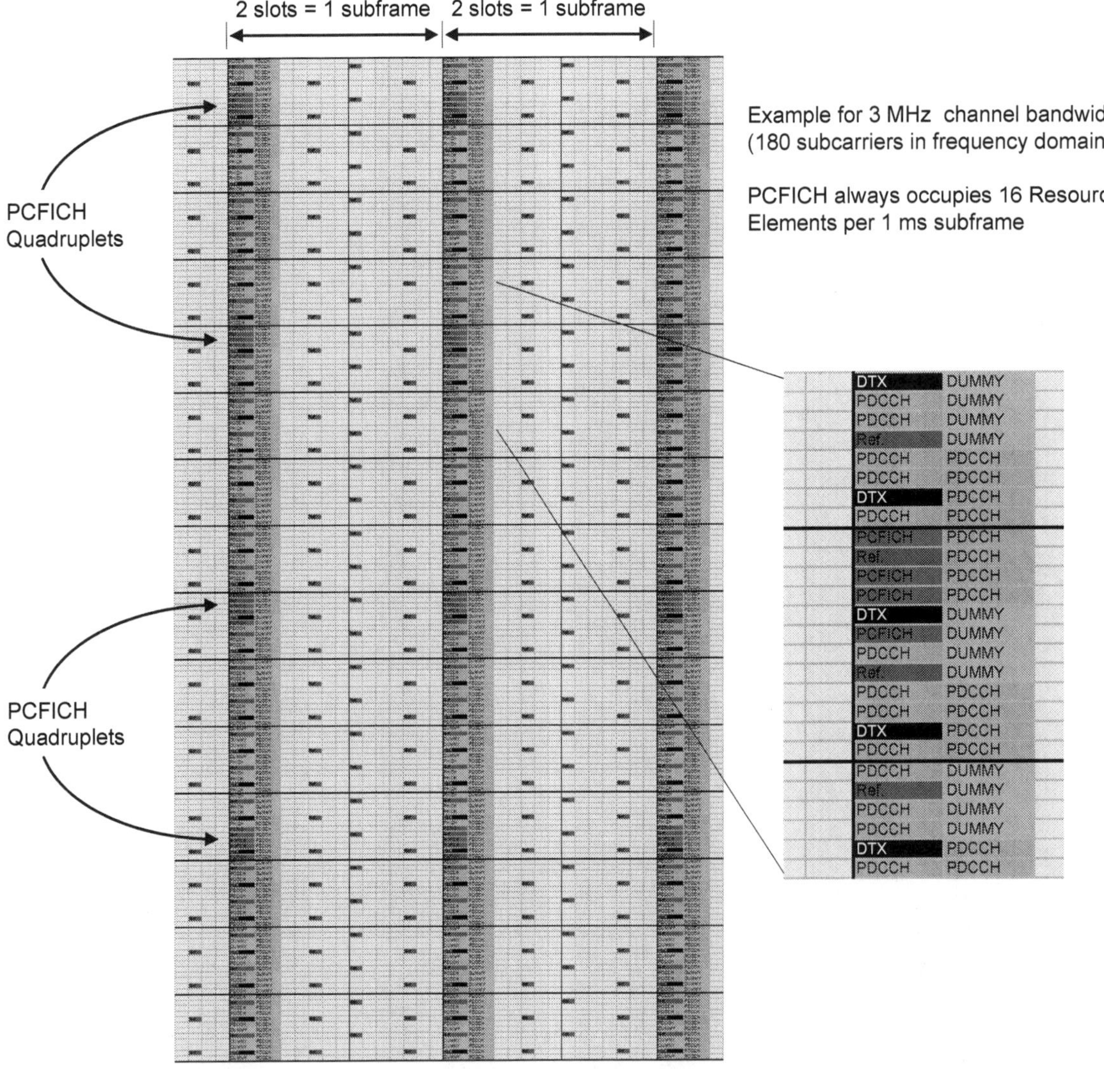

Figure 94 – PCFICH Physical Channel within the first OFDMA symbol of each subframe

★ The PCFICH uses QPSK modulation so the 16 Resource Elements allow 32 bits

★ The PCFICH transfers a Control Format Indicator (CFI) which has a value ranging from 1 to 3

- o actual value = signalled value + 1 for the 1.4 MHz channel bandwidth
- o actual value = signalled value for other channel bandwidths

The CFI is channel coded to 32 bits to occupy the complete PCFICH capacity

★ The example illustrated in Figure 94 shows 2 symbols used by the PDCCH at the start of every subframe. The example is based upon the 3 MHz channel bandwidth so the signalled value is also 2. If the example was based upon the 1.4 MHz channel bandwidth then the signalled value would be 1

★ The range of values signalled by the PCFICH depends upon the scenario and the channel bandwidth, as presented in Table 48. The default values are applicable to the majority of scenarios, e.g. an FDD network without MBSFN nor Positioning Reference Signals

		Channel Bandwidth					
		1.4 MHz	3 MHz	5 MHz	10 MHz	15 MHz	20 MHz
Default values		2, 3, 4	1, 2, 3				
TDD subframes 1 and 6		2	1, 2				
MBSFN subframes on a carrier supporting PDSCH	1, 2 cell specific antenna ports	2	1, 2				
	4 cell specific antenna ports	2	2				
Subframes on a carrier not supporting PDSCH		0	0				
Non-MBSFN subframes configured with Positioning Reference Signals (except TDD subframe 6)		2, 3	1, 2, 3				

Table 48 – Range of PCFICH values as a function of the scenario and channel bandwidth

★ The 1.4 MHz channel bandwidth requires an increased number of symbols in the time domain because there are fewer subcarriers in the frequency domain, i.e. an increased number of symbols in the time domain is required to maintain a reasonable PDCCH capacity

★ The signalled value is determined by the eNode B radio resource management and depends upon the number of active connections, i.e. PDCCH signalling increases as the number of active connections increases

★ The PCFICH represents an overhead which reduces the number of Resource Elements available for user plane data

- o overhead decreases for larger channel bandwidths
- o overhead increases when using the extended cyclic prefix

★ The overhead generated by the PCFICH for FDD is presented in Table 49

	1.4 MHz	3 MHz	5 MHz	10 MHz	15 MHz	20 MHz
PCFICH Res. Elements per 10 ms radio frame	160					
Normal Cyclic Prefix						
Total Resource Elements per 1 ms subframe	1008	2520	4200	8400	12600	16800
Overhead	1.6 %	0.6 %	0.4 %	0.2 %	0.1 %	0.1 %
Extended Cyclic Prefix						
Total Resource Elements per 1 ms subframe	864	2160	3600	7200	10800	14400
Overhead	1.9 %	0.7 %	0.4 %	0.2 %	0.2 %	0.1 %

Table 49 – Overhead generated by the PCFICH (FDD)

★ In the case of TDD, the overhead generated by the PCFICH is the same as that for FDD when calculating within a normal downlink subframe, i.e. 16 Resource Elements relative to the total number of Resource Elements within a normal downlink subframe

★ The overhead for TDD is greater when considering all downlink subframes and accounting for the special subframes which have fewer symbols available for downlink data transmission. This calculation is multi-dimensional because the number of downlink Resource Elements depends upon the cyclic prefix length, the channel bandwidth, the uplink-downlink subframe configuration and the special subframe configuration

★ Table 50 presents the overheads when assuming the uplink-downlink subframe configurations 0 and 5. These configurations represent those with the highest and lowest overheads, i.e. uplink-downlink subframe configuration 0 has the least downlink Resource Elements, while uplink-downlink subframe configuration 5 has the most downlink Resource Elements

★ In the worst case of uplink-downlink subframe configuration 0, using the extended cyclic prefix and special subframe configuration 0, the overhead for TDD reaches 3.0 %

★ 3GPP References: TS 36.211

		1.4 MHz	3 MHz	5 MHz	10 MHz	15 MHz	20 MHz
Normal Cyclic Prefix							
Uplink-Downlink Subframe Config. 0	PCFICH Res. Elements per 10 ms radio frame	64	64	64	64	64	64
	Special SF Config. 0	2.6 %	1.0 %	0.6 %	0.3 %	0.2 %	0.2 %
	Special SF Config. 1	1.9 %	0.8 %	0.5 %	0.2 %	0.2 %	0.1 %
	Special SF Config. 2	1.9 %	0.7 %	0.4 %	0.2 %	0.1 %	0.1 %
	Special SF Config. 3	1.8 %	0.7 %	0.4 %	0.2 %	0.1 %	0.1 %
	Special SF Config. 4	1.7 %	0.7 %	0.4 %	0.2 %	0.1 %	0.1 %
	Special SF Config. 5	2.6 %	1.0 %	0.6 %	0.3 %	0.2 %	0.2 %
	Special SF Config. 6	1.9 %	0.8 %	0.5 %	0.2 %	0.2 %	0.1 %
	Special SF Config. 7	1.9 %	0.7 %	0.4 %	0.2 %	0.1 %	0.1 %
	Special SF Config. 8	1.8 %	0.7 %	0.4 %	0.2 %	0.1 %	0.1 %
Uplink-Downlink Subframe Config. 5	PCFICH Res. Elements per 10 ms radio frame	144	144	144	144	144	144
	Special SF Config. 0	1.7 %	0.7 %	0.4 %	0.2 %	0.1 %	0.1 %
	Special SF Config. 1	1.7 %	0.7 %	0.4 %	0.2 %	0.1 %	0.1 %
	Special SF Config. 2	1.6 %	0.7 %	0.4 %	0.2 %	0.1 %	0.1 %
	Special SF Config. 3	1.6 %	0.7 %	0.4 %	0.2 %	0.1 %	0.1 %
	Special SF Config. 4	1.6 %	0.6 %	0.4 %	0.2 %	0.1 %	0.1 %
	Special SF Config. 5	1.7 %	0.7 %	0.4 %	0.2 %	0.1 %	0.1 %
	Special SF Config. 6	1.7 %	0.7 %	0.4 %	0.2 %	0.1 %	0.1 %
	Special SF Config. 7	1.6 %	0.7 %	0.4 %	0.2 %	0.1 %	0.1 %
	Special SF Config. 8	1.6 %	0.7 %	0.4 %	0.2 %	0.1 %	0.1 %
Extended Cyclic Prefix							
Uplink-Downlink Subframe Config. 0	PCFICH Res. Elements per 10 ms radio frame	64	64	64	64	64	64
	Special SF Config. 0	3.0 %	1.2 %	0.7 %	0.4 %	0.2 %	0.2 %
	Special SF Config. 1	2.2 %	0.9 %	0.5 %	0.3 %	0.2 %	0.1 %
	Special SF Config. 2	2.1 %	0.8 %	0.5 %	0.3 %	0.2 %	0.1 %
	Special SF Config. 3	2.0 %	0.8 %	0.5 %	0.2 %	0.2 %	0.1 %
	Special SF Config. 4	3.0 %	1.2 %	0.7 %	0.4 %	0.2 %	0.2 %
	Special SF Config. 5	2.2 %	0.9 %	0.5 %	0.3 %	0.2 %	0.1 %
	Special SF Config. 6	2.1 %	0.8 %	0.5 %	0.3 %	0.2 %	0.1 %
Uplink-Downlink Subframe Config. 5	PCFICH Res. Elements per 10 ms radio frame	144	144	144	144	144	144
	Special SF Config. 0	2.0 %	0.8 %	0.5 %	0.2 %	0.2 %	0.1 %
	Special SF Config. 1	1.9 %	0.8 %	0.5 %	0.2 %	0.2 %	0.1 %
	Special SF Config. 2	1.9 %	0.8 %	0.5 %	0.2 %	0.2 %	0.1 %
	Special SF Config. 3	1.9 %	0.8 %	0.5 %	0.2 %	0.2 %	0.1 %
	Special SF Config. 4	2.0 %	0.8 %	0.5 %	0.2 %	0.2 %	0.1 %
	Special SF Config. 5	1.9 %	0.8 %	0.5 %	0.2 %	0.2 %	0.1 %
	Special SF Config. 6	1.9 %	0.8 %	0.5 %	0.2 %	0.2 %	0.1 %

Table 50 – Overhead generated by PCFICH Physical Channel (TDD uplink-downlink subframe configurations 0 and 5)

6.3 PHICH

- The Physical Hybrid ARQ Indicator Channel (PHICH) is used to signal positive or negative acknowledgments for uplink data transferred on the PUSCH
- Prior to 3GPP release 10 (LTE Advanced), each connection could transfer a maximum of 1 transport block per subframe on the PUSCH so a maximum of 1 PHICH acknowledgement per subframe per connection was required
- 3GPP release 10 allows connections to transfer 2 transport blocks per subframe when MIMO is used in the uplink direction. This requires 2 PHICH acknowledgements per subframe per connection
- A PHICH acknowledgement is identified by its:
    - PHICH Group
    - PHICH Orthogonal Sequence Index within the PHICH Group
- In the case of FDD, the number of PHICH groups is a function of the downlink channel bandwidth and the PHICH Group Scaling Factor. Both the downlink channel bandwidth and the PHICH Group Scaling Factor are broadcast within the Master Information Block (MIB) on the PBCH. The number of PHICH groups for FDD is presented in Table 51

	1.4 MHz	3 MHz	5 MHz	10 MHz	15 MHz	20 MHz
PHICH Group Scaling Factor	Normal Cyclic Prefix					
1/6	1	1	1	2	2	3
1/2	1	1	2	4	5	7
1	1	2	4	7	10	13
2	2	4	7	13	19	25
PHICH Group Scaling Factor	Extended Cyclic Prefix					
1/6	2	2	2	4	4	6
1/2	2	2	4	8	10	14
1	2	4	8	14	20	26
2	4	8	14	26	38	50

Table 51 – Number of PHICH groups (FDD)

- In the case of TDD, PHICH groups are only included within specific downlink subframes. Other downlink subframes do not include any PHICH groups. The number of PHICH groups within a TDD downlink subframe is given by the equivalent FDD figure multiplied by the factors presented in Table 52

Uplink-Downlink Configuration	Subframe Number									
	0	1	2	3	4	5	6	7	8	9
0	2	1	-	-	-	2	1	-	-	-
1	0	1	-	-	1	0	1	-	-	1
2	0	0	-	1	0	0	0	-	1	0
3	1	0	-	-	-	0	0	0	1	1
4	0	0	-	-	0	0	0	0	1	1
5	0	0	-	0	0	0	0	0	1	0
6	1	1	-	-	-	1	1	-	-	1

Table 52 – Multiplying factors for the number of PHICH groups for TDD

- The number of PHICH groups is doubled when using the extended cyclic prefix because the number of orthogonal sequences within each group is halved
    - normal cyclic prefix has 8 orthogonal sequences per PHICH group
    - extended cyclic prefix has 4 orthogonal sequences per PHICH group
- A UE selects its PHICH group and Orthogonal Sequence Index using:

$$\text{PHICH Group} = \left(I_{lowest} + n_{DMRS}\right) \bmod \left(N_{group}\right)$$

$$\text{PHICH Orthogonal Sequence Index} = \left(\lfloor I_{lowest} / N_{group} \rfloor + n_{DMRS}\right) \bmod \left(2 \times N_{SF}^{PHICH}\right)$$

where,

I_{lowest} is the lowest Physical Resource Block (PRB) index allocated to the first slot of the PUSCH transmission

n_{DMRS} is the Demodulation Reference Signal cyclic shift signalled within DCI format 0 or DCI format 4

N_{group} is the number of PHICH groups

N_{SF}^{PHICH} is the spreading factor size used for PHICH modulation (normal cyclic prefix: 4; extended cyclic prefix: 2)

★ In the case of TDD uplink-downlink subframe configuration 0 with PUSCH transmission during subframe 4 or 9, the PHICH Group number is given by the value calculated above + N_{group}

★ In the case of 3GPP release 10, when uplink MIMO is used to transfer 2 transport blocks within a subframe, the value of I_{lowest} is increased by 1 when calculating the PHICH group and sequence number for the second transport block

★ Table 53 presents the combinations of PHICH group and orthogonal sequence index when assuming the FDD normal cyclic prefix, a 3 MHz channel bandwidth and 2 PHICH groups. There are 8 sequences within each of the 2 groups so this provides 16 combinations of group and sequence, i.e. sufficient to support 15 acknowledgements for 15 connections when each connection is allocated a single Resource Block

	Cyclic Shift for Demodulation Reference Signal							
PUSCH PRB Index	0	1	2	3	4	5	6	7
0	0, 0	1, 1	0, 2	1, 3	0, 4	1, 5	0, 6	1, 7
1	1, 0	0, 1	1, 2	0, 3	1, 4	0, 5	1, 6	0, 7
2	0, 1	1, 2	0, 3	1, 4	0, 5	1, 6	0, 7	1, 0
3	1, 1	0, 2	1, 3	0, 4	1, 5	0, 6	1, 7	0, 0
4	0, 2	1, 3	0, 4	1, 5	0, 6	1, 7	0, 0	1, 1
5	1, 2	0, 3	1, 4	0, 5	1, 6	0, 7	1, 0	0, 1
6	0, 3	1, 4	0, 5	1, 6	0, 7	1, 0	0, 1	1, 2
7	1, 3	0, 4	1, 5	0, 6	1, 7	0, 0	1, 1	0, 2
8	0, 4	1, 5	0, 6	1, 7	0, 0	1, 1	0, 2	1, 3
9	1, 4	0, 5	1, 6	0, 7	1, 0	0, 1	1, 2	0, 3
10	0, 5	1, 6	0, 7	1, 0	0, 1	1, 2	0, 3	1, 4
11	1, 5	0, 6	1, 7	0, 0	1, 1	0, 2	1, 3	0, 4
12	0, 6	1, 7	0, 0	1, 1	0, 2	1, 3	0, 4	1, 5
13	1, 6	0, 7	1, 0	0, 1	1, 2	0, 3	1, 4	0, 5
14	0, 7	1, 0	0, 1	1, 2	0, 3	1, 4	0, 5	1, 6

Table 53 – Example PHICH resource allocation (PHICH group, PHICH Orthogonal Sequence Index)

★ Each PHICH acknowledgement is represented by a bit string of 3 bits: 000 for a negative acknowledgement and 111 for a positive acknowledgement

★ The PHICH acknowledgement is modulated using BPSK to generate 3 modulation symbols. These modulation symbols are then spread using the appropriate PHICH orthogonal sequence. A spreading factor of 4 is used for the normal cyclic prefix and a spreading factor of 2 is used for the extended cyclic prefix

★ This results in PHICH acknowledgments occupying

  - 12 Resource Elements when using the normal cyclic prefix
  - 6 Resource Elements when using the extended cyclic prefix

★ PHICH acknowledgements belonging to the same group occupy the same set of Resource Elements, i.e. they are differentiated by their orthogonal sequences. This allows the multiplexing of multiple transmissions within the same set of Resource Elements

★ In the case of the extended cyclic prefix, PHICH groups are paired and each pair of groups occupies 12 Resource Elements

★ The PHICH is transmitted on the same set of antenna ports as the PBCH. This means that it can be broadcast using only antenna port 0, or transmit diversity can be used to transmit the PHICH using antenna ports {0, 1} or {0, 1, 2, 3}. In the case of using 4 antenna ports, the PHICH has its own precoding matrices which differ to those used for the PDSCH

★ The set of 12 Resource Elements allocated to each PHICH group (or pair of PHICH groups for the extended cyclic prefix) is divided into 3 quadruplets

- The Master Information Block on the PBCH indicates whether the PHICH uses a normal or extended duration. A normal duration means that the PHICH uses the first OFDMA symbol belonging to a subframe. An extended duration means that the PHICH uses the first 3 OFDMA symbols belonging to a subframe
- In the case of TDD using the extended PHICH duration, the PHICH is limited to the first 2 OFDMA symbols within subframes 1 and 6
- The PHICH is also limited to the first 2 OFDMA symbols when the extended PHICH duration is used within MBSFN subframes
- The PHICH duration puts a lower limit on the value signalled by the PCFICH. If the PHICH occupies the first 3 OFDMA symbols belonging to a subframe then the PCFICH must signal a value of at least 3
- The position of the 3 PHICH quadruplets depends upon the Physical layer Cell Identity. An example PHICH allocation is shown in Figure 95. This example is based upon a single PHICH group using the normal PHICH duration and normal cyclic prefix

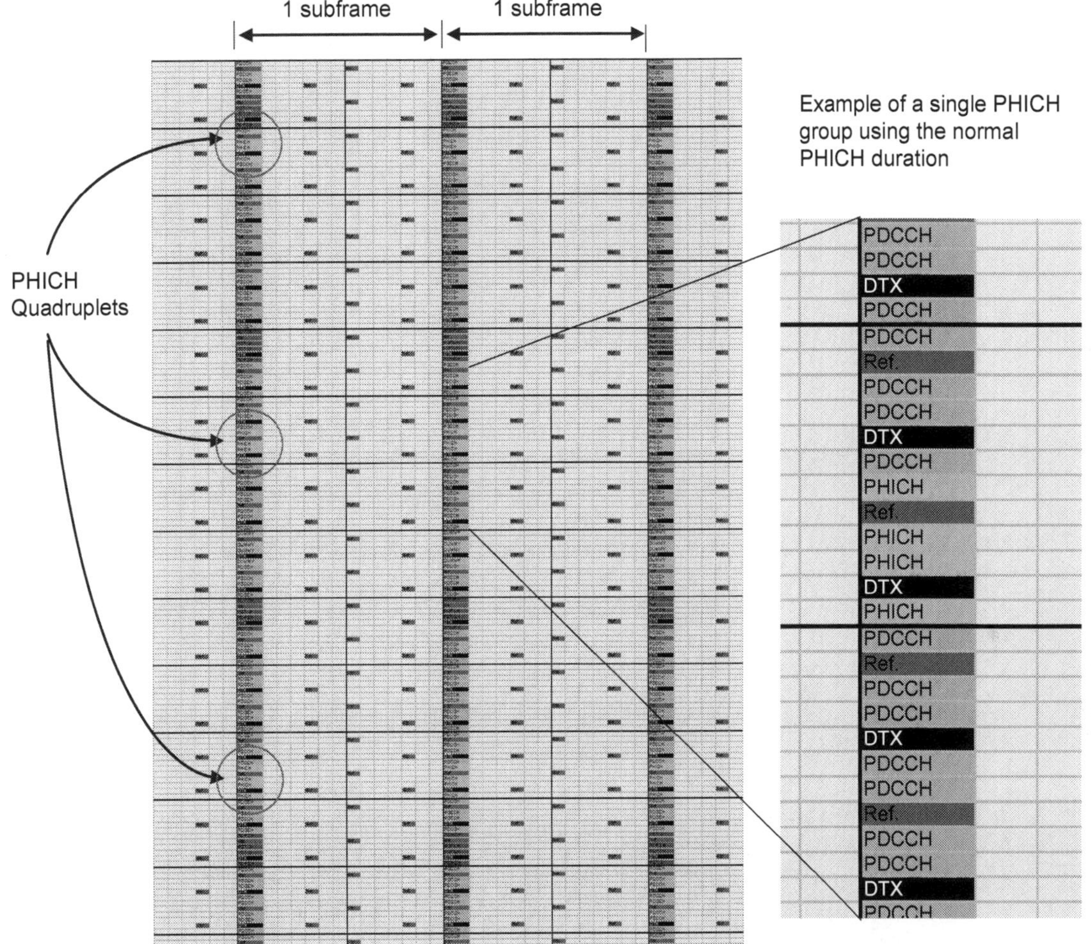

Figure 95 – Single PHICH Group using the normal PHICH duration and normal cyclic prefix

- Figure 96 illustrates an example of a single PHICH group using the extended PHICH duration and normal cyclic prefix. In this case, the PHICH quadruplets are distributed across the first 3 OFDMA symbols belonging to each subframe

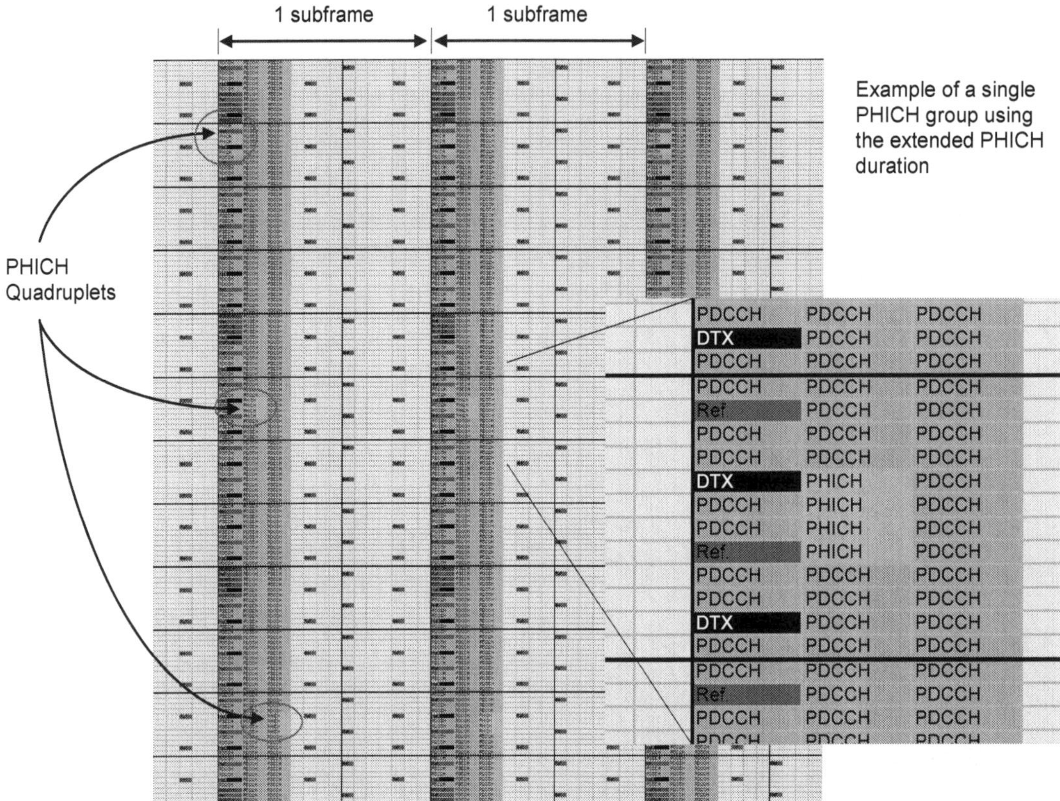

Figure 96 – Single PHICH Group using the extended PHICH duration and normal cyclic prefix

★ Figure 97 illustrates an example of 2 PHICH groups using the extended PHICH duration and normal cyclic prefix. The PHICH quadruplets belonging to each PHICH group are positioned adjacent to each other

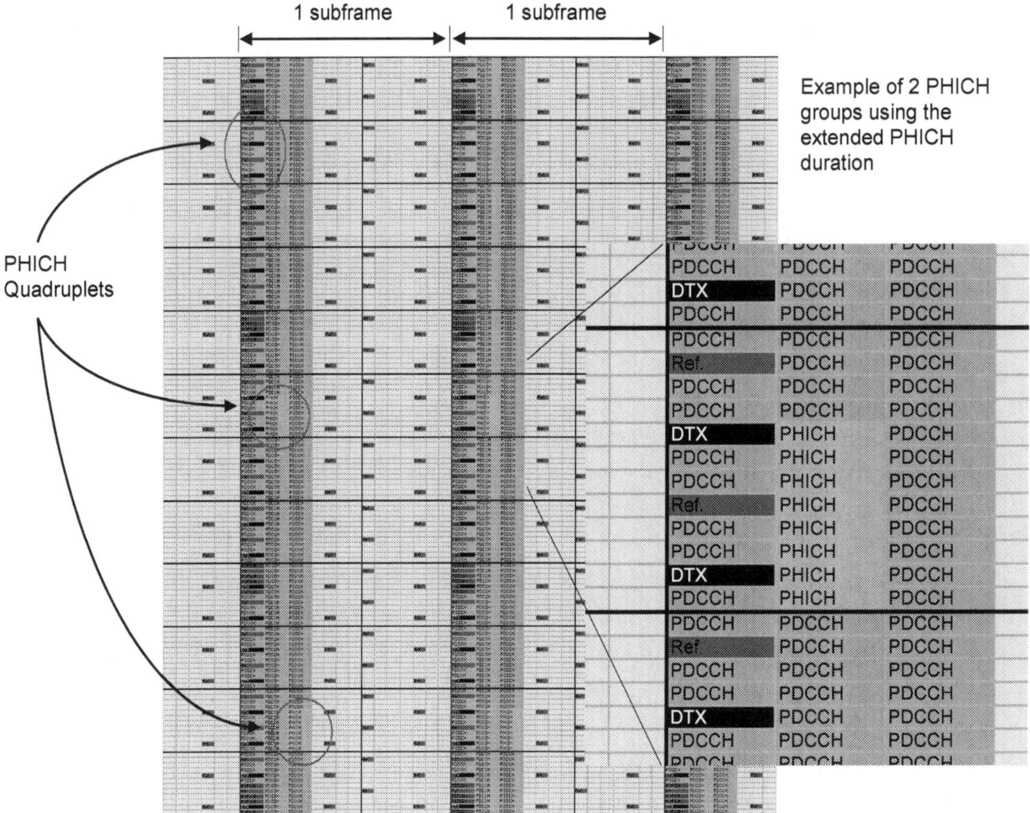

Figure 97 – 2 PHICH Groups using the extended PHICH duration and normal cyclic prefix

★ The PHICH represents an overhead which reduces the number of Resource Elements available for user plane data

- o for a specific channel bandwidth, the overhead increases as the number of PHICH groups increases
- o maximum and minimum overheads remain approximately constant as the channel bandwidth increases because the number of PHICH groups also increases
- o the overhead for the extended cyclic prefix is only slightly greater than that for the normal cyclic prefix despite the number of PHICH groups being double because the number of Resource Elements per group is half that of the normal cyclic prefix

★ The maximum and minimum overheads generated by the PHICH for FDD are presented in Table 54. The minimum figures are generated when assuming a PHICH group scaling factor of 1/6. The maximum figures are generated when assuming a PHICH group scaling factor of 2. All figures are calculated over a 10 ms radio frame

		1.4 MHz	3 MHz	5 MHz	10 MHz	15 MHz	20 MHz
Normal Cyclic Prefix							
Minimum	PHICH Groups	10	10	10	20	20	30
	PHICH Resource Elements	120	120	120	240	240	360
	Total Resource Elements	10080	25200	42000	84000	126000	168000
	Overhead	1.2 %	0.5 %	0.3 %	0.3 %	0.2 %	0.2 %
Maximum	PHICH Groups	20	40	70	130	190	250
	PHICH Resource Elements	240	480	840	1560	2280	3000
	Total Resource Elements	10080	25200	42000	84000	126000	168000
	Overhead	2.4 %	1.9 %	2.0 %	1.9 %	1.8 %	1.8 %
Extended Cyclic Prefix							
Minimum	PHICH Groups	20	20	20	40	40	60
	PHICH Resource Elements	120	120	120	240	240	360
	Total Resource Elements	8640	21600	36000	72000	108000	144000
	Overhead	1.4 %	0.6 %	0.3 %	0.3 %	0.2 %	0.3 %
Maximum	PHICH Groups	40	80	140	260	380	500
	PHICH Resource Elements	240	480	840	1560	2280	3000
	Total Resource Elements	8640	21600	36000	72000	108000	144000
	Overhead	2.8 %	2.2 %	2.3 %	2.2 %	2.1 %	2.1 %

Table 54 – Maximum and minimum overheads generated by the PHICH (FDD)

★ In the case of TDD, the maximum and minimum overheads generated by the PHICH are presented in Table 55. All figures are calculated over a 10 ms radio frame

- o normal cyclic prefix
  - ▪ minimum is generated when assuming uplink-downlink subframe configuration 5, special subframe configuration 4, and PHICH group scaling factor of 1/6
  - ▪ maximum is generated when assuming uplink-downlink subframe configuration 0, special subframe configurations 0 or 5, and PHICH group scaling factor of 2
- o extended cyclic prefix
  - ▪ minimum is generated when assuming uplink-downlink subframe configuration 5, special subframe configuration 4, and PHICH group scaling factor of 1/6
  - ▪ maximum is generated when assuming uplink-downlink subframe configuration 0, special subframe configurations 0 or 4, and PHICH group scaling factor of 2

		1.4 MHz	3 MHz	5 MHz	10 MHz	15 MHz	20 MHz
Normal Cyclic Prefix							
Minimum	PHICH Groups	1	1	1	2	2	3
	PHICH Resource Elements	12	12	12	24	24	36
	Total Resource Elements	3744	9360	15600	31200	46800	62400
	Overhead	0.1 %	0.1 %	< 0.1 %	<0.1 %	< 0.1 %	<0.1 %
Maximum	PHICH Groups	12	24	42	78	114	150
	PHICH Resource Elements	144	288	504	936	1368	1800
	Total Resource Elements	2448	6120	10200	20400	30600	40800
	Overhead	5.9 %	4.7 %	4.9 %	4.6 %	4.5 %	4.4 %
Extended Cyclic Prefix							
Minimum	PHICH Groups	2	2	2	4	4	6
	PHICH Resource Elements	12	12	12	24	24	36
	Total Resource Elements	7128	17820	29700	59400	89100	118800
	Overhead	0.2 %	0.1 %	< 0.1 %	<0.1 %	< 0.1 %	<0.1 %
Maximum	PHICH Groups	24	48	84	156	228	300
	PHICH Resource Elements	144	288	504	936	1368	1800
	Total Resource Elements	2160	5400	9000	18000	27000	36000
	Overhead	6.7 %	5.3 %	5.6 %	5.2 %	5.1 %	5.0 %

Table 55 – Maximum and minimum overheads generated by the PHICH (TDD)

★ 3GPP References: TS 36.211

6.4 PDCCH

- The Physical Downlink Control Channel (PDCCH) is used to transfer Downlink Control Information (DCI). The detailed content of DCI is described in section 9
- The PCFICH signals the number of OFDMA symbols which can be occupied by the PDCCH. These symbols are always at the start of each downlink subframe
- Resource Elements allocated to the PDCCH are grouped into quadruplets (groups of 4 Resource Elements). The number of quadruplets available to the PDCCH is equal to the number of quadruplets within the set of OFDMA symbols signalled by the PCFICH, which have not already been allocated to the PCFICH, PHICH or Reference Signals
- Resource element quadruplets are grouped into Control Channel Elements (CCE). There are 9 quadruplets within a single CCE, i.e. 36 Resource Elements per CCE. The PDCCH uses QPSK modulation so a single CCE can transfer 72 bits
- 3GPP TS 36.211 specifies 4 PDCCH formats. These are presented in Table 56

PDCCH Format	Number of CCE	Number of RE Quadruplets	Number of Bits
0	1	9	72
1	2	18	144
2	4	36	288
3	8	72	576

Table 56 – PDCCH formats

- The PDCCH format is selected according to the size of the DCI. DCI bits have a 16 bit CRC attached prior to rate 1/3 channel coding and rate matching. The PDCCH format must offer sufficient capacity to avoid puncturing the DCI bits too heavily during rate matching
- Table 57 presents some example FDD coding rates for each DCI and PDCCH format. The coding rate is calculated as the ratio of the number of DCI bits after CRC attachment to the capacity of the PDCCH. The entries in bold represent the lowest PDCCH format which avoids the requirement for puncturing (coding rates greater than 0.33 require puncturing). In general, the DCI sizes are different for TDD so the coding rates are also different

			Coding Rate			
DCI Format	Channel Bandwidth	DCI Bits after CRC	PDCCH Format 0	PDCCH Format 1	PDCCH Format 2	PDCCH Format 3
0 / 1A / 3 / 3A	5	41	0.57	**0.28**	0.14	0.07
	10	43	0.60	**0.30**	0.15	0.07
	20	44	0.61	**0.31**	0.15	0.08
1	5	43	0.60	**0.30**	0.15	0.07
	10	47	0.65	**0.33**	0.16	0.08
	20	55	0.76	0.38	**0.19**	0.10
1B	5	43	0.60	**0.30**	0.15	0.07
	10	44	0.61	**0.31**	0.15	0.08
	20	46	0.64	**0.32**	0.16	0.08
1C	5	28	0.39	**0.19**	0.10	0.05
	10	29	0.40	**0.20**	0.10	0.05
	20	31	0.43	**0.22**	0.11	0.05
1D	5	43	0.60	**0.30**	0.15	0.07
	10	44	0.61	**0.31**	0.15	0.08
	20	46	0.64	**0.32**	0.16	0.08
2	5	55	0.76	0.38	**0.19**	0.10
	10	59	0.82	0.41	**0.20**	0.10
	20	67	0.93	0.47	**0.23**	0.12
2A / 2B	5	52	0.72	0.36	**0.18**	0.09
	10	57	0.79	0.40	**0.20**	0.10
	20	64	0.89	0.44	**0.22**	0.11
2C	5	54	0.75	0.38	**0.19**	0.09
	10	58	0.81	0.40	**0.20**	0.10
	20	66	0.92	0.46	**0.23**	0.11
4	5	50	0.69	0.35	**0.17**	0.09
	10	52	0.72	0.36	**0.18**	0.09
	20	54	0.75	0.38	**0.19**	0.09

Table 57 – Coding rates for each DCI and PDCCH format (channel bandwidths of 5, 10 and 20 MHz)

★ Larger coding rates require more puncturing and have less redundancy to protect the DCI bits from the propagation channel. Larger coding rates could be used when coverage conditions are good

★ The number of CCE available depends upon the channel bandwidth and the number of OFDMA symbols allocated to the PDCCH. It also depends upon the number of Resource Element quadruplets used by the PHICH and whether or not the cell specific Reference Signal is broadcast on antenna ports 0 to 3. Table 58 presents the maximum number of available CCE assuming that no quadruplets have been allocated to the PHICH and that the cell specific Reference Signals are not broadcast on all four antenna ports

	Channel Bandwidth					
OFDMA Symbols for PDCCH	1.4 MHz	3 MHz	5 MHz	10 MHz	15 MHz	20 MHz
1	-	2	5	10	16	21
2	2	7	13	27	41	55
3	4	12	21	44	66	88
4	6	-	-	-	-	-

Table 58 – Maximum number of CCE (assuming 1 or 2 transmit antenna ports)

★ The maximum number of CCE is relatively small when only a single OFDMA symbol is allocated to the PDCCH

★ A subset of the CCE shown in Table 58 will be required to broadcast DCI for system information and paging purposes. A further subset can be used to broadcast DCI for power control purposes. This reduces the total number of CCE available for allocating user plane resources

★ Figure 98 illustrates the Resource Elements available to the PDCCH when 2 OFDMA symbols are allocated. The first OFDMA symbol is shared between the Reference Signal, PCFICH, PHICH and PDCCH, whereas the second OFDMA symbol is dedicated to the PDCCH

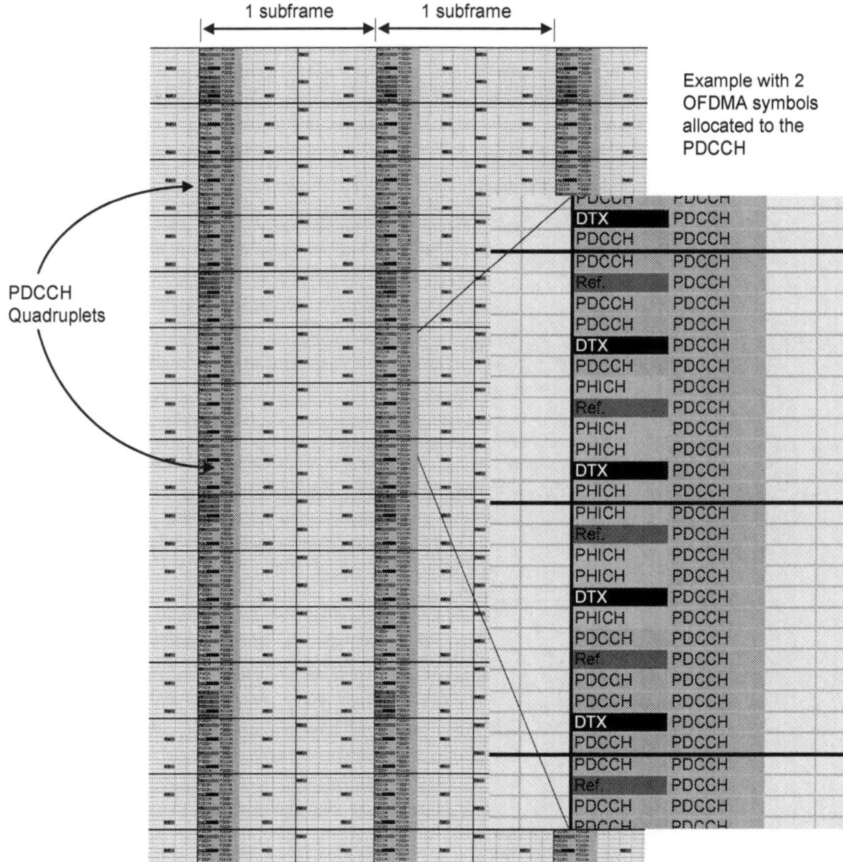

Figure 98 – PDCCH quadruplets with an allocation of 2 OFDMA symbols

★ The example shown in Figure 98 assumes a normal PHICH duration. An extended PHICH duration requires that at least 3 OFDMA symbols are allocated to the PDCCH. In that case, the PHICH would be distributed across the first 3 OFDMA symbols

★ The PDCCH is transmitted on the same set of antenna ports as the PBCH. This means that it can be broadcast using only antenna port 0, or transmit diversity can be used to transmit the PDCCH using antenna ports {0, 1} or {0, 1, 2, 3}

★ Individual UE are not required to decode all PDCCH DCI when searching for paging messages, system information, power control commands or user plane resource allocations. Common and UE specific 'search spaces' are defined to help restrict the quantity of decoding that each UE has to complete. These search spaces are described in section 9.2

★ The PDCCH represents an overhead which reduces the number of Resource Elements available for user plane data

 o the overhead increases as more OFDMA symbols are allocated to the PDCCH

 o the overhead increases when the extended cyclic prefix is used

★ Table 59 presents the combined overhead of the PCFICH, PHICH and PDCCH for FDD as a function of the number of symbols allocated to the PDCCH. These figures exclude the overhead generated by the Reference Signal (this should be added separately using the figures in Table 34). The overheads presented in Table 59 are applicable to either 1 or 2 transmit antenna ports. These overheads decrease when 4×4 MIMO is used but the Reference Signal overheads increase

	1.4 MHz	3 MHz	5 MHz	10 MHz	15 MHz	20 MHz
Normal Cyclic Prefix						
1 OFDMA Symbol	-	4.8 %				
2 OFDMA Symbols	11.9 %					
3 OFDMA Symbols	19.0 %					
4 OFDMA Symbols	26.2 %	-				
Extended Cyclic Prefix						
1 OFDMA Symbols	-	5.6 %				
2 OFDMA Symbols	13.9 %					
3 OFDMA Symbols	22.2 %					
4 OFDMA Symbols	27.8 %	-				

Table 59 – Overheads generated by the combination of PCFICH, PHICH and PDCCH (FDD)

★ In the case of TDD, the overhead calculation has an increased number of dimensions due to its dependence upon the uplink-downlink subframe configuration and special subframe configuration. Table 60 presents the maximum and minimum overheads for each cyclic prefix. These overheads are calculated over the duration of a radio frame

★ The minimum overheads for TDD can be less than that for FDD because subframes 1 and 6 can have a maximum of 2 OFDMA symbols allocated to the PDCCH. Thus, the results for 3 and 4 OFDMA symbols still assume 2 OFDMA symbols within subframes 1 and 6

		UL/DL Config	Spec. SF Config	1.4 MHz	3 MHz	5 MHz	10 MHz	15 MHz	20 MHz
Normal Cyclic Prefix									
Minimum	1 OFDMA Symbol	5	4	-	4.8 %				
	2 OFDMA Symbols	5	4	12.1 %					
	3 OFDMA Symbols	0	4	16.7 %					
	4 OFDMA Symbols	0	4	20.5 %	-				
Maximum	1 OFDMA Symbol	0	0 or 5	-	7.8 %				
	2 OFDMA Symbols	0	0 or 5	19.6 %					
	3 OFDMA Symbols	0	0 or 5	25.5 %					
	4 OFDMA Symbols	0	0 or 5	31.4 %	-				
Extended Cyclic Prefix									
Minimum	1 OFDMA Symbols	5	3	-	5.7 %				
	2 OFDMA Symbols	5	3	14.2 %					
	3 OFDMA Symbols	0	3	19.7 %					
	4 OFDMA Symbols	0	3	22.7 %	-				
Maximum	1 OFDMA Symbols	0	0 or 4	-	8.9 %				
	2 OFDMA Symbols	0	0 or 4	22.2 %					
	3 OFDMA Symbols	0	0 or 4	28.9 %					
	4 OFDMA Symbols	0	0 or 4	33.3 %	-				

Table 60 – Overheads generated by the combination of PCFICH, PHICH and PDCCH (TDD)

6.5 PDSCH

★ The Physical Downlink Shared Channel (PDSCH) is used to transfer:
  - system information
  - paging and other RRC signalling messages
  - application data

★ The set of System Information Blocks (SIB) are broadcast using the PDSCH. This is in contrast to the Master Information Block (MIB) which is broadcast using the PBCH. Both the MIB and SIB are described within section 10

★ Paging messages are broadcast using the PDSCH. UE in RRC Idle mode monitor the PDCCH for paging indications. A positive paging indication triggers the UE to decode the paging message from the relevant PDSCH Resource Blocks

★ Downlink RRC signalling messages are transferred using the PDSCH, i.e. Signalling Radio Bearers (SRB) make use of the PDSCH. Every connection has its own set of SRB

★ System information, paging and other downlink RRC messages represent an additional overhead from the perspective of transferring application data

★ The PDSCH can be modulated using 64QAM, 16QAM or QPSK. The eNode B selects the appropriate modulation scheme according to its link adaptation algorithm. In general, this is based upon the propagation channel and buffer occupancy. QPSK is always used when transferring system information or paging messages. This helps to ensure that these low bit rate messages can be received across the entire cell

★ The application throughputs achievable from the PDSCH are discussed in section 20. The PDSCH is a shared channel so its Resource Blocks and the associated throughputs must be shared between all active connections

★ Downlink Control Information (DCI) formats 1, 1A, 1B, 1C, 1D, 2, 2A, 2B and 2C are used to allocate PDSCH resources to individual UE. DCI are transferred using the PDCCH. They inform UE of which Resource Blocks to decode. The content of the various DCI formats is presented in section 9

★ Figure 99 illustrates the set of Resource Elements available to the PDSCH for FDD after making allocations for the Physical Signals and other Physical Channels. PDSCH Resource Elements are the lightly shaded areas. The example assumes that 2 OFDMA symbols per subframe are allocated to the PDCCH

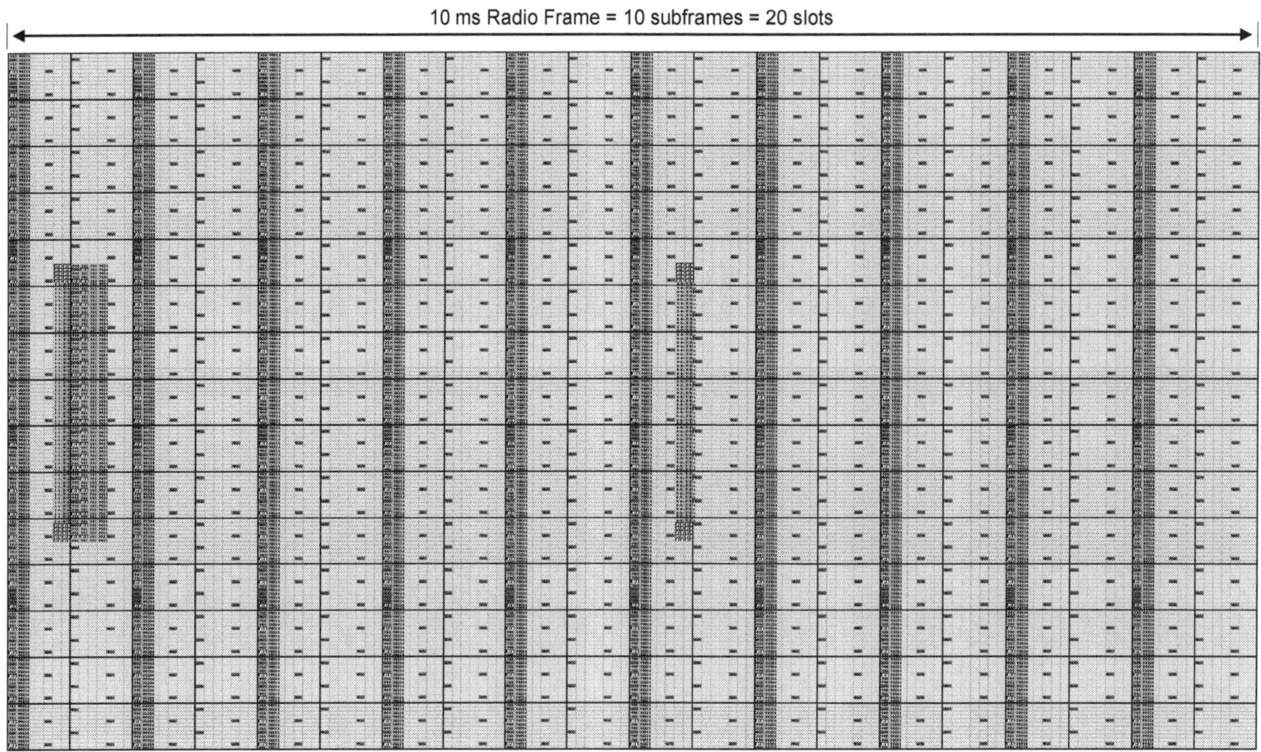

Figure 99 – PDSCH resources (shown as the lightly shaded areas) after allocation of other Physical Signals and Channels

★ Table 61 quantifies the percentage of FDD Resource Elements remaining for the PDSCH after making allocations for the Physical Signals and other Physical Channels. Figures are included for 1 and 2 transmit antenna ports

★ The variable having the greatest impact upon the percentage of Resource Elements available to the PDSCH is the number of OFDMA symbols allocated to the PDCCH

★ The channel bandwidth has relatively little impact upon the percentage of Resource Elements available to the PDSCH

		1.4 MHz	3 MHz	5 MHz	10 MHz	15 MHz	20 MHz
1 Transmit Antenna Port	Normal Cyclic Prefix						
	1 PDCCH Symbol	-	87.0 %	87.9 %	88.6 %	88.8 %	89.0 %
	2 PDCCH Symbols	76.5 %	79.9 %	80.8 %	81.5 %	81.7 %	81.8 %
	3 PDCCH Symbols	69.4 %	72.8 %	73.7 %	74.3 %	74.6 %	74.7 %
	4 PDCCH Symbols	62.3 %	-	-	-	-	-
	Extended Cyclic Prefix						
	1 PDCCH Symbol	-	84.9 %	86.0 %	86.7 %	87.0 %	87.1 %
	2 PDCCH Symbols	72.8 %	76.6 %	77.6 %	78.4 %	78.7 %	78.8 %
	3 PDCCH Symbols	64.4 %	68.3 %	69.3 %	70.1 %	70.3 %	70.5 %
	4 PDCCH Symbols	57.5 %	-	-	-	-	-
2 Transmit Antenna Ports	Normal Cyclic Prefix						
	1 PDCCH Symbol	-	83.5 %	84.4 %	85.1 %	85.3 %	85.4 %
	2 PDCCH Symbols	73.1 %	76.4 %	77.3 %	77.9 %	78.1 %	78.2 %
	3 PDCCH Symbols	66.0 %	69.2 %	70.1 %	70.8 %	71.0 %	71.1 %
	4 PDCCH Symbols	58.8 %	-	-	-	-	-
	Extended Cyclic Prefix						
	1 PDCCH Symbol	-	80.9 %	81.9 %	82.6 %	82.8 %	83.0 %
	2 PDCCH Symbols	68.9 %	72.6 %	73.5 %	74.3 %	74.5 %	74.6 %
	3 PDCCH Symbols	60.6 %	64.2 %	65.2 %	65.9 %	66.2 %	66.3 %
	4 PDCCH Symbols	55.0 %	-	-	-	-	-

Table 61 – Percentage of Resource Elements remaining for PDSCH (FDD)

★ The figures within Table 61 can be translated into application throughputs after accounting for the overheads generated by the system information, RRC signalling, retransmissions and the various protocol stack headers. Application throughputs are discussed and quantified within section 20

★ 3GPP References: TS 36.211

6.6 PMCH

★ The Physical Multicast Channel (PMCH) is used to transfer:
  - Multimedia Broadcast Multicast Service (MBMS) application data on the MTCH logical channel
  - MBMS signalling on the MCCH logical channel

★ Both the MTCH and MCCH are mapped onto the MCH transport channel prior to being mapped onto the PMCH

★ The PMCH is broadcast using antenna port 4. Only single antenna port transmission is supported so transmit diversity and spatial multiplexing are not applicable to the PMCH and MBMS

★ The MBSFN Reference Signal is broadcast with the PMCH on antenna port 4

★ A single MCH transport block is transferred during a single 1 ms subframe

★ UE decode SIB13 to identify the radio frames and subframes during which the MCCH is broadcast. Each MBSFN area has its own MCCH. SIB13 lists the scheduling information for each MCCH

★ The MCCH is broadcast during radio frames which satisfy:

SFN mod MCCH Repetition Period = MCCH Offset

where, the MCCH Repetition Period can be signalled with a value of 32, 64, 128 or 256 radio frames, and the MCCH Offset can be signalled with a value between 0 and 10

★ 3GPP does not allow the MCCH on the PMCH to be broadcast during all subframes:
  - in the case of FDD, the MCCH can be broadcast using subframes 1, 2, 3, 6, 7 and 8
  - in the case of TDD, the MCCH can be broadcast using subframes 3, 4, 7, 8 and 9

Other subframes are excluded to avoid conflict with the PBCH, synchronisation signals and paging messages. In the case of TDD, subframe 2 is excluded because it is always an uplink subframe

★ The allowed subframes for MBMS are illustrated in Figure 100

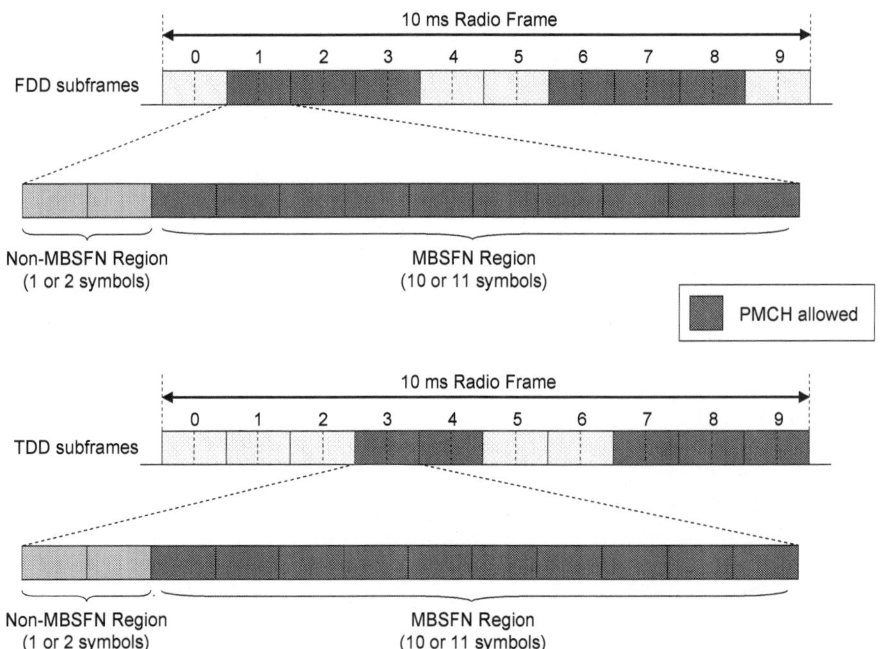

Figure 100 – Subframes and symbols during which PMCH transmission is possible (15 kHz subcarrier spacing)

★ A non-MBSFN region is reserved at the start of each MBMS subframe. This region can have a duration of either 1 or 2 symbols. The choice between these two values is signalled in SIB13. The non-MBSFN region can be used for PDCCH signalling, e.g. uplink resource allocations on the PUSCH. The PMCH can only be broadcast within the MBSFN region

★ The non-MBSFN region uses the same cyclic prefix as used for subframe 0. This could be either the normal or extended cyclic prefix, whereas the MBSFN region always uses the extended cyclic prefix

- ★ Cell specific Reference Signals are only transmitted within the non-MBSFN region of the MBMS subframe. These help non-MBMS UE to decode the PDCCH information within those 1 or 2 OFDMA symbols. MBSFN Reference Signals are only transmitted within the MBSFN region of the MBMS subframes
- ★ The PMCH always uses the extended cyclic prefix but can use either the 15 kHz or 7.5 kHz subcarrier spacing. The 7.5 kHz subcarrier spacing provides twice as many Resource Elements in the frequency domain but half as many Resource Elements in the time domain, i.e. 24 × 3 Resource Elements within a Resource Block rather than 12 × 6
- ★ The PMCH can be modulated using either QPSK, 16QAM or 64QAM although the allocated modulation scheme does not change dynamically over time. The modulation scheme applied during MCCH subframes is signalled within SIB13, whereas the modulation scheme applied during MTCH subframes is signalled within the MCCH. Each PMCH defined within the MCCH can have a different modulation scheme
- ★ The modulation scheme is signalled using a Modulation and Coding Scheme (MCS) index. This also defines the transport block size index. The transport block size index defines the transport block size when combined with the number of allocated Resource Blocks. The PMCH uses the complete set of Resource Blocks during the allocated subframes so the number of Resource Blocks is either 6, 15, 25, 50, 75 or 100 depending upon the channel bandwidth
- ★ Table 62 presents the set of allowed modulation schemes and transport block sizes for the PMCH when transferring MCCH information. The throughputs are calculated by dividing the transport block size by the 1 ms subframe period. Only 4 transport block sizes are allowed for each channel bandwidth

		Channel Bandwidth					
Transport Block Size Index	Modulation Scheme	1.4 MHz	3 MHz	5 MHz	10 MHz	15 MHz	20 MHz
Transport Block Sizes (bits)							
2	QPSK	256	648	1096	2216	3368	4584
7	QPSK	712	1800	3112	6200	9144	12216
12	16QAM	1352	3368	5736	11448	16992	22920
17	64QAM	2152	5352	9144	18336	27376	36696
Equivalent Throughput (Mbps)							
2	QPSK	0.26	0.65	1.10	2.22	3.37	4.58
7	QPSK	0.71	1.80	3.11	6.20	9.14	12.22
12	16QAM	1.35	3.37	5.74	11.45	16.99	22.92
17	64QAM	2.15	5.35	9.14	18.34	27.38	36.70

Table 62 – Modulation schemes, transport block sizes and equivalent throughputs associated with the MCCH over PMCH

- ★ The actual capacity of the PMCH depends upon the number of available Resource Elements. The number of Resource Elements is reduced by the non-MBSFN region and by the MBSFN Reference Signal. Example pairs of Resource Blocks are illustrated in Figure 101. These examples assume that a single OFDMA symbol is allocated to the non-MBSFN region

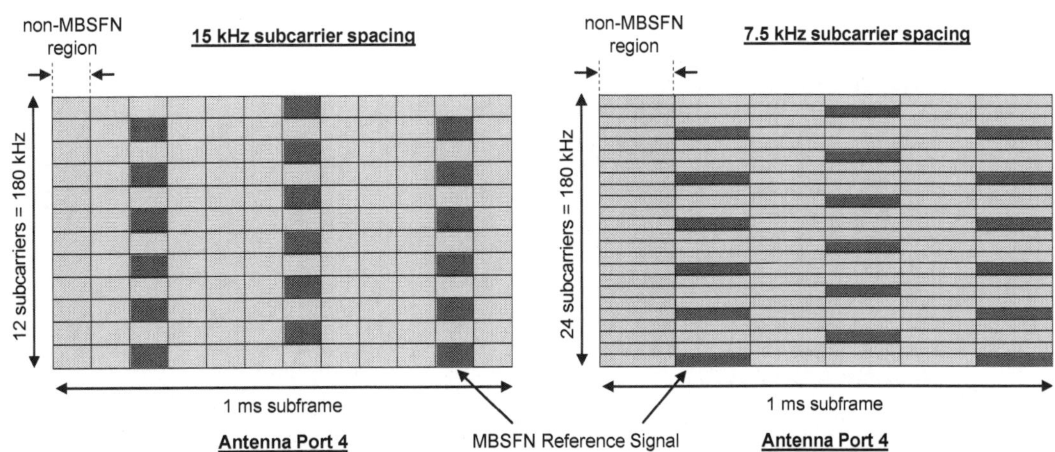

Figure 101 – MBMS Resource Blocks showing the non-MBSFN region and MBSFN Reference Signal

- ★ Table 63 presents the capacity of the PMCH in Mbps assuming a single OFDMA symbol has been allocated for the non-MBSFN region. These figures have been generated by summing all Resource Elements available for the PMCH and multiplying by the number of bits per modulation symbol
- ★ The capacity of the PMCH is sufficient to transfer the throughputs generated by the MCCH although it should be recognised that margin is required between the figures in Table 62 and Table 63 to accommodate the coding gain, i.e. the figures in Table 62 will be increased by a factor of 3 after rate 1/3 Turbo coding has been applied. Rate matching is then used to match the size of the resultant payload to the capacity of the PMCH

Transport Block Size Index	Channel Bandwidth 1.4 MHz	3 MHz	5 MHz	10 MHz	15 MHz	20 MHz
15 kHz subcarrier spacing						
QPSK	1.37	3.42	5.70	11.40	17.10	22.80
16QAM	2.74	6.84	11.40	22.80	34.20	45.60
64QAM	4.10	10.26	17.10	34.20	51.30	68.40
7.5 kHz subcarrier spacing						
QPSK	1.22	3.06	5.10	10.20	15.30	20.40
16QAM	2.45	6.12	10.20	20.40	30.60	40.80
64QAM	3.67	9.18	15.30	30.60	45.90	61.20

Table 63 – Capacity in Mbps of the PMCH assuming a single OFDMA symbol for the non-MBSFN region

★ Table 64 presents the set of allowed modulation schemes and throughputs for the PMCH when transferring MTCH information. These throughputs have been calculated by dividing the transport block size by the 1 ms subframe period

Transport Block Size Index	Modulation Scheme	Channel Bandwidth 1.4 MHz	3 MHz	5 MHz	10 MHz	15 MHz	20 MHz
Equivalent Throughput (Mbps)							
0	QPSK	0.15	0.39	0.68	1.38	2.09	2.79
1	QPSK	0.21	0.52	0.90	1.80	2.73	3.62
2	QPSK	0.26	0.65	1.10	2.22	3.37	4.58
3	QPSK	0.33	0.87	1.42	2.86	4.39	5.74
4	QPSK	0.41	1.06	1.80	3.62	5.35	7.22
5	QPSK	0.50	1.32	2.22	4.39	6.71	8.76
6	QPSK	0.60	1.54	2.60	5.16	7.74	10.30
7	QPSK	0.71	1.80	3.11	6.20	9.14	12.22
8	QPSK	0.81	2.09	3.50	6.97	10.68	14.11
9	QPSK / 16QAM	0.94	2.34	4.01	9.99	11.83	15.84
10	16QAM	1.03	2.66	4.39	8.76	12.96	17.57
11	16QAM	1.19	2.98	4.97	9.91	15.26	19.85
12	16QAM	1.35	3.37	5.74	11.45	16.99	22.92
13	16QAM	1.54	3.88	6.46	12.96	19.08	25.46
14	16QAM	1.74	4.26	7.22	14.11	21.38	28.34
15	16QAM / 64QAM	1.80	4.58	7.74	15.26	22.92	30.58
16	64QAM	1.93	4.97	7.99	16.42	24.50	32.86
17	64QAM	2.15	5.35	9.14	18.34	27.38	36.70
18	64QAM	2.34	5.99	9.91	19.85	29.30	39.23
19	64QAM	2.60	6.46	10.68	21.38	32.86	43.82
20	64QAM	2.79	6.97	11.45	22.92	35.16	46.89
21	64QAM	2.98	7.48	12.58	25.46	37.89	51.02
22	64QAM	3.24	7.99	13.54	27.38	40.58	55.06
23	64QAM	3.50	8.50	14.11	28.34	43.82	57.34
24	64QAM	3.62	9.14	15.26	30.58	45.35	61.66
25	64QAM	3.75	9.53	15.84	31.70	46.89	63.78
26	64QAM	4.39	11.06	18.34	36.70	55.06	75.38

Table 64 – Modulation schemes and equivalent throughputs associated with the MTCH over PMCH

★ Either QPSK or 16QAM can be used in combination with transport block size index 9. Likewise, either 16QAM or 64QAM can be used in combination with transport block size index 15

★ The complete set of transport block sizes is allowed for the PMCH when transferring an MTCH. This provides greater flexibility and allows an increased range of throughputs and BLER targets. It should be possible to receive the MTCH across the entire cell area so the higher throughputs and higher order modulation schemes are only selected when radio conditions within the cell are good

★ Not all of the throughputs shown in Table 64 can be supported by the capacity of the PMCH. Table 63 illustrates that the maximum throughput capacity of the PMCH is 68.4 Mbps, whereas Table 64 shows one transport block size which generates a throughput of 75.38 Mbps

★ The PMCH and MBMS are not supported by Home eNode B (Femto)

★ 3GPP References: TS 36.211, TS 36.212, TS 36.213, TS 36.331

7 DOWNLINK TRANSPORT CHANNELS

7.1 BCH

★ The Broadcast Channel (BCH) transfers the Master Information Block (MIB) using Transparent Mode (TM) RLC and the BCCH logical channel. Section 10.2 describes the content of the MIB

★ The BCH does not transfer System Information Blocks (SIB). These are transferred using the BCCH logical channel and the Downlink Shared Channel (DL-SCH)

★ The BCH uses a fixed Transmission Time Interval (TTI) of 40 ms, i.e. 4 radio frames

★ The total size of the MIB is fixed and is equal to 24 bits. The distribution of those 24 bits between the various MIB fields is presented in Table 65

Information Element	Size
Downlink Bandwidth	3 bits
PHICH Duration	1 bit
PHICH Resource	2 bits
System Frame Number	8 bits
Spare	10 bits

Table 65 – Distribution of the 24 bits belonging to the MIB

★ A 16 bit CRC is added and the result is channel coded using rate 1/3 convolutional coding. This generates a total of (24 + 16) × 3 = 120 bits

★ Rate matching is applied to increase the total number of bits to equal the capacity offered by the PBCH Resource Elements in 4 radio frames. The number of bits offered by the PBCH Resource Elements is shown in Table 66

	Normal Cyclic Prefix	Extended Cyclic Prefix
Number of PBCH Resource Elements per Radio Frame	240	216
Number of PBCH Resource Elements per 40 ms TTI	240 × 4 = 960	216 × 4 = 864
Number of bits available for PBCH (QPSK modulation)	960 × 2 = 1920	864 × 2 = 1728

Table 66 – Capacity available to PBCH across 40 ms TTI

★ These figures indicate that the 120 bits of channel coded BCH information is increased to 1920 bits or 1728 bits for the normal and extended cyclic prefixes respectively. The large quantity of redundancy means that the MIB can be fully decoded from any 10 ms radio frame, i.e. the UE does not have to collect data from the complete 40 ms TTI to decode the content of the MIB

★ 3GPP References: TS 36.212

7.2 PCH

- The Paging Channel (PCH) transfers the RRC Paging message using Transparent Mode (TM) RLC and the PCCH logical channel. The paging procedure and content of the Paging message is presented in section 23.6
- Transport blocks belonging to the PCH have a variable size depending upon the content of the Paging message, i.e. number of paging records and whether or not it includes a system information modification flag or Earthquake and Tsunami Warning System (ETWS) indication
- The physical layer processing applied to each PCH transport block is illustrated in Figure 102. This processing is completed prior to modulation and the subsequent mapping of modulation symbols onto PDSCH Resource Elements

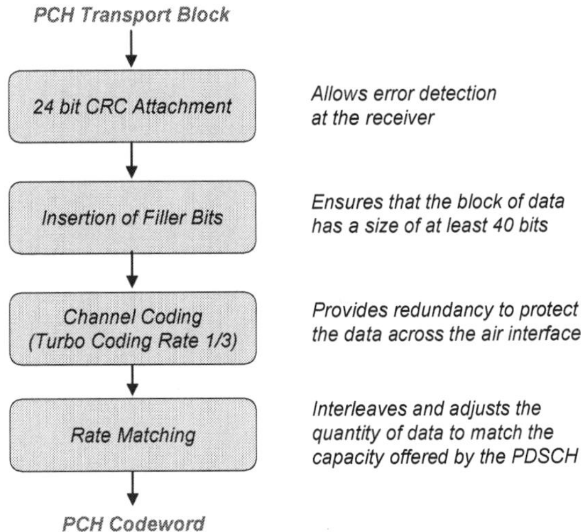

Figure 102 – Physical layer processing for PCH transport block

- A single PCH transport block is generated from each RRC paging message
- A 24 bit CRC is attached to the transport block to allow error detection at the receiver:
    - if a UE successfully receives a PCH transport block without error then it is passed to the higher layers. It is not necessary for the UE to send a positive HARQ acknowledgement to the eNode B
    - if a UE detects an error within the received PCH transport block then it is discarded and the UE waits for the next paging occasion. It is not necessary for the UE to send a negative HARQ acknowledgement to the eNode B
- Filler bits are attached if the total size of the transport block plus CRC is less than 40 bits. This may be the case if the paging message includes only a system information modification flag, or only an ETWS indication (paging messages do not have to include any paging records). Filler bits are necessary because Turbo coding requires a minimum block size of 40 bits
- Rate 1/3 Turbo coding is completed to triple the total number of bits, and to provide redundancy which protects the data when transferred across the air-interface. Turbo coding generates 1 set of Systematic bits and 2 sets of Parity bits. The Systematic bits are a copy of the input bits. The Parity bits provide the redundancy to protect the input bits
- Rate matching starts with an interleaving function. Interleaving is used to help avoid bursts of contiguous bit errors at the input of the Turbo decoder at the receiver. The performance of the decoder is improved when bit errors are distributed at random rather than in contiguous groups. Propagation channel fading tends to generate bursts of contiguous errors but de-interleaving at the receiver randomises the position of those bit errors
- Rate matching then adjusts the total number of bits to match the capacity of the allocated PDSCH Resource Blocks. The total number of bits can be increased using repetition or decreased using puncturing. The PCH uses Redundancy Version 0 which means that puncturing targets the Parity bits rather than the Systematic bits
- The PCH codeword is modulated using QPSK and mapped onto the allocated PDSCH Resource Elements. The Paging message is transferred during a single 1 ms subframe, i.e. during the paging occasion of the relevant DRX cycle
- 3GPP References: TS 36.212

7.3 DL-SCH

- ★ The Downlink Shared Channel (DL-SCH) is used to transfer System Information Blocks (SIB), RRC signalling and application data
- ★ Transport blocks belonging to the DL-SCH have a variable size. The set of allowed Transport Block Sizes (TBS) is presented in section 33.3
- ★ The TBS scheduled during a specific subframe is signalled to the UE within the PDCCH Downlink Control Information (DCI). The Modulation and Coding Scheme (MCS) defines both the modulation scheme and the TBS index. The TBS index can be combined with the number of allocated Resource Blocks to deduce the TBS
- ★ System information and RRC messages are likely to require relatively small transport blocks while application data may require larger transport blocks. Large transport blocks are scheduled by the eNode B when the UE is in good coverage and has sufficient data to be transferred
- ★ The DL-SCH can transfer either 1 or 2 transport blocks of application data per subframe per connection. 2 transport blocks can be transferred when the UE is in good coverage and is configured with 2×2 MIMO, 4×4 MIMO or 8×8 MIMO
- ★ The physical layer processing applied to each DL-SCH transport block is illustrated in Figure 103. This processing is completed prior to modulation and the subsequent mapping of modulation symbols onto PDSCH Resource Elements
- ★ A 24 bit CRC is attached to the transport block to allow error detection at the receiver:
  - o if a UE successfully receives a DL-SCH transport block without error then it is passed to the higher layers. The UE sends a positive HARQ acknowledgement if the transport block includes RRC signalling or application data
  - o if a UE detects an error within the received DL-SCH transport block containing RRC signalling or application data then it is buffered ready to be combined with a retransmission. The UE sends a negative HARQ acknowledgement to trigger the retransmission
  - o if a UE detects an error within the received DL-SCH transport block containing system information then it can be buffered ready to be combined with a retransmission, but without returning a negative HARQ acknowledgement
- ★ Filler bits are attached if the total size of the transport block plus CRC is less than 40 bits. This is necessary because Turbo coding requires at least a minimum block size

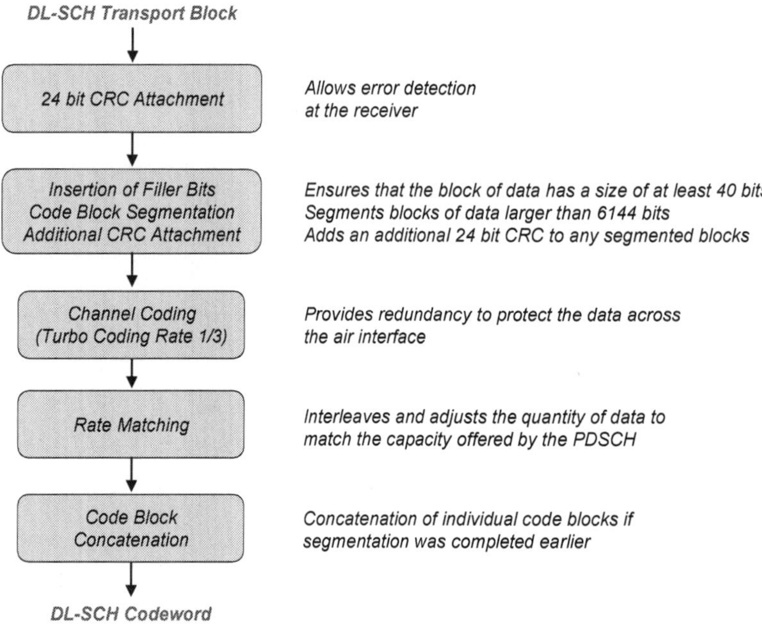

Figure 103 – Physical layer processing for DL-SCH transport block

- ★ If the total size of the transport block plus CRC (and any filler bits) is less than or equal to 6144 bits then it is forwarded for channel coding
- ★ If the code block exceeds 6144 bits then segmentation is required. This is necessary for the purposes of Turbo coding. A code block of size 'Z' is segmented into 'X' sections where X = ROUNDUP(Z / 6120). The denominator is 6120 rather than 6144 to allow for an additional 24 bit CRC which is added to each segment

★ Rate 1/3 Turbo coding is completed to triple the total number of bits, and to provide redundancy which protects the data when transferred across the air-interface. Turbo coding generates 1 set of Systematic bits and 2 sets of Parity bits. The Systematic bits are a copy of the input bits. The Parity bits provide the redundancy to protect the input bits

★ Rate matching starts with an interleaving function. Interleaving is used to help avoid bursts of contiguous bit errors at the input of the Turbo decoder at the receiver. The performance of the decoder is improved when bit errors are distributed at random rather than in contiguous groups. Propagation channel fading tends to generate bursts of contiguous errors but de-interleaving at the receiver randomises the position of those bit errors

★ The concept of interleaving is illustrated in Figure 104. UE travelling at low speeds are more likely to experience fades with long durations (UE travelling at high speeds pass through the fades more rapidly). This means that UE travelling at low speeds may experience wider bursts of errors. This can lead to contiguous errors even after de-interleaving which can subsequently lead to reduced physical layer performance

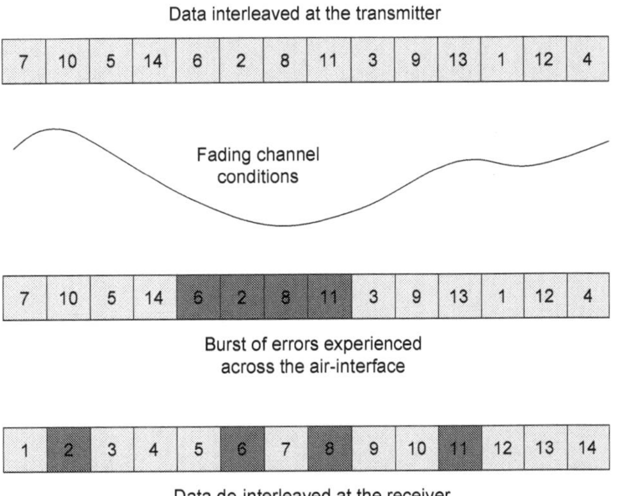

Figure 104 – General concept of interleaving

★ Rate matching then adjusts the total number of bits to match the capacity of the allocated PDSCH Resource Blocks. The total number of bits can be increased using repetition or decreased using puncturing

★ The Redundancy Version (RV) applied to the DL-SCH is signalled within the DCI when transferring signalling messages or application data. In the case of transferring system information, the RV may not be signalled within the DCI, i.e. when using DCI format 1C. In that case the RV is calculated from the SFN when transferring SIB1 and from the subframe number when transferring other SIB

★ Code block concatenation is completed if segmentation was required prior to channel coding. Segments are concatenated to generate a single codeword

★ DL-SCH codewords are modulated using QPSK, 16QAM or 64QAM. QPSK is always used when transferring system information. Signalling messages may only require QPSK if they generate relatively small transport blocks. Application data can benefit from 16QAM or 64QAM when the UE is in good coverage

★ Modulated codewords are mapped onto the allocated PDSCH Resource Elements. Codewords are transferred during a single 1 ms subframe

★ 3GPP References: TS 36.212

7.4 MCH

- The Multicast Channel (MCH) is used to transfer both the MCCH and MTCH logical channels. The MCCH transfers the MBSFN Area Configuration message belonging to the RRC protocol. The MTCH transfers application data belonging to the MBMS service
- Unacknowledged mode RLC is used to transfer MCH transport blocks to the UE
- In the case of the MCCH, there are 4 different Transport Block Sizes (TBS) allowed for each channel bandwidth. These TBS are signalled to the UE using the Modulation and Coding Scheme (MCS) information within SIB13. The MCS defines both the modulation scheme and the TBS index. The TBS index can be combined with the number of Resource Blocks within the total channel bandwidth to deduce the TBS (see tables within section 33.3)
- In the case of the MTCH, there are 26 different TBS allowed for each channel bandwidth. These TBS are signalled to the UE using the MCS information within the MBSFN Area Configuration message. The MCS defines both the modulation scheme and the TBS index. The TBS index can be combined with the number of Resource Blocks within the total channel bandwidth to deduce the TBS
- The MCH can transfer 1 transport block per 1 ms subframe
- The physical layer processing applied to each MCH transport block is illustrated in Figure 105. This processing is completed prior to modulation and the subsequent mapping of modulation symbols onto PMCH Resource Elements

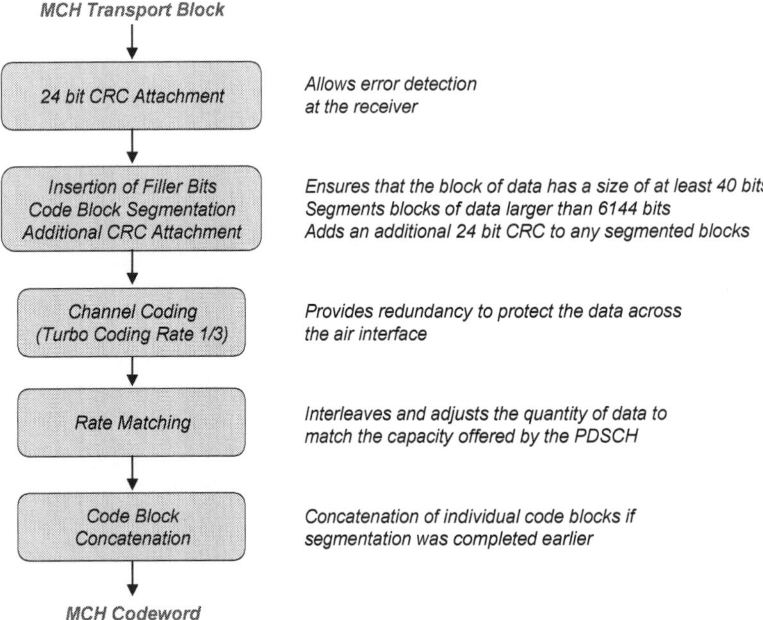

Figure 105 – Physical layer processing for MCH transport block

- Processing of the MCH transport channel is the same as that for the DL-SCH transport channel
- MCH codewords are modulated using QPSK, 16QAM or 64QAM. It should be possible to receive the MCH across the entire cell area so 64QAM is only applied when throughput requirements are high and radio conditions allow
- 3GPP References: TS 36.212

8 DOWNLINK CHANNEL TYPE MAPPINGS

8.1 LOGICAL, TRANSPORT AND PHYSICAL CHANNEL TYPES

- ★ Logical channels transfer data between the RLC and MAC layers
- ★ Transport channels transfer data between the MAC and Physical layers
- ★ Physical channels transfer data across the air-interface
- ★ The mappings between the various channel types are illustrated in Figure 106

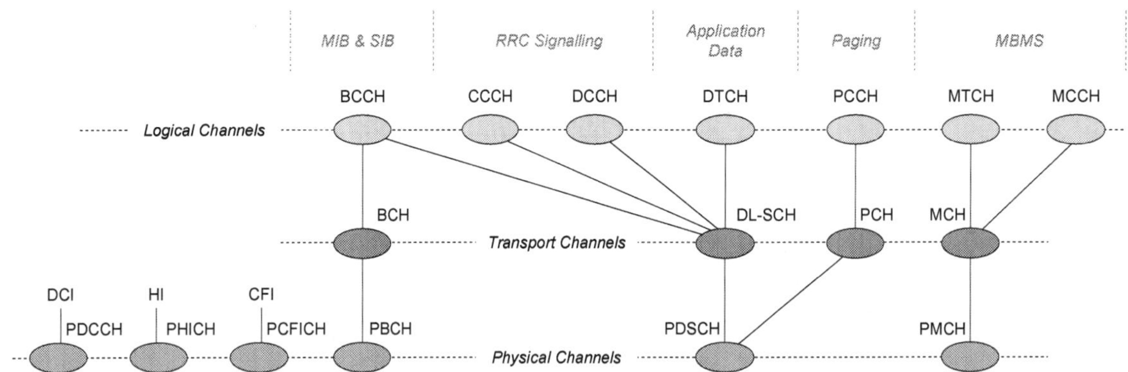

Figure 106 – Mappings between logical, transport and physical channels

- ★ The Broadcast Control Channel (BCCH) is used to transfer both the Master Information Block (MIB) and System Information Blocks (SIB). The MIB is mapped onto the BCH and PBCH, whereas the SIB are mapped onto the DL-SCH and PDSCH
- ★ The Common Control Channel (CCCH) and Dedicated Control Channel (DCCH) are used to transfer RRC messages, i.e. data belonging to the set of Signalling Radio Bearers (SRB). All SRB data is mapped onto the DL-SCH and PDSCH
- ★ The Dedicated Traffic Channel (DTCH) is used to transfer application data. All application data is mapped onto the DL-SCH and PDSCH
- ★ The Paging Control Channel (PCCH) is used to transfer paging messages. All paging messages are mapped onto the PCH and PDSCH
- ★ The Multicast Traffic Channel (MTCH) and Multicast Control Channel (MCCH) are mapped onto the Multicast Channel (MCH) and Physical Multicast Channel (PMCH). The MTCH transfers application data belonging to the MBMS service, whereas the MCCH transfers the MBSFN Area Configuration message
- ★ The PDCCH, PHICH and PCFICH physical channels are not used to transfer higher layer information, so do not have associated logical nor transport channels
  - o PDCCH transfers Downlink Control Information (DCI)
  - o PHICH transfers HARQ Indicators (HI), i.e. Hybrid ARQ acknowledgements for uplink data
  - o PCFICH transfers Control Format Indicators (CFI)
- ★ When a UE receives data on the PDSCH, the PDCCH indicates whether the data belongs to the DL-SCH or the PCH. This is done by using a specific type of Radio Network Temporary Identifier (RNTI), i.e. P-RNTI indicates PCH data whereas C-RNTI and SI-RNTI indicate DL-SCH data
- ★ When a UE receives data on the DL-SCH, the MAC header indicates whether the data belongs to the CCCH, DCCH or DTCH (BCCH data is identified by the SI-RNTI on the PDCCH). The MAC header includes a Logical Channel Identity (LCID) field where a value of 0 corresponds to the CCCH (SRB 0), values of 1 and 2 correspond to the DCCH (SRB1 and SRB2 respectively), and values 3 to 10 correspond to the DTCH
- ★ 3GPP References: TS 36.212, TS 36.321, TS 36.331

9 DOWNLINK CONTROL INFORMATION

9.1 STRUCTURE

★ Downlink Control Information (DCI) is mapped onto the PDCCH physical channel

★ DCI can be used to schedule uplink resources on the PUSCH or downlink resources on the PDSCH. Alternatively, DCI can be used to signal Transmit Power Control (TPC) commands for either the PUSCH or PUCCH

★ 3GPP TS 36.212 specifies the range of DCI formats presented in Table 67

- formats 0 and 4 schedule uplink resources on the PUSCH
- formats 1, 1A, 1B, 1C, 1D, 2, 2A, 2B and 2C schedule downlink resources on the PDSCH
- formats 3 and 3A signal TPC commands for the PUSCH and PUCCH

	DCI Format	Application	RNTI
Uplink Resource Allocation	0	Scheduling of one PUSCH codeword Single antenna port transmission	C-RNTI Temporary C-RNTI SPS-RNTI
	4	Scheduling of one or two PUSCH codewords Closed loop spatial multiplexing	C-RNTI
Downlink Resource Allocation	1	Scheduling of one PDSCH codeword Single antenna port transmission or transmit diversity	C-RNTI Temporary C-RNTI SPS-RNTI
	1A	Compact scheduling of single PDSCH codeword or random access initiation Single antenna port transmission or transmit diversity	P-RNTI SI-RNTI RA-RNTI C-RNTI Temporary C-RNTI SPS-RNTI
	1B	Compact scheduling of single PDSCH codeword with precoding information Closed loop spatial multiplexing using a single layer	C-RNTI
	1C	Very compact Resource Block scheduling of single PDSCH codeword Single antenna port transmission or transmit diversity	P-RNTI SI-RNTI RA-RNTI
	1D	Compact scheduling of single PDSCH codeword with precoding and power offset information Multi-user MIMO	C-RNTI
	2	Scheduling of one or two PDSCH codewords Closed loop spatial multiplexing or transmit diversity	C-RNTI SPS-RNTI
	2A	Scheduling of one or two PDSCH codewords Open loop spatial multiplexing or transmit diversity	C-RNTI SPS-RNTI
	2B	Scheduling of one or two PDSCH codewords Dual or single layer transmission using antenna ports 7 and 8	C-RNTI SPS-RNTI
	2C	Scheduling of one or two PDSCH codewords Up to 8 layer transmission using antenna ports 7 to 14	C-RNTI SPS-RNTI
Transmit Power Control	3	TPC commands for PUCCH and PUSCH with 2 bit power adjustments	TPC-PUSCH-RNTI TPC-PUCCH-RNTI
	3A	TPC commands for PUCCH and PUSCH with 1 bit power adjustments	

Table 67 – Summary of Downlink Control Information (DCI)

★ The majority of DCI formats were introduced within the release 8 version of the 3GPP specifications. DCI format 2B was introduced within the release 9 version, while DCI formats 2C and 4 were introduced within the release 10 version

★ Each DCI format includes a 16 bit CRC which is scrambled by an RNTI. The RNTI is used to address the appropriate UE. The RNTI also provides an indication of the information content of the resource allocation (with the exception of DCI formats 3 and 3A in which case the RNTI indicates whether the TPC command is applicable to the PUSCH or PUCCH)

- DCI whose CRC are scrambled by the P-RNTI allocate PDSCH resources for paging messages. All UE share the same P-RNTI
- DCI whose CRC are scrambled by the SI-RNTI allocate PDSCH resources for system information messages. All UE share the same SI-RNTI

- DCI whose CRC are scrambled by the RA-RNTI allocate PDSCH resources for a random access response. The RA-RNTI is determined from the time-frequency resource used for the random access preamble
- DCI whose CRC are scrambled by the temporary C-RNTI, C-RNTI or SPS-RNTI allocate resources for application data or RRC signalling. These RNTI are allocated on a per UE basis by the eNode B
- DCI whose CRC are scrambled by the TPC-PUSCH-RNTI or TPC-PUCCH-RNTI signal TPC commands. These RNTI are allocated on a per UE basis by the eNode B

The various types of RNTI are described in section 33.2

★ Prior to the release 10 version of the 3GPP specifications, DCI format 0 was the only DCI format available for scheduling resources on the PUSCH. This provided support for single antenna port transmission on the uplink. The release 10 version of the specifications introduced DCI format 4 to provide support for closed loop spatial multiplexing in the uplink

★ Prior to release 10 of the 3GPP specifications, there was only a single approach for signalling an uplink resource allocation within the DCI (uplink resource allocation type 0). The release 10 version of the specifications introduced a second approach (uplink resource allocation type 1). These 2 approaches are described in sections 9.4 and 9.5

★ Both DCI formats 0 and 4 provide support for uplink resource allocation types 0 and 1. This is summarised in Table 68

	Uplink Resource Allocation	
DCI Format	Type 0	Type 1
0	Supported	Supported
4	Supported	Supported

Table 68 – Uplink resource allocation types supported by each DCI format

★ The set of DCI which schedule resources on the PDSCH provide support for single antenna transmission, transmit diversity, closed loop spatial multiplexing, open loop spatial multiplexing and multi-user MIMO

★ 3GPP has defined 3 different approaches for signalling a downlink resource allocation within the DCI. These 3 approaches are described in sections 9.7, 9.8 and 9.9. The DCI format determines which of the 3 approaches can be used. Table 69 presents the resource allocation methods supported by each DCI

	Downlink Resource Allocation		
DCI Format	Type 0	Type 1	Type 2
1	Supported	Supported	-
1A	-	-	Supported
1B	-	-	Supported
1C	-	-	Supported
1D	-	-	Supported
2	Supported	Supported	-
2A	Supported	Supported	-
2B	Supported	Supported	-
2C	Supported	Supported	-

Table 69 – Downlink resource allocation types supported by each DCI format

★ Each DCI format is mapped onto one of the PDCCH formats presented in Table 56. The PDCCH format is selected according to the DCI size and the level of puncturing applied after channel coding

★ 3GPP References: TS 36.212

9.2 SEARCH SPACES

★ UE are not required to decode every PDCCH. UE are only required to decode PDCCH within predefined search spaces during non-DRX subframes. UE decode the PDCCH within:
  - 2 common search spaces
  - 4 UE specific search spaces

★ The 2 common search spaces are the same for all UE, whereas the 4 UE specific search spaces are dependent upon the appropriate RNTI. The UE specific search spaces hop between subframes within a radio frame

★ The sizes of the common and UE specific search spaces are presented in Table 70. The size of each search space is defined in terms of Control Channel Elements (CCE). A single CCE occupies 36 Resource Elements

Type	Size in CCE	Aggregation Level	Number of PDCCH Candidates
Common	16	4	16 / 4 = 4
	16	8	16 / 8 = 2
UE Specific	6	1	6 / 1 = 6
	12	2	12 / 2 = 6
	8	4	8 / 4 = 2
	16	8	16 / 8 = 2

Table 70 – PDCCH candidates monitored by a UE

★ Both of the common search spaces have a size of 16 CCE. The UE attempts to decode 4 PDCCH within the 1st search space and 2 PDCCH within the 2nd search space

★ The UE specific search spaces have sizes of 6, 12, 8 and 16 CCE. The UE attempts to decode 6 PDCCH within the 1st and 2nd search spaces, and 2 PDCCH within the 3rd and 4th search spaces

★ Within each search space, UE do not attempt to decode all DCI but only those applicable to the current transmission mode

★ In some cases, the number of CCE available can be smaller than the size of the search space. For example, the 1.4 MHz channel bandwidth supports a maximum of 2 CCE when 2 OFDMA symbols are allocated to the PDCCH. In this case, the size of the search space reduces to the number of CCE available

★ 3GPP References: TS 36.213

9.3 TIMING FOR RESOURCE ALLOCATIONS

★ In the case of FDD, uplink resource allocations signalled using the PDCCH during subframe 'n' are applicable to subframe 'n + 4', i.e. the UE transmits on the PUSCH during subframe 'n + 4'. This timing relationship is illustrated in Figure 107

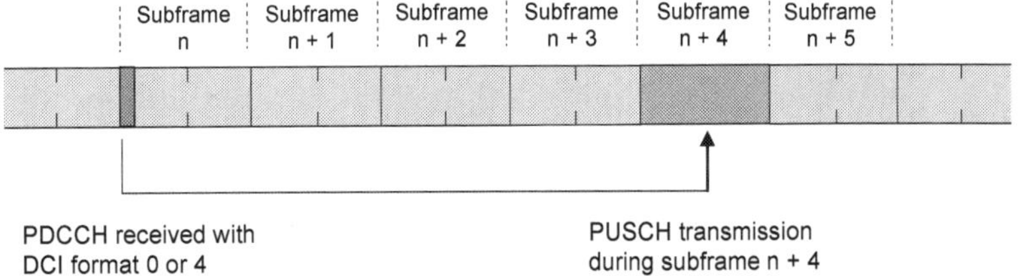

Figure 107 – Timing of the PUSCH transmission relative to the PDCCH uplink resource allocation (FDD)

★ In the case of TDD, the timing relationship is more complicated because the 'n + 4' subframe could be a downlink subframe rather than an uplink subframe

★ The delay between receiving an uplink resource allocation for TDD and transmitting on the PUSCH varies according to the values of 'k' shown in Table 71. Based upon these values, the UE transmits on the PUSCH during subframe 'n + k' after receiving a resource allocation during subframe 'n'

Uplink-Downlink Configuration	Subframe Number 'n'									
	0	1	2	3	4	5	6	7	8	9
1		6			4		6			4
2				4					4	
3	4								4	4
4									4	4
5									4	
6	7	7				7	7			5

Table 71 – Subframe delay 'k' between receiving uplink resource allocation for TDD and transmitting on the PUSCH (TDD uplink-downlink configurations 1 to 6)

★ Figure 108 illustrates the example of uplink-downlink configuration 1. In this case, there are 4 uplink subframes during each radio frame so there are 4 downlink subframes during which uplink resource allocations can be made. Uplink resource allocations can be signalled within the DwPTS section of special subframes, as well as within normal downlink subframes

Uplink-Downlink Configuration 1

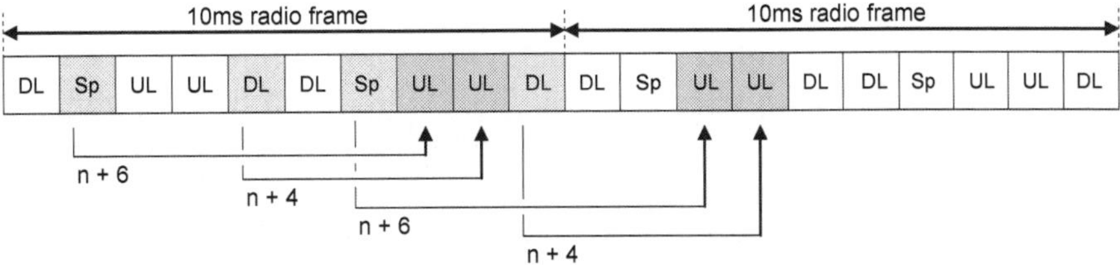

Figure 108 – Timing of the PUSCH transmission relative to the PDCCH uplink resource allocation (TDD uplink-downlink configuration 1)

★ The delays of 'n + 6' and 'n + 4' within Figure 108 follow the values shown in Table 71. Uplink resource allocations are not made during subframes 0 and 5

★ Figure 109 illustrates similar time relationships for TDD uplink-downlink configurations 2 to 6. In each case, the number of downlink and special subframes during which an uplink resource allocation can be made is equal to the number of uplink subframes

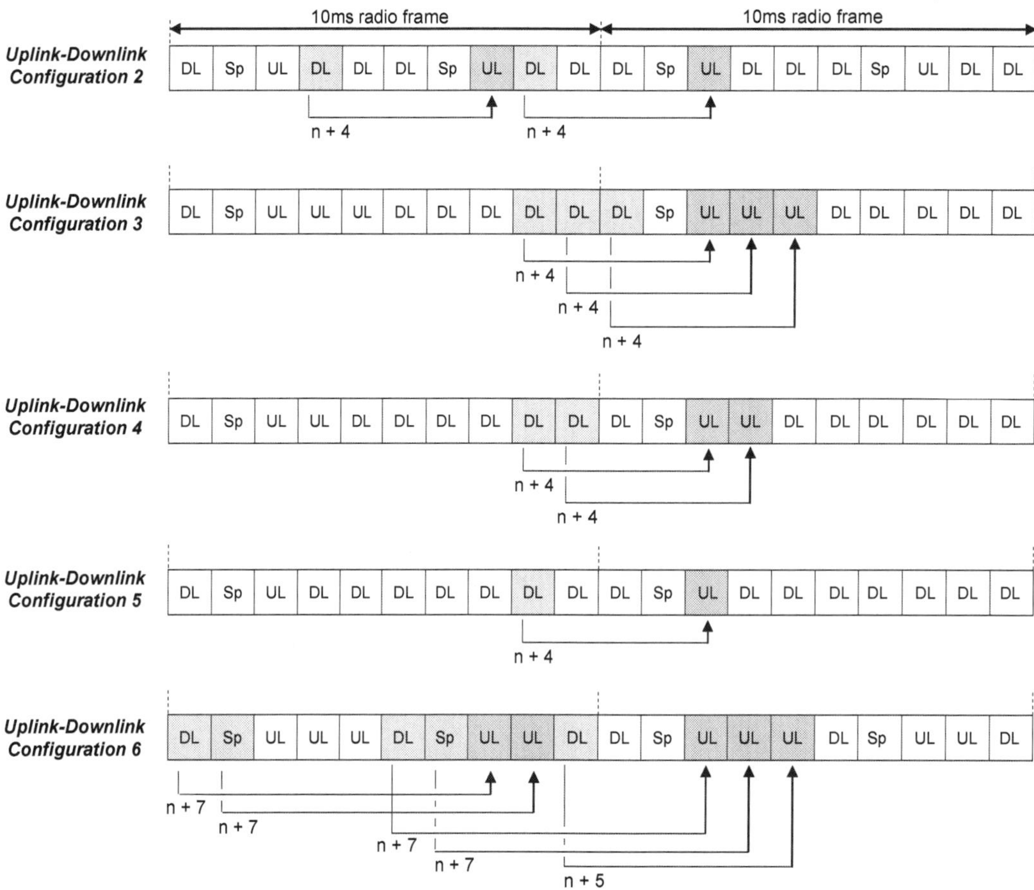

Figure 109 – Timing of the PUSCH transmission relative to the PDCCH uplink resource allocation (TDD uplink-downlink configurations 2 to 6)

★ TDD uplink-downlink subframe configuration 0 is a special case because the number of uplink subframes is greater than the number of downlink and special subframes. This means there are insufficient downlink and special subframes to have a one-to-one mapping with the uplink subframes. 3GPP has overcome this issue using the 'Uplink Index' field within DCI formats 0 and 4

★ The 'Uplink Index' field has a length of 2 bits. If the Most Significant Bit (MSB) is set to 1 while the Least Significant Bit (LSB) is set to 0 then the delay between receiving an uplink resource allocation and transmitting on the PUSCH varies according to the values of 'k' shown in Table 72

Uplink-Downlink Configuration	Subframe Number 'n'									
	0	1	2	3	4	5	6	7	8	9
0	4	6				4	6			

Table 72 – Subframe delay 'k' between receiving uplink resource allocation for TDD and transmitting on the PUSCH (TDD uplink-downlink configuration 0)

★ If the LSB of the 'Uplink Index' is set to 1 while the MSB is set to 0 then the delay between receiving an uplink resource allocation and transmitting on the PUSCH is fixed to 7 subframes

★ If both the LSB and MSB of the 'Uplink Index' are set to 1 then for every resource allocation received on the PDCCH the UE is permitted to transmit on the PUSCH during 2 subframes. The first of these 2 subframes has a delay defined by the values in Table 72. The second of these 2 subframes has a delay of 7 subframes

★ These various scenarios for TDD uplink-downlink configuration 0 are illustrated in Figure 110

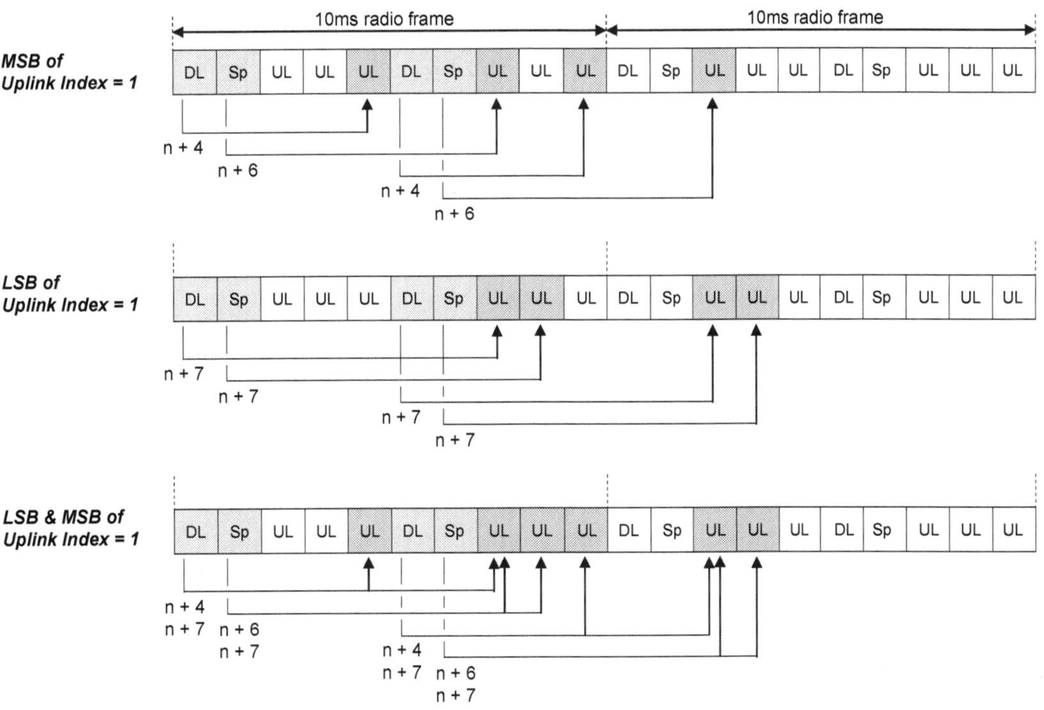

Figure 110 – Timing of the PUSCH transmission relative to the PDCCH uplink resource allocation (TDD uplink-downlink configuration 0)

★ For both FDD and TDD, downlink resource allocations signalled using the PDCCH during subframe ‘n’ are applicable to the same subframe, i.e. the UE receives the PDSCH during the same subframe that it receives the resource allocation on the PDCCH

★ 3GPP References: TS 36.213

9.4 UPLINK RESOURCE ALLOCATION TYPE 0

★ Downlink Control Information (DCI) formats 0 and 4 can be used to signal type 0 uplink Resource Block allocations

★ The DCI allocates a set of virtual Resource Blocks which are subsequently mapped onto a set of physical Resource Blocks. Both the virtual and physical Resource Block allocations are always contiguous

★ If Resource Block hopping is not used, the set of virtual Resource Blocks are the same as the set of physical Resource Blocks. If hopping is used, the virtual Resource Blocks are mapped onto the physical Resource Blocks using a hopping pattern. The Resource Block allocation remains contiguous after the hopping pattern has been applied

★ The number of allocated Resource Blocks is always equal to either 1, or a multiple of 2, 3 or 5, i.e. allowed allocations are 1, 2, 3, 4, 5, 6, 8, 9, 10, 12, 14, etc. The total number of possible Resource Block allocations is presented in Table 73

	1.4 MHz	3 MHz	5 MHz	10 MHz	15 MHz	20 MHz
Total number of possible Resource Blocks allocations	21	117	319	1262	2831	5025
Bits required to signal the complete set of possible allocations	5	7	9	11	12	13
Bits used to signal allocation within DCI (no hopping)	5	7	9	11	12	13
Bits used to signal allocation within DCI (hopping)	4	6	8	9	10	11

Table 73 – Number of allowed combinations of allocated Resource Blocks

★ When hopping is active, the number of bits used to signal the Resource Block allocation is reduced by either 1 or 2 bits. These 1 or 2 bits are used to signal hopping information

★ An uplink resource allocation type 0 is signalled using a Resource Indication Value (RIV). Figure 111 illustrates the RIV calculation for the 1.4 MHz channel bandwidth (total of 6 Resource Blocks in the frequency domain)

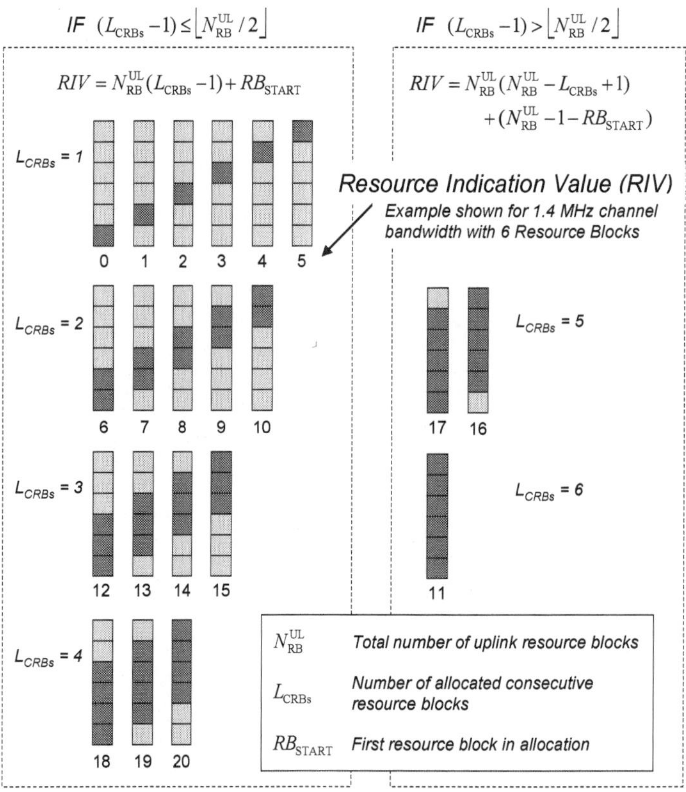

Figure 111 – Uplink Resource Indication Value (RIV) calculations for a 1.4 MHz channel bandwidth

★ The combinations with larger RIV cannot be signalled when hopping is active because the number of bits used to signal the RIV is less. The larger RIV values are less applicable when hopping is active because larger Resource Block allocations offer less flexibility for allocating resources within different parts of the channel bandwidth

★ 3GPP References: TS 36.213

9.5 UPLINK RESOURCE ALLOCATION TYPE 1

★ Downlink Control Information (DCI) formats 0 and 4 can be used to signal type 1 uplink Resource Block allocations

★ Uplink resource allocation type 1 was introduced within the release 10 version of the 3GPP specifications to allow the allocation of non-consecutive Resource Blocks. Allocating non-consecutive Resource Blocks provides the eNode B scheduler with greater flexibility and increases the potential for a frequency selective scheduling gain

★ The drawback of a non-contiguous Resource Block allocation is an increased requirement for power amplifier back-off which leads to reduced efficiency and a reduced maximum transmit power

★ The DCI allocates a set of virtual Resource Blocks which are subsequently mapped onto a set of physical Resource Blocks. Hopping is not supported for uplink resource allocation type 1 so the set of allocated virtual Resource Blocks is the same as the set of allocated physical Resource Blocks

★ The Resource Block allocation defines 2 sets of Resource Blocks, where each set of Resource Blocks includes 1 or more consecutive Resource Block Groups (RBG)

★ The size of the RBG depends upon the channel bandwidth. Smaller channel bandwidths use smaller RBG sizes. The RBG size associated with each channel bandwidth is presented in Table 74. The value 'N' shown in Table 74 is used when signalling the Resource Block allocation to the UE. The use of 'N' is described later in this section.

	1.4 MHz	3 MHz	5 MHz	10 MHz	15 MHz	20 MHz
Total Resource Blocks	6	15	25	50	75	100
Resource Block Group (RBG) Size	1	2	2	3	4	4
N	7	9	14	18	20	26

Table 74 – Resource Block Group (RBG) size associated with each channel bandwidth

★ The Resource Block allocation is defined by 4 indices. Indices s0 and s1 define the first and last RBG within the first set of RBG. Indices s2 and s3 define the first and last RBG within the second set of RBG

★ Some example Resource Block allocations for the 3 MHz channel bandwidth are illustrated in Figure 112. In this case, the RBG size is 2 Resource Blocks. The odd Resource Block at the top of the channel bandwidth also forms an RBG

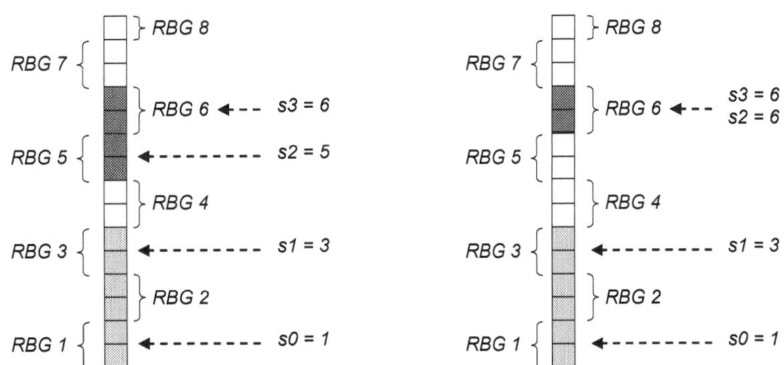

Figure 112 – Example Resource Block allocations for the 3 MHz channel bandwidth

★ The set of 4 indices are combined into a single combinatorial index 'r' before sending to the UE. The value of 'r' appears within the PDCCH DCI formats 0 and 4 when using uplink resource allocation type 1. The value of r is defined as a sum of binomial coefficients using the expression below:

$$r = \binom{N - s0}{4} + \binom{N - (s1 + 1)}{3} + \binom{N - s2}{2} + \binom{N - (s3 + 1)}{1}$$

where, N is dependent upon the channel bandwidth and is given by:

$$N = \left\lceil \frac{\text{Number of uplink Resource Blocks in Channel Bandwidth}}{\text{RBG size}} \right\rceil + 1$$

The value of N for each channel bandwidth is presented in Table 74

★ An example calculation of the combinatorial index 'r' is shown in Figure 113. This example is based upon the 3 MHz channel bandwidth. The first set of RBG includes 3 RBG, whereas the second set of RBG includes 2 RBG

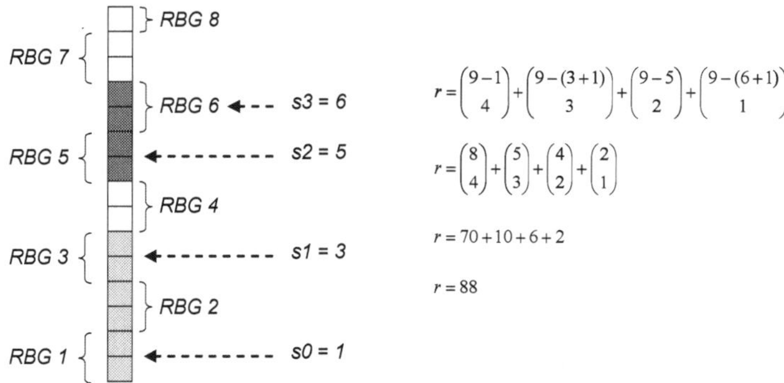

Figure 113 – Example calculation of the combinatorial index 'r'

★ The number of bits required to signal a type 1 uplink resource allocation is given by the expression below:

$$Number\ of\ Bits = \left\lceil LOG_2\left(\binom{\lceil N_{RB}^{UL} / RBG\_Size \rceil + 1}{4}\right)\right\rceil$$

where, N_{RB}^{UL} is the number of uplink Resource Blocks within the channel bandwidth

★ Table 75 presents the result from this equation for each channel bandwidth. The number of bits required is either greater than or equal to the number of bits required for uplink resource allocation type 0

	1.4 MHz	3 MHz	5 MHz	10 MHz	15 MHz	20 MHz
Bits Required to Signal Allocation Type 1	6 bits	7 bits	10 bits	12 bits	13 bits	14 bits

Table 75 – Number of bits required to signal uplink resource allocation type 1

★ 3GPP References: TS 36.213

9.6 UPLINK FREQUENCY HOPPING

★ Uplink Resource Blocks allocated by type 0 resource allocations are always contiguous. This helps to reduce the peak-to-average ratio of the transmitted signal and consequently improves the transmit power amplifier efficiency. A drawback of contiguous allocations is reduced potential for frequency diversity

★ Allocating a small number of Resource Blocks means that the resource allocation spans only a small bandwidth and the propagation channel is relatively well correlated for all Resource Blocks within the allocation. Allocating a large number of Resource Blocks increases the potential for frequency diversity because the resource allocation spans a wider bandwidth

★ Uplink frequency hopping provides frequency diversity while allowing the resource allocations to remain contiguous. This is particularly beneficial to small Resource Block allocations which do not inherently benefit from frequency diversity

★ Uplink frequency hopping is applicable to type 0 resource allocations when the frequency hopping flag within DCI format 0 is set to 1. Frequency hopping is not applied to type 1 resource allocations, nor to any uplink resource allocation made using DCI format 4

★ When hopping is used, the Resource Block allocation field within DCI format 0 includes either 1 or 2 hopping bits. The number of bits is dependent upon the channel bandwidth. The value of the hopping bits determines whether type 1 or type 2 hopping is applied. Table 76 presents the number of hopping bits associated with each channel bandwidth, and their relationship with type 1 and type 2 hopping

Channel Bandwidths	Number of Hopping Bits	Hopping Bit Pattern	PUSCH Hopping
1.4, 3, 5 MHz	1	0	Type 1
		1	Type 2
10, 15, 20 MHz	2	00	Type 1
		01	
		10	
		11	Type 2

Table 76 – Hopping bits within the Resource Block allocation of DCI format 0

9.6.1 TYPE 1 PUSCH HOPPING

★ Type 1 hopping divides the channel bandwidth into a fixed number of sub-bands. Hopping is then supported between those sub-bands. The number of sub-bands is fixed by 3GPP and is dependent upon the channel bandwidth. The number of sub-bands associated with each channel bandwidth is presented in Table 77. This table also presents the maximum number of contiguous Resource Blocks which can be allocated when using type 1 hopping

	1.4 MHz	3 MHz	5 MHz	10 MHz	15 MHz	20 MHz
Total Number of Resource Blocks	6	15	25	50	75	100
Number of Sub-bands	2	2	2	4	4	4
Max. Contiguous Resource Block Allocation	2	4	10	10	13	20

Table 77 – Number of sub-bands and maximum contiguous allocated Resource Blocks for Type 1 hopping

★ Figure 114 illustrates an example for the 5 MHz channel bandwidth. This example assumes that the PUCCH occupies a total of 4 Resource Blocks per time slot at the upper and lower edges of the channel bandwidth. RRC signalling is used to inform the UE that these Resource Blocks are not available for hopping. The PUSCH Hopping Offset (H_{RB}^{HO}) which can be included within an RRC Connection Reconfiguration message is used for this purpose. The signalled value of the hopping offset is rounded up to an even number if an odd number is sent to the UE, e.g. a value of 3 would be rounded up to a value of 4. The value after rounding is known as $\tilde{H}_{RB}^{HO}$

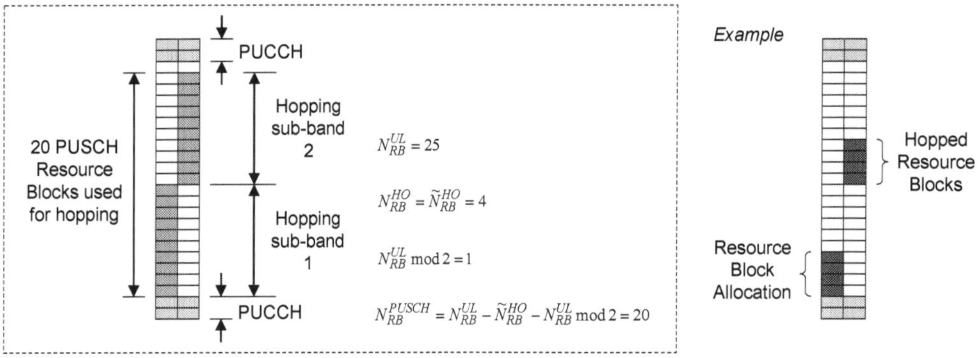

PUCCH
Hopping sub-band 2
20 PUSCH Resource Blocks used for hopping
Hopping sub-band 1
PUCCH

$N_{RB}^{UL} = 25$

$N_{RB}^{HO} = \tilde{N}_{RB}^{HO} = 4$

$N_{RB}^{UL} \bmod 2 = 1$

$N_{RB}^{PUSCH} = N_{RB}^{UL} - \tilde{N}_{RB}^{HO} - N_{RB}^{UL} \bmod 2 = 20$

Example
Hopped Resource Blocks
Resource Block Allocation

Figure 114 – Type 1 hopping for 5 MHz channel bandwidth

★ The example in Figure 114 assumes that DCI format 0 allocates the UE with 4 Resource Blocks at the lowest possible position within hopping sub-band 1. The Resource Blocks move to the lowest possible position within hopping sub-band 2 when hopping is applied. The allocated Resource Blocks are contiguous before and after hopping

★ Figure 115 illustrates an example for the 10 MHz channel bandwidth. This example assumes that the PUCCH occupies a total of 8 Resource Blocks per time slot at the upper and lower edges of the channel bandwidth. This leaves 42 Resource Blocks available for hopping

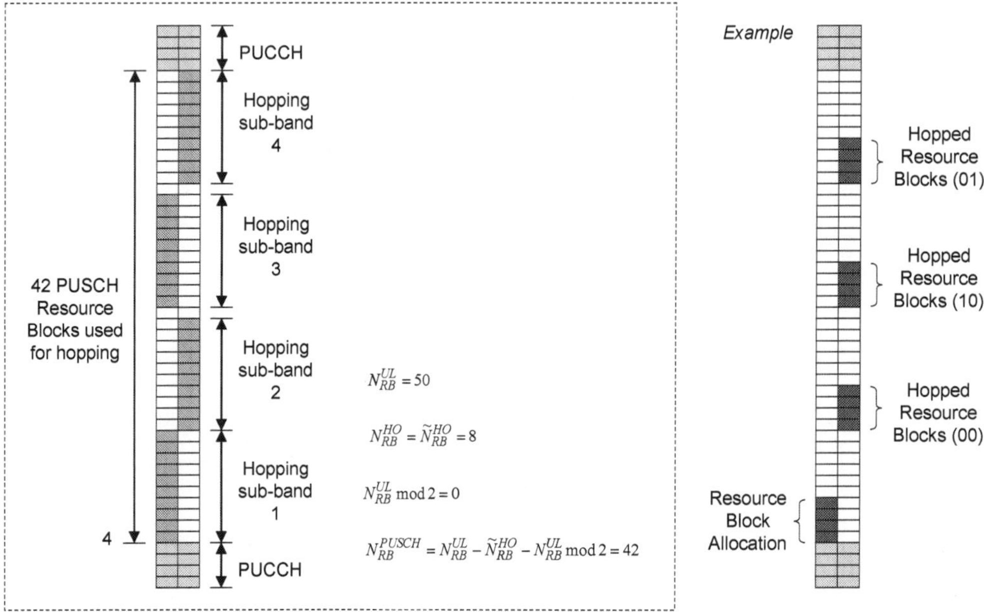

Figure 115 – Type 1 hopping for 10 MHz channel bandwidth

★ In this case, there are 4 hopping sub-bands and the target sub-band is dependent upon the value of the hopping bits within the Resource Block assignment. The example in Figure 115 illustrates all 3 target sub-bands although only one of them would be selected based upon the value of the hopping bits

★ 3GPP has specified 2 hopping modes:

- <u>inter-subframe</u> – the Resource Blocks allocated by DCI format 0 are used for even numbered transmissions of a transport block, while the hopped Resource Blocks are used for odd numbered transmissions of the transport block, i.e. hopping is only applied for retransmissions
- <u>intra and inter-subframe</u> – the Resource Blocks allocated by DCI format 0 are used for the first time slot within the subframe while the hopped Resource Blocks are used for the second time slot within the subframe

The hopping mode is signalled to the UE using RRC signalling, e.g. within the PUSCH configuration of a RRC Connection Reconfiguration message

★ 3GPP References: TS 36.213

9.6.2 TYPE 2 PUSCH HOPPING

- Type 2 hopping allows the number of sub-bands to be configured by the network. The number of sub-bands can be set to a value between 1 and 4, and is signalled to the UE using RRC signalling, e.g. within the PUSCH configuration of an RRC Connection Reconfiguration message
- The maximum number of contiguous Resource Blocks which can be allocated to a UE is dependent upon the channel bandwidth, the number of sub-bands and the number of Resource Blocks reserved for the PUCCH. Some example figures are presented in Table 78. The hopping offset represents the total number of Resource Blocks reserved for the PUCCH within a time slot

	1.4 MHz	3 MHz	5 MHz	10 MHz	15 MHz	20 MHz
Total Resource Blocks	6	15	25	50	75	100
Hopping Offset	2	2	2	4	4	4
1 sub-band	2	4	10	10	13	20
2 sub-bands	2	4	10	10	13	20
3 sub-bands	1	4	7	10	13	20
4 sub-bands	1	3	5	10	13	20
Hopping Offset	-	4	4	8	8	8
1 sub-band	-	4	10	10	13	20
2 sub-bands	-	4	10	10	13	20
3 sub-bands	-	3	7	10	13	20
4 sub-bands	-	2	5	10	13	20

Table 78 – Maximum contiguous allocated Resource Blocks for Type 2 hopping

- The Resource Block allocations of {2, 4, 10, 10, 13, 20} appearing in the first row of results in Table 78 represent the upper limit of the contiguous allocation sizes. These figures decrease as the number of sub-bands increases and as the PUCCH reservation increases
- Configuring a single sub-band represents a special case which generates 'mirroring' of the Resource Block allocation around the center of the channel bandwidth. Examples of mirroring are illustrated in Figure 116

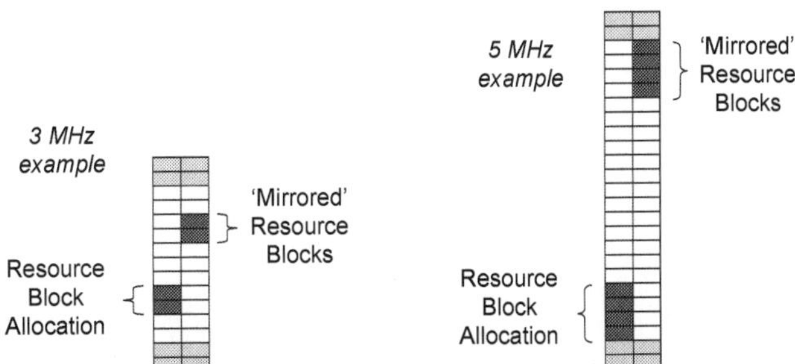

Figure 116 – Type 2 hopping with single sub-band (mirroring)

- Configuring multiple sub-bands leads to hopping based upon pre-defined hopping patterns. These hopping patterns are defined such that the allocations belonging to different UE do not clash. The function used to generate the hopping pattern is presented in Figure 117
- The result from the hopping pattern function is the allocated physical Resource Blocks with indexing which starts at 0 just above the lower PUCCH allocation. For example, if a 5 MHz channel has 2 PUCCH Resource Blocks at its lower edge and 2 PUCCH Resource Blocks at its upper edge then indexing runs from 0 to 20, rather than from 0 to 24, and the modified index equals the original index - 2

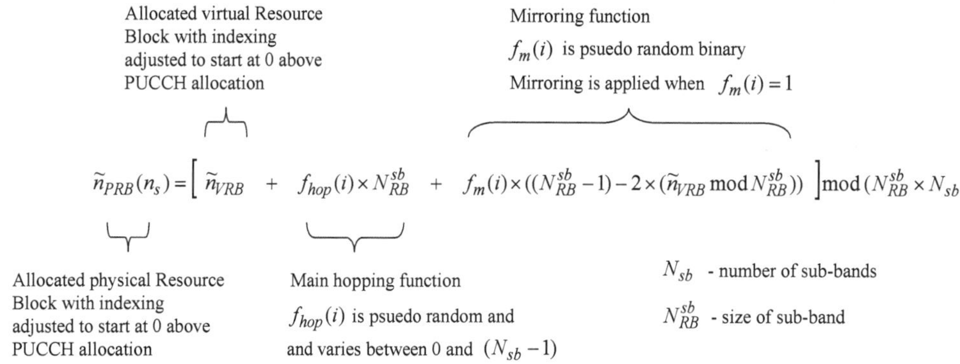

Figure 117 – Function used to generate pre-defined hopping pattern

- The result from the hopping pattern function is the sum of the allocated virtual Resource Block with 2 other components:
    - component 1 - the main hopping pattern function shifts the virtual Resource Block allocation by an integer number of sub-bands. The shift is defined using a pseudo random (deterministic) sequence which can be calculated by both the UE and eNode B. In the case of FDD, the pseudo random sequence is initialised using the Physical layer Cell Identity (PCI). In the case of TDD, the pseudo random sequence is initialised using a combination of the PCI and radio frame number
    - component 2 - a mirroring function which can mirror the virtual Resource Block allocation within a sub-band. Mirroring is applied when the binary pseudo random (deterministic) function, $f_m(i)$ generates a '1'. Mirroring moves a virtual Resource Block allocation which is 'x' Resource Blocks from the bottom of the sub-band to a virtual Resource Block allocation which is 'x' Resource Blocks from the top of the sub-band. Similar to the first component, the pseudo random sequence is initialised by the PCI for FDD, and by the PCI and radio frame number for TDD
- Modular arithmetic is applied to the result of the hopping pattern function to provide wrap around, which ensures that the allocated Resource Blocks after hopping remain within the set of sub-bands
- The two hopping modes available for type 1 hopping are also available for type 2 hopping:
    - inter-subframe
        - 1 sub-band - the Resource Blocks allocated by DCI format 0 are used for even numbered transmissions of a transport block, while the mirrored Resource Blocks are used for odd numbered transmissions of the transport block, i.e. mirroring is applied for retransmissions
        - 2 to 4 sub-bands - the main hopping pattern function and mirroring function are updated at the start of each subframe
    - intra and inter-subframe
        - 1 sub-band - the Resource Blocks allocated by DCI format 0 are used for the first time slot within the subframe while the mirrored Resource Blocks are used for the second time slot within the subframe
        - 2 to 4 sub-bands - the main hopping pattern function and mirroring function are updated at the start of each time slot
- 3GPP References: TS 36.213, TS 36.211

9.7 DOWNLINK RESOURCE ALLOCATION TYPE 0

★ A type 0 downlink Resource Block allocation can be signalled from the eNode B to the UE using Downlink Control Information (DCI) formats 1, 2, 2A, 2B and 2C

★ An allocation received during downlink subframe 'n' defines the allocated Resource Blocks within the same downlink subframe

★ A type 0 Resource Block allocation uses a bitmap to indicate which Resource Block Groups (RBG) are allocated to the UE. A single RBG is a set of consecutive Resource Blocks. The allocated RBG do not need to be contiguous

★ The number of Resource Blocks within an RBG is predetermined and is a function of the channel bandwidth. The RBG size as a function of the channel bandwidth is shown in Table 79

	1.4 MHz	3 MHz	5 MHz	10 MHz	15 MHz	20 MHz
Total Number of Resource Blocks	6	15	25	50	75	100
RBG Size (RB)	1	2	2	3	4	4
Number of complete RBG	6	7	12	16	18	25
Size of remaining RBG (RB)	-	1	1	2	3	-
Total Number of RBG	6	8	13	17	19	25
Size of bitmap (bits)	**6**	**8**	**13**	**17**	**19**	**25**

Table 79 – Resource Block Groups (RBG) for Resource Block allocation type 0

★ Each channel bandwidth includes a number of complete RBG. A partial RBG is also included if the total number of Resource Blocks is not a multiple of the RBG size

★ The bitmap signalled using the Type 0 Resource Block allocation includes a single bit for each RBG. A value of 1 indicates that the RBG has been allocated to the UE

★ Figure 118 illustrates the RBG and corresponding bitmaps for the channel bandwidths of 1.4, 3 and 5 MHz

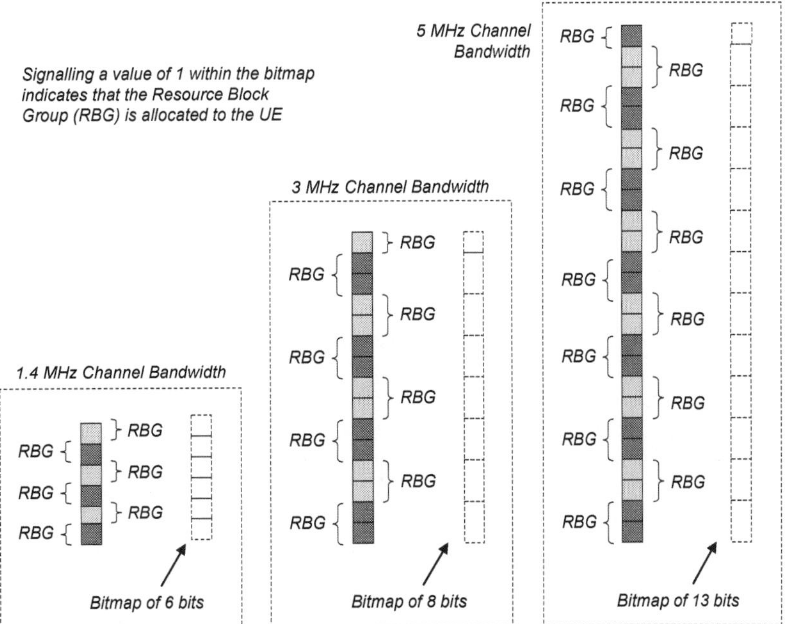

Figure 118 – Type 0 downlink Resource Block allocation

★ 3GPP References: TS 36.213

9.8 DOWNLINK RESOURCE ALLOCATION TYPE 1

★ A type 1 downlink Resource Block allocation can be signalled from the eNode B to the UE using Downlink Control Information (DCI) formats 1, 2, 2A, 2B and 2C. An allocation received during downlink subframe 'n' defines the Resource Block allocation within the same downlink subframe

★ Type 1 Resource Block allocations are not applicable to the 1.4 MHz channel bandwidth. DCI formats 1, 2 and 2A always signal a type 0 Resource Block allocation when the channel bandwidth is 1.4 MHz

★ Type 1 Resource Block allocations are divided into 3 sections:

- Resource Block Group (RBG) subset number
- Resource Block offset flag
- Resource Block bitmap

★ The RBG sizes are the same as those specified for a Type 0 Resource Block allocation. The number of RBG subsets is equal to the RBG size. The RBG size and the number of RBG subsets for each channel bandwidth are shown in Table 80

	1.4 MHz	3 MHz	5 MHz	10 MHz	15 MHz	20 MHz
Total Number of RB	N/A	15	25	50	75	100
RBG Size (RB)	N/A	2	2	3	4	4
Number of RBG subsets	N/A	2	2	3	4	4
RBG subset (bits)	N/A	1	1	2	2	2
Offset flag (bits)	N/A	1	1	1	1	1
Bitmap (bits)	N/A	6	11	14	16	22
Total Bits	N/A	**8**	**13**	**17**	**19**	**25**

Table 80 – RBG sizes and RBG subsets for Resource Block allocation type 1

★ The number of bits used to signal the RBG subset is either 1 or 2 depending upon the number of subsets. The Resource Blocks allocated to a UE always belong to a single RBG subset

★ The Resource Block offset flag indicates whether the subsequent Resource Block bitmap should be aligned with the bottom of the lowest Resource Block within the subset, or aligned with the top of the highest Resource Block within the subset. This offset is necessary because the bitmap is not sufficiently large to include all Resource Blocks within the subset

★ Figure 119 illustrates the concept of RBG subsets and the offset flag for the 3 MHz channel bandwidth

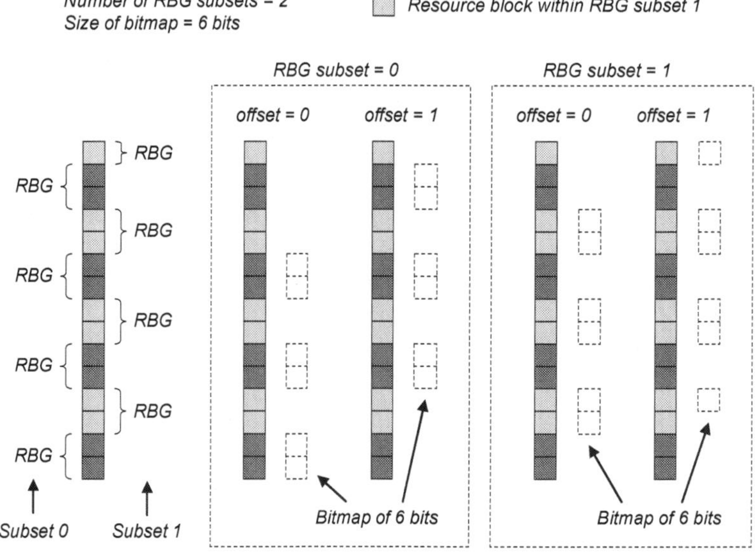

Figure 119 – Type 1 Resource Block allocation for the 3 MHz channel bandwidth

★ Each bit within the bitmap corresponds to a single Resource Block within the RBG subset. A value of 1 indicates that the Resource Block has been allocated to the UE

★ 3GPP References: TS 36.213

9.9 DOWNLINK RESOURCE ALLOCATION TYPE 2

★ A type 2 downlink Resource Block allocation can be signalled from the eNode B to the UE using Downlink Control Information (DCI) formats 1A, 1B, 1C and 1D

★ An allocation received during downlink subframe 'n' defines the allocated Resource Blocks within the same downlink subframe

★ The resource allocation can be made in terms of:

- o contiguous virtual Resource Blocks (also known as localised Resource Blocks), or
- o distributed virtual Resource Blocks

★ DCI 1A, 1B and 1D include a flag to indicate whether the resource allocation is contiguous or distributed. DCI 1C always allocates distributed virtual Resource Blocks

★ The set of allocated virtual Resource Blocks are mapped onto the set of allocated physical Resource Blocks

★ Contiguous virtual Resource Blocks are contiguous both before and after mapping onto their physical Resource Blocks. In this case, the set of allocated physical Resource Blocks is the same as the set of allocated virtual Resource Blocks. In addition, the Resource Block allocation is the same in both time slots belonging to the subframe

★ Contiguous allocations can range from a single virtual Resource Block to the complete set of virtual Resource Blocks spanning the entire channel bandwidth. These Resource Block allocation ranges are shown in Table 81

	1.4 MHz	3 MHz	5 MHz	10 MHz	15 MHz	20 MHz
DCI 1A, 1B, 1D	1 to 6	1 to 15	1 to 25	1 to 50	1 to 75	1 to 100

Table 81 – Range of Resource Block allocations (contiguous allocation)

★ Contiguous virtual Resource Block allocations are signalled using Resource Indication Values (RIV). The calculation of the RIV is the same as when calculating the RIV for type 0 uplink resource allocations. This calculation is presented in section 9.4

★ Distributed Resource Block allocations make use of 'Gap1' and 'Gap2' parameters. These parameters define an offset between the physical Resource Block allocations during the first and second time slots of the subframe

★ The number of Resource Blocks which can be scheduled by a distributed allocation depends upon the DCI format and the choice between Gap1 and Gap2. The smaller bandwidths are always based upon Gap1, while the larger bandwidths can be based upon either Gap1 or Gap2. The choice between Gap1 and Gap2 is signalled as part of the DCI

★ The range of Resource Blocks which can be scheduled by a distributed allocation is presented in Table 82

	1.4 MHz	3 MHz	5 MHz	10 MHz	15 MHz	20 MHz
Distributed Allocation using Gap 1						
DCI 1A (P, RA, SI-RNTI)	1 to 6	1 to 14	1 to 24	1 to 46	1 to 64	1 to 96
DCI 1A (C-RNTI), 1B, 1D	1 to 6	1 to 14	1 to 24	1 to 16	1 to 16	1 to 16
DCI 1C	2 to 6	2 to 14	2 to 24	4 to 44	4 to 64	4 to 96
Distributed Allocation using Gap 2						
DCI 1A (P, RA, SI-RNTI)	-	-	-	1 to 36	1 to 64	1 to 96
DCI 1A (C-RNTI), 1B, 1D	-	-	-	1 to 16	1 to 16	1 to 16
DCI 1C	-	-	-	4 to 36	4 to 64	4 to 96

Table 82 – Range of Resource Block allocations (distributed allocation)

★ Distributed virtual Resource Block allocations are also signalled using Resource Indication Values (RIV)

★ The RIV calculations for DCI formats 1A, 1B and 1D are the same as those used for type 0 uplink Resource Block allocations. However, virtual Resource Blocks cannot be allocated outside the window defined by N_{VRB}, where N_{VRB} is equal to the maximum number of Resource Blocks which can be allocated according to Table 82

★ Figure 120 illustrates this concept for the 3 MHz channel bandwidth. Table 82 indicates that a maximum of 14 Resource Blocks can be allocated so the 15th Resource Block remains outside the allocation window. Figure 120 illustrates RIV values for 2 allocated Resource Blocks. In practice, 1, 2, 3, 4, 5, 6, .. , 14 Resource Blocks can also be allocated generating a total of 105 possible allocations, i.e. requiring 7 bits within the DCI

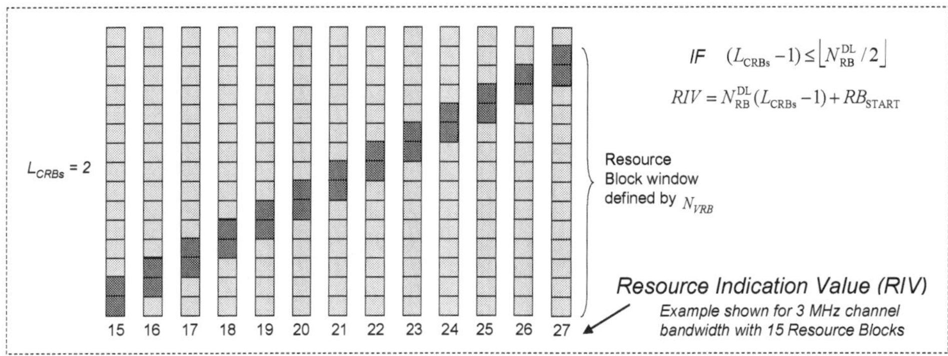

Figure 120 – Resource Block window for a 3 MHz channel bandwidth with a distributed allocation

★ In the case of DCI format 1C, the number of allocated Resource Blocks, and the starting Resource Block are always a multiple of 2 when the channel bandwidth is 1.4, 3 or 5 MHz, and a multiple of 4 when the channel bandwidth is 10, 15 or 20 MHz

★ Figure 121 illustrates the RIV calculations for the 1.4 MHz channel bandwidth when using DCI format 1C. There are 6 different RIV values which can be signalled using only 3 bits. This illustrates the very compact scheduling capability of DCI format 1C

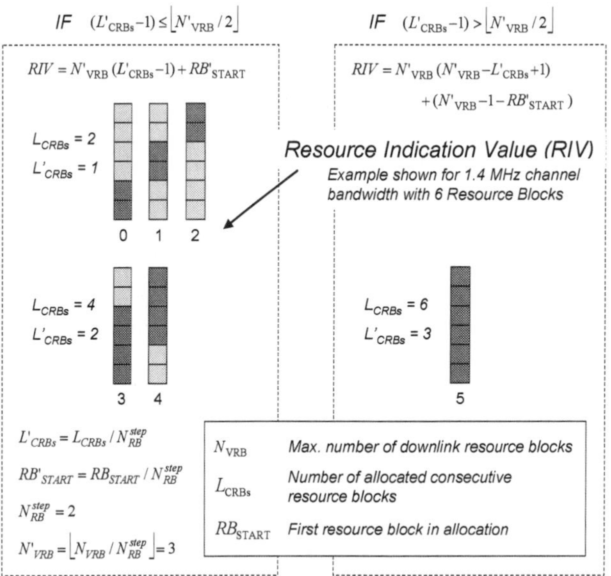

Figure 121 – RIV calculations for a 1.4 MHz channel bandwidth when using DCI format 1C

★ The mapping of distributed virtual Resource Blocks onto physical Resource Blocks uses an interleaving table. The interleaving table always has 4 columns, to provide 4th order diversity after interleaving. The number of rows is given by:

$$Number \text{ of Rows} = \left\lceil \tilde{N}_{VRB}^{DL} / 4P \right\rceil \times P$$

$$\text{where,} \quad \tilde{N}_{VRB}^{DL} = N_{VRB}^{DL} \quad \text{when using Gap1}$$

$$\tilde{N}_{VRB}^{DL} = 2 \times \mathrm{N_{gap}} \quad \text{when using Gap2}$$

The value of N_{VRB}^{DL} is given by the upper limits presented in Table 82, e.g. N_{VRB}^{DL} = 46 when using the 10 MHz channel bandwidth with Gap1. The values of P, Gap1 and Gap2 are presented in Table 83. The value of P corresponds to the Resource Block Group (RBG) size associated with the channel bandwidth

	1.4 MHz	3 MHz	5 MHz	10 MHz	15 MHz	20 MHz
P	1	2	2	3	4	4
Gap1	3	8	12	27	32	48
Gap2	-	-	-	9	16	16

Table 83 – RBG size (P) and Gap values associated with each channel bandwidth

★ In the case of using Gap1 with the 10 MHz channel bandwidth, the interleaving table has 12 rows. Nulls are placed within the interleaving table at the bottom of the 2nd and 4th columns. The total number of nulls is given by the difference between the total number of positions within the interleaving table and the value of $\tilde{N}_{VRB}^{DL}$

★ Figure 122 illustrates the interleaving table for the 10 MHz channel bandwidth when using Gap1. Virtual Resource Block numbers are written into the table row-by-row, and subsequently read out of the table column-by-column. The Nulls within the table are ignored and do not generate outputs

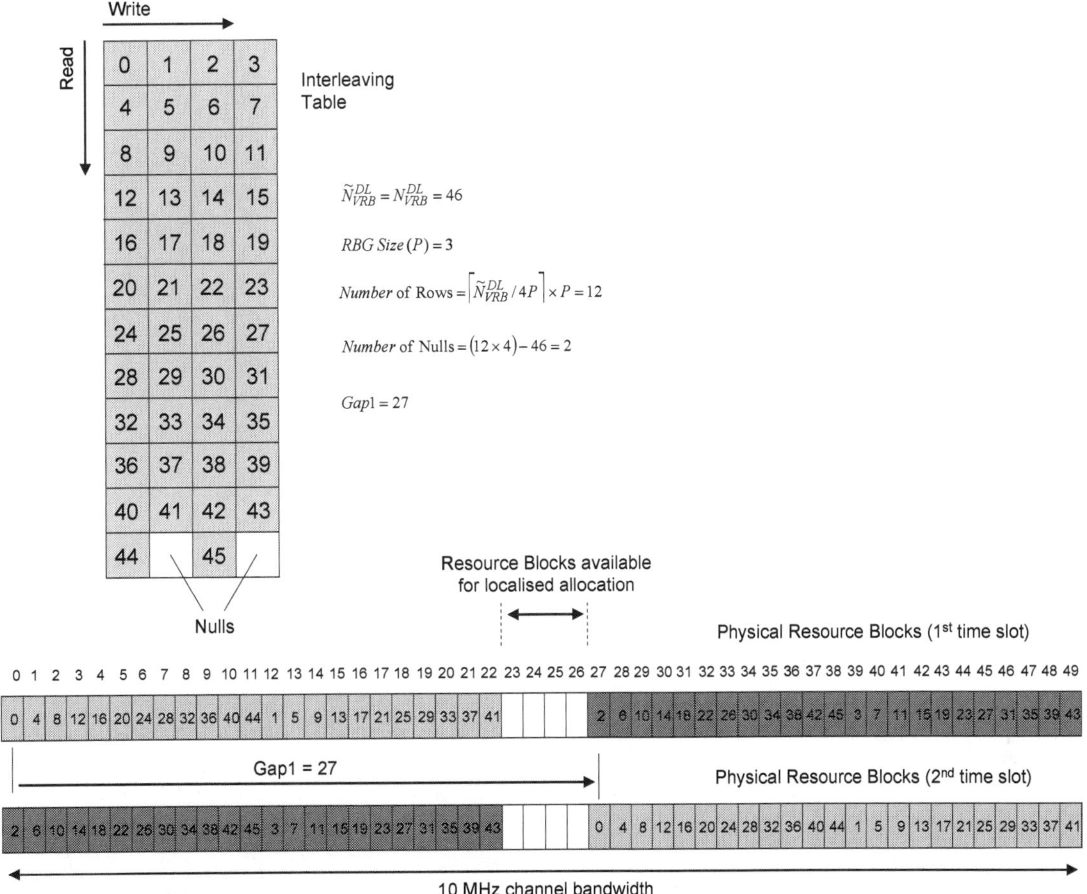

Figure 122 – Interleaving table and virtual to physical Resource Block mapping (10 MHz channel using Gap1)

★ When using Gap1, the first half of the outputs from the interleaving table map onto the lower set of physical Resource Blocks starting from the first physical Resource Block. The second half of the outputs from the interleaving table map onto the upper set of physical Resource Blocks ending at the last physical Resource Block. This generates the virtual to physical Resource Block mapping for the first time slot of the subframe. The mapping is cyclic shifted by the Gap value for the second time slot of the subframe

★ Using a different mapping for the first and second time slots within the subframe represents inter-slot hopping which provides frequency diversity. This is especially beneficial to small Resource Block allocations which do not inherently benefit from frequency diversity

★ The physical Resource Blocks at the center of the channel bandwidth are not used by distributed virtual Resource Blocks but can be used by contiguous virtual Resource Blocks

★ When using Gap2, a pair of interleaving tables are used to define the mapping between the virtual and physical Resource Blocks. In the case of using Gap2 with the 10 MHz channel bandwidth, each interleaving table has 6 rows. Nulls are placed within each interleaving table at the bottom of the 2nd and 4th columns. The total number of nulls within each table is given by the difference between the total number of positions within the interleaving table and the value of $\tilde{N}_{VRB}^{DL}$

★ Figure 123 illustrates the interleaving tables for the 10 MHz channel bandwidth when using Gap2. Virtual Resource Block numbers are written into each table row-by-row, and subsequently read out of each table column-by-column. The Nulls within the table are ignored and do not generate outputs

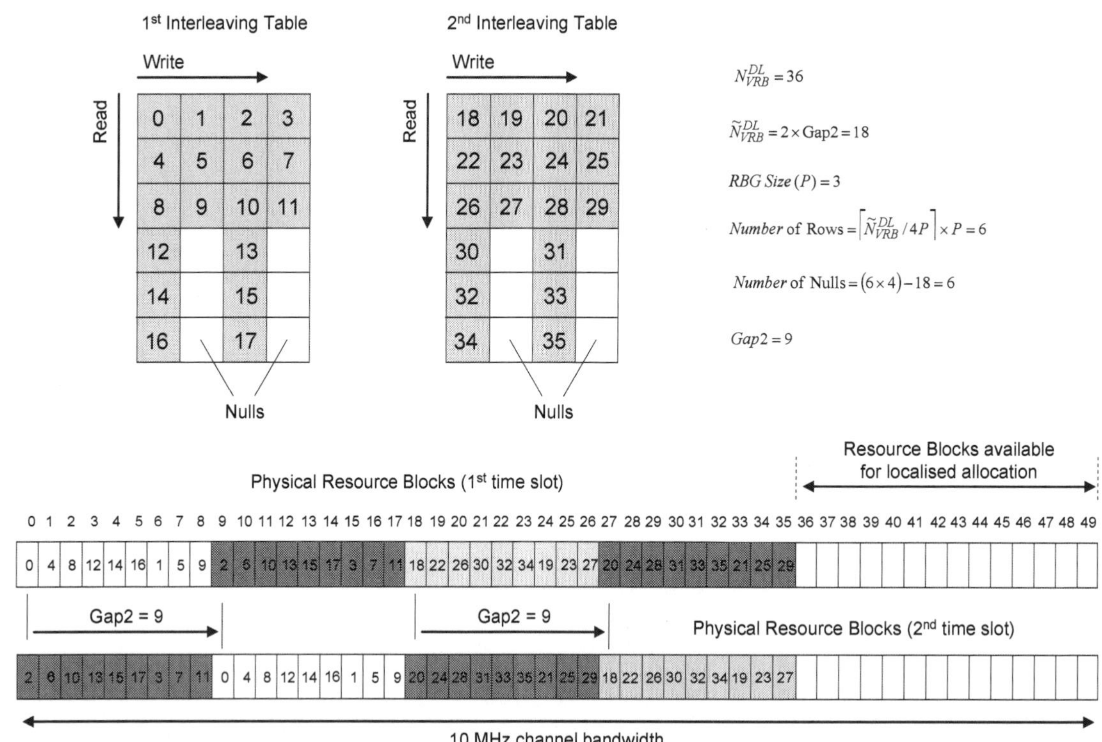

Figure 123 – Interleaving table and virtual to physical Resource Block mapping (10 MHz channel using Gap2)

- ★ When using Gap2, the complete set of outputs from the interleaving tables map onto the lower set of physical Resource Blocks starting from the first physical Resource Block. This generates the virtual to physical Resource Block mapping for the first time slot of the subframe. The mapping from each table is cyclic shifted by the Gap value for the second time slot of the subframe
- ★ The physical Resource Blocks at the upper end of the channel bandwidth are not used by distributed virtual Resource Blocks but can be used by contiguous virtual Resource Blocks
- ★ Type 2 Resource Block allocations (DCI formats 1A, 1B, 1C and 1D) are compact relative to type 0 and type 1 Resource Block allocations (DCI formats 1, 2, 2A, 2B and 2C) because the allocated virtual Resource Blocks have to be contiguous. This results in less flexibility but requires fewer bits
- ★ Table 84 summarises the number of bits used to signal the resource allocation within each DCI. These figures illustrate the difference between standard, compact and very compact signalling. Very compact signalling offers the least flexibility

	Resource Allocation	Signalling Overhead	1.4 MHz	3 MHz	5 MHz	10 MHz	15 MHz	20 MHz
DCI 1, 2, 2A, 2B, 2C	Type 0 or 1	Standard	6	8	13	17	19	25
DCI 1A, 1B, 1D	Type 2	Compact	5	7	9	11	12	13
DCI 1C	Type 2	Very Compact	3	5	7	7	8	9

Table 84 – Number of bits used to signal downlink resource allocations within each DCI

- ★ 3GPP References: TS 36.213

9.10 DCI FORMAT 0

★ Downlink Control Information (DCI) format 0 is used to schedule resources on the PUSCH. The content of DCI format 0 is shown in Table 85

	3GPP Release	1.4 MHz	3 MHz	5 MHz	10 MHz	15 MHz	20 MHz
Carrier Indicator	10	0 or 3 bits					
Format 0 / Format 1A Flag	8	1 bit					
Frequency Hopping Flag	8	1 bit					
Resource Block Allocation	8	5 bits	7 bits	9 bits	11 bits	12 bits	13 bits
MCS and Redundancy Version	8	5 bits					
New Data Indicator	8	1 bit					
TPC Command	8	2 bits					
Cyclic Shift for DM RS	8	3 bits					
Cyclic Shift for DM RS and OCC Index	10	3 bits					
Uplink Index	8	2 bits (TDD only)					
Downlink Assignment Index	8	2 bits (TDD only)					
CQI Request	8	1 bit					
CSI Request	10	1 or 2 bits					
SRS Request	10	0 or 1 bit					
Resource Allocation Type	10	1 bit					

Table 85 – Content of DCI Format 0

★ The Carrier Indicator is only applicable when carrier aggregation is used. This field points towards the RF carrier to be used as the serving cell. A value of 0 means that the primary cell is to be used as the serving cell. Values 1 to 7 mean that a secondary cell is to be used as the serving cell. An RRC Connection Reconfiguration message is sent to the UE before using the Carrier Indicator field. This message informs the UE of which integer value (secondary cell index) to associate with each secondary cell

★ The Format 0 / Format 1A Flag allows the UE to differentiate between the two formats. A value of 0 indicates that the subsequent data belongs to DCI format 0

★ The Frequency Hopping Flag indicates whether or not the UE should apply frequency hopping on the PUSCH. Frequency hopping is only applicable to resource allocation type 0, so this flag is interpreted in a different way when using resource allocation type 1. When using resource allocation type 1 the Frequency Hopping Flag is used as the Most Significant Bit (MSB) of the Resource Block Allocation.

★ The Resource Block Allocation defines the set of Resource Blocks allocated to the UE (see section 9.4). The number of bits required to signal the Resource Block allocation increases for larger channel bandwidths, i.e. there is an increased number of possible allocations

★ The Resource Block Allocation field includes either 1 or 2 hopping bits when hopping is enabled. The channel bandwidths of 1.4, 3 and 5 MHz use 1 hopping bit, whereas channel bandwidths of 10, 15 and 20 MHz use 2 hopping bits. This reduces the number of bits available to signal the actual Resource Block allocation

★ The number of bits belonging to the Resource Block Allocation field has not changed since the release 8 version of the 3GPP specifications. This means that the number of bits is sufficient for an uplink resource allocation type 0, but is not always sufficient for an uplink resource allocation type 1. Table 86 presents the number of bits required to signal uplink resource allocation types 0 and 1

	1.4 MHz	3 MHz	5 MHz	10 MHz	15 MHz	20 MHz
Uplink Resource Allocation Type 0	5 bits	7 bits	9 bits	11 bits	12 bits	13 bits
Uplink Resource Allocation Type 1	6 bits	7 bits	10 bits	12 bits	13 bits	14 bits

Table 86 – Number of bits required to signal uplink resource allocation types 0 and 1

★ In the case of resource allocation type 1, the Frequency Hopping Flag is concatenated with the Resource Block Allocation to generate a bit string of sufficient length to signal the Resource Block allocation. The set of concatenated bits represent the combinatorial index 'r' described in section 9.5

★ The Modulation and Coding Scheme (MCS) and Redundancy Version (RV) bits define a pointer to a row within the MCS and RV table (Table 395 within section 33.4). The set of 5 bits provide a range from 0 to 31, i.e. equal to the number of rows within the MCS and RV table. The MCS and RV table specifies the modulation scheme to be used for the PUSCH. It also specifies the Transport Block Size (TBS) index and the RV

- ★ The New Data Indicator (NDI) informs the UE of whether the uplink resource allocation is for new data or a retransmission. Toggling the NDI from 0 to 1, or from 1 to 0, indicates that the resource allocation is for new data. Leaving the NDI with its previous value indicates that the resource allocation is for a retransmission
- ★ The TPC Command impacts the UE transmit power for the PUSCH. Interpretation of the TPC command depends upon whether accumulated or absolute power control is used. This is configured during connection establishment within the RRC Connection Setup message. Interpretation of the TPC command is described in section 24.4.2
- ★ The Cyclic Shift for the Demodulation Reference Signal (DM-RS) defines a pointer to a look-up table which is used as an input when the UE generates the sequence for its uplink Demodulation Reference Signal. This field is replaced by the 'Cyclic Shift for DM RS and OCC Index' in the release 10 version of the 3GPP specifications
- ★ The 'Cyclic Shift for DM RS and OCC Index' in 3GPP release 10 uses the same look-up table as 'Cyclic Shift for DM RS' in 3GPP releases 8 and 9. The only difference is that the look-up table has been expanded in release 10 to include a set of columns to specify the Orthogonal Covering Code (OCC). The sequence for the uplink Demodulation Reference Signal is then multiplied by the OCC if the UE is configured to do so
- ★ The Uplink Index is only included for TDD uplink-downlink subframe configuration 0. It has an impact upon the timing for power control, aperiodic CQI/PMI/RI reporting and PUSCH transmission:
  - o power control: the uplink index is applicable when the PUSCH is scheduled during subframes 2 or 7. It determines the timing of the power control command to be applied
  - o aperiodic CQI/PMI/RI reporting: the uplink index determines the timing of when the aperiodic report should be sent
  - o PUSCH transmission: the uplink index determines the timing of when to apply the information received in DCI format 0
- ★ The Downlink Assignment Index (DAI) is only included for TDD uplink-downlink configurations 1 to 6. It is used to inform the UE of the number of PDSCH transmissions requiring acknowledgement
- ★ The CQI Request can be used to instruct a UE to provide an aperiodic CQI, Precoding Matrix Indicator (PMI) and Rank Indication (RI) report using the PUSCH. The UE provides a report during uplink subframe 'n+k' after receiving the request during downlink subframe 'n'. The value of 'k' is 4 for FDD and is dependant upon the uplink-downlink subframe configuration for TDD. This field is replaced by the 'CSI Request' field within the release 10 version of the 3GPP specifications
- ★ The CSI Request can have a length of either 1 or 2 bits. The 2 bit length is applicable to UE which are configured with more than a single downlink cell, and when the DCI is mapped onto a PDCCH within the UE specific search space defined by the C-RNTI. When using a single downlink cell, the 1 bit CSI Request can be used to instruct the UE to send a CSI report for that cell. When using multiple downlink cells, the 2 bit CSI Request can be used to instruct the UE to send a CSI report for either a single serving cell, or a set of serving cells. Interpretation of the 2 bit CSI Request is shown in Table 87

CSI Request	Meaning
00	No aperiodic CSI report is triggered
01	Aperiodic CSI report is triggered for serving cell 'c'
10	Aperiodic CSI report is triggered for the 1st set of serving cells
11	Aperiodic CSI report is triggered for the 2nd set of serving cells

Table 87 – Interpretation of the 2 bit CSI Request

The eNode B uses RRC signalling to inform the UE of which serving cells are included within set 1 and which serving cells are included within set 2. This is done using the Trigger1 and Trigger2 information elements within, for example an RRC Connection Reconfiguration message

- ★ The SRS Request can only be included when the DCI is mapped onto a PDCCH within the UE specific search space defined by the C-RNTI. When included, the SRS Request is used to trigger the UE to transmit the Sounding Reference Signal (SRS) in the uplink direction. In the case of DCI format 0, the SRS Request field is only a single bit which triggers the transmission of the SRS using a single SRS configuration. DCI format 4 includes an SRS Request with 2 bits which can be used to select between SRS configurations
- ★ The Resource Allocation Type is used to signal the type of uplink resource allocation. Prior to 3GPP release 10, there was only a single type of uplink resource allocation so this flag was not necessary. Release 10 of the 3GPP specifications introduced a second type of uplink resource allocation. The single bit allocated to the Resource Allocation Type indicates which of the two resource allocations is being used
- ★ When DCI format 0 and DCI format 1A are used to schedule resources from the same serving cell, using the same search space and RNTI, padding is added to DCI format 0 if its payload size is less than that of DCI format 1A. Padding is added until the payload sizes are equal
- ★ Table 88 presents an example of the total number of bits within DCI format 0 when applied to FDD. This example assumes the use of a release 8 or release 9 device so the release 10 fields have been excluded. A 16 bit CRC is added to the payload after padding has been included. The complete set of bits is channel coded using rate 1/3 convolutional coding before rate matching is applied to ensure that the resultant number of bits equals the capacity of the PDCCH

	1.4 MHz	3 MHz	5 MHz	10 MHz	15 MHz	20 MHz
Total Bits prior to Padding	19	21	23	25	26	27
Padding Bits	2	1	2	2	1	1
Total Bits	**21**	**22**	**25**	**27**	**27**	**28**
After inclusion of 16 bit CRC	37	38	41	43	43	44
After Rate 1/3 Channel Coding	111	114	123	129	129	132

Table 88 – Example total number of bits for DCI Format 0 (FDD)

★ Table 89 presents an example of the total number of bits within DCI format 0 when applied to TDD. This example also assumes the use of a release 8 or release 9 device so the release 10 fields have been excluded. The total number of bits prior to padding is 2 bits greater than for FDD due to the inclusion of either the Uplink Index or Downlink Assignment Index

	1.4 MHz	3 MHz	5 MHz	10 MHz	15 MHz	20 MHz
Total Bits prior to Padding	21	23	25	27	28	29
Padding Bits	2	2	2	2	2	2
Total Bits	**23**	**25**	**27**	**29**	**30**	**31**
After inclusion of 16 bit CRC	39	41	43	45	46	47
After Rate 1/3 Channel Coding	117	123	129	135	138	141

Table 89 – Example total number of bits for DCI Format 0 (TDD)

9.11 DCI FORMAT 1

★ Downlink Control Information (DCI) format 1 is used to schedule one PDSCH codeword in one cell. The content of DCI format 1 is shown in Table 90

	3GPP Release	1.4 MHz	3 MHz	5 MHz	10 MHz	15 MHz	20 MHz
Carrier Indicator	10	0 or 3 bits					
Resource Allocation Header	8	0 bits	1 bit				
Resource Block Allocation	8	6 bits	8 bits	13 bits	17 bits	19 bits	25 bits
Modulation and Coding Scheme (MCS)	8	5 bits					
HARQ Process Number	8	3 bits (FDD) / 4 bits (TDD)					
New Data Indicator	8	1 bit					
Redundancy Version	8	2 bits					
TPC Command for PUCCH	8	2 bits					
Downlink Assignment Index	8	2 bits (TDD only)					

Table 90 – Content of DCI Format 1

★ The Carrier Indicator is applicable when carrier aggregation is used. This field points towards the RF carrier to be used as the serving cell. A value of 0 means that the primary cell is to be used as the serving cell, whereas values 1 to 7 mean that a secondary cell is to be used as the serving cell. An RRC Connection Reconfiguration message is sent to the UE before using the Carrier Indicator field. This message informs the UE of which integer value (secondary cell index) to associate with each secondary cell

★ The Resource Allocation Header determines the way in which the UE interprets the subsequent Resource Block Allocation. A value of 0 indicates a type 0 allocation, whereas a value of 1 indicates a type 1 allocation. The 1.4 MHz channel bandwidth does not require a Resource Allocation Header because type 0 is always used

★ The Resource Block Allocation defines the set of Resource Blocks allocated to the UE (see sections 9.7 and 9.8). The number of bits required to signal the Resource Block allocation increases for larger channel bandwidths, i.e. there is an increased number of possible allocations

★ The Modulation and Coding Scheme (MCS) bits define a pointer to a row within the MCS table (Table 387). The set of 5 bits provide a range from 0 to 31, i.e. equal to the number of rows within the MCS table. The MCS table specifies the modulation scheme to be used for the PDSCH. It also specifies the Transport Block Size (TBS) index. In contrast to the uplink, it does not specify a Redundancy Version (RV)

★ The HARQ Process Number specifies the Hybrid Automatic Repeat Request (HARQ) process for the transport block being sent within the allocated Resource Blocks. The set of 3 bits for FDD provide support for a maximum of 8 HARQ processes. The set of 4 bits for

TDD provide support for a maximum of 16 HARQ processes. In practice, the number of downlink HARQ processes for TDD depends upon the uplink-downlink configuration. Uplink-downlink configuration 5 has the greatest requirement and uses a maximum of 15 HARQ processes

★ The New Data Indicator (NDI) informs the UE of whether the downlink resource allocation is for new data or a retransmission. Toggling the NDI from 0 to 1, or from 1 to 0, indicates that the resource allocation is for new data. Leaving the NDI with its previous value indicates that the resource allocation is for a retransmission

★ The Redundancy Version (RV) is signalled separately for the downlink rather than as part of the MCS table. The RV bits inform the UE of the puncturing pattern applied to the channel coded transport block being sent within the allocated Resource Blocks. The RV does not change between transmissions of the same transport block if Chase Combining is used. The RV changes between transmissions of the same transport block if Incremental Redundancy is used

★ The TPC Command impacts the UE transmit power for the PUCCH. The PUCCH uses accumulated rather than absolute transmit power control. Interpretation of the TPC command is presented in section 24.4.3

★ The Downlink Assignment Index (DAI) is applicable to TDD uplink-downlink subframe configurations 1 to 6. It is used by the UE to determine the number of accumulated downlink transmissions for which the eNode B is expecting an acknowledgement. Its value is updated from subframe to subframe as more downlink transmissions are sent. The UE detects that it has missed a downlink transmission if the value deduced from the DAI does not match the actual number of transmissions detected by the UE

★ Padding is added to the payload of DCI format 1 to ensure that the following 2 conditions are satisfied:

  - the total number of bits is different to that belonging to DCI format 0 and DCI format 1A when used to schedule resources for the same serving cell, and using the same search space and RNTI
  - the total number of bits is not equal to {12, 14, 16, 20, 24, 26, 32, 40, 44, 56}

★ Table 91 presents an example of the total number of bits within DCI format 1 when applied to FDD. This example assumes the use of a release 8 or release 9 device so the Carrier Indicator field has been excluded. A 16 bit CRC is added to the payload after padding has been included. The complete set of bits is channel coded using rate 1/3 convolutional coding before rate matching is applied to ensure that the resultant number of bits equals the capacity of the PDCCH

	1.4 MHz	3 MHz	5 MHz	10 MHz	15 MHz	20 MHz
Total Bits prior to Padding	19	22	27	31	33	39
Padding Bits	0	1	0	0	0	0
Total Bits	**19**	**23**	**27**	**31**	**33**	**39**
After inclusion of 16 bit CRC	35	39	43	47	49	55
After Rate 1/3 Channel Coding	105	117	129	141	147	165

Table 91 – Example total number of bits for DCI Format 1 (FDD)

★ Table 92 presents an example of the total number of bits within DCI format 1 when applied to TDD. This example also assumes the use of a release 8 or release 9 device so the Carrier Indicator field has been excluded. The total number of bits prior to padding is 3 bits greater than for FDD due to the inclusion of the Downlink Assignment Index and using 4 bits rather than 3 bits to signal the HARQ process number

	1.4 MHz	3 MHz	5 MHz	10 MHz	15 MHz	20 MHz
Total Bits prior to Padding	22	25	30	34	36	42
Padding Bits	0	2	0	0	0	0
Total Bits	**22**	**27**	**30**	**34**	**36**	**42**
After inclusion of 16 bit CRC	38	43	46	50	52	58
After Rate 1/3 Channel Coding	114	129	138	150	156	174

Table 92 – Example total number of bits for DCI Format 1 (TDD)

9.12 DCI FORMAT 1A

★ Downlink Control Information (DCI) format 1A is used for the compact scheduling of one PDSCH codeword in one cell, or to initiate the random access procedure

★ When DCI format 1A is used for scheduling a PDSCH codeword, some fields have an alternative interpretation when the PDSCH codeword is used to transfer a random access response, paging message or system information. These alternative interpretations are shown in italics within the remainder of this section

★ The content of DCI Format 1A is shown in Table 93

	3GPP Release	1.4 MHz	3 MHz	5 MHz	10 MHz	15 MHz	20 MHz
Carrier Indicator	10	0 or 3 bits					
Format 0 / Format 1A Flag	8	1 bit					
Localised/Distributed VRB Assignment Flag	8	1 bit					
Resource Block Allocation	8	5 bits	7 bits	9 bits	11 bits	12 bits	13 bits
Modulation and Coding Scheme (MCS)	8	5 bits					
HARQ Process Number	8	3 bits (FDD) / 4 bits (TDD)					
New Data Indicator	8	1 bit					
Redundancy Version	8	2 bits					
TPC Command for PUCCH	8	2 bits					
Downlink Assignment Index	8	2 bits (TDD only)					
SRS Request	10	0 or 1 bit					

Table 93 – Content of DCI Format 1A

★ The Carrier Indicator is applicable when carrier aggregation is used. This field points towards the RF carrier to be used as the serving cell. A value of 0 means that the primary cell is to be used as the serving cell, whereas values 1 to 7 mean that a secondary cell is to be used as the serving cell. An RRC Connection Reconfiguration message is sent to the UE before using the Carrier Indicator field. This message informs the UE of which integer value (secondary cell index) to associate with each secondary cell

★ The Format 0 / Format 1A Flag allows the UE to differentiate between the two formats. A value of 1 indicates that the subsequent data belongs to DCI format 1A

★ DCI format 1A always uses downlink resource allocation type 2 (described in section 9.9). This type of resource allocation can schedule either localised or distributed Resource Blocks

★ The Localised/Distributed Virtual Resource Block (VRB) Flag signals the selection between scheduling localised or distributed Resource Blocks. This impacts the mapping between the allocated VRB and the actual physical Resource Blocks

★ The Resource Block Allocation defines the set of Resource Blocks allocated to the UE

- o if the Resource Block allocation is localised then the full set of bits are used to signal the allocation
- o if the Resource Block allocation is distributed:
  - o if the channel bandwidth is 1.4, 3 or 5 MHz then the full set of bits are used to signal the allocation
  - o *if DCI Format 1A is used to schedule resources for a random access response, paging message or system information then the full set of bits are used to signal the allocation (the 'gap' value is signalled using the New Data Indicator bit)*
  - o else, the most significant bit is used to indicate whether 'gap 1' or 'gap 2' has been used when mapping the VRB onto the physical Resource Blocks, while the remaining bits are used to signal the allocation

★ The Modulation and Coding Scheme (MCS) bits define a pointer to a row within the MCS table (Table 387). The set of 5 bits provide a range from 0 to 31, i.e. equal to the number of rows within the MCS table. The MCS table specifies the modulation scheme used for the PDSCH. It also specifies the Transport Block Size (TBS) index. In contrast to the uplink, it does not specify the Redundancy Version (RV). The RV is signalled separately within the DCI

★ *When DCI Format 1A is used to schedule resources for a random access response, paging message or system information then the modulation scheme is always QPSK (the modulation scheme specified within the MCS table is ignored). In addition, the TBS index is set equal to the value of the MCS bits rather than reading its value from the MCS table*

★ The HARQ Process Number specifies the Hybrid Automatic Repeat Request (HARQ) process for the transport block being sent within the allocated Resource Blocks. The set of 3 bits for FDD provide support for a maximum of 8 HARQ processes. The set of 4 bits for TDD provide support for a maximum of 16 HARQ processes. In practice, the number of downlink HARQ processes for TDD depends upon the uplink-downlink configuration. Uplink-downlink configuration 5 has the greatest requirement and uses a maximum of 15 HARQ processes

- ★ *When DCI Format 1A is used to schedule resources for a random access response, paging message or system information then the HARQ Process Number field is reserved. Random access responses, paging messages and system information do not require HARQ acknowledgements*
- ★ The New Data Indicator (NDI) informs the UE of whether the downlink resource allocation is for new data or a retransmission. Toggling the NDI from 0 to 1, or from 1 to 0, indicates that the resource allocation is for new data. Leaving the NDI with its previous value indicates that the resource allocation is for a retransmission
- ★ *When DCI Format 1A is used to schedule resources for a random access response, paging message or system information then the NDI field is ignored when the resource allocation is localised. If the resource allocation is distributed and the channel bandwidth is 10, 15 or 20 MHz then the NDI field indicates the 'gap' value. The normal meaning of the NDI is not relevant because retransmissions are not used for random access responses, paging messages and system information*
- ★ The Redundancy Version (RV) is signalled separately for the downlink rather than as part of the MCS table. The RV bits inform the UE of the puncturing pattern applied to the transport block being sent within the allocated Resource Blocks. The RV does not change between transmissions of the same transport block if Chase Combining is used. In contrast, the RV changes between transmissions of the same transport block if Incremental Redundancy is used
- ★ The TPC Command impacts the UE transmit power for the PUCCH. The PUCCH uses accumulated rather than absolute transmit power control. Interpretation of the TPC command is presented in section 24.4.3
- ★ *When DCI Format 1A is used to schedule resources for a random access response, paging message or system information then the TPC Command is not used for power control but is used to identify the transport block size. The most significant bit is ignored while the least significant bit indicates whether column 2 or column 3 should be selected from within the Transport Block Size table (Table 388), i.e. the number of allocated Resource Blocks should be assumed to be either 2 or 3 for the purposes of identifying the transport block size. A value of 0 corresponds to 2 assumed Resource Blocks, while a value of 1 corresponds to 3 assumed Resource Blocks*
- ★ The Downlink Assignment Index (DAI) is applicable to TDD uplink-downlink subframe configurations 1 to 6. It is used by the UE to determine the number of accumulated downlink transmissions for which the eNode B is expecting an acknowledgement. Its value is updated from subframe to subframe as more downlink transmissions are sent. The UE detects that it has missed a downlink transmission if the value deduced from the DAI does not match the actual number of transmissions detected by the UE
- ★ The SRS Request can only be included when the DCI is mapped onto a PDCCH within the UE specific search space defined by the C-RNTI. When included, the SRS Request is used to trigger the UE to transmit the Sounding Reference Signal (SRS) in the uplink direction. In the case of DCI format 1A, the SRS Request field is only a single bit which triggers the transmission of the SRS using a single SRS configuration. DCI format 4 includes an SRS Request with 2 bits which can be used to select between SRS configurations
- ★ Padding is added to the payload of DCI 1A to ensure that the following 2 conditions are satisfied:
  - o the total number of bits is not less than that of DCI 0 when used to schedule resources for the same serving cell, and using the same search space and RNTI
  - o the total number of bits is not equal to {12, 14, 16, 20, 24, 26, 32, 40, 44, 56}
- ★ DCI format 1A is used to initiate the random access procedure when its CRC bits are scrambled by a C-RNTI and the following fields are configured:
  - o Localised/Distributed VRB Flag set to 0
  - o all bits of the Resource Block Allocation set to 1
  - o a 6 bit Preamble Index is included
  - o a 4 bit PRACH Mask Index is included

  All remaining bits within the DCI are set to 0
- ★ Table 94 presents an example of the total number of bits within DCI format 1A when applied to FDD. This example assumes the use of a release 8 or release 9 device so the Carrier Indicator field has been excluded. A 16 bit CRC is added to the payload after padding has been included. The complete set of bits is channel coded using rate 1/3 convolutional coding before rate matching is applied to ensure that the resultant number of bits equals the capacity of the PDCCH

	1.4 MHz	3 MHz	5 MHz	10 MHz	15 MHz	20 MHz
Total Bits prior to Padding	20	22	24	26	27	28
Padding Bits	1	0	1	1	0	0
Total Bits	**21**	**22**	**25**	**27**	**27**	**28**
After inclusion of 16 bit CRC	37	38	41	43	43	44
After Rate 1/3 Channel Coding	111	114	123	129	129	132

Table 94 – Example total number of bits for DCI Format 1A (FDD)

★ Table 95 presents an example of the total number of bits within DCI format 1A when applied to TDD. This example also assumes the use of a release 8 or release 9 device so the Carrier Indicator field has been excluded. The total number of bits prior to padding is 3 bits greater than for FDD due to the inclusion of the Downlink Assignment Index and using 4 bits rather than 3 bits to signal the HARQ process number

	1.4 MHz	3 MHz	5 MHz	10 MHz	15 MHz	20 MHz
Total Bits prior to Padding	23	25	27	29	30	31
Padding Bits	0	0	0	0	0	0
Total Bits	**23**	**25**	**27**	**29**	**30**	**31**
After inclusion of 16 bit CRC	39	41	43	45	46	47
After Rate 1/3 Channel Coding	117	123	129	135	138	141

Table 95 – Example total number of bits for DCI Format 1A (TDD)

9.13 DCI FORMAT 1B

★ Downlink Control Information (DCI) format 1B is used for the compact scheduling of one PDSCH codeword in one cell. It also includes precoding information so is applicable to closed loop spatial diversity. It includes information regarding only a single transport block so is applicable when transferring a single codeword

★ The content of DCI format 1B is shown in Table 96

	3GPP Release	1.4 MHz	3 MHz	5 MHz	10 MHz	15 MHz	20 MHz
Carrier Indicator	10	0 or 3 bits					
Localised/Distributed VRB Assignment Flag	8	1 bit					
Resource Block Allocation	8	5 bits	7 bits	9 bits	11 bits	12 bits	13 bits
Modulation and Coding Scheme (MCS)	8	5 bits					
HARQ Process Number	8	3 bits (FDD) / 4 bits (TDD)					
New Data Indicator	8	1 bit					
Redundancy Version	8	2 bits					
TPC Command for PUCCH	8	2 bits					
Downlink Assignment Index	8	2 bits (TDD only)					
TPMI Precoding Information	8	2 bits (2 antenna ports) / 4 bits (4 antenna ports)					
PMI Confirmation for Precoding	8	1 bit					

Table 96 – Content of DCI Format 1B

★ The content of DCI format 1B is the same as that for DCI format 1A with the exception that the Format 0 / Format 1A Flag and SRS Request fields are excluded, while the following 2 additional fields are included:

- Transmitted Precoding Matrix Indicator (TPMI) Precoding Information
- Precoding Matrix Indicator (PMI) Confirmation for Precoding

DCI format 1B is not used to allocate resources for random access responses, paging messages nor system information, so the content does not have alternative interpretations like DCI format 1A

★ TPMI Precoding Information specifies the set of weights used to precode the PDSCH data. Table 97 presents the set of weights applicable when using 2 antenna ports

Codebook Index	0	1	2	3
Weight for Antenna 1	$\frac{1}{\sqrt{2}}$	$\frac{1}{\sqrt{2}}$	$\frac{1}{\sqrt{2}}$	$\frac{1}{\sqrt{2}}$
Weight for Antenna 2	$\frac{1}{\sqrt{2}}$	$\frac{-1}{\sqrt{2}}$	$\frac{j}{\sqrt{2}}$	$\frac{-j}{\sqrt{2}}$

Table 97 – Closed loop spatial multiplexing weights (single layer & 2 antenna ports)

★ The number of bits used to signal the TPMI Precoding Information increases from 2 to 4 when 4 antenna ports are used, i.e. the number of weight combinations increases from 4 to 16

- PMI Confirmation for Precoding indicates whether the PDSCH has been precoded using the weights specified by the TPMI Precoding Information field, or using the latest PMI report sent by the UE
- Padding is added to the payload of DCI format 1B to ensure that the total number of bits is not equal to {12, 14, 16, 20, 24, 26, 32, 40, 44, 56}
- Table 98 presents an example of the total number of bits within DCI format 1B when applied to FDD. This example assumes the use of a release 8 or release 9 device so the Carrier Indicator field has been excluded. It also assumes 2 antenna ports are used for transmission so the length of the TPMI Precoding Information is 2 bits. A 16 bit CRC is added to the payload after padding has been included. The complete set of bits is channel coded using rate 1/3 convolutional coding before rate matching is applied to ensure that the resultant number of bits equals the capacity of the PDCCH

	1.4 MHz	3 MHz	5 MHz	10 MHz	15 MHz	20 MHz
Total Bits prior to Padding	22	24	26	28	29	30
Padding Bits	0	1	1	0	0	0
Total Bits	**22**	**25**	**27**	**28**	**29**	**30**
After inclusion of 16 bit CRC	38	41	43	44	45	46
After Rate 1/3 Channel Coding	114	123	129	132	135	138

Table 98 – Example total number of bits for DCI Format 1B (FDD with 2 antenna ports)

- Table 99 presents an example of the total number of bits within DCI format 1B when applied to TDD. This example also assumes the use of a release 8 or release 9 device, and the use of 2 antenna ports. The total number of bits prior to padding is 3 bits greater than for FDD due to the inclusion of the Downlink Assignment Index and using 4 bits rather than 3 bits to signal the HARQ process number

	1.4 MHz	3 MHz	5 MHz	10 MHz	15 MHz	20 MHz
Total Bits prior to Padding	25	27	29	31	32	33
Padding Bits	0	0	0	0	1	0
Total Bits	**25**	**27**	**29**	**31**	**33**	**33**
After inclusion of 16 bit CRC	41	43	45	47	49	49
After Rate 1/3 Channel Coding	123	129	135	141	147	147

Table 99 – Example total number of bits for DCI Format 1B (TDD with 2 antenna ports)

9.14 DCI FORMAT 1C

- Downlink Control Information (DCI) format 1C is used for very compact scheduling of one PDSCH codeword. It is used to allocate resources for random access responses, paging messages and system information. It can also be used to notify UE of a change in the MCCH signalling for MBMS
- The content of DCI format 1C is shown in Table 100. The figures within the table correspond to the number of bits used to represent each field

	3GPP Release	1.4 MHz	3 MHz	5 MHz	10 MHz	15 MHz	20 MHz
Gap Value	8	0 bits			1 bit		
Resource Block Allocation	8	3 bits	5 bits	7 bits	7 bits	8 bits	9 bits
Transport Block Size	8	5 bits					

Table 100 – Content of DCI Format 1C when used for PDSCH scheduling

- DCI format 1C always signals a type 2 resource allocation (see section 9.9) using a distributed mapping between virtual and physical Resource Blocks
- The smaller channel bandwidths always use 'gap 1' as an input parameter for the mapping between the virtual and physical Resource Blocks. DCI format 1C includes a Gap Value for the larger channel bandwidths to indicate whether 'gap 1' or 'gap 2' is used as an input parameter for the mapping between virtual and physical Resource Blocks
- The Resource Block Allocation defines the set of Resource Blocks allocated to the UE. DCI format 1C uses fewer bits to signal the Resource Block allocation by being less flexible in terms of the number of Resource Blocks which can be allocated, i.e. a fixed step size is used
- DCI format 1C does not use the main Transport Block Size (TBS) table presented in section 33.3. DCI format 1C has its own relatively small TBS table, the content of which is presented in Table 101. The TBS index is signalled within DCI format 1C using 5 bits

TBS Index	0	1	2	3	4	5	6	7	8	9	10
TBS Size	40	56	72	120	136	144	176	208	224	256	280

TBS Index	11	12	13	14	15	16	17	18	19	20	21
TBS Size	296	328	336	392	488	552	600	632	696	776	840

TBS Index	22	23	24	25	26	27	28	29	30	31
TBS Size	904	1000	1064	1128	1224	1288	1384	1480	1608	1736

Table 101 – Transport block sizes applicable to DCI Format 1C

★ DCI format 1C does not include any bits to indicate the modulation order. The modulation is always QPSK (modulation order of 2). Random access responses, paging messages and system information do not require higher order modulations

★ DCI format 1C does not include any bits to signal the Redundancy Version (RV)

- if the CRC bits have been scrambled by the P-RNTI or RA-RNTI then RV 0 is used
- if the CRC bits have been scrambled by the SI-RNTI then the RV is calculated from the SFN when transferring SIB1 and from the subframe number when transferring other SIB

★ DCI format 1C does not include any padding

★ Table 102 presents the total number of bits within DCI format 1C. There is no dependency upon FDD and TDD so the total number of bits is the same in both cases. A 16 bit CRC is added to the payload after padding has been included. The complete set of bits is channel coded using rate 1/3 convolutional coding before rate matching is applied to ensure that the resultant number of bits equals the capacity of the PDCCH

	1.4 MHz	3 MHz	5 MHz	10 MHz	15 MHz	20 MHz
Total Bits	**8**	**10**	**12**	**13**	**14**	**15**
After inclusion of 16 bit CRC	24	26	28	29	30	31
After Rate 1/3 Channel Coding	72	78	84	87	90	93

Table 102 – Total number of bits for DCI Format 1C (FDD & TDD)

★ When DCI format 1C is used to notify UE of a change in the content of the MCCH it includes a single 8 bit field – 'Information for MCCH change notification'. Padding is added so the resultant size of DCI format 1C is the same as when used to schedule PDSCH resources, i.e. the same size as presented in Table 102

★ The Information for MCCH change notification defines a bitmap where each bit is associated with a specific MBSFN area. If a bit is set to 1 then it indicates that the MCCH within that MBSFN area will change during the next MCCH modification period. The link between MBSFN area and position within the bitmap is broadcast within SIB13 using the Notification Indicator information element

9.15 DCI FORMAT 1D

★ Downlink Control Information (DCI) format 1D is applicable to multi-user MIMO. It provides compact scheduling of one PDSCH codeword in one cell with precoding and power offset information

★ Each UE using multi-user MIMO receives its own version of DCI format 1D, i.e. each UE can be allocated different transport block sizes and precoding weights

★ The content of DCI format 1D when using 2 antenna ports is shown in Table 103

	3GPP Release	1.4 MHz	3 MHz	5 MHz	10 MHz	15 MHz	20 MHz
Carrier Indicator	10	0 or 3 bits					
Localised/Distributed VRB Assignment Flag	8	1 bit					
Resource Block Allocation	8	5 bits	7 bits	9 bits	11 bits	12 bits	13 bits
Modulation and Coding Scheme (MCS)	8	5 bits					
HARQ Process Number	8	3 bits (FDD) / 4 bits (TDD)					
New Data Indicator	8	1 bit					
Redundancy Version	8	2 bits					
TPC Command for PUCCH	8	2 bits					
Downlink Assignment Index	8	2 bits (TDD only)					
TPMI Precoding Information	8	2 bits (2 antenna ports) / 4 bits (4 antenna ports)					
Downlink Power Offset	8	1 bit					

Table 103 – Content of DCI Format 1D

★ The content of DCI format 1D is the same as that of DCI format 1B with the exception that the PMI Confirmation for Precoding is replaced by a Downlink Power Offset

★ The Downlink Power Offset is necessary for multi-user MIMO to inform the UE of its downlink transmit power relative to the single user downlink transmit power. The mapping between the actual downlink power offset and its signalled value is presented in Table 104

Signalled Downlink Power Offset	Downlink Power Offset
0	- 3 dB
1	0 dB

Table 104 – Mapping between the downlink power offset and its signalled value

★ Padding is added to the payload of DCI format 1D to ensure that the total number of bits is not equal to {12, 14, 16, 20, 24, 26, 32, 40, 44, 56}

★ Table 105 presents an example of the total number of bits within DCI format 1D when applied to FDD. This example assumes the use of a release 8 or release 9 device so the Carrier Indicator field has been excluded. It also assumes 2 antenna ports are used for transmission so the length of the TPMI Precoding Information is 2 bits

	1.4 MHz	3 MHz	5 MHz	10 MHz	15 MHz	20 MHz
Total Bits prior to Padding	22	24	26	28	29	30
Padding Bits	0	1	1	0	0	0
Total Bits	**22**	**25**	**27**	**28**	**29**	**30**
After inclusion of 16 bit CRC	38	41	43	44	45	46
After Rate 1/3 Channel Coding	114	123	129	132	135	138

Table 105 – Example total number of bits for DCI Format 1D (FDD with 2 antenna ports)

★ Table 106 presents an example of the total number of bits within DCI format 1D when applied to TDD. This example also assumes the use of a release 8 or release 9 device, and the use of 2 antenna ports

	1.4 MHz	3 MHz	5 MHz	10 MHz	15 MHz	20 MHz
Total Bits prior to Padding	25	27	29	31	32	33
Padding Bits	0	0	0	0	1	0
Total Bits	**25**	**27**	**29**	**31**	**33**	**33**
After inclusion of 16 bit CRC	41	43	45	47	49	49
After Rate 1/3 Channel Coding	123	129	135	141	147	147

Table 106 – Example total number of bits for DCI Format 1D (TDD with 2 antenna ports)

9.16 DCI FORMAT 2

★ Downlink Control Information (DCI) format 2 is used to schedule one or two PDSCH codewords and is applicable to closed loop spatial multiplexing

★ The content of DCI format 2 is shown in Table 107

		3GPP Release	1.4 MHz	3 MHz	5 MHz	10 MHz	15 MHz	20 MHz
Carrier Indicator		10	0 or 3 bits					
Resource Allocation Header		8	0 bits	1 bit				
Resource Block Allocation		8	6 bits	8 bits	13 bits	17 bits	19 bits	25 bits
TPC Command for PUCCH		8	2 bits					
Downlink Assignment Index		8	2 bits (TDD only)					
HARQ Process Number		8	3 bits (FDD) / 4 bits (TDD)					
Transport Block to Codeword Swap Flag		8	1 bit					
Transport Block 1	Modulation and Coding Scheme	8	5 bits					
	New Data Indicator	8	1 bit					
	Redundancy Version	8	2 bits					
Transport Block 2	Modulation and Coding Scheme	8	5 bits					
	New Data Indicator	8	1 bit					
	Redundancy Version	8	2 bits					
Precoding Information		8	3 bits (2 antenna ports) / 6 bits (4 antenna ports)					

Table 107 – Content of DCI Format 2

★ The Carrier Indicator is applicable when carrier aggregation is used. This field points towards the RF carrier to be used as the serving cell. A value of 0 means that the primary cell is to be used as the serving cell, whereas values 1 to 7 mean that a secondary cell is to be used as the serving cell. An RRC Connection Reconfiguration message is sent to the UE before using the Carrier Indicator field. This message informs the UE of which integer value (secondary cell index) to associate with each secondary cell

★ The Resource Allocation Header determines the way in which the UE interprets the subsequent Resource Block Allocation. A value of 0 indicates a type 0 allocation, whereas a value of 1 indicates a type 1 allocation. The 1.4 MHz channel bandwidth does not require a Resource Allocation Header because type 0 is always used

★ The Resource Block Allocation defines the set of Resource Blocks allocated to the UE (see sections 9.7 and 9.8). The number of bits required to signal the Resource Block allocation increases for larger channel bandwidths

★ The TPC Command impacts the UE transmit power for the PUCCH. The PUCCH uses accumulated rather than absolute transmit power control. Interpretation of the TPC command is presented in section 24.4.3

★ The Downlink Assignment Index (DAI) is applicable to TDD uplink-downlink subframe configurations 1 to 6. It is used by the UE to determine the number of accumulated downlink transmissions for which the eNode B is expecting an acknowledgement. Its value is updated from subframe to subframe as more downlink transmissions are sent. The UE detects that it has missed a downlink transmission if the value deduced from the DAI does not match the actual number of transmissions detected by the UE

★ The HARQ Process Number specifies the Hybrid Automatic Repeat Request (HARQ) process for the transport block being sent within the allocated Resource Blocks. The set of 3 bits for FDD provide support for a maximum of 8 HARQ processes. The set of 4 bits for TDD provide support for a maximum of 16 HARQ processes. In practice, the number of downlink HARQ processes for TDD depends upon the uplink-downlink configuration. Uplink-downlink configuration 5 has the greatest requirement and uses a maximum of 15 HARQ processes

★ The Transport Block to Codeword Swap Flag defines the relationship between transport blocks 1 and 2 and codewords 0 and 1. If the flag is set to 0, codeword 0 is generated from transport block 1, while codeword 1 is generated from transport block 2. The transport blocks are swapped if the flag is set to 1. The flag is only applicable if 2 transport blocks are scheduled. If a single transport block is scheduled then that transport block is used to generate codeword 0

★ Transport block information is provided for either one or two transport blocks (depending upon whether single or dual stream transmission is scheduled). The Modulation and Coding Scheme (MCS) bits define a pointer to a row within the MCS table (Table 387). The New Data Indicator (NDI) informs the UE of whether the downlink resource allocation is for new data or a retransmission. The RV bits inform the UE of the puncturing pattern applied to the transport block being sent within the allocated Resource Blocks

★ In the case of 2 antenna ports, the precoding information points to a row within Table 108. The column is dependent upon whether 1 or 2 codewords are transferred during that subframe. Rows 5 and 6 within the 1st column cater for the scenario where a UE reports a Rank Indication (RI) of 2 while the eNode B schedules a single layer (single rank). In this case, the PMI reported by the UE is a 2×2 matrix (one of the two matrices shown at the top of the second column of Table 108), while precoding requires only a 2×1 matrix so one of the two columns is selected

Precoding Information	1 Codeword	2 Codewords
0	2 Layers: Transmit Diversity	2 Layers: Precoding vector: $\frac{1}{2}\begin{bmatrix}1 & 1\\1 & -1\end{bmatrix}$
1	1 Layer: Precoding vector: $\frac{1}{\sqrt{2}}\begin{bmatrix}1\\1\end{bmatrix}$	2 Layers: Precoding vector: $\frac{1}{2}\begin{bmatrix}1 & 1\\j & -j\end{bmatrix}$
2	1 Layer: Precoding vector: $\frac{1}{\sqrt{2}}\begin{bmatrix}1\\-1\end{bmatrix}$	2 Layers: Precoding vector based upon latest PMI report on PUSCH
3	1 Layer: Precoding vector: $\frac{1}{\sqrt{2}}\begin{bmatrix}1\\j\end{bmatrix}$	*Reserved*
4	1 Layer: Precoding vector: $\frac{1}{\sqrt{2}}\begin{bmatrix}1\\-j\end{bmatrix}$	*Reserved*
5	1 layer: Precoding vector based upon latest PMI report on PUSCH. If RI=2 was reported then use 1st column × $\sqrt{2}$	*Reserved*
6	1 layer: Precoding vector based upon latest PMI report on PUSCH. If RI=2 was reported then use 2nd column × $\sqrt{2}$	*Reserved*
7	*Reserved*	*Reserved*

Table 108 – Precoding information applicable to DCI Format 2 (2 antenna ports)

A similar table is defined for 4 antenna ports but the table includes 64 rows to cater for the larger set of supported combinations

★ Padding is added to the payload of DCI format 2 to ensure that the total number of bits is not equal to {12, 14, 16, 20, 24, 26, 32, 40, 44, 56}

★ Table 109 presents an example of the total number of bits within DCI format 2 when applied to FDD. This example assumes the use of a release 8 or release 9 device so the Carrier Indicator field has been excluded. It also assumes 2 antenna ports are used for transmission so the length of the Precoding Information is 3 bits. A 16 bit CRC is added to the payload after padding has been included. The complete set of bits is channel coded using rate 1/3 convolutional coding before rate matching is applied to ensure that the resultant number of bits equals the capacity of the PDCCH

	1.4 MHz	3 MHz	5 MHz	10 MHz	15 MHz	20 MHz
Total Bits prior to Padding	31	34	39	43	45	51
Padding Bits	0	0	0	0	0	0
Total Bits	**31**	**34**	**39**	**43**	**45**	**51**
After inclusion of 16 bit CRC	47	50	55	59	61	67
After Rate 1/3 Channel Coding	141	150	165	177	183	201

Table 109 – Example total number of bits for DCI Format 2 (FDD with 2 antenna ports)

★ Table 110 presents an example of the total number of bits within DCI format 2 when applied to TDD. This example also assumes the use of a release 8 or release 9 device, and the use of 2 antenna ports. The total number of bits prior to padding is 3 bits greater than for FDD due to the inclusion of the Downlink Assignment Index and using 4 bits rather than 3 bits to signal the HARQ process number

	1.4 MHz	3 MHz	5 MHz	10 MHz	15 MHz	20 MHz
Total Bits prior to Padding	34	37	42	46	48	54
Padding Bits	0	0	0	0	0	0
Total Bits	**34**	**37**	**42**	**46**	**48**	**54**
After inclusion of 16 bit CRC	50	53	58	62	64	70
After Rate 1/3 Channel Coding	150	159	174	186	192	210

Table 110 – Example total number of bits for DCI Format 2 (TDD with 2 antenna ports)

9.17 DCI FORMAT 2A

★ Downlink Control Information (DCI) format 2A is used to schedule one or two PDSCH codewords and is applicable to open loop spatial multiplexing

★ The content of DCI format 2A is shown in Table 111

		3GPP Release	1.4 MHz	3 MHz	5 MHz	10 MHz	15 MHz	20 MHz
Carrier Indicator		10	0 or 3 bits					
Resource Allocation Header		8	0 bits	1 bit				
Resource Block Allocation		8	6 bits	8 bits	13 bits	17 bits	19 bits	25 bits
TPC Command for PUCCH		8	2 bits					
Downlink Assignment Index		8	2 bits (TDD only)					
HARQ Process Number		8	3 bits (FDD) / 4 bits (TDD)					
Transport Block to Codeword Swap Flag		8	1 bit					
Transport Block 1	Modulation and Coding Scheme	8	5 bits					
	New Data Indicator	8	1 bit					
	Redundancy Version	8	2 bits					
Transport Block 2	Modulation and Coding Scheme	8	5 bits					
	New Data Indicator	8	1 bit					
	Redundancy Version	8	2 bits					
Precoding Information		8	0 bits (2 antenna ports) / 2 bits (4 antenna ports)					

Table 111 – Content of DCI Format 2A

★ The content of DCI format 2A is the same as that for DCI format 2 with the exception that the Precoding information refers to open loop spatial multiplexing rather than closed loop spatial multiplexing

★ Precoding information is not necessary for 2×2 open loop spatial multiplexing because the eNode B always sends 2 codewords using 2 layers and the weights are deterministic (section 4.4.2). Transmit diversity is used when only a single codeword can be transferred

★ 2 bits of Precoding information is included within DCI format 2A when 4×4 open loop spatial multiplexing is used. The precoding information points to a row in Table 112. The column is dependent upon whether 1 or 2 codewords are transferred during that subframe

	1 Codeword	2 Codewords
0	4 Layers: Transmit Diversity	2 Layers: Precoding cycling with large delay CDD
1	2 Layers: Precoding cycling with large delay CDD	3 Layers: Precoding cycling with large delay CDD
2	*Reserved*	4 Layers: Precoding cycling with large delay CDD
3	*Reserved*	*Reserved*

Table 112 – Precoding information applicable to DCI Format 2A (4 antenna ports)

★ The Precoding information does not specify any weights because these are deterministic. Instead, the Precoding information signals the number of layers that the eNode B has used when transmitting the 1 or 2 codewords

★ Padding is added to the payload of DCI format 2A to ensure that the total number of bits is not equal to {12, 14, 16, 20, 24, 26, 32, 40, 44, 56}

★ Table 113 presents an example of the total number of bits within DCI format 2A when applied to FDD. This example assumes the use of a release 8 or release 9 device so the Carrier Indicator field has been excluded. It also assumes 2 antenna ports are used for transmission so the length of the Precoding Information is 3 bits. A 16 bit CRC is added to the payload after padding has been included. The complete set of bits is channel coded using rate 1/3 convolutional coding before rate matching is applied to ensure that the resultant number of bits equals the capacity of the PDCCH

	1.4 MHz	3 MHz	5 MHz	10 MHz	15 MHz	20 MHz
Total Bits prior to Padding	28	31	36	40	42	48
Padding Bits	0	0	0	1	0	0
Total Bits	**28**	**31**	**36**	**41**	**42**	**48**
After inclusion of 16 bit CRC	44	47	52	57	58	64
After Rate 1/3 Channel Coding	132	141	156	171	174	192

Table 113 – Example total number of bits for DCI Format 2A (FDD with 2 antenna ports)

★ Table 114 presents an example of the total number of bits within DCI format 2A when applied to TDD. This example also assumes the use of a release 8 or release 9 device, and the use of 2 antenna ports. The total number of bits prior to padding is 3 bits greater than for FDD due to the inclusion of the Downlink Assignment Index and using 4 bits rather than 3 bits to signal the HARQ process number

	1.4 MHz	3 MHz	5 MHz	10 MHz	15 MHz	20 MHz
Total Bits prior to Padding	31	34	39	43	45	51
Padding Bits	0	0	0	0	0	0
Total Bits	**31**	**34**	**39**	**43**	**45**	**51**
After inclusion of 16 bit CRC	47	50	55	59	61	67
After Rate 1/3 Channel Coding	141	150	165	177	183	201

Table 114 – Example total number of bits for DCI Format 2A (TDD with 2 antenna ports)

9.18 DCI FORMAT 2B

★ Downlink Control Information (DCI) format 2B is used to schedule one or two PDSCH codewords and is applicable to transmission mode 8, which supports both single user and multi-user MIMO with beamforming on antenna ports 7 and 8

★ DCI format 2B was introduced within the release 9 version of the 3GPP specifications

★ The content of DCI format 2B is shown in Table 115

		3GPP Release	1.4 MHz	3 MHz	5 MHz	10 MHz	15 MHz	20 MHz
Carrier Indicator		10	0 or 3 bits					
Resource Allocation Header		9	0 bits	1 bit				
Resource Block Allocation		9	6 bits	8 bits	13 bits	17 bits	19 bits	25 bits
TPC Command for PUCCH		9	2 bits					
Downlink Assignment Index		9	2 bits (TDD only)					
HARQ Process Number		9	3 bits (FDD) / 4 bits (TDD)					
Scrambling Identity		9	1 bit					
SRS Request		10	0 or 1 bit (TDD only)					
Transport Block 1	Modulation and Coding Scheme	9	5 bits					
	New Data Indicator	9	1 bit					
	Redundancy Version	9	2 bits					
Transport Block 2	Modulation and Coding Scheme	9	5 bits					
	New Data Indicator	9	1 bit					
	Redundancy Version	9	2 bits					

Table 115 – Content of DCI Format 2B

★ The content of DCI format 2B is the same as that for DCI format 2 with the exception that

  - the Transport Block to Codeword Swap Flag is excluded
  - the Precoding Information is excluded
  - a Scrambling Identity is included
  - an SRS Request can be included

★ The Transport Block to Codeword Swap Flag is not required because 3GPP specifies that when the eNode B transmits 2 codewords then transport block 1 is mapped to codeword 0, and transport block 2 is mapped to codeword 1

★ Precoding Information is not included because the precoding for beamforming is implementation dependent and is not specified by 3GPP. This means that the UE is not provided with any information regarding the precoding weights, i.e. precoding at the eNode B is non-codebook based

★ The Scrambling Identity provides support for multi-user MIMO by allowing separate users to re-use the same antenna port but with a different Scrambling Identity. 2 Scrambling Identities are available to allow 2 users to share the same antenna port

★ The SRS Request within DCI format 2B can only be included for TDD. When included, the SRS Request is used to trigger the UE to transmit the Sounding Reference Signal (SRS) in the uplink direction. In the case of DCI format 2B, the SRS Request field is only a single bit which triggers the transmission of the SRS using a single SRS configuration. DCI format 4 includes an SRS Request with 2 bits which can be used to select between SRS configurations

★ If DCI format 2B is used for the transmission of a single transport block then the New Data Indicator field from the 'disabled' transport block instructs the UE to use either antenna port 7 or antenna port 8

★ Padding is added to the payload of DCI format 2B to ensure that the total number of bits is not equal to {12, 14, 16, 20, 24, 26, 32, 40, 44, 56}

★ Table 116 presents an example of the total number of bits within DCI format 2B when applied to FDD. This example assumes the use of a release 9 device so the Carrier Indicator field has been excluded. A 16 bit CRC is added to the payload after padding has been included. The complete set of bits is channel coded using rate 1/3 convolutional coding before rate matching is applied to ensure that the resultant number of bits equals the capacity of the PDCCH

	1.4 MHz	3 MHz	5 MHz	10 MHz	15 MHz	20 MHz
Total Bits prior to Padding	28	31	36	40	42	48
Padding Bits	0	0	0	1	0	0
Total Bits	**28**	**31**	**36**	**41**	**42**	**48**
After inclusion of 16 bit CRC	44	47	52	57	58	64
After Rate 1/3 Channel Coding	132	141	156	171	174	192

Table 116 – Example total number of bits for DCI Format 2B (FDD)

★ Table 117 presents an example of the total number of bits within DCI format 2B when applied to TDD. This example also assumes the use of a release 9 device. It also assumes that the SRS Request field is included. The total number of bits prior to padding is 4 bits greater than for FDD due to the inclusion of the Downlink Assignment Index and SRS Request, and using 4 bits rather than 3 bits to signal the HARQ process number

	1.4 MHz	3 MHz	5 MHz	10 MHz	15 MHz	20 MHz
Total Bits prior to Padding	32	35	40	44	46	52
Padding Bits	1	0	1	1	0	0
Total Bits	**33**	**35**	**41**	**45**	**46**	**52**
After inclusion of 16 bit CRC	49	51	57	61	62	68
After Rate 1/3 Channel Coding	147	153	171	183	186	204

Table 117 – Example total number of bits for DCI Format 2B (TDD)

9.19 DCI FORMAT 2C

★ Downlink Control Information (DCI) format 2C is used to schedule one or two PDSCH codewords using up to 8 layers of transmission on antenna ports 7 to 14. It is applicable to transmission mode 9

★ DCI format 2C was introduced within the release 10 version of the 3GPP specifications

★ The content of DCI format 2C is shown in Table 118

<table>
<tr><th colspan="2"></th><th>3GPP Release</th><th>1.4 MHz</th><th>3 MHz</th><th>5 MHz</th><th>10 MHz</th><th>15 MHz</th><th>20 MHz</th></tr>
<tr><td colspan="2">Carrier Indicator</td><td>10</td><td colspan="6">0 or 3 bits</td></tr>
<tr><td colspan="2">Resource Allocation Header</td><td>10</td><td>0 bits</td><td colspan="5">1 bit</td></tr>
<tr><td colspan="2">Resource Block Allocation</td><td>10</td><td>6 bits</td><td>8 bits</td><td>13 bits</td><td>17 bits</td><td>19 bits</td><td>25 bits</td></tr>
<tr><td colspan="2">TPC Command for PUCCH</td><td>10</td><td colspan="6">2 bits</td></tr>
<tr><td colspan="2">Downlink Assignment Index</td><td>10</td><td colspan="6">2 bits (TDD only)</td></tr>
<tr><td colspan="2">HARQ Process Number</td><td>10</td><td colspan="6">3 bits (FDD) / 4 bits (TDD)</td></tr>
<tr><td colspan="2">Antenna Ports, Scrambling Identity and Layers</td><td>10</td><td colspan="6">3 bits</td></tr>
<tr><td colspan="2">SRS Request</td><td>10</td><td colspan="6">0 or 1 bit (TDD only)</td></tr>
<tr><td rowspan="3">Transport Block 1</td><td>Modulation and Coding Scheme</td><td>10</td><td colspan="6">5 bits</td></tr>
<tr><td>New Data Indicator</td><td>10</td><td colspan="6">1 bit</td></tr>
<tr><td>Redundancy Version</td><td>10</td><td colspan="6">2 bits</td></tr>
<tr><td rowspan="3">Transport Block 2</td><td>Modulation and Coding Scheme</td><td>10</td><td colspan="6">5 bits</td></tr>
<tr><td>New Data Indicator</td><td>10</td><td colspan="6">1 bit</td></tr>
<tr><td>Redundancy Version</td><td>10</td><td colspan="6">2 bits</td></tr>
</table>

Table 118 – Content of DCI Format 2C

★ The content of DCI format 2C is the same as that for DCI format 2 with the exception that

- o the Transport Block to Codeword Swap Flag is excluded
- o the Precoding Information is excluded
- o Antenna Ports, Scrambling Identity and Layers information is included
- o an SRS Request can be included

★ The Transport Block to Codeword Swap Flag is not required because 3GPP specifies that when the eNode B transmits 2 codewords then transport block 1 is mapped to codeword 0, and transport block 2 is mapped to codeword 1

★ Precoding Information is not included because the precoding for beamforming is implementation dependent and is not specified by 3GPP. This means that the UE is not provided with any information regarding the precoding weights

★ The Antenna Ports, Scrambling Identity and Layers information defines a pointer to a row within the look-up table presented in Table 119. This look-up table includes one set of rows for when a single codeword is transferred, and another set of rows for when 2 codewords are transferred. The Scrambling Identity allows the UE to differentiate between connections when multi-user MIMO is used

★ For single codeword transmission, the 4, 5 and 6 values within Table 119 are only supported for retransmission of a transport block, and if that transport block was previously transmitted using 2, 3 or 4 layers respectively

<table>
<tr><th colspan="4">1 Codeword</th></tr>
<tr><th>Value</th><th>Number of Layers</th><th>Antenna Ports</th><th>Scrambling Identity</th></tr>
<tr><td>0</td><td rowspan="4">1</td><td rowspan="2">7</td><td>0</td></tr>
<tr><td>1</td><td>1</td></tr>
<tr><td>2</td><td rowspan="2">8</td><td>0</td></tr>
<tr><td>3</td><td>1</td></tr>
<tr><td>4</td><td>2</td><td>7 - 8</td><td rowspan="3">-</td></tr>
<tr><td>5</td><td>3</td><td>7 – 9</td></tr>
<tr><td>6</td><td>4</td><td>7 - 10</td></tr>
<tr><td>7</td><td colspan="3">Reserved</td></tr>
</table>

<table>
<tr><th colspan="4">2 Codewords</th></tr>
<tr><th>Value</th><th>Number of Layers</th><th>Antenna Ports</th><th>Scrambling Identity</th></tr>
<tr><td>0</td><td rowspan="2">2</td><td rowspan="2">7 – 8</td><td>0</td></tr>
<tr><td>1</td><td>1</td></tr>
<tr><td>2</td><td>3</td><td>7 – 9</td><td rowspan="6">-</td></tr>
<tr><td>3</td><td>4</td><td>7 – 10</td></tr>
<tr><td>4</td><td>5</td><td>7 – 11</td></tr>
<tr><td>5</td><td>6</td><td>7 – 12</td></tr>
<tr><td>6</td><td>7</td><td>7 – 13</td></tr>
<tr><td>7</td><td>8</td><td>7 - 14</td></tr>
</table>

Table 119 – Antenna Ports, Scrambling Identity and Layers look-up table

★ The SRS Request within DCI format 2C can only be included for TDD. When included, the SRS Request is used to trigger the UE to transmit the Sounding Reference Signal (SRS) in the uplink direction. In the case of DCI format 2C, the SRS Request field is only a single bit which triggers the transmission of the SRS using a single SRS configuration. DCI format 4 includes an SRS Request with 2 bits which can be used to select between SRS configurations

★ Padding is added to the payload of DCI format 2C to ensure that the total number of bits is not equal to {12, 14, 16, 20, 24, 26, 32, 40, 44, 56}

★ Table 120 presents an example of the total number of bits within DCI format 2C when applied to FDD. This example assumes the use of a release 10 device but the Carrier Indicator field has been excluded. A 16 bit CRC is added to the payload after padding has been included. The complete set of bits is channel coded using rate 1/3 convolutional coding before rate matching is applied to ensure that the resultant number of bits equals the capacity of the PDCCH

	1.4 MHz	3 MHz	5 MHz	10 MHz	15 MHz	20 MHz
Total Bits prior to Padding	30	33	38	42	44	50
Padding Bits	0	0	0	0	1	0
Total Bits	**30**	**33**	**38**	**42**	**44**	**50**
After inclusion of 16 bit CRC	46	49	54	58	61	66
After Rate 1/3 Channel Coding	138	147	162	174	183	198

Table 120 – Example total number of bits for DCI Format 2C (FDD)

★ Table 121 presents an example of the total number of bits within DCI format 2C when applied to TDD. This example also assumes the use of a release 10 device. It also assumes that the SRS Request field is included. The total number of bits prior to padding is 4 bits greater than for FDD due to the inclusion of the Downlink Assignment Index and SRS Request, and using 4 bits rather than 3 bits to signal the HARQ process number

	1.4 MHz	3 MHz	5 MHz	10 MHz	15 MHz	20 MHz
Total Bits prior to Padding	34	37	42	46	48	54
Padding Bits	0	0	0	0	0	0
Total Bits	**34**	**37**	**42**	**46**	**48**	**54**
After inclusion of 16 bit CRC	50	53	58	62	64	70
After Rate 1/3 Channel Coding	150	159	174	186	192	210

Table 121 – Example total number of bits for DCI Format 2C (TDD)

9.20 DCI FORMAT 3

★ Downlink Control Information (DCI) format 3 is used for the transmission of TPC commands for the PUCCH and PUSCH with 2 bit power adjustments. The content of DCI format 3 is shown in Table 122. The figures within the table correspond to the number of bits used to represent each field

	3GPP Release	1.4 MHz	3 MHz	5 MHz	10 MHz	15 MHz	20 MHz
TPC Commands	8	20	22	24	26	26	28
Padding		1	0	1	1	1	0
Total		**21**	**22**	**25**	**27**	**27**	**28**
After inclusion of 16 bit CRC		37	38	41	43	43	44
After Rate 1/3 Channel Coding		111	114	123	129	129	132

Table 122 – Content of DCI Format 3

★ The Transmit Power Control (TPC) commands are signalled using a pair of bits, e.g. DCI 3 for the 1.4 MHz channel bandwidth uses 20 bits to signal 10 TPC commands

★ The RNTI used to scramble the CRC bits determines whether the TPC commands are applicable to the PUSCH or PUCCH. The TPC-PUSCH-RNTI indicates that the TPC commands are applicable to the PUSCH. Whereas, the TPC-PUCCH-RNTI indicates that the TPC commands are applicable to the PUCCH

★ UE extract a single TPC command from the string of TPC commands based upon the TPC Index provided by the eNode B. The TPC index can be signalled using the RRC Connection Setup, RRC Connection Reconfiguration or RRC Connection Re-establishment messages. Accumulative power control is applied to both the PUSCH and PUCCH when TPC commands are received from DCI format 3

★ Padding is added to ensure that the size of DCI 3 is equal to the size of DCI 0

9.21 DCI FORMAT 3A

★ Downlink Control Information (DCI) format 3A is used for the transmission of TPC commands for the PUCCH and PUSCH with 1 bit power adjustments. The content of DCI format 3A is shown in Table 123. The figures within the table correspond to the number of bits used to represent each field

	3GPP Release	1.4 MHz	3 MHz	5 MHz	10 MHz	15 MHz	20 MHz
TPC Commands	8	21	22	25	27	27	28
Total		**21**	**22**	**25**	**27**	**27**	**28**
After inclusion of 16 bit CRC		37	38	41	43	43	44
After Rate 1/3 Channel Coding		111	114	123	129	129	132

Table 123 – Content of DCI Format 3A

★ The Transmit Power Control (TPC) commands are signalled using a single bit, e.g. DCI 3A for the 1.4 MHz channel bandwidth uses 21 bits to signal 21 TPC commands

★ The number of TPC commands within DCI format 3A is specified to equal the size of DCI format 0, i.e. the size of DCI format 3A equals the size of DCI format 0

★ The RNTI used to scramble the CRC bits determines whether the TPC commands are applicable to the PUSCH or PUCCH. The TPC-PUSCH-RNTI indicates that the TPC commands are applicable to the PUSCH. Whereas, the TPC-PUCCH-RNTI indicates that the TPC commands are applicable to the PUCCH

★ UE extract a single TPC command from the string of TPC commands based upon the TPC Index provided by the eNode B. The TPC index can be signalled using the RRC Connection Setup, RRC Connection Reconfiguration or RRC Connection Re-establishment messages. Accumulative power control is applied to both the PUSCH and PUCCH when TPC commands are received from DCI format 3A

9.22 DCI FORMAT 4

★ Downlink Control Information (DCI) format 4 is used to schedule one or two PUSCH codewords for multi-antenna port transmission in the uplink. It is used by uplink transmission mode 2

★ The content of DCI format 4 is shown in Table 124

		3GPP Release	1.4 MHz	3 MHz	5 MHz	10 MHz	15 MHz	20 MHz
Carrier Indicator		10	0 or 3 bits					
Resource Block Allocation		10	6 bits	7 bits	10 bits	12 bits	13 bits	14 bits
TPC Command for Scheduled PUSCH		10	2 bits					
Cyclic Shift for DM RS and OCC		10	3 bits					
Uplink Index		10	2 bits (TDD only)					
Downlink Assignment Index		10	2 bits (TDD only)					
CSI Request		10	1 or 2 bits					
SRS Request		10	2 bits					
Resource Allocation Type		10	1 bit					
Transport Block 1	Modulation and Coding Scheme	10	5 bits					
	New Data Indicator	10	1 bit					
Transport Block 2	Modulation and Coding Scheme	10	5 bits					
	New Data Indicator	10	1 bit					
Precoding Information and Number of Layers		10	3 bits (2 antenna ports) / 6 bits (4 antenna ports)					

Table 124 – Content of DCI Format 4

★ The Carrier Indicator is only applicable when carrier aggregation is used. This field points towards the RF carrier to be used as the serving cell. A value of 0 means that the primary cell is to be used as the serving cell. Values 1 to 7 mean that a secondary cell is to be used as the serving cell. An RRC Connection Reconfiguration message is sent to the UE before using the Carrier Indicator field. This message informs the UE of which integer value (secondary cell index) to associate with each secondary cell

★ The Resource Block Allocation defines the set of Resource Blocks allocated to the UE. These Resource Blocks can be allocated using either uplink resource allocation type 0 or type 1 (see sections 9.4 and 9.5). The number of bits required to signal the Resource Block allocation increases for larger channel bandwidths, i.e. there is an increased number of possible allocations

★ The number of bits allocated to the Resource Block Allocation field is greater than that for DCI format 0. The increased number of bits is intended to accommodate the requirements of uplink resource allocation type 1. DCI format 0 was first specified before uplink resource allocation type 1 was defined so only accounts for uplink resource allocation type 0

★ Resource Block hopping is not supported when allocating resources using DCI format 4

★ The TPC Command impacts the UE transmit power for the PUSCH. Interpretation of the TPC command depends upon whether accumulated or absolute power control is used. This is configured during connection establishment within the RRC Connection Setup message. Interpretation of the TPC command is described in section 24.4.2

★ The 'Cyclic Shift for DM RS and OCC Index' defines a pointer to a look-up table which is used as an input when the UE generates the sequence for its uplink Demodulation Reference Signal. The cyclic shift is used when generating the sequence itself while the Orthogonal Covering Code (OCC) is an additional multiplying factor

★ The Uplink Index is only included for TDD uplink-downlink configuration 0. It has an impact upon the timing for power control, aperiodic CQI/PMI/RI reporting and PUSCH transmission:

- o power control: the uplink index is applicable when the PUSCH is scheduled during subframes 2 or 7. It determines the timing of the power control command to be applied
- o aperiodic CQI/PMI/RI reporting: the uplink index determines the timing of when the aperiodic report should be sent
- o PUSCH transmission: the uplink index determines the timing of when to apply the information received in DCI format 4

★ The Downlink Assignment Index (DAI) is only included for TDD uplink-downlink configurations 1 to 6. It is used to inform the UE of the number of PDSCH transmissions requiring acknowledgement

★ The CSI Request can have a length of either 1 or 2 bits. The 2 bit length is applicable to UE which are configured with more than a single downlink cell. When using a single downlink cell, the 1 bit CSI Request can be used to instruct the UE to send a CSI report for that cell. When using multiple downlink cells, the 2 bit CSI Request can be used to instruct the UE to send a CSI report for either a single serving cell, or a set of serving cells. Interpretation of the 2 bit CSI Request is shown in Table 125

CSI Request	Meaning
00	No aperiodic CSI report is triggered
01	Aperiodic CSI report is triggered for serving cell 'c'
10	Aperiodic CSI report is triggered for the 1st set of serving cells
11	Aperiodic CSI report is triggered for the 2nd set of serving cells

Table 125 – Interpretation of the 2 bit CSI Request

- ★ The eNode B uses RRC signalling to inform the UE of which serving cells are to be included within set 1 and which serving cells are to be included within set 2. This is done using the Trigger1 and Trigger2 information elements within, for example an RRC Connection Reconfiguration message
- ★ The SRS Request is used to trigger the UE to transmit the Sounding Reference Signal (SRS) in the uplink direction. In the case of DCI format 4, the SRS Request field has 2 bits which allows the eNode B to select between SRS configurations
- ★ The Resource Allocation Type is used to signal the type of uplink resource allocation, i.e. type 0 or type 1
- ★ The Modulation and Coding Scheme (MCS) bits define a pointer to a row within the MCS and RV table (Table 395 within section 33.4). The set of 5 bits provide a range from 0 to 31, i.e. equal to the number of rows within the MCS and RV table. The MCS and RV table specifies the modulation scheme to be used for the PUSCH. It also specifies the Transport Block Size (TBS) index and the RV
- ★ The New Data Indicator (NDI) informs the UE of whether the uplink resource allocation is for new data or a retransmission. Toggling the NDI from 0 to 1, or from 1 to 0, indicates that the resource allocation is for new data. Leaving the NDI with its previous value indicates that the resource allocation is for a retransmission
- ★ The Precoding Information and Number of Layers field has a length of 3 bits when 2 antenna ports are used and a length of 6 bits when 4 antenna ports are used. This field defines a pointer to a row within a look-up table which specifies the number of layers and a Transmitted Precoding Matrix Indicator (TPMI). The TPMI defines a pointer to a row in another look-up table which specifies the precoding weights for uplink spatial multiplexing. The Precoding Information and Number of Layers look-up table for 2 antenna ports is presented in Table 126

1 Codeword		
Value	Number of Layers	TPMI
0	1	0
1	1	1
2	1	2
3	1	3
4	1	4
5	1	5
6	Reserved	
7		

2 Codewords		
Value	Number of Layers	TPMI
0	2	0
1	Reserved	
2		
3		
4		
5		
6		
7		

Table 126 – Look-up table for Precoding Information and Number of Layers (2 antenna ports)

- ★ A single bit of padding is added to the content of DCI format 4 if the size of its payload equals the payload size of DCI formats 1, 2, 2A, 2B or 2C when they are used in the same serving cell and for the same downlink transmission mode
- ★ Table 127 presents an example of the total number of bits within DCI format 4 when applied to FDD. This example assumes the use of a release 10 device configured with a single RF carrier and 2 uplink antenna ports. It also assumes a single bit CSI Request. A 16 bit CRC is added to the payload after padding has been included. The complete set of bits is channel coded using rate 1/3 convolutional coding before rate matching is applied to ensure that the resultant number of bits equals the capacity of the PDCCH

	1.4 MHz	3 MHz	5 MHz	10 MHz	15 MHz	20 MHz
Total Bits prior to Padding	30	31	34	36	37	38
Padding Bits	0	1	0	0	0	0
Total Bits	**30**	**32**	**34**	**36**	**37**	**38**
After inclusion of 16 bit CRC	46	48	50	52	53	54
After Rate 1/3 Channel Coding	138	144	150	156	159	162

Table 127 – Example total number of bits for DCI Format 4 (FDD with 2 antenna ports)

★ Table 128 presents an example of the total number of bits within DCI format 4 when applied to TDD. This example also assumes the use of a release 10 device configured with a single RF carrier and 2 uplink antenna ports. The total number of bits prior to padding is 2 bits greater than for FDD due to the inclusion of either the Uplink Index or Downlink Assignment Index

	1.4 MHz	3 MHz	5 MHz	10 MHz	15 MHz	20 MHz
Total Bits (prior to padding)	32	33	36	38	39	40
Padding	0	0	0	0	0	0
Total Bits	**32**	**33**	**36**	**38**	**39**	**40**
After inclusion of 16 bit CRC	48	49	52	54	55	56
After Rate 1/3 Channel Coding	144	147	156	162	165	168

Table 128 – Example total number of bits for DCI Format 4 (TDD with 2 antenna ports)

10 SYSTEM INFORMATION

10.1 STRUCTURE

★ UE read system information in RRC Idle mode to acquire the parameters necessary to complete cell selection and cell reselection. System information also provides the parameters necessary to access the network and detect paging messages

★ Paging messages can be used to inform the UE that there has been a change to the system information and so trigger the UE to re-acquire the system information. Paging messages can also be used to alert the UE of Earthquake and Tsunami Warning System (ETWS) notifications which are broadcast on the system information

★ System information is broadcast using a Master Information Block (MIB) and a series of System Information Blocks (SIB). Figure 124 illustrates the logical, transport and physical channels used to transfer the MIB and SIB. The MIB is the only system information transferred using the BCH and PBCH. SIB are transferred using the DL-SCH and PDSCH. SIB1 has its own RRC message whereas SIB2 to SIB13 are encapsulated within the more general System Information RRC message

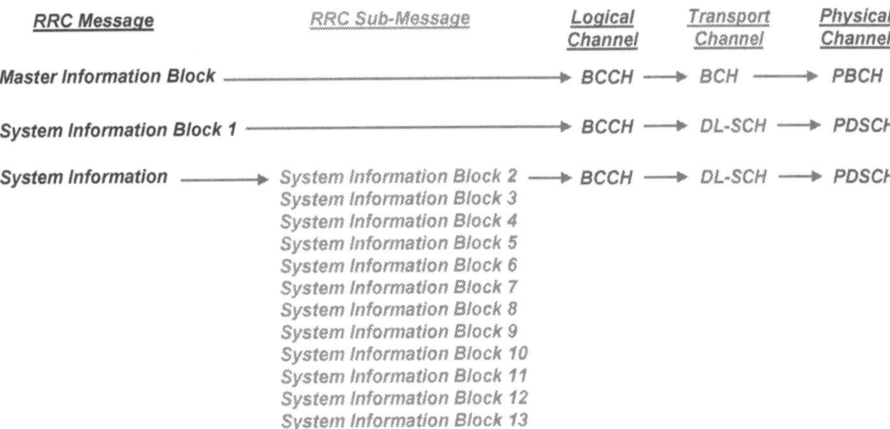

Figure 124 – System Information messages mapped onto Logical, Transport and Physical Channels

★ The set of Resource Elements used by the MIB on the PBCH is standardised by 3GPP so does not require any additional signalling, i.e. the PBCH occupies the central 72 subcarriers within the first 4 OFDMA symbols of the second slot of a radio frame. The same allocation is made for both FDD and TDD

★ The set of Resource Elements used by the SIB on the PDSCH is not standardised by 3GPP so requires additional signalling to inform the UE of where to look. The PDCCH is used to provide this additional signalling

★ The PDCCH includes a CRC which is scrambled by the System Information RNTI (SI-RNTI) if it includes resource allocation information relevant to the SIB. The SI-RNTI has been standardised to have a single fixed value of FFFF

★ Downlink Control Information (DCI) formats 1A and 1C can be used to signal PDSCH resource allocations for the SIB, i.e. DCI formats 1A and 1C can have their CRC scrambled by the SI-RNTI

★ The eNode B is responsible for scheduling the Resource Blocks used to transfer the SIB. The MIB and SIB1 are broadcast at a rate which is specified by 3GPP. The rate at which SIB2 to SIB11 are broadcast is implementation dependent

★ UE start by reading the MIB and this provides sufficient information to read SIB1. SIB1 provides scheduling information for the remaining SIB. This hierarchy of reading system information is shown in Figure 125

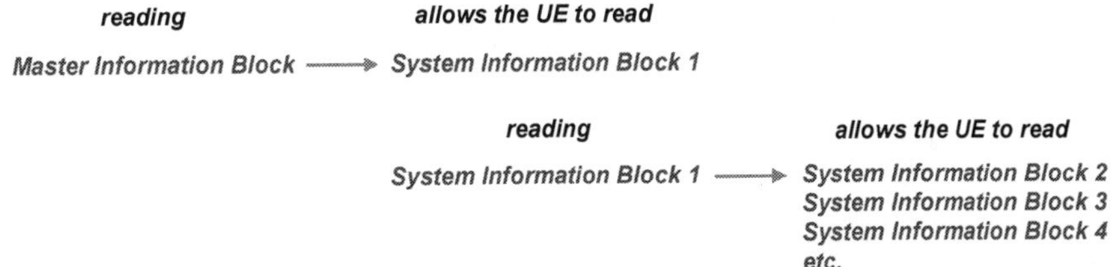

Figure 125 – Hierarchy of reading System Information

★ A summary of the information included within the MIB and each of the SIB is provided in Table 129. More detailed information is provided within the following sections

System Information	3GPP Release	Content
Master Information Block	8	Downlink channel bandwidth, PHICH configuration, SFN
System Information Block 1	8	PLMN Id, tracking area code, cell selection parameters, frequency band, cell barring, scheduling information for other SIB
System Information Block 2	8	Access class barring, RACH, BCCH, PCCH, PRACH, PDSCH, PUSCH, PUCCH parameters, UE timers and constants, uplink carrier frequency
System Information Block 3	8	Cell reselection parameters
System Information Block 4	8	Intra-frequency neighbouring cell information for cell reselection
System Information Block 5	8	Inter-frequency neighbouring cell information for cell reselection
System Information Block 6	8	UMTS neighbouring cell information for cell reselection
System Information Block 7	8	GERAN neighbouring cell information for cell reselection
System Information Block 8	8	CDMA2000 neighbouring cell information for cell reselection
System Information Block 9	8	Home eNode B name
System Information Block 10	8	Earthquake and Tsunami Warning System primary notification
System Information Block 11	8	Earthquake and Tsunami Warning System secondary notification
System Information Block 12	9	Commercial Mobile Alert Service (CMAS) notification
System Information Block 13	9	MBMS Single Frequency Network (MBSFN) configuration information

Table 129 – Summary of System Information content

★ The release 9 version of the 3GPP specifications introduced SIB12 for Commercial Mobile Alert Service (CMAS) parameters. This required an extension to the paging message to alert UE to when they should read CMAS messages from SIB12

★ The release 9 version of the specifications also introduced SIB13 for the support of Multimedia Broadcast Multicast Services on a Single Frequency Network (MBSFN). SIB13 includes information which allows UE to subsequently acquire MBMS control information associated with one or more MBSFN areas

★ The MIB and all SIB are transferred using transparent mode RLC

★ 3GPP References: TS 36.331

10.2 MASTER INFORMATION BLOCK

★ The Master Information Block (MIB) can be fully decoded from the PBCH during any 10 ms radio frame. Section 7.1 explains that the physical layer adds a large quantity of redundancy to the MIB to help ensure reliable reception by the UE. The content of the MIB is shown in Table 130

Information Elements	
Downlink channel bandwidth	
PHICH configuration	PHICH duration
	PHICH resource
System Frame Number (SFN)	

Table 130 – Content of Master Information Block (MIB)

★ The content of the MIB is the same for both FDD and TDD

★ The downlink channel bandwidth is signalled in terms of the number of Resource Blocks, i.e. 6, 15, 25, 50, 75 or 100. This information allows the UE to proceed to decode the PCFICH and PDCCH, i.e. the UE can then identify Resource Blocks allocated to the System Information Blocks (SIB)

★ The PHICH configuration defines:

- o the PHICH duration which can be either 'normal' or 'extended'
- o the PHICH resource (PHICH Group Scaling Factor) which can be 1/6, 1/2, 1, 2

Section 6.3 explains the use of these PHICH parameters

★ The SFN field defines the 8 most significant bits of the SFN. The SFN requires 10 bits in total to provide a range from 0 to 1023. The 2 least significant bits are deduced from the 4 radio frame cycle used to transmit the complete BCH Transmission Time Interval (TTI), i.e. the first radio frame within the cycle corresponds to 00, the second to 01, the third to 10 and the forth to 11

10.3 SYSTEM INFORMATION BLOCK 1

★ System Information Block 1 (SIB1) can be decoded from the PDSCH once every second 10 ms radio frame. A SIB1 transmission is sent whenever SFN mod 8 = 0. Each transmission of SIB1 is followed by 3 retransmissions when SFN mod 2 = 0. This transmission timing for SIB1 is illustrated in Figure 126

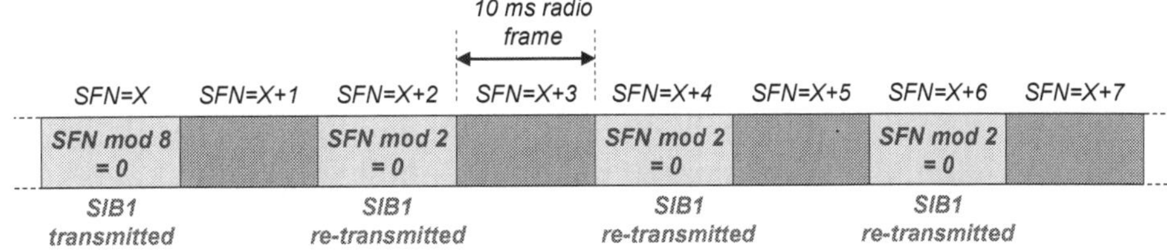

Figure 126 – Timing for the transmission and retransmission of SIB1

★ The content of SIB1 is shown in Table 131

★ The TDD Configuration information is only included by TDD networks. The IMS Emergency Support and Cell Selection Information at the end of SIB1 were introduced within the release 9 version of the specifications

Information Elements		
Cell Access Information	PLMN Identity List (1 to 6 instances)	PLMN Identity
		Cell Reserved for Operator Use
	Tracking Area Code	
	Cell Identity	
	Cell Barred	
	Intra-Frequency Cell Reselection Allowed	
	CSG Indication	
	CSG Identity	
Cell Selection Information	Qrxlevmin	
	Qrxlevmin Offset	
Pmax		
Frequency Band Indicator		
Scheduling Information List (1 to 32 instances)	SI Periodicity (8, 16, 32, 64, 128, 256, 512 radio frames)	
	SIB Mapping (1 to 32 instances)	SIB Type
TDD Configuration	Subframe Assignment (uplink-downlink configuration)	
	Special Subframe Pattern (special subframe configuration)	
System Information Window Length (1, 2, 5, 10, 15, 20, 40 ms)		
System Information Value Tag		
IMS Emergency Support (rel. 9)		
Release 9 Extension	Cell Selection Information	Qqualmin
		Qqualmin Offset

Table 131 – Content of System Information Block 1 (SIB1)

★ The PLMN identity list can specify up to 6 PLMN identities, i.e. a single cell can be shared between PLMN. The first PLMN within the list is the Primary PLMN. A flag is included for each PLMN to indicate whether or not the cell is reserved for operator use

★ The Tracking Area Code (TAC) is applicable to all PLMN within the PLMN identity list, i.e. all PLMN share a common TAC. The TAC has a range from 0 to 65536. Tracking area planning is described in section 31.9

★ The Cell Identity has a length of 28 bits and encapsulates the eNode B identity. The eNode B identity can have a length of either 20 or 28 bits, allowing either 256 cells or 1 cell per eNode B respectively. The cell identity is applicable to all PLMN within the PLMN identity list but can be concatenated with the PLMN identity to generate a unique global cell identity. The cell and eNode B identities are presented in section 31.8

★ The Cell Barred flag is included to inform the UE of whether or not the UE is allowed to select or reselect the cell in RRC Idle mode. It is applicable to all PLMN within the PLMN identity list

★ The Intra-Frequency Cell Reselection Allowed flag is used to indicate whether or not the UE is allowed to reselect other cells on the same RF carrier when the cell is barred

★ If the Closed Subscriber Group (CSG) indication is set to TRUE then the UE is only allowed to access the cell if the subsequent CSG Identity matches an entry within the UE's own list of allowed CSG. The CSG identity is applicable to the primary PLMN

★ The minimum RSRP requirement for cell selection is defined in terms of Qrxlevmin. The actual value of Qrxlevmin = (signalled value × 2) + 1

★ Qrxlevmin Offset is subtracted from Qrxlevmin when the UE is camped on a Visited PLMN and cell selection is triggered due to a periodic search for a higher priority PLMN. The offset reduces the value of Qrxlevmin so allows cell selection to become easier. The actual value of the offset = (signalled value × 2)

★ Pmax defines the maximum allowed uplink transmit power for the cell. This also impacts the cell selection criteria if the UE transmit power capability is less than the maximum allowed for the cell

★ The operating band is specified by the Frequency Band Indicator because the downlink channel upon which the UE is receiving the system information does not uniquely identify the operating band, i.e. some operating bands overlap and have different duplex spacings

★ SIB1 includes scheduling information for SIB2 to SIB13. Information is provided in terms of a System Information Periodicity and a System Information Window. The Periodicity is specific to groups of the SIB whereas the Window is common to all SIB

★ SIB2 is not explicitly signalled within the scheduling information list but is always associated with the first of the signalled Periodicity values

★ Use of the Periodicity and Window values is illustrated in Figure 127

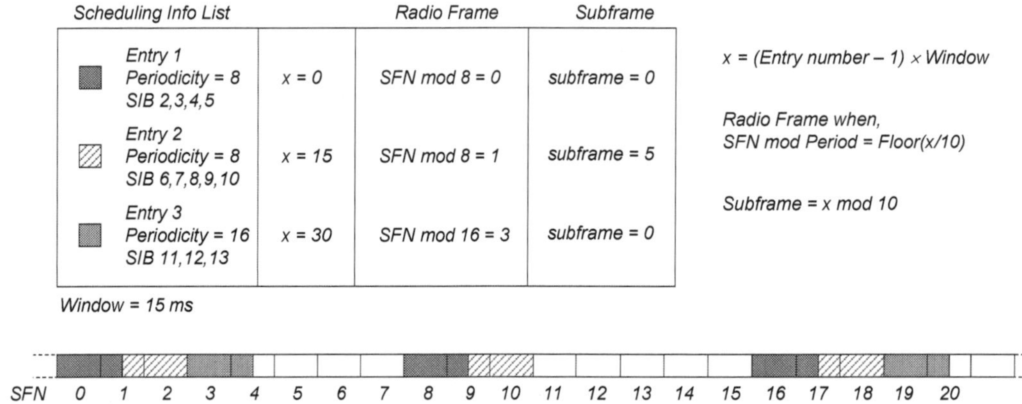

Figure 127 – Periodicity and Window used to schedule SIB2 to SIB13

★ The TDD Configuration specifies the Subframe Assignment and the Special Subframe Pattern. The Subframe Assignment represents the uplink-downlink subframe configuration, i.e. the number and pattern of subframes allocated to the uplink and downlink. The Special Subframe Pattern represents the special subframe configuration, i.e. the duration of the DwPTS, guard period and UpPTS. These parameters are described in section 3.2

★ A System Information Value Tag is included within SIB1 to provide an indication of whether or not the content of SIB2 to SIB9 or SIB 13 has changed and needs to be re-acquired. The value tag is not applicable to the MIB, SIB1, SIB10, SIB11 nor SIB12

★ The IMS Emergency Support flag was introduced within the release 9 version of the 3GPP specifications. It is used to indicate whether or not the cell supports IMS emergency bearer services for UE in Limited Service Mode (LSM). UE are in LSM when they are unable to authenticate with the MME

★ Qqualmin and Qqualmin Offset were introduced within the release 9 version of the 3GPP specifications. The release 8 version of the specifications defines its cell selection criteria based upon RSRP only. The release 9, and newer versions of the specifications use both RSRP and RSRQ as criteria for cell selection. The actual value equals the signalled value for both Qqualmin and Qqualmin Offset. The cell selection procedure is described in section 23.2

10.4 SYSTEM INFORMATION BLOCK 2

★ The first part of the content of SIB2 is shown in Table 132

Information Elements		
AC Barring for Emergency Calls		
AC Barring for MO Signalling	Barring Factor (0 to 0.95, step 0.05)	
	Barring Time (4, 8, 16, 32, 64, 128, 256, 512 seconds)	
	Barring for Special AC	
AC Barring for MO Data	Barring Factor (0 to 0.95, step 0.05)	
	Barring Time (4, 8, 16, 32, 64, 128, 256, 512 seconds)	
	Barring for Special AC	
Radio Resource Configuration	RACH Configuration	Number of RA Preambles
		Size of RA Preamble Group A
		Message Size Group A
		Message Power Offset Group B
		Power Ramping Step
		Preamble Initial Received Target Power
		Preamble Trans Max
		RA Response Window Size
		MAC Contention Resolution Timer
		Max HARQ Msg3Tx
	BCCH Configuration	Modification Period Coefficient

Table 132 – Content of System Information Block 2 (SIB2) - Part 1

★ Access Class (AC) barring for emergency calls determines whether or not UE with AC 0 to 9 should treat the cell as barred for emergency calls. The same rule is applicable to UE without a USIM (AC information is stored on the USIM). UE with AC 11 to 15 are able to make emergency calls unless the cell is both AC class barred for emergency calls and AC class barred for mobile originating data calls

★ Access Class (AC) barring for Mobile Originated (MO) signalling determines whether or not UE should treat the cell as barred when initiating a MO signalling connection. UE with AC 0 to 9 generate a random number between 0 and 1. If it is less than the Barring Factor then the cell is treated as not barred. UE with AC 11 to 15 check the Barring for Special AC to determine whether or not it is necessary to generate a random number with which to compare against the Barring Factor

★ If the cell is to be treated as barred for MO signalling connections then the Barring Time is used to define a penalty time during which the UE is not allowed to re-attempt access. The guard time is defined as T305 = (0.7 + 0.6 × RAND) × Barring Time, where RAND is a random number between 0 and 1

★ Similar AC barring information is provided for UE establishing MO data connections. In this case, if the cell is to be treated as barred, the penalty time is defined as T303 = (0.7 + 0.6 × RAND) × Barring Time, where RAND is a random number between 0 and 1

★ The RACH configuration is specified in terms of:

- total number of random access preambles (4 to 64, step 4)
- number of random access preambles within Group A (4 to 60, step 4)
- message size threshold for selecting preamble Group A (56, 144, 208, 256 bits)
- power offset for selecting preamble Group B (-∞, 0, 5, 8, 10, 12, 15, 18 dB)
- power ramping step size (0, 2, 4, 6 dB)
- preamble initial received target power (-120 to -90, step 2 dBm)
- maximum number of preamble transmissions (3, 4, 5, 6, 7, 8, 10, 20, 50, 100, 200)
- random access response window size (2, 3, 4, 5, 6, 7, 8, 10 subframes)
- MAC contention resolution timer (8, 16, 24, 32, 40, 48, 56, 64 subframes)
- maximum number of HARQ transmissions (1 to 8, step 1)

Use of these RACH parameters is described in section 24.1

★ The BCCH configuration is specified in terms of a modification period coefficient (2, 4, 8, 16). This coefficient is multiplied by the default DRX cycle (also broadcast within SIB2) to generate the BCCH modification period

★ Changes to the content of the BCCH can only occur at modification period boundaries (with the exception of ETWS information). Boundaries occur at SFN for which SFN mod m = 0, where m is the number of radio frames within the modification period

★ UE are informed that system information content is going to change during the preceding modification period. This is done using the modification flag within an RRC: Paging message. Figure 128 illustrates the concept of BCCH modification

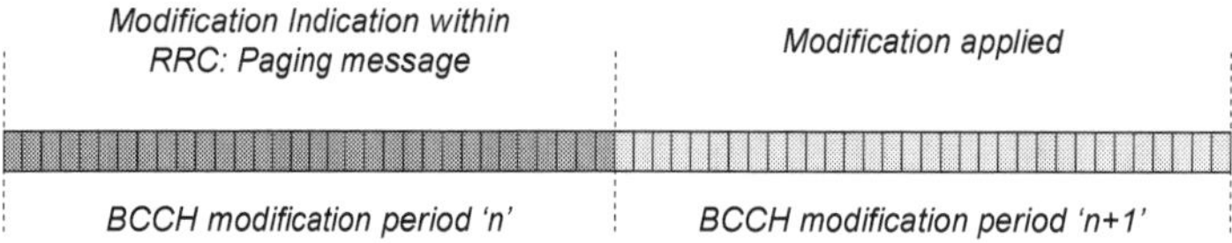

BCCH modification period = Default DRX Cyle Length × Modification Period Coefficient

Figure 128 – BCCH modification period

★ The second part of the content of SIB2 is shown in Table 133

Information Elements			
Radio Resource Configuration (continued)	PCCH Configuration	Default DRX Cycle	
		nB	
	PRACH Configuration	Root Sequence Index	
		PRACH Configuration Index	
		High Speed Flag	
		Zero Correlation Zone Configuration	
		PRACH Frequency Offset	
	PDSCH Configuration	Reference Signal Power	
		p-b	
	PUSCH Configuration	n-SB	
		Hopping Mode	
		PUSCH Hopping Offset	
		Enable 64QAM	
		Uplink Reference Signals	Group Hopping Enabled
			Group Assignment
			Sequence Hopping Enabled
			Cyclic Shift
	PUCCH Configuration	Delta PUCCH Shift	
		nRB CQI	
		nCS AN	
		nlPUCCH AN	

Table 133 – Content of System Information Block 2 (SIB2) - Part 2

★ The PCCH configuration is specified in terms of the Default DRX Cycle (32, 64, 128, 256 radio frames) and the 'nB' variable (4T, 2T, T, T/2, T/4, T/8, T/16, T/32). Use of these parameters is described in section 23.7

★ The PRACH configuration is specified in terms of:

- Random access root sequence index (0 to 837)
- PRACH configuration index (0 to 63)
- High speed flag (true/false)
- Zero correlation zone configuration (0 to 15)
- PRACH frequency offset (0 to 94 Resource Blocks)

Use of these PRACH parameters is described in section 24.1

- ★ The Reference Signal Power is specified as part of the PDSCH configuration. This specifies the Energy Per Resource Element (EPRE) for the cell specific Reference Signal using a range from -60 to 50 dBm. The Reference Signal Power is used to calculate the power difference between the PDSCH and cell specific Reference Signal
- ★ The 'p-b' variable (P_B) is also used as an input when calculating the power difference between the PDSCH and the cell specific Reference Signal. It has a range from 0 to 3. The ratio between the PDSCH and cell specific Reference Signal powers can depend upon whether or not the ratio is being calculated for an OFDMA symbol which includes Reference Signal Resource Elements. The P_B variable is used to quantify this dependence
- ★ The PUSCH configuration is specified in terms of:
    - N_{sb} variable (1 to 4)
    - Hopping mode (inter-subframe, intra and inter-subframe)
    - Hopping offset (0 to 98)
    - Enable 64QAM (true/false)
    - Uplink Reference Signal
        - group hopping enabled (true/false)
        - group assignment (0 to 29)
        - frequency hopping enabled (true/false)
        - cyclic shift (0 to 7)
- ★ The PUCCH configuration is specified in terms of:
    - Δ_{shift}^{PUCCH} variable (1 to 3)
    - $N_{CS}^{(1)}$ variable (0 to 7)
    - $N_{RB}^{(2)}$ variable (0 to 98)
    - $N_{PUCCH}^{(1)}$ variable (0 to 2047)

  Use of these PUCCH parameters is described in section 14.2
- ★ The third part of the content of SIB2 is shown in Table 134
- ★ The uplink Sounding Reference Signal (SRS) is configured in terms of its bandwidth. The SRS bandwidth configuration is signalled using a value between 0 and 7. This defines a pointer to a row within a look-up table specified by 3GPP TS 36.211. The look-up table to which the pointer is applied depends upon the uplink channel bandwidth.
- ★ The uplink SRS is also configured in terms of subframes. The allocated subframes are signalled using a value between 0 and 15. This defines a pointer to a row within a look-up table specified by 3GPP TS 36.211
- ★ The Ack/Nack SRS Simultaneous Transmission parameter defines whether or not UE transmit both ACK/NACK on the PUCCH and SRS during a single subframe
- ★ The Max UpPTS parameter is only applicable to TDD. It determines whether or not the bandwidth of the Sounding Reference Signal is reconfigured when transmitted within the UpPTS field of a TDD special subframe
- ★ The signalled pO Nominal PUSCH corresponds to the $P_{O\_NOMINAL\_PUSCH}(1)$ parameter. This parameter is only applicable to non-persistent scheduling and impacts the calculation of the PUSCH transmit power. It has a range from -126 to 24 dBm
- ★ The alpha parameter also impacts the calculation of the PUSCH transmit power. It is used to scale the contribution of the path loss. Alpha can have values of 0, 0.4, 0.5, 0.6, 0.7, 0.8, 0.9 and 1.0. The signalled value of alpha is not applied when the PUSCH is being used to respond to a random access response grant. Power control is described in section 24.4
- ★ The signalled pO Nominal PUCCH corresponds to the $P_{O\_NOMINAL\_PUCCH}$ parameter. This parameter impacts the calculation of the PUCCH transmit power. It has a range from -127 to -96 dBm
- ★ The signalled DeltaF List values correspond to the $\Delta_{F\_PUCCH}(F)$ parameters, where F represents the PUCCH format. These parameters impact the calculation of the PUCCH transmit power. Each value corresponds to a power offset relative to the power used for PUCCH format 1a. The release 10 version of the 3GPP specifications added two additional power offsets (shown slightly lower down Table 134). These additional power offsets apply to PUCCH format 3 and PUCCH format 1b with Channel Selection
- ★ The signalled Delta Preamble Msg3 corresponds to the $\Delta_{PREAMBLE\_Msg3}$ parameter. The actual value of the parameter equals the signalled value × 2. The signalled value can have values between -1 and 6 dB. $\Delta_{PREAMBLE\_Msg3}$ impacts the transmit power of the PUSCH when responding to a random access response grant
- ★ The length of the uplink cyclic prefix is signalled as len1 (normal) or len2 (extended)
- ★ T300 is used during the RRC connection establishment procedure described in section 26.1. It defines the period of time during which the UE waits for a response to the RRC Connection Request message

<table>
<tr><th colspan="4">Information Elements</th></tr>
<tr><td rowspan="17">Radio Resource Configuration (continued)</td><td rowspan="4">Uplink Sounding Reference Signal Configuration</td><td rowspan="4">setup</td><td>SRS Bandwidth Configuration</td></tr>
<tr><td>SRS Subframe Configuration</td></tr>
<tr><td>Ack/Nack SRS Simul. Trans.</td></tr>
<tr><td>Max UpPTS (TDD)</td></tr>
<tr><td rowspan="9">Uplink Power Control</td><td colspan="2">pO Nominal PUSCH</td></tr>
<tr><td colspan="2">Alpha</td></tr>
<tr><td colspan="2">pO Nominal PUCCH</td></tr>
<tr><td rowspan="5">DeltaF List (power offsets in dB)</td><td>PUCCH Format 1 (-2, 0, 2)</td></tr>
<tr><td>PUCCH Format 1b (1, 3, 5)</td></tr>
<tr><td>PUCCH Format 2 (-2, 0, 1, 2)</td></tr>
<tr><td>PUCCH Format 2a (-2, 0, 2)</td></tr>
<tr><td>PUCCH Format 2b (-2, 0, 2)</td></tr>
<tr><td colspan="2">Delta Preamble Msg3</td></tr>
<tr><td colspan="3">Uplink Cyclic Prefix Length</td></tr>
<tr><td rowspan="2">Uplink Power Control Extension (rel. 10)</td><td colspan="2">DeltaF PUCCH Format 3 (-1, 0, 1, 2, 3, 4, 5, 6)</td></tr>
<tr><td colspan="2">DeltaF PUCCH Format 1b CS (1, 2)</td></tr>
<tr><td rowspan="6">UE Timers and Constants</td><td colspan="3">T300 (100, 200, 300, 400, 600, 1000, 1500, 2000 ms)</td></tr>
<tr><td colspan="3">T301 (100, 200, 300, 400, 600, 1000, 1500, 2000 ms)</td></tr>
<tr><td colspan="3">T310 (0, 50, 100, 200, 500, 1000, 2000 ms)</td></tr>
<tr><td colspan="3">N310 (1, 2, 3, 4, 6, 8, 10, 20)</td></tr>
<tr><td colspan="3">T311 (1, 3, 5, 10, 15, 20, 30 seconds)</td></tr>
<tr><td colspan="3">N311 (1, 2, 3, 4, 5, 6, 8, 10)</td></tr>
<tr><td rowspan="3">Frequency Information</td><td colspan="3">Uplink Carrier Frequency</td></tr>
<tr><td colspan="3">Uplink Bandwidth</td></tr>
<tr><td colspan="3">Additional Spectrum Emission</td></tr>
<tr><td rowspan="3">MBSFN Subframe Configuration List (1 to 8 instances)</td><td colspan="3">Radio Frame Allocation Period</td></tr>
<tr><td colspan="3">Radio Frame Allocation Offset</td></tr>
<tr><td colspan="3">Subframe Allocation – one frame or four frames</td></tr>
<tr><td colspan="4">Time Alignment Timer Common</td></tr>
<tr><td colspan="2" rowspan="3">SSAC Barring for MMTEL Voice (rel. 9)</td><td colspan="2">Barring Factor (0 to 0.95, step 0.05)</td></tr>
<tr><td colspan="2">Barring Time (4, 8, 16, 32, 64, 128, 256, 512 seconds)</td></tr>
<tr><td colspan="2">Barring for Special AC</td></tr>
<tr><td colspan="2" rowspan="3">SSAC Barring for MMTEL Video (rel. 9)</td><td colspan="2">Barring Factor (0 to 0.95, step 0.05)</td></tr>
<tr><td colspan="2">Barring Time (4, 8, 16, 32, 64, 128, 256, 512 seconds)</td></tr>
<tr><td colspan="2">Barring for Special AC</td></tr>
<tr><td colspan="2" rowspan="3">AC Barring for CS Fallback (rel. 10)</td><td colspan="2">Barring Factor (0 to 0.95, step 0.05)</td></tr>
<tr><td colspan="2">Barring Time (4, 8, 16, 32, 64, 128, 256, 512 seconds)</td></tr>
<tr><td colspan="2">Barring for Special AC</td></tr>
</table>

Table 134 – Content of System Information Block 2 (SIB2) - Part 3

★ T301 is started after sending the RRC Connection Re-establishment Request message. If T301 expires prior to receiving an RRC Connection Re-establishment or RRC Connection Re-establishment Reject message, or if the selected cell becomes unsuitable, the UE moves to RRC Idle mode

★ T310 is started after receiving N310 consecutive out-of-sync indications from the lower layers. If T310 expires prior to receiving N311 consecutive in-sync indications then the UE either moves to RRC Idle mode or initiates connection re-establishment

★ T311 is started after initiating the connection re-establishment procedure. If T311 expires before the UE finds a suitable cell then the UE moves to RRC Idle mode

★ The uplink carrier frequency can be specified in terms of the EARFCN but is optional. The default duplex spacing is assumed if the uplink carrier frequency is not specified

- ★ The uplink bandwidth can be specified in terms of Resource Blocks but is optional. The uplink bandwidth is assumed to be equal to the downlink bandwidth (specified in the MIB) if the uplink bandwidth is not specified
- ★ The additional spectrum emission figure defines a value between 1 and 32. This figure refers to a set of requirements within 3GPP TS 36.101. This allows the spectrum emission limits to be configured according to local requirements
- ★ The optional MBMS Single Frequency Network (MBSFN) subframe configuration list specifies the set of downlink subframes reserved for MBSFN. Radio frames which include MBSFN subframes occur when SFN mod Radio Frame Allocation Period = Radio Frame Allocation Offset. The MBSFN subframes themselves are signalled using the subframe allocation which can be a bit string for either one frame or four frames
- ★ MBSFN subframe information can be included within SIB2 to inform release 8 UE that reception of the cell specific Reference Signal and the PDSCH should not be attempted during those subframes. The release 9 version of the specifications introduced SIB13 to provide more detailed scheduling information for release 9, and newer UE supporting MBMS
- ★ The time alignment timer is used to control how long a UE considers itself uplink time aligned. It can be allocated values of 500, 750, 1280, 1920, 2560, 5120, 10240 and infinity subframes. Uplink timing is controlled using the timing advance procedure described in 24.2
- ★ The release 9 version of the 3GPP specifications introduced Service Specific Access Class (SSAC) barring for Multimedia Telephony (MMTEL) voice and video calls. The parameters in SIB2 are used to determine whether or not UE should treat the cell as barred when initiating MMTEL voice and data calls.
    - o UE with AC 0 to 9 generate a random number between 0 and 1. If it is less than the Barring Factor then the cell is treated as not barred. UE with AC 11 to 15 check the Barring for Special AC to determine whether or not it is necessary to generate a random number with which to compare against the Barring Factor
    - o If the cell is to be treated as barred then the Barring Time is used to define a penalty time during which the UE is not allowed to re-attempt access. The penalty time is defined as $(0.7 + 0.6 \times \text{RAND}) \times$ Barring Time, where RAND is a random number between 0 and 1
- ★ The release 10 version of the 3GPP specifications introduced a similar set of parameters for UE establishing mobile originating CS Fallback connections

10.5 SYSTEM INFORMATION BLOCK 3

★ SIB3 provides parameters for cell reselection. The content of SIB3 is shown in Table 135

Information Elements		
Cell Reselection Common Information	Qhyst	
	Mobility State Parameters	Tevaluation
		T Hyst Normal
		N Cell Change Medium
		N Cell Change High
	Qhyst Scaling Factors	Scaling Factor Medium
		Scaling Factor High
Cell Reselection Serving Frequency Information	Snon-intrasearch	
	Thresh Serving Low	
	Cell Reselection Priority	
Intra Frequency Cell Reselection Information	Qrxlevmin	
	Pmax	
	Sintrasearch	
	Allowed Measurement Bandwidth	
	Presence of Antenna Port 1	
	Neighbour Cell Configuration	
	Treselection EUTRA	
	Treselection EUTRA Scaling Factors	Scaling Factor Medium
		Scaling Factor High
Release 9 Extension	SintraSearchP	
	SintraSearchQ	
	Snon-intrasearchP	
	Snon-intrasearchQ	
	Qqualmin	
	Thresh Serving Low Q	

Table 135 – Content of System Information Block 3 (SIB3)

★ Qhyst represents the hysteresis added to the serving cell RSRP measurement during cell reselection ranking. It can be configured with a value between 0 and 24 dB. Cell reselection is described in section 23.3

★ The mobility state parameters are used by the UE to categorise itself as having normal, medium or high mobility

- Tevaluation corresponds to T_{CRmax} within TS 36.304. It is used to define the time interval during which the number of cell reselections is counted. It can have values of 30, 60, 120, 180 and 240 seconds

- T Hyst Normal corresponds to $T_{CRmaxHyst}$ within TS 36.304. A UE categorises itself as having normal mobility if neither the criteria for medium nor high mobility are achieved during the time interval defined by $T_{CRmaxHyst}$. It can have values of 30, 60, 120, 180 and 240 seconds

- N Cell Change Medium corresponds to N_{CR_M} within TS 36.304. It defines the minimum number of cell reselections within the time interval T_{CRmax} for a UE to categorise itself as having medium mobility. It can have values between 1 and 16

- N Cell Change High corresponds to N_{CR_H} within TS 36.304. It defines the minimum number of cell reselections within the time interval T_{CRmax} for a UE to categorise itself as having high mobility. It can have values between 1 and 16

★ The Qhyst Scaling Factors are added to Qhyst if either medium or high mobility is detected. The scaling factors are negative so tend to reduce the ranking of the serving cell and allow reselection to occur more readily. Both the medium and high scaling factors can be allocated values of -6, -4, -2 and 0 dB

★ Snon-intrasearch is used to trigger inter-frequency and inter-RAT measurements for cell reselection when the target inter-frequency layer has an equal or lower priority, and when the target inter-RAT layer has a lower priority. Snon-intrasearch defines a threshold based upon RSRP. The signalled value can range from 0 to 31, but the actual value = signalled value × 2 dB

- ★ Thresh Serving Low represents an RSRP based threshold, which the serving cell must fall below before cell reselection to a lower priority layer is allowed. The signalled value can range from 0 to 31, but the actual value = signalled value × 2 dB
- ★ The Cell Reselection Priority defines the absolute priority of the frequency layer to which the serving cell belongs. It can be allocated a value between 0 and 7, where 0 represents the lowest priority
- ★ Qrxlevmin defines the minimum RSRP for cell reselection. This differs from the Qrxlevmin in SIB1 which is used for cell selection. The signalled value of Qrxlevmin ranges from -70 to -22, but the actual value = signalled value × 2 dBm
- ★ Pmax defines the maximum allowed uplink transmit power for intra-frequency neighbouring cells. This impacts the cell reselection criteria if the UE transmit power capability is less than the maximum allowed for the cell. It is in contrast to Pmax for cell selection which is included in SIB1
- ★ Sintrasearch is an RSRP based threshold used to trigger intra-frequency measurements for cell reselection. The signalled value can range from 0 to 31, but the actual value = signalled value × 2 dB
- ★ The allowed measurement bandwidth defines the maximum allowed measurement bandwidth. It is defined in terms of the Resource Blocks associated with a specific channel bandwidth, i.e. 6, 15, 25, 50, 75 or 100 Resource Blocks. If this parameter is excluded, the UE assumes a value equal to the downlink channel bandwidth specified within the MIB. This parameter is useful when neighbouring cells have different bandwidths to the serving cell
- ★ The presence of antenna port 1 parameter informs the UE of whether or not it can assume that all neighbouring cells have at least two cell specific antenna ports, i.e. at least ports 0 and 1. RSRP measurements can be based upon transmissions from both antenna ports when available
- ★ The neighbour cell configuration provides information regarding the MBSFN subframe allocations in neighbouring cells. In the case of TDD, it is also used to provide information regarding the TDD uplink-downlink subframe configuration in neighbouring cells
- ★ Treselection EUTRA defines the time-to-trigger for cell reselection. It can be allocated a value between 0 and 7 seconds. Its value can be scaled by the associated scaling factors if medium or high mobility is detected. The scaling factors can have values of 0.25, 0.5, 0.75 and 1.0, i.e. they typically decrease the value of Treselection to allow more rapid cell reselections
- ★ The release 9 version of the 3GPP specifications introduces an extension to SIB3 which defines additional parameters for cell reselection
- ★ SintraSearchP overwrites the value of Sintrasearch for 3GPP release 9, and newer UE. Similar to Sintrasearch, the signalled value of SintraSearchP can range from 0 to 31, while the actual value = signalled value × 2 dB
- ★ SintraSearchQ defines an RSRQ based threshold for triggering intra-frequency cell reselection measurements. Prior to 3GPP release 9, intra-frequency cell reselection measurements were only triggered by the SintraSearch threshold based upon RSRP. SintraSearchQ can be signalled using a value between 0 and 31 dB. The actual value equals the signalled value
- ★ Snon-intraSearchP overwrites the value of Snon-intrasearch for 3GPP release 9, and newer UE. The signalled value of Snon-intraSearchP can range from 0 to 31, while the actual value = signalled value × 2 dB
- ★ Snon-intraSearchQ defines an RSRQ based threshold for triggering inter-frequency and inter-RAT cell reselection measurements when the target inter-frequency layer has an equal or lower priority, and when the target inter-RAT layer has a lower priority. Prior to 3GPP release 9, inter-frequency and inter-RAT cell reselection measurements were only triggered by the Snon-intraSearch threshold based upon RSRP. Snon-intraSearchQ can be signalled using a value between 0 and 31, and the actual value equals the signalled value
- ★ Qqualmin defines a minimum RSRQ threshold for cell reselection. 3GPP release 9, and newer UE check both Qrxlevmin and Qqualmin when completing cell reselection. Qqualmin can be signalled using a value between -34 and -3, and the actual value equals the signalled value
- ★ Thresh Serving Low Q represents an RSRQ based threshold, which the serving cell must fall below before cell reselection to a lower priority layer is allowed. The signalled value can range from 0 to 31 dB, and the actual value = signalled value

10.6 SYSTEM INFORMATION BLOCK 4

★ SIB4 provides cell reselection information for intra-frequency neighbours. The content of SIB4 is shown in Table 136

<table>
<tr><th colspan="3">Information Elements</th></tr>
<tr><td rowspan="2">Intra-Frequency Neighbour Cell List
(1 to 16 instances)</td><td rowspan="2">Intra-Frequency Neighbour Cell Information</td><td>Physical layer Cell Identity</td></tr>
<tr><td>Qoffset</td></tr>
<tr><td rowspan="2">Intra-Frequency Black Cell List
(1 to 16 instances)</td><td rowspan="2">Physical Cell Identity Range</td><td>Start</td></tr>
<tr><td>Range</td></tr>
<tr><td rowspan="2">CSG Physical Cell Identity Range</td><td rowspan="2">Physical Cell Identity Range</td><td>Start</td></tr>
<tr><td>Range</td></tr>
</table>

Table 136 – Content of System Information Block 4 (SIB4)

★ All 3 of the main fields within SIB4 are optional. It is not mandatory to specify any intra-frequency neighbours because UE are able to complete cell reselection without a neighbour list

★ RSRP measurement offsets can be specified for up to 16 intra-frequency neighbours. Neighbours are identified by their Physical layer Cell Identity (PCI) while the measurement offset is specified by Qoffset. Qoffset is applicable to cell reselection and has a range from -24 to +24 dB. Negative values make neighbours appear more attractive

★ A maximum of 16 groups of cells can be blacklisted. Blacklisted cells are not considered as candidates for cell reselection. Cells belonging to a specific group have consecutive PCI. The number of cells within a group can range from 1 to 504

★ A single group of PCI can be specified as being reserved for Closed Subscriber Group (CSG) cells, i.e. Home eNode B. This group is defined in terms of a range of consecutive PCI. The number of cells within the group can range from 1 to 504

10.7 SYSTEM INFORMATION BLOCK 5

★ SIB5 provides cell reselection information for inter-frequency neighbours. The content of SIB5 is shown in Table 137

<table>
<tr><th colspan="4">Information Elements</th></tr>
<tr><td rowspan="17">Inter-Frequency Carrier Frequency List
(1 to 8 instances)</td><td colspan="3">Downlink Carrier Frequency</td></tr>
<tr><td colspan="3">Qrxlevmin</td></tr>
<tr><td colspan="3">Pmax</td></tr>
<tr><td colspan="3">Treselection EUTRA</td></tr>
<tr><td colspan="2" rowspan="2">Treselection EUTRA Scaling Factors</td><td>SF Medium</td></tr>
<tr><td>SF High</td></tr>
<tr><td colspan="3">ThreshX-High</td></tr>
<tr><td colspan="3">ThreshX-Low</td></tr>
<tr><td colspan="3">Allowed Measurement Bandwidth</td></tr>
<tr><td colspan="3">Presence Antenna Port 1</td></tr>
<tr><td colspan="3">Priority</td></tr>
<tr><td colspan="3">Neighbour Cell Configuration</td></tr>
<tr><td colspan="3">QoffsetFreq</td></tr>
<tr><td rowspan="2">Inter-Frequency Neighbour Cell List
(1 to 16 instances)</td><td rowspan="2">Inter-Frequency Neighbour Cell Information</td><td>Physical Cell Identity</td></tr>
<tr><td>Qoffset</td></tr>
<tr><td rowspan="2">Inter-Frequency Black Cell List
(1 to 16 instances)</td><td rowspan="2">Physical Cell Identity Range</td><td>Start</td></tr>
<tr><td>Range</td></tr>
<tr><td rowspan="3">Release 9 Extension</td><td colspan="3">Qqualmin</td></tr>
<tr><td colspan="3">ThreshX-HighQ</td></tr>
<tr><td colspan="3">ThreshX-LowQ</td></tr>
</table>

Table 137 – Content of System Information Block 5 (SIB5)

★ Inter-frequency neighbouring cells use different center frequencies relative to the cell upon which the UE is camped, i.e. they have different EARFCN

★ Information can be specified for up to 8 LTE frequencies. It is not mandatory to specify individual cells on each frequency but the RF carrier itself must be specified. Specifying the RF carrier provides UE with information on where to search for inter-frequency neighbouring cells

★ The minimum RSRP requirement for inter-frequency cell reselection is specified in terms of Qrxlevmin. The signalled value of Qrxlevmin ranges from -70 to -22, but the actual value = signalled value × 2 dBm. Qrxlevmin for intra-frequency cell reselection is specified in SIB3, while Qrxlevmin for the serving cell is specified in SIB1

★ Pmax defines the maximum allowed uplink transmit power for inter-frequency neighbouring cells. This has an impact upon the cell reselection criteria if the UE transmit power capability is less than the maximum allowed for the cell. Pmax for intra-frequency cell reselection is specified in SIB3, while Pmax for the serving cell is specified in SIB1

★ Treselection EUTRA defines the time-to-trigger for cell reselection. It can be allocated a value between 0 and 7 seconds. Its value can be scaled by the associated scaling factors if medium or high mobility is detected. The scaling factors can have values of 0.25, 0.5, 0.75 and 1.0, i.e. they typically decrease the value of Treselection to allow more rapid cell reselection

★ ThreshX-High is used for cell reselection to a higher priority RF carrier. Cell reselection to a higher priority RF carrier is allowed if Srxlev for the candidate cell is greater than the value of ThreshX-High. The signalled value of ThreshX-High has a range from 0 to 31 but the actual value = signalled value × 2 dB

★ ThreshX-Low is used for cell reselection to a lower priority RF carrier. Cell reselection to a lower priority RF carrier is allowed if Srxlev for the candidate cell is greater than ThreshX-Low, while Srxlev for the serving cell is less than the value of Thresh Serving Low signalled within SIB3. The signalled value of ThreshX-Low has a range from 0 to 31 but the actual value = signalled value × 2 dB

★ The allowed measurement bandwidth defines the maximum allowed measurement bandwidth. It is defined in terms of the Resource Blocks associated with a specific channel bandwidth, i.e. 6, 15, 25, 50, 75 or 100 Resource Blocks. This parameter is useful when neighbouring cells have different bandwidths to the serving cell

★ The presence of the antenna port 1 parameter informs the UE of whether or not it can assume that all neighbouring cells have at least two cell specific antenna ports, i.e. at least ports 0 and 1. RSRP measurements can be based upon transmissions from both antenna ports when available

★ The cell reselection priority defines the absolute priority of the frequency layer. It can be allocated a value between 0 and 7, where 0 represents the lowest priority

★ The neighbour cell configuration provides information regarding the MBSFN subframe allocations in neighbouring cells. In the case of TDD, it is also used to provide information regarding the TDD uplink-downlink subframe configuration in neighbouring cells

★ QoffsetFreq defines an RSRP measurement offset which is applied to all cells on the specified RF carrier. It is applicable to inter-frequency cell reselection towards a frequency layer which has equal priority. QoffsetFreq has a range from -24 to +24 dB. Negative values make cells on a specific RF carrier appear more attractive

★ Cell specific RSRP measurement offsets can be defined for up to 16 inter-frequency neighbours on each of the specified RF carriers. Neighbours are identified by their Physical layer Cell Identity (PCI) while the measurement offset is specified by Qoffset. Qoffset is applicable to inter-frequency cell reselection to a frequency layer which has equal priority. It has a range from -24 to +24 dB. Negative values make neighbours appear more attractive

★ A maximum of 16 groups of cells can be blacklisted on each of the specified RF carriers. Blacklisted cells are not considered as candidates for cell reselection. Cells belonging to a specific group have consecutive PCI. The number of cells within a group can range from 1 to 504

★ The release 9 version of the 3GPP specifications introduces an extension to SIB5 which defines additional cell reselection parameters. These cell reselection parameters are based upon RSRQ whereas the equivalent 3GPP release 8 parameters are based upon RSRP

★ Qqualmin defines a minimum RSRQ threshold for cell reselection. 3GPP release 9, and newer UE check both Qrxlevmin and Qqualmin when completing cell reselection. Qqualmin can be signalled using a value between -34 and -3 dB, and the actual value equals the signalled value

★ ThreshX-HighQ is an RSRQ based threshold used for cell reselection to a higher priority RF carrier. Cell reselection to a higher priority RF carrier is allowed if Squal for the candidate cell is greater than the value of ThreshX-HighQ. The signalled value of ThreshX-HighQ has a range from 0 to 31 dB, and the actual value = signalled value

★ ThreshX-LowQ is an RSRQ based threshold used for cell reselection to a lower priority RF carrier. Cell reselection to a lower priority RF carrier is allowed if Squal for the candidate cell is greater than ThreshX-LowQ, while Squal for the serving cell is less than the value of ThreshServingLowQ signalled within SIB3. The signalled value of ThreshX-LowQ has a range from 0 to 31 dB, and the actual value = signalled value

10.8 SYSTEM INFORMATION BLOCK 6

★ SIB6 provides cell reselection information for UMTS neighbours. The content of SIB6 is shown in Table 138

<table>
<tr><th colspan="3">Information Elements</th></tr>
<tr><td rowspan="9">FDD UMTS Inter-RAT Carrier Frequency List
(1 to 16 instances)</td><td colspan="2">Downlink Carrier Frequency</td></tr>
<tr><td colspan="2">Priority</td></tr>
<tr><td colspan="2">ThreshX-High</td></tr>
<tr><td colspan="2">ThreshX-Low</td></tr>
<tr><td colspan="2">Qrxlevmin</td></tr>
<tr><td colspan="2">Pmax</td></tr>
<tr><td colspan="2">Qqualmin</td></tr>
<tr><td rowspan="2">Release 9 Extension</td><td>ThreshX-HighQ</td></tr>
<tr><td>ThreshX-LowQ</td></tr>
<tr><td rowspan="6">TDD UMTS Inter-RAT Carrier Frequency List
(1 to 16 instances)</td><td colspan="2">Downlink Carrier Frequency</td></tr>
<tr><td colspan="2">Priority</td></tr>
<tr><td colspan="2">ThreshX-High</td></tr>
<tr><td colspan="2">ThreshX-Low</td></tr>
<tr><td colspan="2">Qrxlevmin</td></tr>
<tr><td colspan="2">Pmax</td></tr>
<tr><td colspan="3">Treselection UTRA</td></tr>
<tr><td rowspan="2">Treselection UTRA Scaling Factors</td><td colspan="2">SF Medium</td></tr>
<tr><td colspan="2">SF High</td></tr>
</table>

Table 138 – Content of System Information Block 6 (SIB6)

★ Information can be specified for up to 16 FDD UMTS RF carriers. Individual cells cannot be specified, i.e. there is no mechanism to signal a scrambling code. The RF carrier itself is specified to inform the UE of where to search for a neighbouring UMTS cell

★ The cell reselection priority defines the absolute priority of the UMTS layer. It can be allocated a value between 0 and 7, where 0 represents the lowest priority

★ ThreshX-High is used for cell reselection to a higher priority UMTS layer. Cell reselection to a higher priority layer is allowed if Srxlev for the candidate cell is greater than the value of ThreshX-High. The signalled value of ThreshX-High has a range from 0 to 31, but the actual value = signalled value × 2 dB

★ ThreshX-Low is used for cell reselection to a lower priority UMTS layer. Cell reselection to a lower priority layer is allowed if Srxlev for the candidate cell is greater than ThreshX-Low, while Srxlev for the serving cell is less than the value of Thresh Serving Low signalled within SIB3. The signalled value of ThreshX-Low has a range from 0 to 31, but the actual value = signalled value × 2 dB

★ The minimum CPICH RSCP requirement for candidate UMTS cells is specified in terms of Qrxlevmin. The signalled value of Qrxlevmin ranges from -60 to -13, but the actual value = signalled value × 2 + 1 dBm

★ Pmax defines the maximum allowed uplink transmit power for the specified UMTS layer. This impacts the cell reselection criteria if the UE transmit power capability is less than the maximum allowed for the cell

★ The minimum CPICH Ec/Io requirement for candidate UMTS cells is specified in terms of Qqualmin. It ranges from -24 to 0 dB

★ The release 9 version of the 3GPP specifications introduces some additional cell reselection parameters within SIB6. These cell reselection parameters are based upon CPICH Ec/Io whereas the equivalent 3GPP release 8 parameters are based upon CPICH RSCP

★ ThreshX-HighQ is a CPICH Ec/Io based threshold used for cell reselection to a higher priority UMTS layer. Cell reselection to a higher priority UMTS layer is allowed if Squal for the candidate cell is greater than the value of ThreshX-HighQ. The signalled value of ThreshX-HighQ has a range from 0 to 31 dB, and the actual value = signalled value

★ ThreshX-LowQ is a CPICH Ec/Io based threshold used for cell reselection to a lower priority UMTS layer. Cell reselection to a lower priority UMTS layer is allowed if Squal for the candidate cell is greater than ThreshX-LowQ, while Squal for the serving cell is less than the value of ThreshServingLowQ signalled within SIB3. The signalled value of ThreshX-LowQ has a range from 0 to 31 dB, and the actual value = signalled value

★ Measurement offsets cannot be specified for UMTS neighbouring cells

★ Similar information is specified for both FDD and TDD UMTS technologies although Qqualmin is not applicable to TDD

★ Treselection UTRA defines the time-to-trigger for cell reselection. It can be allocated a value between 0 and 7 seconds. Its value can be scaled by the associated scaling factors if medium or high mobility is detected. The scaling factors can have values of 0.25, 0.5, 0.75 and 1.0, i.e. they typically decrease the value of Treselection to allow more rapid cell reselections

10.9 SYSTEM INFORMATION BLOCK 7

★ SIB7 provides cell reselection information for GERAN neighbours. The content of SIB7 is shown in Table 139

Information Elements			
Treselection GERAN			
Treselection GERAN Scaling Factors	Medium Mobility		
	High Mobility		
Carrier Frequencies Information List (1 to 16 instances)	Carrier Frequencies	Starting ARFCN	
		Band Indicator	
		CHOICE	
		1. Explicit List of ARFCN	
		2. Equally Spaced ARFCN	ARFCN Spacing
			Number of ARFCN
		3. Bit Map of ARFCN	
	Priority		
	NCC Permitted		
	Qrxlevmin		
	Pmax		
	ThreshX-High		
	ThreshX-Low		

Table 139 – Content of System Information Block 7 (SIB7)

★ Treselection GERAN defines the time-to-trigger for cell reselection. It can be allocated a value between 0 and 7 seconds. Its value can be scaled by the associated scaling factors if medium or high mobility is detected. The scaling factors can have values of 0.25, 0.5, 0.75 and 1.0, i.e. they typically decrease the value of Treselection to allow more rapid cell reselections

★ Information can be specified for up to 16 groups of GERAN RF carriers. Each group is specified in terms of its starting Absolute Radio Frequency Channel Number (ARFCN) and band indicator. The ARFCN is a value between 0 and 1023 whereas the band indicator can be either DCS1800 or PCS1900. The band indicator determines the interpretation of the ARFCN. It is only applicable to the DCS1800 and PCS1900 bands

★ Each group of ARFCN is then specified using one of three different approaches. The list of ARFCN which follow the starting ARFCN can be listed explicitly. Alternatively, they can be defined as a set of equally spaced ARFCN. Otherwise a bitmap can be specified to indicate which ARFCN belong to the group.

★ The cell reselection priority defines the absolute priority of the GERAN layer. It can be allocated a value between 0 and 7, where 0 represents the lowest priority

★ The Network Colour Code (NCC) field defines which NCC can be considered for cell reselection. The field is encoded as a bitmap. If the nth bit = 1 then the BCCH carrier with NCC = n-1 can be considered as a candidate

★ The minimum RSSI requirement for candidate GERAN cells is specified in terms of Qrxlevmin. The signalled value of Qrxlevmin ranges from 0 to 45, but the actual value = signalled value × 2 – 115 dBm

★ Pmax defines the maximum allowed uplink transmit power for the specified GERAN RF carrier group. This impacts the cell reselection criteria if the UE transmit power capability is less than the maximum allowed for the cell

★ ThreshX-High is used for cell reselection to a higher priority GERAN RF carrier group. Cell reselection to a higher priority group is allowed if Srxlev for the candidate cell is greater than the value of ThreshX-High. The signalled value of ThreshX-High has a range from 0 to 31 but the actual value = signalled value× 2 dB

★ ThreshX-Low is used for cell reselection to a lower priority GERAN RF carrier group. Cell reselection to a lower priority group is allowed if Srxlev for the candidate cell is greater than ThreshX-Low, while Srxlev for the serving cell is less than the value of ThreshServingLow signalled within SIB 3. The signalled value of ThreshX-Low has a range from 0 to 31 but the actual value = signalled value× 2 dB

10.10 SYSTEM INFORMATION BLOCK 8

★ SIB8 provides cell reselection information for CDMA2000 neighbours. The first part of the content of SIB8 is shown in Table 140

<table>
<tr><th colspan="5">Information Elements</th></tr>
<tr><td rowspan="4">System Time Info</td><td colspan="4">CDMA-EUTRA Synchronisation</td></tr>
<tr><td rowspan="3">CDMA System Time</td><td colspan="3">CHOICE</td></tr>
<tr><td colspan="3">1.Synchronous System Time</td></tr>
<tr><td colspan="3">2. Asynchronous System Time</td></tr>
<tr><td colspan="5">Search Window Size (0 to 15)</td></tr>
<tr><td rowspan="13">HRPD Parameters</td><td rowspan="3">Pre-Registration Information HRPD</td><td colspan="3">Pre-Registration Allowed</td></tr>
<tr><td colspan="3">Pre-Registration Zone Id</td></tr>
<tr><td colspan="2">Secondary Pre-Registration Zone Id List (1 to 2 instances)</td><td>Pre-Registration Zone Id</td></tr>
<tr><td rowspan="10">Cell Reselection Parameters</td><td rowspan="4">Band Class List (1 to 32 instances)</td><td colspan="2">Band Class</td></tr>
<tr><td colspan="2">Priority</td></tr>
<tr><td colspan="2">ThreshX-High</td></tr>
<tr><td colspan="2">ThreshX-Low</td></tr>
<tr><td rowspan="3">Neighbour Cell List (1 to 16 instances)</td><td colspan="2">Band Class</td></tr>
<tr><td rowspan="2">Neighbour Cells per Band List (1 to 16 instances)</td><td>ARFCN</td></tr>
<tr><td>Physical Cell Id List (1 to 16 instances)</td></tr>
<tr><td colspan="3">Treselection</td></tr>
<tr><td rowspan="2">Speed State Scale Factors</td><td colspan="2">SF Medium</td></tr>
<tr><td colspan="2">SF High</td></tr>
</table>

Table 140 – Content of System Information Block 8 (SIB8) – Part 1

★ The CDMA-EUTRA synchronisation indication informs the UE of whether or not the CDMA2000 and LTE networks are synchronised. If synchronised, there is no drift in the timing between the two networks

★ The choice between including the synchronous or asynchronous system time depends upon whether or not the two networks are synchronised. In both cases, the system time defines the CDMA2000 system time corresponding to the 10 ms radio frame boundary of the SI window in which SIB8 is broadcast. The units are 10 ms when the networks are synchronous, and 8 chips when the systems are asynchronous

★ The search window size helps UE find neighbouring pilots by defining a range of PN sequence offsets. The signalled value between 0 and 15 represents a pointer to a look-up table specified by 3GPP2. A signalled value of 0 points to a window size of 4 PN chips whereas a signalled value of 15 points to a window size of 452 PN chips

★ The HRPD pre-registration information indicates whether or not UE should pre-register with the CDMA2000 network. If pre-registration is required then a pre-registration zone is provided. Secondary pre-registration zones can also be defined

★ The HRPD cell reselection information defines

- o a list of up to 32 CDMA2000 frequency bands. Each frequency band can be allocated an absolute priority for cell reselection as well as the ThreshX-high and ThreshX-Low thresholds
- o up to 16 neighbour cell lists where each list is associated with one of the CDMA2000 frequency bands. Each neighbour cell list can include up to 16 RF carriers (defined by their ARFCN) and up to 16 physical cell identities can be associated with each RF carrier
- o Treselection CDMA2000 which defines the time-to-trigger for cell reselection. It can be allocated a value between 0 and 7 seconds. Its value can be scaled by the associated scaling factors if medium or high mobility is detected. The scaling factors can have values of 0.25, 0.5, 0.75 and 1.0, i.e. they typically decrease the value of Treselection to allow more rapid cell reselections

★ The second part of the content of SIB8 is shown in Table 141

Information Elements				
1xRTT Parameters	CSFB Registration Parameters	SID, NID, Multiple SID, Multiple NID, Home Reg, Foreign SID Reg, Foreign NID Reg, Parameter Reg, Power Up Reg, Registration Period, Registration Zone, Total Zone, Zone Timer		
	Long Code State			
	Cell Reselection Parameters	Band Class List (1 to 32 instances)	Band Class	
			Priority	
			ThreshX-High	
			ThreshX-Low	
		Neighbour Cell List (1 to 16 instances)	Band Class	
			Neighbour Cells per Band List (1 to 16 instances)	ARFCN
				Physical Cell Id List (1 to 16 instances)
		Treselection		
		Speed State Scale Factors	SF Medium	
			SF High	
Release 9 Extension	CSFB Support for Dual Rx UE			
	Cell Reselection Parameters for HRPD	Neighbour Cell List (1 to 16 instances)	Neighbour Cells per Band List (1 to 16 instances)	Physical Cell Id List (1 to 24 instances)
	Cell Reselection Parameters for 1xRTT	Neighbour Cell List (1 to 16 instances)	Neighbour Cells per Band List (1 to 16 instances)	Physical Cell Id List (1 to 24 instances)
	CSFB Registration Parameters 1xRTT	Power Down Registration Indicator		
	Access Class Barring Configuration 1xRTT	Access Class Barring 0 to 9		
		Access Class Barring 10		
		Access Class Barring 11		
		Access Class Barring 12		
		Access Class Barring 13		
		Access Class Barring 14		
		Access Class Barring 15		
		Access Class Barring Message		
		Access Class Barring Registration		
		Access Class Barring Emergency Calls		
Release 10 Extension	CSFB Dual Rx Tx Support			

Table 141 – Content of System Information Block 8 (SIB8) – Part 2

★ Single Carrier Radio Transmission Technology (1xRTT) CS Fallback (CSFB) registration information is included if CSFB to CDMA2000 is supported

★ The long code state is included if Single Radio Voice Call Continuity (SRVCC) handover to CDMA2000 is supported. Its value is based upon the CDMA system time

★ The cell reselection parameters for 1xRTT are the same as those for HRPD

★ The release 9 version of the 3GPP specifications introduces some additional CDMA2000 parameters to SIB8

★ A flag is used to indicate whether or not the network supports dual receiver CS Fallback, i.e. a CS Fallback procedure which can be used when the UE has two receivers

★ Extended neighbour lists are provided for both HRPD and 1xRTT. The physical cell identities at a specific position within the extended neighbour lists are associated with the operating band and ARFCN defined for the same position within the original neighbour lists

★ An additional CS Fallback Registration parameter is defined for 1xRTT. The Power Down Registration Indicator instructs UE that are pre-registered with CDMA2000 1xRTT to complete a registration procedure when switched off

- ★ A set of Access Class barring parameters are included for 1xRTT. These parameters are used when calculating the access class barring factor for access overload. They correspond to the PSIST parameter within the CDMA2000 specifications. Access classes 0 to 9 use a range from 0 to 63, whereas access classes 10 to 15 use a range from 0 to 7
- ★ The Access Class Barring Message parameter is used to modify the access class barring factor for message transmissions. It corresponds to the MSG_PSIST parameter within the CDMA2000 specifications. It has a range from 0 to 7
- ★ The Access Class Barring Registration parameter is used to modify the access class barring factor for autonomous registrations. It corresponds to the REG_PSIST parameter within the CDMA2000 specifications. It has a range from 0 to 7
- ★ The Access Class Barring Emergency Calls parameter is used to modify the access class barring factor for emergency calls and emergency message transmissions for access overload classes 0 to 9. It corresponds to the PSIST_EMG parameter within the CDMA2000 specifications. It has a range from 0 to 7
- ★ The release 10 version of the 3GPP specifications introduces an additional parameter for CS fallback. The CS fallback dual Rx Tx Support flag indicates whether or not the network supports dual Rx Tx enhanced CS fallback for 1xRTT. This capability allows UE which are capable of dual Rx Tx enhanced 1xCSFB to switch off their 1xRTT receiver/transmitter while camped on LTE

10.11 SYSTEM INFORMATION BLOCK 9

- ★ The content of SIB9 is shown in Table 142

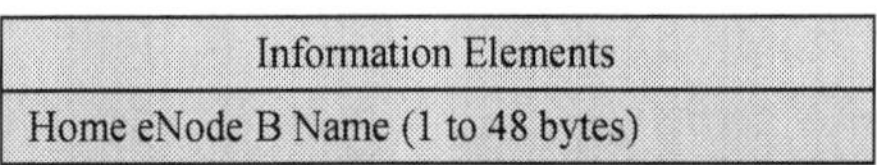

Information Elements
Home eNode B Name (1 to 48 bytes)

Table 142 – Content of System Information Block 9 (SIB9)

- ★ SIB9 is associated with Home eNode B
- ★ Home eNode B provide services to Closed Subscriber Groups (CSG)
- ★ The Home eNode B name is coded using UTF-8 to provide flexibility for using different languages. It allows a maximum of 48 characters when coded using 1 byte per character, 24 characters when coded using 2 bytes per character, etc
- ★ The Home eNode B name helps the subscriber to choose the correct CSG identity when performing a manual selection

10.12 SYSTEM INFORMATION BLOCK 10

- SIB10 provides support for the Earthquake and Tsunami Warning System (ETWS). ETWS is an example of a Public Warning System (PWS)
- ETWS warning notifications can be either primary (short notifications delivered within 4 seconds) or secondary (more detailed information). SIB10 broadcasts primary notifications whereas SIB11 broadcasts secondary notifications
- The LTE radio access network performs scheduling and broadcasting of the warning message content received from the Cell Broadcast Center (CBC). This content is received from the CBC via the MME. LTE is responsible for paging UE to inform them that a warning notification is being broadcast
- The content of SIB10 is shown in Table 143

Information Elements
Message Identifier (16 bits)
Serial Number (16 bits)
Warning Type (16 bits)
Warning Security Information (50 bytes)

Table 143 – Content of System Information Block 10 (SIB10)

- The Message Identifier is a string of 16 bits which defines the type of ETWS notification. These are specified within 3GPP TS 23.041. Examples include earthquake warning, tsunami warning, earthquake and tsunami warning, other emergency, or test message
- The Serial Number is a string of 16 bits used to track any changes in the ETWS notification. It includes:
  - 1 bit emergency user alert flag to instruct the UE to alert the end-user, e.g. by playing a tone or vibrating
  - 1 bit popup flag to instruct the UE to display a message on its screen
  - 8 bit message code used to further specify the type of ETWS notification
  - 2 bit geographical scope to indicate the area over which the message is applicable, i.e. cell, location area, or PLMN. It also indicates the display mode, i.e. whether or not the warning is to be displayed at all times
  - 4 bit update number used to indicate a change in the message content. The update number differentiates between older and newer versions of the same message
- The Warning Type specifies whether the warning is for a tsunami, earthquake or tsunami and earthquake. It also includes user alert and popup information
- The Warning Security Information is optional and is only included when security is applied. It includes a 7 byte time stamp and a 43 byte digital signature

10.13 SYSTEM INFORMATION BLOCK 11

- SIB11 also provides support for the Earthquake and Tsunami Warning System (ETWS)
- ETWS warning notifications can be either primary (short notifications delivered within 4 seconds) or secondary (more detailed information). SIB10 broadcasts primary notifications whereas SIB11 broadcasts secondary notifications
- The content of SIB11 is shown in Table 144

Information Elements
Message Identifier (16 bits)
Serial Number (16 bits)
Warning Message Segment Type (last / not last)
Warning Message Segment Number (0 to 63)
Warning Message Segment
Data Coding Scheme (8 bits)

Table 144 – Content of System Information Block 11 (SIB11)

- The Message Identifier is a string of 16 bits which defines the type of ETWS notification. These are specified within 3GPP TS 23.041. Examples include earthquake warning, tsunami warning, earthquake and tsunami warning, other emergency, or test message

- The Serial Number is a string of 16 bits used to track any changes in the ETWS notification. It includes:
  - 1 bit emergency user alert flag to instruct the UE to alert the end-user, e.g. by playing a tone or vibrating
  - 1 bit popup flag to instruct the UE to display a message on its screen
  - 8 bit message code used to further specify the type of ETWS notification
  - 2 bit geographical scope to indicate the area over which the message is applicable, i.e. cell, location area, or PLMN. It also indicates the display mode, i.e. whether or not the warning is on the display at all times
  - 4 bit update number used to indicate a change in the message content. The update number differentiates between older and newer versions of the same message
- The Warning Message Segment Type indicates whether or not the included message segment is the last segment of the complete message
- The Warning Message Segment Number represents the number of the subsequent Message Segment. This number is used when reconstructing the complete message. It can be allocated values between 0 and 63
- The Warning Message Segment provides an actual segment of the message
- The Data Coding Scheme identifies the alphabet/coding and language used for the ETWS message. The allowed values are specified within 3GPP TS 23.038

10.14 SYSTEM INFORMATION BLOCK 12

- SIB12 was introduced within the release 9 version of the 3GPP specifications
- SIB12 is used to provide a Commercial Mobile Alert Service (CMAS) notification. CMAS is an example of a Public Warning System (PWS) which is able to deliver multiple, concurrent warning text messages
- The Federal Communications Commission (FCC) established CMAS to allow service providers to send emergency alerts as text messages to their users who have CMAS capable UE
- The content of SIB12 is shown in Table 145

Information Elements
Message Identifier
Serial Number
Message Segment Type
Message Segment Number
Message Segment
Data Coding Scheme

Table 145 – Content of System Information Block 12

- The Message Identifier is a string of 16 bits which defines the type of CMAS message broadcast by SIB12. These are specified within 3GPP TS 23.041. Examples include 'Presidential Alerts', 'Extreme Alerts with Severity of Extreme, Urgency of Immediate, and Certainty of Observed' and 'Severe Alerts with Severity of Severe, Urgency of Expected, and Certainty of Likely'
- The Serial Number is a string of 16 bits used to track any changes in the CMAS notification. The serial number is updated every time a CMAS message with a specific Message Identifier is updated
- The Message Segment Type is used to indicate whether or not the subsequent Message Segment is the last segment belonging to the message
- The Message Segment Number represents the number of the subsequent Message Segment. This number is used when reconstructing the complete message. It can be allocated values between 0 and 63
- The Message Segment is a section of the message itself
- The Data Coding Scheme identifies the alphabet/coding and language used for the CMAS message. The allowed values are specified within 3GPP TS 23.038

10.15 SYSTEM INFORMATION BLOCK 13

★ SIB13 was introduced within the release 9 version of the 3GPP specifications

★ SIB13 supports the transmission of MBMS using a Single Frequency Network (SFN). It includes information which allows UE to subsequently acquire MBMS control information associated with one or more MBSFN areas

★ The content of SIB13 is shown in Table 146

Information Elements		
MBSFN Area Information List (1 to 8 instances)	MBSFN Area Identity	
	Non-MBSFN Region Length	
	Notification Indicator	
	MCCH Configuration	MCCH Repetition Period
		MCCH Offset
		MCCH Modification Period
		Subframe Allocation Information
		Signalling MCS
Notification Configuration	Notification Repetition Coefficient	
	Notification Offset	
	Notification Subframe Index	

Table 146 – Content of System Information Block 13

★ The first part of SIB13 provides MBSFN Area Information. This part of the SIB can be repeated for up to 8 MBSFN areas

★ The MBSFN Area Identity is signalled as an integer within the range 0 to 255. This identifies the MBSFN area to which the subsequent information applies

★ The Non-MBSFN Region Length defines the number of symbols at the start of a subframe which cannot be used by the PMCH. This region can be either 1 or 2 symbols and is used by the PDCCH and cell specific Reference Signals

★ The Notification Indicator informs the UE of which PDCCH bit is used to indicate a change of the MCCH. Values within the range 0 to 7 can be signalled, where a value of 0 indicates the least significant bit. This is applicable to Downlink Control Information (DCI) format 1C

★ The MCCH Repetition Period defines the time interval between consecutive transmissions of the MCCH. This can be signalled using values of 32, 64, 128 or 256 radio frames. There is an MCCH for each MBSFN area

★ The MCCH Offset combined with the MCCH Repetition Period defines the specific radio frame during which the MCCH is scheduled. The MCCH is scheduled when the System Frame Number (SFN) mod MCCH Repetition Period = MCCH Offset. The MCCH Offset can be configured with values between 0 and 10

★ The MCCH Modification Period defines the number of radio frames during which the MCCH content cannot change. The MCCH content can only change when a boundary defined by SFN mod MCCH Modification Period = 0 is passed. The Modification Period can be configured with a value of 512 or 1024 radio frames

★ The Subframe Allocation Information indicates the subframes within the radio frames indicated by the MCCH Repetition Period and MCCH Offset which transmit the MCCH. The allocation information is signalled as a bitmap of length 6. This bitmap is mapped onto 6 of the 10 subframes within the radio frame. This mapping differs for FDD and TDD. The mapping is presented in Table 147

	Bit 1	Bit 2	Bit 3	Bit 4	Bit 5	Bit 6
FDD	SF 1	SF 2	SF 3	SF 6	SF 7	SF 8
TDD	SF 3	SF 4	SF 7	SF 8	SF 9	Not Used

Table 147 – Mapping between Subframe Allocation Information bits and subframe (SF) numbers

★ The Signalling MCS defines the Modulation and Coding Scheme (MCS) used for the subframes indicated by the Subframe Allocation Information, and for the first subframe of each MCH scheduling period. MCS indices of 2 (QPSK, transport block size index = 2), 7 (QPSK, transport block size index = 7), 13 (16QAM, transport block size index = 12), and 19 (64QAM, transport block size index = 17) can be signalled

★ The Notification Configuration within SIB13 is applicable to all MBSFN areas

- The Notification Repetition Coefficient can be configured with a value of either 2 or 4. It defines the actual change notification repetition period which is common for all MCCH. It is defined by the shortest result from the calculation of (MCCH Modification Period) / (Notification Repetition Coefficient) for all MCCH
- The Notification Offset combined with the Notification Repetition Coefficient defines the specific radio frames during which the MCCH information change notification is scheduled. The MCCH information change notification is scheduled in radio frames for which SFN mod notification repetition period = Notification Offset. The Notification Offset can be configured with values between 0 and 10
- The Notification Subframe Index indicates the subframes within the radio frames indicated by the Notification Repetition Period and Notification Offset which are used to transmit MCCH change notifications on the PDCCH. The Notification Subframe Index is signalled using an integer value between 1 and 6, and uses the same mapping as presented in Table 147, e.g. a signalled value with the 4th bit set to '1' would indicate subframe 6 for FDD, or subframe 8 for TDD

11 UPLINK AIR-INTERFACE

11.1 FRAME STRUCTURE AND TIMING

★ In the case of FDD, the uplink radio frame structure is the same as the downlink radio frame structure

- frame structure type 1 is applicable to FDD (both full duplex and half duplex)

★ In the case of TDD, the uplink and downlink share the same radio frame structure

- frame structure type 2 is applicable to TDD

★ 3GPP TS 36.211 specifies frame structures 'type 1' and 'type 2'. Both have 10 ms durations and are presented in section 3.2

★ The timing of the uplink radio frame relative to the downlink radio frame is defined at the UE according to Figure 129. The uplink radio frame timing precedes the downlink radio frame timing at the UE to allow for the round trip time propagation delay. The uplink and downlink are then synchronised at the eNode B

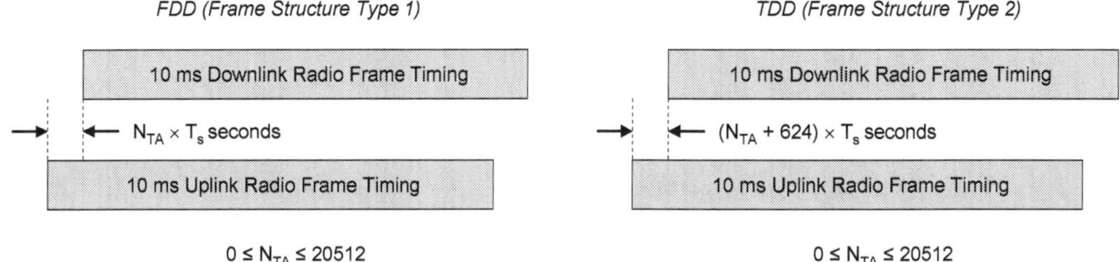

Figure 129 – Uplink frame timing relative to downlink frame timing at the UE

★ In the case of FDD, the start of an uplink subframe transmission can overlap with the end of a previous downlink subframe transmission. This is not an issue for full duplex FDD because UE can transmit and receive simultaneously. In the case of half duplex FDD, the UE prioritises transmission over reception and does not receive the end of the downlink subframe (this assumes that the eNode B scheduler has allocated an uplink subframe directly after an allocated downlink subframe)

★ In the case of TDD, the Guard Period within the special subframe avoids any overlap between downlink reception and uplink transmission (illustrated in Figure 17 within section 3.2.2)

★ N_{TA} can range from 0 to 20512 and is multiplied by $T_S = 1/30.72$ μs. This provides a maximum timing advance of 0.67 ms, which is equivalent to 200 km. This represents the round trip distance so the maximum cell range is 100 km

★ The TDD case includes a fixed additional 624 T_S which is equivalent to 20 μs. This additional offset allows time for the eNode B to switch from receiving to transmitting. This additional guard period is illustrated in Figure 17 within section 3.2.2. The timing advance instructed by the eNode B corresponds to N_{TA} so the UE adds this additional offset of 624 T_S when defining its transmission timing

★ The eNode B generates timing advance instructions for each UE such that all uplink transmissions are synchronous when they arrive at the eNode B. UE towards the cell edge are instructed to use a large timing advance while UE close to the eNode B are instructed to use a small timing advance

★ Timing advance is initialised during the random access procedure. The MAC payload belonging to the Random Access Response includes an 11 bit Timing Advance Command which corresponds to T_A within the equation:

$$N_{TA} = T_A \times 16$$

★ The set of 11 bits allows a range of T_A from 0 to 1282. The maximum value of 1282 does not fully utilise the complete range offered by 11 bits but corresponds to the maximum value of N_{TA}, i.e. 1282 × 16 = 20512

★ Subsequent to the random access procedure, timing advance changes are signalled using the Timing Advance Command MAC Control Element. The Timing Advance Command MAC Control Element forms part of the MAC PDU payload. It has a size of 1 byte from which 6 bits are used to signal the timing advance command, i.e. a range from 0 to 63

★ The timing advance command within the MAC Control Element corresponds to T_A within the equation:

$$N_{TAnew} = N_{TAold} + (T_A - 31) \times 16$$

Subtracting 31 from T_A allows the eNode B to shift the timing advance in both positive and negative directions

★ Timing advance commands received during downlink subframe 'n' are applied to uplink subframe 'n+6'

★ When a timing advance command causes subframe 'm+1' to overlap with subframe 'm' the UE transmits all of subframe 'm' but does not transmit the overlapping part of subframe 'm+1'

★ Timing advance is further described in section 24.2

★ 3GPP References: TS 36.211, TS 36.213

11.2 RESOURCE BLOCKS

- ★ The eNode B scheduler allocates Resource Blocks to UE when allowing data transfer. Uplink Resource Blocks have the same structure as downlink Resource Blocks using the 15 kHz subcarrier spacing
  - a single 0.5 ms slot in the time domain
  - 12 × 15 kHz subcarriers in the frequency domain, i.e. 180 kHz

  The 7.5 kHz subcarrier spacing is not applicable to the uplink direction
- ★ Uplink Resource Blocks are also the same as downlink Resource Blocks in terms of using either the normal or extended cyclic prefix, i.e. providing support for either 7 or 6 time domain symbols per Resource Block respectively
- ★ Resource Blocks based upon normal and extended cyclic prefixes are illustrated in Figure 19. Fewer symbols can be accommodated when using the extended cyclic prefix because the cyclic prefix occupies a greater percentage of the slot
- ★ The grid generated by subcarriers in the frequency domain and symbols in the time domain defines a set of Resource Elements. The number of uplink frequency domain Resource Blocks equals the number of downlink frequency domain Resource Blocks. The number of Resource Blocks and subcarriers for each channel bandwidth are shown in Table 148

Channel Bandwidth		1.4 MHz	3 MHz	5 MHz	10 MHz	15 MHz	20 MHz
Resource Blocks		6	15	25	50	75	100
Subcarriers		72	180	300	600	900	1200
Total Subcarrier Bandwidth (MHz)		1.08	2.70	4.50	9.00	13.50	18.00
Modulation Symbol Rate (Msps)	Normal Cyclic Prefix	1.008	2.52	4.2	8.4	12.6	16.8
	Extended Cyclic Prefix	0.864	2.16	3.6	7.2	10.8	14.4

Table 148 – Resource Block characteristics as a function of the channel bandwidth

- ★ In contrast to the downlink, the uplink does not include a 'null' subcarrier at the center of the channel bandwidth. This reduces the total subcarrier bandwidth by 15 kHz. A 'null' subcarrier is not required because all subcarriers are offset by 7.5 kHz so the subcarrier at 0 Hz is avoided
- ★ A single Resource Block provides a modulation symbol rate of 168 ksps when using the normal cyclic prefix and 144 ksps when using the extended cyclic prefix. The modulation symbol rates allow an initial estimate of the LTE maximum throughput capability, e.g. 16.8 Msps corresponds to 100.8 Mbps when using 64QAM. This bit rate figure assumes that all Resource Elements are available for transferring data. In practice, not all Resource Elements are available
- ★ The uplink differs from the downlink in terms of mapping the modulation symbols onto Resource Elements. The difference for single antenna transmission is illustrated in Figure 130

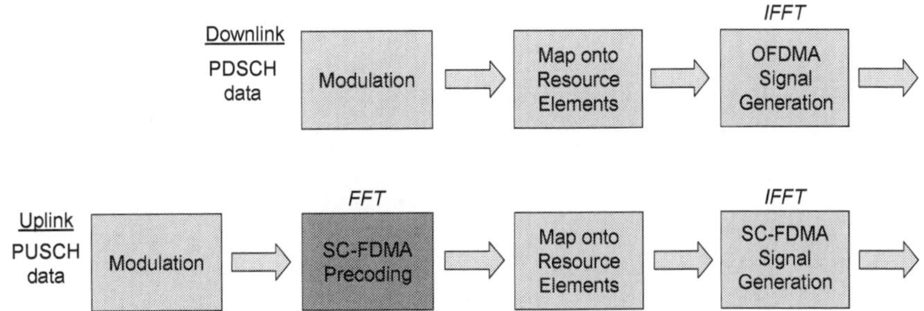

Figure 130 – Single antenna transmission mapping of modulation symbols onto Resource Elements

- ★ In the downlink, a single modulation symbol (QPSK, 16QAM, 64QAM) is mapped directly onto each Resource Element prior to generating the OFDMA signal
- ★ In the uplink, a set of modulation symbols are processed by taking their Fast Fourier Transform (FFT) prior to mapping onto a set of Resource Elements. The combination of FFT precoding with the subsequent IFFT generates a Single Carrier Frequency Division Multiple Access (SC-FDMA) signal
- ★ Figure 131 makes the same comparison for multiple antenna transmission. Multiple antenna transmission adds the layer mapping and precoding procedures. Multiple antenna transmission for the uplink was introduced within the release 10 version of the 3GPP specifications

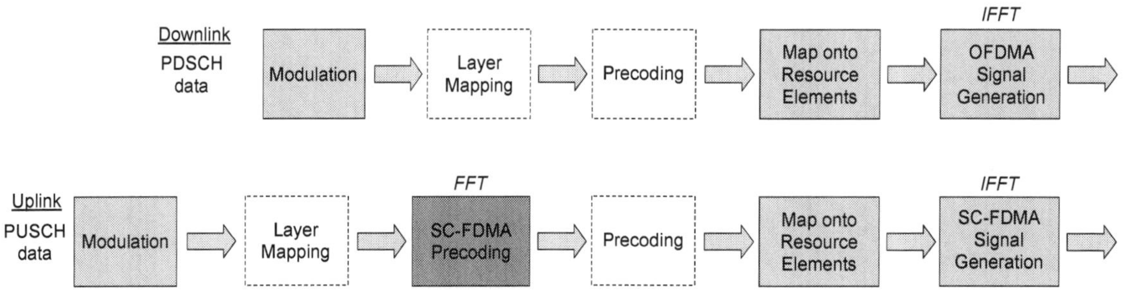

Figure 131 – Multiple antenna transmission mapping of modulation symbols onto Resource Elements

- ★ In the downlink, the layer mapping and precoding procedures are adjacent to one another
- ★ In the Uplink direction, the layer mapping and precoding procedures are separated by the FFT SC-FDMA precoding
- ★ 3GPP References: TS 36.211

11.3 MULTIPLE ACCESS

★ Multiple access techniques allow resources to be shared between a group of users

★ As presented in section 3.1, the downlink of LTE uses Orthogonal Frequency Division Multiple Access (OFDMA)

★ A drawback of OFDMA is its relatively high Peak to Average Power Ratio (PAPR), generated by summing large numbers of subcarriers. High PAPR means that power amplifiers have to operate with increased back-off to remain within the linear section of the amplifier characteristic and to avoid clipping. This leads to reduced efficiency of the amplifier

★ Reduced power efficiency has an impact upon battery powered handheld devices. 3GPP selected a different multiple access scheme for the uplink of LTE to help avoid the negative impacts upon power efficiency and battery life

★ The uplink of LTE uses a multiple access scheme known as Single Carrier Frequency Division Multiple Access (SC-FDMA)

★ SC-FDMA reduces the Peak to Average Power Ratio associated with OFDMA but requires additional baseband processing. As illustrated in Figure 130, SC-FDMA requires an FFT Precoding stage prior to mapping the modulation symbols onto the set of Resource Elements

★ Once the FFT Precoding has been completed, the processing for SC-FDMA is similar to that of OFDMA. The similarity with OFDMA allows the benefits of OFDMA to be retained, e.g. SC-FDMA is also resilient to delay spread

★ Both OFDMA and SC-FDMA are tolerant to delay spread because their time domain symbols have relatively long durations. The reduced impact of delay spread means that fading is flat in the frequency domain and receiver equalization becomes simpler

★ Table 149 compares SC-FDMA with other multiple access schemes. Although the name implies that SC-FDMA is similar to FDMA it has more in common with OFDMA

	FDMA	TDMA	CDMA	OFDMA	SC-FDMA
Resource Allocation	RF Carrier	Time Slot	Code	Subcarriers	Subcarriers
Example System	AMPS	GSM	UMTS	LTE Downlink	LTE Uplink

Table 149 – Range of multiple access techniques

★ Uplink resource allocation type 0 (described in section 9.4) is used to allocate a set of contiguous Resource Blocks to an individual UE. Each set of contiguous Resource Blocks can be thought of as an FDMA RF carrier although the lower level processing is based upon a group of subcarriers

★ This is in contrast to the downlink, in which case the allocated Resource Blocks do not need to be contiguous

★ The release 10 version of the 3GPP specifications introduced uplink resource allocation type 1 (described in section 9.5) which can be used to allocate non-contiguous Resource Blocks. Allocating non-contiguous Resource Blocks provides the eNode B scheduler with greater flexibility and increases the potential for a frequency selective scheduling gain. The drawback of a non-contiguous Resource Block allocation is an increase in the Peak to Average Power Ratio which leads to reduced efficiency and a reduced maximum transmit power

★ SC-FDMA is applied to the PUSCH physical channel but not to the PRACH and PUCCH physical channels. Neither the PRACH nor PUCCH physical channels have FFT Precoding completed prior to mapping onto Resource Elements

★ The PUSCH represents the main uplink physical channel so it is reasonable to state that the uplink of LTE uses SC-FDMA

★ When using the release 8 and 9 versions of the 3GPP specifications, the PUCCH physical channel is only transmitted if the UE needs to transfer physical layer control information (HARQ acknowledgements, CQI or a scheduling request) when there is nothing else to transfer. Physical layer control information is transferred on the PUSCH if there is also RRC signalling or application data to transfer. This means that from the perspective of a single UE, the PUCCH is time multiplexed with the PUSCH

★ The release 8 and 9 versions of the 3GPP specifications do not support simultaneous transmission of the PUCCH and PUSCH to avoid non-contiguous transmissions in the frequency domain (the PUCCH is allocated Resource Blocks towards each edge of the channel bandwidth, whereas the PUSCH is allocated Resource Blocks towards the center of the channel bandwidth). Non-contiguous transmissions in the frequency domain would result in an increased Peak to Average Power Ratio

★ The release 10 version of the specifications introduces support for the simultaneous transmission of the PUCCH and PUSCH, i.e. non-contiguous transmissions in the frequency domain are allowed. This is consistent with the introduction of uplink resource allocation type 1 which allows non-contiguous Resource Block allocations for the PUSCH

★ 3GPP References: TS 36.211, TS 36.213

11.4 MODULATION

★ Table 150 presents the modulation schemes applied to each uplink Physical Channel

Physical Channel		Modulation Scheme
PRACH		-
PUCCH	Format 1	-
	Format 1a	1 BPSK symbol
	Format 1b	1 QPSK symbol
	Format 2	10 QPSK symbols
	Format 2a	10 QPSK symbols + 1 BPSK symbol
	Format 2b	10 QPSK symbols + 1 QPSK symbol
	Format 3	24 QPSK symbols
PUSCH		QPSK, 16QAM, 64QAM

Table 150 – Modulation schemes applied to each uplink Physical Channel

★ The PRACH is not associated with a modulation scheme because it is only responsible for transmitting the random access preambles which are represented by a sequence of complex numbers, i.e. there are no binary information bits to modulate

★ PUCCH format 1 does not use a modulation scheme. Information is transferred by the presence/absence of the PUCCH. When PUCCH format 1 is present, a value of '1' is provided as an input to generate the complex sequence to be transmitted

★ PUCCH format 1a transfers a single BPSK symbol, while format 1b transfers a single QPSK symbol. The single modulation symbol is provided as an input to generate the complex sequence to be transmitted

★ PUCCH formats 2, 2a and 2b transfer 10 QPSK symbols which are channel coded prior to mapping onto Resource Elements (see Figure 171 in section 14.2.2). Format 2a transfers an additional BPSK symbol, while format 2b transfers an additional QPSK symbol. These additional symbols are used when generating the PUCCH Demodulation Reference Signal

★ PUUCH format 3 was introduced within the release 10 version of the 3GPP specifications to support the increased number of HARQ acknowledgements associated with Carrier Aggregation. It allows the transfer of 24 QPSK symbols which are channel coded prior to mapping onto Resource Elements (see Figure 172 in section 14.2.3)

★ The PUSCH supports QPSK, 16QAM and 64QAM. The modulation scheme selected depends upon the RF channel conditions and the quantity of data to be transferred

★ Similar to the downlink, Gray coding is used to minimise the number of bit errors when the receiver misinterprets a modulation constellation point for one of its neighbours. The principle of Gray coding is illustrated in Figure 20

★ 3GPP References: TS 36.211

11.5 SC-FDMA SIGNAL GENERATION

★ The first stage of generating an SC-FDMA signal from a stream of modulation symbols is illustrated in Figure 132. This is applicable to the PUSCH physical channel and represents the FFT Precoding stage

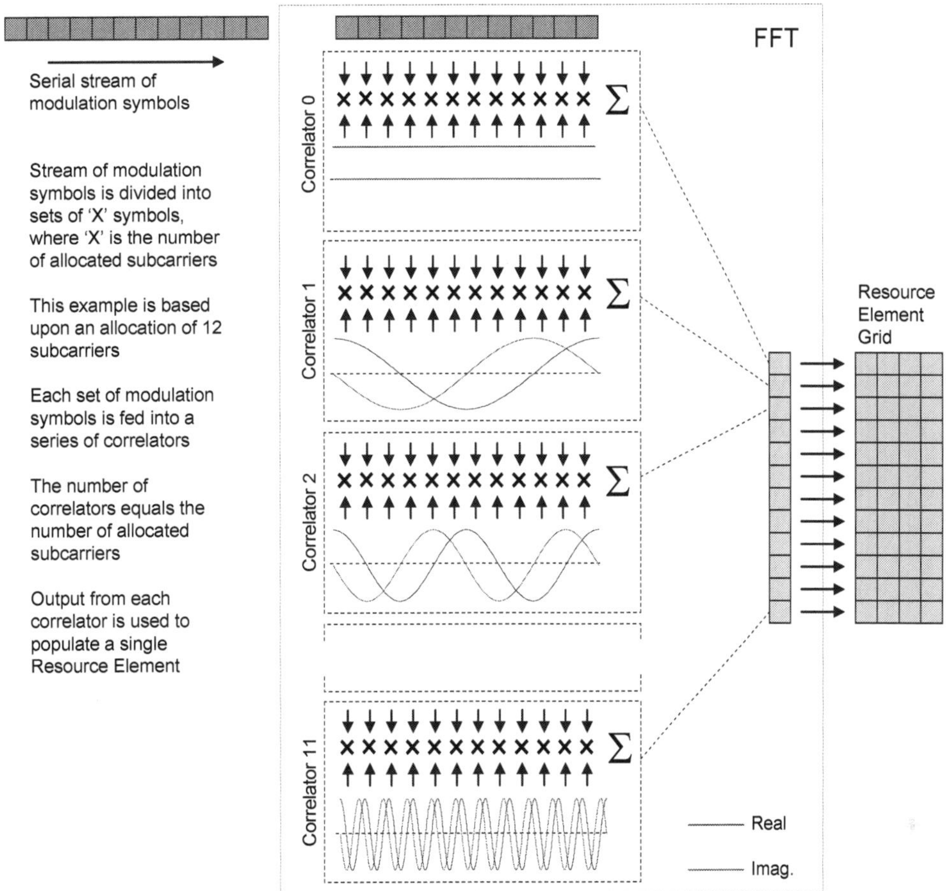

Figure 132 – Generation of the PUSCH SC-FDMA signal (part 1)

★ The stream of modulation symbols (QPSK, 16QAM or 64QAM) is divided into sets of 'X' symbols, where X is the number of allocated subcarriers. If a single Resource Block has been allocated then X=12, if two Resource Blocks have been allocated then X=24, etc.

★ Each set of modulation symbols is processed by an FFT function which generates a frequency domain representation of the symbols. The FFT function is a block of correlators which extracts a set of frequency components from the set of modulation symbols

★ The first correlator (shown at the top of Figure 132) extracts the DC component by taking an average of the modulation symbols. The second correlator extracts the frequency component defined by a single cycle of a sine/cosine. The final correlator extracts the frequency component defined by 'X-1' cycles of a sine/cosine. The number of correlators is equal to the number of allocated subcarriers so there is a single correlator result for each subcarrier

★ The output from each correlator is used to populate a single Resource Element. The resulting modulation symbol rate is the same as if the modulation symbols were mapped directly onto the set of Resource Elements

★ 3GPP TS 36.211 expresses the FFT Precoding function using the following expression:

$$z(l \times M_{sc}^{PUSCH} + k) = \frac{1}{\sqrt{M_{sc}^{PUSCH}}} \sum_{i=0}^{M_{sc}^{PUSCH}-1} d(l \times M_{sc}^{PUSCH} + i) \times e^{-j\frac{2\pi ik}{M_{sc}^{PUSCH}}}$$

$$k = 0,\ldots, M_{sc}^{PUSCH} - 1 \qquad l = 0,\ldots,(M_{symb} / M_{sc}^{PUSCH}) - 1$$

Where, M_{sc}^{PUSCH} is the number of allocated subcarriers

M_{symb} is the total number of modulation symbols to be processed

'l' is incremented for each set of 'X' modulation symbols

★ The second stage of generating an SC-FDMA signal is illustrated in Figure 133. This second stage is applicable to both the PUSCH and PUCCH physical channels

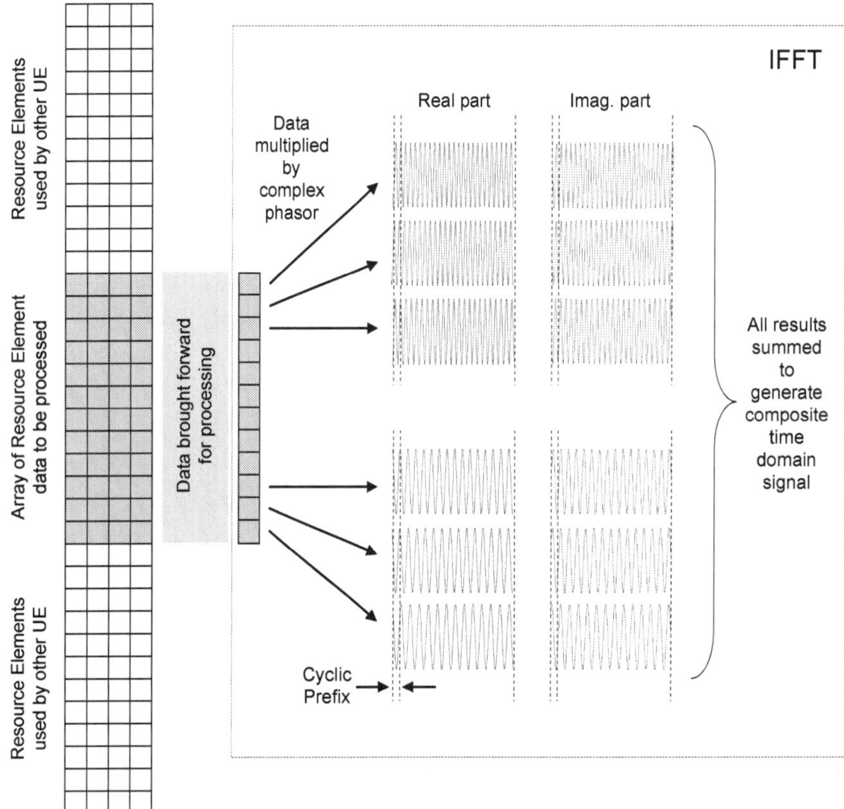

Figure 133 – Generation of the PUSCH SC-FDMA signal (part 2 – example A)

★ The second stage of generating an SC-FDMA signal is similar to the procedure used to generate an OFDMA signal (compare with Figure 21 in section 3.5). Each column of Resource Elements generates a single SC-FDMA symbol. Each column of Resource Elements is multiplied by a series of complex phasors. A single complex phasor is applied to each Resource Element

★ The frequencies of the complex phasors depend upon the allocated Resource Blocks. Resource Blocks allocated towards the center of the channel bandwidth are processed using low frequencies while Resource Blocks away from the center of the channel bandwidth are processed using high frequencies. Resource Blocks allocated in the lower half of the channel bandwidth are processed using negative frequencies (complex phasors rotating in a clockwise direction), while Resource Blocks allocated in the upper half of the channel bandwidth are processed using positive frequencies. (complex phasors rotating in an anti-clockwise direction)

★ Once the complete column of Resource Elements has been multiplied by the appropriate complex phasors, the results are summed to generate the SC-FDMA symbol, i.e. the SC-FDMA symbol is a sum of complex phasors

★ A UE only has to process the Resource Blocks which it has been allocated. This is in contrast to the downlink where the eNode B has to process the Resource Blocks allocated to all UE

★ Figure 134 illustrates a second example of the second stage of generating an SC-FDMA signal. This example is applicable to a UE which has been allocated a Resource Block at the center of the channel bandwidth

★ The first complex phasor above the center of the channel bandwidth has a frequency of (0.5 × 15 kHz) = 7.5 kHz. This results in half a cycle of the complex phasor within the duration of the SC-FDMA symbol payload. The second complex phasor above the center of the channel bandwidth has a frequency of (1.5 × 15 kHz) = 22.5 kHz. This results in one and a half cycles of the complex phasor within the duration of the SC-FDMA symbol payload

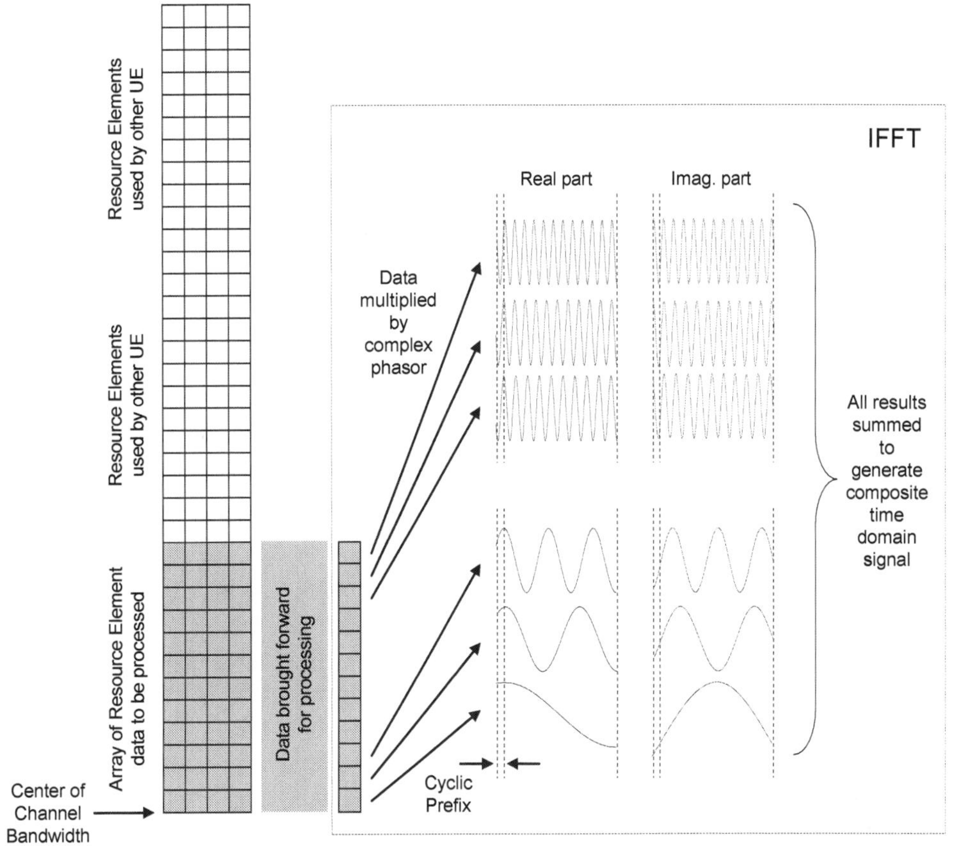

Figure 134 – Generation of the PUSCH SC-FDMA signal (part 2 – example B)

★ There is no null subcarrier in the uplink because the first Resource Element within the upper half of the channel bandwidth uses a subcarrier frequency of 7.5 kHz while the first Resource Element within the lower half of the channel bandwidth uses a subcarrier frequency of -7.5 kHz

★ Similar to an OFDMA symbol, the duration of an SC-FDMA symbol is divided into 2 parts: a relatively short first part which corresponds to the cyclic prefix and a longer second part which corresponds to the main body (payload) of the SC-FDMA symbol. The durations of the cyclic prefix and main body of the symbol are the same as those for the OFDMA symbols in the downlink, i.e.

Normal Cyclic Prefix

- $160 / 3.072 \times 10^7 = 5.2$ μs in the first SC-FDMA symbol of a 0.5 ms slot
- $144 / 3.072 \times 10^7 = 4.7$ μs in the remaining SC-FDMA symbols of a 0.5 ms slot

Extended Cyclic Prefix

- $512 / 3.072 \times 10^7 = 16.7$ μs in all SC-FDMA symbols of a 0.5 ms slot

Main body of Symbol

- $2048 / 3.072 \times 10^7 = 66.67$ μs

★ 3GPP TS 36.211 expresses the second stage of generating an SC-FDMA signal as:

$$s_l(t) = \sum_{k=-\lfloor 12\times N_{RB}^{UL}/2 \rfloor}^{\lceil 12\times N_{RB}^{UL}/2 \rceil - 1} a_{k,l} \times e^{j2\pi(k+1/2)\times 15000 \times (t - N_{CP}T_s)}$$

$$0 \le t < (N_{CP} + 2048) \times T_s$$

★ This expression is similar to that used to generate an OFDMA signal in section 3.5. The expression for generating an OFDMA signal is divided into two halves as a result of the null subcarrier at the center of the channel bandwidth

- Similar to the OFDMA downlink, the SC-FDMA uplink uses the variable T_s as a minimum unit of time
  - the main body of an OFDMA symbol includes 2048 samples when sampling at a rate defined by T_s
  - $T_s = 66.67\ \mu s / 2048 = 32.55$ ns
- 3GPP References: TS 36.211

11.6 SC-FDMA SYMBOL

- Uplink SC-FDMA symbols have the same structures as downlink OFDMA symbols
- There are 7 SC-FDMA symbols per time slot when using the normal Cyclic Prefix, and 6 SC-FDMA symbols per time slot when using the extended Cyclic Prefix. These symbols are illustrated in Figure 135
- The 7.5 kHz subcarrier scenario is not defined for the uplink because it is only used for downlink MBMS transmissions

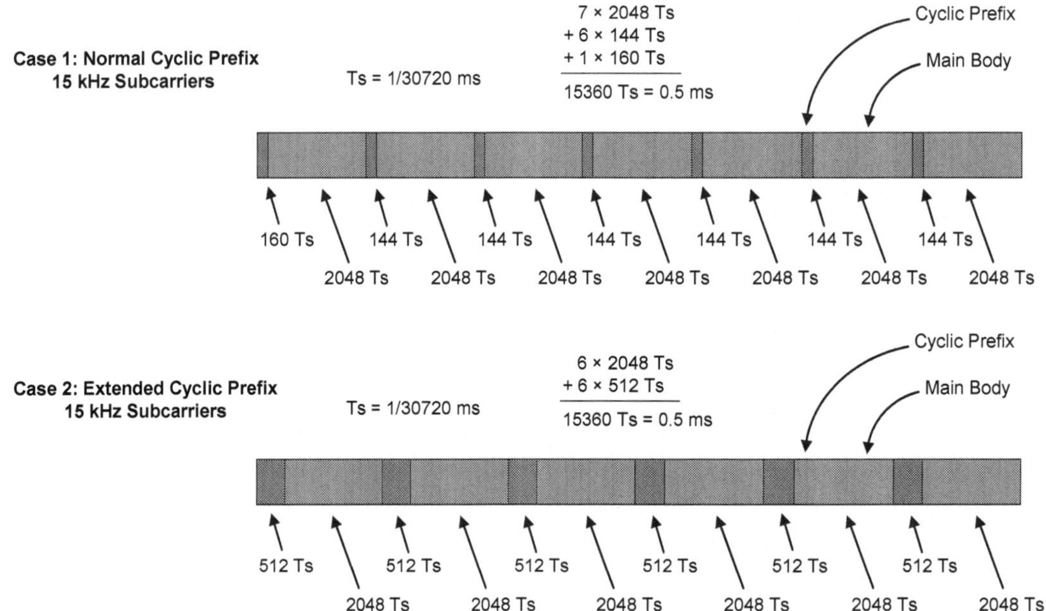

Figure 135 – Structure of the symbols within a 0.5 ms slot (normal and extended cyclic prefixes)

- Each SC-FDMA symbol has a Cyclic Prefix followed by the main body of the symbol
- When using the normal Cyclic Prefix, the duration of the first prefix within a time slot is 160 × Ts, whereas the duration of subsequent prefixes within a time slot is 144 × Ts
- When using the extended Cyclic Prefix, the duration of the prefix is always 512 × Ts
- The duration of the main body of the symbol is always 2048 × Ts = 66.67 μs, which is equal to the period of a 15 kHz cosine, i.e. 1/15000 = 66.67 μs
- 3GPP References: TS 36.211

11.7 CYCLIC PREFIX AND WINDOWING

★ The concepts of Cyclic Prefix and Windowing are the same for both the uplink and downlink. These concepts are described in sections 3.7 and 3.8. Some of the main points are repeated below

★ The cyclic prefix represents a guard band at the start of each symbol which provides protection against multi-path delay spread. The cyclic prefix also represents an overhead which should be minimised

★ The duration of the cyclic prefix should be greater than the multi-path delay spread

★ LTE specifies both normal and extended cyclic prefix lengths. The normal cyclic prefix is intended to be sufficient for the majority of scenarios while the extended cyclic prefix is intended for scenarios with particularly high delay spread. Normal and extended cyclic prefixes are summarised in Table 151

	Normal		Extended
	160 Ts	144 Ts	512 Ts
Duration	5.2 μs	4.7 μs	16.7 μs
Equivalent Distance	1.6 km	1.4 km	5 km
Overhead	160 / 2048 = 7.8 %	144 / 2048 = 7.0 %	512 / 2048 = 25 %

Table 151 – Cyclic prefix lengths for the uplink of LTE

★ 3GPP has not specified the use of windowing for LTE. Equipment vendors may introduce windowing to help achieve the spectrum emission requirements

★ The signal is likely to be discontinuous at the boundary between SC-FDMA symbols. Discontinuities generate increased spectrum emissions outside the channel bandwidth. The impact of discontinuities can be reduced by using a windowing function to smooth the signal at inter-symbol boundaries

★ Windowing reduces the levels of unwanted spectrum emissions but also reduces the effectiveness of the cyclic prefix because the signal is distorted to some extent

11.8 TRANSMITTER AND RECEIVER CHAIN

★ The functions provided by the receiver are the inverse of those provided by the transmitter

★ In the case of the PUSCH, the transmitter combines the FFT Precoding function with the IFFT to generate the SC-FDMA symbols. The PUSCH receiver combines the FFT function with IFFT Decoding to extract the stream of modulation symbols. A cyclic prefix is added at the transmitter and removed at the receiver

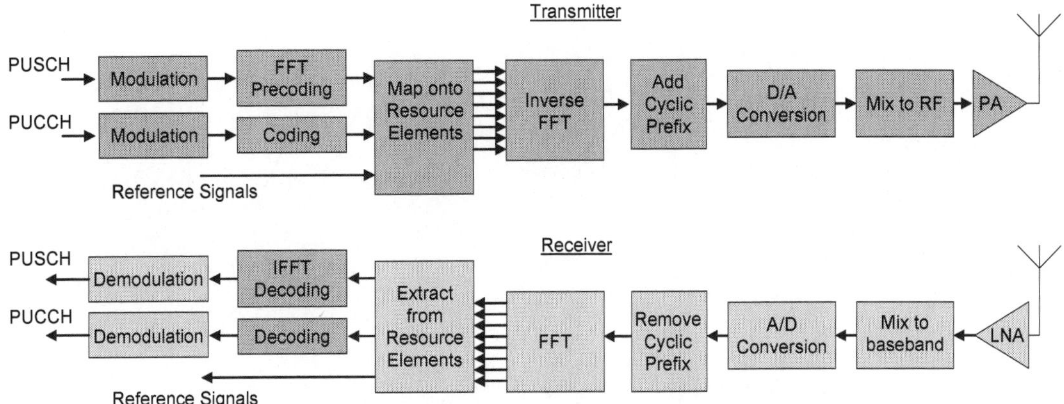

Figure 136 – Transmitter and receiver chain for LTE uplink

★ The transmitter includes a Power Amplifier (PA) to increase the signal strength to a level which can be transmitted across the target coverage area. The receiver includes a Low Noise Amplifier (LNA) to help improve the signal to noise ratio

12 UPLINK MULTIPLE ANTENNA TECHNOLOGIES

12.1 ANTENNA PORTS

★ 3GPP uses the concept of 'antenna ports'. It is important to differentiate between the concept of 'antenna ports' and physical antenna elements. Antenna ports map onto physical antenna elements

★ Prior to the release 10 version of the 3GPP specifications, it was not necessary to specify antenna ports in the uplink direction because only single antenna port transmission was used

★ The release 10 version of the specifications requires multiple antenna ports to provide support for:

- spatial multiplexing for the PUSCH (2×2 or 4×4 MIMO)
- transmit diversity for the PUCCH (dual antenna port)

★ The complete set of antenna ports and their associated physical channels and Reference Signals are presented in Table 152

Physical Channels and Signals	Index $\tilde{p}$	Antenna Port (p) as a function of the number of antenna ports		
		1	2	4
PUSCH and Demodulation Reference Signal for the PUSCH	0	10	20	40
	1	-	21	41
	2	-	-	42
	3	-	-	43
Sounding Reference Signal (SRS)	0	10	20	40
	1	-	21	41
	2	-	-	42
	3	-	-	43
PUCCH and Demodulation Reference Signal for the PUCCH	0	100	200	-
	1	-	201	-

Table 152 – Antenna ports and their associated physical channels and Reference Signals

★ The PUSCH and its associated demodulation Reference Signal use antenna ports {10}, {20, 21} and {40, 41, 42, 43}

★ The Sounding Reference Signal (SRS) uses the same antenna ports as the PUSCH so any channel estimates derived from the SRS are directly applicable to the PUSCH

★ The PUCCH and its associated demodulation Reference Signal use antenna ports {100} and {200, 201}

★ The index $\tilde{p}$ is used as a reference to each antenna port within the 3GPP specifications

★ Each antenna port has its own grid of Resource Elements

★ The physical antenna elements at the UE are shared between the antenna ports used by the PUSCH, SRS and PUCCH

★ An example mapping for 2 physical antenna elements at the UE is illustrated in Figure 137

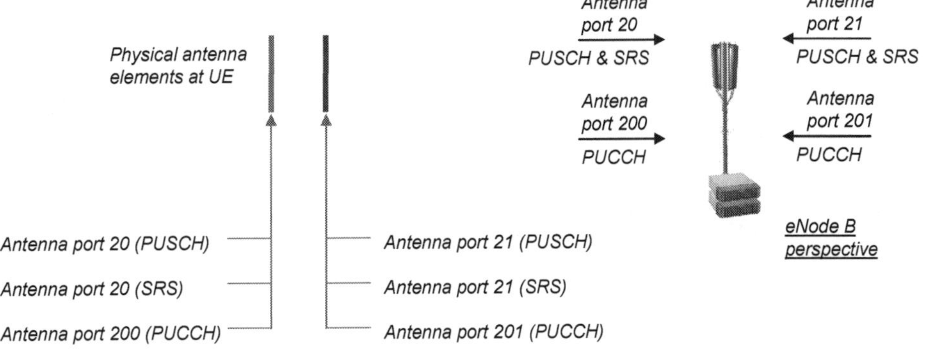

Figure 137 – Example of one-to-one mapping between antenna port and physical antenna elements

★ Use of antenna ports 40, 41, 42 and 43 requires at least 4 physical antenna elements at the UE. This is likely to be more practical for larger devices, e.g. laptop or tablet, rather than smaller handheld devices

★ 3GPP References: TS 36.211

12.2 TRANSMISSION MODES

★ The release 8 and 9 versions of the 3GPP specifications do not specify a set of uplink transmission modes because only single antenna port transmission is supported

★ The release 10 version of the specifications introduces transmission modes 1 and 2 to differentiate between single antenna port transmission and closed loop spatial multiplexing

★ The eNode B signals the uplink transmission mode to the UE within an RRC Connection Setup, RRC Connection Reconfiguration or RRC Connection Re-establishment message

★ Transmission mode 2 allows dynamic switching between single antenna port transmission and closed loop spatial multiplexing. This switching can be completed without any RRC signalling

★ The set of uplink transmission modes is summarised in Table 153. These are applicable when the UE is addressed by its C-RNTI. Closed loop spatial multiplexing is not supported when addressing the UE by an SPS-RNTI

★ The Search Space defines the set of PDCCH within which the UE checks for a resource allocation. This avoids the UE having to decode all PDCCH. Search Spaces are described in greater detail in section 9.2

★ The Downlink Control Information (DCI) format defines the structure and content of the resource allocation on the PDCCH. DCI formats are described in greater detail in section 9

Mode	PUSCH Transmission Scheme	Antenna Ports	DCI Format	Search Space
1	Single Antenna Port, port 10	{10}	0	Common and UE Specific
2	Single Antenna Port, port 10	{10}	0	Common and UE Specific
	Closed Loop Spatial Multiplexing	{20, 21}, {40, 41, 42, 43}	4	UE Specific

Table 153 – PUSCH transmission modes when using C-RNTI to address the UE

★ Transmission mode 1 provides support for single antenna port transmission. DCI format 0 provides support for either contiguous Resource Block allocations (resource allocation type 0) or non-contiguous Resource Block allocations (resource allocation type 1). Neither layer mapping nor precoding are applied at the UE

★ Transmission mode 2 provides support for both 2×2 and 4×4 closed loop spatial multiplexing. DCI format 4 provides support for signalling the use of either 1 or 2 transport blocks. It also provides support for signalling the number of layers and the precoding to be applied by the UE. Similar to DCI format 0, DCI format 4 provides support for either contiguous Resource Block allocations (resource allocation type 0) or non-contiguous Resource Block Allocations (resource allocation type 1)

★ 3GPP References: TS 36.213

12.3 LTE TECHNOLOGIES

12.3.1 TRANSMIT ANTENNA SELECTION

★ It is mandatory for all UE categories to support downlink receive diversity. This means that UE are already equipped with multiple antenna for the reception of downlink transmissions

★ Transmit antenna selection allows the uplink to take advantage of these multiple antenna with minimal changes to the implementation. The concept of transmit antenna selection is shown in Figure 138

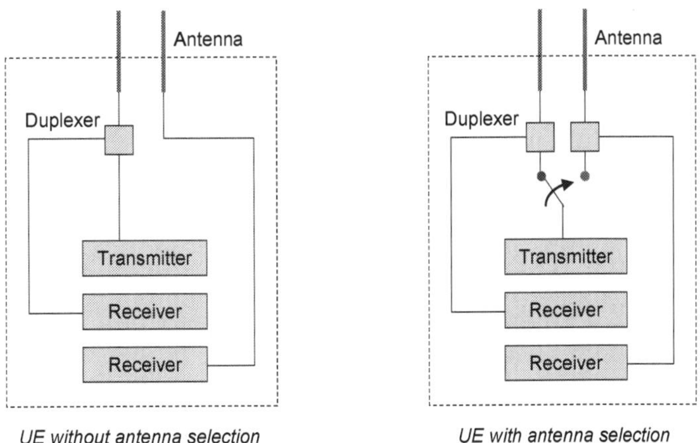

Figure 138 – UE with and without transmit antenna selection

★ Transmit antenna selection can be useful if the end-user holds the UE in such a way that one antenna is covered by his or her hand, i.e. one antenna is shielded so experiences increased link loss

★ The inclusion of a switch to allow the transmitter to use either antenna is likely to introduce a small insertion loss

★ Transmit antenna selection is introduced within the release 8 version of the 3GPP specifications

★ It is not mandatory for UE to support transmit antenna selection. The UE Capability Information message is used by the UE to inform the network of whether or not transmit antenna selection is supported

★ The network can instruct the UE to use transmit antenna selection in either open loop or closed loop mode. This instruction can be sent using an RRC Connection Setup, RRC Connection Reconfiguration or RRC Connection Re-establishment message

★ The open loop mode allows the UE to select the antenna for uplink transmissions

★ The closed loop mode involves the eNode B instructing the UE to use one of two antenna. The instructions are sent using DCI format 0. The instructions do not explicitly appear within the content of DCI format 0 but are used in combination with the RNTI to scramble the CRC bits, i.e. UE can deduce the instruction when decoding the CRC bits. The instruction simply tells the UE to use either antenna 0 or antenna 1

★ 3GPP References: TS 36.331, TS 36.212, TS 36.213

12.3.2 MULTI-USER MIMO

★ Multi-User MIMO (MU-MIMO) refers to the eNode B allocating the same time and frequency resources to more than a single UE, i.e. multiple UE are allocated the same Resource Blocks during the same subframe. The general concept of uplink multi-user MIMO is illustrated in Figure 139

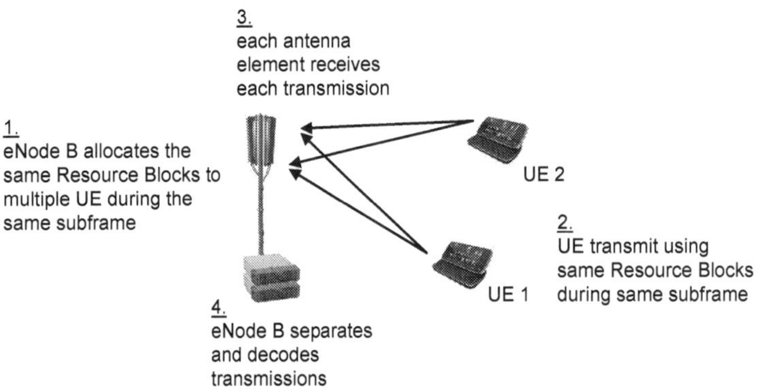

Figure 139 – Concept of uplink multi-user MIMO

★ The main objective of multi-user MIMO is to increase the overall system throughput, i.e. improve spectrum efficiency. Multi-user MIMO represents a form of Space Division Multiple Access (SDMA)

★ 3GPP has not explicitly specified support for uplink multi-user MIMO. Nevertheless, it has been possible to support uplink multi-user MIMO since the release 8 version of the 3GPP specifications

★ The general concept is similar to Single User MIMO (SU-MIMO) except that each layer of transmission originates from a different UE. Only a single transmit antenna is required at each UE so there is no impact upon UE implementation complexity

★ Uplink multi-user MIMO is transparent to the UE. None of the UE are aware that any other UE has been allocated the same set of Resource Blocks

★ The network vendor is responsible for implementing functionality for multi-user MIMO within the eNode B:

- o selecting the UE to share the same Resource Block allocation
- o making the uplink resource allocations on the PDCCH
- o separating and decoding the uplink transmissions from each UE

★ Each multi-user MIMO UE is allocated a different cyclic shift for its Demodulation Reference Signal (DM-RS). This cyclic shift is allocated with the Resource Block allocation within either DCI format 0 or DCI format 4

★ The use of different cyclic shifts provides orthogonality between the UE and helps the eNode B to separate them. Larger cyclic shifts provide greater orthogonality. Orthogonality is achieved when each UE is allocated the same number of Resource Blocks. Orthogonality is lost if each UE is allocated a different number of Resource Blocks. This constraint reduces the flexibility of Resource Block scheduling

★ The release 8 and 9 versions of the 3GPP specifications define only a cyclic shift to differentiate between Demodulation Reference Signals. The release 10 version of the specifications introduces the use of an Orthogonal Cover Code (OCC) to allow an additional dimension of differentiation. The OCC allows orthogonality to be maintained when multi-user MIMO UE are allocated different numbers of Resource Blocks

★ 3GPP References: TS 36.211

12.4 LTE ADVANCED TECHNOLOGIES

- ★ LTE Advanced has an objective to increase the peak uplink spectrum efficiency to 15 bps/Hz. Carrier Aggregation has limited impact upon spectrum efficiency because increased throughputs are achieved by increasing the bandwidth. Multiple antenna transmission technologies allow the throughput to increase without increasing the bandwidth
- ★ The release 8 and 9 versions of the 3GPP specifications do not support spatial multiplexing in the uplink
- ★ The release 10 version of the 3GPP specifications introduces the concept of transmission modes for the uplink. Transmission mode 2 is introduced to support the requirements of multiple antenna transmission for LTE Advanced. Transmission mode 2 supports:
    - o spatial multiplexing with 4 layers of parallel data transfer which allows a single UE to benefit from 4×4 MIMO. Antenna ports 40, 41, 42 and 43 are used to transfer these 4 layers
    - o spatial multiplexing with 2 layers of parallel data transfer which allows a single UE to benefit from 2×2 MIMO. Antenna ports 20 and 21 are used to transfer these 2 layers
    - o dynamic switching between single antenna port transmission and either 4×4 MIMO or 2×2 MIMO. This switching can be completed without any RRC signalling
- ★ The introduction of spatial multiplexing in the uplink direction generates the requirement for:
    - o Demodulation Reference Signals (DM-RS) for each transmission layer
    - o Sounding Reference Signals (SRS) for each antenna port
- ★ DM-RS are generated for each transmission layer. The UE applies the same precoding to both the DM-RS and the PUSCH to generate the signals to transmit from each antenna port. The eNode B can then use the DM-RS to estimate the channel response associated with each layer and consequently help to decode the uplink data
- ★ The DM-RS belonging to each layer is differentiated by its cyclic shift. Orthogonal Covering Codes (OCC) can also be applied to provide additional differentiation. DM-RS are described in greater detail in section 13.1
- ★ The release 8 and 9 versions of the 3GPP specifications are limited to SRS transmission using a single antenna port. 4×4 MIMO requires transmission of the SRS on 4 antenna ports. Likewise, 2×2 MIMO requires transmission of the SRS on 2 antenna ports. The release 10 version of the specifications allows the SRS to be transmitted on the same antenna ports as the PUSCH
- ★ Triggering SRS transmission is also made more flexible in the release 10 version of the specifications. Rather than relying upon configuring SRS transmissions with RRC signalling, the release 10 version of the specifications also allows SRS transmissions to be triggered using flags within Downlink Control Information (DCI)
    - o trigger type 0: SRS transmission triggered by RRC signalling configuration
    - o trigger type 1: SRS transmissions triggered by a DCI flag
- ★ The following DCI provide support for SRS transmissions using trigger type 1:
    - o DCI Format 0: single bit flag
    - o DCI Format 1A: single bit flag
    - o DCI Format 2B: single bit flag for TDD only
    - o DCI Format 2C: single bit flag for TDD only
    - o DCI Format 4: 2 bits used to select between SRS parameter sets configured by RRC signalling
- ★ The content of each DCI format is presented in section 9. SRS are described in greater detail in section 13.2
- ★ Uplink Multi-User MIMO is enhanced by the release 10 version of the specifications by defining Orthogonal Cover Codes (OCC) as an additional method to differentiate between the uplink DM-RS transmitted by each co-scheduled UE. These OCC allow the DM-RS to remain orthogonal when different but overlapping bandwidths are allocated to each UE. The release 8 and 9 versions of the specifications define cyclic shifts to differentiate between DM-RS. These only remain orthogonal when equal bandwidths are allocated to each of the co-scheduled UE
- ★ 3GPP References: TS 36.211, TS 36.213

12.4.1 CLOSED LOOP SPATIAL MULTIPLEXING

- LTE Advanced introduces closed loop spatial multiplexing for the uplink within the release 10 version of the 3GPP specifications:
    - 2×2 closed loop spatial multiplexing, requiring at least 2 physical antenna elements at the UE
    - 4×4 closed loop spatial multiplexing, requiring at least 4 physical antenna elements at the UE
- The antenna elements required at the UE may already be present for downlink spatial multiplexing, e.g. 4×4 closed loop spatial multiplexing in the downlink direction requires the UE to have at least 4 physical antenna elements
- However, uplink spatial multiplexing also requires the UE to support multiple transmit paths within its transmitter implementation. This requires multiple amplifiers and multiple RF stages running in parallel. These additional transmit paths will have an impact upon UE power consumption
- The eNode B provides the UE with instructions regarding:
    - the number of layers to transmit
    - the precoding to apply

  These instructions are provided in Downlink Control Information (DCI) format 4 when the uplink resources are allocated to the UE. The precoding to be applied is specified using the Transmitted Precoding Matrix Indicator (TPMI)
- The eNode B selects a TPMI to maximise the signal to noise ratio at its receiver. Applying the set of precoding weights at the UE represents a form of maximum ratio combining at the transmitter
- Layer mapping and precoding for uplink spatial multiplexing are separated by the SC-FDMA precoding procedure. This is in contrast to the downlink where the precoding procedure immediately follows layer mapping. Figure 140 illustrates the case for uplink layer mapping and precoding when transmitting a single layer

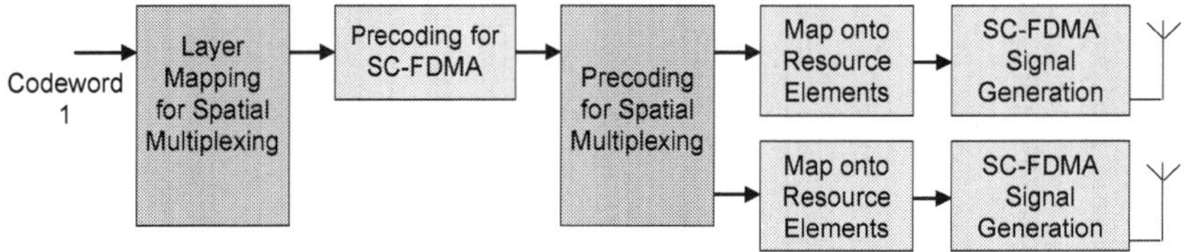

Figure 140 – Layer mapping and precoding for uplink spatial multiplexing (2×2 MIMO using 1 layer)

- Transmitting a single layer is applicable to both 2×2 and 4×4 closed loop spatial multiplexing
- Figure 141 illustrates the equivalent case when transmitting 2 layers. The SC-FDMA precoding is completed for each layer prior to the precoding for spatial multiplexing

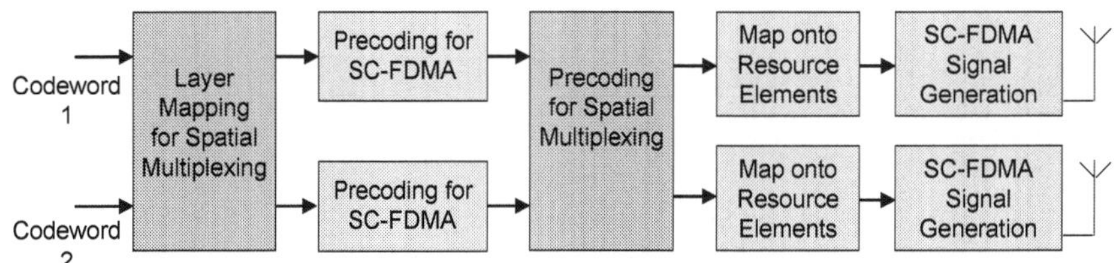

Figure 141 – Layer mapping and precoding for uplink spatial multiplexing (2×2 MIMO using 2 layers)

- Transmitting 2 layers is also applicable to both 2×2 and 4×4 closed loop spatial multiplexing
- When transmitting 4 layers, SC-FDMA precoding is completed in 4 parallel streams prior to the precoding for spatial multiplexing. The transmission of 4 layers is only applicable to 4×4 closed loop spatial multiplexing

2×2 CLOSED LOOP SPATIAL MULTIPLEXING

★ Table 154 presents the number of codewords and layers which are supported by 2×2 closed loop spatial multiplexing:

- either 1 or 2 codewords can be transferred during each 1 ms subframe
- the number of layers is always equal to the number of codewords

Number of Codewords	Number of Layers	Number of Antenna Ports
1	1	2
2	2	

Table 154 – Codewords, layers and antenna elements for 2×2 closed loop spatial multiplexing

★ The layer mapping process for the transfer of 1 and 2 codewords is shown in Figure 142. In both cases, layer mapping simply transfers the codeword for precoding

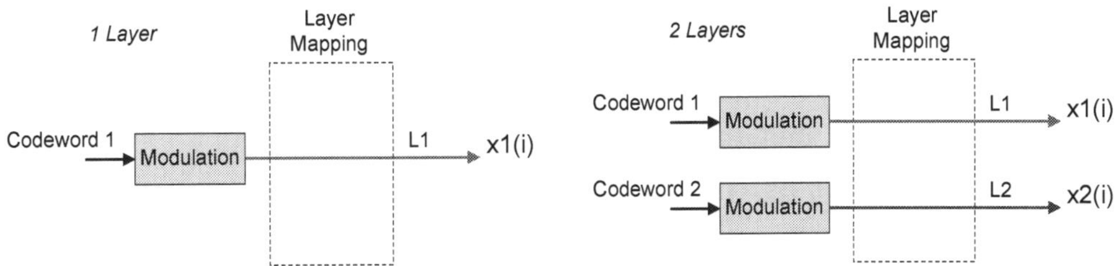

Figure 142 – Layer mapping for 2×2 closed loop spatial multiplexing

★ Layer mapping is followed by FFT precoding for SC-FDMA. This is described in section 11.5. Precoding for spatial multiplexing follows the precoding for SC-FDMA

★ The structure of precoding for 2×2 spatial multiplexing is shown in Figure 143

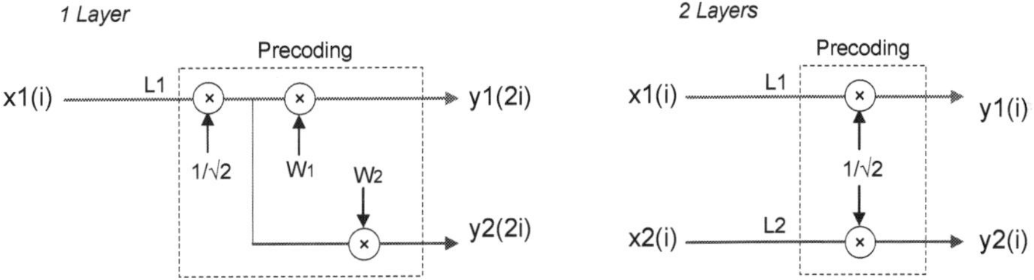

Figure 143 – Precoding for 2×2 closed loop spatial multiplexing

★ In the case of a single codeword and a single layer, precoding results in:

- the 1st antenna transmitting, (codeword × 1/√2 × precoding weight 1)
- the 2nd antenna transmitting (codeword × 1/√2 × precoding weight 2)

★ The precoding weights for transmission of a single codeword are selected from the set of values presented in Table 155

Codebook Index	Precoding Weight 1	Precoding Weight 2
0	1	1
1	1	-1
2	1	j
3	1	-j
4	1	0
5	0	1

Table 155 – Codebook for 2×2 closed loop spatial multiplexing with 1 codeword

★ Codebook indices 0 to 3 apply a fixed weight of 1 to the first antenna and a configurable weight to the second antenna. The set of weights are the same as those applied to the downlink for 2x2 closed loop spatial multiplexing. They are based upon the QPSK alphabet and provide a trade-off between performance and complexity

★ It is not necessary to consider the impact of precoding upon the peak-to-average power ratio when using a single transmission layer (rank 1) because the peak-to-average power ratio remains the same after precoding (measured at each individual antenna port)

★ Codebook indices 4 and 5 have been included to provide support for transmit antenna selection, i.e. one of the two weights is zero. Transmit antenna selection may be used if the eNode B detects that the end-user is holding the UE in such a way that one antenna is covered by his or her hand, i.e. one antenna is shielded so experiences increased link loss. This helps to save UE battery power when the antenna is having limited impact

★ The codebook index to be applied during a specific subframe is signalled to the UE within Downlink Control Information (DCI) format 4 on the PDCCH physical channel

★ In the case of two codewords and two layers, precoding results in:

  o the 1st antenna transmitting, (codeword 1 × 1/√2)

  o the 2nd antenna transmitting (codeword 2 × 1/√2)

  In this case, there is no precoding other than the amplitude scaling. It is assumed that the gain from precoding is smaller for full-rank transmission when using advanced receivers at the eNode B. Full-rank transmission means that the number of layers equals the number of antenna ports

★ 3GPP References: TS 36.211, TS 36.213

4×4 CLOSED LOOP SPATIAL MULTIPLEXING

★ Table 156 presents the number of codewords and layers which are supported by 4×4 closed loop spatial multiplexing.

  o either 1 or 2 codewords can be transferred during each 1 ms subframe

  o the number of layers is always greater than or equal to the number of codewords

Number of Codewords	Number of Layers	Number of Antenna Elements
1	1	4
	2	
2	2	
	3	
	4	

Table 156 – Codewords, layers and antenna elements for 4×4 closed loop spatial multiplexing

★ The layer mapping process is illustrated in Figure 144. Layer mapping provides serial to parallel conversion when 2 layers are generated from a single codeword

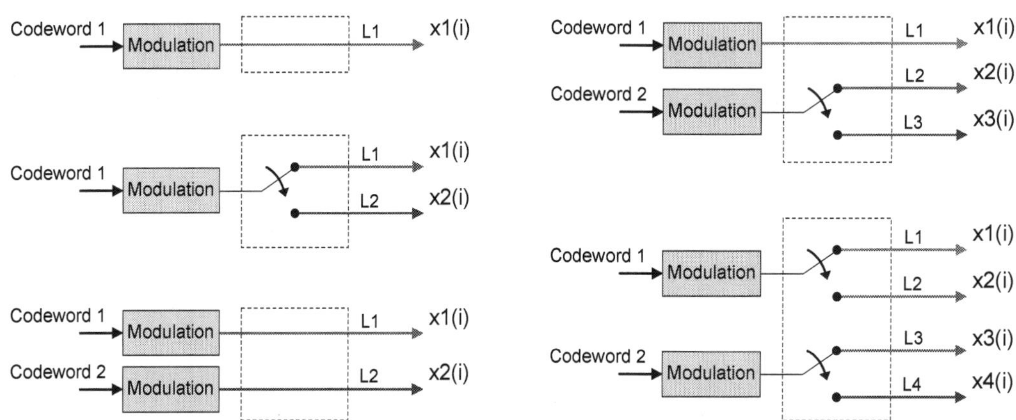

Figure 144 – Layer mapping for 4×4 closed loop spatial multiplexing

★ The number of layers should match the current channel conditions between the set of transmit antenna and the set of receive antenna. Low correlation propagation paths between the transmit and receive antenna allows increased numbers of layers. This corresponds to a propagation channel with increased rank

★ The mapping of a single codeword onto 2 layers is included to provide support for retransmissions. Retransmissions are completed using the same layer mapping as the original transmission. For example, if 2 codewords were originally used to generate 4 layers (2 layers per codeword), but only 1 codeword was received successfully then the other codeword is retransmitted after serial to parallel conversion to generate 2 layers

★ Layer mapping is followed by FFT precoding for SC-FDMA. This is completed independently for each layer, and is described in section 11.5. Precoding for spatial multiplexing follows the precoding for SC-FDMA

★ The codebooks used for each rank of uplink spatial multiplexing are independent, e.g. the codebook used for 2-layer transmission is not a subset of the codebook used for 3 or 4 layer transmission

★ The structure of precoding for 4×4 spatial multiplexing when using a single layer is shown in Figure 145

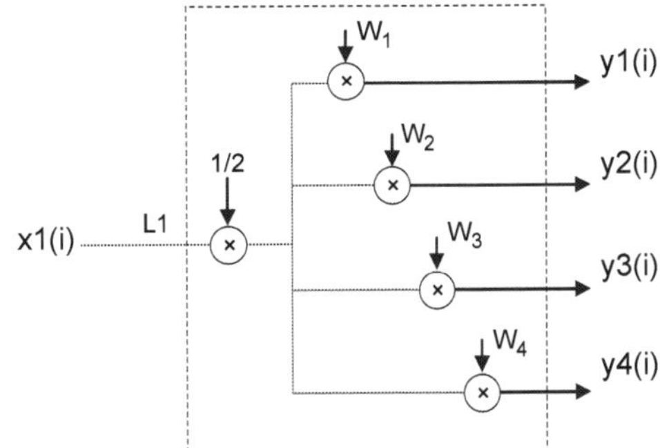

Figure 145 – Precoding for 4×4 closed loop spatial multiplexing (1 layer)

★ The precoding weights, w_1 to w_4 are selected from the set of values presented in Table 157

Codebook Index	Precoding Weight			
	W1	W2	W3	W4
0	1	1	1	-1
1			j	j
2			-1	1
3			-j	-j
4		j	1	j
5			j	1
6			-1	-j
7			-j	-1
8	1	-1	1	1
9			j	-j
10			-1	-1
11			-j	j
12		-j	1	-j
13			j	-1
14			-1	j
15			-j	1
16	1	0	1	0
17			-1	
18			j	
19			-j	
20	0	1	0	1
21				-1
22				j
23				-j

Table 157 – Codebook for 4×4 closed loop spatial multiplexing (1 layer)

★ The codebook is based upon the QPSK alphabet to provide a trade-off between performance and complexity

★ It is not necessary to consider the impact of precoding upon the peak-to-average power ratio when using a single transmission layer (rank 1) because the peak-to-average power ratio remains the same after precoding (measured at each individual antenna port), i.e. the peak-to-average power ratio experienced by each amplifier within the UE remains unchanged

★ Codebook indices 0 to 15 utilise all 4 antenna elements. Codebook indices 16 to 23 utilise only 2 of the 4 antenna elements. These last 8 codebook entries provide support for antenna selection when the UE is experiencing antenna gain imbalance, potentially caused by the end-user's hand wrapped around the device. It is assumed that the UE would have 2 pairs of cross polar elements so a hand would cover 2 elements at a time rather than a single element

★ The use of antenna selection helps to conserve UE power consumption which is more likely to be important when the UE is using single layer transmission because the UE is more likely to be towards the cell edge and transmitting at a relatively high power

★ The structure of precoding for 4×4 spatial multiplexing when using 2 layers is shown in Figure 146

★ The precoding weights, w_{1a}, w_{1b} to w_{4a}, w_{4b} are selected by the eNode B from the set of values presented in Table 158. The upper value in each entry represents the 'a' weight, whereas the lower value in each entry represents the 'b' weight

★ Similar to single layer transmission, the codebook is based upon the QPSK alphabet to provide a trade-off between performance and complexity

★ An important characteristic is that for each codebook entry, only one of the two values is non-zero. This means that each antenna transmits a precoded version of layer 1, or a precoded version of layer 2, but does not transmit a combination of both layers. This type of codebook design preserves the peak-to-average power ratio of the signal, i.e. the precoding does not change the peak-to-average power ratio measured at each individual antenna port

★ The codebook for 2 layer transmission does not support antenna selection because one of the two weights associated with each antenna element is always non-zero, i.e. the antenna port does not switch off

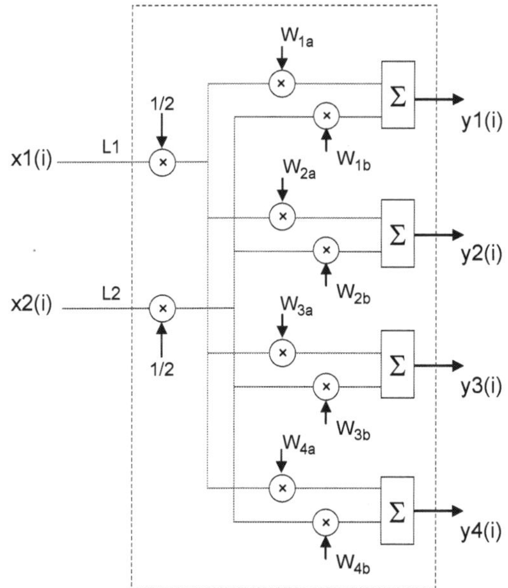

Figure 146 – Precoding for 4×4 closed loop spatial multiplexing (2 layers)

Codebook Index	Precoding Weight			
	W1	W2	W3	W4
0	1 0	1 0	0 1	0 -j
1				0 j
2		-j 0	0 1	0 1
3				0 -1

Codebook Index	Precoding Weight			
	W1	W2	W3	W4
4	1 0	-1 0	0 1	0 -j
5				0 j
6		j 0		0 1
7				0 -1

Codebook Index	Precoding Weight			
	W1	W2	W3	W4
8	1 0	0 1	1 0	0 1
9				0 -1
10			-1 0	0 1
11				0 -1

Codebook Index	Precoding Weight			
	W1	W2	W3	W4
12	1 0	0 1	0 1	1 0
13			0 -1	
14			0 1	-1 0
15			0 -1	

Table 158 – Codebook for 4×4 closed loop spatial multiplexing (2 layers)

- ★ The structure of precoding for 4×4 spatial multiplexing when using 3 layers is shown in Figure 147
- ★ The precoding weights, w_{1a}, w_{1b}, w_{1c} to w_{4a}, w_{4b}, w_{4c} are selected by the eNode B from the set of values presented in Table 159. The upper value in each entry represents the 'a' weight, whereas the lower value in each entry represents the 'c' weight
- ★ In this case, the codebook is based upon the BPSK alphabet rather than the QPSK alphabet. This is possible because the number of codebook entries is relatively small, and the impact upon performance is expected to be minor
- ★ Each entry within the codebook for 3 layer transmission preserves the SC-FDMA peak-to-average power ratio by defining only a single non-zero value for each antenna. The total downlink transmit power is equally distributed across the set of 4 antenna because each codebook entry has a single equal amplitude value associated with each antenna. The total downlink transmit power is divided between the 3 layers in the ratio of 2:1:1. This ratio is used to provide the 2 codewords with equal transmit power (recall that codeword 1 forms layer 1, while serial-to-parallel conversion results in codeword 2 forming layers 2 and 3)
- ★ The codebook for 3 layer transmission does not support antenna selection because one of the three weights associated with each antenna element is always non-zero, i.e. the antenna port does not switch off

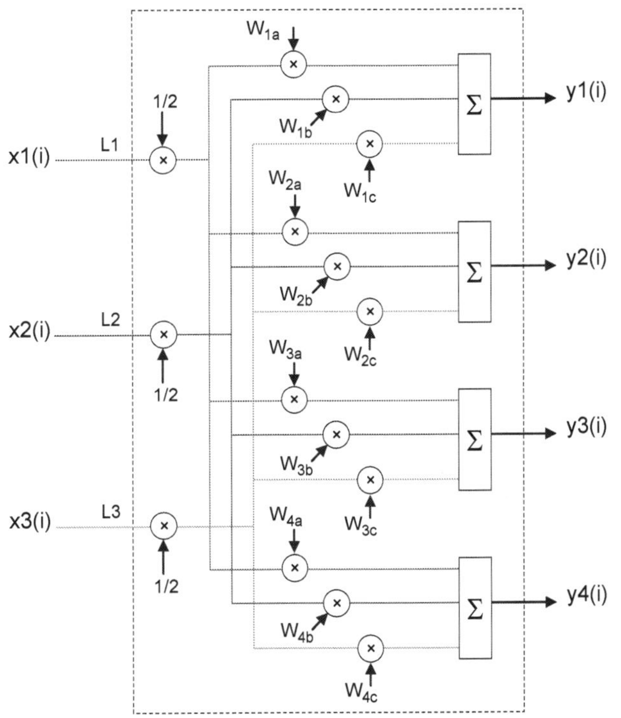

Figure 147 – Precoding for 4×4 closed loop spatial multiplexing (3 layers)

Codebook Index	Antenna 1	Antenna 2	Antenna 3	Antenna 4
0	1 0 0	1 0 0	0 1 0	0 0 1
1		-1 0 0		
2		0 1 0	1 0 0	
3			-1 0 0	

Codebook Index	Antenna 1	Antenna 2	Antenna 3	Antenna 4
4	1 0 0	0 1 0	0 0 1	1 0 0
5				-1 0 0
6	0 1 0	1 0 0	1 0 0	0 0 1
7			-1 0 0	

Codebook Index	Antenna 1	Antenna 2	Antenna 3	Antenna 4
8	0 1 0	1 0 0	0 0 1	1 0 0
9				-1 0 0
10		0 0 1	1 0 0	1 0 0
11				-1 0 0

Table 159 – Codebook for 4×4 closed loop spatial multiplexing (3 layers)

★ The structure of precoding for 4×4 spatial multiplexing when using 4 layers is shown in Figure 148

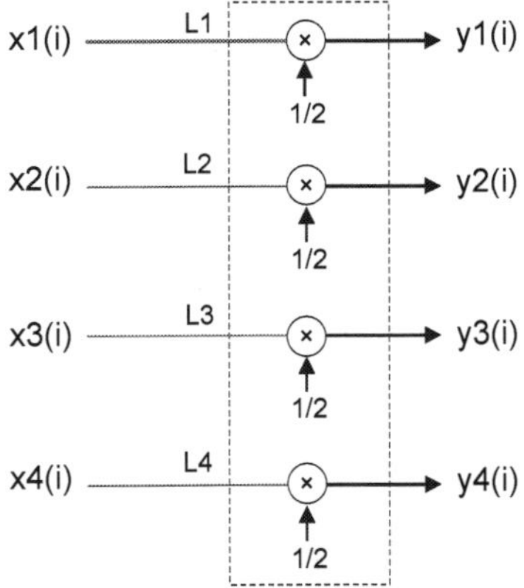

Figure 148 – Precoding for 4×4 closed loop spatial multiplexing (4 layers)

★ In this case, there is no precoding other than the amplitude scaling. Similar to full-rank transmission for 2×2 MIMO, it is assumed that the gain from precoding is small when using advanced receivers at the eNode B

★ 3GPP References: TS 36.211, TS 36.213

12.4.2 TRANSMIT DIVERSITY

★ 3GPP discussed the following types of transmit diversity for the PUSCH:

  - Frequency Switch Transmit Diversity (FSTD)
  - Cyclic Delay Diversity (CDD)
  - Space-Frequency Block Coding (SFBC)
  - Space-Time Block Coding (STBC)

  However, none of these have been included within the release 8, 9 nor 10 versions of the 3GPP specifications

★ The release 10 version of the 3GPP specifications introduces transmit diversity for the PUCCH

  - transmit diversity for the PUCCH has been specified for 2 antenna ports {200, 201}
  - these 2 antenna ports map onto 2, or more physical antenna elements at the UE

★ Transmit diversity for the PUCCH is based upon Spatial Orthogonal Resource Transmit Diversity (SORTD). This involves transmitting the same Uplink Control Information (UCI) from 2 different antenna ports. Each antenna port uses a different set of PUCCH resources, so SORTD doubles the utilisation of resources. This represents the main drawback of using SORTD. From the perspective of the eNode B, the two transmissions are similar to those which would be received from 2 separate UE

★ The general concept of SORTD is illustrated in Figure 149

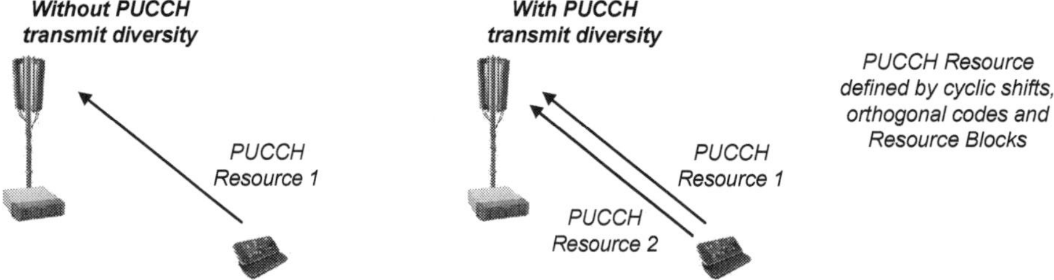

Figure 149 – General concept of PUCCH transmit diversity using SORTD

★ The high level processing used to generate the SORTD signals is presented in Figure 150. The 2 PUCCH transmissions transfer the same control information (scheduling request, HARQ acknowledgements, channel state information) but are generated independently in 2 parallel streams. This illustrates how SORTD consumes twice as many PUCCH resources

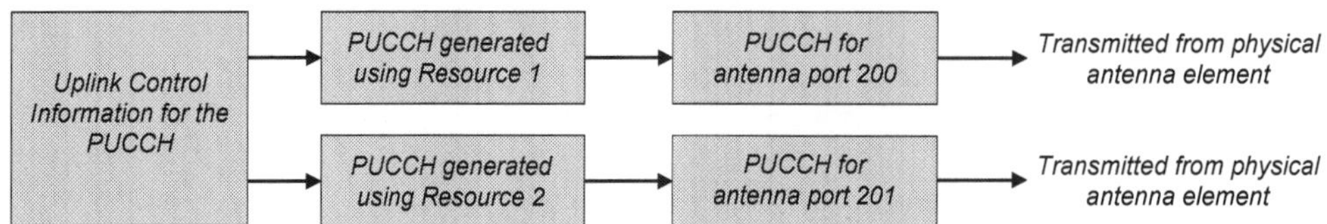

Figure 150 – General processing flow used to generate SORTD transmissions for the PUCCH

★ The method used to select the pair of PUCCH resources for SORTD depends upon the scenario:

  <u>PUCCH Formats 1, 1a and 1b (Scheduling Request, HARQ Acknowledgements)</u>:

  - The eNode B uses RRC signalling to provide the UE with 2 Scheduling Request (SR) resource indices, i.e. there is one resource index per antenna port. Each index can be within the range 0 to 2047. The UE transmits the PUCCH using Scheduling Request resources whenever it has a Scheduling Request to send. The 2 Scheduling Request resource indices are referenced as:
    - $n^{(1,p0)}_{\text{PUCCH,SRI}}$ and $n^{(1,p1)}_{\text{PUCCH,SRI}}$
  - The eNode B uses RRC signalling to provide the UE with a single HARQ acknowledgement resource index. This index can range from 0 to 2047. The UE uses this resource index to derive the resource indices for both antenna ports. The UE transmits the

PUCCH using the HARQ acknowledgement resource indices when it has HARQ acknowledgments to send without a Scheduling Request. The pair of resource indices are defined as:

- $n^{(1,p0)}_{\text{PUCCH}} = n_{\text{CCE}} + N^{(1)}_{\text{PUCCH}}$ and $n^{(1,p1)}_{\text{PUCCH}} = n_{\text{CCE}} + 1 + N^{(1)}_{\text{PUCCH}}$

  where, n_{CCE} is the index of the lowest Control Channel Element (CCE) used by the eNode B to transmit the PDCCH which allocated the downlink PDSCH transmission which the UE is acknowledging. $N^{(1)}_{\text{PUCCH}}$ is the single resource index signaled by the eNode B

- In the case of Semi Persistent Scheduling (SPS) (described in section 27.6.2), the eNode B uses RRC signalling to provide the UE with 2 sets of resource indices, i.e. there is one set of resource indices per antenna port. Each index can be within the range 0 to 2047. The UE transmits the PUCCH from 2 antenna ports using one resource index from each set. Selection of a resource index from within each set is based upon instructions from the eNode B within the PDCCH
- In the case of PUCCH Formats 1, 1a and 1b, the resource indices are used to determine the cyclic shift and orthogonal codes applied during the processing of the PUCCH signal. Processing of the PUCCH signal is illustrated in Figure 169 (section 14.2.1). The resource indices also determine the Resource Blocks used for transmission of the PUCCH
- The release 10 version of the 3GPP specifications does not define transmit diversity for PUCCH format 1b when using channel selection. Format 1b with channel selection can be used to increase the number of HARQ acknowledgements signalled by a single PUCCH transmission

<u>PUCCH Formats 2, 2a and 2b (Channel State Information, HARQ Acknowledgements)</u>:

- The eNode B uses RRC signalling to provide the UE with 2 resource indices, i.e. there is one resource index per antenna port. Each index can be within the range 0 to 1184. The 2 resource indices are referenced as:
  - $n^{(2,p0)}_{\text{PUCCH}}$ and $n^{(2,p1)}_{\text{PUCCH}}$
- In the case of PUCCH Formats 2, 2a and 2b, the resource indices are used to determine the cyclic shift applied during the processing of the PUCCH signal. Processing of the PUCCH signal is illustrated in Figure 171 (section 14.2.2). The resource indices also determine the Resource Blocks used for transmission of the PUCCH

<u>PUCCH Format 3 (Scheduling Request, HARQ Acknowledgements)</u>:

- PUCCH Format 3 can be used when carrier aggregation is configured
- The eNode B uses RRC signalling to provide the UE with 2 sets of resource indices, i.e. there is one set of resource indices per antenna port. Each index can be within the range 0 to 549
- When data is transferred from a secondary cell, the UE transmits PUCCH format 3 from 2 antenna ports using one resource index from each set of resource indices. Selection of a resource index from within each set is based upon instructions from the eNode B within the PDCCH. The 2 resource indices selected from the 2 sets of resource indices are referenced as:
  - $n^{(3,p0)}_{\text{PUCCH}}$ and $n^{(3,p1)}_{\text{PUCCH}}$
- In the case of PUCCH Format 3, the resource indices are used to determine the cyclic shift and orthogonal codes applied during the processing of the PUCCH signal. Processing of the PUCCH signal is illustrated in Figure 172 (section 14.2.3). The resource indices also determine the Resource Blocks used for transmission of the PUCCH
- When data is scheduled from only the primary cell then PUCCH formats 1a and 1b are used to send acknowledgements

★ 3GPP References: TS 36.211, TS 36.213, TS 36.331

13 UPLINK SIGNALS

13.1 DEMODULATION REFERENCE SIGNAL

★ Demodulation Reference Signals (DM-RS) are used for channel estimation and synchronisation during demodulation. The eNode B can also use the timing of the Demodulation Reference Signal to generate timing advance instructions for the UE

★ There are two types of Demodulation Reference Signal:

- PUSCH Demodulation Reference Signal
- PUCCH Demodulation Reference Signal

★ Both types of Demodulation Reference Signal are generated from a common set of base sequences. 3GPP TS 36.211 specifies 30 groups of base sequences, where the number of sequences within each group depends upon the uplink channel bandwidth. Each group includes

- 5 base sequences which have lengths of {12, 24, 36, 48, 60}
- 2 × y base sequences which have lengths of {72, 72, 84, 84, 96, 96, 108, 108, ...}, where y = maximum number of uplink Resource Blocks – 5

All base sequences have lengths which are multiples of 12, i.e. the number of subcarriers within a Resource Block. Figure 151 illustrates the 30 groups of sequences for the 3 MHz channel bandwidth (15 Resource Blocks)

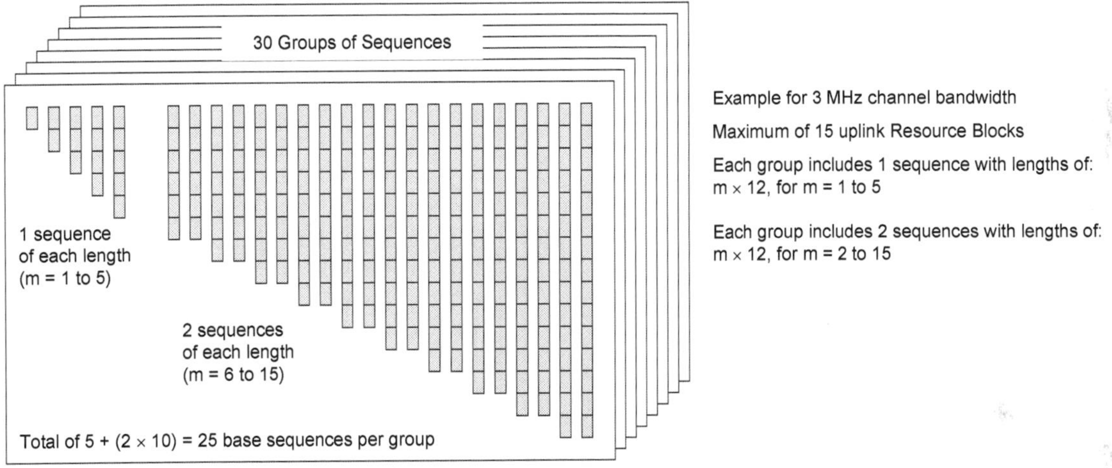

Figure 151 – Groups of base sequences used by the PUSCH and PUCCH Demodulation Reference Signals

★ The UE selects a group of base sequences using a Group Hopping calculation. This calculation (presented in the following sections) is dependent upon whether the sequence is going to be used for the PUSCH or PUCCH Reference Signal

★ The UE selects a sequence length depending upon the number of Resource Blocks being transmitted. For example, if the UE is transmitting the PUSCH using 10 Resource Blocks then it will select a sequence with a length of 10 × 12 = 120. Similarly, if the UE is transmitting the PUCCH using 1 Resource Block then it well select a sequence with a length of 1 × 12 = 12

★ When the sequence length is greater than 60, the UE also has to select which sequence to use from the pair of sequences with the appropriate length. This is only applicable to the PUSCH Reference Signal because the PUCCH Reference Signal always has a length of 12, i.e. a single Resource Block is used within each time slot when transmitting the PUCCH. The sequence selection procedure for the PUSCH is presented in the following section

★ The eNode B applies the same selection procedures as the UE so it knows which Reference Signal sequence to expect

★ 3GPP References: TS 36.211

13.1.1 PUSCH DEMODULATION REFERENCE SIGNAL

★ The PUSCH Demodulation Reference Signal is included within every Resource Block allocated to the PUSCH. Figure 152 illustrates the Resource Elements allocated to the PUSCH Demodulation Reference Signal , i.e. a single column of Resource Elements within each Resource Block

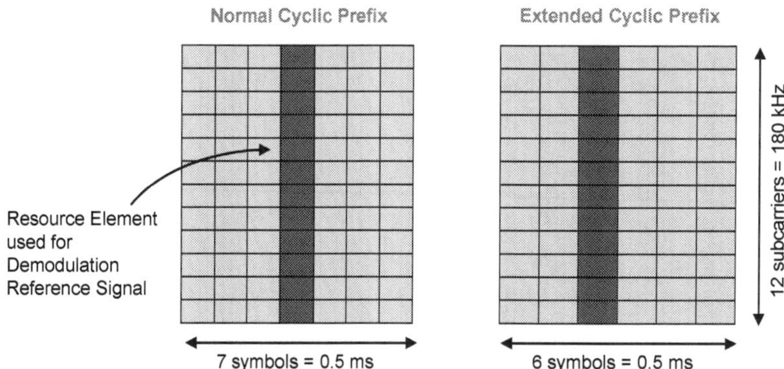

Figure 152 – Resource Elements used by the PUSCH Demodulation Reference Signal

★ Recall that in the downlink direction, Cell Specific Reference Signals are distributed in both the time and frequency domains. This approach allows channel estimation to generate a result which is representative of the complete Resource Block. In the uplink direction, PUSCH Demodulation Reference Signals are distributed in only the frequency domain. This is necessary to preserve the reduced peak-to-average power ratio provided by SC-FDMA

★ The release 8 and 9 versions of the 3GPP specifications limit the PUSCH to transmission using a single antenna port. In this case, it is only necessary to generate a single PUSCH Demodulation Reference Signal sequence for each time slot. This sequence is given by:

$$r_{u,v}^{(\alpha)}(n)$$

where,

α is the cyclic shift applied to the base sequence (12 different cyclic shifts are available)

u is the base sequence group number (0 to 29)

v is the index of the base sequence with an appropriate length (0 or 1)

n ranges from 0 to (number of allocated subcarriers – 1)

★ The base sequence group number, 'u' is selected using the equation:

$$u = (\mathrm{f_{gh}} + \mathrm{f_{ss}}) \bmod 30$$

where,

$\mathrm{f_{gh}} = 0$ if group hopping is disabled (configuration is cell specific and is broadcast in SIB2)

$\mathrm{f_{gh}}$ is a pseudo-random number between 0 and 29 if group hopping is enabled. This pseudo-random number is dependent upon the time slot number and the Physical layer Cell Identity (PCI)

$\mathrm{f_{ss}} = ((\mathrm{PCI} \bmod 30) + \Delta_{\mathrm{ss}}) \bmod 30$

Δ_{ss} is cell specific and is broadcast in SIB2. It defines an offset between 0 and 29 which can be used to select different sequence groups for cells with equal PCI

★ The equation above indicates that the sequence group is cell specific rather than UE specific. Planning of sequence groups is discussed in section 31.7

★ The index of the sequence, 'v' is always 0 when the sequence length belongs to the set {12, 24, 36, 48, 60} because there is only 1 sequence of each length. The index can have values of 0 or 1 for longer sequences because there are 2 sequences of each length

- index 0 is used by default
- if group hopping is disabled and sequence hopping is enabled then the index is based upon a pseudo-random sequence which is dependent upon the PCI and the value of $\mathrm{f_{ss}}$. SIB2 includes a flag to indicate whether or not sequence hopping is enabled for the cell. If enabled for the cell, it can be disabled for a specific UE using the 'Disable Sequence Group Hopping' parameter within an RRC Connection Setup, RRC Connection Reconfiguration or RRC Connection Re-establishment message

★ Sequence hopping helps to reduce the probability of clashes between the sequences used by the population of UE

★ The cyclic shift applied to the base sequence is given by:

$$\alpha = 2\pi\, n_{cs} / 12$$

where,

$n_{cs} = (n^{(1)}_{DMRS} + n^{(2)}_{DMRS} + n_{PRS}) \bmod 12$

$n^{(1)}_{DMRS}$ is cell specific, and is determined from a look-up table indexed using the cyclic shift broadcast in SIB2

$n^{(2)}_{DMRS}$ is UE specific, and is determined from a look-up table indexed using the cyclic shift within the DCI

n_{PRS} is generated using a pseudo-random sequence which is dependent upon the time slot number, the Physical layer Cell Identity (PCI) and the value of f_{ss}

★ The cyclic shift is applied to the appropriate base sequence using the equation:

$$r^{(\alpha)}_{u,v}(n) = e^{j\alpha n} \times \bar{r}_{u,v}(n) \quad \text{where} \quad \bar{r}_{u,v}(n) \text{ is the base sequence}$$

★ In general, users within a cell are allocated different Resource Blocks so do not need to be differentiated by their cyclic shifts. Multiple UE share the same uplink Resource Blocks if multi-user MIMO is used. In this case, users are differentiated by their cyclic shifts which provide orthogonality as long as the number of Resource Blocks allocated to each user is equal

★ The processing used to generate the Demodulation Reference Signal for the PUSCH is shown in Figure 153

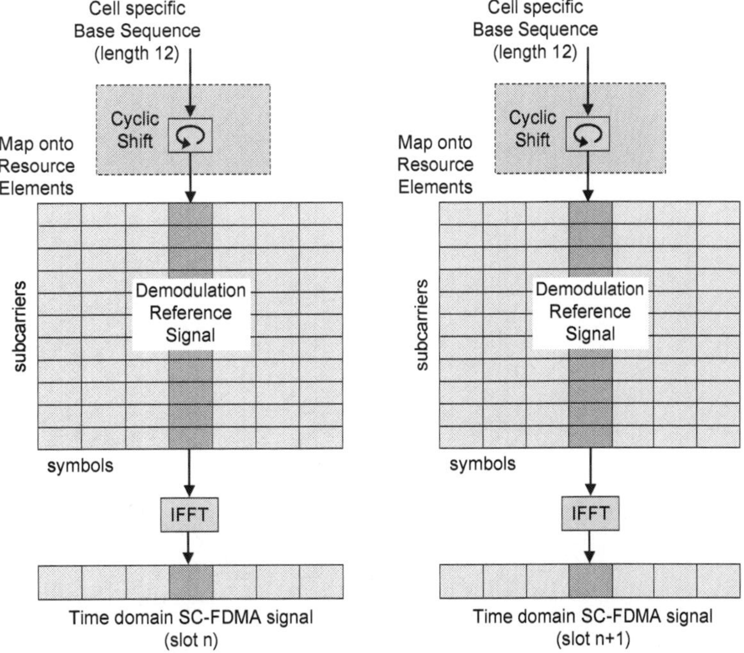

Figure 153 – Processing to generate Demodulation Reference Signal for the PUSCH (3GPP releases 8 and 9, normal cyclic prefix)

★ The release 10 version of the specifications introduces the use of an Orthogonal Cover Code (OCC) to allow an additional dimension of differentiation. The OCC allows orthogonality to be maintained when multi-user MIMO UE are allocated different numbers of Resource Blocks

★ The release 10 version of the specifications provides support for uplink MIMO and allows transmission using up to 4 layers. In this case, a PUSCH Demodulation Reference Signal is generated for each layer. These Demodulation Reference Signals are differentiated using a combination of OCC and cyclic shifts. The 3GPP release 10, PUSCH Demodulation Reference Signal sequence is defined as:

$$w^{(\lambda)}(m) \times r^{(\alpha(\lambda))}_{u,v}(n)$$

where,

$w^{(\lambda)}(m)$ is the OCC which is a function of the layer number 'λ'

m corresponds to the time slot within the subframe (0 for the first time slot and 1 for the second time slot)

$\alpha(\lambda)$ is the cyclic shift which is a function of the layer number 'λ'

★ The use of an OCC can be enabled/disabled on a per user basis using the 'dmrs-WithOCC-Activated' flag within an RRC Connection Setup, RRC Connection Reconfiguration or RRC Connection Re-establishment message

★ The OCC is applied in the time domain so an OCC $[w^{(\lambda)}(0), w^{(\lambda)}(1)] = [1, -1]$ means that the Demodulation Reference Signal belonging to the first time slot is multiplied by 1, while the Demodulation Reference Signal belonging to the second time slot is multiplied by -1

★ The OCC is determined from a look-up table indexed using the cyclic shift within the DCI. This means that both $n^{(2)}_{DMRS}$ and the OCC are determined by the cyclic shift value within the DCI. This approach has been adopted to avoid the requirement for additional signalling

★ The cyclic shift becomes a function of the layer number by making the value of $n^{(2)}_{DMRS}$ a function of the layer number. The look-up table which defines $n^{(2)}_{DMRS}$ as a function of the cyclic shift within the DCI, specifies a $n^{(2)}_{DMRS}$ value for each layer. There are 12 possible cyclic shifts so the values of $n^{(2)}_{DMRS}$ are separated by 3 to provide support for up to 4 layers. For example, when the cyclic shift within the DCI is signaled as 000 then the $n^{(2)}_{DMRS}$ values for layers 1 to 4 and 0, 6, 3 and 9

★ The processing used to generate the Demodulation Reference Signal for the PUSCH when using OCC is shown in Figure 154

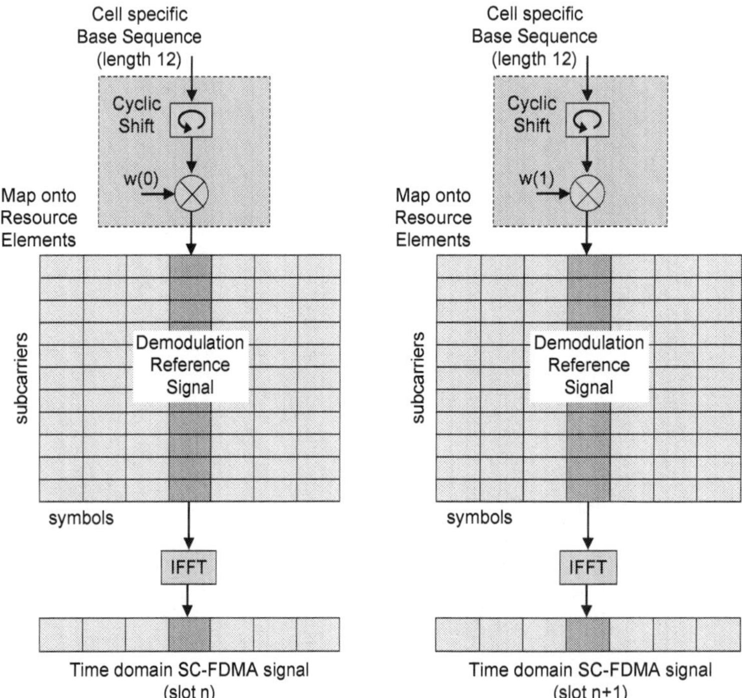

Figure 154 – Processing to generate Demodulation Reference Signal for the PUSCH (3GPP release 10 normal cyclic prefix)

★ If uplink MIMO is used, the set of layer specific Demodulation Reference Signals are precoded using the same matrix as used for the PUSCH data. This allows the eNode B to use the Demodulation Reference Signals to estimate the propagation channel for each layer of PUSCH data. The concept of applying the same precoding to both the PUSCH data and the Demodulation Reference Signals is illustrated in Figure 155. This example is based upon 2×2 MIMO and assumes that 2 layers of data are being transferred. The precoded data and the precoded Reference Signals are subsequently mapped onto their Resource Elements ready for SC-FDMA signal generation

★ 12 Resource Elements per Resource Block are allocated to the PUSCH Demodulation Reference Signal. These 12 Resource Elements represent an overhead from the perspective of transferring data on the PUSCH. The overheads generated by the PUSCH Demodulation Reference Signal are presented in Table 160

★ 3GPP References: TS 36.211, TS 36.331

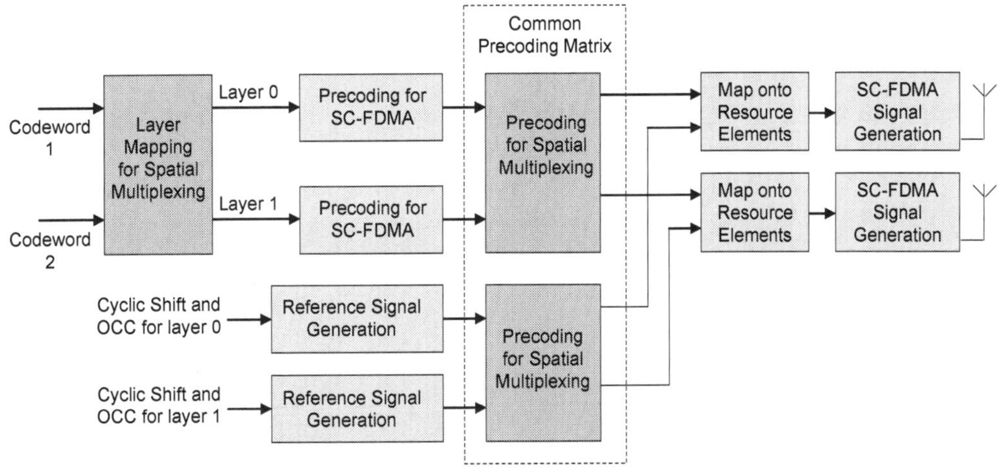

Figure 155 – Common precoding applied to both PUSCH data and Demodulation Reference Signal when using uplink MIMO

	Normal Cyclic Prefix	Extended Cyclic Prefix
Total Number of Resource Elements	84	72
Reference Signal Resource Elements	12	12
Reference Signal Overhead	12 / 84 = 14.3 %	12 / 72 = 16.7 %

Table 160 – Overheads generated by the PUSCH Demodulation Reference Signal

13.1.2 PUCCH DEMODULATION REFERENCE SIGNAL

★ The PUCCH Demodulation Reference Signal is included within every Resource Block allocated to the PUCCH. The PUCCH occupies 1 Resource Block within each time slot of a subframe

★ The number of Resource Elements occupied by the PUCCH Demodulation Reference Signal depends upon the PUCCH format and whether a normal or extended cyclic prefix is used. Figure 156 shows the Resource Elements used by the PUCCH Demodulation Reference Signal. One Resource Block is shown for each scenario. Each PUCCH transmission uses two Resource Blocks but the allocation of Resource Elements is the same for both. PUCCH formats 2a and 2b are not applicable to the extended cyclic prefix

★ The Demodulation Reference Signal for PUCCH formats 1, 1a and 1b occupies a relatively large number of Resource Elements. This is possible because these PUCCH formats transfer only 1 or 2 bits of information. 96 Resource Elements are sufficient to reliably signal this small quantity of information

★ Increasing the number of Resource Elements allocated to the Demodulation Reference Signal helps to improve the performance of channel estimation

★ PUCCH formats 2, 2a, 2b and 3 transfer greater quantities of information so require a larger number of Resource Elements. Formats 2a and 2b transfer an additional 1 and 2 bits of information respectively using the Demodulation Reference Signal sequence, i.e. the receiver can deduce these bits of information by identifying which sequence has been used for the Demodulation Reference Signal

★ PUCCH format 1b supports channel selection which means that additional information can be signalled by selecting a specific PUCCH resource for transmission, i.e. allocating the UE with 4 PUCCH resources allows an additional 2 bits of information to be signalled when the UE selects 1 of those 4 resources

★ PUCCH formats 1, 1a 1b, 2, 2a and 2b were introduced within the release 8 version of the 3GPP specifications, whereas format 3 was introduced within the release 10 version

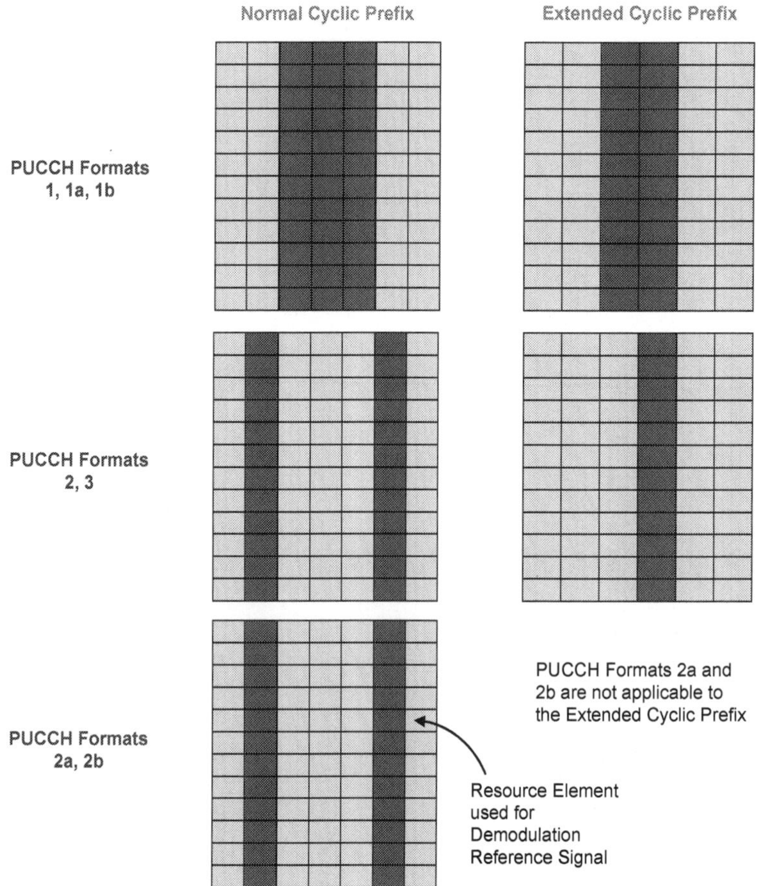

Figure 156 – Resource Elements used by the PUCCH Demodulation Reference Signal

★ The release 8 and 9 versions of the 3GPP specifications limit the PUCCH to transmission using a single antenna port. The PUCCH Demodulation Reference Signal for release 8 and 9 is given by:

$$\overline{w}(m) \times \mathrm{z(m)} \times r_{u,v}^{(\alpha)}(n)$$

where,

$\overline{w}(m)$ is only applicable to PUCCH formats 1, 1a and 1b. It represents a time domain orthogonal code. The length of the code is equal to the number of SC-FDMA symbols allocated to the Demodulation Reference Signal within a time slot, i.e. the length is 3 for the normal cyclic prefix and 2 for the extended cyclic prefix. These orthogonal codes are UE specific and provide one mechanism for differentiating between multiple users sharing the same PUCCH Resource Blocks. Similar orthogonal codes are applied to the PUCCH data as illustrated in Figure 169 (section 14.2.1)

z(m) is set equal to 1 by default. In the case of PUCCH formats 2a and 2b, z(m) for the second SC-FDMA symbol within each time slot (m = 1) is set equal to the 11th modulation symbol generated from the PUCCH information bits. The first 10 modulation symbols are transferred using the PUCCH itself

'm' references the SC-FDMA symbols for the PUCCH Demodulation Reference Signal within a time slot, e.g. m ranges from 0 to 2 for PUCCH formats 1, 1a and 1b when using the normal cyclic prefix

$r_{u,v}^{(\alpha)}(n)$ represents the Demodulation Reference Signal sequence generated from the appropriate base sequence by having a cyclic shift applied

α is the cyclic shift applied to the base sequence (12 different cyclic shifts are available). The cyclic shift is UE specific so can be used in combination with the orthogonal codes to differentiate between multiple users sharing the same PUCCH Resource Blocks (the cyclic shift is applicable to all PUCCH formats, whereas the use of orthogonal codes is only applicable to PUCCH formats 1, 1a and 1b within the release 8 and 9 versions of the specifications)

u is the base sequence group number (0 to 29)

v is the index of the sequence with an appropriate length (0 or 1)

n ranges from 0 to 11, i.e. a value for each subcarrier within the PUCCH Resource Block

★ The base sequence group number, 'u' is selected using the equation:

$$u = (f_{gh} + f_{ss}) \bmod 30$$

where,

$f_{gh} = 0$ if group hopping is disabled (configuration is cell specific and is broadcast in SIB2)

f_{gh} is a pseudo-random number between 0 and 29 if group hopping is enabled. This pseudo-random number is dependent upon the time slot number and the Physical layer Cell Identity (PCI)

$f_{ss} = \text{PCI} \bmod 30$

★ The expression for f_{gh} is the same for both PUSCH and PUCCH Demodulation Reference Signals, whereas the expression for f_{ss} is different. The PUCCH expression for f_{ss} does not include Δ_{ss}

★ The index of the base sequence, 'v' is always 0 for the PUCCH because there is only 1 sequence of length 12

★ The processing used to generate the Demodulation Reference Signal for PUCCH when using OCC (applicable to PUCCH formats 1, 1a and 1b) is shown in Figure 157

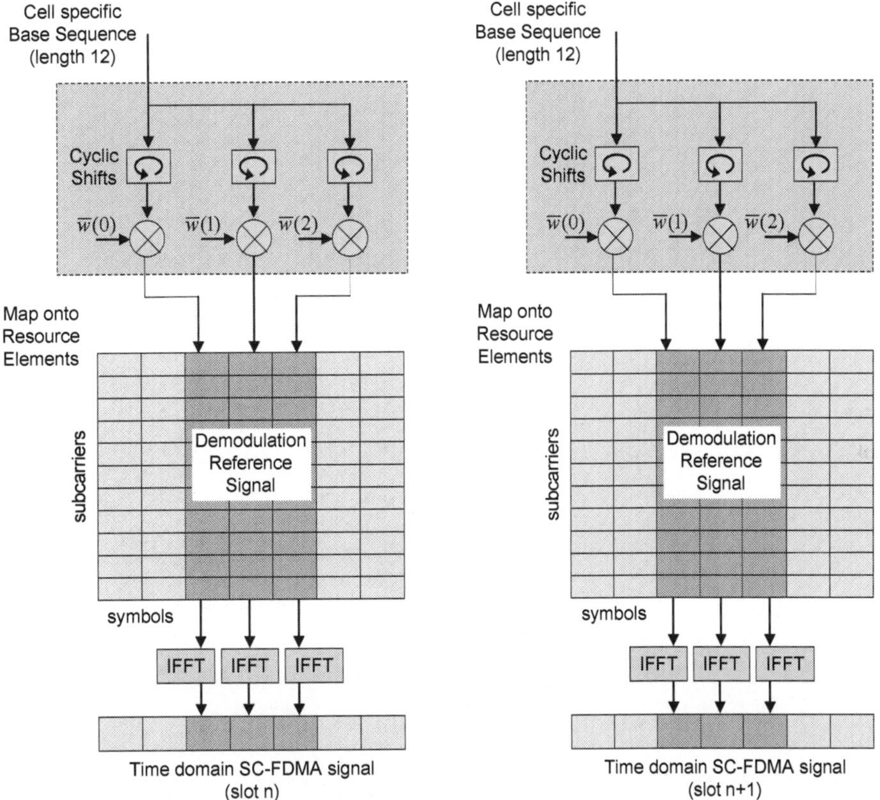

Figure 157 – Processing to generate Demodulation Reference Signal for PUCCH formats 1, 1a and 1b (normal cyclic prefix)

★ This figure illustrates the use of the time domain orthogonal codes. In the case of the normal cyclic prefix, the orthogonal codes have a length of 3 and there are 3 codes available to provide differentiation between UE. In the case of the extended cyclic prefix, the orthogonal codes have a length of 2 and there are 2 codes available to provide differentiation between UE

★ The processing used to generate the Demodulation Reference Signal for PUCCH formats 2a and 2b is shown in Figure 157. This case uses the 11th modulation symbol belonging to the PUCCH information bits to mask the Reference Signal sequence belonging to the second Reference Signal SC-FDMA symbol within each time slot

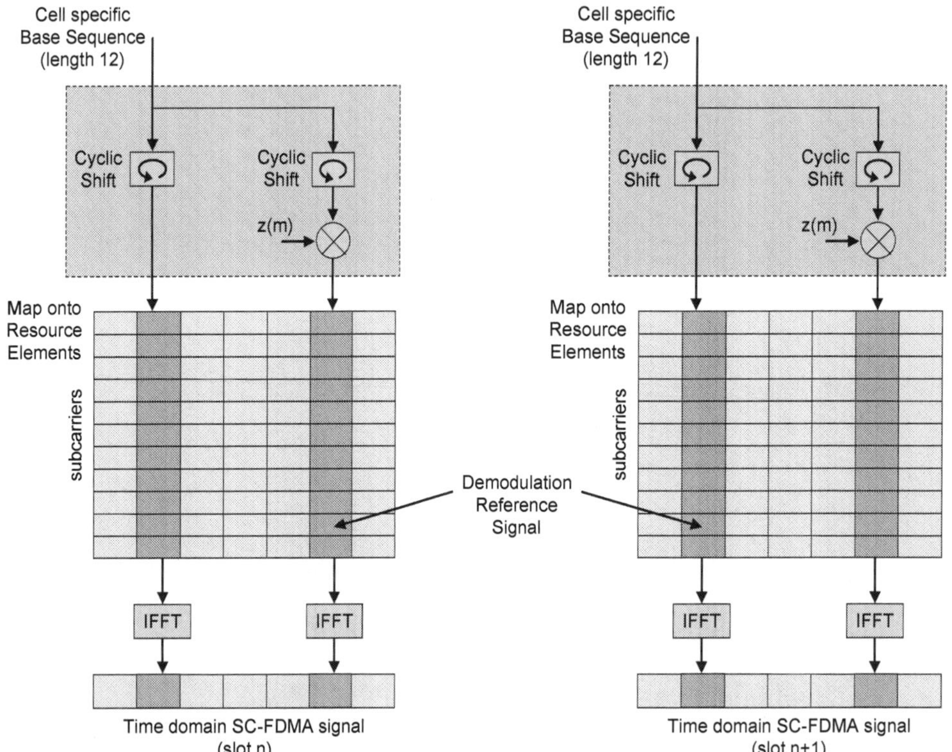

Figure 158 – Processing to generate Demodulation Reference Signal for PUCCH formats 2a and 2b (normal cyclic prefix)

★ The release 10 version of the specifications provides support for PUCCH transmit diversity using 2 antenna ports. In this case, a PUCCH Demodulation Reference Signal is generated for each antenna port. This is in contrast to the PUSCH Demodulation Reference Signals for MIMO which are generated for each layer, rather than each antenna port. The PUCCH Demodulation Reference Signals are generated directly for the antenna ports because transmit diversity does not involve any precoding across layers

★ These Demodulation Reference Signals are differentiated using a combination of OCC and cyclic shifts. The 3GPP release 10, PUSCH Demodulation Reference Signal sequence is defined as:

$$1/\sqrt{P} \times \overline{w}^{(\tilde{p})}(m) \times z(m) \times r_{u,v}^{(\alpha(\tilde{p}))}(n)$$

★ This is similar to the expression used by the release 8 and 9 versions of the specifications but now includes an initial scaling factor to keep the total transmit power independent of the number of antenna ports ('P' is the number of antenna ports). The release 10 expression also introduces an antenna port dependency upon the orthogonal codes and cyclic shift

★ Use of the orthogonal codes remains limited to PUCCH formats 1, 1a and 1b. Different orthogonal codes can be allocated to each antenna port when using transmit diversity. Similarly, different cyclic shifts can be applied to each antenna port when using transmit diversity

★ Table 161 presents the Resource Element figures for each PUCCH Demodulation Reference Signal scenario. These figures can be deduced from Figure 156

PUCCH Formats	1, 1a, 1b		2, 3		2a, 2b	
Cyclic Prefix	Normal	Extended	Normal	Extended	Normal	Extended
Total Resource Elements Available	168	144	168	144	168	-
Reference Signal Resource Elements	72	48	48	24	48	-
PUCCH Resource Elements	96		120		120	-

Table 161 – Resource Elements occupied by the PUCCH Demodulation Reference Signal

★ Both the PUCCH and the PUCCH Demodulation Reference Signal represent overheads from the perspective of transferring PUSCH data. This means that Resource Blocks allocated to the PUCCH are 100% overhead from the perspective of transferring PUSCH data

★ 3GPP References: TS 36.211

13.2 SOUNDING REFERENCE SIGNAL

★ The Sounding Reference Signal (SRS) can be used to measure the uplink channel quality over a section of the channel bandwidth. The eNode B can use this information for uplink frequency selective scheduling and link adaptation. When uplink/downlink channel reciprocity is assumed, measurements from the SRS can also be used to support downlink transmissions, e.g. the SRS can be used to support Angle of Arrival (AoA) measurements for downlink beamforming. Channel reciprocity is most applicable to TDD in which case the same RF carrier is used for uplink and downlink transmissions. Angle of Arrival measurements from the SRS can also be used to support location based services

★ The SRS was introduced within the release 8 version of the 3GPP specifications, and was subsequently enhanced within the release 10 version. Enhancements provide support for uplink MIMO and rapid triggering of SRS transmissions using a flag within the DCI

★ An example of the Sounding Reference Signal is illustrated in Figure 159. This example is based upon the 3 MHz channel bandwidth and a Sounding Reference Signal which spans 8 Resource Blocks

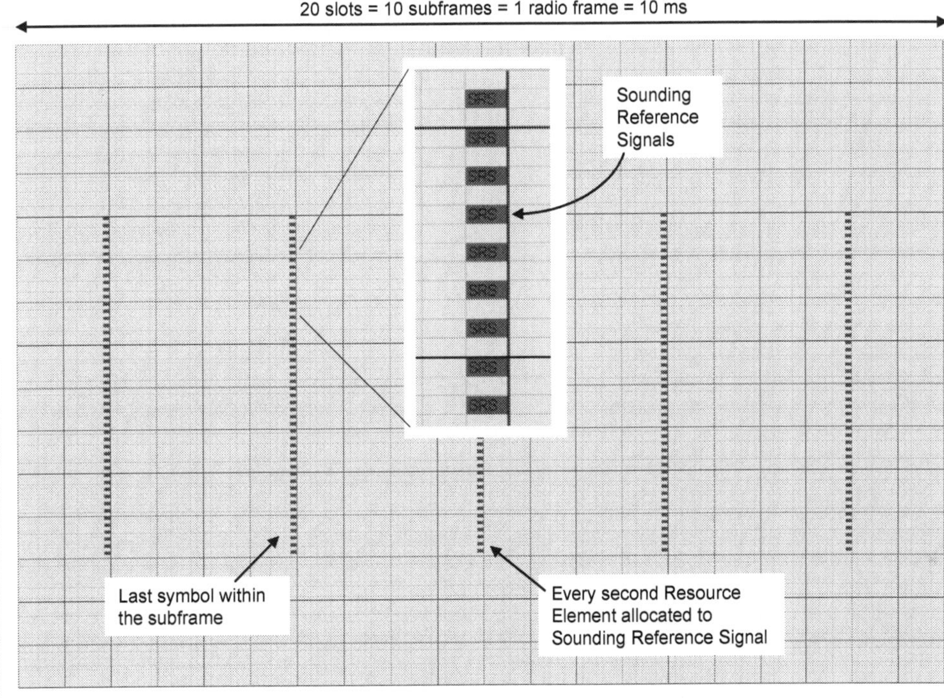

Figure 159 – Example Sounding Reference Signal (SRS)

★ The eNode B instructs the UE to transmit the Sounding Reference Signal across a specific section of the channel bandwidth using a combination of common information in SIB2, and dedicated information within an RRC Connection Setup, RRC Connection Reconfiguration or RRC Connection Re-establishment message

★ The SRS information broadcast by SIB2 is presented in Table 162, while the 3GPP release 8 and 9 information transmitted within an RRC Connection Setup, RRC Connection Reconfiguration or RRC Connection Re-establishment message is presented in Table 163

★ SIB2 broadcasts the 'SRS Bandwidth Configuration', which can have a value from 0 to 7, while the dedicated signalling message specifies an 'SRS Bandwidth', which can have a value from 0 to 3. Theses two parameters define a row and column within a look-up table specified by 3GPP TS 36.211

★ An example of the SRS Bandwidth Configuration / SRS Bandwidth look-up table is shown in Table 164. This example is applicable to the 1.4, 3 and 5 MHz channel bandwidths. 3GPP TS 36.211 also includes similar tables for the larger channel bandwidths

★ The first parameter defined by the look-up table, m_{SRS} defines the number of Resource Blocks over which the SRS is transmitted. The 1.4 MHz channel bandwidth has 6 Resource Blocks in total so channel sounding is always completed across 4 Resource Blocks. The 3 MHz channel bandwidth has 15 Resource Blocks in total so channel sounding can be across 4, 8 or 12 Resource Blocks

★ The UE is never instructed to transmit the SRS across the entire channel bandwidth. It is not necessary to transmit the SRS within the Resource Blocks reserved for the PUCCH. These Resource Blocks are located at the two edges of the channel bandwidth. The SRS is used for frequency selective scheduling of the PUSCH rather than frequency selective scheduling of the PUCCH

★ The m_{SRS} values have been defined in a tree structure. The 'SRS Bandwidth Configuration' is cell specific so a single cell operates using a single row from Table 164 (assuming a channel bandwidth of 1.4, 3 or 5 MHz). For example, a cell using the 5 MHz channel bandwidth (25 Resource Blocks) could use 'SRS Bandwidth Configuration' 3 so an individual UE within the cell could be instructed to transmit the SRS across either 20 Resource Blocks or 4 Resource Blocks. This example is illustrated in Figure 160

Information Elements		
Uplink Sounding Reference Signal Configuration Common	Setup	SRS Bandwidth Configuration
		SRS Subframe Configuration
		Ack/Nack + SRS Simultaneous Transmission
		SRS Max UpPTS (TDD)

Table 162 – Common SRS configuration information from SIB2

Information Elements		
Uplink Sounding Reference Signal Configuration Dedicated	Setup	SRS Bandwidth
		SRS Hopping Bandwidth
		Frequency Domain Position
		Duration
		SRS Configuration Index
		Transmission Comb
		Cyclic Shift

Table 163 – UE specific SRS configuration information from dedicated signalling

SRS Bandwidth Configuration	SRS Bandwidth = 0		SRS Bandwidth = 1		SRS Bandwidth = 2		SRS Bandwidth = 3	
	$m_{SRS,0}$	N_0	$m_{SRS,1}$	N_1	$m_{SRS,2}$	N_2	$m_{SRS,3}$	N_3
0	36	1	12	3	4	3	4	1
1	32	1	16	2	8	2	4	2
2	24	1	4	6	4	1	4	1
3	20	1	4	5	4	1	4	1
4	16	1	4	4	4	1	4	1
5	12	1	4	3	4	1	4	1
6	8	1	4	2	4	1	4	1
7	4	1	4	1	4	1	4	1

Table 164 – Look-up table for SRS Bandwidth Configuration and Bandwidth parameters (channel bandwidths of 1.4, 3 and 5 MHz)

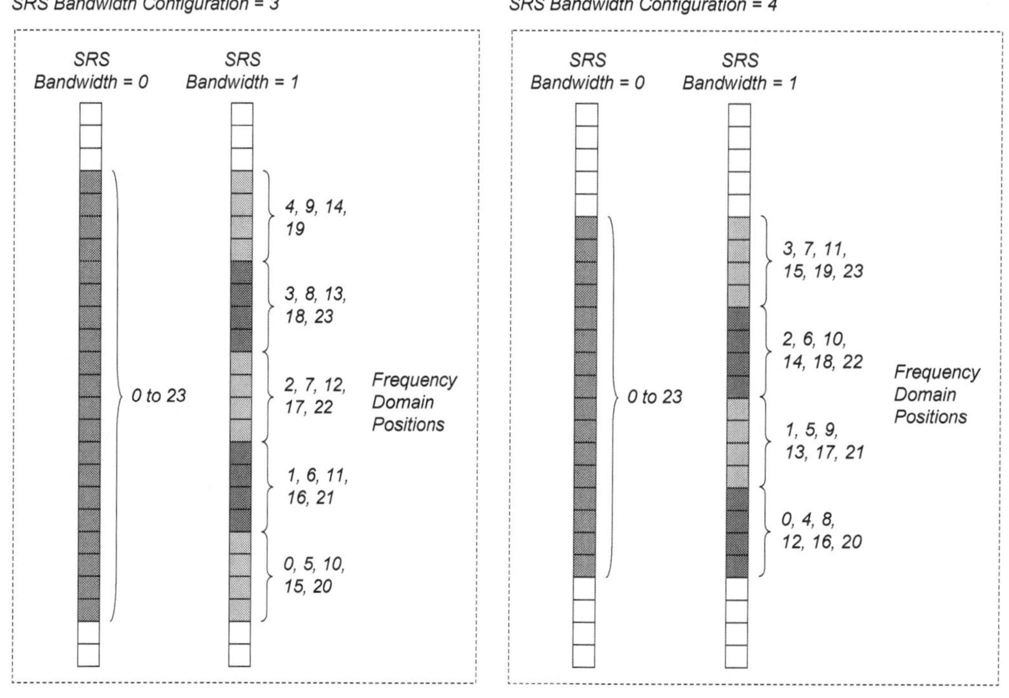

Figure 160 – Example SRS transmissions for the 5 MHz channel bandwidth (25 Resource Blocks)

★ The tree structure is visible in Figure 160 as there is support for SRS transmission across a single set of 20 Resource Blocks, and 5 sets of 4 Resource Blocks. The set of 20 Resource Blocks is used when a UE is allocated an 'SRS Bandwidth' of 0, whereas a set of 4 Resource Blocks is used when a UE is allocated an 'SRS Bandwidth' of 1. In the latter case, selection between the 5 sets of 4 Resource Blocks is dependent upon the 'Frequency Domain Position' parameter signalled to the UE within an RRC Connection Setup, RRC Connection Reconfiguration or RRC Connection Re-establishment message. The 'Frequency Domain Position' parameter can range from 0 to 23, and is shown in Table 163

★ The N_0, N_1, N_2 and N_3 parameters shown in Table 164 are used in combination with the m_{SRS} and 'Frequency Domain Position' parameters when calculating the starting position of the SRS in the frequency domain. The result of the calculation is a pointer towards the lowest subcarrier to be used by the SRS

★ Figure 160 shows a second example for the 5 MHz channel bandwidth based upon a cell configured with an 'SRS Bandwidth Configuration' of 4. In this case, the maximum bandwidth for SRS transmission is 16 Resource Blocks, while the tree structure also supports SRS transmission across 4 Resource Blocks. The 'Frequency Domain Position' parameter determines which set of 4 Resource Blocks is used

★ The tree structures in Figure 160 have depths of 2 because they are based upon a relatively small channel bandwidth. The larger channel bandwidths allow increased flexibility and the tree structures have depths of up to 4. The tree structures for each combination of channel bandwidth and 'SRS Bandwidth Configuration' are presented in Table 165. This table presents the number of Resource Blocks × the number of possible allocations. The layer within the tree is determined by the 'SRS Bandwidth' parameter

		Channel Bandwidth					
		1.4	3	5	10	15	20
		Total Resource Blocks					
SRS Bandwidth Configuration (cell specific)	SRS Bandwidth (UE specific)	6	15	25	50	75	100
0	0	-	-	-	48 RB × 1	72 RB × 1	96 RB × 1
	1				24 RB × 2	24 RB × 3	48 RB × 2
	2				12 RB × 4	12 RB × 6	24 RB × 4
	3				4 RB × 12	4 RB × 18	4 RB × 24
1	0	-	-	-	48 RB × 1	64 RB × 1	96 RB × 1
	1				16 RB × 3	32 RB × 2	32 RB × 2
	2				8 RB × 6	16 RB × 4	16 RB × 4
	3				4 RB × 12	4 RB × 16	4 RB × 24
2	0	-	-	24 RB × 1	40 RB × 1	60 RB × 1	80 RB × 1
	1			4 RB × 6	20 RB × 2	20 RB × 3	40 RB × 2
	2				4 RB × 10	4 RB × 15	20 RB × 4
	3						4 RB × 20
3	0	-	-	20 RB × 1	36 RB × 1	48 RB × 1	72 RB × 1
	1			4 RB × 5	12 RB × 3	24 RB × 2	24 RB × 3
	2				4 RB × 9	12 RB × 4	12 RB × 6
	3					4 RB × 12	4 RB × 18
4	0	-	-	16 RB × 1	32 RB × 1	48 RB × 1	64 RB × 1
	1			4 RB × 4	16 RB × 2	16 RB × 3	32 RB × 2
	2				8 RB × 4	8 RB × 6	16 RB × 4
	3				4 RB × 8	4 RB × 12	4 RB × 16
5	0	-	12 RB × 1	12 RB × 1	24 RB × 1	40 RB × 1	60 RB × 1
	1		4 RB × 3	4 RB × 3	4 RB × 6	20 RB × 2	20 RB × 3
	2					4 RB × 10	4 RB × 15
	3						
6	0	-	8 RB × 1	8 RB × 1	20 RB × 1	36 RB × 1	48 RB × 1
	1		4 RB × 2	4 RB × 2	4 RB × 5	12 RB × 3	24 RB × 2
	2					4 RB × 9	12 RB × 4
	3						4 RB × 12
7	0	4 RB × 1	4 RB × 1	4 RB × 1	16 RB × 1	32 RB × 1	48 RB × 1
	1				4 RB × 4	16 RB × 2	16 RB × 3
	2					8 RB × 4	8 RB × 6
	3					4 RB × 8	4 RB × 12

Table 165 – Tree structures for SRS transmission (Number of Resource Blocks × Number of Possibilities)

★ The minimum bandwidth for SRS transmission is 4 Resource Blocks. Smaller bandwidths provide the eNode B with less information regarding the propagation channel but allow the UE to transmit the SRS towards cell edge. UE at cell edge may not have sufficient transmit power for a wideband SRS transmission. When transmitting the SRS using a small bandwidth, multiple transmissions can be used to allow channel estimation across the same section of the channel bandwidth as the wideband SRS. This concept is illustrated in Figure 161

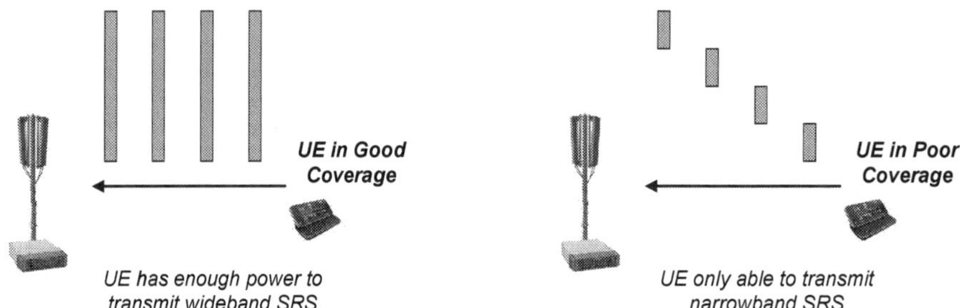

Figure 161 – Wideband SRS transmissions in good coverage, and narrowband SRS transmissions in poor coverage

★ Frequency hopping is used to change the set of Resource Blocks between one narrowband SRS transmission and the next. If frequency hopping is disabled, the Resource Blocks used for SRS transmission remain unchanged unless they are reconfigured by dedicated signalling (RRC Connection Setup, RRC Connection Reconfiguration or RRC Connection Re-establishment message)

★ The 'SRS Hopping Bandwidth' parameter shown in Table 163 is used to determine whether or not frequency hopping is enabled. This UE specific parameter can be configured with values of 0, 1, 2 and 3. Frequency hopping is enabled if the value of the 'SRS Hopping Bandwidth' is less than the value of the 'SRS Bandwidth'. Otherwise, hopping is disabled

★ Transmitting the SRS with a wide bandwidth reduces the number of UE which can simultaneously transmit the SRS during a specific SC-FDMA symbol. The 'Transmission Comb' parameter presented in Table 163 has been introduced to increase the number of UE which can be multiplexed during the same SC-FDMA symbol. The transmission comb parameter is signalled to individual UE using a value of either 0 or 1. The value of 1 instructs the UE to apply a 1 subcarrier offset when allocating Resource Elements to the SRS. The SRS is transmitted on every second subcarrier so this allows 2 transmissions to be interleaved (frequency multiplexing is applied)

★ Figure 162 illustrates how the transmission comb parameter can be combined with the tree structure to multiplex the SRS transmissions from different users. This example is based upon the 5 MHz channel bandwidth with a total of 25 Resource Blocks and an 'SRS Bandwidth Configuration' of 4 (16 Resource Blocks × 1, and 4 Resource Blocks × 4)

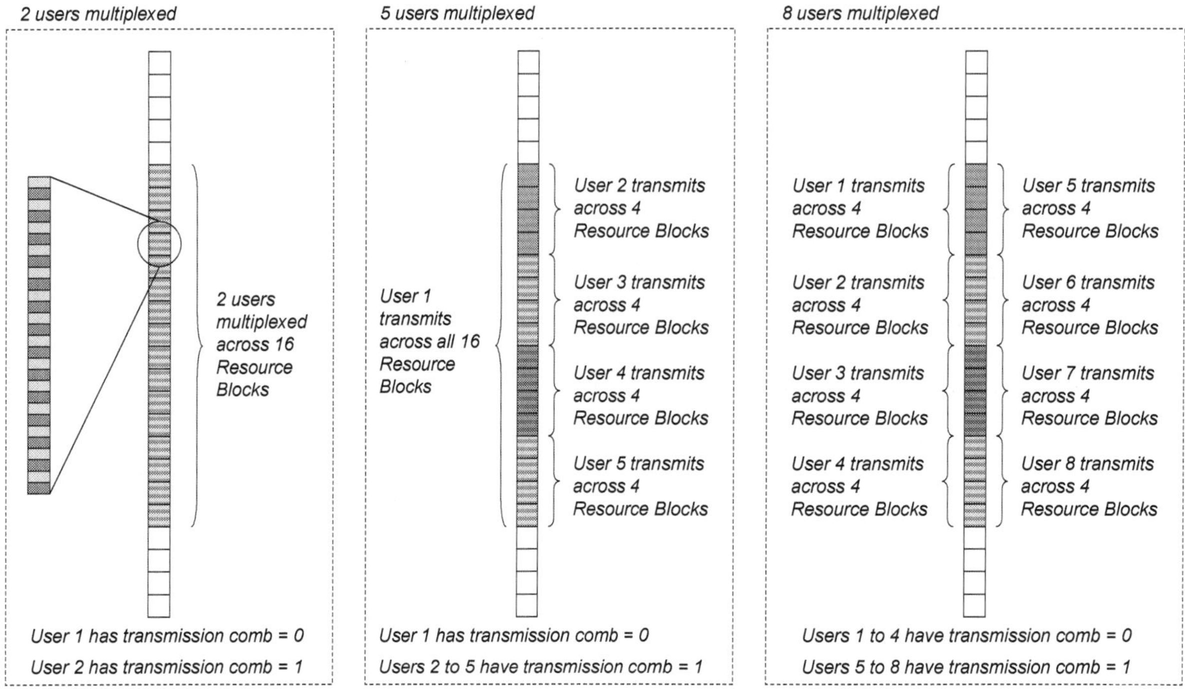

Figure 162 – Use of the comb parameter and tree structure to multiplex SRS transmissions from different users

★ Additional multiplexing capacity can be achieved by allocating different cyclic shifts to each user. The UE specific 'cyclic shift' parameter shown in Table 163 can be allocated values between 0 and 7 (8 different cyclic shifts are defined). The cyclic shift is applied to the sequence used for SRS transmission. The sequence used for SRS transmission is given by:

$$r_{u,v}^{(\alpha)}(n)$$

where,

α is the cyclic shift applied to the base sequence
u is the base sequence group number (0 to 29)
v is the index of the base sequence with an appropriate length (0 or 1)
n ranges from 0 to (number of allocated subcarriers – 1)

★ The SRS uses the same 30 groups of base sequences as the Demodulation Reference Signals. These 30 groups of base sequences are illustrated in Figure 151 within section 13.1. The value of 'u' is selected using the same calculation as for the PUCCH Demodulation Reference Signal. The value of 'v' is selected using the same calculation as for the PUSCH ('v' is always 0 for the PUCCH)

★ The cyclic shift applied to the base sequence is defined as:

$$\alpha = 2\pi \, n_{SRS}^{cs} / 8$$

where,

n_{SRS}^{cs} is the 'cyclic shift' parameter shown in Table 163 (0 to 7)

★ The minimum number of Resource Blocks for SRS transmission (4 Resource Blocks) has been selected to ensure that the SRS sequence length is always a multiple of 8, i.e. 4 Resource Blocks include 48 subcarriers in total, with 24 of those subcarriers used by an individual UE for SRS transmission. Using sequences which are a multiple of 8 helps to improve the performance of the eNode B receiver when differentiating between the SRS transmissions sharing the same subcarriers with different cyclic shifts

★ With the exception of TDD special subframes, the SRS is always transmitted during the last symbol of the subframe. In the case of TDD special subframes, the SRS can be transmitted during the UpPTS field. If the UpPTS field includes 2 SC-FDMA symbols then both symbols can be used for SRS transmission, and both symbols can be allocated to a single UE

★ The set of subframes during which the Sounding Reference Signal is transmitted is determined by combining the cell specific 'SRS Subframe Configuration', with the UE specific 'SRS Configuration Index'

★ The cell specific 'SRS Subframe Configuration' can be allocated values between 0 and 15, although the value of 15 is currently reserved and is not used. This parameter is broadcast in SIB2 and is shown in Table 162. The FDD subframes during which the SRS can be active for each 'SRS Subframe Configuration' are presented in Table 166

SRS Subframe Configuration	Subframe Number									
	0	1	2	3	4	5	6	7	8	9
0	✓	✓	✓	✓	✓	✓	✓	✓	✓	✓
1	✓		✓		✓		✓		✓	
2		✓		✓		✓		✓		✓
3	✓					✓				
4		✓					✓			
5			✓					✓		
6				✓					✓	
7	✓	✓				✓	✓			
8			✓	✓				✓	✓	
9	✓									
10		✓								
11			✓							
12				✓						
13	✓	✓	✓	✓	✓		✓		✓	
14	✓	✓	✓	✓	✓	✓	✓		✓	

Table 166 – Subframes during which Sounding Reference Signal can be transmitted (FDD)

★ The equivalent information for TDD is presented in Table 167. In the case of TDD, SRS transmissions are not permitted during subframes 0 and 5 because these are always downlink subframes. Subframe 1 is always a special subframe so SRS transmission during subframe 1 is within the UpPTS field. Subframe 6 may, or may not be a special subframe depending upon the TDD uplink-downlink subframe configuration. In the case of TDD, 'SRS Subframe Configuration' values of 14 and 15 are reserved, and are not used

★ The cell specific 'SRS Subframe Configuration' defines the FDD and TDD subframes which are available for SRS transmission within a cell, but individual UE are limited to using the subframes defined by their 'SRS Configuration Index' value. This parameter can be allocated values between 0 and 1023, although values 637 to 1023 are not used by FDD, and values 645 to 1023 are not used by TDD. The 'SRS Configuration Index' is used to determine the SRS Periodicity and SRS Subframe Offset. Table 168 presents the relationships for both FDD and TDD

SRS Subframe Configuration	Subframe Number									
	0	1	2	3	4	5	6	7	8	9
0		✓					✓			
1		✓	✓				✓	✓		
2		✓		✓			✓		✓	
3		✓			✓		✓			✓
4		✓	✓	✓			✓	✓	✓	
5		✓	✓		✓		✓	✓		✓
6		✓		✓	✓		✓		✓	✓
7		✓	✓	✓	✓		✓	✓	✓	✓
8		✓	✓				✓			
9		✓		✓			✓			
10		✓					✓	✓		
11		✓	✓				✓		✓	
12		✓		✓			✓			✓
13		✓			✓		✓	✓		

Table 167 – Subframes during which Sounding Reference Signal can be transmitted (TDD)

FDD		
SRS Configuration Index (I_{SRS})	SRS Periodicity (T_{SRS})	SRS Subframe Offset (T_{offset})
0 – 1	2 ms	I_{SRS}
2 – 6	5 ms	I_{SRS} - 2
7 - 16	10 ms	I_{SRS} - 7
17 – 36	20 ms	I_{SRS} - 17
37 – 76	40 ms	I_{SRS} - 37
77 – 156	80 ms	I_{SRS} - 77
157 – 316	160 ms	I_{SRS} - 157
317 - 636	320 ms	I_{SRS} - 317

TDD		
SRS Configuration Index (I_{SRS})	SRS Periodicity (T_{SRS})	SRS Subframe Offset (T_{offset})
0	2 ms	0, 1
1	2 ms	0, 2
2	2 ms	1, 2
3	2 ms	0, 3
4	2 ms	1, 3
5	2 ms	0, 4
6	2 ms	1, 4
7	2 ms	2, 3
8	2 ms	2, 4
9	2 ms	3, 4
10 - 14	5 ms	I_{SRS} - 10
15 - 24	10 ms	I_{SRS} - 15
25 - 44	20 ms	I_{SRS} - 25
45 - 84	40 ms	I_{SRS} - 45
85 - 164	80 ms	I_{SRS} - 85
165 - 324	160 ms	I_{SRS} - 165
325 - 644	320 ms	I_{SRS} - 325

Table 168 – Look-up tables for UE specific Sounding Reference Signal subframe parameters (FDD and TDD)

★ In the case of FDD, and TDD when the SRS Periodicity > 2 ms, the SRS subframes for a specific UE are those which satisfy the expression:

$$(10 \cdot n_f + k_{SRS} - T_{offset}) \bmod T_{SRS} = 0$$

where,

n_f is the System Frame Number (0 to 1023)

k_{SRS} for FDD is the subframe number within the radio frame (0 to 9)

k_{SRS} for TDD is defined using the look-up table presented as Table 169

★ For example, if an FDD user is allocated an SRS Configuration Index of 4 then T_{SRS} is 5 ms and T_{offset} is 2. SRS transmissions are consequently allocated during subframes 2 and 7 of all radio frames

★ Similarly, if a TDD user is allocated an SRS Configuration Index of 12 then T_{SRS} is 5 ms and T_{offset} is 2. SRS transmissions are consequently allocated during subframes 2 and 7 of all radio frames (assuming those subframes are uplink rather than downlink)

	Subframe Number											
	0	1		2	3	4	5	6		7	8	9
		1st Symbol	2nd Symbol					1st Symbol	2nd Symbol			
UpPTS with 2 SC-FDMASymbols		0	1	2	3	4		5	6	7	8	9
UpPTS with 1 SC-FDMA Symbol		1						6				

Table 169 – Look-up table for the value of k_{SRS} when using TDD

- ★ If a TDD user is allocated an SRS Configuration Index of 11 then T_{SRS} is 5 ms and T_{offset} is 1. SRS transmissions are then allocated during subframes 1 and 6 of all radio frames. Subframe 1 is always a special subframe and the T_{offset} value of 1 means that the 2nd SC-FDMA symbol is used for SRS transmission when the UpPTS field includes 2 symbols, and the 1st SC-FDMA symbol is used for SRS transmission when the UpPTS field includes only 1 symbol. Subframe 6 is always either a special subframe or a downlink subframe. When subframe 6 is a special subframe, the selection between SC-FDMA symbols is the same as for subframe 1
- ★ In the case of TDD when the SRS Periodicity = 2 ms, the SRS subframes for a specific UE are those which satisfy the expression:

$$(k_{\text{SRS}} - T_{offset}) \bmod 5 = 0$$

- ★ For example, if a TDD user is allocated an SRS Configuration Index of 5 then T_{SRS} is 2 ms and T_{offset} is 0, 4. SRS transmissions are consequently allocated for k_{SRS} values of 0, 4, 5 and 9. The values of 0 and 5 are applicable to special subframes and are only used when there are 2 SC-FDMA symbols within the UpPTS field. The values of 4 and 9 are applicable to subframes 4 and 9 when those subframes are uplink rather than downlink
- ★ Table 163 also shows that the UE is provided with a 'Duration' parameter which can be signalled using values of either True or False. A value of True indicates that the UE should continue transmitting the SRS during the allocated subframes until instructed otherwise. A value of False indicates that the UE should complete only a single transmission
- ★ When transmit antenna selection is enabled for a UE (described in section 12.3.1) transmission of the SRS switches between the antenna so the eNode B is able to measure the propagation channel from each. The eNode B can then use this information for the closed loop mode of operation to provide the UE with instructions regarding the best antenna from which to transmit
- ★ The release 10 version of 3GPP TS 36.213 defines two types of triggering for the SRS:
  - o triggering type 0: based upon RRC signalling (similar to the 3GPP release 8 and 9 mechanism for transmitting the SRS)
  - o triggering type 1: aperiodic transmission triggered by information within the Downlink Control Information (DCI)
- ★ When using triggering type 1, the UE is initially provided with aperiodic SRS configuration information using RRC signalling but the UE does not transmit an aperiodic SRS until instructed by a DCI. The UE completes a single SRS transmission after receiving an instruction from a DCI. This allows the eNode B to rapidly request an SRS transmission as and when required
  - o DCI formats 0, 4 and 1A can be used to trigger an SRS transmission for both FDD and TDD
  - o DCI formats 2B and 2C can be used to trigger an SRS transmission for TDD
- ★ The UE can be provided with 3 aperiodic SRS configurations to be triggered by DCI format 4. The 'SRS Request' field within DCI format 4 has a length of 2 bits so is able to select between 4 different instructions. An 'SRS Request' signalled using the bit combination '00' indicates that the UE should not transmit an aperiodic SRS, whereas the remaining bit combinations point towards the set of 3 aperiodic SRS configurations
- ★ DCI formats 0, 4 and 1A each use separate aperiodic SRS configurations, whereas DCI formats 2B and 2C share the same configuration information
- ★ In the case of triggering type 1, the 'SRS Configuration Index' is limited to a range between 0 and 31, and additional look-up tables are specified in 3GPP TS 36.213. These additional look-up tables replace the release 8 and 9 tables presented in Table 168 when using aperiodic SRS transmission
- ★ The release 10 version of the 3GPP specifications supports SRS transmission on multiple antenna ports. This allows the eNode B to complete SRS based channel estimation for each antenna port when the UE is configured to use uplink MIMO. The SRS transmissions on each antenna port occur during the same SC-FDMA symbol and occupy the same Resource Blocks but are differentiated by their combination of Cyclic Shift and Transmission Comb
- ★ The release 10 version of the 3GPP specifications also supports SRS transmission on multiple RF carriers. This allows the eNode B to complete SRS based channel estimation for each RF carrier when the UE is configured to use Carrier Aggregation
- ★ The Sounding Reference Signal represents an overhead from the perspective of transferring data on the PUSCH so its periodicity should be carefully controlled
- ★ 3GPP References: TS 36.211, TS 36.213, TS 36.331

14 UPLINK PHYSICAL CHANNELS

14.1 PRACH

★ The Physical Random Access Channel (PRACH) is used to transfer the random access preambles used to initiate the random access procedure. The complete random access procedure is described in section 24.1

★ The PRACH does not transfer any RRC messages nor any application data. This differs from UMTS where the PRACH is used to transfer both RRC messages, e.g. the RRC Connection Request and Cell Update messages, and application data

★ The general structure of a random access preamble is illustrated in Figure 163. It includes a cyclic prefix, a sequence and a guard time

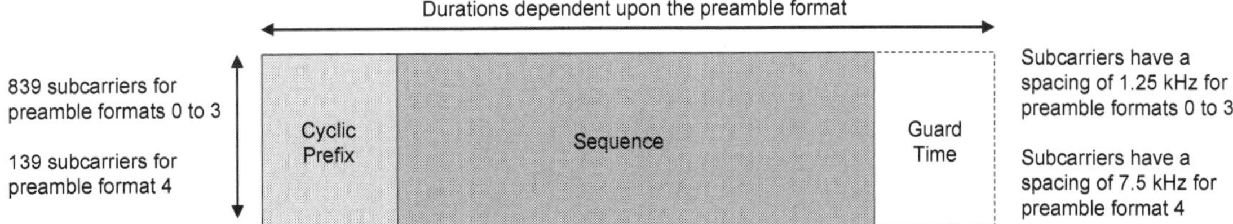

Figure 163 – Structure of random access preamble

★ A cyclic prefix is usually required to account for the maximum delay spread, i.e. the maximum delay spread should not exceed the duration of the cyclic prefix. This principle is shown in section 3.7. In the case of the PRACH, the cyclic prefix has to account for both the maximum delay spread and the maximum cell range. This additional requirement is caused by the UE transmitting the PRACH before any Timing Advance instructions have been provided, i.e. the PRACH is transmitted using a Timing Advance of 0

★ The requirement for the cyclic prefix to account for both the maximum delay spread and the maximum cell range is illustrated in Figure 164. This figure illustrates the PRACH received by an eNode B from 2 different UE:

- the 1st UE is assumed to be at the same location as the eNode B so there is negligible propagation delay and negligible delay spread. The received PRACH is aligned with the eNode B subframe timing
- the 2nd UE is assumed to be at cell edge so experiences the maximum propagation delay and an assumed maximum delay spread
    - the first delay spread component arrives with a delay equal to the round trip time between UE and eNode B. The delay is equal to the round trip time rather than the one-way delay because the subframe timing at the UE (based upon downlink measurements) is already delayed by the one-way delay before the UE transmits the PRACH
    - the last delay spread component arrives with a delay equal to the round trip time + the maximum delay spread. The eNode B observation window uses the cyclic prefix to replace the section of the sequence lost outside the window

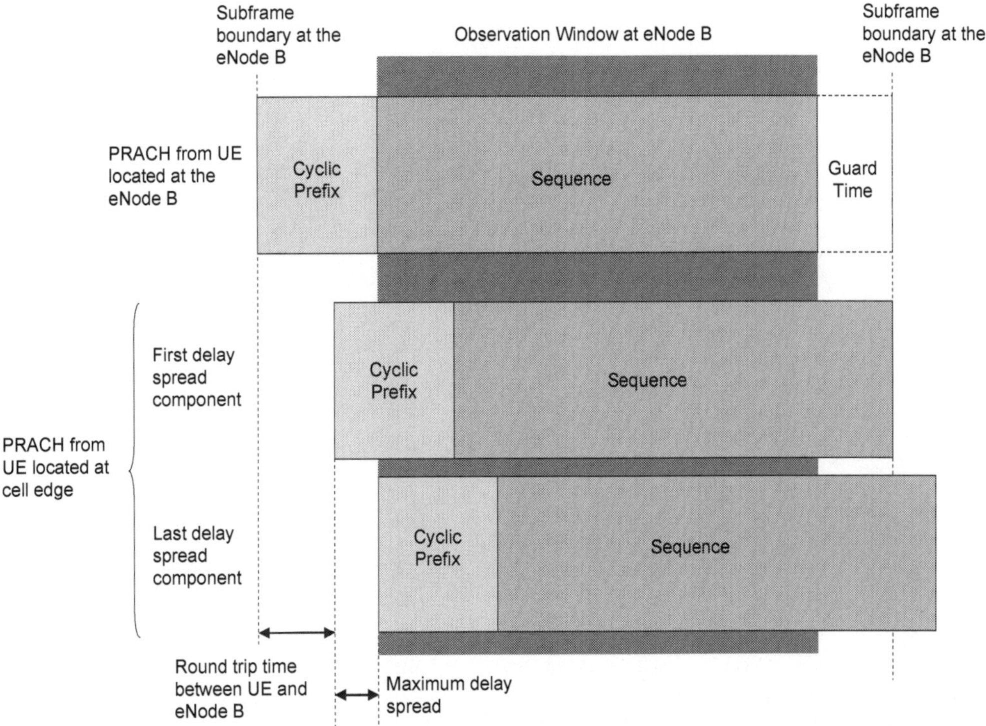

Figure 164 – Cyclic Prefix allowing for both the maximum round trip time and the maximum delay spread

★ The guard time duration should be long enough to accommodate the round trip time. This limits any overlap into the subsequent subframe, i.e. only delay spread components overlap into the subsequent subframe. Figure 164 illustrates that the PRACH received from a UE at cell edge and with maximum delay spread overlaps into the subsequent subframe. This is deemed to be acceptable because the overlap is into the cyclic prefix region of the subsequent subframe

★ 3GPP TS 36.211 specifies the set of 5 preamble formats presented in Table 170. These are all based upon the structure illustrated in Figure 163 but have different durations for the cyclic prefix, sequence and guard time

- Formats 0 to 3 can be used by either FDD or TDD, and are based upon relatively long preamble sequences
- Format 4 can only be used by TDD when using special subframe configurations with UpPTS lengths of 4384 Ts (2 symbols with normal cyclic prefix) and 5120 Ts (2 symbols with extended cyclic prefix). This format has a short preamble sequence

Preamble Format	Application	Cyclic Prefix Duration	Sequence Duration	Guard Time	Total Length	Typical Max. Cell Range
0	FDD & TDD	103.13 μs	800 μs	96.88 μs	1 ms	14.5 km
1	FDD & TDD	684.38 μs	800 μs	515.63 μs	2 ms	77.3 km
2	FDD & TDD	203.13 μs	1600 μs	196.88 μs	2 ms	29.5 km
3	FDD & TDD	684.38 μs	1600 μs	715.63 μs	3 ms	100.2 km
4	TDD only	14.58 μs	133 μs	9.38 μs	0.16 ms	1.4 km

Table 170 – PRACH format parameters

★ Preamble formats 0 to 3 occupy an integer number of subframes (1, 2 or 3 subframes) after allowing for the guard time

★ Preamble format 4 occupies only part of a subframe because it is transmitted during TDD special subframes which are divided into a downlink section (DwPTS), a guard period and an uplink section (UpPTS)

- when using the normal cyclic prefix in the uplink direction, the PRACH fully occupies the duration of the UpPTS and also occupies some of the guard period. This effectively reduces the duration of the guard period and so has an impact upon the cell range figures presented in Table 14 (section 3.2.2). This is acceptable because PRACH format 4 is limited to supporting small cell ranges
- when using the extended cyclic prefix in the uplink direction, the PRACH occupies part of the duration of the UpPTS

These PRACH format 4 timings relative to the UpPTS are illustrated in Figure 165

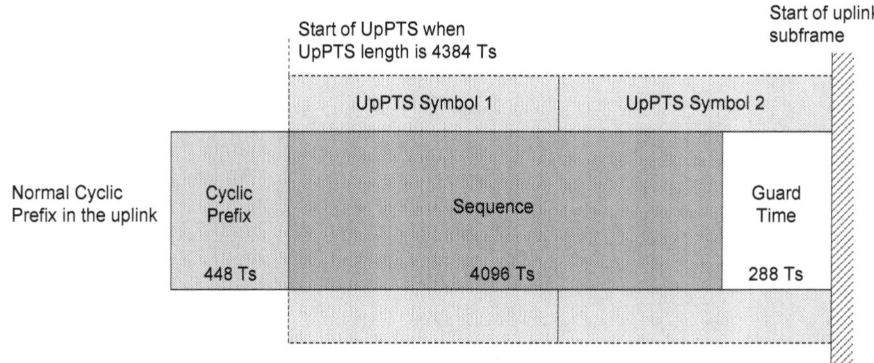

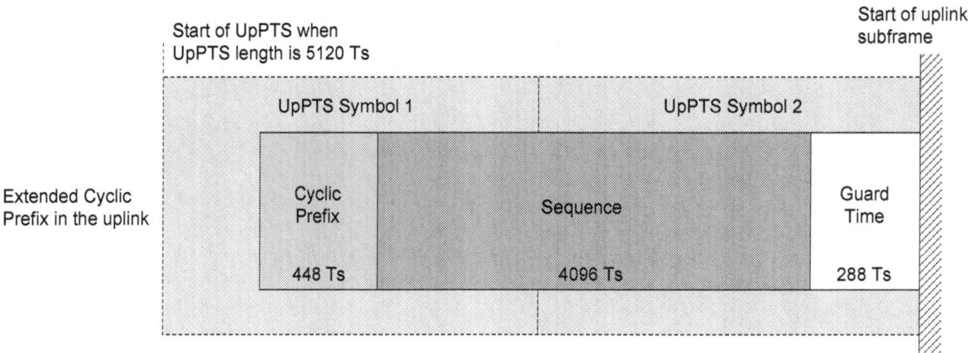

Figure 165 – Timing of PRACH format 4 relative to symbols within UpPTS

★ Section 31.6.1 discusses the selection between PRACH preamble formats, and provides the delay spread assumptions used to calculate the maximum cell ranges. In general, the radio network planner selects the appropriate preamble format based upon the cell range. Formats with longer cyclic prefixes and longer guard times are suitable for larger cell ranges. The drawback associated with using longer PRACH durations is an increased PRACH overhead, i.e. an increased number of Resource Blocks are allocated to the PRACH

★ The preamble format to be used within a specific cell is signalled using the PRACH Configuration Index which is broadcast in SIB2. The PRACH configuration index also defines the System Frame Numbers (SFN) and the subframes during which the PRACH can be transmitted. Table 171 presents the PRACH Configuration Indices for FDD

Config. Index	Preamble Format	SFN	Subframe Numbers
0	0	Even	1
1	0	Even	4
2	0	Even	7
3	0	Any	1
4	0	Any	4
5	0	Any	7
6	0	Any	1, 6
7	0	Any	2, 7
8	0	Any	3, 8
9	0	Any	1, 4, 7
10	0	Any	2, 5, 8
11	0	Any	3, 6, 9
12	0	Any	0, 2 , 4, 6, 8
13	0	Any	1, 3, 5, 7, 9
14	0	Any	0 to 9
15	0	Even	9
16	1	Even	1
17	1	Even	4
18	1	Even	7
19	1	Any	1
20	1	Any	4
21	1	Any	7
22	1	Any	1, 6
23	1	Any	2, 7
24	1	Any	3, 8
25	1	Any	1, 4, 7
26	1	Any	2, 5, 8
27	1	Any	3, 6, 9
28	1	Any	0, 2 , 4, 6, 8
29	1	Any	1, 3, 5, 7, 9
30			
31	1	Even	9
32	2	Even	1
33	2	Even	4
34	2	Even	7
35	2	Any	1
36	2	Any	4
37	2	Any	7
38	2	Any	1, 6
39	2	Any	2, 7
40	2	Any	3, 8
41	2	Any	1, 4, 7
42	2	Any	2, 5, 8
43	2	Any	3, 6, 9
44	2	Any	0, 2 , 4, 6, 8
45	2	Any	1, 3, 5, 7, 9
46			
47	2	Even	9
48	3	Even	1
49	3	Even	4
50	3	Even	7
51	3	Any	1
52	3	Any	4
53	3	Any	7
54	3	Any	1, 6
55	3	Any	2, 7
56	3	Any	3, 8
57	3	Any	1, 4, 7
58	3	Any	2, 5, 8
59	3	Any	3, 6, 9
60			
61			
62			
63	3	Even	9

Table 171 – PRACH Configuration Indices for FDD

★ Section 31.6.2 discusses the selection of an appropriate PRACH configuration index and also presents the equivalent table for TDD. The PRACH configuration index is signalled to the UE within SIB2

★ Preamble formats 0 to 3 use a subcarrier spacing of 1.25 kHz rather than the usual uplink subcarrier spacing of 15 kHz. These preamble formats use 839 subcarriers so occupy 1.05 MHz in the frequency domain. These dimensions are illustrated in Figure 166. There is a 15 kHz guard band either side of the random access preamble so a total of 1.08 MHz is used (6 Resource Blocks)

★ Preamble format 4 uses 139 subcarriers with a spacing of 7.5 kHz, so occupies 1.04 MHz in the frequency domain. Similar to preamble formats 0 to 3, a total of 6 Resource Blocks are reserved for the PRACH so there is a small guard band either side of the PRACH subcarriers

★ In the case of FDD, there is a maximum of 1 PRACH position within a single subframe. The position of this PRACH is defined by the 'PRACH Frequency Offset' parameter broadcast in SIB2. This parameter has a range from 0 to 94, and specifies the first Resource Block within which the PRACH is located. The maximum value of 94 is applicable to the 20 MHz channel bandwidth. Smaller channel bandwidths will use smaller maximum values

★ In the case of TDD, there can be up to 6 PRACH positions within a single subframe. This compensates for the reduced number of uplink subframes available for TDD, and helps to maintain the PRACH capacity. The position of each PRACH is defined using a combination of the 'PRACH Frequency Offset' parameter broadcast in SIB2, and predefined 'f_{RA}' values which are dependent upon the PRACH configuration index. Examples of the 'f_{RA}' parameter are presented in Table 350 (section 31.6.2)

★ Section 31.6.5 discusses the selection of the PRACH frequency offset. In general, the frequency offset should be selected such that the PRACH is located adjacent to the PUCCH. This avoids fragmentation of the Resource Blocks available to the PUSCH

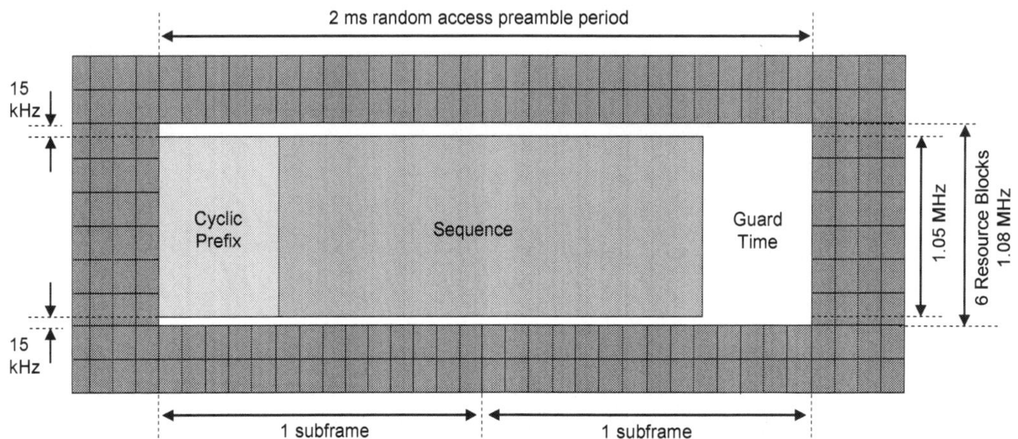

Figure 166 – Random access preamble amongst other Resource Blocks (preamble format 2)

★ Multiple UE are able to simultaneously transmit their PRACH using the same time and frequency domain resources. This is made possible by using different preamble sequences. Each cell has 64 preamble sequences

★ 3GPP TS 36.211 specifies that preamble sequences are generated from a set of 838 root sequences when using preamble formats 0 to 3, and from a set of 138 root sequences when using preamble format 4. Each preamble sequence is generated from its root sequence by applying a cyclic shift. The zero correlation zone parameter determines the size of the cyclic shift and the number of preamble sequences which can be generated from each root sequence Sections 31.6.3 and 31.6.4 discuss the selection of a zero correlation zone and root sequence index

★ The set of 64 preamble sequences within a cell are divided into two groups:

- a group for the contention based random access procedure. UE are responsible for randomly selecting a sequence from the group of sequences advertised by SIB2. These sequences are shared by all UE and there is a danger that multiple UE select the same sequence. This group of sequences can be further divided into 2 sub-groups:
  - 'Group A' sequences selected by the UE when the uplink data quantity to be sent is relatively small, or when the UE is in poor coverage
  - 'Group B' sequences selected by the UE when the uplink data quantity to be sent is relatively large, and the UE is in good coverage
- a group for the non-contention based random access procedure. The eNode B uses dedicated signalling to allocate specific sequences to individual UE. During the handover procedure, the target eNode B can allocate a specific sequence to a UE to ensure that the UE does not experience contention when accessing the target cell, and this helps to make the handover procedure seamless

★ Both the contention and non-contention based random access procedures are described in section 24.1

★ The division of PRACH sequences is illustrated in Figure 167. The number of 'Group B' preamble sequences can be set to 0 by configuring the value of 'Size of RA Preambles in Group A' to equal the value of 'Number of RA Preambles'. The maximum allowed value for 'Number of RA Preambles' is 64 so it is possible to use all of the sequences for contention based random access, and not reserve any sequences for the non-contention based procedure

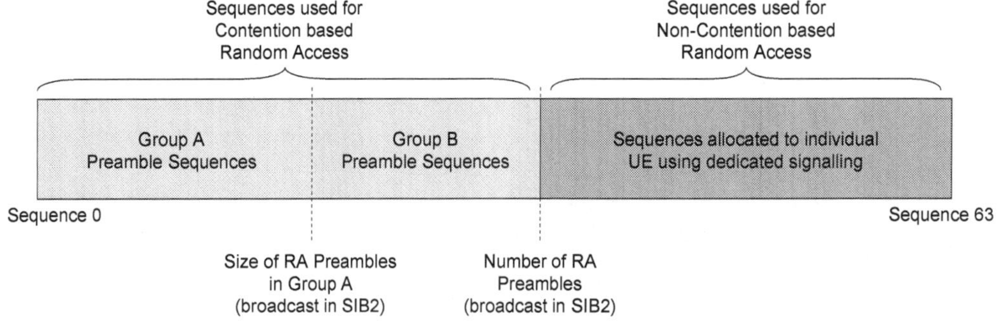

Figure 167 – Division of random access preamble sequences

★ Group A sequences are indexed from 0 to 'Size of RA Preambles in Group A' - 1, where the 'Size of RA Preambles in Group A' parameter ranges from 4 to 60 and is broadcast in SIB2

★ Group B sequences are indexed from 'Size of RA Preambles in Group A' to 'Number of RA Preambles' - 1, where the 'Number of RA Preambles' parameter ranges from 4 to 64 and is broadcast in SIB2

★ Defining groups A and B provides the eNode B with information regarding the quantity of data to be transferred and the coverage conditions experienced by the UE. This information can be used by the eNode B when allocating resources

★ The UE selects a group B sequence if both of the following conditions are satisfied:

Message Size > *Group A Message Size Threshold* AND

Path Loss < P_{MAX} – *Target Rx Power* – *Preamble to Msg Delta* – *Group B Offset*

where, 'Group A Message Size Threshold', 'Target Rx Power', 'Preamble to Msg Delta' and 'Group B Offset' are broadcast in SIB2

★ Once the UE has selected either Group A or Group B, a preamble is selected at random from within that group

★ The length of the preamble sequence is equal to the number of subcarriers used by the preamble format:

  o cells configured to use preamble formats 0 to 3 have sequences with a length of 839

  o cells configured to use preamble format 4 have sequences with a length of 139

★ The processing used to generate the PRACH signal from the preamble sequence is illustrated in Figure 168. This processing follows the same pattern as the processing used to generate the SC-FDMA symbols for the PUSCH

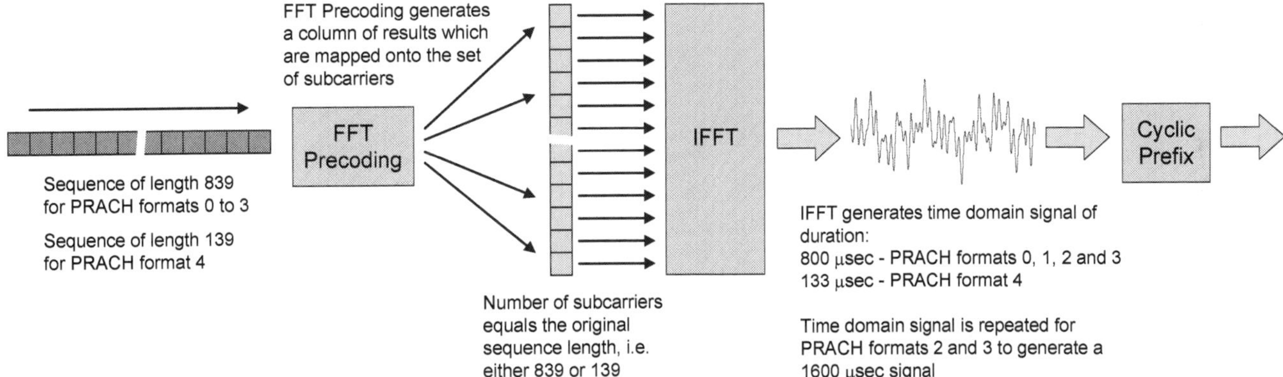

Figure 168 – Processing used to generate PRACH signal from PRACH sequence

★ The sequence is precoded using a Fast Fourier Transform (FFT) function. This generates a set of outputs which are mapped onto the set of subcarriers reserved for the PRACH transmission. The frequency of these subcarriers depends upon the position of the PRACH allocation within the channel bandwidth

★ An Inverse Fast Fourier Transform (IFFT) function is then used to generate a time domain signal from the set of subcarriers. The duration of the time domain signal is equal to the period of the frequency used to define the subcarrier spacing:

  o preamble formats 0 to 3 use a subcarrier spacing of 1.25 kHz so the period of the output signal is 1 / 1250 = 800 μsec

  o preamble format 4 uses a subcarrier spacing of 7.5 kHz so the period of the output signal is 1 / 7500 = 133 μsec

★ Table 170 indicates that the signal duration for PRACH formats 2 and 3 is 1600 μsec. This duration is generated by duplicating and concatenating the 800 μsec output from the IFFT function

★ The bandwidth of this time domain signal is defined by the subcarrier spacing and the number of subcarriers (recalling that the IFFT is just a sum of the subcarriers after modulating the subcarriers with the output from the FFT precoding)

  o preamble formats 0 to 3 use a subcarrier spacing of 1.25 kHz and 839 subcarriers so generate a bandwidth of 1.05 MHz

  o preamble format 4 uses a subcarrier spacing of 7.5 kHz and 139 subcarriers so generates a bandwidth of 1.04 MHz

★ The cyclic prefix is generated by copying the end of the time domain signal and attaching it to the beginning

★ 3GPP References: TS 36.211, TS 36.321, TS 36.331

14.2 PUCCH

★ The Physical Uplink Control Channel (PUCCH) is used to transfer Uplink Control Information (UCI). UCI can also be transferred using the PUSCH. The detailed content of UCI is described in section 17

★ The release 8 and 9 versions of the 3GPP specifications do not allow an individual UE to transmit both the PUCCH and PUSCH during the same subframe. If a release 8 or 9 UE has application data or RRC signalling to send then UCI is transferred using the PUSCH. A release 8 or 9 UE transfers UCI using the PUCCH if it does not have any application data nor RRC signalling to transfer

★ The release 10 version of the 3GPP specifications introduces the option to support simultaneous PUCCH and PUSCH transmission. The UE signals its capability for simultaneous transmission using a combination of information elements:

- o baseband support is signalled using the simultaneousPUCCH-PUSCH information element
- o if baseband support is indicated then RF support can be signaled on a per operating band basis using the nonContiguousUL-RA-WithinCC-Info information element
- o if baseband support is indicated and the UE supports uplink Carrier Aggregation then simultaneous PUCCH and PUSCH transmission is supported for any Component Carrier which the UE is capable of aggregating

All of the above capability information is signalled within the UE EUTRA Capability section of the UE Capability Information message. Support for uplink Carrier Aggregation is deduced from the supported band combinations information

★ The release 10 version of the 3GPP specifications also introduces the possibility to use transmit diversity for the PUCCH. This involves the UE transmitting the PUCCH from 2 antenna ports. This capability is optional and is signalled to the eNode B within the UE EUTRA Capability section of the UE Capability Information message. Transmit diversity for the PUCCH is described in section 12.4.2

★ In the case of TDD, the PUCCH is not transmitted within the UpPTS field of special subframes

★ 3GPP TS 36.211 and TS 36.213 specify the 7 PUCCH formats presented in Table 172. PUCCH formats 2a and 2b are not applicable when using the extended cyclic prefix.

<table>
<tr><th>PUCCH Format</th><th>Number of Bits per Subframe</th><th>FDD / TDD</th><th>Normal Cyclic Prefix</th><th>Extended Cyclic Prefix</th></tr>
<tr><td>1</td><td>-</td><td>FDD & TDD</td><td colspan="2">Scheduling Request (SR)</td></tr>
<tr><td rowspan="2">1a</td><td rowspan="2">1</td><td>FDD & TDD</td><td colspan="2">1 × HARQ-ACK</td></tr>
<tr><td>FDD only</td><td colspan="2">1 × HARQ-ACK + SR</td></tr>
<tr><td rowspan="2">1b</td><td rowspan="2">2</td><td>FDD & TDD</td><td colspan="2">2 × HARQ-ACK, or
2 × HARQ-ACK + SR, or
up to 4 × HARQ-ACK with channel selection when using Carrier Aggregation</td></tr>
<tr><td>TDD only</td><td colspan="2">up to 4 × HARQ-ACK with channel selection</td></tr>
<tr><td>2</td><td>20</td><td>FDD & TDD</td><td>CSI report</td><td>CSI report, or
CSI report + up to 2 × HARQ-ACK</td></tr>
<tr><td>2a</td><td>21</td><td>FDD & TDD</td><td>CSI report + 1 × HARQ-ACK</td><td rowspan="2">Not Applicable</td></tr>
<tr><td>2b</td><td>22</td><td>FDD & TDD</td><td>CSI report + 2 × HARQ-ACK</td></tr>
<tr><td rowspan="2">3</td><td rowspan="2">48</td><td>FDD</td><td colspan="2">up to 10 × HARQ-ACK, or
up to 10 × HARQ-ACK + SR</td></tr>
<tr><td>TDD</td><td colspan="2">up to 20 × HARQ-ACK, or
up to 20 × HARQ-ACK + SR</td></tr>
</table>

Table 172 – Information transferred by each PUCCH format

★ The PUCCH is able to transfer various combinations of Scheduling Requests (SR), Hybrid Automatic Repeat reQuest (HARQ) acknowledgements and Channel State Information (CSI) reports. CSI reports can include Channel Quality Indicators (CQI), Precoding Matrix Indicators (PMI), Precoding Type Indicators (PTI) and Rank Indications (RI). In summary,

- o PUCCH formats 1, 1a and 1b transfer HARQ acknowledgments and scheduling requests
- o PUCCH formats 2, 2a and 2b transfer HARQ acknowledgements and CSI reports
- o PUCCH format 3 transfers HARQ acknowledgments for carrier aggregation and scheduling requests

★ PUCCH formats 1, 1a, 1b, 2, 2a and 2b were introduced within the release 8 version of the 3GPP specifications. PUCCH format 3 was introduced within the release 10 version of the specifications to handle the increased number of HARQ acknowledgements associated with Carrier Aggregation. PUCCH format 1b was also updated for the release 10 version of the specifications to support the HARQ acknowledgements associated with Carrier Aggregation

★ Table 172 also presents the number of bits associated with each PUCCH format. These bits are used to generate modulation symbols prior to further baseband processing and mapping onto Resource Elements. The number of bits is small when HARQ acknowledgments and scheduling requests are transferred

★ The modulation scheme and the number of Resource Elements occupied by each PUCCH format is presented in Table 173

<table>
<tr><th rowspan="2">PUCCH Format</th><th rowspan="2">Number of Bits per Subframe</th><th rowspan="2">Modulation Scheme</th><th colspan="2">Number of Resource Elements Occupied</th></tr>
<tr><th>Normal Cyclic Prefix</th><th>Extended Cyclic Prefix</th></tr>
<tr><td>1</td><td>-</td><td>-</td><td colspan="2" rowspan="3">48 + 48 = 96 or 48 + 36 = 84</td></tr>
<tr><td>1a</td><td>1</td><td>BPSK</td></tr>
<tr><td>1b</td><td>2</td><td>QPSK</td></tr>
<tr><td>2</td><td>20</td><td>QPSK</td><td colspan="2">60 + 60 = 120</td></tr>
<tr><td>2a</td><td>21</td><td>QPSK + BPSK</td><td rowspan="2">60 + 60 = 120</td><td rowspan="2">Not Applicable</td></tr>
<tr><td>2b</td><td>22</td><td>QPSK + QPSK</td></tr>
<tr><td>3</td><td>48</td><td>QPSK</td><td colspan="2">60 + 60 = 120 or 60 + 48 = 108</td></tr>
</table>

Table 173 – Modulation scheme and number of Resource Elements for each PUCCH format

★ A single PUCCH transmission always occupies 2 Resource Blocks which are distributed across the 2 time slots belonging to a subframe

★ Each pair of Resource Blocks allocated to the PUCCH can be used simultaneously by multiple UE. The use of different cyclic shifts and different orthogonal spreading codes allows the eNode B to differentiate the PUCCH transmissions from multiple UE sharing the same pair of Resource Blocks

14.2.1 FORMATS 1, 1a, 1b

★ PUCCH format 1 is used to signal a positive Scheduling Request, i.e. a UE sends format 1 when it would like the eNode B to allocate some uplink resources for the PUSCH

★ PUCCH format 1a is used to signal 1 HARQ acknowledgement for downlink data on the PDSCH. In the case of FDD, it can also be used to combine 1 HARQ acknowledgement with a positive Scheduling Request. The HARQ acknowledgement is transferred using the single bit of information, while the Scheduling Request is signalled by transmitting the PUCCH on the Scheduling Request PUCCH resource rather than on the standard PUCCH resource

- o the Scheduling Request PUCCH resource is defined by 'sr-PUCCH-ResourceIndex' within the range 0 to 2047
- o the standard PUCCH resource is defined by 'n1PUCCH-AN' within the range 0 to 2047

★ PUCCH format 1b is used to signal 2 HARQ acknowledgements for downlink data on the PDSCH. It can also be used to combine 2 HARQ acknowledgements with a Scheduling Request. Similar to format 1a, the Scheduling Request is signalled by transmitting the PUCCH on the Scheduling Request PUCCH resource rather than on the standard PUCCH resource

★ In the case of TDD, PUCCH format 1b can also be used to transfer up to 4 HARQ acknowledgements with channel selection. Transferring 4 HARQ acknowledgements would normally require 4 bits of information but in this case only 2 bits are available. The additional 2 bits of information are signalled by selecting 1 of 4 PUCCH resources. The set of 4 PUCCH resources are derived from the value of 'n1PUCCH-AN'. This approach is used for the TDD ACK/NACK multiplexing described in section 24.3.4

★ The use of format 1b for Carrier Aggregation is described in section 18.11

★ PUCCH format 1 does not have an associated modulation scheme. Information is transferred by the presence/absence of the PUCCH. When present, d(0) = 1 is used as an input when generating the sequence to occupy the PUCCH Resource Elements

★ PUCCH formats 1a and 1b use BPSK and QPSK respectively. In both cases, a single modulation symbol is generated. d(0) is set equal to that modulation symbol and is used as an input when generating the sequence to occupy the PUCCH Resource Elements

★ Figure 169 illustrates the baseband processing applied to d(0) prior to transmission by the UE

★ d(0) is multiplied by a cell specific base sequence of length 12. This generates a new sequence of length 12. The cell specific base sequence is a function of the Physical layer Cell Identity (PCI). If group hopping is enabled (see section 31.7) the base sequence changes between time slots. Otherwise, the same base sequence is used for all time slots. Group hopping helps to randomise intercell interference

★ The new sequence of length 12 then has a series of time domain cyclic shifts applied. These cyclic shifts are a function of both the time slot and the SC-FDMA symbol. Changing the cyclic shift between time slots and SC-FDMA symbols helps to randomise intercell interference

★ The time domain cyclic shift is applied by modifying the sequence in the frequency domain. The sequence is already in the frequency domain because it will be mapped onto the 12 subcarriers belonging to a Resource Block. Applying a phase ramp in the frequency domain generates the same result as a cyclic shift in the time domain. Thus, a set of different phase ramps are applied where each phase ramp corresponds to a different cyclic shift. An example phase ramp would rotate the first number in the sequence by 30°, the second number by 60°, the third number by 90°, etc

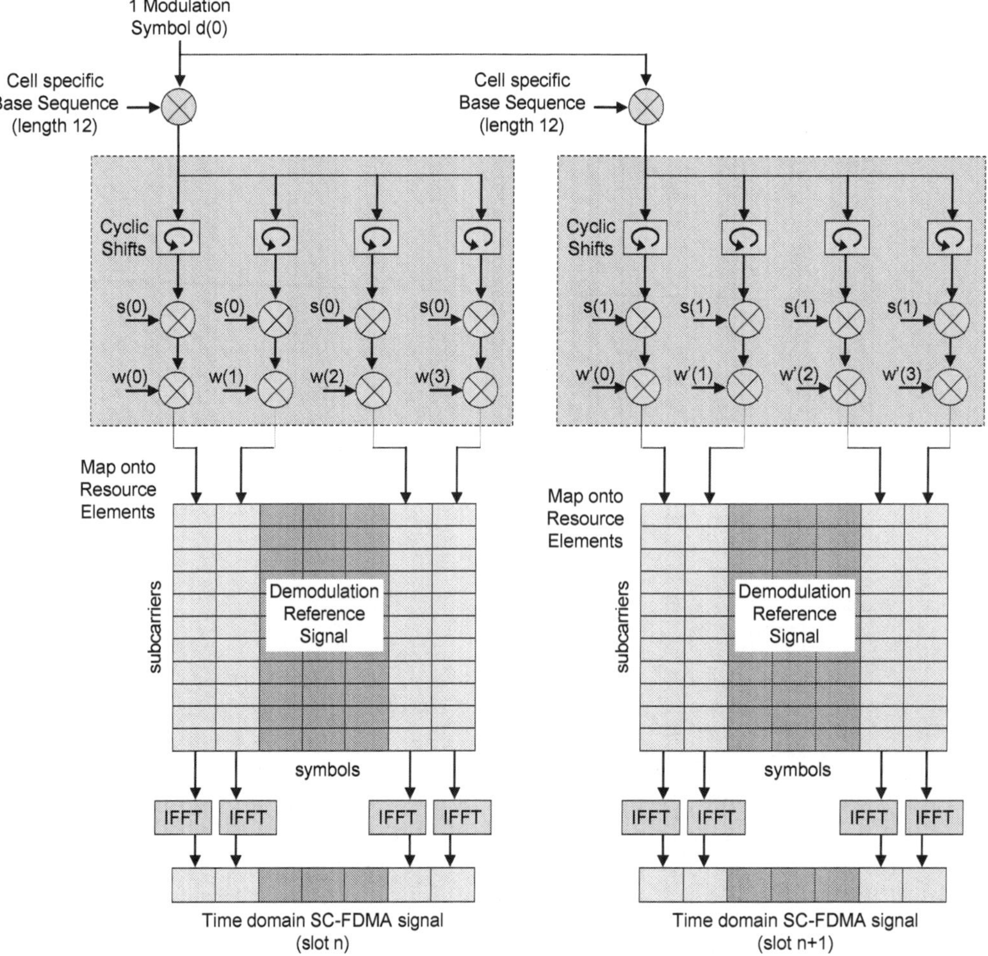

Figure 169 – Processing to generate PUCCH formats 1, 1a and 1b (normal cyclic prefix and without SRS)

★ Figure 170 illustrates the time domain impact of an example phase ramp applied in the frequency domain

Figure 170 – Concept of cyclic shifts

★ The set of cyclic shifts are UE specific and each UE sharing the same pair of Resource Blocks for PUCCH transmission uses a different combination of cyclic shifts and orthogonal codes (orthogonal codes are described below). This allows the eNode B to differentiate between the transmissions from each UE and represents a form of multiplexing

★ The number of cyclic shifts available is dependent upon the 'delta PUCCH shift' parameter signalled by the eNode B within an RRC Connection Setup, RRC Connection Reconfiguration or RRC Connection Re-establishment message. This parameter can be allocated values of 1, 2 or 3. The number of cyclic shift possibilities for each value is presented in Table 174

Delta PUCCH Shift	Cyclic Shift Possibilities
1	12
2	6
3	4

Table 174 – Number of cyclic shift possibilities for each Delta PUCCH Shift value

★ Allowing an increased number of cyclic shifts increases the number of UE which can share the same pair of PUCCH Resource Blocks. However, it can also have a negative impact upon the air-interface performance. Assuming the normal cyclic prefix, the main body of each SC-FDMA symbol has a duration of 66.67 μs. When 12 cyclic shifts are permitted, each cyclic shift is equivalent to 66.67 / 12 = 5.56 μs. Air-interface performance is likely to degrade if the propagation channel delay spread exceeds this value. Halving the number of allowed cyclic shifts doubles the maximum allowed delay spread

- ★ Time slot specific scrambling is applied after the cyclic shifts by multiplying each sequence of length 12 by s(n), where s(n) is given by either a 0° phase shift which does not change any of the sequences in the time slot, or a 90° phase shift which rotates each entry within each sequence in the time slot by 90°. Time slot specific scrambling helps to reduce the impact of inter-code interference
- ★ After scrambling, a pair of time domain orthogonal codes are applied. The first orthogonal code is applied to the sequences belonging to the first time slot, and the second orthogonal code is applied to the sequences belonging to the second time slot. Each orthogonal code has a length of 4, i.e. a single entry for each SC-FDMA symbol within the time slot
- ★ The orthogonal codes are UE specific and each UE sharing the same pair of Resource Blocks for PUCCH transmission uses a different combination of cyclic shifts and orthogonal codes. The set of allowed orthogonal codes is presented in Table 175

Code Index	w(0)	w(1)	w(2)	w(3)
0	1	1	1	1
1	1	-1	1	-1
2	1	-1	-1	1

Table 175 – PUCCH format 1, 1a and 1b orthogonal codes of length 4

- ★ The orthogonal codes have a length of 4 so in principle it is possible to define a set of 4 orthogonal codes, i.e. code 1 1 -1 -1 is missing. This is similar to the concept of spreading codes in UMTS where the number of spreading codes with a specific spreading factor (code length) is equal to the spreading factor, e.g. there are 16 codes with a spreading factor of 16
- ★ In this case, the 4th orthogonal code is not included in the set because the PUCCH demodulation Reference Signal limits the number of codes which can be used. The PUCCH demodulation Reference Signal is also generated using a UE specific orthogonal code but there are only 3 SC-FDMA symbols so only 3 orthogonal codes can be defined
- ★ The combination of time domain cyclic shifts and time domain orthogonal codes define the number of UE which can be multiplexed within a single pair of PUCCH Resource Blocks. The total number of multiplexing possibilities is defined by the product of the number of cyclic shifts and the number of orthogonal codes. These figures are presented in Table 176

Delta PUCCH Shift	Cyclic Shift Possibilities	Orthogonal Codes	Multiplexing Possibilities
1	12	3	$12 \times 3 = 36$
2	6	3	$6 \times 3 = 18$
3	4	3	$4 \times 3 = 12$

Table 176 – Multiplexing possibilities for each Delta PUCCH Shift value

- ★ The number of multiplexing possibilities does not directly translate to the number of UE which can share the same pair of Resource Blocks because a single UE is allocated more than a single PUCCH resource. UE are allocated a standard PUCCH resource and a Scheduling Request (SR) PUCCH resource. The standard PUCCH resource is used when signalling HARQ acknowledgements without a SR, whereas the SR PUCCH resource is used when signalling a SR either with or without HARQ acknowledgements. TDD UE can be allocated 4 PUCCH resources when using format 1b to send 4 HARQ acknowledgements with channel selection (described in section 24.3.4). In this case, the selection of the PUCCH resource (channel selection) provides the eNode B with 2 additional bits of information
- ★ After multiplication by the orthogonal codes, each resultant sequence of length 12 is mapped onto a column of Resource Elements. Each column of Resource Elements is then processed in the normal way using an Inverse Fast Fourier Transform (IFFT) to generate the time domain SC-FDMA symbols with their cyclic prefixes
- ★ As shown in Figure 169, the PUCCH demodulation Reference Signal for formats 1, 1a and 1b occupies 3 out of the 7 symbols per time slot when using the normal cyclic prefix. When using the extended cyclic prefix, the demodulation Reference Signal occupies 2 out of 6 symbols per time slot. In both cases, there are 4 symbols available for PUCCH transmission. This corresponds to $4 \times 12 = 48$ Resource Elements per time slot
- ★ When transmitting the Sounding Reference Signal (SRS), the final symbol in the subframe becomes unavailable for the PUCCH. This decreases the number of available Resource Elements within the second time slot to $3 \times 12 = 36$. In this case, a length 3 orthogonal code is applied to the sequences belonging to the second time slot. Similar to the length 4 orthogonal sequences, there are 3 possibilities available so it does not impact the number of UE which can be multiplexed into the same pair of Resource Blocks

14.2.2 FORMATS 2, 2a, 2b

- ★ PUCCH format 2 can be used to transfer a Channel State Information (CSI) report. This report can include Channel Quality Indicator (CQI), Precoding Matrix Indication (PMI), Precoding Type Indication (PTI) and Rank Indication (RI) information. The content of CSI reports is described in section 17. In all cases, the combinations of CQI, PMI, PTI and RI are coded to a total of 20 bits

★ In the case of the extended cyclic prefix, PUCCH format 2 can also be used to signal the combination of a CSI report and either 1 or 2 HARQ acknowledgements. The combination of CSI report and HARQ acknowledgement(s) is coded to a total of 20 bits so the payload size for PUCCH format 2 remains constant irrespective of whether or not HARQ acknowledgements are included

★ PUCCH formats 2a and 2b are only applicable when using the normal cyclic prefix. PUCCH format 2a provides support for a CSI report and 1 HARQ acknowledgement, whereas PUCCH format 2b provides support for a CSI report and 2 HARQ acknowledgements. In both cases, the CSI report is coded to a total of 20 bits and the 1 or 2 HARQ acknowledgements are concatenated without additional coding, i.e. generating 21 bits for PUCCH format 2a, and 22 bits for PUCCH format 2b

★ Figure 171 illustrates the baseband processing applied to the 20 bits associated with the CSI report. The additional 1 bit belonging to the HARQ acknowledgement for PUCCH format 2a is BPSK modulated to generate a single modulation symbol. Likewise, the additional 2 bits belonging to the HARQ acknowledgements for PUCCH format 2b are QPSK modulated to generate a single modulation symbol. In each case, the modulation symbol is used as an input when generating the PUCCH demodulation Reference Signal, i.e. the eNode B receiver can deduce the additional bits of information when detecting the Reference Signal sequence

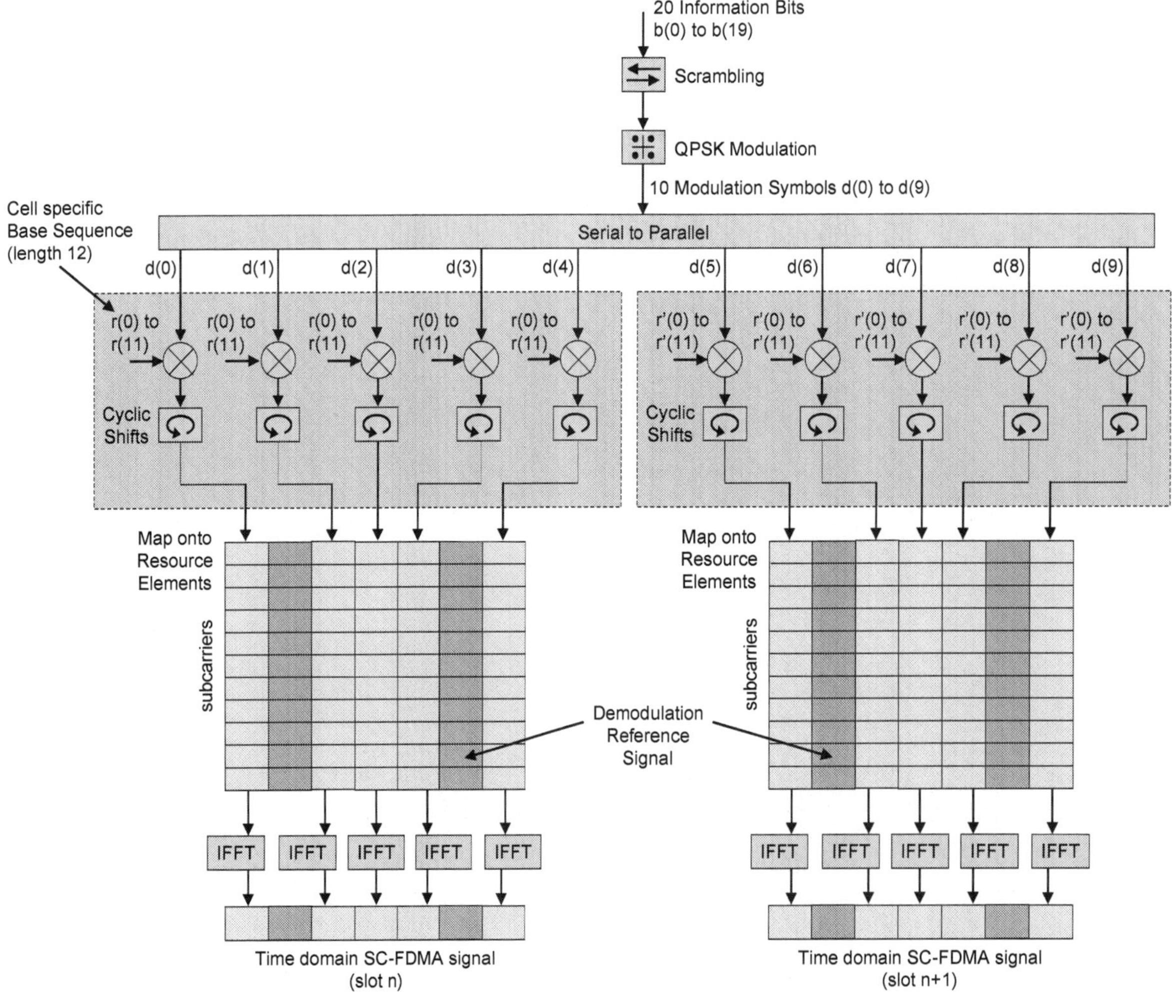

Figure 171 – Processing to generate PUCCH formats 2, 2a and 2b (normal cyclic prefix)

★ The set of 20 bits is scrambled with a UE specific scrambling sequence. The scrambling sequence is dependent upon both the Physical layer Cell Identity (PCI) which has been allocated to the cell, and the C-RNTI which has been allocated to the UE

★ The set of 20 scrambled bits are then QPSK modulated to generate a set of 10 modulation symbols

★ The set of 10 modulation symbols are separated into 10 parallel streams. These 10 parallel streams are used to generate the sequences which are mapped onto the 10 SC-FDMA symbols within the pair of PUCCH Resource Blocks

★ Each modulation symbol is multiplied by a cell specific base sequence of length 12. This generates a new sequence of length 12. The cell specific base sequence is a function of the Physical layer Cell Identity (PCI). If group hopping is enabled (see section 31.7) the base sequence changes between time slots. Otherwise, the same base sequence is used for all time slots. Group hopping helps to randomise intercell interference

★ Each new sequence of length 12 then has a time domain cyclic shift applied. These cyclic shifts are a function of both the time slot and the SC-FDMA symbol. Changing the cyclic shift between time slots and SC-FDMA symbols helps to randomise intercell interference

- The time domain cyclic shift is applied by modifying the sequence in the frequency domain. The sequence is already in the frequency domain because it will be mapped onto the 12 subcarriers belonging to a Resource Block. Applying a phase ramp in the frequency domain generates the same result as a cyclic shift in the time domain. Thus, a set of different phase ramps are applied where each phase ramp corresponds to a different cyclic shift
- The set of cyclic shifts are UE specific and each UE sharing the same pair of Resource Blocks for PUCCH transmission uses a different set of cyclic shifts. In contrast to PUCCH formats 1, 1a and 1b, orthogonal time domain codes are not applied to PUCCH formats 2, 2a and 2b. This means that the number of UE which can share the same pair of Resource Blocks is determined only by the number of available cyclic shifts
- PUCCH formats 2, 2a and 2b do not use the 'delta PUCCH shift' parameter to control the number of possible cyclic shifts so the full set of 12 possible cyclic shifts are available to the eNode B when allocating PUCCH resources
- After applying the time domain cyclic shifts, each resultant sequence of length 12 is mapped onto a column of Resource Elements. Each column of Resource Elements is then processed in the normal way using an Inverse Fast Fourier Transform (IFFT) to generate the time domain SC-FDMA symbols with their cyclic prefixes
- As shown in Figure 171, the PUCCH demodulation Reference Signal for formats 2, 2a and 2b occupies 2 out of the 7 symbols per time slot when using the normal cyclic prefix. When using the extended cyclic prefix, the demodulation Reference Signal occupies 1 out of 6 symbols per time slot. In both cases, there are 5 symbols available for PUCCH transmission. This corresponds to 5 × 12 = 60 Resource Elements per time slot
- The demodulation Reference Signal for PUCCH formats 2, 2a and 2b occupies fewer SC-FDMA symbols then for PUCCH formats 1, 1a and 1b. This helps to increase the number of Resource Elements available for the PUCCH payload and decreases the coding rate, i.e. it allows increased redundancy to help transfer the payload across the air-interface. The payload is smaller for PUCCH formats 1, 1a and 1b so the corresponding requirement for Resource Elements is less
- PUCCH formats 2, 2a and 2b are never transferred during the same subframe as the Sounding Reference Signal. PUCCH formats 2, 2a and 2b have priority if they coincide with the Sounding Reference Signal

14.2.3 FORMAT 3

- PUCCH format 3 was introduced within the release 10 version of the 3GPP specifications to support the increased number of HARQ acknowledgements associated with Carrier Aggregation
  - in the case of FDD, there can be up to 5 Component Carriers and each Component Carrier can generate up to 2 HARQ acknowledgments during each subframe. This generates the requirement to support up to 10 HARQ acknowledgments
  - in the case of TDD, there can also be up to 5 Component Carriers and each Component Carrier can generate up to 2 HARQ acknowledgments during each subframe. In the case of TDD, a single uplink subframe may have to acknowledge the data sent during multiple downlink subframes (depending upon the uplink-downlink subframe configuration)
    - for example, uplink-downlink subframe configuration 4 includes 2 uplink subframes, 7 downlink subframes and 1 special subframe. Within the special subframe, downlink data can be transferred during the DwPTS field, but the PUCCH cannot be transmitted during the UpPTS field. This means that each uplink subframe has to acknowledge 4 downlink subframes and the PUCCH must be capable of transferring up to 40 HARQ acknowledgements
  - the TDD version of PUCCH format 3 has been designed to support up to 20 HARQ acknowledgements. When there is a requirement to transfer more than 20 HARQ acknowledgements, a logical AND operation is applied to all pairs of acknowledgements belonging to the same subframe and the same Component Carrier, i.e. transferring 2 transport blocks with spatial multiplexing generates a resultant single HARQ acknowledgement instead of the usual 2 HARQ acknowledgements
  - TDD uplink-downlink subframe configuration 5 which includes 1 uplink subframe, 8 downlink subframes and 1 special subframe is only supported with PUCCH format 3 when the UE is configured with either 1 or 2 Component Carriers. 3 Component Carriers would generate the requirement to support 3 × 9 × 2 = 54 HARQ acknowledgements which would reduce to 3 × 9 = 27 HARQ acknowledgements after applying the logical AND operation. This is greater than the capacity of PUCCH format 3
- PUCCH format 3 can also be used to transfer a Scheduling Request (SR) when transmission coincides with a SR subframe. In contrast to PUCCH formats 1, 1a and 1b where the SR is signalled by using the SR PUCCH resource rather than the standard PUCCH resource, PUCCH format 3 allocates a single bit of information to signal the SR. This additional bit of information is concatenated with the HARQ acknowledgements
- The total set of bits (HARQ acknowledgements and potentially a SR) has a maximum length of 11 bits for FDD, and a maximum length of 21 bits for TDD. In both cases, the set of bits is coded to generate a resultant set of 48 bits
- Figure 172 illustrates the baseband processing applied to the 48 bits generated for PUCCH format 3
- The set of 48 bits is scrambled with a UE specific scrambling sequence. The scrambling sequence is dependent upon both the Physical layer Cell Identity (PCI) which has been allocated to the cell, and the C-RNTI which has been allocated to the UE
- The set of 48 scrambled bits are then QPSK modulated to generate a set of 24 modulation symbols

★ The set of 24 modulation symbols is divided into 2 halves. The first half of 12 symbols is used as an input for the first PUCCH time slot, while the second half of 12 symbols is used as an input for the second PUCCH time slot

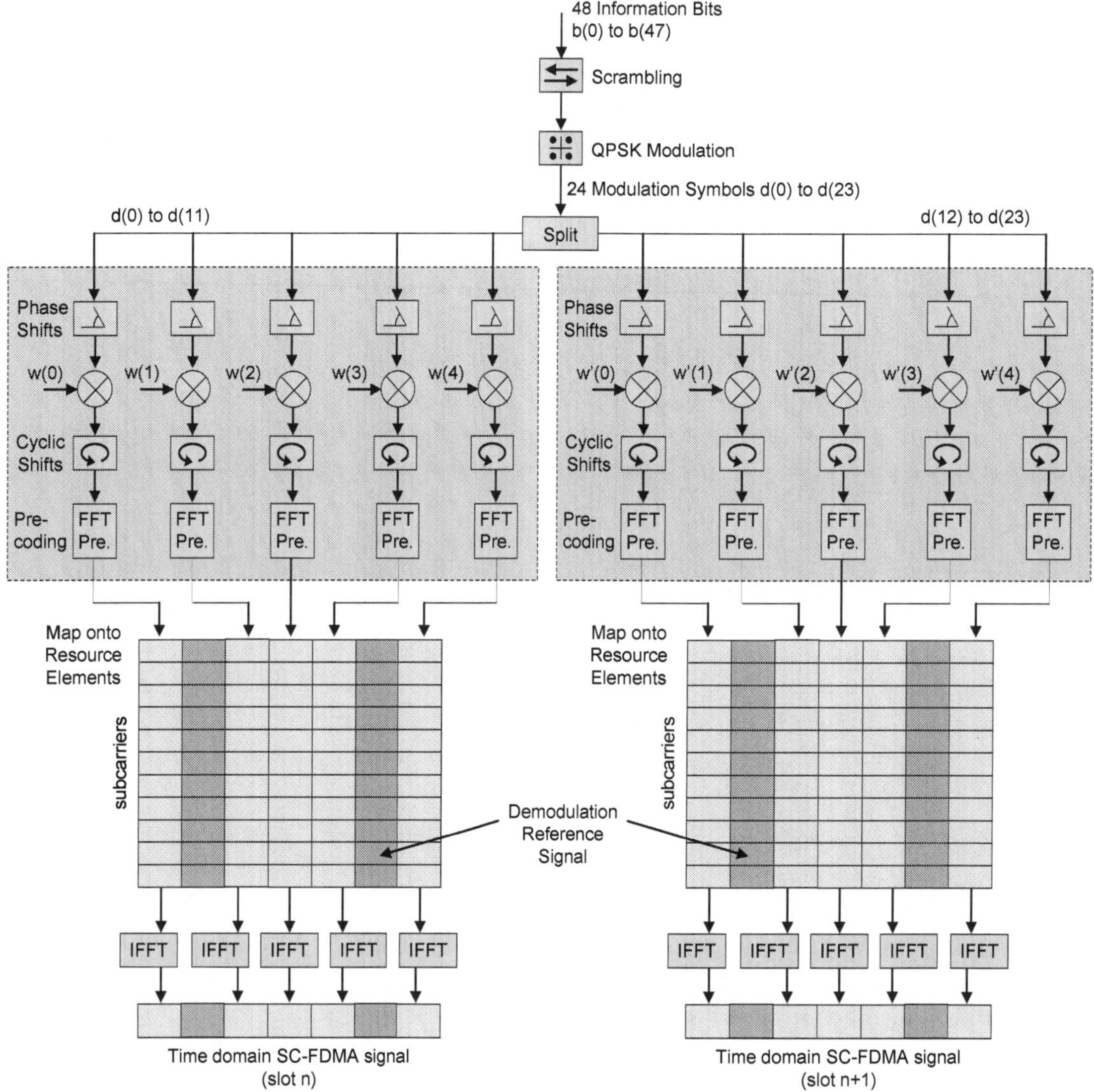

Figure 172 – Processing to generate PUCCH format 3 (normal cyclic prefix and without SRS)

★ 5 duplicates of the 12 symbols are generated for each time slot and a phase shift is applied to each duplicate. The phase shifts are a function of the cell's Physical layer Cell Identity (PCI), as well as the SC-FDMA symbol number and time slot number. The objective of these phase shifts is to help randomise intercell interference

★ The 10 sequences of length 12 are then multiplied by 2 time domain orthogonal codes of length 5. The first orthogonal code (shown as w(0) to w(4)) is applied to the sequences belonging to the first time slot, while the second orthogonal code is applied to the sequences belonging to the second time slot (shown as w'(0) to w'(4)). A single entry from each orthogonal code is multiplied by each sequence so the resultant 10 sequences remain with a length of 12

★ The orthogonal codes are UE specific and each UE sharing the same pair of Resource Blocks for PUCCH transmission uses a different pair of orthogonal codes. 5 different orthogonal codes have been defined because each code has a length of 5. This allows 5 UE to share the same pair of Resource Blocks. This differs from PUCCH formats 1, 1a and 1b where only 3 of the possible 4 orthogonal codes are used, i.e. all of the available orthogonal codes are used for PUCCH format 3. This is because the demodulation Reference Signal for PUCCH format 3 does not use orthogonal codes so there is no limitation in terms of the number of codes supported by the Reference Signal

★ A set of time domain cyclic shifts are then applied to the 10 sequences. These cyclic shifts are a function of the cell's PCI, as well as the SC-FDMA symbol number and time slot number

★ Fast Fourier Transform (FFT) precoding is applied in a similar way to precoding for the PUSCH. This generates the results which are mapped onto the set of Resource Elements. Each sequence of length 12 is mapped onto a single column of Resource Elements. The Inverse Fast Fourier Transform (IFFT) is then applied to generate the time domain SC-FDMA symbols

- As shown in Figure 172, the PUCCH demodulation Reference Signal for format 3 occupies 2 out of the 7 symbols per time slot when using the normal cyclic prefix. When using the extended cyclic prefix, the demodulation Reference Signal occupies 1 out of 6 symbols per time slot. In both cases, there are 5 symbols available for PUCCH transmission. This corresponds to 5 × 12 = 60 Resource Elements per time slot
- When transmitting the Sounding Reference Signal (SRS), the final symbol in the subframe becomes unavailable for the PUCCH. This decreases the number of available Resource Elements within the second time slot to 4 × 12 = 48. In this case, a length 4 orthogonal code is applied to the sequences belonging to the second time slot. A set of 4 orthogonal codes has been specified for this purpose. This reduces the multiplexing capacity of PUCCH format 3 from 5 UE per Resource Block to 4 UE per Resource Block

14.2.4 RESOURCE ALLOCATION

- The PUCCH is allocated Resource Blocks at the 2 edges of the channel bandwidth. Each PUCCH transmission uses 1 Resource Block on each side of the channel bandwidth and these 2 Resource Blocks are distributed across 2 time slots
- Figure 173 illustrates examples based upon a single UE and a group of UE. Resource Block numbering for the PUCCH starts on the outside edges of the channel bandwidth and increases inwards

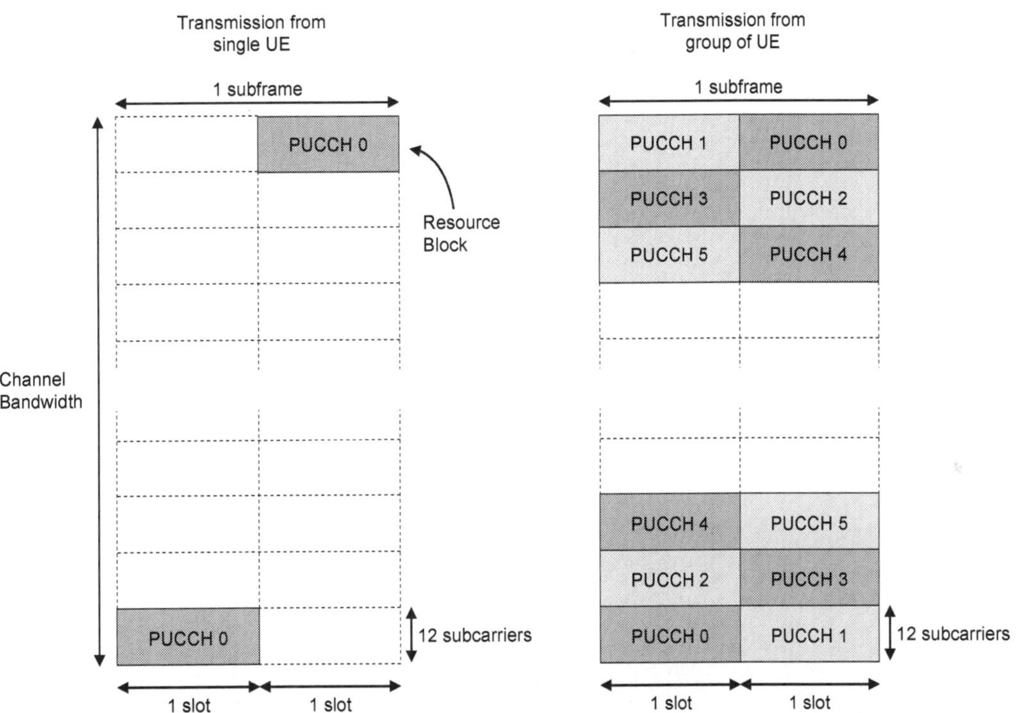

Figure 173 – PUCCH Resource Blocks at the edges of the channel bandwidth

- The PUCCH is allocated Resource Blocks at the edge of the channel bandwidth to avoid fragmenting the Resource Blocks available to the PUSCH. Avoiding fragmentation improves the efficiency with which the PUSCH scheduler within the eNode B can operate. Allocating PUCCH Resource Blocks at the 2 channel edges also provides frequency diversity
- Resource Block allocations are required for PUCCH formats 1, 1a, 1b, 2, 2a and 2b when Carrier Aggregation for LTE Advanced is not used, i.e. PUCCH format 3 is only required when Carrier Aggregation is used
- PUCCH formats 1, 1a and 1b share one set of Resource Blocks, while PUCCH formats 2, 2a and 2b share another set of Resource Blocks. A single Resource Block within each time slot can be shared between PUCCH formats 1, 1a, 1b, 2, 2a and 2b if the $N_{CS}^{(1)}$ parameter (nCS-An) is configured with a value greater than 0. This parameter is broadcast in SIB2 and can be configured with values between 0 and 7. The value of $N_{CS}^{(1)}$ defines the number of cyclic shifts available to PUCCH formats 1, 1a and 1b within the shared Resource Block
- PUCCH formats 2, 2a and 2b have Resource Blocks reserved on the outside edges of the channel bandwidth. The $N_{RB}^{(2)}$ parameter (nRB-CQI) defines the number of Resource Blocks within each time slot reserved for PUCCH formats 2, 2a and 2b. This parameter is broadcast in SIB2 and can be configured with values between 0 and 98
- PUCCH formats 1, 1a and 1b are allocated the next available Resource Blocks after accounting for those which have been reserved for PUCCH formats 2, 2a and 2b

★ If the $N_{CS}^{(1)}$ parameter has been configured with a value greater than 0, the shared Resource Block is positioned between the format 2 and format 1 allocations

★ Figure 174 illustrates 3 example Resource Block allocations to demonstrate the impact of the $N_{RB}^{(2)}$ and $N_{CS}^{(1)}$ parameters

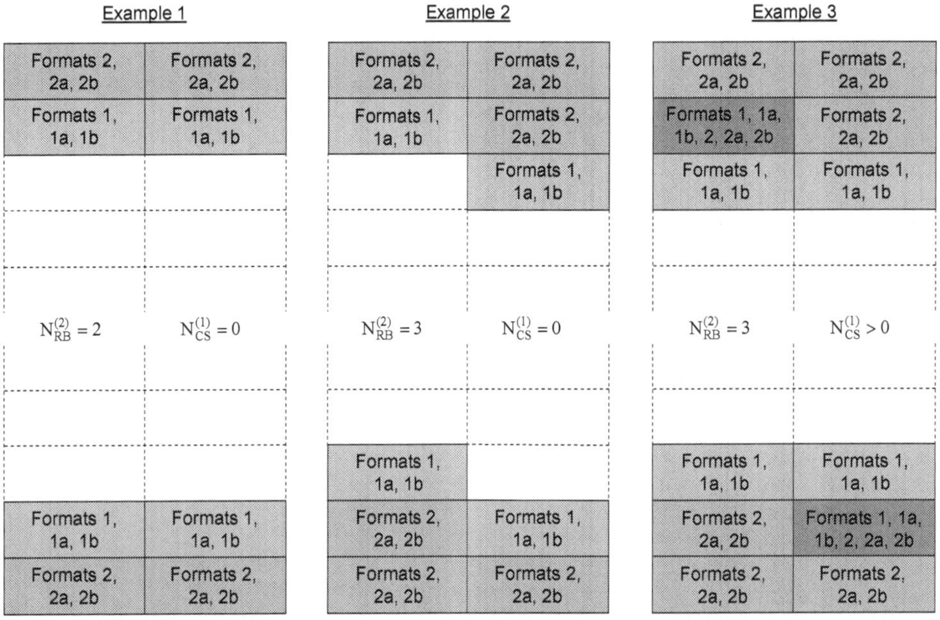

Example 1		Example 2		Example 3	
Formats 2, 2a, 2b	Formats 2, 2a, 2b	Formats 2, 2a, 2b	Formats 2, 2a, 2b	Formats 2, 2a, 2b	Formats 2, 2a, 2b
Formats 1, 1a, 1b	Formats 1, 1a, 1b	Formats 1, 1a, 1b	Formats 2, 2a, 2b	Formats 1, 1a, 1b, 2, 2a, 2b	Formats 2, 2a, 2b
			Formats 1, 1a, 1b	Formats 1, 1a, 1b	Formats 1, 1a, 1b
$N_{RB}^{(2)} = 2$	$N_{CS}^{(1)} = 0$	$N_{RB}^{(2)} = 3$	$N_{CS}^{(1)} = 0$	$N_{RB}^{(2)} = 3$	$N_{CS}^{(1)} > 0$
		Formats 1, 1a, 1b		Formats 1, 1a, 1b	Formats 1, 1a, 1b
Formats 1, 1a, 1b	Formats 1, 1a, 1b	Formats 2, 2a, 2b	Formats 1, 1a, 1b	Formats 2, 2a, 2b	Formats 1, 1a, 1b, 2, 2a, 2b
Formats 2, 2a, 2b	Formats 2, 2a, 2b	Formats 2, 2a, 2b	Formats 2, 2a, 2b	Formats 2, 2a, 2b	Formats 2, 2a, 2b

Figure 174 – Resource Block allocations for PUCCH formats 1, 1a, 1b, 2, 2a and 2b

★ When transmitting PUCCH format 1, 1a or 1b, the PUCCH resource used by a specific UE is determined by $n_{PUCCH}^{(1)}$

- in the case of FDD, $n_{PUCCH}^{(1)} = n_{CCE} + N_{PUCCH}^{(1)}$, where n_{CCE} is the number of the first Control Channel Element (CCE) used for transmission of the corresponding DCI, and $N_{PUCCH}^{(1)}$ is common to all UE and is broadcast in SIB2 using a range from 0 to 2047
- $n_{PUCCH}^{(1)}$ has a second value when using PUCCH formats 1, 1a and 1b to transfer a Scheduling Request. An RRC Connection Reconfiguration message is used to signal the value of $n_{PUCCH,SR}^{(1)}$ which can also range from 0 to 2047
- in the case of TDD when up to 4 HARQ acknowledgements are transferred with channel selection, up to 4 values of $n_{PUCCH}^{(1)}$ are calculated. The UE then selects the resource to use based upon the combination of positive and negative acknowledgements

★ In the case of formats 1, 1a and 1b, the PUCCH resource defines the allocated Resource Blocks, the time domain cyclic shifts and the time domain orthogonal codes

★ When transmitting PUCCH format 2, 2a or 2b, the PUCCH resource used by a specific UE is determined by the $n_{PUCCH}^{(2)}$ parameter (cqi-PUCCH-ResourceIndex) signalled to the UE within an RRC Connection Reconfiguration message. This parameter ranges from 0 to 1185

★ In practice, the range of $n_{PUUCH}^{(2)}$ is limited by the number of Resource Blocks allocated to PUCCH formats 2, 2a and 2b. A maximum of 12 $n_{PUCCH}^{(2)}$ values can be associated with a single Resource Block due to the maximum of 12 time domain cyclic shifts

★ In the case of formats 2, 2a and 2b, the PUCCH resource defines the allocated Resource Blocks and the time domain cyclic shifts

★ When carrier aggregation is used for LTE Advanced, PUCCH format 3 can be allocated a set of Resource Blocks

★ PUCCH format 3 uses Resource Blocks from within the PUCCH format 2 allocation, i.e. the $N_{RB}^{(2)}$ parameter then defines the number of Resource Blocks within each time slot reserved for PUCCH formats 2, 2a, 2b and 3

★ Figure 175 illustrates some example Resource Block allocations for PUCCH format 3

★ It is the responsibility of the eNode B scheduler to ensure that resource allocations for PUCCH format 3 do not coincide with resource allocations for PUCCH formats 2, 2a and 2b, i.e. they use separate Resource Blocks

★ When transmitting PUCCH format 3, the PUCCH resource used by a specific UE is determined by the $n^{(3)}_{PUCCH}$ parameter (n3PUCCH-AN-List) signalled by the eNode B. This parameter can have a range from 0 to 549 to account for the possibility of multiplexing up to 5 UE within each Resource Block, i.e. the set of 550 values corresponds to 110 Resource Blocks

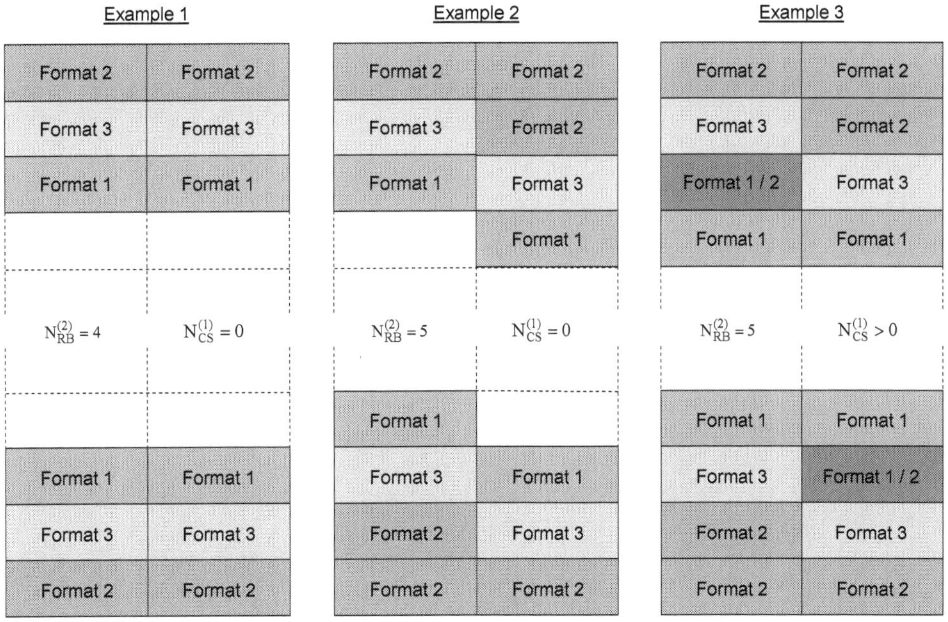

Figure 175 – Resource Block allocations for PUCCH formats 1, 1a, 1b, 2, 2a, 2b and 3

★ The UE is provided with a list of 4 values for the $n^{(3)}_{PUCCH}$ parameter. This is done semi-statically using RRC signalling. The UE is then dynamically instructed which of the 4 values to use during a specific subframe by the 'TPC command for PUCCH' bits within the Downlink Control Information (DCI), i.e. the TPC commands are interpreted in a different way when using PUCCH format 3

★ The set of 4 values for the $n^{(3)}_{PUCCH}$ parameter provides the eNode B scheduler with increased flexibility to avoid resource allocation collisions. This can be necessary when the Sounding Reference Signal (SRS) is transmitted and the PUCCH format 3 multiplexing capacity within a Resource Block reduces from 5 to 4

★ The value of $n^{(3)}_{PUCCH}$ determines the allocated Resource Blocks and the time domain orthogonal codes for PUCCH format 3

★ 3GPP References: TS 36.211, TS 36.212, TS 36.213, TS 36.331

14.3 PUSCH

★ The Physical Uplink Shared Channel (PUSCH) is used to transfer:

  - RRC signalling messages
  - Uplink Control Information (UCI)
  - application data

★ Uplink RRC signalling messages are transferred using the PUSCH, i.e. Signalling Radio Bearers (SRB) use the PUSCH and every connection has its own set of SRB

★ UCI can be transferred using either the PUSCH or the PUCCH. 3GPP release 8 and 9 devices do not support simultaneous transmission of the PUSCH and PUCCH so UCI is transferred using the PUSCH whenever there is any RRC signalling or application data to transfer. The release 10 version of the specifications allows UE to support simultaneous transmission of the PUSCH and PUCCH so UCI can be transferred using the PUCCH while RRC signalling or application data is transferred using the PUSCH

★ RRC signalling and UCI represent additional overheads from the perspective of transferring application data (additional to the overheads generated by the Reference Signals, PRACH and PUCCH)

★ The PUSCH can be modulated using 64QAM, 16QAM or QPSK. The eNode B selects the appropriate modulation scheme according to its link adaptation algorithm

★ The modulation scheme is signalled to the UE within the Modulation and Coding Scheme (MCS) information belonging to PDCCH Downlink Control Information (DCI) formats 0 and 4. These DCI are used to signal the uplink resource allocations to the population of UE. The content of the various DCI formats is presented in section 9

★ The PUSCH always uses QPSK when TTI Bundling is enabled. If the eNode B instructs a UE to use 64QAM but the UE does not support 64QAM then 16QAM is selected

★ The application throughputs achievable from the PUSCH are discussed in section 20. The PUSCH is a shared channel so the total set of Resource Blocks available to the PUSCH, and the associated throughputs must be shared between all active connections

★ Figure 176 illustrates an example of the Resource Elements available to the PUSCH after making allocations for the Reference Signals and other Physical Channels

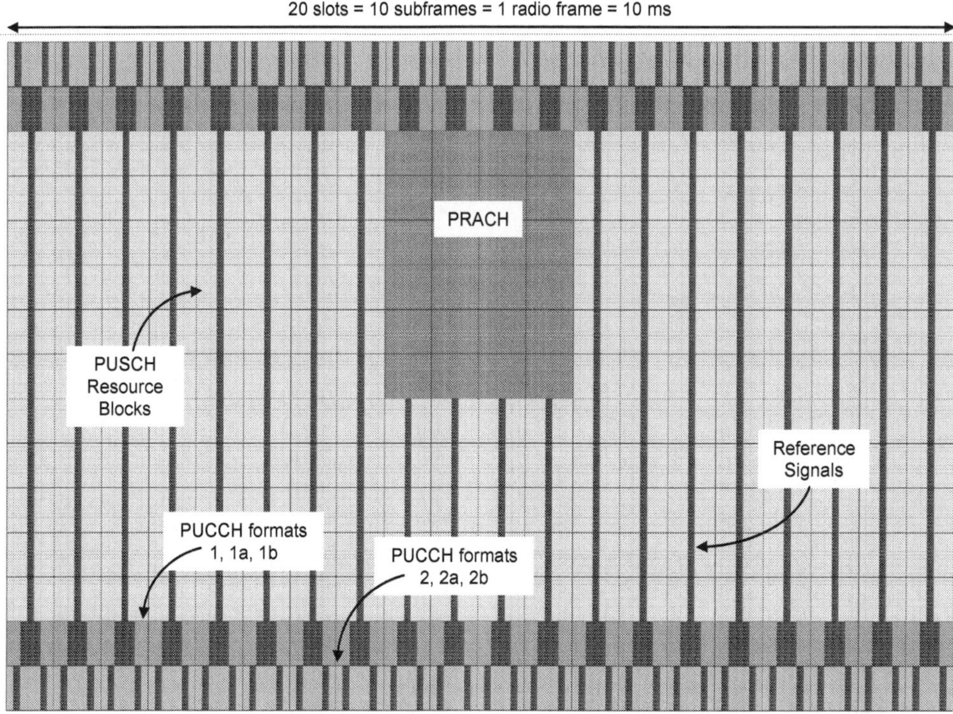

Figure 176 – Example Resource Elements available to the PUSCH

★ The example shown in Figure 176 is based upon the 3 MHz channel bandwidth, i.e. 15 Resource Blocks in the frequency domain. It is assumed that 4 of the 15 Resource Blocks are allocated to the PUCCH. It is also assumed that the radio frame includes a single 2ms PRACH transmission. The eNode B is responsible for sharing the set of PUSCH Resource Blocks shown in Figure 176 between the population of UE

★ Table 177 quantifies the percentage of Resource Elements remaining for the PUSCH after making allocations for the Reference Signals and other Physical Channels

		1.4 MHz	3 MHz	5 MHz	10 MHz	15 MHz	20 MHz
Without PRACH	Normal Cyclic Prefix						
	2 PUCCH Res. Blks.	57 %	74 %	79 %	82 %	83 %	84 %
	4 PUCCH Res. Blks.	29 %	63 %	72 %	79 %	81 %	82 %
	6 PUCCH Res. Blks.	-	51 %	65 %	75 %	79 %	81 %
	8 PUCCH Res. Blks.	-	40 %	58 %	72 %	77 %	79 %
	Extended Cyclic Prefix						
	2 PUCCH Res. Blks.	56 %	72 %	77 %	80 %	81 %	82 %
	4 PUCCH Res. Blks.	28 %	61 %	70 %	77 %	79 %	80 %
	6 PUCCH Res. Blks.	-	50 %	63 %	73 %	77 %	78 %
	8 PUCCH Res. Blks.	-	39 %	57 %	70 %	74 %	77 %
With 1 PRACH	Normal Cyclic Prefix						
	2 PUCCH Res. Blks.	46 %	67 %	75 %	80 %	82 %	83 %
	4 PUCCH Res. Blks.	23 %	56 %	68 %	77 %	80 %	81 %
	6 PUCCH Res. Blks.	-	45 %	61 %	73 %	77 %	80 %
	8 PUCCH Res. Blks.	-	33 %	54 %	70 %	75 %	78 %
	Extended Cyclic Prefix						
	2 PUCCH Res. Blks.	44 %	66 %	73 %	78 %	80 %	81 %
	4 PUCCH Res. Blks.	22 %	54 %	66 %	75 %	78 %	79 %
	6 PUCCH Res. Blks.	-	43 %	59 %	71 %	75 %	77 %
	8 PUCCH Res. Blks.	-	32 %	53 %	68 %	73 %	76 %

Table 177 – Percentage of Resource Elements remaining for PUSCH

★ These figures exclude the impact of the Sounding Reference Signal. The results which include a single PRACH transmissions (lower half of the table), are calculated over a single radio frame. With the exception of the 1.4 MHz channel bandwidth, it has been assumed that the PRACH transmission does not overlap with the PUCCH Resource Blocks

★ The variable having the greatest impact upon the percentage of Resource Elements available to the PUSCH is the number of Resource Blocks allocated to the PUCCH. In general, the number of Resource Blocks allocated to the PUCCH will increase for the larger channel bandwidths

★ The figures within Table 177 can be translated into application throughputs after accounting for the overheads generated by RRC signalling, UCI on the PUSCH and the various protocol stack headers. Bit rates are discussed in section 20

★ 3GPP References: TS 36.211, TS 36.212, TS 36.213

15 UPLINK TRANSPORT CHANNELS

15.1 UL-SCH

★ The Uplink Shared Channel (UL-SCH) is used to transfer RRC signalling and application data. Uplink Control Information (UCI) can also be added during physical layer processing prior to mapping onto the PUSCH physical channel. UCI is transferred using the PUCCH rather than the PUSCH when there is no RRC signalling nor application data to transfer

★ The release 8 and 9 versions of the 3GPP specifications support the transfer of 1 uplink transport block per subframe for each connection with a resource allocation. The release 10 version of the 3GPP specifications increases this to 2 uplink transport blocks per subframe when using uplink MIMO. The release 10 version of the specifications further increases this when using Carrier Aggregation, i.e. there can be up to 2 transport blocks per Component Carrier

★ Transport blocks belonging to the UL-SCH have a variable size. The set of allowed Transport Block Sizes (TBS) is presented in section 33.4. The set of allowed TBS is the same as that for the DL-SCH (excluding the larger sizes which are applicable to the downlink when a single transport block is mapped onto 3 or 4 layers - Table 393 and Table 394)

★ The TBS scheduled during a specific subframe is signalled to the UE within PDCCH Downlink Control Information (DCI) formats 0 and 4. These DCI specify the Modulation and Coding Scheme (MCS) which defines the modulation scheme, the TBS index and the Redundancy Version (RV). The TBS index can be combined with the number of allocated Resource Blocks to deduce the TBS

★ An UL-SCH transport block includes the RRC signalling and application data to be transferred. UCI is multiplexed separately as part of the physical layer processing. The physical layer processing applied to an UL-SCH transport block is illustrated in Figure 177. This processing is completed prior to modulation and the subsequent generation of the uplink SC-FDMA signal

★ A 24 bit CRC is attached to the transport block to allow error detection at the eNode B receiver. The eNode B uses the PHICH to provide acknowledgements to the UE based upon the result of the CRC check. Transport block retransmissions are completed if the UE does not receive a positive acknowledgement

★ Filler bits are attached if the total size of the transport block plus CRC is less than 40 bits. This is necessary because Turbo coding requires at least a minimum block size

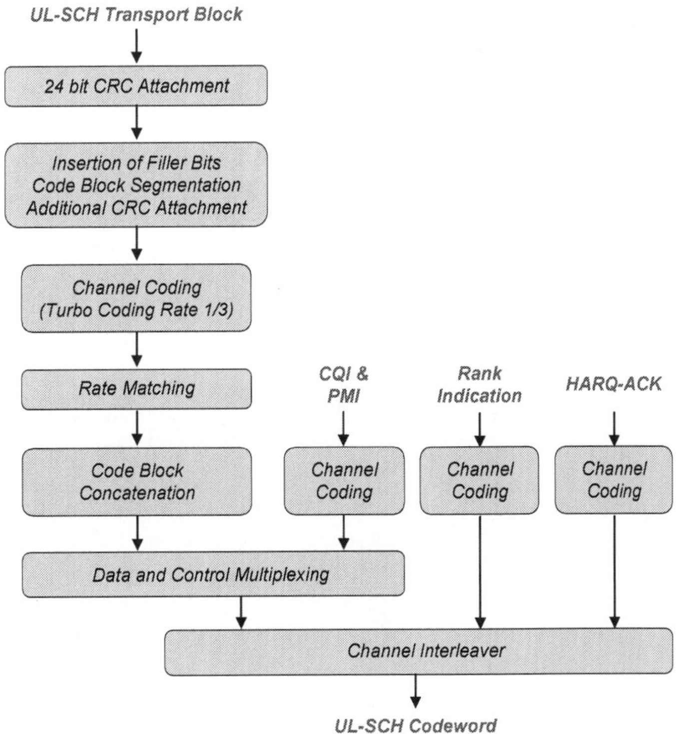

Figure 177 – Physical layer processing for UL-SCH transport block

★ If the total size of the transport block plus CRC (and any filler bits) is less than or equal to 6144 bits then it is forwarded for channel coding

★ If the code block exceeds 6144 bits then segmentation is required. This is necessary for the purposes of Turbo coding. A code block of size 'Z' is segmented into 'X' sections where X = ROUNDUP(Z / 6120). The denominator is 6120 rather than 6144 to allow for an additional 24 bit CRC which is added to each segment

★ Rate 1/3 Turbo coding is completed to triple the total number of bits, and to provide redundancy which protects the data when transferred across the air-interface. Turbo coding generates 1 set of Systematic bits and 2 sets of Parity bits. The Systematic bits are a copy of input bits. The Parity bits provide the redundancy to protect the input bits

★ Rate matching starts with an interleaving function. Interleaving is used to help avoid bursts of contiguous bit errors at the input of the Turbo decoder at the receiver. The performance of the decoder is improved when bit errors are distributed at random rather than in contiguous groups. Propagation channel fading tends to generate bursts of contiguous errors but de-interleaving at the receiver randomises the position of those bit errors

★ The concept of interleaving is illustrated in Figure 178. UE travelling at low speeds are more likely to experience fades with long durations (UE travelling at high speeds pass through the fades more rapidly). This means that UE travelling at low speeds may experience wider bursts of errors. This can lead to contiguous errors even after de-interleaving which can subsequently lead to reduced physical layer performance

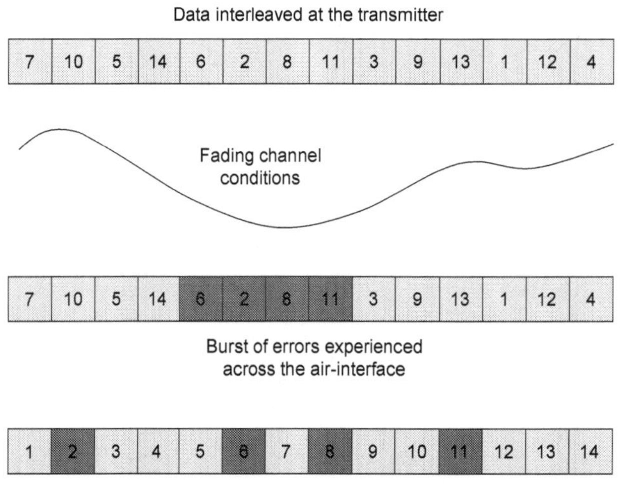

Figure 178 – General concept of interleaving

★ Rate matching then adjusts the total number of bits to match the capacity of the allocated PUSCH Resource Blocks. The total number of bits can be increased using repetition or decreased using puncturing

★ The Redundancy Version (RV) is used as an input during rate matching, i.e. it determines the puncturing pattern when puncturing is applied. The RV is signalled using the MCS Index within DCI formats 0 and 4

★ Code block concatenation is completed if segmentation was required prior to channel coding. Segments are concatenated to generate a single codeword

★ Data and control multiplexing is used to combine the code block (which contains RRC signalling and/or application data) with CQI and PMI control information. The CQI and PMI control information is coded into a block of 32 bits prior to multiplexing with the data

★ A channel interleaving stage is then used to include Rank Indication (RI) and HARQ acknowledgement information. Interleaving is completed such that HARQ acknowledgments occupy the Resource Elements adjacent to the PUSCH Demodulation Reference Signal. RI information is allocated Resource Elements adjacent to the HARQ Acknowledgements. Keeping the control information close to the Demodulation Reference Signal helps to improve its reliability

★ The resulting UL-SCH codeword is modulated and subsequently used to generate the SC-FDMA signal. Each UL-SCH codeword is transferred during a single 1 ms subframe

★ 3GPP References: TS 36.212

15.2 RACH

★ The RACH transport channel is used to transfer random access preamble control information between the MAC and Physical layers. This concept is illustrated in Figure 179

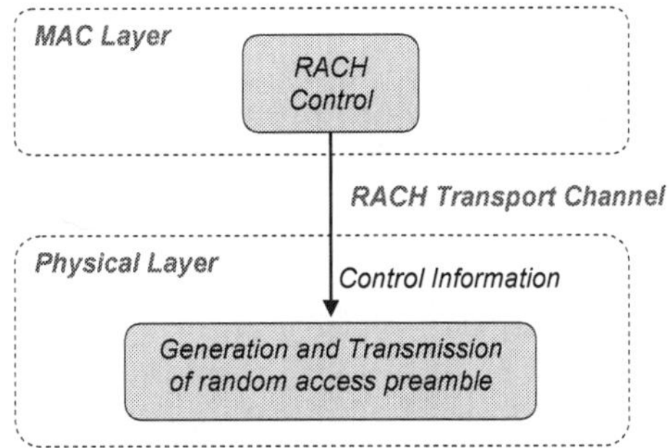

Figure 179 – RACH transport channel transferring control information

★ The RACH transport channel for LTE does not transfer any higher layer messages. This differs from the RACH transport channel for UMTS which can be used to transfer both RRC signalling and application data. In the case of LTE, the UL-SCH and DL-SCH are used to transfer RRC signalling and application data

★ The Physical layer is responsible for using the control information to calculate the PRACH transmit power, selecting a preamble sequence and transmitting the preamble using the indicated PRACH resources

★ The structure of the random access preamble is described in section 14.1, while the complete random access procedure is described in section 24.1

★ 3GPP References: TS 36.321

16 UPLINK CHANNEL TYPE MAPPINGS

16.1 LOGICAL, TRANSPORT AND PHYSICAL CHANNEL TYPES

★ Logical channels transfer data between the RLC and MAC layers

★ Transport channels transfer data between the MAC and Physical layers

★ Physical channels transfer data across the air-interface

★ The mappings between the various channel types are illustrated in Figure 180

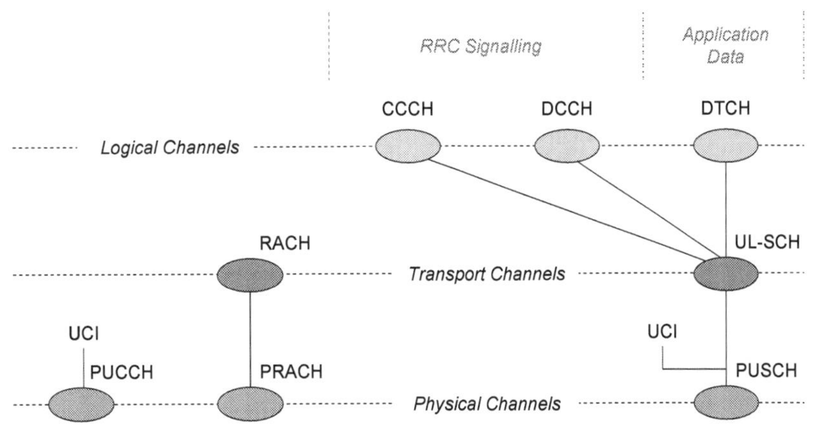

Figure 180 – Mappings between logical, transport and physical channels

★ The Common Control Channel (CCCH) and Dedicated Control Channel (DCCH) are used to transfer RRC messages, i.e. data belonging to the set of Signalling Radio Bearers (SRB). Table 5 in section 2.9 associates the various RRC messages with the CCCH and DCCH logical channels. All SRB data is mapped onto the UL-SCH and PUSCH

★ The Dedicated Traffic Channel (DTCH) is used to transfer application data. The DTCH is always mapped onto the UL-SCH and PUSCH

★ Uplink Control Information (UCI) can be added to the data from the UL-SCH during physical layer processing. This allows UCI to be transferred using the PUSCH when there is RRC signalling or application data to send

★ The PUCCH and PRACH physical channels are not used to transfer higher layer information. The PUCCH is not associated with any logical channels nor any transport channels. The PRACH is associated with the RACH transport channel but this transport channel is only used to transfer random access preamble control information from the MAC layer to the Physical layer

★ The PRACH is used to transfer the random access preambles used to initiate the random access procedure

★ The PUCCH is used to transfer UCI. UCI can include Channel Quality Indicators (CQI) , Precoding Matrix Indicators (PMI), Precoding Type Indicators (PTI), Rank Indicators (RI), HARQ acknowledgements and Scheduling Requests

★ 3GPP References: TS 36.212, TS 36.321, TS 36.331

17 UPLINK CONTROL INFORMATION

17.1 INTRODUCTION

★ The UE uses Uplink Control Information (UCI) to provide the eNode B with feedback. It includes:
  - Channel State Information (CSI)
    - Channel Quality Indicators (CQI)
    - Precoding Matrix Indicators (PMI)
    - Precoding Type Indicators (PTI)
    - Rank Indicators (RI)
  - Scheduling Requests (SR)
  - HARQ Acknowledgements

★ The UE sends uplink control information to the eNode B using either the PUCCH or the PUSCH

★ The release 8 and 9 versions of the 3GPP specifications do not support simultaneous transmission of the PUCCH and PUSCH. In this case, the physical channels used for the transfer of uplink control information are presented in Table 178

	No Data to send on the PUSCH	Data to send on the PUSCH
SR	PUCCH format 1	PUSCH
HARQ ACK	PUCCH format 1a/1b/3	
SR + HARQ ACK	PUCCH format 1a/1b/3	
Periodic CSI	PUCCH format 2	
Aperiodic CSI	PUSCH	
Periodic CSI + HARQ ACK	PUCCH format 2/2a/2b	
Aperiodic CSI + HARQ ACK	PUSCH	

Table 178 – Physical Channels for Uplink Control Information when simultaneous PUCCH and PUSCH transmission is not supported

★ In general, the PUCCH is used to transfer uplink control information, unless the PUSCH is already being used to transfer uplink data (RRC signalling messages or application data). Uplink control information can then be multiplexed with the data and transferred on the PUSCH

★ The various PUCCH formats are presented in section 14.2

★ As will be explained in the following sections, CSI reporting is divided into periodic and aperiodic categories. Periodic CSI reporting uses the PUCCH unless there is data being transferred on the PUSCH. Aperiodic CSI reporting is always transferred on the PUSCH. Aperiodic reporting can be triggered using a flag within the DCI used to allocate PUSCH resources. Thus, the PUSCH resources used to transfer the aperiodic CSI information can be allocated at the same time as triggering the report. The capacity of the PUSCH is greater than the capacity of the PUCCH so greater quantities of information can be included within aperiodic CSI reports

★ Scheduling Requests (SR) are not explicitly transferred on the PUSCH. Instead, the UE can include Buffer Status Report MAC control elements within the uplink MAC PDU. This control element provides the eNode B with information regarding the quantity of data that the UE has within its transmit buffers

★ The release 10 version of the 3GPP specifications introduces the capability to transmit both the PUCCH and PUSCH during the same subframe. This is an optional capability for the UE. When supported, it changes the mapping of uplink control information onto the PUCCH and PUSCH

★ In this case, uplink control information can be sent on the PUCCH while data is transferred on the PUSCH, e.g. HARQ acknowledgements or periodic CSI reports can be sent on the PUCCH while data is transferred on the PUSCH. When there is a combination of CSI reports and HARQ acknowledgements to send, the CSI reports can be sent on the PUSCH while the HARQ acknowledgements are sent on the PUCCH

17.2 CHANNEL STATE INFORMATION

★ Channel State Information (CSI) includes the following 4 components:
  - o Channel Quality Indicators (CQI)
  - o Precoding Matrix Indicators (PMI)
  - o Precoding Type Indicators (PTI)
  - o Rank Indicators (RI)

★ The requirement for each of these components depends upon the downlink transmission mode. The set of downlink transmission modes is described in section 4.2. Table 20 presents the dependency between transmission mode and channel state information

★ Figure 181 indicates that CSI reports can be either aperiodic or periodic. Aperiodic reports are always transferred using the PUSCH. Periodic reports are transferred using the PUCCH unless a report coincides with a PUSCH transmission (when simultaneous PUCCH and PUSCH transmission is supported, periodic reports can be transferred on the PUCCH while data is transferred on the PUSCH)

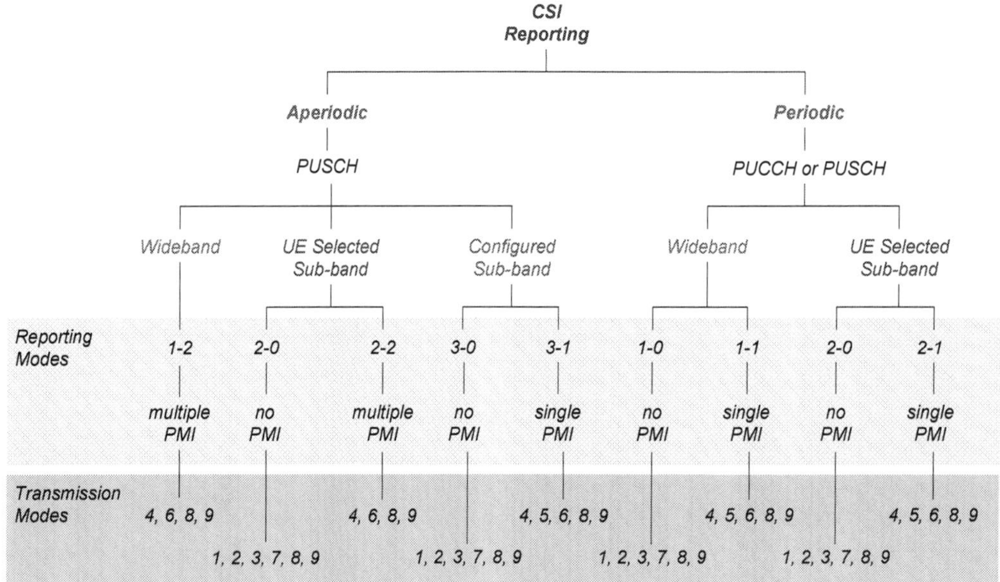

Figure 181 – Types of CSI reporting

★ Aperiodic and periodic CSI reporting are described in sections 17.3 and 17.4. Aperiodic reporting provides support for wideband, UE selected sub-band and higher layer configured sub-band reporting. Periodic reporting is limited to wideband and UE selected sub-band reporting

★ Wideband reporting generates a CSI result based upon the complete channel bandwidth.

★ UE selected sub-band reporting generates a CSI result based upon a set of one or more preferred sub-bands. The details of this approach differ between aperiodic and periodic reporting (described in sections 17.3 and 17.4)

★ Higher layer configured sub-band reporting generates a series of CSI results where each CSI result is based upon a specific sub-band. This approach generates a relatively large quantity of data (multiple CSI results) so is limited to aperiodic reporting on the PUSCH

★ CSI reporting modes are defined to specify the information reported by the UE

★ Figure 181 also shows which transmission modes are applicable to each reporting mode. Transmission modes 1, 2, 3 and 7 do not require PMI feedback, i.e. they are open loop transmission modes. Transmission modes 4 and 6 are variants of closed loop spatial multiplexing while transmissions mode 5 corresponds to multi-user MIMO. Transmission modes 8 and 9 can be configured with or without PMI reporting

★ PMI reports provide feedback in terms of the preferred set of weights to be used by the eNode B during closed loop spatial multiplexing and multi-user MIMO. The UE identifies the set of weights which maximise the downlink signal to noise ratio

★ RI reports provide feedback in terms of the preferred number of layers to be sent by the eNode B. RI reports are applicable to both open and closed loop spatial multiplexing. RI reports are not applicable to closed loop spatial multiplexing when using transmission mode 6 because transmission mode 6 is limited to using a single layer and transferring a single transport block per subframe

17.3 APERIODIC CSI REPORTING

★ Aperiodic reporting is always transferred using the PUSCH. It is not applicable to the 1.4 MHz channel bandwidth

★ Aperiodic reporting can be triggered by:

- a 'CSI Request' within Downlink Control Information (DCI) formats 0 and 4 on the PDCCH
- a 'CSI Request' within the Random Access Response Grant belonging to a Random Access Response message on the PDSCH

★ 3GPP releases 8 and 9 use a single bit within DCI format 0 for the purposes of the 'CSI Request'. This single bit triggers a CSI report for the serving cell. 3GPP release 10 introduces the ability to use either 1 or 2 bits within DCI formats 0 and 4. The use of 2 bits is applicable when Carrier Aggregation is enabled. Interpretation of these 2 bits is presented in Table 179

CSI Request Field	Description
00	No CSI report is triggered
01	CSI report is triggered for the serving cell
10	CSI report is triggered for serving cell set 1
11	CSI report is triggered for serving cell set 2

Table 179 – Interpretation of 2 bit CSI Request field

★ Serving cell sets 1 and 2 are signalled to the UE within the CQI Report Configuration belonging to an RRC Connection Setup, RRC Connection Reconfiguration or RRC Connection Re-establishment message. Set 1 is signalled using a bitmap of length 8 bits and is referenced as 'trigger 1'. Similarly, set 2 is signalled using a bitmap of length 8 bits and is referenced as 'trigger 2'

★ The 'CSI Request' within a Random Access Response Grant is limited to 1 bit and is used to trigger a CSI report for the serving cell

★ When a 'CSI Request' is received within DCI formats 0 or 4:

- in the case of FDD, the CSI report is sent 4 subframes after the trigger has been received
- in the case of TDD uplink-downlink subframe configurations 1 to 6, the CSI report is sent 'k' subframes after the trigger has been received, where 'k' is specified in Table 71 (section 9.3)
- in the case of TDD uplink-downlink subframe configuration 0, the timing of the CSI report is dependent upon the value of the 'Uplink Index' within the DCI. Values of '10' and '11' indicate that the CSI report should be sent 'k' subframes after the trigger has been received, where 'k' is specified in Table 72 (section 9.3). A value of '01' indicates that the CSI report should be sent 7 subframes after the trigger has been received

★ When a 'CSI Request' is received within a Random Access Response Grant:

- if the uplink delay field within the Random Access Response Grant is set to 0, the CSI report is sent during the first available uplink subframe for PUSCH transmission after at least 6 subframes from when the trigger was received
- if the uplink delay field within the Random Access Response Grant is set to 1, the CSI report is sent during the second available uplink subframe for PUSCH transmission after at least 6 subframes from when the trigger was received

★ An RRC Connection Setup, RRC Connection Reconfiguration or RRC Connection Re-establishment message is used to configure the aperiodic reporting mode from the set of modes listed in Table 180

Reporting Mode	Type of Reporting	Applicable Downlink Transmission Modes	PMI Reported
1-2	Wideband	4 and 6 8 when PMI/RI reporting is configured 9 when PMI/RI reporting is configured and the number of CSI RS ports > 1	Yes, multiple PMI
2-0	UE Selected Sub-band	1, 2, 3 and 7 8 when PMI/RI reporting is not configured 9 when PMI/RI reporting is not configured or the number of CSI RS ports is not > 1	No
2-2		4 and 6 8 when PMI/RI reporting is configured 9 when PMI/RI reporting is configured and the number of CSI RS ports > 1	Yes, multiple PMI
3-0	Configured Sub-band	1, 2, 3 and 7 8 when PMI/RI reporting is not configured 9 when PMI/RI reporting is not configured or the number of CSI RS ports is not > 1	No
3-1		4, 5 and 6 8 when PMI/RI reporting is configured 9 when PMI/RI reporting is configured and the number of CSI RS ports > 1	Yes, single PMI

Table 180 – Reporting modes for aperiodic CSI reporting on the PUSCH

★ The general concepts for the three types of reporting are presented in Figure 182. Wideband CQI reports are based upon the complete channel bandwidth. Higher layer configured sub-band reporting generates multiple CQI results, i.e. a CQI result is generated for each sub-band. UE selected sub-band reporting generates a single CQI result based upon a set of preferred sub-bands which have been selected by the UE

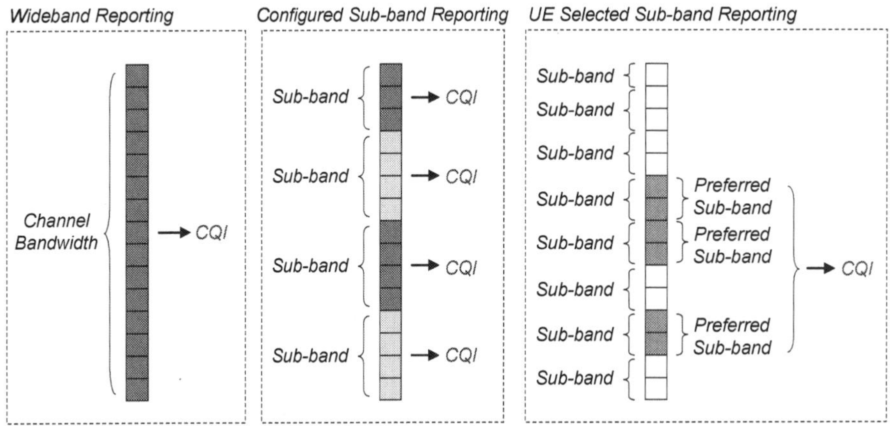

Figure 182 – Concepts for wideband, higher layer configured sub-band and UE selected sub-band reporting

★ The sub-bands applicable to wideband and higher layer configured sub-band reporting are defined by the parameters presented in Table 181. Sub-bands are applicable to wideband reporting because PMI are reported on a per sub-band basis

Channel Bandwidth	Resource Blocks	Sub-band Size (Resource Blocks)	Number of Sub-bands
1.4 MHz	6	Not Applicable	Not Applicable
3 MHz	15	4	4
5 MHz	25	4	7
10 MHz	50	6	9
15 MHz	75	8	10
20 MHz	100	8	13

Table 181 – Sub-band parameters for wideband and higher layer configured sub-band reporting

★ Sub-bands are numbered from the bottom of the channel bandwidth upwards. For each channel bandwidth, the final sub-band is smaller than the sub-band size because the channel bandwidths are not integer multiples of the sub-band size, e.g. the 20 MHz channel bandwidth has 12 sub-bands of 8 Resource Blocks and 1 sub-band of 4 Resource Blocks

★ For reporting mode 1-2 (wideband reporting), the UE:

- o selects a preferred precoding matrix for each sub-band
- o reports a wideband CQI for each codeword, while assuming that the preferred precoding matrices are used for each sub-band and that transmission is across the complete channel bandwidth
- o reports the preferred Precoding Matrix Indicator (PMI) for each sub-band. In the case of transmission mode 9 with 8 CSI Reference Signal antenna ports, the UE reports a first PMI value for the complete channel bandwidth (PMI_1) and a second PMI value for each sub-band (PMI_2) (section 4.5.1 describes the concept of 2-stage PMI reporting for transmission mode 9, i.e. PMI_1 providing wideband information and PMI_2 providing frequency selective information)
- o reports a Rank Indication (RI) when using transmission modes 4, 8 and 9
- o assumes that the reported RI is applied when generating its CQI and PMI values for transmission modes 4, 8 and 9. The UE assumes a rank of 1 for transmission mode 6

★ For reporting mode 3-0 (higher layer configured sub-band reporting without PMI), the UE:

- o reports a wideband CQI for the first codeword, while assuming that transmission is across the complete channel bandwidth
- o reports a sub-band CQI for each sub-band, while assuming transmission within each sub-band
- o reports a Rank Indication (RI) when using transmission mode 3
- o assumes that the reported RI is being applied when generating its CQI values for transmission mode 3. The UE assumes a rank of 1 for transmission modes 1, 2, 7, 8 and 9

★ For reporting mode 3-1 (higher layer configured sub-band reporting with PMI), the UE:
 - o selects a preferred precoding matrix for the complete channel bandwidth
 - o reports a wideband CQI for each codeword, while assuming that the preferred precoding matrix is used and that transmission is across the complete channel bandwidth
 - o reports a sub-band CQI for each sub-band and for each codeword, while assuming that the preferred precoding matrix is used for each sub-band and that transmission is within each sub-band
 - o reports the preferred Precoding Matrix Indicator (PMI). In the case of transmission mode 9 with 8 CSI Reference Signal antenna ports, the UE reports the first and second PMI values corresponding to the preferred precoding matrix (section 4.5.1 describes the concept of 2-stage PMI reporting for transmission mode 9, i.e. PMI_1 providing wideband information and PMI_2 providing frequency selective information)
 - o reports a Rank Indication (RI) when using transmission modes 4, 8 and 9
 - o assumes that the reported RI is being applied when generating its CQI and PMI values for transmission modes 4, 8 and 9. The UE assumes a rank of 1 for transmission modes 5 and 6

★ Sub-band CQI values are reported using differential coding which defines the sub-band CQI relative to the corresponding wideband CQI. The offset is defined as the 'sub-band CQI' – 'wideband CQI'. The mapping between the signalled value and the actual offset is presented in Table 182

Reported Sub-band Differential CQI	Actual Offset
0	0
1	1
2	≥2
3	≤-1

Table 182 – Differential coding used for sub-band CQI aperiodic reporting modes 3-0 and 3-1

★ There is slightly more scope for signalling positive offsets because the eNode B is interested in being able to select the sub-bands which provide better channel conditions than the 'average' wideband channel conditions

★ The sub-bands applicable to UE selected sub-band reporting are defined by the parameters presented in Table 183

Channel Bandwidth	Resource Blocks	Sub-band Size (Resource Blocks)	Number of Preferred Sub-bands (M)
1.4 MHz	6	Not Applicable	Not Applicable
3 MHz	15	2	3
5 MHz	25	2	3
10 MHz	50	3	5
15 MHz	75	4	6
20 MHz	100	4	6

Table 183 – Sub-band parameters for UE selected sub-band reporting

★ In this case, the number of preferred sub-bands (M) multiplied by the sub-band size is much less than the total number of Resource Blocks belonging to the channel bandwidth. The UE is responsible for selecting the set of 'M' preferred sub-bands

★ For reporting mode 2-0 (UE selected sub-band reporting without PMI), the UE:
 - o selects a set of preferred sub-bands. The number of sub-bands within this set is specified in Table 183
 - o reports a wideband CQI for the first codeword, while assuming that transmission is across the complete channel bandwidth
 - o reports a single CQI for the first codeword, assuming transmission across the preferred sub-bands
 - o reports the positions of the preferred sub-bands
 - o reports a Rank Indication (RI) when using transmission mode 3
 - o assumes that the reported RI is being applied when generating its CQI values for transmission mode 3. The UE assumes a rank of 1 for transmission modes 1, 2, 7, 8 and 9

★ For reporting mode 2-2 (UE selected sub-band reporting with PMI), the UE:

  - selects a set of preferred sub-bands and a single preferred precoding matrix for transmission across those preferred sub-bands. The number of sub-bands within the set is specified in Table 183
  - reports the preferred Precoding Matrix Indicator (PMI) for transmission across the preferred sub-bands. In the case of transmission mode 9 with 8 CSI Reference Signal antenna ports, the UE reports a first PMI value (PMI_1) for the complete channel bandwidth, and a second PMI value (PMI_2) for the complete channel bandwidth, plus an additional second PMI value (PMI_2) for the preferred sub-bands (section 4.5.1 describes the concept of 2-stage PMI reporting for transmission mode 9, i.e. PMI_1 providing wideband information and PMI_2 providing frequency selective information)
  - reports a single CQI for each codeword, assuming that the preferred precoding matrix is used and that transmission is across the preferred sub-bands
  - reports the positions of the preferred sub-bands
  - selects a preferred precoding matrix assuming transmission across the complete channel bandwidth
  - reports the preferred Precoding Matrix Indicator (PMI) assuming transmission across the complete channel bandwidth
  - reports a single CQI for each codeword, assuming that the preferred precoding matrix is used and that transmission is across the complete channel bandwidth
  - reports a Rank Indication (RI) when using transmission modes 4, 8 and 9
  - assumes that the reported RI is being applied when generating its CQI and PMI values for transmission modes 4, 8 and 9. The UE assumes a rank of 1 for transmission mode 6

★ Similar to higher layer configured sub-band reporting, UE selected sub-band reporting signals sub-band CQI values using differential values which define the sub-band CQI relative to the corresponding wideband CQI. The offset is defined as the 'sub-band CQI' – 'wideband CQI'. The mapping between the signalled values and the actual offsets is presented Table 184

Reported Sub-band Differential CQI	Actual Offset
0	≤1
1	2
≤1	3
3	≥4

Table 184 – Differential coding used for sub-band CQI aperiodic reporting modes 2-0 and 2-2

★ 3GPP References: TS 36.212, TS 36.213, TS 36.331

17.4 PERIODIC CSI REPORTING

★ Periodic reporting can be used to provide the eNode B with Channel State Information (CSI) reports which include Channel Quality Indicators (CQI), Precoding Matrix Indicators (PMI), Precoding Type Indicators (PTI) and Rank Indicators (RI)

★ 3GPP TS 36.213 specifies the set of periodic CSI reporting modes presented in Table 185. The 2 types of periodic reporting are wideband and UE selected sub-band

★ The eNode B uses an RRC Connection Setup, RRC Connection Reconfiguration, or RRC Connection Re-establishment message to provide the UE with instructions regarding the reporting mode to use. In contrast to aperiodic reporting, the 'reporting mode' is not signalled explicitly. Instead, the eNode B signals the 'type of reporting' and the 'transmission mode' The UE can then deduce the reporting mode from these 2 inputs

Reporting Mode	Type of Reporting	Applicable Transmission Modes	PMI Reported
1-0	Wideband	1, 2, 3, 7 8 when PMI/RI reporting is not configured 9 when PMI/RI reporting is not configured or the number of CSI RS ports is not > 1	No
1-1		4, 5, 6 8 when PMI/RI reporting is configured 9 when PMI/RI reporting is configured and the number of CSI RS ports > 1	Yes, single PMI
2-0	UE Selected Sub-band	1, 2, 3, 7 8 when PMI/RI reporting is not configured 9 when PMI/RI reporting is not configured or the number of CSI RS ports is not > 1	No
2-1		4, 5, 6 8 when PMI/RI reporting is configured 9 when PMI/RI reporting is configured and the number of CSI RS ports > 1	Yes, single PMI

Table 185 – Reporting modes for periodic CSI reporting

★ 2 submodes have been specified for transmission mode 9 when using reporting mode 1-1:

- submode 1 allows the UE to report the Rank Indicator (RI) and PMI_1 within one type of report, and wideband CQI and PMI_2 in another type of report (section 4.5.1 describes the concept of 2-stage PMI reporting for transmission mode 9)
- submode 2: allows the UE to report the wideband CQI, PMI_1 and PMI_2 within a single report

★ The eNode B provides the UE with instructions regarding the submode to use. This is done using the 'PUCCH format 1-1 CSI reporting mode' parameter

★ The general concepts for wideband and sub-band periodic reporting are shown in Figure 183. The concept of periodic wideband reporting is the same as that for aperiodic wideband reporting. However, UE selected sub-band reporting uses a different concept for periodic reporting

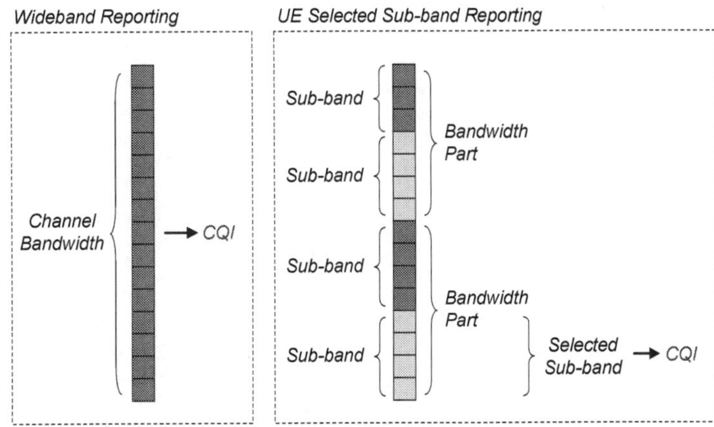

Figure 183 – Periodic CSI reporting concepts for wideband and UE selected sub-band reporting

★ The channel bandwidth is divided into 'bandwidth parts' for UE selected sub-band reporting. Each bandwidth part includes a number of sub-bands. The number of bandwidth parts and sub-bands for each channel bandwidth are defined by the parameters in Table 186

★ The UE considers a single bandwidth part each time it generates a UE selected sub-band report. The UE selects the preferred sub-band from within the bandwidth part and generates a CQI result based upon that sub-band. The UE then moves onto the next bandwidth part when generating the next UE selected sub-band report

Channel Bandwidth	Resource Blocks	Sub-band Size (Resource Blocks)	Bandwidth Parts (J)
1.4 MHz	6	Not Applicable	Not Applicable
3 MHz	15	4	2
5 MHz	25	4	2
10 MHz	50	6	3
15 MHz	75	8	4
20 MHz	100	8	4

Table 186 – Sub-band parameters for periodic CSI reporting

- ★ Sub-band reporting is not applicable to the 1.4 MHz channel bandwidth, i.e. reporting modes 2-0 and 2-1 are not used
- ★ UE send both wideband and sub-band reports when configured with UE selected sub-band reporting. A pre-defined pattern is used to determine which type of report to send during a specific subframe
- ★ Figure 184 illustrates the general concepts for sending wideband reports alone, and sending a combination of wideband and UE selected sub-band reports. Sub-band reporting uses the concept of 'bandwidth parts'. At least one sub-band report from each bandwidth part is sent between every two wideband reports

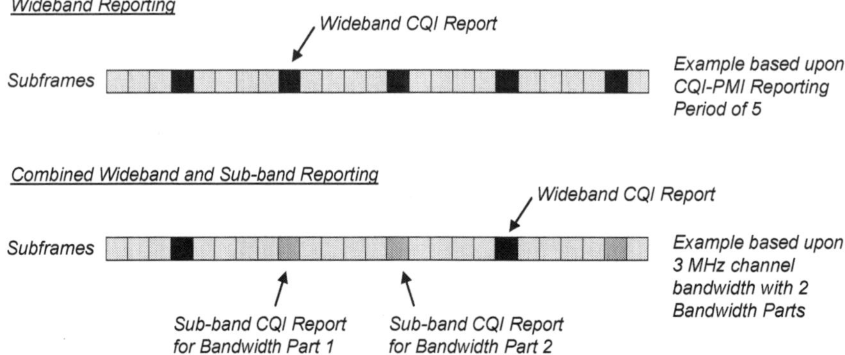

Figure 184 – Timing concepts for wideband reporting and combined wideband and sub-band reporting

- ★ The rate at which a UE sends reports is dependent upon the 'CQI-PMI Configuration Index' which can be signalled to the UE using an RRC Connection Setup, RRC Connection Reconfiguration or RRC Connection Re-establishment message. The mapping between this configuration index and the period between consecutive reports is presented in Table 187. The reporting period and reporting offset are quantified in terms of subframes

CQI-PMI Configuration Index $I_{CQI/PMI}$	CQI-PMI Reporting Period N_{pd}	Reporting Offset $N_{OFFSET,CQI}$
0 ≤ Index ≤ 1	2	Index
2 ≤ Index ≤ 6	5	Index - 2
7 ≤ Index ≤ 16	10	Index – 7
17 ≤ Index ≤ 36	20	Index – 17
37 ≤ Index ≤ 76	40	Index – 37
77 ≤ Index ≤ 156	80	Index – 77
157 ≤ Index ≤ 316	160	Index – 157
Index =317	Reserved	
318 ≤ Index ≤ 349	32	Index – 318
350 ≤ Index ≤ 413	64	Index – 350
414 ≤ Index ≤ 541	128	Index - 414
542 ≤ Index ≤ 1023	Reserved	

Table 187 – Reporting period and reporting offset as a function of the CQI-PMI Configuration Index (FDD)

- ★ Table 187 is applicable to FDD but 3GPP TS 36.213 also includes a similar table for TDD. In the case of TDD, the set of allowed reporting periods has a dependency upon the uplink-downlink subframe configuration. Reporting periods which are a multiple of 10 subframes are applicable to all uplink-downlink subframe configurations

★ When only wideband reporting is configured, CQI / PMI reports are sent during subframes that satisfy the following expression:

$$\left(10 \times n_f + k - N_{OFFSET,CQI}\right) \bmod N_{pd} = 0 \qquad \textit{equation 1}$$

where, n_f is the System Frame Number (0 to 1023)

k is the subframe number within a radio frame (0 to 9)

$N_{OFFSET,CQI}$ is the reporting offset defined in Table 187

N_{pd} is the reporting period defined in Table 187

★ When both wideband and sub-band reporting are configured:

- if PTI values are not used, or the most recent PTI value was 1,
  - wideband CQI / wideband PMI (wideband CQI / wideband PMI_2 for transmission mode 9) reports are sent during subframes that satisfy the following expression:

$$\left(10 \times n_f + k - N_{OFFSET,CQI}\right) \bmod \left(H \times N_{pd}\right) = 0$$

where, H = (J × K) + 1

J is the number of sub-band parts associated with the channel bandwidth presented in Table 186

K is signalled by the eNode B using a value from 1 to 4

  - sub-band CQI (and sub-band PMI_2 for transmission mode 9) reports are sent during subframes that satisfy equation 1 if the subframe is not already occupied by a wideband report:

- if the most recent PTI value was 0,
  - wideband PMI_1 reports are sent during subframes that satisfy the following expression:

$$\left(10 \times n_f + k - N_{OFFSET,CQI}\right) \bmod \left(H' \times N_{pd}\right) = 0$$

where, H' is signalled by the eNode B using the 'periodicity factor' parameter with a value of 2 or 4

  - wideband PMI_2 and wideband CQI reports are sent during subframes that satisfy equation 1 if the subframe is not already occupied by a PMI_1 report

★ Rank Indicators (RI) are reported using their own timing rules. The rate at which a UE sends RI reports is dependent upon the RI Configuration Index signalled by the eNode B. The mapping between this configuration index and the various timing parameters is presented in Table 188. This table is applicable to both FDD and TDD

RI Configuration Index	RI Reporting Period Multiplier, M_{RI}	RI Reporting Offset $N_{OFFSET,RI}$
$0 \leq$ Index ≤ 160	1	-Index
$161 \leq$ Index ≤ 321	2	-(Index - 161)
$322 \leq$ Index ≤ 482	4	-(Index - 322)
$483 \leq$ Index ≤ 643	8	-(Index - 483)
$644 \leq$ Index ≤ 804	16	-(Index - 644)
$805 \leq$ Index ≤ 965	32	-(Index - 805)
$966 \leq$ Index ≤ 1023	Reserved	

Table 188 – Reporting period and reporting offset as a function of the RI Configuration Index

★ When only wideband reporting is configured, RI reports are sent during subframes that satisfy the following expression:

$$\left(10 \times n_f + k - N_{OFFSET,CQI} - N_{OFFSET,RI}\right) \bmod \left(N_{pd} \times M_{RI}\right) = 0$$

where, $N_{OFFSET,RI}$ is the reporting offset defined in Table 188

M_{RI} is the reporting period multiplier defined in Table 188

★ When UE selected sub-band reporting is configured, RI reports are sent during subframes that satisfy the following expression:

$$\left(10 \times n_f + k - N_{OFFSET,CQI} - N_{OFFSET,RI}\right) \bmod \left(H \times N_{pd} \times M_{RI}\right) = 0$$

★ 3GPP TS 36.213 defines the set of 'Report Types' presented in Table 189 for use with periodic CSI reporting

	Reported Quantities						
Report Type	Sub-band CQI	Wideband CQI	PMI	1st PMI	2nd PMI	RI	PTI
1	✓						
1a	✓				✓		
2		✓	✓				
2a				✓			
2b		✓			✓		
2c		✓		✓	✓		
3						✓	
4		✓					
5				✓		✓	
6						✓	✓

Table 189 – Content of each Report Type for periodic CSI reporting

★ For reporting mode 1-0 (wideband reporting without PMI),

- o during RI subframes (applicable to transmission mode 3) the UE reports a RI using a 'type 3' report
- o during wideband CQI subframes the UE reports a single wideband CQI for the first codeword using a 'type 4' report. This CQI is generated assuming transmission across the complete channel bandwidth and a rank of 1, with the exception of transmission mode 3 for which the UE assumes the last reported rank

★ For reporting mode 1-1 (wideband reporting with PMI),

- o during RI subframes (applicable to transmission modes 4, 8 and 9) the UE reports a RI using a 'type 3' report
- o during RI and PMI_1 subframes (applicable to transmission mode 9 submode 1 when using 8 CSI RS antenna ports) the UE reports a RI and a wideband PMI_1 using a 'type 5' report. The PMI_1 is generated assuming transmission across the complete channel bandwidth and use of the reported RI
- o during CQI and PMI subframes (applicable to all transmission modes except transmission mode 9 when using 8 CSI RS antenna ports) the UE reports a single wideband CQI and a single wideband PMI using a 'type 2' report. This report also includes a spatial differential CQI when the RI > 1. The wideband CQI is generated assuming use of the reported PMI and transmission across the complete channel bandwidth. Both CQI and PMI are generated assuming a rank of 1, with the exception of transmission modes 4, 8 and 9 for which the UE assumes the last reported rank
- o during CQI and PMI_2 subframes (applicable to transmission mode 9 submode 1 when using 8 CSI RS antenna ports) the UE reports a single wideband CQI and a single wideband PMI_2 using a 'type 2b' report. This report also includes a spatial differential CQI when the RI > 1. The wideband CQI is generated assuming use of the selected PMI matrix and transmission across the complete channel bandwidth. Both CQI and PMI are generated assuming the last reported rank
- o during CQI, PMI_1 and PMI_2 subframes (applicable to transmission mode 9 submode 2 when using 8 CSI RS antenna ports) the UE reports a single wideband CQI, a single wideband PMI_1 and a single wideband PMI_2 using a 'type 2c' report. This report also includes a spatial differential CQI when the RI > 1. The wideband CQI is generated assuming use of the selected PMI matrix and transmission across the complete channel bandwidth. Both CQI and PMI are generated assuming the last reported rank

★ For reporting mode 2-0 (UE selected sub-band reporting without PMI),

- o during RI subframes (applicable to transmission mode 3) the UE reports a RI using a 'type 3' report
- o during wideband CQI subframes the UE reports a single wideband CQI for the first codeword using a 'type 4' report. This CQI is generated assuming transmission across the complete channel bandwidth and a rank of 1, with the exception of transmission mode 3 for which the UE assumes the last reported rank
- o during sub-band CQI subframes the UE reports a single sub-band CQI for the first codeword using a 'type 1' report. The report also includes a label to specify the preferred sub-band within the relevant bandwidth part. The CQI is generated assuming transmission across only the preferred sub-band. The preferred sub-band is selected and the corresponding CQI is generated assuming a rank of 1, with the exception of transmission mode 3 for which the UE assumes the last reported rank

★ For reporting mode 2-1 (UE selected sub-band reporting with PMI),

- during RI subframes (applicable to transmission modes 4, 8 and 9 if the number of CSI RS antenna ports is 2 or 4) the UE reports a RI using a 'type 3' report
- during RI subframes (applicable to transmission mode 9 if the number of CSI RS antenna ports is 8) the UE reports a RI and PTI using a 'type 6' report
- during wideband CQI and PMI subframes (applicable to all transmission modes except transmission mode 9 when using 8 CSI RS antenna ports) the UE reports a single wideband CQI and a single wideband PMI using a 'type 2' report. This report also includes a spatial differential CQI when the RI > 1. The wideband CQI is generated assuming use of the reported PMI and transmission across the complete channel bandwidth. Both CQI and PMI are generated assuming a rank of 1, with the exception of transmission modes 4, 8 and 9 for which the UE assumes the last reported rank
- during PMI_1 subframes with the last reported PTI = 0 (applicable to transmission mode 9 when using 8 CSI RS antenna ports) the UE reports a wideband PMI_1 using a 'type 2a' report. The PMI_1 is generated assuming transmission across the complete channel bandwidth and use of the last reported RI
- during wideband CQI and PMI_2 subframes with the last reported PTI = 1 (applicable to transmission mode 9 when using 8 CSI RS antenna ports) the UE reports a wideband CQI and a wideband PMI_2 using a 'type 2b' report. This report also includes a spatial differential CQI when the RI > 1. The wideband CQI is generated assuming use of the selected precoding matrix and transmission across the complete channel bandwidth. Both CQI and PMI_2 are generated assuming the last reported rank
- during sub-band CQI subframes (applicable to all transmission modes except transmission mode 9 when using 8 CSI RS antenna ports) the UE reports a single sub-band CQI for the first codeword using a 'type 1' report. The report also includes a label to specify the preferred sub-band within the relevant bandwidth part. This report also includes a spatial differential CQI when the RI > 1. The CQI is generated assuming transmission across only the preferred sub-band. The preferred sub-band is selected and the corresponding CQI is generated assuming a rank of 1, with the exception of transmission modes 4, 8 and 9 for which the UE assumes the last reported rank
- during wideband CQI and PMI_2 subframes with the last reported PTI = 0 (applicable to transmission mode 9 when using 8 CSI RS antenna ports) the UE reports a single wideband CQI and a single wideband PMI_2 using a 'type 2b' report. This report also includes a spatial differential CQI when the RI > 1. The wideband CQI is generated assuming use of the selected PMI matrix and transmission across the complete channel bandwidth. Both CQI and PMI are generated assuming the last reported rank
- during sub-band CQI and PMI_2 subframes with the last reported PTI = 1 (applicable to transmission mode 9 when using 8 CSI RS antenna ports) the UE reports a single sub-band CQI and a single sub-band PMI_2 using a 'type 1a' report. The report also includes a label to specify the preferred sub-band within the relevant bandwidth part. This report also includes a spatial differential CQI when the RI > 1. The sub-band CQI is generated assuming use of the selected PMI matrix and transmission across the selected sub-band. Both CQI and PMI_2 are generated assuming the last reported rank

★ A spatial differential CQI defines the CQI of the second codeword relative to the CQI of the first codeword. It is defined by an offset equal to 'CQI for codeword 0' – 'CQI for codeword 1'. The mapping between the signalled spatial differential CQI and the actual offset is presented in Table 190

Signalled Spatial Differential CQI	Offset	Signalled Spatial Differential CQI	Offset
0	0	4	≤ -4
1	1	5	-3
2	2	6	-2
3	≥ 3	7	-1

Table 190 – Mapping between signalled spatial differential CQI and actual offset

★ 3GPP References: TS 36.212, TS 36.213, TS 36.331

17.5 PRECODING MATRIX INDICATOR

- Precoding Matrix Indicators (PMI) are applicable to closed loop transmission modes:
  - transmission mode 4: closed loop spatial multiplexing
  - transmission mode 5: single layer multi-user MIMO
  - transmission mode 6: closed loop spatial multiplexing using a single layer
  - transmission mode 8: dual layer multi-user MIMO (3GPP release 9)
  - transmission mode 9: closed loop spatial multiplexing with up to 8 layers (3GPP release 10)
- The UE uses PMI to signal the preferred set of weights to be applied during precoding. Precoding is described in section 4
- The preferred set of weights are selected by the UE to maximise the downlink signal to noise ratio
- PMI can be transferred using either the PUSCH or PUCCH
- Transmission modes 4, 5 and 6 use codebook based precoding. This means that the UE selects the preferred precoding matrix from the set of matrices available for downlink transmission at the eNode B, i.e. the eNode B can directly apply the precoding matrix selected by the UE
- Transmission modes 8 and 9 use non-codebook based precoding. This means that the UE selects a preferred precoding matrix from a set of matrices defined by 3GPP, but the eNode B is not constrained to using the same set of matrices. In this case, precoding at the eNode B is implementation dependent and the network vendor is free to chose the precoding algorithm. When channel reciprocity is assumed, the downlink precoding can be based upon uplink channel measurements, e.g. from the Sounding Reference Signal (SRS). Precoding typically aims to achieve beamforming and is transparent to the UE. Precoding for beamforming is presented in section 4
- In the case of transmission modes 8 and 9, the precoding algorithms used by the eNode B may not require PMI feedback from the UE. These transmission modes can be configured without PMI/RI reporting. The eNode B then relies upon uplink measurements to support its downlink transmissions
- Non-codebook based precoding can also be applied to transmission mode 7 for the purposes of beamforming. This transmission mode does not support PMI/RI reporting from the UE, i.e. the eNode B can apply implementation dependent downlink precoding without PMI reports from the UE
- In the case of transmission mode 9, PMI reporting is based upon the 2-stage approach described in section 4.5.1. The first stage corresponds to reporting wideband longer term channel properties, e.g. the general direction of the UE relative to the eNode B. The second stage corresponds to reporting frequency selective short term channel properties, e.g. the precise direction of the UE relative to the eNode B, and the correlation between cross polar antenna elements
- Codebook sizes can be limited using the codebook subset restriction capability defined by 3GPP. RRC signalling can be used to exclude specific precoding matrices. The eNode B may wish to restrict the set of matrices if it decides that some matrices are not useful for the current channel conditions or antenna configuration. Codebook subset restrictions are signalled to the UE using a bitmap which flags the subset of precoding matrices to exclude
- Codebook subset restriction can be applied to transmission modes 4, 5, 6, 8 and 9. It can also be applied to transmission mode 3 although transmission mode 3 does not involve the reporting of PMI values. In this case, the codebook subset restriction limits the set of ranks which can be reported by the UE
- Codebook sub-sampling is also specified by 3GPP to help reduce the signalling overhead generated by PMI reporting. Sub-sampling the codebook limits the number of choices available to the UE when selecting the preferred precoding matrix. This reduces the resolution of the reporting accuracy but also reduces the number of bits required to signal the resultant choice of PMI to the eNode B. This is useful when reporting PMI values on the PUCCH which has limited capacity. Transmission mode 9 provides support for codebook sub-sampling
- 3GPP References: TS 36.211, TS 36.213, TS 36.331

17.6 PRECODING TYPE INDICATOR

- ★ Precoding Type Indicators (PTI) were introduced by the release 10 version of the 3GPP specifications
- ★ PTI values allow the UE to dynamically switch between the type of PMI reports provided to the eNode B
- ★ They are applicable to transmission mode 9 which uses the 2-stage approach to PMI reporting described in section 4.5.1. The first stage corresponds to reporting wideband longer term channel properties (PMI_1). The second stage corresponds to reporting frequency selective short term channel properties (PMI_2)
- ★ A PTI value of '0' indicates that the UE has detected that PMI_1 is changing so it needs to be signalled to the eNode B. A PTI value of '1' indicates that the UE has detected that PMI_1 is not changing so reporting can focus on PMI_2. Focusing on PMI_2 provides scope for reporting a series of sub-band PMI_2 values rather than a single wideband value
- ★ Examples of PTI reporting are presented in section 4.5.1
- ★ 3GPP References: TS 36.213

17.7 RANK INDICATOR

- ★ Rank Indicators (RI) are applicable to the open and closed loop transmission modes which are able to use more than a single layer between the layer mapping and precoding functions:
  - o transmission mode 3: open loop spatial multiplexing
  - o transmission mode 4: closed loop spatial multiplexing
  - o transmission mode 8: dual layer multi-user MIMO (3GPP release 9)
  - o transmission mode 9: closed loop spatial multiplexing with up to 8 layers (3GPP release 10)
- ★ The UE uses the RI to provide the eNode B with a suggestion regarding the number of layers to be generated during layer mapping
- ★ The number of useful layers depends upon the propagation channel conditions. Transferring a high number of layers requires uncorrelated propagation channels between each transmit antenna and each receive antenna
- ★ In the case of transmission mode 3, the RI is used to select between transmit diversity and open loop spatial multiplexing
- ★ In the case of transmission modes 8 and 9, the precoding algorithms used by the eNode B may not require RI feedback from the UE. These transmission modes can be configured without PMI/RI reporting. The eNode B then relies upon uplink measurements to support its downlink transmissions
- ★ RI can be transferred using either the PUSCH or PUCCH
- ★ The timing of RI reports is dependent upon the RI Configuration Index signalled to the UE within an RRC Connection Setup, RRC Connection Reconfiguration or RRC Re-establishment message. The mapping between this configuration index and the various timing parameters is presented in Table 188 (section 17.4). This table is applicable to both FDD and TDD
- ★ 3GPP References: TS 36.212, TS 36.213, TS 36.331

17.8 SCHEDULING REQUEST

★ A Scheduling Request (SR) is a single bit flag used to request PUSCH resources from the eNode B

★ Scheduling Requests are always transferred using the PUCCH

★ UE are provided with configuration information regarding Scheduling Requests within an RRC Connection Setup, RRC Connection Reconfiguration or RRC Connection Re-establishment message. This configuration information is presented in Table 191

Information Elements		
Scheduling Request Configuration	Setup	Scheduling Request PUCCH Resource Index
		Scheduling Request Configuration Index
		DSR Transmission Maximum
		Scheduling Request PUCCH Resource Index P1 (3GPP release 10)

Table 191 – Scheduling Request configuration information

★ The Scheduling Request PUCCH Resource Index can be signalled using a value between 0 and 2047. This parameter is applicable when sending an SR using PUCCH formats 1, 1a or 1b. In these cases, the Resource Index defines the allocated Resource Blocks, cyclic shifts and orthogonal codes which are used when sending an SR (see section 14.2.1)

★ SR can also be transferred using PUCCH format 3. In contrast to PUCCH formats 1, 1a and 1b where the SR is signalled by using the SR PUCCH resource rather than the standard PUCCH resource, PUCCH format 3 allocates a single bit of information to signal the SR. This additional bit of information is concatenated with the HARQ acknowledgements

★ The Scheduling Request Configuration Index can be signalled with a value between 0 and 157. This parameter defines the set of SR transmission instances. The Configuration Index is used in combination with the look-up table shown in Table 192 to determine values for the Scheduling Request Periodicity and the Scheduling Request Offset

SR Configuration Index I_{SR}	SR Periodicity (ms)	SR Subframe Offset
0 to 4	5	I_{SR}
5 to 14	10	I_{SR} - 5
15 to 34	20	I_{SR} - 15
35 to 74	40	I_{SR} - 35
75 to 154	80	I_{SR} - 75
155 to 156	2	I_{SR} - 155
157	1	I_{SR} - 157

Table 192 – Scheduling Request (SR) Periodicity and Subframe Offset as a function of the SR Configuration Index

★ SR transmission instances are then defined as the subframes which satisfy the following condition:

$$(10 \cdot n_f + k - SR_{offset}) \bmod SR_{\text{periodicity}} = 0$$

where,

n_f is the System Frame Number (0 to 1023)

k is the subframe number within the radio frame (0 to 9)

SR_{offset} is the Scheduling Request Offset from Table 192

$SR_{periodicity}$ is the Scheduling Request Periodicity from Table 192

★ The release 10 version of the 3GPP specifications provides support for PUCCH transmit diversity using 2 antenna ports. The Scheduling Request PUCCH Resource Index P1 defines the PUCCH resources for the second antenna port when sending an SR using PUCCH formats 1, 1a or 1b with transmit diversity. Similar to the resource index for the first antenna port, this parameter can range from 0 to 2047

★ After sending a first SR, the UE initialises an SR counter with a value of 0 and starts an SR Prohibit Timer. The Prohibit Timer is signalled to the UE by the RRC layer and can be allocated values from 0 to 7. A value of 1 means that the UE must wait 1 SR period before being able to send another SR. Likewise, a value of 2 means that the UE must wait 2 SR periods before being able to send another SR. A value of 0 means that the Prohibit Timer is not used

★ If the Prohibit Timer expires and the UE has not received an uplink resource allocation able to accommodate all pending data available for uplink transmission, then the UE is able to send another SR to the eNode B

- The SR counter is incremented by 1, and the Prohibit Timer is reset after every SR transmission
- The UE can continue sending SR until either a resource allocation is received from the eNode B which is able to accommodate all pending uplink data, or the SR counter reaches the value of the DSR Transmission Maximum parameter shown in Table 191. This parameter represents the maximum allowed number of SR transmissions and can be configured with values of 4, 8, 16, 32 or 64
- Once the maximum allowed number of transmissions has been reached, the UE initiates the random access procedure and uses that procedure as a way to request an uplink resource allocation
- 3GPP References: TS 36.213, TS 36.321, TS 36.331

17.9 HARQ ACK/NACK

- Uplink Hybrid Automatic Repeat Request (HARQ) ACK/NACK are used to acknowledge downlink data transferred on the PDSCH
- Prior to Carrier Aggregation for LTE Advanced, it is necessary to transfer 1 or 2 acknowledgments for each FDD downlink transmission on the PDSCH, i.e. one acknowledgement for each transport block
- Carrier Aggregation increases the number of transport blocks which can be sent to an individual UE during a single subframe (up to 10 transport blocks when using 5 Component Carriers). This increases the requirement for HARQ acknowledgements in the uplink
- The number of HARQ acknowledgments to be sent during a single uplink subframe can be greater for TDD because there can be multiple downlink subframes before the UE is able to return any acknowledgements in the uplink
- HARQ acknowledgment bundling and multiplexing can be applied when there are large numbers of acknowledgments to transfer
- The procedures for sending HARQ acknowledgments are described in section 24.3
- HARQ acknowledgements can be transferred using either the PUSCH or PUCCH
- The mapping of HARQ acknowledgements onto the PUCCH is described in section 14.2. The mapping of HARQ acknowledgements onto the UL-SCH / PUSCH is illustrated in section 15.1
- When using TTI bundling, each bundle of 4 TTI requires a single HARQ acknowledgement from the UE (the set of 4 TTI provide support for an original transmission and 3 autonomous retransmissions. TTI bundling is described in section 27.6.1
- 3GPP References: TS 36.212, TS 36.213, TS 36.321

18 CARRIER AGGREGATION

18.1 INTRODUCTION

★ LTE-Advanced has peak data rate requirements of 1 Gbps in the downlink, and 500 Mbps in the uplink

★ These requirements can be achieved by increasing the channel bandwidth

★ Carrier Aggregation (CA) increases the channel bandwidth by combining multiple RF carriers. Application data can then be sent and received using multiple RF carriers rather than a single RF carrier. Each individual RF carrier is known as a Component Carrier (CC)

★ Carrier Aggregation is applicable to both the uplink and downlink directions. A Component Carrier can be either uplink and downlink, or downlink only, but cannot be uplink only. UE signal their support for Carrier Aggregation independently for each direction, i.e. some UE may support Carrier Aggregation in the downlink only, while other UE may support Carrier Aggregation in both the uplink and downlink

★ Carrier Aggregation is applicable to both FDD and TDD. In the case of TDD, the uplink-downlink subframe configuration must be the same for all Component Carriers

★ Figure 185 illustrates two examples of Carrier Aggregation – the first based upon 5 Component Carriers, and the second based upon 2 Component Carriers. The bandwidth of each Component Carrier is scenario dependent and does not have to be the maximum channel bandwidth of 20 MHz

★ All Component Carriers belong to the same eNode B and are synchronised on the air-interface. This means that a single set of timing advance commands are used for all Component Carriers

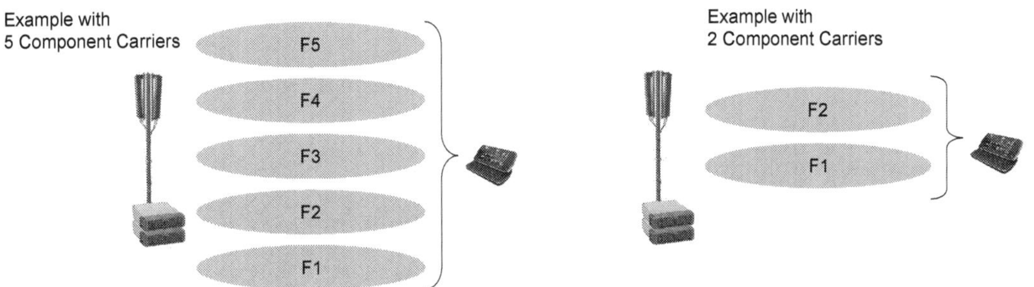

Figure 185 – Examples of Carrier Aggregation

★ 3 general types of Carrier Aggregation scenario have been defined by 3GPP:

- o intra-band contiguous
- o intra-band non-contiguous
- o inter-band non-contiguous

These 3 scenarios are illustrated in Figure 186. In the case of the inter-band non-contiguous scenario, the Component Carriers within an individual operating band can be either contiguous or non-contiguous (Figure 186 illustrates only a single Component Carrier within each operating band)

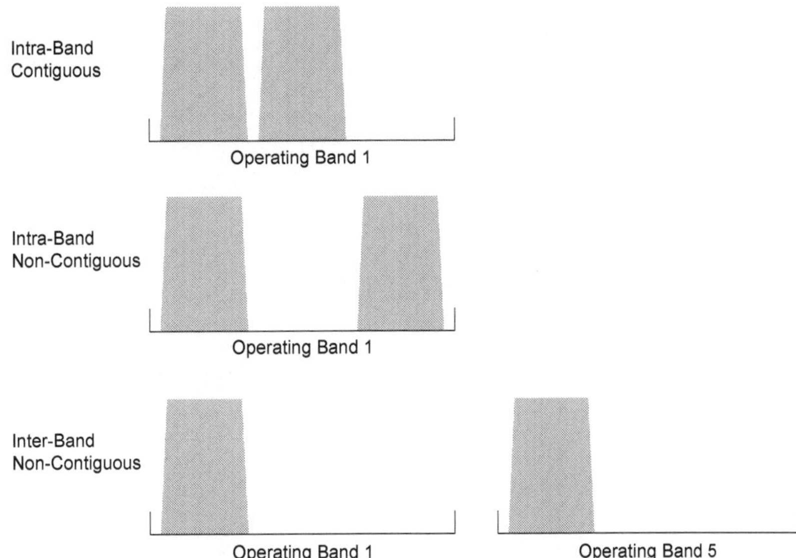

Figure 186 – Carrier Aggregation scenarios

- ★ The intra-band scenarios provide benefits in terms of implementation effort. A single transceiver can transmit and receive multiple RF carriers when they are positioned within the same operating band
- ★ The inter-band scenario provides benefits in terms of spectrum availability. An operator's spectrum is likely to be distributed across multiple operating bands rather than located within a single band
- ★ The release 10 version of the 3GPP specifications has initially focused upon introducing the intra-band contiguous and inter-band non-contiguous scenarios
- ★ The 3GPP release 10 version of Carrier Aggregation has been specified such that Component Carriers are backwards compatible for release 8 and release 9 devices, i.e. older devices can use individual Component Carriers in the normal way
- ★ The release 10 version of the 3GPP specifications defines signalling to support up to 5 Component Carriers, i.e. a maximum combined channel bandwidth of 100 MHz (RRC signalling is specified in 3GPP TS 36.331). From the RF perspective, a maximum of 2 Component Carriers have been defined initially within the release 10 version of the specifications (Component Carrier combinations are specified in 3GPP TS 36.101)
- ★ Carrier Aggregation categorises cells as:
  - o Primary Cell:
    - ▪ the cell upon which the UE performs initial connection establishment
    - ▪ the primary cell can be changed during the handover procedure
    - ▪ each connection has a single primary cell
    - ▪ the primary cell uses the primary RF carrier
    - ▪ a PDCCH order to initiate the random access procedure can only be received on the primary cell
    - ▪ PRACH preambles can only be sent on the primary cell
    - ▪ the primary cell is used to generate inputs during security procedures
    - ▪ the primary cell is used to define NAS mobility information, e.g. Tracking Area Identity
  - o Secondary Cell:
    - ▪ a cell which has been configured to provide additional radio resources after connection establishment
    - ▪ a secondary cell uses a secondary RF carrier
    - ▪ each connection can have multiple secondary cells
  - o Serving Cell:
    - ▪ both primary and secondary cells are categorised as serving cells
    - ▪ there is one HARQ entity per serving cell at the UE
- ★ Component Carriers belonging to a specific eNode B may have different coverage, e.g. a Component Carrier in the 900 MHz operating band may have greater coverage than a Component Carrier in the 2.6 GHz operating band. The eNode B can use measurement reporting events to trigger the release of a specific Component Carrier as its coverage becomes weak, e.g. measurement reporting event A2 (serving cell becomes worse than a threshold) can be configured to trigger separately for each Component Carrier
- ★ Uplink power control is applied independently for each Component Carrier. The path loss estimate used within the open loop section of the power control calculation can be derived from the relevant Component Carrier. Alternatively, the path loss estimate can be derived from a 'reference' Component Carrier but the 'reference' Component Carrier should always be within the same operating band. Inner loop power control commands are signalled for each Component Carrier
- ★ The PUSCH and Sounding Reference Signal (SRS) can be transmitted on each serving cell with an uplink (both primary and secondary), but the PUCCH can only be transmitted on the primary serving cell
- ★ The Carrier Indicator Field (CIF) can be included as part of the PDCCH Downlink Control Information (DCI) within the UE specific search space. The CIF can be used for cross carrier scheduling, i.e. allocating resources on serving cell 'x' by sending the PDCCH DCI on serving cell 'y'. The CIF is not included within DCI when using the common search space
- ★ The increased number of transport blocks transferred when using Carrier Aggregation require an increased number of HARQ acknowledgements. Acknowledgements for the PUSCH are sent on the PHICH using the Component Carrier which allocated the PUSCH resources. Acknowledgements for the PDSCH are either sent on the PUCCH belonging to the primary serving cell, or on the PUSCH belonging to either the primary or a secondary serving cell
- ★ MAC control elements can be used for fast activation and deactivation of secondary cells after they have been configured by the RRC layer. The primary cell is always activated and cannot be deactivated
- ★ 3GPP References: TS 36.101, TS 36.104, TS 36.213, TS 36.306, TS 36.321, TS 36.331

18.2 CARRIER ORGANISATION

18.2.1 INTRA-BAND

★ The release 10 version of the 3GPP specifications defines intra-band contiguous Carrier Aggregation for FDD within operating band 1, and for TDD within operating band 40. These bands are presented in Table 193

Carrier Aggregation Band	E-UTRA Operating Band	Duplex Technology	Uplink Operating Band	Downlink Operating Band
CA_1	1	FDD	1920 – 1980 MHz	2110 – 2170 MHz
CA_40	40	TDD	2300 – 2400 MHz	

Table 193 – Intra-band contiguous carrier aggregation operating bands

★ Within each of these operating bands, a range of channel bandwidths can be used for Carrier Aggregation. The channel bandwidths specified for intra-band Carrier Aggregation within operating bands 1 and 40 are presented in Table 194

Carrier Aggregation Configuration	E-UTRA Operating Band	10 MHz + 20 MHz	15 MHz + 15 MHz	20 MHz + 20 MHz
CA_1C	1	-	Yes	Yes
CA_40C	40	Yes	Yes	Yes

Table 194 – Channel bandwidths for intra-band contiguous carrier aggregation within operating bands 1 and 40

★ Intra-band Carrier Aggregation focuses upon the larger channel bandwidths and a total channel bandwidth which is greater than 20 MHz, i.e. total channel bandwidths of 30 MHz and 40 MHz are shown in Table 194. Smaller total channel bandwidths are less likely to require Carrier Aggregation, e.g. a single 20 MHz channel can be used instead of two 10 MHz channels (assuming the two 10 MHz channels are adjacent to each other)

★ The letter 'C' appended to the Carrier Aggregation Configuration refers to the 'Carrier Aggregation Bandwidth Class'.

★ The bandwidth class defines the maximum number of component carriers, the range of aggregated Resource Blocks and the nominal guard band. Bandwidth classes 'A' and 'C' have been finalised within 3GPP TS 36.101, while RRC signalling specified within TS 36.331 provides support for bandwidth classes 'A' to 'F'. UE signal their support for specific bandwidth classes as part of their UE capability information

★ The characteristics associated with bandwidth classes 'A' and 'C' are presented in Table 195

Carrier Aggregation Bandwidth Class	Maximum Number of Component Carriers	Aggregated Transmission Bandwidth Configuration	Nominal Guard Band (BW_{GB})
A	1	$N_{RB,agg} \leq 100$ Res. Blks	$0.05 \times BW_{channel(1)}$
C	2	100 Res. Blks $< N_{RB,agg} \leq 200$ Res. Blks	$0.05 \times Max(BW_{channel(1)}, BW_{channel(2)})$

Table 195 – Carrier Aggregation bandwidth classes

★ Both of the intra-band Carrier Aggregation configurations have a bandwidth class of 'C' so a maximum of 2 Component Carriers are supported, i.e. a maximum aggregated channel bandwidth of 40 MHz and a maximum of 200 Resource Blocks

★ Bandwidth class 'A' is used for inter-band Carrier Aggregation and a single Component Carrier is used from each operating band

★ The nominal guard band shown in Table 195 quantifies the guard bands at the upper and lower edges of the aggregated channel bandwidth. $BW_{channel}$ is the channel bandwidth of the Component Carrier in MHz, i.e. 1.4, 3, 5, 10, 15 or 20 MHz

★ The aggregated channel bandwidth and the pair of guard bands are illustrated in Figure 187. The aggregated channel bandwidth includes the guard bands at the upper and lower edges of the Component Carriers and is defined as:

$$BW_{channel\_CA} = F_{edge,high} - F_{edge,low}$$

where, $F_{edge,low} = F_{c,low} - F_{offset,low}$ and $F_{edge,high} = F_{c,high} + F_{offset,high}$

$$F_{offset,low} = 0.18 \times N_{RB,low} / 2 + BW_{GB} \quad \text{and} \quad F_{offset,high} = 0.18 \times N_{RB,high} / 2 + BW_{GB}$$

The factor of 0.18 corresponds to the 180 kHz bandwidth of a Resource Block, i.e. 12 × 15 kHz

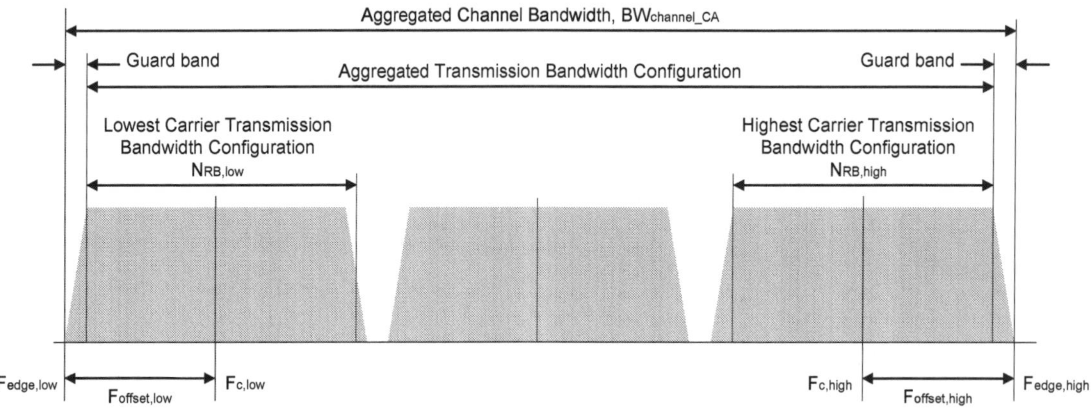

Figure 187 – Bandwidth definitions for intra-band contiguously aggregated Component Carriers

- ★ The channel raster for each Component Carrier is 100 kHz, i.e. the normal channel raster for LTE
- ★ The spacing between the center frequencies of intra-band contiguously aggregated RF carriers is specified to be a multiple of 300 kHz
  - o compatible with the 100 kHz raster
  - o multiple of 15 kHz to preserve subcarrier orthogonality
- ★ The nominal channel spacing between two contiguously aggregated Component Carriers is defined as:

Nominal Channel Spacing =

$$\text{ROUNDDOWN}[\ (BW_{channel(1)} + BW_{channel(2)} - 0.1 \times \text{ABS}(BW_{channel(1)} - BW_{channel(2)})\ ) / (2 \times 0.3)\ ] \times 0.3$$

The nominal channel spacings for the channel bandwidth combinations specified in Table 194 are presented in Table 196

$BW_{channel(1)}$	$BW_{channel(2)}$	Nominal Channel Spacing
20 MHz	20 MHz	19.8 MHz
15 MHz	15 MHz	15.0 MHz
20 MHz	10 MHz	14.4 MHz

Table 196 – Nominal Channel Spacings for intra-band Component Carriers

- ★ 3GPP uses the nominal channel spacing when specifying RF requirements for the UE and eNode B. In practice, the actual channel spacing for intra-band contiguous carrier allocation can be reduced to any multiple of 300 kHz less than the nominal channel spacing. This provides some flexibility for optimisation

18.2.2 INTER-BAND

- ★ The release 10 version of the 3GPP specifications defines inter-band Carrier Aggregation for FDD within operating bands 1 and 5. These bands are presented in Table 197

Carrier Aggregation Band	E-UTRA Operating Band	Duplex Technology	Uplink Operating Band	Downlink Operating Band
CA_1-5	1	FDD	1920 – 1980 MHz	2110 – 2170 MHz
	5	FDD	824 – 849 MHz	869 – 895 MHz

Table 197 – Inter-band carrier aggregation operating bands

- ★ Inter-band carrier aggregation within operating bands 1 and 5 is currently specified to use the 10 MHz channel bandwidth. The same channel bandwidth is used in both operating bands. This is presented in Table 198

Carrier Aggregation Configuration	E-UTRA Operating Band	10 MHz
CA_1A-5A	1	Yes
	5	Yes

Table 198 – Channel bandwidths for inter-band carrier aggregation within operating bands 1 and 5

★ The letter 'A' appended to the operating band number within the Carrier Aggregation Configuration refers to the 'Carrier Aggregation Bandwidth Class' as presented in Table 195

★ Bandwidth class 'A' defines a single Component Carrier so a total of 2 Component Carriers is supported, i.e. one from each operating band. Bandwidth class 'A' also specifies a maximum of 100 Resource Blocks within each operating band. This is consistent with the 10 MHz channel bandwidth which supports 50 Resource Blocks per RF carrier

★ 3GPP References: TS 36.101, TS 36.104

18.3 UE CAPABILITY

★ The UE signals its capability to the eNode B using a UE Capability Information message

★ From the perspective of Carrier Aggregation, the most important information within this message is:

- UE Category
- RF Parameters – Supported Band Combinations
- Physical Layer Parameters
- Feature Group Indicators

★ The release 10 version of the 3GPP specifications defines 8 UE categories. Categories 1 to 5 were introduced within the release 8 version of the specifications. Nevertheless, 3GPP release 10 UE which are category 1 to 5 may still support Carrier Aggregation

★ Table 199 presents some example downlink Carrier Aggregation scenarios for each UE category

UE Category	Maximum Throughput Capability	Example Carrier Aggregation Throughput	Component Carriers	Layers for Spatial Multiplexing	Total Transport Blocks	Bits per Transport Blk
1	10.3 Mbps	8.8 Mbps	2 × 10 MHz	1	2	4392
2	51.0 Mbps	48.9 Mbps	2 × 10 MHz	2	4	12 216
3	102.0 Mbps	101.8 Mbps	2 × 10 MHz	2	4	25 456
4	150.8 Mbps	146.8 Mbps	2 × 10 MHz	2	4	36 696
5	299.6 Mbps	255.1 Mbps	2 × 10 MHz	4	4	63 776
6	301.5 Mbps	301.5 Mbps	2 × 20 MHz	2	4	75 376
7	301.5 Mbps	301.5 Mbps	2 × 20 MHz	2	4	75 376
8	3.0 Gbps	3.0 Gbps	5 × 20 MHz	8	10	299 856

Table 199 – Example downlink Carrier Aggregation scenarios for each UE category

★ A category 8 UE is able to support 8×8 MIMO across 5 × 20 MHz Component Carriers. Each Component Carrier can transfer a maximum of 2 transport blocks per TTI so there can be a maximum of 10 transport blocks. The set of 10 transport blocks transfer a maximum of 10 × 299 856 bits, which corresponds to a maximum throughput of 3.0 Gbps

★ These figures represent the maximum capability from the perspective of the UE category. The bandwidth class and MIMO capability must also be checked for each supported combination of Component Carriers. For example, a category 8 UE will not support 3 Gbps if it is specified to use bandwidth class 'C' which limits it to 2 Component Carriers

★ From a downlink perspective, UE categories 6 and 7 offer the same capability. A category 6 or 7 UE is able to support 2×2 MIMO across 2 × 20 MHz Component Carriers. Each Component Carrier can transfer a maximum of 2 transport blocks per TTI so there can be a maximum of 4 transport blocks. The set of 4 transport blocks transfer a maximum of 4 × 75 376 bits, which corresponds to a maximum throughput of 301.5 Mbps

★ UE categories 1 to 5 are able to achieve their maximum throughput capabilities using a single 20 MHz channel (as defined for the release 8 version of the 3GPP specifications). Alternatively, these UE categories could achieve their maximum throughput capabilities using 2 × 10 MHz with Carrier Aggregation

★ Table 200 presents some example uplink Carrier Aggregation scenarios for each UE category

★ A category 8 UE is able to support 4×4 MIMO across 5 × 20 MHz Component Carriers. Each Component Carrier can transfer a maximum of 2 transport blocks per TTI so there can be a maximum of 10 transport blocks. The set of 10 transport blocks transfer a maximum of 10 × 149 776 bits, which corresponds to a maximum throughput of 1.5 Gbps

- note: the transport block size of 149 776 bits requires an allocation of 100 Resource Blocks for the PUSCH. In practice, this is not achievable when the Resource Blocks towards the edge of the channel bandwidth are allocated to the PUCCH, i.e. the 20 MHz channel bandwidth has less than 100 Resource Blocks available to the PUSCH

UE Category	Maximum Throughput Capability	Example Carrier Aggregation Throughput	Component Carrriers	Layers for Spatial Multiplexing	Total Transport Blocks	Bits per Transport Blk
1	5.2 Mbps	5.1 Mbps	2 × 10 MHz	1	2	2536
2	25.5 Mbps	24.4 Mbps	2 × 10 MHz	1	2	12 216
3	51.0 Mbps	50.9 Mbps	2 × 10 MHz	1	2	25 456
4	51.0 Mbps	50.9 Mbps	2 × 10 MHz	1	2	25 456
5	75.4 Mbps	68.0 Mbps	2 × 10 MHz	1	2	34 008
6	51.0 Mbps	50.9 Mbps	2 × 20 MHz	1	2	25 456
7	102.0 Mbps	102.0 Mbps	2 × 20 MHz	1	2	51 024
8	1.5 Gbps	1.5 Gps	5 × 20 MHz	4	10	149 776

Table 200 – Example uplink Carrier Aggregation scenarios for each UE category

★ Category 6 and 7 UE are differentiated by their maximum throughput capabilities in the uplink. Neither UE category supports 64QAM. A Category 7 UE can achieve its maximum throughput capability using either Carrier Aggregation or MIMO. UE categories 1 to 6 can achieve their maximum throughput capabilities without having to use either Carrier Aggregation nor MIMO. The use of Carrier Aggregation is shown in Table 200 to illustrate some examples

★ The content of the Supported Band Combinations information element from within the UE Capability Information message is presented in Table 201. The message structure allows a UE to signal up to 128 different band combinations, and include up to 64 bands within each combination. Each band included within a combination is identified by its band number, ranging from 1 to 64. Each band can have up to 16 sets of uplink parameters and up to 16 sets of downlink parameters

Information Elements					
Supported Band Combinations	Band Combination Parameters (1 to 128)	Band Parameters (1 to 64)	E-UTRA Band		
			Uplink Band Parameters (1 to 16)	Carrier Aggregation Bandwidth Class	
				Uplink MIMO Capability	
			Downlink Band Parameters (1 to 16)	Carrier Aggregation Bandwidth Class	
				Downlink MIMO Capability	

Table 201 – Content of Supported Band Combinations information element

★ Each set of uplink parameters associated with a specific band defines the Carrier Aggregation bandwidth class (bandwidth classes of A, B, C, D, E or F). The uplink MIMO capability is also defined as either 2-layers or 4-layers

★ Each set of downlink parameters associated with a specific band defines the Carrier Aggregation bandwidth class (bandwidth classes of A, B, C, D, E or F). The downlink MIMO capability is also defined as either 2-layers, 4-layers or 8-layers

★ The 3GPP release 10 physical layer UE capability parameters are presented in Table 202. Not all of these parameters are associated with Carrier Aggregation, e.g. 'Two antenna ports for PUCCH' indicates whether or not the UE supports transmit diversity for the PUCCH

Information Elements	
Physical Layer Parameters	Two antenna ports for PUCCH
	Transmission mode 9 with 8 Tx FDD
	PMI disabling
	Cross Carrier Scheduling
	Simultaneous PUCCH and PUSCH
	Multi-cluster PUSCH within a Component Carrier
	Non-contiguous uplink resource allocation within a Component Carrier

Table 202 – 3GPP release 10 physical layer UE capability parameters

★ The cross carrier scheduling parameter indicates whether or not cross carrier scheduling is supported by the UE. Cross carrier scheduling is described in section 18.7. This parameter also indicates that the UE supports the use of the primary cell as a path loss reference for a secondary cell. This is relevant to power control. It is mandatory for UE to support the use of a secondary cell as a path loss reference for that secondary cell

★ If a UE supports Carrier Aggregation in the uplink (indicated by the supported band combinations information element) then inclusion of the 'Simultaneous PUCCH and PUSCH' parameter indicates that the UE supports simultaneous PUCCH and PUSCH transmission on any Component Carrier that can be aggregated

- If a UE supports uplink Carrier Aggregation then it is mandatory for the UE to support PUSCH transmission using non-contiguous Resource Block allocations (PUSCH resource allocation type 1). This is irrespective of whether or not the 'Multi-cluster PUSCH within a Component Carrier' parameter is included within the UE capability information. Otherwise, if a UE does not support uplink Carrier Aggregation then inclusion of the 'Multi-cluster PUSCH within a Component Carrier' parameter indicates that the UE supports PUSCH transmission using non-contiguous Resource Block allocations within the operating bands specified by the 'Non-contiguous uplink resource allocation within a Component Carrier' parameter set
- A subset of the 3GPP release 10 Feature Group Indicators (FGI) within the UE Capability Information message are directly applicable to Carrier Aggregation:
    - bit 11, Measurement Reporting Trigger Event A6 (neighbouring cell becomes better than a secondary cell by an offset)
    - bit 12, Secondary Cell Addition during the Handover to E-UTRA procedure
    - bit 13, Periodic transmission of Sounding Reference Signal (SRS) on each Component Carrier within a band combination

  These bits can only be set to '1' if the UE supports Carrier Aggregation. Each bit indicates whether or not the UE supports the associated functionality
- 3GPP References: TS 36.306, TS 36.331

18.4 MEASUREMENTS

- Section 22 provides a description of measurement reporting. Measurement reporting uses a combination of Measurement Objects, Measurement Identities and Reporting Configurations
- Measurement reporting was modified for the release 10 version of the 3GPP specifications to account for the possibility of multiple serving cells. For example, reporting event A3: 'Neighbour becomes Offset better than Serving Cell' required clarification regarding the choice of serving cell to compare against the neighbour
- Table 203 compares the definition of measurement reporting events in 3GPP release 10 with those in 3GPP releases 8 and 9

Measurement Reporting Event	3GPP Release 10	3GPP Releases 8 and 9
A1	Serving Cell becomes better than threshold	
A2	Serving Cell becomes worse than threshold	
A3	Neighbour becomes offset better than Primary Cell	Neighbour becomes offset better than Serving Cell
A4	Neighbour becomes better than threshold	
A5	Primary Cell becomes worse than threshold 1 and neighbour becomes better than threshold 2	Serving Cell becomes worse than threshold 1 and neighbour becomes better than threshold 2
A6	Neighbour becomes offset better than Secondary Cell	Does not exist within releases 8 and 9
B1	Inter RAT neighbour becomes better than threshold	
B2	Primary Cell becomes worse than threshold 1 and inter-RAT neighbour becomes better than threshold 2	Serving Cell becomes worse than threshold 1 and inter-RAT neighbour becomes better than threshold 2

Table 203 – Comparison of measurement reporting events in 3GPP release 10 with those in 3GPP releases 8 and 9

- In the case of reporting events A1 and A2, the serving cell is the primary or secondary cell on the RF carrier specified by the Measurement Object (there can be a single Measurement Object per RF carrier)
- Reporting events A3 and A5 have been re-worded for the release 10 version of the specifications to indicate that the serving cell to be considered is the primary cell
- The neighbouring cell associated with reporting events A3, A4 and A5 must be on the RF carrier specified by the Measurement Object
- Reporting event A6 was introduced for Carrier Aggregation within the release 10 version of the 3GPP specifications. It is triggered by a neighbouring cell on the same RF carrier as the secondary cell. The secondary cell is identified by the RF carrier specified within the Measurement Object
- Reporting events B1 and B2 are triggered by a neighbouring cell on the RF carrier specified within the Measurement Object. In the case of GERAN, the Measurement Object can define a group of RF carriers and reporting events B1 and B2 can be triggered by any of the RF carriers within that group
- Reporting event B2 has been re-worded for the release 10 version of the specifications to indicate that the serving cell to be considered is the primary cell

- ★ Intra-frequency measurements are categorised as measurements at frequencies that belong to any of the serving cells, i.e. measurements on RF carriers belonging to both the primary and secondary cells are categorised as intra-frequency
- ★ Inter-frequency measurements are categorised as measurements at frequencies that differ from those that belong to any of the serving cells, i.e. measurements on an RF carrier that belongs to a secondary cell are not categorised as inter-frequency
- ★ 3GPP References: TS 36.331

18.5 UPLINK POWER CONTROL

- ★ Section 24.4 provides a description of uplink power control for the PUSCH, PUCCH and Sounding Reference Signal (SRS). The PUSCH and SRS are power controlled on each Component Carrier. The PUCCH is only transmitted on the primary cell and its power control is handled in the same way as without Carrier Aggregation
- ★ The power control equations presented in section 24.4 remain applicable when Carrier Aggregation is configured. These equations are applied separately for each Component Carrier, so each Component Carrier has its own independent power control
- ★ Uplink power control uses a combination of open loop and inner loop components. The open loop component includes an estimate of the path loss, whereas the inner loop component is derived from eNode B power control commands
    - o path loss estimates can be based upon either the downlink of the primary serving cell, or the downlink of the secondary serving cell to which the power control is being applied
    - o power control commands for the PUSCH and SRS are received within the PDCCH Downlink Control Information (DCI)
        - ▪ commands received on DCI formats 3 and 3A are applicable to the uplink of the cell from which they were received
        - ▪ commands received on DCI formats 0 and 4 are applicable to cell where the uplink resources are being allocated (can be different to the cell from which they were received if cross carrier scheduling is configured)
- ★ Some power control parameters are signalled to the UE on a per carrier basis. The parameters presented in Table 204 are signalled for each secondary carrier. These are divided into common and dedicated parameters

UplinkPowerControlCommonSCell-r10
p0-NominalPUSCH-r10
alpha-r10

UplinkPowerControlDedicatedSCell-r10
p0-UE-PUSCH-r10
deltaMCS-Enabled-r10
accumulationEnabled-r10
pSRS-Offset-r10
pSRS-OffsetAp-r10
filterCoefficient-r10
pathlossReferenceLinking-r10

Table 204 – Secondary cell power control parameters

- ★ The p0-NominalPUSCH and p0-UE-PUSCH parameters are summed by the UE to generate the $P_{O\_PUSCH}(j)$ variable within the PUSCH power control calculation
- ★ The alpha parameter ($\alpha(j)$) is used to configure the use of fractional power control. A value of 1 disables fractional power control, while a value < 1 means that the UE does not fully compensate path loss increases with transmit power increases, i.e. the received signal to noise ratio decreases as the path loss increases
- ★ The deltaMCS-Enabled parameter determines whether or not the UE increases its transmit power when transferring a large number of bits per Resource Element, i.e. it links the uplink transmit power to the Modulation and Coding Scheme (MCS). A value of 0 indicates that no transmit power increase is applied, whereas a value of 1 indicates that an MCS dependent transmit power increase is applied
- ★ The accumulationEnabled parameter indicates whether or not TPC command accumulation is to be enabled. TPC commands are specific to the Component Carrier and are sent within the PDCCH allocating the uplink resources, or in DCI formats 3 and 3A
- ★ The filterCoefficient defines the filtering to be applied to the RSRP measurements prior to calculating the downlink path loss
- ★ The pathlossReferenceLinking parameter identifies the downlink RF carrier from which the RSRP is measured to calculate the path loss. The UE can be instructed to use either the primary downlink RF carrier, or the secondary downlink RF carrier which is linked to the uplink RF carrier
- ★ If a UE reaches its maximum transmit power capability, the PUCCH is provided with priority and is transmitted with its required power. The transmit power of the PUSCH is then reduced equally on all Component Carriers
- ★ 3GPP References: TS 36.213, TS 36.331

18.6 RRC SIGNALLING

★ Secondary serving cells are added by the eNode B after RRC connection establishment using an RRC Connection Reconfiguration message. This message can includes sections to add, modify or release secondary serving cells. The section used to add or modify secondary serving cells is presented in Table 205

Information Elements				
Secondary Cells to Add or Modify List	Secondary Cells to Add or Modify (1 to 4)	Secondary Cell Index		
		Cell Identification	Physical Cell Identity	
			Downlink Carrier Frequency	
		Common Radio Resource Configuration	Non-UL Configuration	Downlink Bandwidth
				Antenna Information
				MBSFN Subframe Configuration
				PHICH Configuration
				PDSCH Configuration
				TDD Configuration
			UL Configuration	UL Frequency Information
				Pmax
				Uplink Power Control
				SRS Configuration
				Uplink Cyclic Prefix Length
				PRACH Configuration
				PUSCH Configuration
		Dedicated Radio Resource Configuration	Non-UL Configuration	Antenna Information
				Cross Carrier Scheduling Configuration
				CSI Reference Signal Configuration
				PDSCH Configuration
			UL Configuration	Antenna Information
				PUSCH Configuration
				Uplink Power Control
				CQI Report Configuration
				SRS Configuration
				Aperiodic SRS Configuration

Table 205 – Section of RRC Connection Reconfiguration message used to add or modify secondary serving cells

★ Between 1 and 4 secondary serving cells can be added by the RRC Connection Reconfiguration message. This provides support for Carrier Aggregation across 5 serving cells after accounting for the primary serving cell

★ Each secondary serving cell is allocated an index within the range 1 to 7. The secondary cell is linked to its index by specifying its Physical layer Cell Identity (PCI) and its downlink E-UTRA Absolute Radio Frequency Channel Number (EARFCN)

★ The UE is then provided with a section of common radio resource configuration information. This type of information is present within the Master Information Block (MIB) and the set of System Information Blocks (SIB). For example, the PHICH configuration is present within the MIB, while the MBSFN Subframe Configuration is present within SIB2

★ The first part of the common radio resource configuration is referenced as 'Non-UL Configuration'. This refers to information which is applicable to the downlink, or both the uplink and downlink

- the downlink bandwidth is specified in terms of Resource Blocks (6, 15, 25, 50, 75 or 100)
- the number of antenna ports used by the cell specific Reference Signal is defined (1, 2 or 4)
- the MBSFN subframe configuration is used to specify which subframes (if any) are allocated to MBSFN
- the PHICH configuration defines both the PHICH duration (normal or extended) and the PHICH resource, also known as the PHICH Group Scaling Factor (1/6, 1/2, 1 or 2)
- the PDSCH configuration defines the Reference Signal power in terms of the Energy Per Resource Element (EPRE) using a range from -60 to 50 dBm. It also defines the 'p-b' variable (P_B) which has a range from 0 to 3. Both of these variables are used as inputs when calculating the power difference between the PDSCH and the cell specific Reference Signal

  - the TDD configuration specifies the uplink-downlink subframe configuration (0 to 6) which defines the number and pattern of uplink and downlink subframes. It also specifies the special subframe configuration (0 to 8) which defines the duration of the DwPTS, guard period and UpPTS

★ The second part of the common radio resource configuration is referenced as 'UL Configuration'. This refers to information which is only applicable to the uplink

  - the uplink frequency information defines the E-UTRA Absolute Radio Frequency Channel Number (EARFCN), the uplink bandwidth in terms of Resource Blocks and any additional spectrum emission requirements
  - Pmax is specified to define the maximum permitted uplink transmit power within the secondary cell
  - the uplink power control section includes the p0-NominalPUSCH and alpha parameters presented in section 18.5
  - the Sounding Reference Signal (SRS) information includes the bandwidth configuration and the subframe configuration. These parameters define the Resource Blocks and subframes used for SRS transmission (see section 13.2). This section of SRS information also includes a flag to indicate whether or not SRS transmission is permitted simultaneously with HARQ acknowledgements on the PUCCH. However, the UE ignores this information element because the PUCCH is not transmitted on secondary serving cells. This section of SRS information also includes the Max UpPTS parameter which determines whether or not the bandwidth of the SRS is reconfigured when transmitted within the UpPTS field of a TDD special subframe
  - the uplink cyclic prefix length is signaled as either 'normal' or 'extended'
  - the PRACH configuration defines the PRACH Configuration Index (0 to 63) used to specify the PRACH preamble format and the set of subframes available for PRACH preamble transmission. In the case of TDD, the configuration index also defines the frequency resources available for PRACH preamble transmission
  - the PUSCH configuration defines the frequency hopping mode (inter-subframe, or intra and inter-subframe) and the frequency hopping offset (0 to 98), as well as the number of sub-bands used for frequency hopping. It also specifies whether or not 64QAM can be used in the uplink. Information regarding the uplink Reference Signal is also provided

★ The UE is then provided with a section of dedicated radio resource configuration information. Similar to the common configuration information, the first part is referenced as 'Non-UL Configuration'

  - the antenna information defines the downlink transmission mode (1 to 9). It also specifies any codebook subset restrictions to be applied when generating feedback for spatial multiplexing, i.e. entries within the precoding codebook which are to be excluded. Information regarding transmit antenna selection is also included. The UE can be instructed to use either open loop or closed loop transmit antenna selection
  - the cross carrier scheduling information (see Table 206 in section 18.7) indicates whether or not cross carrier scheduling is to be used. If resources on the secondary cell are to be allocated by the same secondary cell then a value of 'own' is signalled. In that case, a flag is also included to indicate whether or not the Carrier Indicator Field (CIF) is included within the PDCCH belonging to this secondary serving cell to allocate resources for other serving cells. If resources on the secondary cell are to be allocated by another serving cell then a value of 'other' is signalled, and the scheduling cell identity is specified. In that case, the PDSCH start symbol is also provided to avoid the UE having to rely upon the PCFICH to determine the first OFDMA symbol for the PDSCH
  - The CSI Reference Signal configuration can be used to setup CSI Reference Signal transmissions with both non-zero and zero transmit powers (a zero transmit power is used to create instances of DTX which allow interference measurements). The number of antenna ports used for the CSI Reference Signal is configured as 1, 2, 4 or 8. The resource configuration (0 to 31) defines the CSI Reference Signal Configuration and points towards a row within a look-up table that specifies a 'reference' Resource Element. The actual Resource Elements used by the CSI Reference Signal are derived from this 'reference' Resource Element using antenna specific offsets. The Subframe Configuration (0 to 154) is used to define both a periodicity and subframe offset for the timing of the CSI Reference Signal. The variable Pc informs the UE of the ratio between the PDSCH Energy per Resource Element (EPRE) and the CSI Reference Signal EPRE. The value of Pc can be signalled with values between -8 and 15 dB
  - the PDSCH configuration defines the 'p-a' variable (P_A) used as an input when calculating the power difference between the PDSCH and the cell specific Reference Signal

★ The second part of the dedicated radio resource configuration is referenced as 'UL Configuration'. This refers to information which is only applicable to the uplink

  - the antenna information defines the uplink transmission mode (transmission mode 1 or transmission mode 2). If transmission mode 2 is signalled then the eNode B can also include an instruction to use 4 antenna ports for uplink transmission
  - the PUSCH configuration indicates whether or not group hopping is enabled, and also whether or not Orthogonal Covering Codes (OCC) are to be enabled for the demodulation Reference Signal
  - the uplink power control information for a secondary serving cell is presented in Table 204 (section 18.5)
  - the CQI report configuration defines the aperiodic CSI reporting mode (1-2, 2-0, 2-2, 3-0 or 3-1) and all of the information necessary for periodic CSI reporting. A flag is also included to indicate whether or not the UE should provide Precoding Matrix Indicator (PMI) and Rank Indicator (RI) feedback
  - The SRS configuration information includes the SRS bandwidth, hopping bandwidth, frequency domain position, duration, configuration index, transmission comb offset and cyclic shift. It also defines the number of antenna ports to be used for the SRS

- The aperiodic SRS configuration defines the SRS configuration index for aperiodic transmissions. It also defines 1 to 3 SRS configurations which can be triggered by DCI format 4, and configurations which can be triggered by DCI formats 0, 1A, 2B or 2C. In each case, the number of antenna ports, the SRS bandwidth, the frequency domain position, the transmission comb offset and cyclic shift are specified

★ 3GPP References: TS 36.331

18.7 CROSS CARRIER SCHEDULING

★ Cross carrier scheduling refers to the network using PDCCH signalling on RF carrier 'x' to allocate resources on RF carrier 'y'. It avoids the UE having to check the PDCCH transmissions on every Component Carrier

★ Cross carrier scheduling also provides benefits in terms of Inter-Cell Interference Coordination (ICIC), i.e. the OFDMA symbols normally allocated to the PDCCH can be allowed to DTX in one cell to avoid interference towards another cell. Similarly, if one cell is experiencing relatively high levels of downlink interference then the PDCCH can be sent on another cell to help improve reception reliability

★ Cross carrier scheduling is applicable to secondary RF carriers rather than the primary RF carrier. Resources on the primary RF carrier are always allocated using the PDCCH belonging to that carrier

★ Cross carrier scheduling is applicable to application data but not paging messages, system information nor random access messages. The UE checks the common search space on the primary RF carrier for PDCCH transmissions with the CRC bits scrambled using the P-RNTI, SI-RNTI or RA-RNTI

★ The general concept of cross carrier scheduling is illustrated in Figure 188. This figure compares scenarios with and without cross carrier scheduling. Two cross carrier scheduling scenarios are illustrated – the first with all scheduling provided by the PDCCH belonging to the primary RF carrier, and the second with scheduling provided by a combination of the PDCCH belonging to the primary RF carrier and the PDCCH belonging to a secondary RF carrier

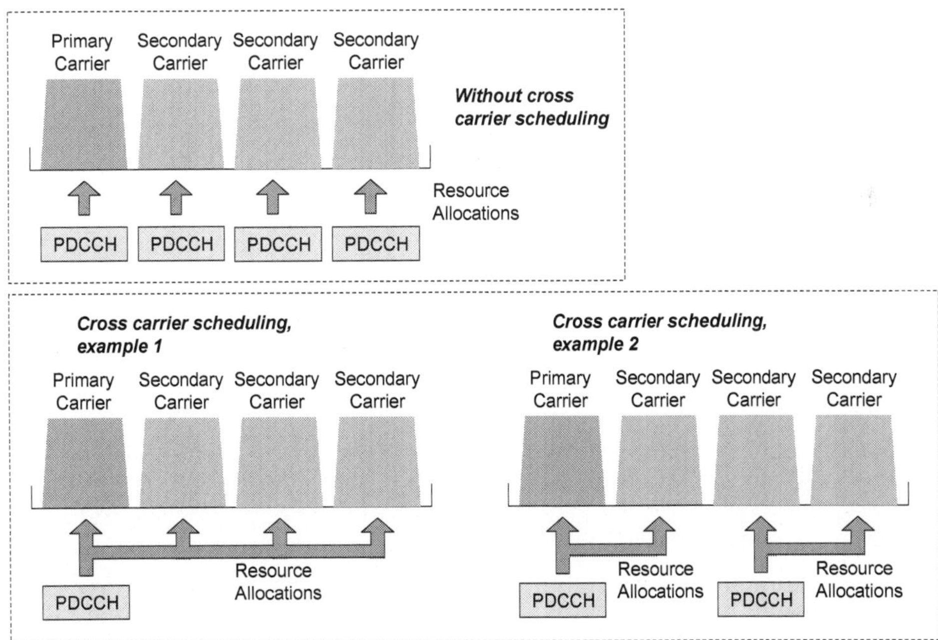

Figure 188 – Concept of cross carrier scheduling

★ Cross carrier scheduling configurations are UE specific rather than cell specific so it is possible to distribute the PDCCH load across multiple RF carriers

★ It is not mandatory for UE to support cross carrier scheduling. A UE signals its cross carrier scheduling capability within the 3GPP release 10 physical layer parameters of the UE Capability Information message

★ An eNode B enables the use of cross carrier scheduling for a specific connection using an RRC Connection Reconfiguration message. This is done using the cross carrier scheduling configuration information element shown in Table 206. This information element is associated with a specific secondary serving cell, i.e. one of these information elements can be sent for each secondary serving cell

★ If a secondary serving cell is to have its resources allocated by its own PDCCH then the cross carrier scheduling configuration is set to 'own'. In addition, the Carrier Indication Field (CIF) Presence flag can be set to 'true' to inform the UE that the PDCCH belonging to that RF carrier may be used to allocate resources on a different RF carrier

Information Elements		
Scheduling Cell Information	CHOICE	
	Own	CIF Presence
	Other	Scheduling Cell Identity
		PDSCH Start

Table 206 – Content of Cross Carrier Scheduling Configuration information element

- ★ If the secondary RF carrier is to have its resources allocated by the PDCCH belonging to a different RF carrier then the cross carrier scheduling configuration is set to 'other'. The Scheduling Cell Identity informs the UE of which RF carrier the UE needs to monitor to receive its resource allocations. The PDSCH Start information signals the OFDMA symbol within which the PDSCH transmission starts. This avoids the UE having to read the PCFICH which normally indicates the number of OFDMA symbols reserved at the start of each subframe for the PDCCH and PHICH
- ★ PDCCH Downlink Control Information (DCI) formats 0, 1, 1A, 1B, 1D, 2, 2A, 2B, 2C and 4 are able to accommodate the Carrier Indicator Field (CIF). DCI format 1C which is intended for very compact scheduling of resource allocations does not support inclusion of the CIF. Likewise the DCI used to signal power control commands (3 and 3A) do not support inclusion of the CIF
- ★ 3GPP References: TS 36.213, TS 36.331

18.8 FAST ACTIVATION AND DEACTIVATION

- ★ Fast activation and deactivation of secondary cells helps to conserve UE battery power, i.e. deactivating a secondary cell means that the UE no longer has to monitor that cell for resource allocations and no longer has to complete measurements for CQI reporting
- ★ Fast activation and deactivation is applicable to the uplink and downlink of secondary cells. The primary cell is always activated and cannot be deactivated. Secondary cells are initially deactivated after their addition and after handover so require a fast activation command before use
- ★ Fast activation and deactivation is completed by the MAC layer after the RRC layer has already added one or more secondary cells. A MAC control element is used for the activation and deactivation procedure. MAC control elements are sent to the UE as part of the MAC payload within a MAC PDU
- ★ The MAC control element used for fast activation and deactivation of secondary cells has a length of 1 byte and is identified within the MAC header using a Logical Channel Identity (LCID) of 11011. It is illustrated in Figure 189

Figure 189 – MAC control element used for fast activation and deactivation of secondary cells

- ★ The values of C1 to C7 indicate which secondary cells are to be activated and which secondary cells are to be deactivated. A value of 1 for Ci indicates that the secondary cell with index 'i' is to be activated, whereas a value of 0 for Ci indicates that the secondary cell with index 'i' is to be deactivated
- ★ A secondary cell can also be deactivated by an inactivity timer (sCellDeactivationTimer). This timer is optional and a value of infinity is assumed if the eNode B does not signal a value to the UE
  - o the timer can be included within the RRC Connection Setup, RRC Connection Reconfiguration or RRC Connection Re-establishment messages. Its value can be set to 2, 4, 8, 16, 32, 64 or 128 radio frames
  - o the timer runs independently for each secondary cell and is started when a secondary cell is activated
  - o the timer is reset if the UE receives an uplink or downlink resource allocation from the PDCCH on that secondary cell. The resource allocation could be for that secondary cell, or for another secondary cell if cross carrier scheduling is used
  - o the timer is also reset if cross carrier scheduling is used and the UE receives an uplink or downlink resource allocation for that secondary cell from the PDCCH belonging to a different serving cell
  - o the secondary cell is deactivated if the timer expires. An activation command from the MAC control element is then required before being able to use the secondary cell again
- ★ When a UE receives an activation command during subframe 'n', the actions associated with activation are applied during subframe 'n+8'. When a UE receives a deactivation command during subframe 'n', or the inactivity timer expires during subframe 'n', the actions associated with deactivation are applied no later than during subframe 'n+8', with the exception that actions associated with the reporting of channel state information are applied during subframe 'n+8'
- ★ The UE actions associated with an active secondary cell are:
  - o PDCCH monitoring 'on' the secondary cell (which could be used for resource allocations on different secondary cells if cross carrier scheduling is used)

- PDCCH monitoring 'for' the secondary cell (which could involve monitoring the PDCCH on a different serving cell if cross carrier scheduling is used)
- transmission of the uplink Sounding Reference Signal (SRS), and reporting of channel state information, including Channel Quality Indicator (CQI), Precoding Matrix Indicator (PMI), Rank Indicator (RI) and Precoding Type Indicator (PTI)

★ 3GPP References: TS 36.213, TS 36.321

18.9 DATA FLOW

★ Carrier Aggregation does not have any impact upon the functionality of the PDCP and RLC layers. These layers just have to be capable of handling the higher throughputs associated with Carrier Aggregation. The number of data flows through these layers depends upon the number of radio bearers

★ The main impact of Carrier Aggregation functionality is within the MAC layer. The MAC layer at the transmitting side is responsible for distributing the data across the set of Component Carriers according to resources allocated by the scheduler. The MAC layer at the receiving side is responsible for aggregating the data received from the set of Component Carriers

★ Each Component Carrier has its own Hybrid Automatic Repeat reQuest (HARQ) entity within the MAC layer. Each HARQ entity supports its own set of parallel HARQ processes based upon the Stop And Wait (SAW) protocol, i.e. each HARQ entity operates in the normal manner

★ A maximum of 2 transport blocks can be scheduled per Component Carrier during each subframe. This results in a maximum of 10 transport blocks when 5 Component Carriers are configured

★ In terms of transport channels, there is 1 DL-SCH and 1 UL-SCH for the primary cell. There is also 1 DL-SCH and either 1 or 0 UL-SCH for each secondary cell

★ Figure 190 illustrates an example flow of downlink data within an eNode B for 2 UE with 2 Component Carriers. The first UE is assumed to have a single radio bearer, while the second UE is assumed to have a pair of radio bearers. The scheduler allocates resources for both UE on both of their Component Carriers

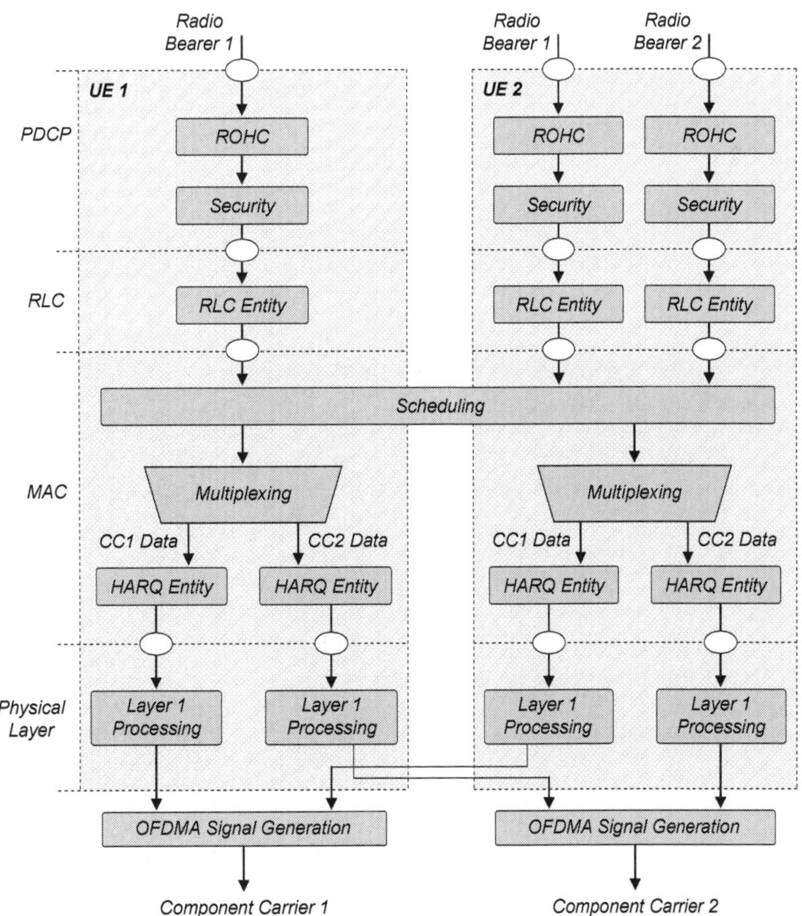

Figure 190 – Example flow of downlink data within an eNode B for 2 UE using 2 Component Carriers

★ Physical layer processing is completed independently for each flow of data towards each Component Carrier. This processing includes the addition of CRC bits, channel coding and rate matching

★ OFDMA signals are then generated for each Component Carrier using the appropriate data being sent to each UE. Both UE will receive data from both Component Carriers assuming the scheduler has shared the total set of Resource Blocks on each Component Carrier between the two UE

★ Figure 191 illustrates a set of example transmitter implementations for both intra-band and inter-band Carrier Aggregation

★ If the two Component Carriers are contiguous then a single Inverse Fast Fourier Transform (IFFT) function can be used to generate both OFDMA signals, a single mixer stage can be used to translate the baseband signal to RF, and a single power amplifier can be used to increase the strength of the transmitted signal. This represents the simplest scenario to implement

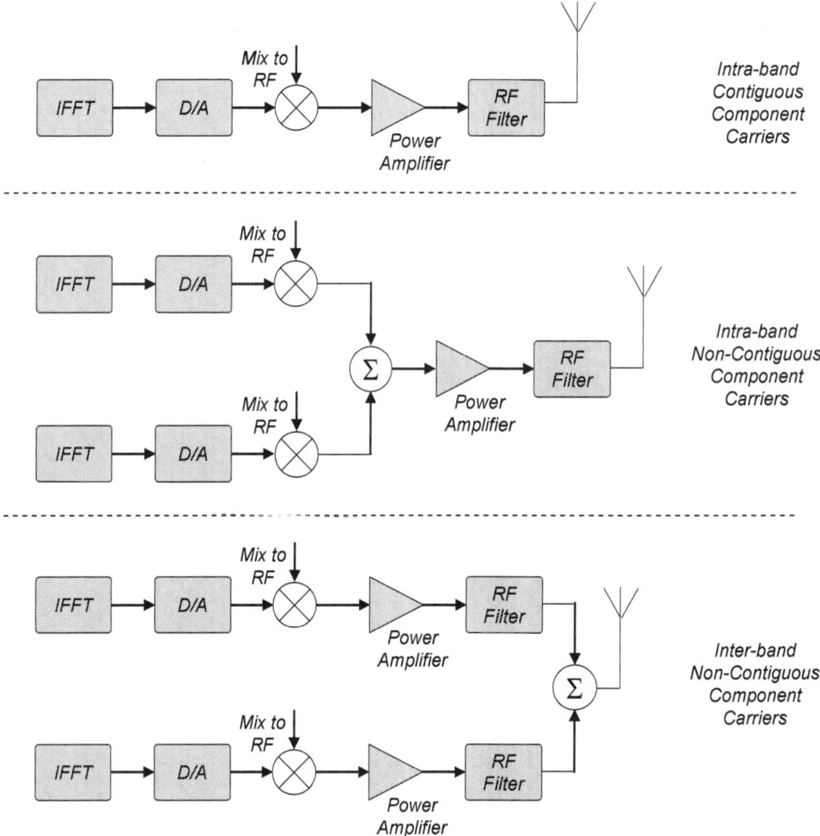

Figure 191 – Example transmitter implementations for intra-band and inter-band Carrier Aggregation

★ If the two Component Carriers are non-contiguous but within the same operating band then separate IFFT functions are used to generate the OFDMA signals, separate mixing would also be required to translate the baseband signals to RF, but a single power amplifier could be used to increase the strength of the transmitted signal. The efficiency of the power amplifier is likely to depend upon its transmit bandwidth which would have a minimum requirement determined by the spacing of the Component Carriers

★ If the two Component Carriers are non-contiguous and in different operating bands then separate IFFT functions are used to generate the OFDMA signals, separate mixing would also be required to translate the baseband signals to RF, and separate power amplifiers would be required to increase the strength of the transmitted signal (unless a multi-band power amplifier is available). This represents the most expensive scenario to implement due to the requirement for multiple power amplifiers

★ Data flow in the uplink direction follows a similar pattern, i.e. the PDCP and RLC layers have no additional functionality while the MAC layer is responsible for distributing the uplink data between Component Carriers. There is also a maximum of 2 transport blocks per Component Carrier and a single HARQ entity per Component Carrier. The transmitter designs shown in Figure 191 are also applicable to the uplink although their implementation is more challenging due to the requirement to keep devices low cost, physically small and with a low power consumption

★ 3GPP References: TS 36.321

18.10 PHICH

- The PHICH is used to acknowledge uplink data transfer on the PUSCH. Acknowledgements are sent using the PHICH belonging to the serving cell which allocated the PUSCH resources
- The PUSCH transmissions and PHICH acknowledgements are sent on different serving cells when cross carrier scheduling is used. Cross carrier scheduling is not applied to the primary serving cell so PUSCH transmissions on the primary serving cell are always acknowledged on the primary serving cell
- A PHICH acknowledgement is identified by its:
  - PHICH Group
  - PHICH Orthogonal Sequence Index within the PHICH Group
- As presented in section 6.3, the PHICH group and Orthogonal Sequence Index are a function of:
  - the lowest Physical Resource Block (PRB) index allocated to the first slot of the PUSCH transmission
  - the transport block number when uplink MIMO is used to transfer 2 transport blocks within a subframe
  - the demodulation Reference Signal cyclic shift signalled within DCI format 0 or DCI format 4
  - the number of PHICH groups
  - the uplink-downlink subframe configuration when using TDD
  - spreading factor size used for PHICH modulation
- Cross carrier scheduling increases the potential for PHICH collisions because scheduling multiple serving cells from a single serving cell increases the number of PHICH acknowledgments from that serving cell. Collisions can be avoided by using different demodulation Reference Signal cyclic shifts, or by scheduling different lowest Physical Resource Block indices
- 3GPP References: TS 36.213

18.11 UPLINK CONTROL SIGNALLING

- Uplink control signalling can be transferred using either the PUCCH belonging to the primary serving cell, or the PUSCH belonging to either the primary or a secondary serving cell. The PUCCH is not transmitted from secondary serving cells
- Uplink control signalling includes HARQ acknowledgments, Channel State Information (CSI) and Scheduling Requests (SR)
- PUCCH format 3 was introduced to support HARQ acknowledgments and Scheduling Requests for Carrier Aggregation:
  - up to 10 HARQ acknowledgements for FDD (with or without a Scheduling Request)
  - up to 20 HARQ acknowledgements for TDD (with or without a Scheduling Request)
- In the case of FDD, 10 HARQ acknowledgements are sufficient to support 5 Component Carriers, each transferring 2 transport blocks during each subframe
- In the case of TDD, the requirement for HARQ acknowledgments is increased when there are multiple downlink subframes associated with each uplink subframe. 20 HARQ acknowledgments support 5 Component Carriers, each transferring 2 transport blocks, and 2 downlink subframes associated with the uplink subframe. HARQ-ACK bundling and multiplexing (see section 24.3.4) can be used when more than 20 HARQ acknowledgments are generated
- PUCCH format 1b has also been adapted to support Carrier Aggregation. PUCCH format 1b with channel selection is able to transfer 4 HARQ Acknowledgments. This is sufficient to support 2 Component Carriers, each transferring 2 transport blocks. HARQ-ACK bundling and multiplexing can be used by TDD when more than a single downlink subframe is associated with each uplink subframe
- CSI reports are only transferred for activated serving cells, i.e. CSI reports are not transferred for serving cells which have been configured by the RRC layer but which are currently deactivated (MAC control element required to activate serving cells)
- Periodic CSI reporting for Carrier Aggregation can be transferred using the existing 3GPP release 8 PUCCH formats. The periodic reporting patterns for each Component Carrier can be interleaved so the UE does not have to transfer CSI information for more than a single Component Carrier during a single subframe. 3GPP has specified a set of prioritisation rules to handle scenarios where periodic reporting patterns for different Component Carriers coincide
- When periodic CSI reporting coincides with the transfer of uplink data on the PUSCH, the periodic CSI is transferred using the PUCCH while the uplink data is transferred using the PUSCH (assuming simultaneous PUSCH and PUCCH transmission is supported and configured). When simultaneous PUSCH and PUCCH transmission is not configured, the UE sends the periodic CSI report on the allocated PUSCH with the lowest serving cell index

- ★ When periodic CSI reports coincide with the transfer of HARQ acknowledgments, the periodic CSI is transferred using the PUSCH while the HARQ acknowledgments are transferred using the PUCCH (assuming simultaneous PUSCH and PUCCH transmission is supported and configured)
- ★ Aperiodic CSI reporting is always transferred using the PUSCH and can be triggered for one or more serving cells
- ★ 3GPP release 10 introduces the ability to trigger aperiodic CSI reports using 2 bits within DCI formats 0 and 4. Interpretation of these 2 bits is presented in Table 179 (section 17.3). A value of '01' indicates that an aperiodic CSI report is triggered for the serving cell. Values of '10' and '11' indicate that aperiodic CSI reports are triggered for serving cell sets 1 and 2 respectively. Serving cell sets 1 and 2 are configured by the RRC layer using bitmaps to indicate which serving cells are to be included
- ★ 3GPP References: TS 36.213

19 UE CAPABILITIES

19.1 UE CATEGORIES

★ The capabilities associated with each UE category are specified by 3GPP TS 36.306

★ A single UE category defines both the uplink and downlink capabilities. This differs from UMTS which uses separate UE categories for HSDPA and HSUPA

★ Table 207 presents the most important capabilities associated with each UE category. Categories 1 to 5 were introduced within the release 8 version of the 3GPP specifications. Categories 6 to 8 were introduced within the release 10 version of the specifications

		Downlink			Uplink		
UE Category	3GPP Release	Max. Total Bits per TTI	Max. Bits per Transport Blk	Max. Layers for Spatial Mux	64QAM Supported	Max. Total Bits per TTI	Max. Bits per Transport Blk
1	8	10 296	10 296	1	No	5 160	5 160
2	8	51 024	51 024	2	No	25 456	25 456
3	8	102 048	75 376	2	No	51 024	51 024
4	8	150 752	75 376	2	No	51 024	51 024
5	8	299 552	149 776	4	Yes	75 376	75 376
6	10	301 504	149 776 (4 layers) 75 376 (2 layers)	2 or 4	No	51 024	51 024
7	10	301 504	149 776 (4 layers) 75 376 (2 layers)	2 or 4	No	102 048	51 024
8	10	2 998 560	299 856	8	Yes	1 497 760	149 776

Table 207 – Capabilities associated with each UE category

★ A 3GPP release 8 or 9 UE can be implemented to support a UE category between 1 and 5, whereas a 3GPP release 10 UE can be implemented to support a UE category between 1 and 8, i.e. 3GPP release 10 UE can be based upon the UE categories introduced within the release 8 version of the specifications

★ The maximum total bits per Transmission Time Interval (TTI) in the downlink defines the maximum downlink throughput. A single TTI corresponds to the 1 ms subframe duration

- o the maximum downlink throughput specified for the release 8 and 9 versions of the 3GPP specifications is 300 Mbps. This is supported when transferring 2 transport blocks per subframe on a single RF carrier
- o the maximum downlink throughput specified for the release 10 version of the 3GPP specifications is 3 Gbps. This is supported when transferring 10 transport blocks per subframe across 5 Component Carriers

★ Similarly, the maximum total bits per TTI in the uplink defines the maximum uplink throughput:

- o the maximum uplink throughput specified for the release 8 and 9 versions of the 3GPP specifications is 75 Mbps. This is supported when transferring 1 transport block per subframe on a single RF carrier
- o the maximum uplink throughput specified for the release 10 version of the 3GPP specifications is 1.5 Gbps. This is supported when transferring 10 transport blocks per subframe across 5 Component Carriers

★ The release 8 and 9 versions of the 3GPP specifications allow only a single transport block to be transferred in the uplink direction during each TTI so the maximum total bits per TTI equals the maximum bits per transport block

★ All UE categories support:

- o the 20 MHz channel bandwidth
- o 64QAM in the downlink
- o downlink transmit diversity with up to 4 transmit antenna elements

★ UE performance requirements have been specified based upon the assumption that all UE support downlink receive diversity

★ UE categories 2, 3 and 4 support 2×2 MIMO in the downlink whereas UE category 5 can also support 4×4 MIMO in the downlink (defined by the number of layers for spatial multiplexing). UE category 8 is required to support 8×8 MIMO in the downlink

★ A UE signals its category (or categories) within the UE Capability Information message

★ UE which are category 6 or 7, also signal support for category 4. This allows the network to treat these UE as category 4 devices when the network does not support categories 6 or 7. Likewise, UE which are category 8, also signal support for category 5

★ 3GPP References: TS 36.306

19.2 OTHER CAPABILITIES

★ UE signal their support for optional capabilities within the UE Capability Information message. This message can include capability information relevant to each supported radio access technology. Information relevant to LTE capabilities is signalled using the 'UE EUTRA Capability' information element

★ Table 208 presents the general UE capabilities included within the 'UE EUTRA Capability' information element

UE Capabilities	3GPP Release
Access Stratum Release	8
Device Type	9

Table 208 – General capabilities

★ The Access Stratum Release specifies the version of the 3GPP specifications upon which the UE implementation is based, i.e. 3GPP release 8, 9 or 10

★ The Device Type indicates whether or not the UE benefits from battery power consumption optimisation. Handheld devices are likely to rely upon batteries as a source of power, whereas data dongles are more likely to obtain their power from a laptop which could be connected to the mains electricity supply. Differentiating between these types of devices provides the eNode B with the potential to apply different radio resource management to devices which would benefit from reduced power consumption, i.e. battery life could be prolonged. For example, UE which benefit from reduced power consumption could be allocated reduced inactivity timers so the DRX mode of operation is entered sooner and more frequently

★ Table 209 presents the Packet Data Convergence Protocol (PDCP) UE capabilities included within the 'UE EUTRA Capability' information element

UE Capabilities	3GPP Release
Supported ROHC Profiles	8
Maximum Number of ROHC Context Sessions	8

Table 209 – Optional PDCP capabilities

★ Robust Header Compression (ROHC) can be used to reduce protocol stack overheads when transmitting user plane data across the air-interface. This is particularly important for services where the payload is relatively small compared to the protocol stack headers. Voice over IP (VoIP) is an example service which requires the use ROHC to ensure that protocol stack overheads are manageable. Header compression for VoIP is described in section 27.3.2

★ The UE can signal support for the following Robust Header Compression (ROHC) profiles:

- 0x0001: for RTP/UDP/IP header compression
- 0x0002: for UDP/IP header compression
- 0x0003: for ESP/IP header compression
- 0x0004: for IP header compression
- 0x0006: for TCP/IP header compression
- 0x0101: for RTP/UDP/IP updated header compression
- 0x0102: for UDP/IP updated header compression
- 0x0103: for ESP/IP updated header compression
- 0x0104: for IP updated header compression

The VoIP service uses the RTP/UDP/IP protocol stack for the speech data and the UDP/IP protocol stack for the associated RTCP signalling

★ The UE also signals its capability in terms of the maximum number of supported ROHC context sessions, i.e. the number of parallel streams of data being compressed/decompressed. The UE can specify values between 2 and 16384

★ Table 210 presents the physical layer capabilities included within the 'UE EUTRA Capability' information element

★ The UE sends a flag when the following capabilities are supported:

- UE Transmit Antenna Selection, i.e. the UE has the ability to switch between multiple antenna for its uplink transmissions. This can be useful if the end-user's hand is shielding one antenna (see section 12.3.1)
- UE specific Reference Signals refers to support for downlink transmission mode 7, i.e. single layer beamforming. Transmission mode 7 uses antenna port 5 to transmit a UE specific Reference Signal. Beamforming can be applied to the UE specific Reference Signal and to the set of allocated Resource Blocks. This directs them towards the appropriate UE (see section 4.4.4)
- Enhanced Dual Layer (FDD) refers to support for downlink transmission mode 8. Transmission mode 8 allows 2 UE to simultaneously benefit from dual layer beamforming, i.e. a total of 4 layers can be transferred. Alternatively, those 4 layers can be used for single layer beamforming towards 4 UE. Dual layer beamforming allows throughputs to be doubled relative to single layer beamforming (see section 4.4.4)
- Enhanced Dual Layer (TDD) refers to the equivalent support for TDD. 3GPP TS 36.306 specifies that release 9 and 10 UE which support TDD, should also support downlink transmission mode 8

- 2 Antenna Ports for PUCCH Transmission refers to PUCCH transmit diversity (see section 12.4.2)
- Transmission Mode 9 with 8 CSI Reference Signal Antenna Ports (FDD) refers to support for single user beamforming with up to 8×8 MIMO (see section 4.5.1)
- PMI Disabling indicates whether or not PMI reporting can be disabled for a UE using transmission mode 9. If PMI reporting is disabled, CQI reports are derived from the cell specific Reference Signals. Otherwise, CQI reports are derived from the CSI Reference Signals
- Cross Carrier Scheduling is applicable to Carrier Aggregation. This capability allows the PDCCH on serving cell 'x' to allocate resources on serving cell 'y' (see section 18.7). When Carrier Aggregation is supported in the uplink direction, this capability also allows the UE to use the primary serving cell as a path loss reference for a secondary serving cell. This is applicable to uplink power control calculations
- Simultaneous PUCCH and PUSCH Transmission refers to capability of the baseband processing within the UE. The eNode B interprets this capability in combination with the RF capability of the UE. The baseband capability is independent of the operating band
  - if a UE indicates baseband support, and if the UE indicates that it supports non-contiguous uplink resource allocation at the RF level for a specific RF carrier within a specific operating band, then the UE supports simultaneous PUCCH and PUSCH transmission for all RF carriers within that operating band
  - if a UE indicates baseband support, and if the UE indicates that it supports uplink Carrier Aggregation, then the UE supports simultaneous PUCCH and PUSCH transmission for any uplink Component Carrier which the UE can aggregate
- Multi-Cluster PUSCH within a Component Carrier refers to the capability of the baseband processing within the UE. Multi-cluster PUSCH transmission corresponds to non-contiguous Resource Block allocations provided by the PDCCH Downlink Control Information (DCI), i.e. PUSCH resource allocation type 1 (see section 9.5). The baseband capability is independent of the operating band
  - if a UE indicates baseband support, and if the UE indicates that it supports non-contiguous uplink resource allocation at the RF level for a specific RF carrier within a specific operating band, then the UE supports multi-cluster PUSCH transmission for all RF carriers within that operating band
  - if a UE indicates that it supports uplink Carrier Aggregation, then the UE supports multi-cluster PUSCH transmission for any uplink Component Carrier which the UE can aggregate (irrespective of whether or not baseband support is indicated)

★ Non-Contiguous Uplink Resource Allocation within a Component Carrier refers the RF capability of the UE. This information is interpreted in combination with the baseband capability of the UE. The RF capability for non-contiguous uplink resource allocation is specified for each supported operating band. The operating bands themselves are not listed explicitly as part of this capability information, but the list of indications are ordered using the same operating band sequence as used for the 'Supported E-UTRA Band List' within the RF capabilities information

★ Release 10 UE can signal their uplink MIMO capability using values of '2 layers' or '4 layers' corresponding to support for 2×2 MIMO or 4×4 MIMO. This capability information is signalled for each Carrier Aggregation bandwidth class that the UE specifies for each operating band within the set of supported band combinations (see Table 201 within section 18.3)

★ Release 10 UE can signal their downlink MIMO capability using values of '2 layers', '4 layers' or '8 layers' corresponding to support for 2×2 MIMO, 4×4MIMO or 8×8 MIMO. This capability information is signalled for each Carrier Aggregation bandwidth class that the UE specifies for each operating band within the set of supported band combinations (see Table 201 within section 18.3)

UE Capabilities	3GPP Release
UE Transmit Antenna Selection	8
UE Specific Reference Signals Supported	8
Enhanced Dual Layer (FDD)	9
Enhanced Dual Layer (TDD)	9
2 Antenna Ports for PUCCH Transmission	10
Transmission Mode 9 with 8 CSI Reference Signal Antenna Ports (FDD)	10
PMI Disabling	10
Cross Carrier Scheduling	10
Simultaneous PUCCH and PUSCH Transmission	10
Multi-Cluster PUSCH within a Component Carrier	10
Non-Contiguous Uplink Resource Allocation within a Component Carrier	10
Uplink MIMO Capability	10
Downlink MIMO Capability	10

Table 210 – Optional Physical Layer capabilities

★ Table 211 presents the RF capabilities included within the 'UE EUTRA Capability' information element

UE Capabilities	3GPP Release
Supported E-UTRA Band List	8
Supported Band Combinations	10

Table 211 – Optional RF capabilities

★ The Supported E-UTRA Band List specifies a list of the supported FDD and TDD operating bands using the numbering scheme presented in section 2.6. A flag can be included to indicate that only half duplex transmission/reception is supported within a specific operating band. Otherwise, support for full duplex transmission/reception is assumed

★ The structure of the Supported Band Combinations information is presented in Table 201 (section 18.3). The message structure allows a UE to signal up to 128 different band combinations, and include up to 64 bands within each combination. Each band included within a combination is identified by its band number, ranging from 1 to 64. Each band can have up to 16 sets of uplink parameters and up to 16 sets of downlink parameters

- each set of uplink parameters associated with a specific band defines the Carrier Aggregation bandwidth class (bandwidth classes of A, B, C, D, E or F). The uplink MIMO capability is also defined as either 2-layers or 4-layers
- each set of downlink parameters associated with a specific band defines the Carrier Aggregation bandwidth class (bandwidth classes of A, B, C, D, E or F). The downlink MIMO capability is also defined as either 2-layers, 4-layers or 8-layers

★ Table 212 presents the measurement capabilities included within the 'UE EUTRA Capability' information element

UE Capabilities	3GPP Release
Inter-Frequency Need for Gaps	8
Inter-RAT Need for Gaps	8

Table 212 – Optional Measurement capabilities

★ These parameters indicate whether or not the UE requires measurement gaps to be configured when completing inter-frequency or inter-RAT measurements. Measurement gaps may not be necessary if the UE has multiple receivers and is able to measure one RF carrier while receiving on another RF carrier

★ The Inter-Frequency Need for Gaps is specified for each supported operating band. The operating bands themselves are not listed explicitly as part of this capability information, but the list of indications are ordered using the same operating band sequence as used for the 'Supported E-UTRA Band List' within the RF capabilities information

★ The Inter-RAT Need for Gaps is specified for each supported operating band and for each supported radio access technology. The operating bands themselves are not listed explicitly as part of this capability information, but the list of indications are ordered using the same operating band sequence as provided within the inter-RAT capabilities information

★ Release 10 UE also provide the inter-frequency and inter-RAT 'need for gaps' information for each supported band combination

★ Table 213 presents the inter-RAT capabilities included within the 'UE EUTRA Capability' information element

★ The UE can signal its support for the following technologies:

- UTRA FDD
- UTRA TDD 1.28 Mcps
- UTRA TDD 3.84 Mcps
- UTRA TDD 7.68 Mcps
- GERAN
- CDMA2000 HRPD
- CDMA2000 1xRTT

★ UE do not use dedicated flags to signal their support for these technologies. Instead, UE include a list of supported operating bands for each supported technology. This allows the eNode B to deduce which technologies are supported

★ UE which support FDD or TDD UTRA also support packet switched inter-system handover from LTE to UTRAN

★ UE which support GERAN use a separate flag to indicate whether or not packet switched inter-system handover from LTE to GERAN is supported

★ UE which support CDMA2000 HRPD or 1xRTT include additional flags to indicate support for either a single transmitter or dual transmitter. UE equipped with a dual transmitter are able to simultaneously transmit on LTE and CDMA2000. Similarly, UE which support CDMA2000 HRPD or 1xRTT include an additional flag to indicate support for either a single receiver or dual receiver. UE equipped with a dual receiver are able to simultaneously receive on LTE and CDMA2000

★ In the case of CDMA2000 1xRTT, the UE can signal its support for the enhanced CS Fallback procedure. This procedure makes CS fallback relatively seamless and avoids the requirement for an RRC connection release with re-direction. The enhanced CS fallback procedure tunnels handover signalling between the UE and the 1xRTT network. This allows the transition between radio access technologies to be handled as a handover rather than a release with re-direction

UE Capabilities	3GPP Release
Support of UTRA FDD	8
Supported Band List (UTRA FDD)	8
Support of UTRA TDD 1.28 Mcps	8
Supported Band List (UTRA TDD 1.28 Mcps)	8
Support of UTRA TDD 3.84 Mcps	8
Supported Band List (UTRA TDD 3.84 Mcps)	8
Support of UTRA TDD 7.68 Mcps	8
Supported Band List (UTRA TDD 7.68 Mcps)	8
Support of GERAN	8
Supported Band List (GERAN)	8
Inter-RAT PS Handover to GERAN	8
Support of CDMA2000 HRPD	8
Supported Band List (CDMA2000 HRPD)	8
Transmit Configuration HRPD	8
Receive Configuration HRPD	8
Support of CDMA2000 1xRTT	8
Supported Band List (CDMA2000 1xRTT)	8
Transmit Configuration 1xRTT	8
Receive Configuration 1xRTT	8
Enhanced CS Fallback to CDMA2000 1xRTT	9
Enhanced CS Fallback to CDMA2000 1xRTT with concurrent PS Mobility to CDMA2000 HRPD	9
Enhanced Redirection to UTRA	9
Enhanced Redirection to GERAN	9
Dual Transfer Mode (DTM)	9
Enhanced CS Fallback to CDMA2000 1xRTT with Dual Transmitter/Receiver	10
Enhanced Redirection to UTRA TDD	10

Table 213 – Optional inter-RAT capabilities

★ If the UE supports CDMA2000 1xRTT and HRPD simultaneously, then it can signal its support for enhanced CS fallback to CDMA2000 1xRTT in combination with a concurrent packet switched handover to CDMA2000 HRPD

★ Enhanced Redirection to UTRA refers to the ability of the UE to receive FDD or TDD UTRA system information within the RRC Connection Release message sent from the LTE network. This avoids the UE having to acquire the system information after synchronising with the UTRA network, i.e. the UE can immediately establish its connection after being re-directed for CS fallback

★ The concept of enhanced re-direction can also be applied to GERAN, i.e. the UE is provided with GERAN system information within the RRC Connection Release message sent from the LTE network. 3GPP TS 36.306 specifies that UE which support CS fallback to GERAN should also support enhanced re-direction to GERAN

★ The Dual Transfer Mode (DTM) parameter specifies whether or not the UE supports DTM for GERAN. DTM allows the simultaneous transfer of CS and PS data on the same GERAN RF carrier

★ The Enhanced CS Fallback to CDMA2000 1xRTT with Dual Transmitter/Receiver is only applicable if the UE supports dual transmitter and dual receiver. This parameter indicates that the UE can complete the enhanced CS fallback procedure using its dual transmitter/receiver

★ The Enhanced Redirection to UTRA TDD introduced within the release 10 version of the 3GPP specifications allows the LTE network to include a list of UTRA TDD RF carriers within the RRC Connection Release message for re-direction purposes. In addition, the system information belonging to each of those UTRA TDD RF carriers can be included

★ Table 214 presents the Closed Subscriber Group (CSG) proximity capabilities included within the 'UE EUTRA Capability' information element

UE Capabilities	3GPP Release
Intra-Frequency Proximity Indication	9
Inter-Frequency Proximity Indication	9
UTRAN Proximity Indication	9

Table 214 – Optional CSG Proximity capabilities

★ Proximity detection allows the UE to inform the eNode B when it enters or leaves the coverage of a CSG cell included within its whitelist. The UE uses a Proximity Indication message to specify the RF carrier belonging to the CSG cell and also to specify whether the UE is entering or leaving the coverage area

★ UE are able to signal their support for intra-frequency, inter-frequency and UTRAN proximity detection. Support relies upon the UE having the ability to complete autonomous searches for CSG cells in its whitelist

★ Table 215 presents the neighbour system information acquisition capabilities included within the 'UE EUTRA Capability' information element

UE Capabilities	3GPP Release
Intra-Frequency System Information Acquisition for Handover	9
Inter-Frequency System Information Acquisition for Handover	9
UTRAN System Information Acquisition for Handover	9

Table 215 – Optional Neighbour System Information acquisition capabilities

★ Neighbour system information acquisition involves the UE generating autonomous gaps to decode the system information from a neighbouring cell. If supported, the eNode B can request the UE to decode and report the Cell Global Identity (CGI) belonging to neighbouring cells. This information can be used when adding neighbour relations (see section 32.4)

★ UE are able to signal their support for intra-frequency, inter-frequency and UTRAN system information acquisition

★ Table 216 presents the Self Organising Network (SON) capabilities included within the 'UE EUTRA Capability' information element

UE Capabilities	3GPP Release
RACH Report	9

Table 216 – Optional SON capabilities

★ RACH reports can be included within the UE Information Response message. They allow UE to report the number of PRACH preambles transmitted for the most recent successfully completed random access procedure. They also allow the UE to report whether or not contention resolution was successful

★ UE use a single flag to indicate their support for RACH reporting

★ Table 217 presents the UE based network performance measurement capabilities included within the 'UE EUTRA Capability' information element

UE Capabilities	3GPP Release
Logged Measurements Idle	10
Standalone GNSS Location	10

Table 217 – Optional UE based network performance measurement capabilities

★ The Logged Measurements Idle capability can be used by the network for the Minimisation of Drive Tests (MDT) (see section 32.15). MDT functionality helps to reduce the requirement for drive testing by collecting measurements from the population of subscribers. Measurements for 'Logged MDT' are completed while the UE is in RRC Idle mode

★ UE use a single flag to indicate their support for recording measurements in RRC Idle mode. 3GPP TS 36.306 specifies that UE which support these measurements must have a minimum of 64 kBytes of memory to record the measurements

★ UE also use a single flag to indicate whether or not they are equipped with a Global Navigation Satellite System (GNSS) receiver that can be used to provide detailed location information when reporting measurements for MDT

★ 3GPP References: TS 36.306, TS 36.331

19.3 FEATURE GROUP INDICATORS

★ Feature Group Indicators (FGI) are used for mandatory features which may not have been implemented and tested in early devices

★ FGI are included within the 'UE EUTRA Capability' information element of the UE Capability Information message

★ FGI are signalled using a bitmap where each bit corresponds to specific area of functionality. The inclusion of FGI within the UE Capability Information message is optional. Excluding the FGI bitmap indicates that the UE supports all of the functionality associated with the set of supported radio access technologies

★ 3GPP TS 36.331 specifies a first FGI bitmap for the release 8 version of the specifications and a second FGI bitmap for the release 10 version of the specifications. An overview of the 3GPP release 8 FGI bitmap is provided in Table 218

Index	Functionality
1	Intra-subframe frequency hopping for PUSCH scheduled by uplink grant DCI format 3a PDSCH transmission mode 5 Aperiodic CSI reporting on the PUSCH: Mode 2-0 UE selected sub-band CQI without PMI Aperiodic CSI reporting on the PUSCH: Mode 2-2 UE selected sub-band CQI with multiple PMI
2	Simultaneous CQI and ACK/NACK on PUCCH (PUCCH formats 2a and 2b) Absolute TPC command for PUSCH Resource Allocation Type 1 for PDSCH Periodic CSI reporting on the PUCCH: Mode 2-0 UE selected sub-band CQI without PMI Periodic CSI reporting on the PUCCH: Mode 2-1 UE selected sub-band CQI with single PMI
3	5 bit RLC Unacknowledged Mode Sequence Number 7 bit PDCP Sequence Number
4	Short DRX cycle
5	Long DRX cycle DRX Command MAC control element
6	Prioritised Bit Rate
7	RLC Unacknowledged Mode
8	EUTRA RRC Connected to UTRA CELL_DCH PS Inter-System Handover
9	EUTRA RRC Connected to GERAN GSM Dedicated Inter-System Handover
10	EUTRA RRC Connected to GERAN Idle Cell Change Order EUTRA RRC Connected to GERAN Idle Cell Change Order with Network Assisted Cell Change (NACC)
11	EUTRA RRC Connected to CDMA2000 1xRTT CS Active Inter-System Handover
12	EUTRA RRC Connected to CDMA2000 HRPD Active Inter-System Handover
13	Inter-Frequency Handover within FDD or TDD
14	Measurement Reporting Event A4 Measurement Reporting Event A5
15	Measurement Reporting Event B1
16	Intra-frequency periodic measurement reporting when trigger type is periodic and purpose is report strongest cells Inter-frequency periodic measurement reporting when trigger type is periodic and purpose is report strongest cells Inter-RAT periodic measurement reporting when trigger type is periodic and purpose is report strongest cells
17	Intra-frequency ANR features
18	Inter-frequency ANR features
19	Inter-RAT ANR features
20	If bit 7 = 0 then SRB1 and SRB2 for DCCH + 8 Acknowledged Mode RLC Data Radio Bearers If bit 7 = 1 then SRB1 and SRB2 for DCCH + 8 Acknowledged Mode RLC Data Radio Bearers If bit 7 = 1 then SRB1 and SRB2 for DCCH + 5 Acknowledged Mode RLC + 3 Unacknowledged Mode RLC Data Radio Bearers
21	Predefined intra and inter subframe frequency hopping for PUSCH with N_sb > 1 Predefined inter subframe frequency hopping for PUSCH with N_sb > 1
22	UTRAN measurements, reporting and measurement reporting event B2 in EUTRA RRC Connected
23	GERAN measurements, reporting and measurement reporting event B2 in EUTRA RRC Connected
24	1xRTT measurements, reporting and measurement reporting event B2 in EUTRA RRC Connected
25	Inter-frequency measurements and reporting in EUTRA RRC Connected
26	HRPD measurements, reporting and measurement reporting event B2 in EUTRA RRC Connected
27	EUTRA RRC Connected to UTRA CELL_DCH CS Inter-System Handover
28	TTI Bundling
29	Semi-Persistent Scheduling
30	Handover between FDD and TDD

Table 218 – 3GPP release 8 Feature Group Indicators (FGI)

★ The release 8 FGI bitmap has a length of 32 bits but only the first 30 bits have been specified for use. Bits 31 and 32 should always be signalled with a value of 0. Other bits are signalled with a value of 1 when the associated feature has been implemented and tested

★ An overview of the 3GPP release 10 FGI bitmap is provided in Table 219

Index	Functionality
101	Demodulation Reference Signal with Orthogonal Cover Code and Sequence Group Hopping disabling
102	Trigger type 1 SRS (aperiodic SRS)
103	PDSCH transmission mode 9 when up to 4 CSI Reference Signal ports are configured
104	PDSCH transmission mode 9 for TDD when 8 CSI Reference Signal ports are configured
105	Periodic CSI reporting on PUCCH : Mode 2-0 UE selected sub-band CQI without PMI, with transmission mode 9 Periodic CSI reporting on PUCCH : Mode 2-1 UE selected sub-band CQI with single PMI, with transmission mode 9 and up to 4 CSI Reference Signal ports
106	Periodic CSI reporting on PUCCH : Mode 2-1 UE selected sub-band CQI with single PMI, with transmission mode 9 and 8 CSI Reference Signal ports
107	Aperiodic CSI reporting on PUSCH : Mode 2-0 UE selected sub-band CQI without PMI, with transmission mode 9 Aperiodic CSI reporting on PUSCH : Mode 2-2 UE selected sub-band CQI with multiple PMI, with transmission mode 9 and up to 4 CSI Reference Signal ports
108	Aperiodic CSI reporting on PUSCH : Mode 2-2 UE selected sub-band CQI with multiple PMI, with transmission mode 9 and 8 CSI Reference Signal ports
109	Periodic CSI reporting on PUCCH Mode 1-1, submode 1
110	Periodic CSI reporting on PUCCH Mode 1-1, submode 2
111	Measurement Reporting trigger Event A6
112	Secondary Cell addition within the Handover to EUTRA procedure
113	Trigger type 0 SRS (periodic SRS) transmission on X serving cells
114	Reporting of both UTRA CPICH RSCP and Ec/Io in a Measurement Report
115	Time domain ICIC RLM/RRM measurement subframe restriction for the serving cell Time domain ICIC RRM measurement subframe restriction for neighbouring cells Time domain ICIC CSI measurement subframe restriction
116	Relative transmit phase continuity for spatial multiplexing in the uplink

Table 219 – 3GPP release 10 Feature Group Indicators (FGI)

★ The release 10 FGI bitmap also has a length of 32 bits but only the first 16 bits have been specified for use. Bits 17 to 32 should always be signalled with a value of 0. Other bits are signalled with a value of 1 when the associated feature has been implemented and tested. Bits belonging to the release 10 FGI bitmap have been allocated indices which are numbered from 101 to 132

★ 3GPP References: TS 36.331

20 BIT RATES

20.1 DOWNLINK BIT RATES

20.1.1 FDD

★ The Physical Downlink Shared Channel (PDSCH) is used to transfer application data. The throughput achieved by the PDSCH depends upon:

- the number of Resource Elements allocated to the PDSCH
- the modulation scheme applied to each Resource Element
- the quantity of redundancy included by physical layer processing
- the use of multiple antenna transmission schemes
- the use of Carrier Aggregation

★ The number of PDSCH Resource Elements depends upon the channel bandwidth and the choice between the normal and extended cyclic prefix. It also depends upon the overheads generated by the other physical channels and physical signals

★ The modulation scheme and quantity of redundancy depend upon the RF channel conditions. UE experiencing good channel conditions are more likely to be allocated higher order modulation schemes with less redundancy

★ Multiple antenna transmission schemes increase the throughput achieved by the PDSCH. 2×2 MIMO using antenna ports 0 and 1 approximately doubles the peak throughput whereas 4×4 MIMO using antenna ports 0, 1, 2 and 3 approximately quadruples the peak throughput. Cell specific Reference Signal overheads increase when using MIMO with antenna ports 0, 1, 2 and 3 so the throughputs are less than double and quadruple the single antenna case

★ The use of antenna ports 7 to 14 for 2×2 MIMO, 4×4 MIMO or 8×8 MIMO is discussed towards the end of this section. Likewise, the use of Carrier Aggregation is discussed towards the end of this section. The throughput figures in Table 220, Table 221 and Table 222 assume the use of antenna ports 0 to 3

★ The PDSCH is a shared channel so its throughput capability has to be shared between all users. Increasing the number of users reduces the throughput per user. Users experiencing poor channel conditions will reduce the total cell throughput

★ Table 220 presents a set of theoretical absolute maximum physical layer throughputs which could be achieved if all Resource Elements were allocated to the PDSCH and the physical layer did not add any redundancy. These figures are not achievable in practice but provide a starting point from which to derive the maximum expected throughputs

★ The non-MIMO throughputs in Table 220 have been generated by multiplying the modulation symbol rate by the number of bits per symbol. For example, the 20 MHz channel bandwidth has 100 Resource Blocks providing 1200 subcarriers in the frequency domain. When using the normal cyclic prefix there are 14 OFDMA symbols during each 1 ms subframe so the modulation symbol rate is given by 1200 × 14 / 0.001 = 16.8 Msps. The bit rate when using 64QAM is then given by 16.8 Msps × 6 bits per symbol = 100.8 Mbps

★ The MIMO throughputs in Table 220 have been generated by multiplying the 64QAM throughputs by the relevant MIMO rank, i.e. the throughputs have been doubled for 2×2 MIMO and quadrupled for 4×4 MIMO

	Channel Bandwidth	1.4 MHz	3 MHz	5 MHz	10 MHz	15 MHz	20 MHz
	Resource Blocks in the frequency domain	6	15	25	50	75	100
Normal Cyclic Prefix	OFDMA symbols per 1 ms	14					
	Modulation symbol rate (Msps)	1.0	2.5	4.2	8.4	12.6	16.8
	QPSK Bit Rate (Mbps)	2.0	5.0	8.4	16.8	25.2	33.6
	16QAM Bit Rate (Mbps)	4.0	10.1	16.8	33.6	50.4	67.2
	64QAM Bit Rate (Mbps)	6.1	15.1	25.2	50.4	75.6	100.8
	2×2 MIMO 64QAM Bit Rate (Mbps)	12.1	30.2	50.4	100.8	151.2	201.6
	4×4 MIMO 64QAM Bit Rate (Mbps)	24.2	60.5	100.8	201.6	302.4	403.2
Extended Cyclic Prefix	OFDMA symbols per 1 ms	12					
	Modulation symbol rate (Msps)	0.9	2.2	3.6	7.2	10.8	14.4
	QPSK Bit Rate (Mbps)	1.7	4.3	7.2	14.4	21.6	28.8
	16QAM Bit Rate (Mbps)	3.5	8.6	14.4	28.8	43.2	57.6
	64QAM Bit Rate (Mbps)	5.2	13.0	21.6	43.2	64.8	86.4
	2×2 MIMO 64QAM Bit Rate (Mbps)	10.4	25.9	43.2	86.4	129.6	172.8
	4×4 MIMO 64QAM Bit Rate (Mbps)	20.7	51.8	86.4	172.8	259.2	345.6

Table 220 – Absolute maximum FDD physical layer throughputs if all Resource Elements were allocated to the PDSCH

★ The first step in deriving the maximum expected throughput is to remove the overheads generated by the other physical channels and physical signals, i.e. the PCFICH, PDCCH, PHICH, PBCH, Primary and Secondary Synchronisation Signals, and the Cell Specific Reference Signal. Table 221 presents a set of maximum physical layer throughputs with these overheads removed. The results still assume a coding rate of 1, i.e. the physical layer has not introduced any redundancy

★ Table 221 illustrates the relatively significant impact of the number of OFDMA symbols allocated to the PDCCH, PCFICH and PHICH. These physical channels can be allocated 2, 3 or 4 symbols when using the 1.4 MHz channel bandwidth, and 1, 2 or 3 symbols when using the other channel bandwidths

		Channel Bandwidth	1.4 MHz	3 MHz	5 MHz	10 MHz	15 MHz	20 MHz
Normal Cyclic Prefix	1 PDCCH Sym.	QPSK Bit Rate (Mbps)	-	4.4	7.4	14.9	22.4	29.9
		16QAM Bit Rate (Mbps)	-	8.8	14.8	29.8	44.8	59.8
		64QAM Bit Rate (Mbps)	-	13.2	22.2	44.7	67.1	89.7
		2x2 MIMO 64QAM Bit Rate (Mbps)	-	25.3	42.5	85.8	129.0	172.2
		4x4 MIMO 64QAM Bit Rate (Mbps)	-	47.7	80.3	161.9	243.4	325.0
	2 PDCCH Sym.	QPSK Bit Rate (Mbps)	1.5	4.0	6.8	13.7	20.6	27.5
		16QAM Bit Rate (Mbps)	3.1	8.1	13.6	27.4	41.2	55.0
		64QAM Bit Rate (Mbps)	4.6	12.1	20.4	41.1	61.8	82.5
		2x2 MIMO 64QAM Bit Rate (Mbps)	8.8	23.1	39.0	78.5	118.1	157.7
		4x4 MIMO 64QAM Bit Rate (Mbps)	17.2	44.8	75.5	152.4	229.2	306.0
	3 PDCCH Sym.	QPSK Bit Rate (Mbps)	1.4	3.7	6.2	12.5	18.8	25.1
		16QAM Bit Rate (Mbps)	2.8	7.3	12.4	25.0	37.6	50.2
		64QAM Bit Rate (Mbps)	4.2	11.0	18.6	37.4	56.4	75.3
		2x2 MIMO 64QAM Bit Rate (Mbps)	8.0	20.9	35.3	71.4	107.4	143.3
		4x4 MIMO 64QAM Bit Rate (Mbps)	15.4	40.5	68.3	137.9	207.4	277.0
	4 PDCCH Sym.	QPSK Bit Rate (Mbps)	1.3	-	-	-	-	-
		16QAM Bit Rate (Mbps)	2.5	-	-	-	-	-
		64QAM Bit Rate (Mbps)	3.8	-	-	-	-	-
		2x2 MIMO 64QAM Bit Rate (Mbps)	7.1	-	-	-	-	-
		4x4 MIMO 64QAM Bit Rate (Mbps)	13.7	-	-	-	-	-
Extended Cyclic Prefix	1 PDCCH Sym.	QPSK Bit Rate (Mbps)	-	3.7	6.2	12.5	18.8	25.1
		16QAM Bit Rate (Mbps)	-	7.3	12.4	25.0	37.6	50.2
		64QAM Bit Rate (Mbps)	-	11.0	18.6	37.5	56.4	75.3
		2x2 MIMO 64QAM Bit Rate (Mbps)	-	21.0	35.4	71.4	107.3	143.4
		4x4 MIMO 64QAM Bit Rate (Mbps)	-	39.1	66.0	133.2	200.4	267.5
	2 PDCCH Sym.	QPSK Bit Rate (Mbps)	1.3	3.3	5.6	11.3	17.0	22.7
		16QAM Bit Rate (Mbps)	2.5	6.6	11.2	22.6	34.0	45.4
		64QAM Bit Rate (Mbps)	3.8	9.9	16.8	33.9	51.0	68.1
		2x2 MIMO 64QAM Bit Rate (Mbps)	7.1	18.8	31.8	64.2	96.6	128.9
		4x4 MIMO 64QAM Bit Rate (Mbps)	13.8	36.2	61.2	123.6	186.1	248.5
	3 PDCCH Sym.	QPSK Bit Rate (Mbps)	1.1	3.0	5.0	10.1	15.2	20.3
		16QAM Bit Rate (Mbps)	2.2	5.9	10.0	20.2	30.4	40.6
		64QAM Bit Rate (Mbps)	3.3	8.9	15.0	30.3	45.6	60.9
		2x2 MIMO 64QAM Bit Rate (Mbps)	6.3	16.6	28.2	56.9	85.8	114.6
		4x4 MIMO 64QAM Bit Rate (Mbps)	12.0	31.9	54.0	109.2	164.3	219.5
	4 PDCCH Sym.	QPSK Bit Rate (Mbps)	1.0	-	-	-	-	-
		16QAM Bit Rate (Mbps)	2.0	-	-	-	-	-
		64QAM Bit Rate (Mbps)	3.0	-	-	-	-	-
		2x2 MIMO 64QAM Bit Rate (Mbps)	5.7	-	-	-	-	-
		4x4 MIMO 64QAM Bit Rate (Mbps)	10.9	-	-	-	-	-

Table 221 – Maximum FDD physical layer throughputs based upon Resource Elements available to the PDSCH (coding rate = 1, no retransmissions, no SRB, paging nor SIB overheads, no protocol stack overheads)

★ In practice, the number of symbols allocated to the PCFICH, PDCCH and PHICH depends upon the quantity of traffic loading the cell. There will be a requirement for an increased number of symbols as the quantity of traffic increases, i.e. the maximum throughput capability will decrease as the traffic and associated overheads increase

★ The figures in Table 221 are significantly less than those in Table 220. For example, the maximum throughput associated with the 20 MHz channel bandwidth, the normal cyclic prefix and 4×4 MIMO decreases from 403 Mbps to 325, 306 or 277 Mbps (depending upon the number of symbols allocated to the PDCCH, PCFICH and PHICH). This demonstrates the impact of the overheads generated by the physical channels and physical signals which do not transfer any application data

★ Redundancy added by the physical layer further reduces the throughputs measured at the top of the physical layer. The PDSCH uses a combination of rate 1/3 Turbo coding and rate matching to generate redundancy. In general, the quantity of redundancy is large when UE experience poor channel conditions and small when UE experience good channel conditions

★ Figure 192 illustrates an example link adaptation strategy which defines the physical layer coding rate as a function of the channel conditions and modulation scheme. The coding rate reflects the quantity of redundancy added by the physical layer. A low coding rate indicates a large quantity of redundancy while a high coding rate reflects a small quantity of redundancy. A coding rate of 1 corresponds to no redundancy

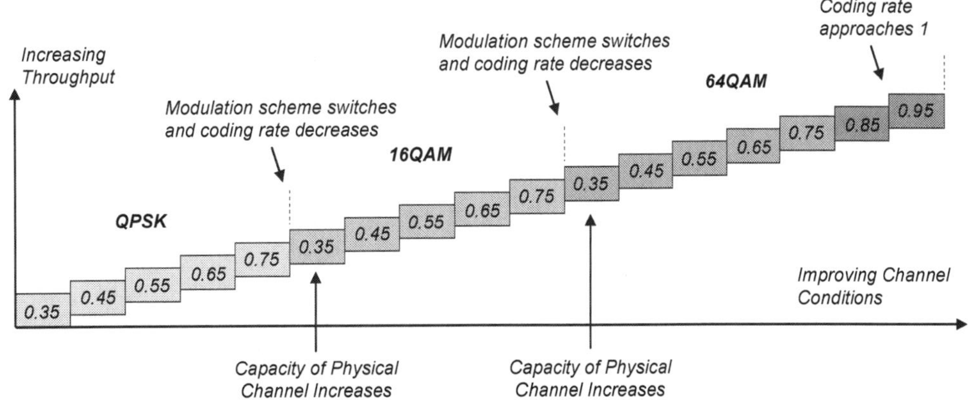

Figure 192 – Physical layer coding rate as a function of channel conditions and modulation scheme

★ QPSK and a low coding rate are associated with poor channel conditions. Link adaptation allocates larger transport block sizes as the channel conditions improve but the modulation scheme is kept as QPSK. This forces the quantity of redundancy to decrease (and the coding rate to increase), i.e. larger quantities of data are transferred without increasing the capacity of the physical channel

★ In this example, the modulation scheme is switched from QPSK to 16QAM once the channel conditions have improved sufficiently to allow the coding rate to increase to 0.75. Switching the modulation scheme increases the capacity of the physical channel so the quantity of redundancy can be increased. Link adaptation then continues to allocate larger transport block sizes as the channel conditions improve. The modulation scheme is switched from 16QAM to 64QAM once the channel conditions have improved enough to allow the coding rate to again reach 0.75

★ Once 64QAM has been allocated, link adaptation continues to allocate larger transport block sizes as the channel conditions improve. In this case, there is no option to switch to a higher order modulation scheme once the coding rate reaches 0.75. Instead, link adaptation continues to allocate larger transport block sizes and the coding rate approaches 1

★ System Information Blocks (SIB), paging messages and RRC signalling are transferred using the PDSCH. This reduces the PDSCH capacity available for application data. The overhead generated by the SIB, paging messages and RRC signalling will depend upon the quantity of traffic loading the cell but is likely to be relatively small, i.e. less than 100 kbps

★ Retransmissions reduce the higher layer throughputs. Hybrid Automatic Repeat Request (HARQ) retransmissions from the MAC layer reduce the throughputs measured from above the MAC layer. Automatic Repeat Request (ARQ) retransmissions from the RLC layer reduce the throughputs measured from above the RLC layer. Likewise TCP retransmissions reduce the throughput measured from above the TCP layer

★ Protocol stack headers also reduce the higher layer throughputs. The MAC, RLC, PDCP and IP layers add headers to the application data. The PDCP layer provides header compression for IP data streams so is able to reduce the impact of the IP header. The TCP and UDP layers also add their own headers when using TCP or UDP applications

★ Table 222 presents a set of example application throughputs assuming physical layer coding rates of 0.75 and 0.95. The coding rate of 0.95 is only shown for the 64QAM modulation scheme to remain consistent with the link adaptation strategy shown in Figure 192. Table 222 assumes the normal cyclic prefix, a 10 % retransmission rate and an additional 5 % overhead generated by a combination of the SIB, paging, RRC signalling and protocol stack headers

★ These throughput figures are comparable to those provided by UMTS High Speed Packet Access (HSPA) when making similar assumptions for both technologies. For example, UMTS HSPA with a 5 MHz channel bandwidth and 64QAM can achieve an application layer throughput of approximately 18 Mbps. LTE offers the same throughput capability when using 5 MHz and 64QAM if a single OFDMA symbol is allocated to the PDCCH, PCFICH and PHICH. Similar comparisons can be made when MIMO is applied to HSPA, and when HSPA is allocated 2 or 4 RF carriers to generate effective UMTS channel bandwidths of 10 or 20 MHz

		Coding Rate	1.4 MHz	3 MHz	5 MHz	10 MHz	15 MHz	20 MHz
1 PDCCH Symbol	QPSK Bit Rate (Mbps)	0.75	-	2.8	4.7	9.6	14.4	19.2
	16QAM Bit Rate (Mbps)	0.75	-	5.6	9.5	19.1	28.7	38.3
	64QAM Bit Rate (Mbps)	0.75	-	8.5	14.2	28.7	43.0	57.5
		0.95	-	10.7	18.0	36.3	54.5	72.9
	2×2 MIMO 64QAM Bit Rate (Mbps)	0.75	-	16.2	27.3	55.0	82.7	110.4
		0.95	-	20.5	34.5	69.7	104.8	139.9
	4×4 MIMO 64QAM Bit Rate (Mbps)	0.75	-	30.6	51.5	103.8	156.1	208.4
		0.95	-	38.7	65.2	131.5	197.7	264.0
2 PDCCH Symbols	QPSK Bit Rate (Mbps)	0.75	1.0	2.6	4.4	8.8	13.2	17.6
	16QAM Bit Rate (Mbps)	0.75	2.0	5.2	8.7	17.6	26.4	35.3
	64QAM Bit Rate (Mbps)	0.75	2.9	7.8	13.1	26.4	39.6	52.9
		0.95	3.7	9.8	16.6	33.4	50.2	67.0
	2×2 MIMO 64QAM Bit Rate (Mbps)	0.75	5.6	14.8	25.0	50.3	75.7	101.1
		0.95	7.1	18.8	31.7	63.8	95.9	128.1
	4×4 MIMO 64QAM Bit Rate (Mbps)	0.75	11.0	28.7	48.4	97.7	147.0	196.2
		0.95	14.0	36.4	61.3	123.8	186.2	248.5
3 PDCCH Symbols	QPSK Bit Rate (Mbps)	0.75	0.9	2.4	4.0	8.0	12.1	16.1
	16QAM Bit Rate (Mbps)	0.75	1.8	4.7	8.0	16.0	24.1	32.2
	64QAM Bit Rate (Mbps)	0.75	2.7	7.1	11.9	24.0	36.2	48.3
		0.95	3.4	8.9	15.1	30.4	45.8	61.2
	2×2 MIMO 64QAM Bit Rate (Mbps)	0.75	5.1	13.4	22.6	45.8	68.9	91.9
		0.95	6.5	17.0	28.7	58.0	87.2	116.4
	4×4 MIMO 64QAM Bit Rate (Mbps)	0.75	9.9	26.0	43.8	88.4	133.0	177.6
		0.95	12.5	32.9	55.5	112.0	168.5	225.0
4 PDCCH Symbols	QPSK Bit Rate (Mbps)	0.75	0.8	-	-	-	-	-
	16QAM Bit Rate (Mbps)	0.75	1.6	-	-	-	-	-
	64QAM Bit Rate (Mbps)	0.75	2.4	-	-	-	-	-
		0.95	3.1	-	-	-	-	-
	2×2 MIMO 64QAM Bit Rate (Mbps)	0.75	4.6	-	-	-	-	-
		0.95	5.8	-	-	-	-	-
	4×4 MIMO 64QAM Bit Rate (Mbps)	0.75	8.8	-	-	-	-	-
		0.95	11.1	-	-	-	-	-

Table 222 – FDD application layer throughputs based upon Resource Elements available to the PDSCH (normal cyclic prefix, 10% retransmissions, 5% additional overheads)

★ Reference Signal overheads change for the release 10 version of the 3GPP specifications when using antenna ports 7 to 14 for 2×2 MIMO, 4×4 MIMO or 8×8 MIMO. In these cases, the cell specific Reference Signal can be transmitted on antenna port 0, while the PDSCH and UE specific Reference Signals are transmitted on antenna ports 7 to 14, and CSI Reference Signals are transmitted on antenna ports 15 to 22. In addition, some subframes can be configured as MBSFN subframes. This allows transmission of the PDSCH to 3GPP release 10, and newer devices while avoiding the requirement to transmit the cell specific Reference Signal. These subframes are known as 'LTE Advanced' subframes. Configuring them as MBSFN subframes is a workaround solution to avoid older devices from expecting to receive the cell specific Reference Signal (MBSFN data is not actually transmitted during those subframes)

★ 3GPP TR 36.912 specifies an LTE Advanced peak spectral efficiency of 30.6 bps/Hz when using 8×8 MIMO with FDD. This figure is calculated as follows:

- total number of Resource Elements per antenna port within a 10 ms radio frame when using the normal cyclic prefix and the 20 MHz channel bandwidth is given by 12 × 14 × 100 × 10 = 168 000 Resource Elements
- accounting for 8 antenna ports increases the total number of Resource Elements to 8 × 168 000 = 1 344 000 Resource Elements
- assuming that the PCFICH, PDCCH and PHICH occupy only the first OFDMA symbol within each subframe generates an overhead of 1 / 14 = 7.14 % (normal cyclic prefix is used)

- assuming that the cell specific Reference Signal is only transmitted on antenna port 0, and assuming that 6 subframes within the radio frame are configured as MBSFN subframes then it is only necessary to transfer cell specific Reference Signals on antenna port 0 during the remaining 4 subframes. The overhead generated by the cell specific Reference Signal transmissions outside the first OFDMA symbol is then given by 6 × 100 × 4 / 168 000 = 1.43 % (the overhead generated by cell specific Reference Signal transmissions during the first OFDMA symbol has already been captured within the PCFICH, PDCCH and PHICH overhead calculation)
- assuming that the UE specific Reference Signal occupies 24 Resource Elements per Resource Block pair, then the corresponding overhead is given by 24 / (12 × 14) = 14.29 %
- assuming that the CSI Reference Signal is transmitted with a periodicity of 10 ms then the corresponding overhead for 8 antenna ports is given by (2 × 8) / (12 × 14 × 10) = 0.95 % (at the time of writing 3GPP TR 36.912 it was assumed that there would be 2 Resource Elements per antenna port for the CSI Reference Signal. The subsequent standardisation process resulted in the use of code division multiplexing so a total of 8 rather than 16 Resource Elements are required for transmission on 8 antenna ports – see Figure 90 in section 5.2.5. This halves the actual overhead generated by the CSI Reference Signal)
- assuming that the PBCH and Synchronisation Signals occupy 564 Resource Elements per radio frame, then the corresponding overhead is given by 564 / 168 000 = 0.34 %
- summing the set of overheads generates a total overhead of (7.14 + 1.43 + 14.29 + 0.95 + 0.34) = 24.15 %
- this overhead reduces the number of Resource Elements available to the PDSCH to 1 344 000 × (1 – 0.2415) = 1 019 424 Resource Elements. Assuming that 64QAM is applied to each Resource Element generates a physical layer throughput which is given by (1 019 424 × 6) / 0.010 = 611.654 Mbps
- the spectral efficiency is then obtained after dividing by the channel bandwidth of 20 MHz, i.e. (611.654 / 20) = 30.6 bps/Hz

★ In practice, the spectral efficiency will be less than 30.6 bps/Hz if more than a single OFDMA symbol is allocated to the PCFICH, PDCCH and PHICH, or if subframes are not configured as MBSFN subframes for PDSCH transmission to release 10 or newer devices

★ Table 223 presents a set of theoretical absolute maximum physical layer throughputs for a selection of LTE Advanced scenarios including both 8×8 MIMO and Carrier Aggregation. These throughputs are the equivalent of those presented in Table 220. They assume that all Resource Elements are allocated to the PDSCH and that the physical layer does not add any redundancy. These figures are not achievable in practice but provide a starting point from which to derive the maximum expected throughputs

★ Figures are only presented for the normal cyclic prefix because the UE specific Reference Signal is not supported on antenna ports 9 to 14 when using the extended cyclic prefix. The scenarios presented in Table 223 assume that the PDSCH is transmitted on antenna ports 7 to 10 when using 4×4 MIMO, and on antenna ports 7 to 14 when using 8×8 MIMO

	Channel Bandwidth	1.4 MHz	3 MHz	5 MHz	10 MHz	15 MHz	20 MHz
	Resource Blocks in the frequency domain	6	15	25	50	75	100
Normal Cyclic Prefix	4×4 MIMO 64QAM Bit Rate (Mbps)	24.2	60.5	100.8	201.6	302.4	403.2
	8×8 MIMO 64QAM Bit Rate (Mbps)	48.4	121.0	201.6	403.2	604.8	806.4
	8×8 MIMO 64QAM, 2 Component Carriers Bit Rate (Mbps)	96.8	241.9	403.2	806.4	1209.6	1612.8
	8×8 MIMO 64QAM, 5 Component Carriers Bit Rate (Mbps)	241.9	604.8	1008.0	2016.0	3024.0	4032.0

Table 223 – Absolute maximum FDD physical layer throughputs if all Resource Elements were allocated to the PDSCH (LTE Advanced)

★ Deriving the maximum expected throughput requires the overheads generated by the other physical channels and physical signals to be taken into account. The following assumptions are made for this analysis:

- the PCFICH, PDCCH and PHICH occupy either 1, 2, 3 or 4 OFDMA symbols per subframe (see section 6.2)
- the cell specific Reference Signal is only transmitted from antenna port 0 (see section 5.2.1)
- no MBSFN subframes are configured so the cell specific Reference Signal is transmitted during all subframes
- the UE specific Reference Signal occupies 24 Resource Elements within each Resource Block pair (see section 5.2.3)
- in the case of 4×4 MIMO, the CSI Reference Signal occupies 4 Resource Elements within each Resource Block pair, during subframes that include CSI Reference Signal transmissions. The CSI Reference Signal periodicity is assumed to be 10 ms (see section 5.2.5)
- in the case of 8×8 MIMO, the CSI Reference Signal occupies 8 Resource Elements within each Resource Block pair, during subframes that include CSI Reference Signal transmissions. The CSI Reference Signal periodicity is assumed to be 10 ms (see section 5.2.5)
- the PBCH and Synchronisation Signals occupy a total of 564 Resource Elements per radio frame (see sections 6.1 and 5.1)

★ These assumptions lead to the physical layer throughputs presented in Table 224. These throughputs do not yet account for the redundancy added by the physical layer. Nor do they account for the overheads generated by the various protocol stack headers and retransmissions. Nor do they account for the overheads generated by RRC signalling

	Channel Bandwidth	1.4 MHz	3 MHz	5 MHz	10 MHz	15 MHz	20 MHz
1 PDCCH Sym.	4×4 MIMO 64QAM Bit Rate (Mbps)	-	44.2	74.4	149.7	225.1	300.4
	8×8 MIMO 64QAM Bit Rate (Mbps)	-	88.1	148.2	298.5	448.7	598.9
	8×8 MIMO 64QAM Bit Rate, 2 Component Carriers (Mbps)	-	176.3	296.5	597.0	897.4	1197.9
	8×8 MIMO 64QAM Bit Rate, 5 Component Carriers (Mbps)	-	440.7	741.1	1492.4	2243.6	2994.7
2 PDCCH Sym.	4×4 MIMO 64QAM Bit Rate (Mbps)	15.3	39.9	67.2	135.3	203.5	271.6
	8×8 MIMO 64QAM Bit Rate (Mbps)	30.6	79.5	133.8	269.7	405.5	541.3
	8×8 MIMO 64QAM Bit Rate, 2 Component Carriers (Mbps)	61.2	159.0	267.7	539.4	811.0	1082.7
	8×8 MIMO 64QAM Bit Rate, 5 Component Carriers (Mbps)	152.9	397.5	669.1	1348.4	2027.6	2706.7
3 PDCCH Sym.	4×4 MIMO 64QAM Bit Rate (Mbps)	13.6	35.6	60.0	120.9	181.9	242.8
	8×8 MIMO 64QAM Bit Rate (Mbps)	27.1	70.9	119.4	240.9	362.3	483.7
	8×8 MIMO 64QAM Bit Rate, 2 Component Carriers (Mbps)	54.3	141.7	238.9	481.8	724.6	967.5
	8×8 MIMO 64QAM Bit Rate, 5 Component Carriers (Mbps)	135.6	354.3	597.1	1204.4	1811.6	2418.7
4 PDCCH Sym.	4×4 MIMO 64QAM Bit Rate (Mbps)	11.9	-	-	-	-	-
	8×8 MIMO 64QAM Bit Rate (Mbps)	23.7	-	-	-	-	-
	8×8 MIMO 64QAM Bit Rate, 2 Component Carriers (Mbps)	47.3	-	-	-	-	-
	8×8 MIMO 64QAM Bit Rate, 5 Component Carriers (Mbps)	118.4	-	-	-	-	-

Table 224 – Maximum FDD physical layer throughputs based upon Resource Elements available to the PDSCH (LTE Advanced) (coding rate = 1, no retransmissions, no SRB, paging nor SIB overheads, no protocol stack overheads)

★ Table 224 illustrates the relatively significant impact of the number of OFDMA symbols allocated to the PDCCH, PCFICH and PHICH. These physical channels can be allocated 2, 3 or 4 symbols when using the 1.4 MHz channel bandwidth, and 1, 2 or 3 symbols when using the other channel bandwidths

★ The 4×4 MIMO throughputs in Table 224 can be compared with those in Table 221. The 'LTE Advanced' throughputs for 4×4 MIMO (using antenna ports 7 to 10) are less than those based upon the release 8 and 9 versions of the specifications (using antenna ports 0 to 3). This is a result of the relative overheads generated by the cell specific Reference Signal and UE specific Reference Signal

  - when using 4×4 MIMO with antenna ports 0 to 3, the cell specific Reference Signal occupies 16 Resource Elements per Resource Block pair (assuming 2 or more OFDMA symbols are allocated to the PCFICH, PDCCH and PHICH, and counting only the Resource Elements outside those OFDMA symbols)
  - when using 4×4 MIMO with antenna ports 7 to 10, the cell specific Reference Signal occupies 6 Resource Elements per Resource Block pair, while the UE specific Reference Signal occupies 24 Resource Elements per Resource Block pair, i.e. a total of 30 Resource Elements are used outside the OFDMA symbols used for the PCFICH, PDCCH and PHICH (ignoring the relatively small impact of the CSI Reference Signal for the purposes this comparison)

★ These figures illustrate that there is an increased overhead when using antenna ports 7 to 10 for 4×4 MIMO. This overhead could be reduced if subframes were configured as MBSFN subframes. This would remove the requirement to transmit the cell specific Reference Signal so the overhead would reduce from 30 to 24 Resource Elements when using antenna ports 7 to 10. Nevertheless, the overhead remains greater than the 16 Resource Elements when using antenna ports 0 to 3

★ The throughputs shown in Table 224 increase in direct proportion to the number of Component Carriers because it is assumed that each Component Carrier has the same configuration. This does not have to be the case in practice, e.g. one Component Carrier could have 2 OFDMA symbols reserved for the PCFICH, PDCCH and PHICH, while another Component Carrier could have 3 OFDMA symbols reserved

★ The figures in Table 224 are significantly less than those in Table 223. For example, the maximum throughput associated with the 20 MHz channel bandwidth and 8×8 MIMO with 5 Component Carriers decreases from 4032 Mbps to 2994.7, 2706.7 or 2418.7 Mbps (depending upon the number of symbols allocated to the PDCCH, PCFICH and PHICH). This demonstrates the impact of the overheads generated by the physical channels and physical signals which do not transfer any application data

★ Table 225 presents a set of example application throughputs assuming physical layer coding rates of 0.75 and 0.95, as well as a 10 % retransmission rate and an additional 5 % overhead generated by a combination of the SIB, paging, RRC signalling and protocol stack headers

★ The maximum throughput generated by 8×8 MIMO with 5 Component Carriers is 2432.5 Mbps. This is obtained when assuming a single OFDMA symbol is used for the PCFICH, PDCCH and PHICH. The figure decreases to 1964.6 Mbps when assuming 3 OFDMA symbols are used for the PCFICH, PDCCH and PHICH. These throughput figures are high but require 100 MHz of spectrum. The equivalent figures for 40 MHz of spectrum are 973.0 and 785.8 Mbps

★ These throughput figures are peak connection throughputs rather than average cell throughputs. They assume that the UE is benefiting from 64QAM and full rank MIMO transmission while being scheduled resources on all Component Carriers

		Coding Rate	1.4 MHz	3 MHz	5 MHz	10 MHz	15 MHz	20 MHz
1 PDCCH Symbol	4×4 MIMO 64QAM Bit Rate (Mbps)	0.75	-	28.4	47.7	96.0	144.3	192.7
		0.95	-	35.9	60.4	121.6	182.8	244.0
	8×8 MIMO 64QAM Bit Rate (Mbps)	0.75	-	56.5	95.1	191.4	287.7	384.1
		0.95	-	71.6	120.4	242.4	364.5	486.5
	8×8 MIMO 64QAM, 2 Component Carriers Bit Rate (Mbps)	0.75	-	113.0	190.1	382.8	575.5	768.1
		0.95	-	143.2	240.8	484.9	728.9	973.0
	8×8 MIMO 64QAM, 5 Component Carriers Bit Rate (Mbps)	0.75	-	282.6	475.3	957.0	1438.7	1920.4
		0.95	-	357.9	602.0	1212.2	1822.3	2432.5
2 PDCCH Symbols	4×4 MIMO 64QAM Bit Rate (Mbps)	0.75	9.8	25.6	43.1	86.8	130.5	174.2
		0.95	12.5	32.4	54.5	109.9	165.3	220.6
	8×8 MIMO 64QAM Bit Rate (Mbps)	0.75	19.6	51.0	85.8	172.9	260.0	347.1
		0.95	24.8	64.6	108.7	219.0	329.4	439.7
	8×8 MIMO 64QAM, 2 Component Carriers Bit Rate (Mbps)	0.75	39.2	102.0	171.6	345.9	520.1	694.3
		0.95	49.7	129.1	217.4	438.1	658.8	879.4
	8×8 MIMO 64QAM, 5 Component Carriers Bit Rate (Mbps)	0.75	98.1	254.9	429.1	864.7	1300.2	1735.7
		0.95	124.2	322.8	543.5	1095.2	1646.9	2198.5
3 PDCCH Symbols	4×4 MIMO 64QAM Bit Rate (Mbps)	0.75	8.7	22.8	38.4	77.5	116.6	155.7
		0.95	11.1	28.9	48.7	98.2	147.7	197.2
	8×8 MIMO 64QAM Bit Rate (Mbps)	0.75	17.4	45.4	76.6	154.5	232.3	310.2
		0.95	22.0	57.6	97.0	195.7	294.3	392.9
	8×8 MIMO 64QAM, 2 Component Carriers Bit Rate (Mbps)	0.75	34.8	90.9	153.2	308.9	464.7	620.4
		0.95	44.1	115.1	194.0	391.3	588.6	785.8
	8×8 MIMO 64QAM, 5 Component Carriers Bit Rate (Mbps)	0.75	87.0	227.2	382.9	772.3	1161.7	1551.0
		0.95	110.2	287.8	485.0	978.3	1471.4	1964.6
4 PDCCH Symbols	4×4 MIMO 64QAM Bit Rate (Mbps)	0.75	7.6	-	-	-	-	-
		0.95	9.7	-	-	-	-	-
	8×8 MIMO 64QAM Bit Rate (Mbps)	0.75	15.2	-	-	-	-	-
		0.95	19.2	-	-	-	-	-
	8×8 MIMO 64QAM, 2 Component Carriers Bit Rate (Mbps)	0.75	30.4	-	-	-	-	-
		0.95	38.5	-	-	-	-	-
	8×8 MIMO 64QAM, 5 Component Carriers Bit Rate (Mbps)	0.75	75.9	-	-	-	-	-
		0.95	96.1	-	-	-	-	-

Table 225 – FDD application layer throughputs based upon Resource Elements available to the PDSCH (LTE Advanced) (10% retransmissions, 5% additional overheads)

★ 3GPP References: TS 36.211, TR 36.912

20.1.2 TDD

- ★ The analysis of bit rates for TDD is complicated by the large number of combinations of uplink-downlink subframe configurations and special subframe configurations:
  - o the set of 7 uplink-downlink subframe configurations is presented in Table 12 within section 3.2.2
  - o the set of 9 special subframe configurations is presented in Table 13 within section 3.2.2
- ★ For the purposes of the analysis within this section:
  - o uplink-downlink subframe configurations 0 and 5 have been selected because these have the least and the most number of downlink subframes
  - o special subframe configurations 0 and 4 have been selected because these have the least and the most number of OFDMA symbols within the DwPTS field of the special subframe
- ★ Similar to FDD, the Physical Downlink Shared Channel (PDSCH) is used to transfer application data, and the throughput achieved by the PDSCH depends upon the number of Resource Elements, the modulation scheme applied to each Resource Element, the quantity of redundancy included by physical layer processing, the use of multiple antenna transmission schemes and the use of Carrier Aggregation
- ★ Table 226 presents a set of theoretical absolute maximum physical layer throughputs which could be achieved if all Resource Elements were allocated to the PDSCH and the physical layer did not add any redundancy. These figures are not achievable in practice but provide a starting point from which to derive the maximum expected throughputs
- ★ All throughputs are averaged across a 10 ms radio frame so the impact of the uplink subframes is captured. Only figures for the normal cyclic prefix are presented

		Channel Bandwidth	1.4 MHz	3 MHz	5 MHz	10 MHz	15 MHz	20 MHz
		Resource Blocks	6	15	25	50	75	100
Uplink-Downlink Subframe Configuration 0	QPSK Bit Rate (Mbps)	Special Subframe Config 0	0.5	1.2	2.0	4.1	6.1	8.2
		Special Subframe Config 4	0.7	1.9	3.1	6.2	9.4	12.5
	16QAM Bit Rate (Mbps)	Special Subframe Config 0	1.0	2.4	4.1	8.2	12.2	16.3
		Special Subframe Config 4	1.5	3.7	6.2	12.5	18.7	25.0
	64QAM Bit Rate (Mbps)	Special Subframe Config 0	1.5	3.7	6.1	12.2	18.4	24.5
		Special Subframe Config 4	2.2	5.6	9.4	18.7	28.1	37.4
	2x2 MIMO 64QAM Bit Rate (Mbps)	Special Subframe Config 0	2.9	7.3	12.2	24.5	36.7	49.0
		Special Subframe Config 4	4.5	11.2	18.7	37.4	56.2	74.9
	4x4 MIMO 64QAM Bit Rate (Mbps)	Special Subframe Config 0	5.9	14.7	24.5	49.0	73.4	97.9
		Special Subframe Config 4	9.0	22.5	37.4	74.9	112.3	149.8
Uplink-Downlink Subframe Configuration 5	QPSK Bit Rate (Mbps)	Special Subframe Config 0	1.7	4.1	6.9	13.8	20.7	27.6
		Special Subframe Config 4	1.8	4.5	7.4	14.9	22.3	29.8
	16QAM Bit Rate (Mbps)	Special Subframe Config 0	3.3	8.3	13.8	27.6	41.4	55.2
		Special Subframe Config 4	3.6	8.9	14.9	29.8	44.6	59.5
	64QAM Bit Rate (Mbps)	Special Subframe Config 0	5.0	12.4	20.7	41.4	62.1	82.8
		Special Subframe Config 4	5.4	13.4	22.3	44.6	67.0	89.3
	2x2 MIMO 64QAM Bit Rate (Mbps)	Special Subframe Config 0	9.9	24.8	41.4	82.8	124.2	165.6
		Special Subframe Config 4	10.7	26.8	44.6	89.3	133.9	178.6
	4x4 MIMO 64QAM Bit Rate (Mbps)	Special Subframe Config 0	19.9	49.7	82.8	165.6	248.4	331.2
		Special Subframe Config 4	21.4	53.6	89.3	178.6	267.8	357.1

Table 226 – Absolute maximum TDD physical layer throughputs if all Resource Elements were allocated to the PDSCH

- ★ These throughputs can be compared with the equivalent FDD figures in Table 220. Throughputs are significantly lower when using TDD uplink-downlink subframe configuration 0 because there are only 2 downlink subframes and 2 special subframes. Uplink-downlink subframe configuration 5 has 8 downlink subframes and 1 special subframe so the throughputs are closer to the equivalent FDD figures
- ★ TDD throughputs are lower than the equivalent FDD throughputs but only a single RF carrier is used rather than separate uplink-downlink RF carriers
- ★ Similar to FDD, the first step in deriving the maximum expected throughput is to remove the overheads generated by the other physical channels and physical signals, i.e. the PCFICH, PDCCH, PHICH, PBCH, Primary and Secondary Synchronisation Signals, and the cell specific Reference Signal

★ Table 227 presents a set of maximum physical layer throughputs with these overheads removed. The results still assume a coding rate of 1. Results are presented for both the maximum and minimum number of PDCCH symbols

- subframes 1 and 6: the maximum is 2 symbols for all channel bandwidths, while the minimum is 2 symbols for a channel bandwidth of 1.4 MHz, and 1 symbol for larger channel bandwidths
- other subframes: the maximum is 4 symbols for a channel bandwidth of 1.4 MHz, and 3 symbols for larger channel bandwidths, while the minimum is 2 symbols for a channel bandwidth of 1.4 MHz, and 1 symbol for larger channel bandwidths

			Channel Bandwidth	1.4 MHz	3 MHz	5 MHz	10 MHz	15 MHz	20 MHz
			Resource Blocks	6	15	25	50	75	100
Uplink-Downlink Subframe Configuration 0	Minimum PDCCH Symbols	QPSK Bit Rate (Mbps)	Special Subframe Config 0	0.2	0.9	1.8	3.8	5.8	7.7
			Special Subframe Config 4	0.5	1.5	2.6	5.4	8.2	10.9
		16QAM Bit Rate (Mbps)	Special Subframe Config 0	0.5	1.9	3.7	7.6	11.5	15.5
			Special Subframe Config 4	1.0	3.1	5.3	10.8	16.3	21.9
		64QAM Bit Rate (Mbps)	Special Subframe Config 0	0.7	2.8	5.5	11.4	17.3	23.2
			Special Subframe Config 4	1.5	4.6	7.9	16.2	24.5	32.8
		2×2 MIMO 64QAM Bit Rate (Mbps)	Special Subframe Config 0	1.4	5.4	10.7	22.1	33.5	44.9
			Special Subframe Config 4	2.8	8.8	15.2	31.0	46.9	62.7
		4×4 MIMO 64QAM Bit Rate (Mbps)	Special Subframe Config 0	2.8	10.0	20.1	41.5	62.8	84.2
			Special Subframe Config 4	5.4	16.6	28.5	58.3	88.0	117.8
	Maximum PDCCH Symbols	QPSK Bit Rate (Mbps)	Special Subframe Config 0	0.2	0.7	1.3	2.6	4.0	5.4
			Special Subframe Config 4	0.4	1.3	2.3	4.7	7.1	9.5
		16QAM Bit Rate (Mbps)	Special Subframe Config 0	0.4	1.4	2.5	5.3	8.1	10.8
			Special Subframe Config 4	0.9	2.7	4.6	9.4	14.2	19.0
		64QAM Bit Rate (Mbps)	Special Subframe Config 0	0.6	2.1	3.8	7.9	12.1	16.2
			Special Subframe Config 4	1.3	4.0	6.9	14.1	21.3	28.5
		2×2 MIMO 64QAM Bit Rate (Mbps)	Special Subframe Config 0	1.1	4.1	7.3	15.2	23.1	31.0
			Special Subframe Config 4	2.4	7.5	13.0	26.7	40.4	54.1
		4×4 MIMO 64QAM Bit Rate (Mbps)	Special Subframe Config 0	2.1	7.9	14.1	29.5	44.8	60.2
			Special Subframe Config 4	4.7	14.6	25.1	51.5	77.9	104.3
Uplink-Downlink Subframe Configuration 5	Minimum PDCCH Symbols	QPSK Bit Rate (Mbps)	Special Subframe Config 0	1.2	3.6	6.0	12.1	18.2	24.4
			Special Subframe Config 4	1.3	3.9	6.5	13.1	19.8	26.4
		16QAM Bit Rate (Mbps)	Special Subframe Config 0	2.5	7.1	12.0	24.3	36.5	48.7
			Special Subframe Config 4	2.7	7.7	13.0	26.3	39.6	52.8
		64QAM Bit Rate (Mbps)	Special Subframe Config 0	3.7	10.7	18.0	36.4	54.7	73.1
			Special Subframe Config 4	4.0	11.6	19.6	39.4	59.3	79.2
		2×2 MIMO 64QAM Bit Rate (Mbps)	Special Subframe Config 0	7.0	20.5	34.6	69.9	105.2	140.5
			Special Subframe Config 4	7.7	22.2	37.5	75.7	113.8	152.0
		4×4 MIMO 64QAM Bit Rate (Mbps)	Special Subframe Config 0	13.7	38.6	65.2	131.7	198.2	264.7
			Special Subframe Config 4	15.0	41.9	70.7	142.7	214.7	286.7
	Maximum PDCCH Symbols	QPSK Bit Rate (Mbps)	Special Subframe Config 0	1.0	3.0	5.0	10.2	15.4	20.5
			Special Subframe Config 4	1.1	3.3	5.6	11.2	16.9	22.6
		16QAM Bit Rate (Mbps)	Special Subframe Config 0	2.0	6.0	10.1	20.4	30.7	41.1
			Special Subframe Config 4	2.3	6.6	11.1	22.5	33.8	45.1
		64QAM Bit Rate (Mbps)	Special Subframe Config 0	3.1	8.9	15.1	30.6	46.1	61.6
			Special Subframe Config 4	3.4	9.9	16.7	33.7	50.7	67.7
		2×2 MIMO 64QAM Bit Rate (Mbps)	Special Subframe Config 0	5.8	17.0	28.9	58.4	87.9	117.4
			Special Subframe Config 4	6.5	18.8	31.7	64.1	96.5	128.9
		4×4 MIMO 64QAM Bit Rate (Mbps)	Special Subframe Config 0	11.2	33.0	55.9	113.0	170.1	227.2
			Special Subframe Config 4	12.6	36.3	61.4	124.0	186.7	249.3

Table 227 – Maximum TDD physical layer throughputs based upon Resource Elements available to the PDSCH (coding rate = 1, no retransmissions, no SRB, paging nor SIB overheads, no protocol stack overheads)

★ The figures in Table 227 are significantly less than those in Table 226. For example, the maximum throughput associated with the 20 MHz channel bandwidth, uplink-downlink subframe configuration 5, special subframe configuration 4 and 4×4 MIMO decreases from 357 Mbps to 287 or 249 Mbps (depending upon the number of symbols allocated to the PDCCH, PCFICH and PHICH). This demonstrates the impact of the overheads generated by the physical channels and physical signals which do not transfer any application data

★ In practice, the number of symbols allocated to the PCFICH, PDCCH and PHICH depends upon the quantity of traffic loading the cell. There will be a requirement for an increased number of symbols as the quantity of traffic increases, i.e. the maximum throughput capability will decrease as the traffic and associated overheads increase

★ As described in section 20.1.1, redundancy added by the physical layer further reduces the throughputs measured at the top of the physical layer. The PDSCH uses a combination of rate 1/3 Turbo coding and rate matching to generate redundancy. In general, the quantity of redundancy is large when UE experience poor channel conditions and small when UE experience good channel conditions

★ System Information Blocks (SIB), paging messages and RRC signalling are transferred using the PDSCH. This reduces the PDSCH capacity available for application data. The overhead generated by the SIB, paging messages and RRC signalling will depend upon the quantity of traffic loading the cell but is likely to be relatively small, i.e. less than 100 kbps

★ Retransmissions reduce the higher layer throughputs. Hybrid Automatic Repeat Request (HARQ) retransmissions from the MAC layer reduce the throughputs measured from above the MAC layer. Automatic Repeat Request (ARQ) retransmissions from the RLC layer reduce the throughputs measured from above the RLC layer. Likewise TCP retransmissions reduce the throughput measured from above the TCP layer

★ Protocol stack headers also reduce the higher layer throughputs. The MAC, RLC, PDCP and IP layers add headers to the application data. The PDCP layer provides header compression for IP data streams so is able to reduce the impact of the IP header. The TCP and UDP layers also add their own headers when using TCP or UDP applications

★ Table 228 presents a set of example application throughputs assuming physical layer coding rates of 0.75 and 0.95. The coding rate of 0.95 is only shown for the 64QAM modulation scheme to remain consistent with the link adaptation strategy shown in Figure 192. Table 228 assumes uplink-downlink subframe configuration 5, special subframe configuration 4, a 10 % retransmission rate and an additional 5 % overhead generated by a combination of the SIB, paging, RRC signalling and protocol stack headers

			Channel Bandwidth	1.4 MHz	3 MHz	5 MHz	10 MHz	15 MHz	20 MHz
			Resource Blocks	6	15	25	50	75	100
Uplink-Downlink Subframe Configuration 5 Special Subframe Configuration 4	Minimum PDCCH Symbols	QPSK Bit Rate (Mbps)	Coding Rate = 0.75	0.87	2.48	4.18	8.43	12.68	16.94
		16QAM Bit Rate (Mbps)	Coding Rate = 0.75	1.73	4.96	8.36	16.86	25.36	33.86
		64QAM Bit Rate (Mbps)	Coding Rate = 0.75	2.60	7.44	12.54	25.29	38.05	50.80
			Coding Rate = 0.95	3.29	9.42	15.88	32.04	48.19	64.35
		2x2 MIMO 64QAM Bit Rate (Mbps)	Coding Rate = 0.75	4.95	14.25	24.05	48.52	72.99	97.46
			Coding Rate = 0.95	6.27	18.06	30.46	61.45	92.45	123.45
		4x4 MIMO 64QAM Bit Rate (Mbps)	Coding Rate = 0.75	9.61	26.89	45.36	91.53	137.70	183.87
			Coding Rate = 0.95	12.17	34.06	57.45	115.93	174.41	232.90
	Maximum PDCCH Symbols	QPSK Bit Rate (Mbps)	Coding Rate = 0.75	0.74	2.11	3.57	7.20	10.84	14.47
		16QAM Bit Rate (Mbps)	Coding Rate = 0.75	1.47	4.22	7.12	14.40	21.67	28.94
		64QAM Bit Rate (Mbps)	Coding Rate = 0.75	2.21	6.33	10.69	21.60	32.50	43.41
			Coding Rate = 0.95	2.79	8.02	13.54	27.36	41.17	54.99
		2x2 MIMO 64QAM Bit Rate (Mbps)	Coding Rate = 0.75	4.17	12.04	20.35	41.13	61.91	82.68
			Coding Rate = 0.95	5.29	15.25	25.78	52.10	78.41	104.73
		4x4 MIMO 64QAM Bit Rate (Mbps)	Coding Rate = 0.75	8.05	23.29	39.35	79.52	119.69	159.86
			Coding Rate = 0.95	10.20	29.50	49.85	100.73	151.61	202.49

Table 228 – TDD application layer throughputs based upon Resource Elements available to the PDSCH (normal cyclic prefix, 10% retransmissions, 5% additional overheads)

★ Reference Signal overheads change for the release 10 version of the 3GPP specifications when using antenna ports 7 to 14 for 2×2 MIMO, 4×4 MIMO or 8×8 MIMO. In these cases, the cell specific Reference Signal can be transmitted on antenna port 0, while the PDSCH and UE specific Reference Signals are transmitted on antenna ports 7 to 14, and CSI Reference Signals are transmitted on antenna ports 15 to 22

★ In addition, some subframes can be configured as MBSFN subframes. This allows transmission of the PDSCH to 3GPP release 10, and newer devices while avoiding the requirement to transmit the cell specific Reference Signal. These subframes are known as 'LTE Advanced' subframes. Configuring them as MBSFN subframes is a workaround solution to avoid older devices from expecting to receive the cell specific Reference Signal (MBSFN data is not actually transmitted during those subframes)

- 3GPP TR 36.912 specifies an LTE Advanced peak spectral efficiency of 30.0 bps/Hz when using 8×8 MIMO with TDD. This figure is calculated as follows:
    - total number of Resource Elements per antenna port within a 10 ms radio frame when using the normal cyclic prefix, the 20 MHz channel bandwidth, uplink-downlink subframe configuration 2 and special subframe configuration 4 is given by (12 × 14 × 100 × 4) + (12 × 12 × 100 × 2) = 96 000 Resource Elements
    - accounting for 8 antenna ports increases the total number of Resource Elements to 8 × 96 000 = 768 000 Resource Elements
    - assuming that the PCFICH, PDCCH and PHICH occupy only the first OFDMA symbol within each subframe generates an overhead of 6 / (14 × 4 + 12 × 2) = 7.5 % (normal cyclic prefix is used)
    - assuming that the cell specific Reference Signal is only transmitted on antenna port 0, and assuming that 2 subframes within the radio frame are configured as MBSFN subframes then it is only necessary to transfer cell specific Reference Signals on antenna port 0 during the remaining 2 subframes and 2 special subframes. The overhead generated by the cell specific Reference Signal transmissions outside the first OFDMA symbol is then given by 6 × 100 × 4 / 96 000 = 2.5 % (the overhead generated by cell specific Reference Signal transmissions during the first OFDMA symbol has already been captured within the PCFICH, PDCCH and PHICH overhead calculation)
    - assuming that the UE specific Reference Signal occupies 24 Resource Elements per Resource Block pair, then the corresponding overhead is given by 24 × 6 × 100 / 96 000 = 15 %
    - assuming that the PBCH and Synchronisation Signals occupy 564 Resource Elements per radio frame, then the corresponding overhead is given by 564 / 96 000 = 0.6 %
    - summing the set of overheads generates a total overhead of (7.5 + 2.5 + 15 + 0.6) = 25.6 %
    - this overhead reduces the number of Resource Elements available to the PDSCH to 768 000 × (1 – 0.256) = 571 392 Resource Elements. Assuming that 64QAM is applied to each Resource Element generates a physical layer throughput which is given by (571 392 × 6) / 0.010 = 342.84 Mbps
    - within a 10 ms radio frame the TDD downlink activity factor is defined by (14 × 4 + 12 × 2) / (14 × 10) = 57.1 %
    - the spectral efficiency is then obtained by including the activity factor and dividing by the channel bandwidth of 20 MHz, i.e. 342.84 / (20 × 0.571) = 30.0 bps/Hz
- In practice, the spectral efficiency will be less than 30.0 bps/Hz if more than a single OFDMA symbol is allocated to the PCFICH, PDCCH and PHICH, or if subframes are not configured as MBSFN subframes for PDSCH transmission to release 10 or newer devices
- Table 229 presents a set of theoretical absolute maximum physical layer throughputs for a selection of LTE Advanced scenarios including both 8×8 MIMO and Carrier Aggregation. These throughputs are the equivalent of those presented in Table 226. They assume that all Resource Elements are allocated to the PDSCH and that the physical layer does not add any redundancy. These figures are not achievable in practice but provide a starting point from which to derive the maximum expected throughputs

		Channel Bandwidth	1.4 MHz	3 MHz	5 MHz	10 MHz	15 MHz	20 MHz
		Resource Blocks	6	15	25	50	75	100
Uplink-Downlink Subframe Configuration 0	4×4 MIMO 64QAM Bit Rate (Mbps)	Special Subframe Config 0	5.9	14.7	24.5	49.0	73.4	97.9
Uplink-Downlink Subframe Configuration 0	4×4 MIMO 64QAM Bit Rate (Mbps)	Special Subframe Config 4	9.0	22.5	37.4	74.9	112.3	149.8
Uplink-Downlink Subframe Configuration 0	8×8 MIMO 64QAM Bit Rate (Mbps)	Special Subframe Config 0	11.8	29.4	49.0	97.9	146.9	195.8
Uplink-Downlink Subframe Configuration 0	8×8 MIMO 64QAM Bit Rate (Mbps)	Special Subframe Config 4	18.0	44.9	74.9	149.8	224.6	299.5
Uplink-Downlink Subframe Configuration 0	8×8 MIMO 64QAM 2 Component Carriers Bit Rate (Mbps)	Special Subframe Config 0	23.5	58.8	97.9	195.8	293.8	391.7
Uplink-Downlink Subframe Configuration 0	8×8 MIMO 64QAM 2 Component Carriers Bit Rate (Mbps)	Special Subframe Config 4	35.9	89.9	149.8	299.5	449.3	599.0
Uplink-Downlink Subframe Configuration 0	8×8 MIMO 64QAM 5 Component Carriers Bit Rate (Mbps)	Special Subframe Config 0	58.8	146.9	244.8	489.6	734.4	979.2
Uplink-Downlink Subframe Configuration 0	8×8 MIMO 64QAM 5 Component Carriers Bit Rate (Mbps)	Special Subframe Config 4	89.9	224.6	374.4	748.8	1123.2	1497.6
Uplink-Downlink Subframe Configuration 5	4×4 MIMO 64QAM Bit Rate (Mbps)	Special Subframe Config 0	19.9	49.7	82.8	165.6	248.4	331.2
Uplink-Downlink Subframe Configuration 5	4×4 MIMO 64QAM Bit Rate (Mbps)	Special Subframe Config 4	21.4	53.6	89.3	178.6	267.8	357.1
Uplink-Downlink Subframe Configuration 5	8×8 MIMO 64QAM Bit Rate (Mbps)	Special Subframe Config 0	39.7	99.4	165.6	331.2	496.8	662.4
Uplink-Downlink Subframe Configuration 5	8×8 MIMO 64QAM Bit Rate (Mbps)	Special Subframe Config 4	42.9	107.1	178.6	357.1	535.7	714.2
Uplink-Downlink Subframe Configuration 5	8×8 MIMO 64QAM 2 Component Carriers Bit Rate (Mbps)	Special Subframe Config 0	79.5	198.7	331.2	662.4	993.6	1324.8
Uplink-Downlink Subframe Configuration 5	8×8 MIMO 64QAM 2 Component Carriers Bit Rate (Mbps)	Special Subframe Config 4	85.7	214.3	357.1	714.2	1071.4	1428.5
Uplink-Downlink Subframe Configuration 5	8×8 MIMO 64QAM 5 Component Carriers Bit Rate (Mbps)	Special Subframe Config 0	198.7	496.8	828.0	1656.0	2484.0	3312.0
Uplink-Downlink Subframe Configuration 5	8×8 MIMO 64QAM 5 Component Carriers Bit Rate (Mbps)	Special Subframe Config 4	214.3	535.7	892.8	1785.6	2678.4	3571.2

Table 229 – Absolute maximum TDD physical layer throughputs if all Resource Elements were allocated to the PDSCH (LTE Advanced)

- Figures are only presented for the normal cyclic prefix because the UE specific Reference Signal is not supported on antenna ports 9 to 14 when using the extended cyclic prefix. The scenarios presented in Table 229 assume that the PDSCH is transmitted on antenna ports 7 to 10 when using 4×4 MIMO, and on antenna ports 7 to 14 when using 8×8 MIMO

★ Deriving the maximum expected throughput requires the overheads generated by the other physical channels and physical signals to be taken into account. The following assumptions are made for this analysis:

- the PCFICH, PDCCH and PHICH occupy either the minimum or maximum allowed symbols per subframe (see section 6.2)
- the cell specific Reference Signal is only transmitted from antenna port 0 (see section 5.2.1)
- no MBSFN subframes are configured so the cell specific Reference Signal is transmitted during all subframes
- the UE specific Reference Signal occupies 24 Resource Elements within each Resource Block pair (see section 5.2.3)
- in the case of 4×4 MIMO, the CSI Reference Signal occupies 4 Resource Elements within each Resource Block pair, during subframes that include CSI Reference Signal transmissions. The periodicity is assumed to be 10 ms (see section 5.2.5)
- in the case of 8×8 MIMO, the CSI Reference Signal occupies 8 Resource Elements within each Resource Block pair, during subframes that include CSI Reference Signal transmissions. The CSI Reference Signal periodicity is assumed to be 10 ms
- the PBCH and Synchronisation Signals occupy a total of 564 Resource Elements per radio frame (see sections 6.1 and 5.1)

★ These assumptions lead to the physical layer throughputs presented in Table 230. These throughputs do not yet account for the redundancy added by the physical layer. Nor do they account for the overheads generated by the various protocol stack headers and retransmissions. Nor do they account for the overheads generated by RRC signalling

			Channel Bandwidth	1.4 MHz	3 MHz	5 MHz	10 MHz	15 MHz	20 MHz
			Resource Blocks	6	15	25	50	75	100
Uplink-Downlink Subframe Configuration 0	Minimum PDCCH Symbols	4×4 MIMO 64QAM Bit Rate (Mbps)	Special SF Config 0	2.4	9.5	16.6	34.3	52.1	69.9
			Special SF Config 4	4.8	15.3	26.1	53.2	80.4	107.5
		8×8 MIMO 64QAM Bit Rate (Mbps)	Special SF Config 0	4.7	18.7	32.7	67.7	102.8	137.8
			Special SF Config 4	9.5	30.2	51.7	105.5	159.3	213.0
		8×8 MIMO 64QAM 2 Component Carriers Bit Rate (Mbps)	Special SF Config 0	9.3	37.3	65.4	135.4	205.5	275.6
			Special SF Config 4	19.0	60.5	103.5	211.0	318.5	426.0
		8×8 MIMO 64QAM 5 Component Carriers Bit Rate (Mbps)	Special SF Config 0	23.3	93.3	163.4	338.6	513.8	689.0
			Special SF Config 4	46.9	149.7	256.2	522.4	788.7	1055.0
	Maximum PDCCH Symbols	4×4 MIMO 64QAM Bit Rate (Mbps)	Special SF Config 0	1.7	6.9	12.3	25.7	39.1	52.6
			Special SF Config 4	4.1	12.7	21.8	44.6	67.4	90.2
		8×8 MIMO 64QAM Bit Rate (Mbps)	Special SF Config 0	3.3	13.5	24.0	50.4	76.8	103.2
			Special SF Config 4	8.1	25.1	43.1	88.2	133.3	178.5
		8×8 MIMO 64QAM 2 Component Carriers Bit Rate (Mbps)	Special SF Config 0	6.6	27.0	48.1	100.9	153.7	206.5
			Special SF Config 4	16.2	50.1	86.2	176.4	266.7	356.9
		8×8 MIMO 64QAM 5 Component Carriers Bit Rate (Mbps)	Special SF Config 0	16.4	67.4	120.2	252.2	384.2	516.2
			Special SF Config 4	40.6	125.3	215.5	441.1	666.7	892.3
Uplink-Downlink Subframe Configuration 5	Minimum PDCCH Symbols	4×4 MIMO 64QAM Bit Rate (Mbps)	Special SF Config 0	12.1	35.8	60.5	122.2	183.9	245.5
			Special SF Config 4	13.3	38.7	65.3	131.6	198.0	264.3
		8×8 MIMO 64QAM Bit Rate (Mbps)	Special SF Config 0	24.0	71.4	120.5	243.4	366.3	489.2
			Special SF Config 4	26.4	77.2	130.1	262.3	394.5	526.8
		8×8 MIMO 64QAM 2 Component Carriers Bit Rate (Mbps)	Special SF Config 0	48.0	142.7	241.0	486.8	732.6	978.3
			Special SF Config 4	52.9	154.3	260.1	524.6	789.1	1053.5
		8×8 MIMO 64QAM 5 Component Carriers Bit Rate (Mbps)	Special SF Config 0	120.1	356.8	602.6	1217.0	1831.4	2445.8
			Special SF Config 4	132.2	385.8	650.3	1311.5	1972.7	2633.9
	Maximum PDCCH Symbols	4×4 MIMO 64QAM Bit Rate (Mbps)	Special SF Config 0	9.6	28.9	49.0	99.1	149.3	199.5
			Special SF Config 4	10.9	28.8	48.7	98.5	148.3	198.1
		8×8 MIMO 64QAM Bit Rate (Mbps)	Special SF Config 0	19.2	57.5	97.5	197.3	297.2	397.0
			Special SF Config 4	21.6	57.3	96.9	196.1	295.2	394.3
		8×8 MIMO 64QAM 2 Component Carriers Bit Rate (Mbps)	Special SF Config 0	38.4	115.1	195.0	394.6	594.3	794.0
			Special SF Config 4	43.2	114.6	193.9	392.1	590.3	788.6
		8×8 MIMO 64QAM 5 Component Carriers Bit Rate (Mbps)	Special SF Config 0	95.9	287.7	487.4	986.6	1485.8	1985.0
			Special SF Config 4	108.0	286.4	484.7	980.3	1475.9	1971.5

Table 230 – Maximum TDD physical layer throughputs based upon Resource Elements available to the PDSCH (LTE Advanced) (coding rate = 1, no retransmissions, no SRB, paging nor SIB overheads, no protocol stack overheads)

★ The 4×4 MIMO throughputs in Table 230 can be compared with those in Table 227. The 'LTE Advanced' throughputs for 4×4 MIMO (using antenna ports 7 to 10) are less than those based upon the release 8 and 9 versions of the specifications (using antenna ports 0 to 3). This is a result of the relative overheads generated by the cell specific Reference Signal and UE specific Reference Signal

★ The throughputs shown in Table 230 increase in direct proportion to the number of Component Carriers because it is assumed that each Component Carrier has the same configuration. This does not have to be the case in practice, e.g. one Component Carrier could have 2 OFDMA symbols reserved for the PCFICH, PDCCH and PHICH, while another Component Carrier could have 3 OFDMA symbols reserved

★ Table 231 presents a set of example application throughputs assuming physical layer coding rates of 0.75 and 0.95, as well as a 10 % retransmission rate and an additional 5 % overhead generated by a combination of the SIB, paging, RRC signalling and protocol stack headers

★ The maximum throughput generated by 8×8 MIMO with 5 Component Carriers is 2139.3 Mbps. This is obtained when assuming the minimum OFDMA symbol allocation for the PCFICH, PDCCH and PHICH. The figure decreases to 1601.3 Mbps when assuming the maximum OFDMA symbol allocation for the PCFICH, PDCCH and PHICH. These throughput figures are high but require 100 MHz of spectrum. The equivalent figures for 40 MHz of spectrum are 855.7 and 640.5 Mbps

★ These throughput figures are peak connection throughputs rather than average cell throughputs. They assume that the UE is benefiting from 64QAM and full rank MIMO transmission while being scheduled resources on all Component Carriers

			Channel Bandwidth	1.4 MHz	3 MHz	5 MHz	10 MHz	15 MHz	20 MHz
			Resource Blocks	6	15	25	50	75	100
Uplink-Downlink Subframe Configuration 5 Special Subframe Configuration 4	Minimum PDCCH Symbols	4x4 MIMO 64QAM Bit Rate (Mbps)	Coding Rate = 0.75	8.5	24.8	41.9	84.4	127.0	169.5
			Coding Rate = 0.95	10.8	31.5	53.0	106.9	160.8	214.7
		8x8 MIMO 64QAM Bit Rate (Mbps)	Coding Rate = 0.75	17.0	49.5	83.4	168.2	253.0	337.8
			Coding Rate = 0.95	21.5	62.7	105.6	213.0	320.5	427.9
		8x8 MIMO 64QAM 2 Component Carriers Bit Rate (Mbps)	Coding Rate = 0.75	33.9	99.0	166.8	336.4	506.0	675.6
			Coding Rate = 0.95	42.9	125.3	211.3	426.1	640.9	855.7
		8x8 MIMO 64QAM 5 Component Carriers Bit Rate (Mbps)	Coding Rate = 0.75	84.8	247.4	417.0	841.0	1265.0	1689.0
			Coding Rate = 0.95	107.4	313.3	528.2	1065.2	1602.3	2139.3
	Maximum PDCCH Symbols	4x4 MIMO 64QAM Bit Rate (Mbps)	Coding Rate = 0.75	7.0	18.5	31.2	63.2	95.1	127.0
			Coding Rate = 0.95	8.8	23.4	39.6	80.0	120.5	160.9
		8x8 MIMO 64QAM Bit Rate (Mbps)	Coding Rate = 0.75	13.9	36.7	62.2	125.7	189.3	252.8
			Coding Rate = 0.95	17.5	46.5	78.7	159.2	239.8	320.3
		8x8 MIMO 64QAM 2 Component Carriers Bit Rate (Mbps)	Coding Rate = 0.75	27.7	73.5	124.3	251.4	378.6	505.7
			Coding Rate = 0.95	35.1	93.1	157.5	318.5	479.5	640.5
		8x8 MIMO 64QAM 5 Component Carriers Bit Rate (Mbps)	Coding Rate = 0.75	69.3	183.7	310.8	628.6	946.4	1264.2
			Coding Rate = 0.95	87.7	232.6	393.7	796.2	1198.8	1601.3

Table 231 – TDD application layer throughputs based upon Resource Elements available to the PDSCH (LTE Advanced) (10% retransmissions, 5% additional overheads)

★ 3GPP References: TS 36.211, TR 36.912

20.2 UPLINK BIT RATES

20.2.1 FDD

★ The Physical Uplink Shared Channel (PUSCH) is used to transfer application data. The throughput achieved by the PUSCH depends upon the:

- o the number of Resource Elements allocated to the PUSCH
- o the modulation scheme applied to each Resource Element
- o the quantity of redundancy included by physical layer processing
- o the use of multiple antenna transmission schemes
- o the use of Carrier Aggregation

★ The number of Resource Elements allocated to the PUSCH depends upon the channel bandwidth and the choice between the normal and extended cyclic prefix. It also depends upon the overheads generated by the other physical channels and physical signals

★ The modulation scheme and quantity of redundancy depend upon the RF channel conditions. UE experiencing good channel conditions are more likely to be allocated higher order modulation schemes with less redundancy

★ Multiple antenna transmission schemes increase the throughput achieved by the PUSCH. 2×2 MIMO doubles the peak throughput whereas 4×4 MIMO quadruples the peak throughput. In the case of the uplink, MIMO forms part of LTE Advanced introduced within the release 10 version of the 3GPP specifications

★ The PUSCH is a shared channel so its throughput capability has to be shared between all users. Increasing the number of users reduces the throughput per user. Users experiencing poor channel conditions will reduce the total cell throughput

★ Table 232 presents a set of theoretical figures which represent the absolute maximum physical layer throughputs which could be achieved if all Resource Elements were allocated to the PUSCH and the physical layer did not add any redundancy. These figures are not achievable in practice but provide a starting point from which to derive the maximum expected throughputs

	Channel Bandwidth	1.4 MHz	3 MHz	5 MHz	10 MHz	15 MHz	20 MHz
	Resource Blocks in the frequency domain	6	15	25	50	75	100
Normal Cyclic Prefix	SC-FDMA symbols per 1 ms	14					
	Modulation symbol rate (Msps)	1.0	2.5	4.2	8.4	12.6	16.8
	QPSK Bit Rate (Mbps)	2.0	5.0	8.4	16.8	25.2	33.6
	16QAM Bit Rate (Mbps)	4.0	10.1	16.8	33.6	50.4	67.2
	64QAM Bit Rate (Mbps)	6.0	15.1	25.2	50.4	75.6	100.8
Extended Cyclic Prefix	SC-FDMA symbols per 1 ms	12					
	Modulation symbol rate (Msps)	0.9	2.2	3.6	7.2	10.8	14.4
	QPSK Bit Rate (Mbps)	1.7	4.3	7.2	14.4	21.6	28.8
	16QAM Bit Rate (Mbps)	3.5	8.6	14.4	28.8	43.2	57.6
	64QAM Bit Rate (Mbps)	5.2	13.0	21.6	43.2	64.8	86.4

Table 232 – Absolute maximum FDD physical layer throughputs if all Resource Elements were allocated to the PUSCH

★ The figures presented in Table 232 are equal to those presented in Table 220 for the PDSCH. The impact of MIMO has been excluded at this stage to focus upon the throughput which can be achieved when using the release 8 and 9 versions of the 3GPP specifications

★ The first step to deriving the maximum expected throughput is to remove the overheads generated by the other physical channels and physical signals. These include the PUCCH, PRACH, Demodulation Reference Signal for the PUSCH, Demodulation Reference Signal for the PUCCH and Sounding Reference Signal

★ Table 233 presents a set of maximum physical layer throughputs with the overheads generated by the PUCCH and the Demodulation Reference Signals taken into account. The PUCCH is assumed to occupy 2, 4, 6 or 8 Resource Blocks per timeslot. The PUCCH cannot occupy 6 or 8 Resource Blocks when the channel bandwidth is 1.4 MHz because the total number of Resource Blocks in the frequency domain is only 6. The overheads generated by the PRACH and Sounding Reference Signal have been ignored in this table. The results assume a coding rate of 1, i.e. the physical layer has not introduced any redundancy

★ Table 233 illustrates that when using the 20 MHz channel bandwidth with the normal cyclic prefix and 64QAM modulation, the maximum physical layer throughput decreases from 100.8 Mbps to between 84.7 and 79.5 depending upon the number of Resource Blocks allocated to the PUCCH (the throughput could decrease further if more than 8 Resource Blocks were allocated to the PUCCH)

★ The overhead generated by the PRACH depends upon the random access preamble configuration. The minimum configuration allows a single 1 ms PRACH preamble every 2 radio frames. The maximum configuration allows ten 1 ms PRACH preambles every radio frame (or five 2 ms PRACH preambles every radio frame)

		Channel Bandwidth	1.4 MHz	3 MHz	5 MHz	10 MHz	15 MHz	20 MHz
Normal Cyclic Prefix	2 PUCCH Res. Blks	QPSK Bit Rate (Mbps)	1.2	3.7	6.6	13.8	21.0	28.2
		16QAM Bit Rate (Mbps)	2.3	7.5	13.2	27.6	42.0	56.4
		64QAM Bit Rate (Mbps)	3.5	11.2	19.9	41.5	63.1	84.7
	4 PUCCH Res. Blks	QPSK Bit Rate (Mbps)	0.6	3.2	6.0	13.2	20.4	27.6
		16QAM Bit Rate (Mbps)	1.2	6.3	12.1	26.5	40.9	55.3
		64QAM Bit Rate (Mbps)	1.7	9.5	18.1	39.7	61.3	82.9
	6 PUCCH Res. Blks	QPSK Bit Rate (Mbps)	-	2.6	5.5	12.7	19.9	27.1
		16QAM Bit Rate (Mbps)	-	5.2	10.9	25.3	39.7	54.1
		64QAM Bit Rate (Mbps)	-	7.8	16.4	38.0	59.6	81.2
	8 PUCCH Res. Blks	QPSK Bit Rate (Mbps)	-	2.0	4.9	12.1	19.3	26.5
		16QAM Bit Rate (Mbps)	-	4.0	9.8	24.2	38.6	53.0
		64QAM Bit Rate (Mbps)	-	6.0	14.7	36.3	57.9	79.5
Extended Cyclic Prefix	2 PUCCH Res. Blks	QPSK Bit Rate (Mbps)	1.0	3.1	5.5	11.5	17.5	23.5
		16QAM Bit Rate (Mbps)	1.9	6.2	11.0	23.0	35.0	47.0
		64QAM Bit Rate (Mbps)	2.9	9.4	16.6	34.6	52.6	70.6
	4 PUCCH Res. Blks	QPSK Bit Rate (Mbps)	0.5	2.6	5.0	11.0	17.0	23.0
		16QAM Bit Rate (Mbps)	1.0	5.3	10.1	22.1	34.1	46.1
		64QAM Bit Rate (Mbps)	1.4	7.9	15.1	33.1	51.1	69.1
	6 PUCCH Res. Blks	QPSK Bit Rate (Mbps)	-	2.2	4.6	10.6	16.6	22.6
		16QAM Bit Rate (Mbps)	-	4.3	9.1	21.1	33.1	45.1
		64QAM Bit Rate (Mbps)	-	6.5	13.7	33.1	49.7	67.7
	8 PUCCH Res. Blks	QPSK Bit Rate (Mbps)	-	1.7	4.1	10.1	16.1	22.1
		16QAM Bit Rate (Mbps)	-	3.4	8.2	20.2	32.2	44.2
		64QAM Bit Rate (Mbps)	-	5.0	12.2	30.2	48.2	66.2

Table 233 – Maximum FDD physical layer throughputs based upon Resource Elements available to the PUSCH (coding rate = 1, no retransmissions, no SRB or protocol stack overheads, no PRACH or Sounding Reference Signal overheads)

★ Table 234 presents a set of maximum physical layer throughputs assuming a single 2ms PRACH transmission per radio frame. Similar to Table 233, the overheads generated by the PUCCH and Demodulation Reference Signals have been taken into account but the Sounding Reference Signal has been ignored and a coding rate of 1 has been assumed

★ The overhead generated by a single 2 ms PRACH transmission per radio frame has a more significant impact upon the throughputs associated with the smaller channel bandwidths. The impact is relatively small for the larger channel bandwidths

★ Redundancy added by the physical layer further reduces the throughputs measured at the top of the physical layer. Similar to the PDSCH, the PUSCH uses a combination of rate 1/3 Turbo coding and rate matching to generate redundancy. In general, the quantity of physical layer redundancy is relatively large when UE experience poor channel conditions and relatively small when UE experience good channel conditions

★ The uplink link adaptation strategy is likely to be similar to the downlink link adaptation strategy. An example is shown in Figure 192 (within section 20.1.1). The coding rate reflects the quantity of redundancy added by the physical layer. A low coding rate indicates a large quantity of redundancy while a high coding rate reflects a small quantity of redundancy. A coding rate of 1 corresponds to no redundancy

★ QPSK and a low coding rate are associated with poor channel conditions. Link adaptation allocates larger transport block sizes as the channel conditions improve but the modulation scheme is kept as QPSK. This forces the quantity of redundancy to decrease (and the coding rate to increase), i.e. larger quantities of data are transferred without increasing the capacity of the physical channel

★ In the Figure 192 example, the modulation scheme is switched from QPSK to 16QAM once the channel conditions have improved sufficiently to allow the coding rate to increase to 0.75. Switching the modulation scheme increases the capacity of the physical channel so the quantity of redundancy can be increased. Link adaptation then continues to allocate larger transport block sizes as the channel conditions improve. The modulation scheme is switched from 16QAM to 64QAM once the channel conditions have improved sufficiently to allow the coding rate to again reach 0.75

★ Once 64QAM has been allocated, link adaptation continues to allocate larger transport block sizes as the channel conditions improve. In this case, there is no option to switch to a higher order modulation scheme once the coding rate reaches 0.75. Instead, link adaptation continues to allocate larger transport block sizes and the coding rate approaches 1

		Channel Bandwidth	1.4 MHz	3 MHz	5 MHz	10 MHz	15 MHz	20 MHz
Normal Cyclic Prefix	2 PUCCH Res. Blks	QPSK Bit Rate (Mbps)	0.9	3.4	6.3	13.5	20.7	27.9
		16QAM Bit Rate (Mbps)	1.8	6.8	12.6	27.0	41.4	55.8
		64QAM Bit Rate (Mbps)	2.8	10.2	18.8	40.4	62.0	83.6
	4 PUCCH Res. Blks	QPSK Bit Rate (Mbps)	0.5	2.8	5.7	12.9	20.1	27.3
		16QAM Bit Rate (Mbps)	0.9	5.6	11.4	25.8	40.2	54.6
		64QAM Bit Rate (Mbps)	1.4	8.5	17.1	38.7	60.3	81.9
	6 PUCCH Res. Blks	QPSK Bit Rate (Mbps)	-	2.2	5.1	12.3	19.5	26.7
		16QAM Bit Rate (Mbps)	-	4.5	10.3	24.7	39.1	53.5
		64QAM Bit Rate (Mbps)	-	6.7	15.4	37.0	58.6	80.2
	8 PUCCH Res. Blks	QPSK Bit Rate (Mbps)	-	1.7	4.6	11.8	19.0	26.2
		16QAM Bit Rate (Mbps)	-	3.3	9.1	23.5	37.9	52.3
		64QAM Bit Rate (Mbps)	-	5.0	13.7	35.3	56.9	78.5
Extended Cyclic Prefix	2 PUCCH Res. Blks	QPSK Bit Rate (Mbps)	0.8	2.8	5.2	11.2	17.2	23.2
		16QAM Bit Rate (Mbps)	1.5	5.7	10.5	22.5	34.5	46.5
		64QAM Bit Rate (Mbps)	2.3	8.5	15.7	33.7	51.7	69.7
	4 PUCCH Res. Blks	QPSK Bit Rate (Mbps)	0.4	2.4	4.8	10.8	16.8	22.8
		16QAM Bit Rate (Mbps)	0.8	4.7	9.5	21.5	33.5	45.5
		64QAM Bit Rate (Mbps)	1.2	7.1	14.3	32.3	50.3	68.3
	6 PUCCH Res. Blks	QPSK Bit Rate (Mbps)	-	1.9	4.3	10.3	16.3	22.3
		16QAM Bit Rate (Mbps)	-	3.7	8.5	20.5	32.5	44.5
		64QAM Bit Rate (Mbps)	-	5.6	12.8	30.8	48.8	66.8
	8 PUCCH Res. Blks	QPSK Bit Rate (Mbps)	-	1.4	3.8	9.8	15.8	21.8
		16QAM Bit Rate (Mbps)	-	2.8	7.6	19.6	31.6	43.6
		64QAM Bit Rate (Mbps)	-	4.2	11.4	29.4	47.4	65.4

Table 234 – Maximum FDD physical layer throughputs based upon Resource Elements available to the PUSCH (coding rate = 1, no retransmissions, no SRB or protocol stack overheads, no Sounding Reference Signal overhead)

- ★ RRC signalling is transferred using the PUSCH and this reduces the PUSCH capacity available for application data. The overhead will depend upon the quantity of traffic loading the cell but is likely to be relatively small, i.e. less than 50 kbps
- ★ Retransmissions reduce the higher layer throughputs. Hybrid Automatic Repeat Request (HARQ) retransmissions from the MAC layer reduce the throughputs measured from above the MAC layer. Automatic Repeat Request (ARQ) retransmissions from the RLC layer reduce the throughputs measured from above the RLC layer. Likewise TCP retransmissions reduce the throughput measured from above the TCP layer
- ★ Protocol stack headers also reduce the higher layer throughputs. The MAC, RLC, PDCP and IP layers add headers to the application data. The PDCP layer provides header compression for IP data streams so is able to reduce the impact of the IP header. The TCP layer also adds its own header when using TCP applications
- ★ Table 235 presents a set of example application layer throughputs assuming physical layer coding rates of 0.75 and 0.95. The coding rate of 0.95 is only shown for the 64QAM modulation scheme to remain consistent with the link adaptation strategy shown in Figure 192. Table 235 assumes a single 2 ms PRACH preamble per radio frame, 10 % retransmission rate and an additional 5 % overhead generated by a combination of RRC signalling and protocol stack headers
- ★ The maximum uplink throughput within Table 235 is 67.9 Mbps. This throughput is provided by the 20 MHz channel bandwidth when 2 Resource Blocks are allocated to the PUCCH and the coding rate is 0.95. Allocating additional Resource Blocks to the PUCCH has relatively little impact when the channel bandwidth is 20 MHz

			Coding Rate	1.4 MHz	3 MHz	5 MHz	10 MHz	15 MHz	20 MHz
Normal Cyclic Prefix	2 PUCCH Res. Blks	QPSK Bit Rate (Mbps)	0.75	0.6	2.2	4.0	8.6	13.3	17.9
		16QAM Bit Rate (Mbps)	0.75	1.2	4.4	8.1	17.3	26.5	35.8
		64QAM Bit Rate (Mbps)	0.75	1.8	6.5	12.1	25.9	39.8	53.6
			0.95	2.2	8.3	15.3	32.8	50.4	67.9
	4 PUCCH Res. Blks	QPSK Bit Rate (Mbps)	0.75	0.3	1.8	3.7	8.3	12.9	17.5
		16QAM Bit Rate (Mbps)	0.75	0.6	3.6	7.3	16.5	25.8	35.0
		64QAM Bit Rate (Mbps)	0.75	0.9	5.4	11.0	24.8	38.7	52.5
			0.95	1.1	6.9	13.9	31.4	49.0	66.5
	6 PUCCH Res. Blks	QPSK Bit Rate (Mbps)	0.75	-	1.4	3.3	7.9	12.5	17.1
		16QAM Bit Rate (Mbps)	0.75	-	2.9	6.6	15.8	25.0	34.3
		64QAM Bit Rate (Mbps)	0.75	-	4.3	9.9	23.7	37.6	51.4
			0.95	-	5.5	12.5	30.0	47.6	65.1
	8 PUCCH Res. Blks	QPSK Bit Rate (Mbps)	0.75	-	1.1	2.9	7.5	12.2	16.8
		16QAM Bit Rate (Mbps)	0.75	-	2.1	5.8	15.1	24.3	33.5
		64QAM Bit Rate (Mbps)	0.75	-	3.2	8.8	22.6	36.5	50.3
			0.95	-	4.1	11.1	28.6	46.2	63.7
Extended Cyclic Prefix	2 PUCCH Res. Blks	QPSK Bit Rate (Mbps)	0.75	0.5	1.8	3.4	7.2	11.1	14.9
		16QAM Bit Rate (Mbps)	0.75	1.0	3.6	6.7	14.4	22.1	29.8
		64QAM Bit Rate (Mbps)	0.75	1.5	5.4	10.1	21.6	33.2	44.7
			0.95	1.9	6.9	12.7	27.4	42.0	56.6
	4 PUCCH Res. Blks	QPSK Bit Rate (Mbps)	0.75	0.2	1.5	3.0	6.9	10.7	14.6
		16QAM Bit Rate (Mbps)	0.75	0.5	3.0	6.1	13.8	21.5	29.2
		64QAM Bit Rate (Mbps)	0.75	0.7	4.5	9.1	20.7	32.2	43.8
			0.95	0.9	5.7	11.6	26.2	40.8	55.4
	6 PUCCH Res. Blks	QPSK Bit Rate (Mbps)	0.75	-	1.2	2.7	6.6	10.4	14.3
		16QAM Bit Rate (Mbps)	0.75	-	2.4	5.5	13.2	20.9	28.6
		64QAM Bit Rate (Mbps)	0.75	-	3.6	8.2	19.8	31.3	42.8
			0.95	-	4.6	10.4	25.0	39.7	54.3
	8 PUCCH Res. Blks	QPSK Bit Rate (Mbps)	0.75	-	0.9	2.4	6.3	10.1	14.0
		16QAM Bit Rate (Mbps)	0.75	-	1.8	4.9	12.6	20.3	27.9
		64QAM Bit Rate (Mbps)	0.75	-	2.7	7.3	18.8	30.4	41.9
			0.95	-	3.4	9.2	23.9	38..5	53.1

Table 235 – FDD application layer throughputs based upon Resource Elements available to the PUSCH (single 2 ms PRACH preamble per radio frame, 10% retransmissions, 5% additional overheads)

★ 3GPP TR 36.912 specifies an LTE Advanced peak spectral efficiency of 16.8 bps/Hz when using 4×4 MIMO with FDD. This figure is calculated as follows:

- o total number of Resource Elements per antenna port within a 10 ms radio frame when using the normal cyclic prefix and the 20 MHz channel bandwidth is given by 12 × 14 × 100 × 10 = 168 000 Resource Elements
- o accounting for 4 antenna ports increases the total number of Resource Elements to 4 × 168 000 = 672 000 Resource Elements
- o assuming that the PUCCH and PUCCH Demodulation Reference Signal occupy only a single Resource Block at each edge of the channel bandwidth generates an overhead of 2 / 100 = 2 %
- o assuming that there is a single PRACH time and frequency resource allocation within each radio frame, and that the PRACH occupies 6 Resource Blocks in the frequency domain and 1 ms in the time domain, generates an overhead of 6 × 12 × 14 / 168 000 = 0.6 %
- o assuming that the PUSCH Demodulation Reference Signal occupies 1 column of Resource Elements within each PUSCH Resource Block generates an overhead of (12 × (98 × 10 × 2 - 6 × 2)) / 168 000 = 13.9 %
- o summing the set of overheads generates a total overhead of (2 + 0.6 + 13.9) = 16.5 %

- this overhead reduces the number of Resource Elements available to the PUSCH to 672 000 × (1 – 0.165) = 561 120 Resource Elements. Assuming that 64QAM is applied to each Resource Element generates a physical layer throughput which is given by (561 120 × 6) / 0.010 = 336.672 Mbps
- the spectral efficiency is then obtained after dividing by the channel bandwidth of 20 MHz, i.e. (336.672 / 20) = 16.8 bps/Hz

★ In practice, the spectral efficiency will be less than 16.8 bps/Hz when using larger PUCCH and PRACH resource allocations

★ Table 236 presents a set of theoretical absolute maximum physical layer throughputs for a selection of LTE Advanced scenarios including both 4×4 MIMO and Carrier Aggregation. Figures are only presented for the normal cyclic prefix. These throughputs are the equivalent of those presented in Table 232. They assume that all Resource Elements are allocated to the PUSCH and that the physical layer does not add any redundancy. These figures are not achievable in practice but provide a starting point from which to derive the maximum expected throughputs

	Channel Bandwidth	1.4 MHz	3 MHz	5 MHz	10 MHz	15 MHz	20 MHz
	Resource Blocks in the frequency domain	6	15	25	50	75	100
Normal Cyclic Prefix	2×2 MIMO 64QAM Bit Rate (Mbps)	12.1	30.2	50.4	100.8	151.2	201.6
	4×4 MIMO 64QAM Bit Rate (Mbps)	24.2	60.5	100.8	201.6	302.4	403.2
	4×4 MIMO 64QAM, 2 Component Carriers Bit Rate (Mbps)	48.4	121.0	201.6	403.2	604.8	806.4
	4×4 MIMO 64QAM, 5 Component Carriers Bit Rate (Mbps)	121.0	302.4	504.0	1008.0	1512.0	2016.0

Table 236 – Absolute maximum FDD physical layer throughputs if all Resource Elements were allocated to the PUSCH (LTE Advanced)

★ The figures presented in Table 236 are half of the equivalent downlink figures presented in Table 223 because the uplink supports a maximum of 4×4 MIMO, whereas the downlink supports a maximum of 8×8 MIMO

★ Table 237 presents the set of LTE Advanced throughputs after accounting for the various overheads: PUCCH, Demodulation Reference Signal for PUCCH, Demodulation Reference Signal for PUSCH, PRACH, 10 % HARQ retransmission rate and an additional 5 % overhead generated by a combination of RRC signalling and protocol stack headers. These assumptions are the same as those used for the 3GPP release 8 and 9 throughputs presented in Table 235

★ The throughput figures in Table 237 are multiples of the throughputs presented in Table 235 because the use of 2×2 MIMO doubles the peak connection throughput and the use of 4×4 MIMO quadruples the peak connection throughput. It is also assumed that each Component Carrier has the same configuration so the throughput increases in direct proportion to the number of RF carriers

			Coding Rate	1.4 MHz	3 MHz	5 MHz	10 MHz	15 MHz	20 MHz
Normal Cyclic Prefix	2 PUCCH Res. Blocks	2×2 MIMO 64QAM Bit Rate (Mbps)	0.75	3.5	13.1	24.2	51.9	79.6	107.3
			0.95	4.5	16.6	30.6	65.7	100.8	135.9
		4×4 MIMO 64QAM Bit Rate (Mbps)	0.75	7.1	26.2	48.3	103.7	159.1	214.5
			0.95	9.0	33.1	61.2	131.4	201.6	271.7
		4×4 MIMO 64QAM, 2 Component Carriers Bit Rate (Mbps)	0.75	14.2	52.3	96.6	207.4	318.2	429.0
			0.95	18.0	66.2	122.4	262.7	403.1	543.5
		4×4 MIMO 64QAM, 5 Component Carriers Bit Rate (Mbps)	0.75	35.5	130.8	241.6	518.6	795.6	1072.6
			0.95	44.9	165.6	306.0	656.9	1007.8	1358.7
	4 PUCCH Res. Blocks	2×2 MIMO 64QAM Bit Rate (Mbps)	0.75	1.8	10.9	21.9	49.6	77.3	105.0
			0.95	2.2	13.8	27.8	62.9	98.0	133.1
		4×4 MIMO 64QAM Bit Rate (Mbps)	0.75	3.5	21.7	43.9	99.3	154.7	210.1
			0.95	4.5	27.5	55.6	125.8	195.9	266.1
		4×4 MIMO 64QAM, 2 Component Carriers Bit Rate (Mbps)	0.75	7.1	43.4	87.8	198.6	309.4	420.2
			0.95	9.0	55.0	111.2	251.5	391.9	532.2
		4×4 MIMO 64QAM, 5 Component Carriers Bit Rate (Mbps)	0.75	17.7	108.6	219.4	496.4	773.4	1050.5
			0.95	22.5	137.5	277.9	628.8	979.7	1330.6

Table 237 – FDD application layer throughputs based upon Resource Elements available to the PUSCH (LTE Advanced)
(single 2 ms PRACH preamble per radio frame, 10% retransmissions, 5% additional overheads)

★ These throughput figures are peak connection throughputs rather than average cell throughputs. They assume that the UE is benefiting from 64QAM and full rank MIMO transmission while being scheduled resources on all Component Carriers

★ 3GPP References: TS 36.211, TR 36.912

20.2.2 TDD

★ The analysis of uplink bit rates for TDD is complicated by the set of uplink-downlink subframe configurations:

  ○ the set of 7 uplink-downlink subframe configurations is presented in Table 12 within section 3.2.2

★ For the purposes of the analysis within this section:

  ○ uplink-downlink subframe configurations 0 and 5 have been selected because these have the most and the least number of uplink subframes. This remains consistent with the analysis of downlink bit rates for TDD presented in section 20.1.2

★ In the case of the uplink, the analysis is not dependent upon the special subframe configuration because the PUSCH cannot be transmitted during the UpPTS field within a special subframe, i.e. application data cannot be transferred during the UpPTS field

★ Similar to FDD, the Physical Uplink Shared Channel (PUSCH) is used to transfer application data, and the throughput achieved depends upon the number of Resource Elements, the modulation scheme, the quantity of redundancy included by physical layer processing, the use of multiple antenna transmission schemes and the use of Carrier Aggregation

★ Table 238 presents a set of theoretical absolute maximum physical layer throughputs which could be achieved if all Resource Elements were allocated to the PUSCH and the physical layer did not add any redundancy. These figures are not achievable in practice but provide a starting point from which to derive the maximum expected throughputs

★ All throughputs are averaged across a 10 ms radio frame so the impact of the downlink and special subframes is captured. Only figures for the normal cyclic prefix are presented

	Channel Bandwidth	1.4 MHz	3 MHz	5 MHz	10 MHz	15 MHz	20 MHz
	Resource Blocks	6	15	25	50	75	100
Uplink-Downlink Subframe Configuration 0	QPSK Bit Rate (Mbps)	1.2	3.0	5.0	10.1	15.1	20.2
	16QAM Bit Rate (Mbps)	2.4	6.0	10.1	20.2	30.2	40.3
	64QAM Bit Rate (Mbps)	3.6	9.1	15.1	30.2	45.4	60.5
Uplink-Downlink Subframe Configuration 5	QPSK Bit Rate (Mbps)	0.2	0.5	0.8	1.7	2.5	3.4
	16QAM Bit Rate (Mbps)	0.4	1.0	1.7	3.4	5.0	6.7
	64QAM Bit Rate (Mbps)	0.6	1.5	2.5	5.0	7.6	10.1

Table 238 – Absolute maximum TDD physical layer throughputs if all Resource Elements were allocated to the PUSCH

★ These throughputs can be compared with the equivalent FDD figures in Table 232. Throughputs are significantly lower when using TDD uplink-downlink subframe configuration 5 because there is only 1 uplink subframe. Uplink-downlink subframe configuration 0 has 6 uplink subframes so the throughputs are closer to the equivalent FDD figures

★ TDD throughputs are lower than the equivalent FDD throughputs but only a single RF carrier is used rather than separate uplink-downlink RF carriers

★ The first step to deriving the maximum expected throughput is to remove the overheads generated by the other physical channels and physical signals. These include the PUCCH, PRACH, Demodulation Reference Signal for the PUSCH, Demodulation Reference Signal for the PUCCH and Sounding Reference Signal

★ Table 239 presents a set of maximum physical layer throughputs with the overheads generated by the PUCCH and the Demodulation Reference Signals taken into account. The PUCCH is assumed to occupy 2, 4, 6 or 8 Resource Blocks per timeslot. The PUCCH cannot occupy 6 or 8 Resource Blocks when the channel bandwidth is 1.4 MHz because the total number of Resource Blocks in the frequency domain is only 6. The overheads generated by the PRACH and Sounding Reference Signal have been ignored in this table. The results assume a coding rate of 1, i.e. the physical layer has not introduced any redundancy

★ Table 239 illustrates that when using the 20 MHz channel bandwidth with uplink-downlink subframe configuration 0 and 64QAM, the maximum physical layer throughput decreases from 60.5 Mbps to between 50.8 and 47.7 depending upon the number of Resource Blocks allocated to the PUCCH (the throughput could decrease further if more than 8 Resource Blocks were allocated to the PUCCH)

★ The overhead generated by the PRACH depends upon the random access preamble configuration. In the case of TDD, the short preamble format can be used to transmit the PRACH during the UpPTS field of special subframes. This can avoid having any impact upon the Resource Blocks available to the PUSCH

★ Redundancy added by the physical layer further reduces the throughputs measured at the top of the physical layer. Similar to the PDSCH, the PUSCH uses a combination of rate 1/3 Turbo coding and rate matching to generate redundancy. In general, the quantity of physical layer redundancy is relatively large when UE experience poor channel conditions and relatively small when UE experience good channel conditions

		Channel Bandwidth	1.4 MHz	3 MHz	5 MHz	10 MHz	15 MHz	20 MHz
Uplink-Downlink Subframe Configuration 0	2 PUCCH Res. Blks	QPSK Bit Rate (Mbps)	0.7	2.2	4.0	8.3	12.6	16.9
		16QAM Bit Rate (Mbps)	1.4	4.5	7.9	16.6	25.2	33.9
		64QAM Bit Rate (Mbps)	2.1	6.7	11.9	24.9	37.8	50.8
	4 PUCCH Res. Blks	QPSK Bit Rate (Mbps)	0.3	1.9	3.6	7.9	12.3	16.6
		16QAM Bit Rate (Mbps)	0.7	3.8	7.3	15.9	24.5	33.2
		64QAM Bit Rate (Mbps)	1.0	5.7	10.9	23.8	36.8	49.8
	6 PUCCH Res. Blks	QPSK Bit Rate (Mbps)	-	1.6	3.3	7.6	11.9	16.2
		16QAM Bit Rate (Mbps)	-	3.1	6.6	15.2	23.8	32.5
		64QAM Bit Rate (Mbps)	-	4.7	9.8	22.8	35.8	48.7
	8 PUCCH Res. Blks	QPSK Bit Rate (Mbps)	-	1.2	2.9	7.3	11.6	15.9
		16QAM Bit Rate (Mbps)	-	2.4	5.9	14.5	23.2	31.8
		64QAM Bit Rate (Mbps)	-	3.6	8.8	21.8	34.7	47.7
Uplink-Downlink Subframe Configuration 5	2 PUCCH Res. Blks	QPSK Bit Rate (Mbps)	0.1	0.4	0.7	1.4	2.1	2.8
		16QAM Bit Rate (Mbps)	0.2	0.7	1.3	2.8	4.2	5.6
		64QAM Bit Rate (Mbps)	0.3	1.1	2.0	4.1	6.3	8.5
	4 PUCCH Res. Blks	QPSK Bit Rate (Mbps)	0.1	0.3	0.6	1.3	2.0	2.8
		16QAM Bit Rate (Mbps)	0.1	0.6	1.2	2.6	4.1	5.5
		64QAM Bit Rate (Mbps)	0.2	1.0	1.8	4.0	6.1	8.3
	6 PUCCH Res. Blks	QPSK Bit Rate (Mbps)	-	0.3	0.5	1.3	2.0	2.7
		16QAM Bit Rate (Mbps)	-	0.5	1.1	2.5	4.0	5.4
		64QAM Bit Rate (Mbps)	-	0.8	1.6	3.8	6.0	8.1
	8 PUCCH Res. Blks	QPSK Bit Rate (Mbps)	-	0.2	0.5	1.2	1.9	2.6
		16QAM Bit Rate (Mbps)	-	0.4	1.0	2.4	3.9	5.3
		64QAM Bit Rate (Mbps)	-	0.6	1.5	3.6	5.8	7.9

Table 239 – Maximum TDD physical layer throughputs based upon Resource Elements available to the PUSCH (coding rate = 1, no retransmissions, no SRB or protocol stack overheads, no PRACH or Sounding Reference Signal overheads)

★ The uplink link adaptation strategy is likely to be similar to the downlink link adaptation strategy. An example is shown in Figure 192 (within section 20.1.1). The coding rate reflects the quantity of redundancy added by the physical layer. A low coding rate indicates a large quantity of redundancy while a high coding rate reflects a small quantity of redundancy. A coding rate of 1 corresponds to no redundancy

★ QPSK and a low coding rate are associated with poor channel conditions. Link adaptation allocates larger transport block sizes as the channel conditions improve but the modulation scheme is kept as QPSK. This forces the quantity of redundancy to decrease (and the coding rate to increase), i.e. larger quantities of data are transferred without increasing the capacity of the physical channel

★ RRC signalling is transferred using the PUSCH and this reduces the PUSCH capacity available for application data. The overhead will depend upon the quantity of traffic loading the cell but is likely to be relatively small, i.e. less than 50 kbps

★ Retransmissions reduce the higher layer throughputs. Hybrid Automatic Repeat Request (HARQ) retransmissions from the MAC layer reduce the throughputs measured from above the MAC layer. Automatic Repeat Request (ARQ) retransmissions from the RLC layer reduce the throughputs measured from above the RLC layer. Likewise TCP retransmissions reduce the throughput measured from above the TCP layer

★ Protocol stack headers also reduce the higher layer throughputs. The MAC, RLC, PDCP and IP layers add headers to the application data. The PDCP layer provides header compression for IP data streams so is able to reduce the impact of the IP header. The TCP layer also adds its own header when using TCP applications

★ Table 240 presents a set of example application throughputs assuming physical layer coding rates of 0.75 and 0.95. The coding rate of 0.95 is only shown for the 64QAM modulation scheme to remain consistent with the link adaptation strategy shown in Figure 192. Table 240 assumes a 10 % retransmission rate and an additional 5 % overhead generated by a combination of RRC signalling and protocol stack headers. The overhead generated by the PRACH has been excluded based upon the assumption that the PRACH is transmitted during the UpPTS field of the special subframes. In practice, the PRACH may have an allocation within the normal uplink subframes

★ The maximum uplink throughput within Table 240 is 41.3 Mbps. This throughput is provided by the 20 MHz channel bandwidth when 2 Resource Blocks are allocated to the PUCCH and the coding rate is 0.95. Allocating additional Resource Blocks to the PUCCH has relatively little impact when the channel bandwidth is 20 MHz

			Coding Rate	1.4 MHz	3 MHz	5 MHz	10 MHz	15 MHz	20 MHz
Uplink-Downlink Subframe Configuration 0	2 PUCCH Res. Blks	QPSK Bit Rate (Mbps)	0.75	0.4	1.4	2.5	5.3	8.1	10.9
		16QAM Bit Rate (Mbps)	0.75	0.9	2.9	5.1	10.6	16.2	21.7
		64QAM Bit Rate (Mbps)	0.75	1.3	4.3	7.6	16.0	24.3	32.6
			0.95	1.7	5.5	9.7	20.2	30.7	41.3
	4 PUCCH Res. Blks	QPSK Bit Rate (Mbps)	0.75	0.2	1.2	2.3	5.1	7.9	10.6
		16QAM Bit Rate (Mbps)	0.75	0.4	2.4	4.7	10.2	15.7	21.3
		64QAM Bit Rate (Mbps)	0.75	0.7	3.7	7.0	15.3	23.6	31.9
			0.95	0.8	4.6	8.8	19.4	29.9	40.4
	6 PUCCH Res. Blks	QPSK Bit Rate (Mbps)	0.75	-	1.0	2.1	4.9	7.6	10.4
		16QAM Bit Rate (Mbps)	0.75	-	2.0	4.2	9.8	15.3	20.8
		64QAM Bit Rate (Mbps)	0.75	-	3.0	6.3	14.6	22.9	31.2
			0.95	-	3.8	8.0	18.5	29.1	39.6
	8 PUCCH Res. Blks	QPSK Bit Rate (Mbps)	0.75	-	0.8	1.9	4.7	7.4	10.2
		16QAM Bit Rate (Mbps)	0.75	-	1.6	3.8	9.3	14.8	20.4
		64QAM Bit Rate (Mbps)	0.75	-	2.3	5.7	14.0	22.3	30.6
			0.95	-	2.9	7.2	17.7	28.2	38.7
Uplink-Downlink Subframe Configuration 5	2 PUCCH Res. Blks	QPSK Bit Rate (Mbps)	0.75	0.1	0.2	0.4	0.9	1.3	1.8
		16QAM Bit Rate (Mbps)	0.75	0.1	0.5	0.8	1.8	2.7	3.6
		64QAM Bit Rate (Mbps)	0.75	0.2	0.7	1.3	2.7	4.0	5.4
			0.95	0.3	0.9	1.6	3.4	5.1	6.9
	4 PUCCH Res. Blks	QPSK Bit Rate (Mbps)	0.75	< 0.1	0.2	0.4	0.8	1.3	1.8
		16QAM Bit Rate (Mbps)	0.75	0.1	0.4	0.8	1.7	2.6	3.5
		64QAM Bit Rate (Mbps)	0.75	0.1	0.6	1.2	2.5	3.9	5.3
			0.95	0.1	0.8	1.5	3.2	5.0	6.7
	6 PUCCH Res. Blks	QPSK Bit Rate (Mbps)	0.75	-	0.2	0.4	0.8	1.3	1.7
		16QAM Bit Rate (Mbps)	0.75	-	0.3	0.7	1.6	2.5	3.5
		64QAM Bit Rate (Mbps)	0.75	-	0.5	1.1	2.4	3.8	5.2
			0.95	-	0.6	1.3	3.1	4.8	6.6
	8 PUCCH Res. Blks	QPSK Bit Rate (Mbps)	0.75	-	0.1	0.3	0.8	1.2	1.7
		16QAM Bit Rate (Mbps)	0.75	-	0.3	0.6	1.6	2.5	3.4
		64QAM Bit Rate (Mbps)	0.75	-	0.4	0.9	2.3	3.7	5.1
			0.95	-	0.5	1.2	2.9	4.7	6.5

Table 240 – TDD application layer throughputs based upon Resource Elements available to the PUSCH (10% retransmissions, 5% additional overheads, PRACH assumed to be within UpPTS)

★ 3GPP TR 36.912 specifies an LTE Advanced peak spectral efficiency of 16.1 bps/Hz when using 4×4 MIMO with TDD. This figure is calculated as follows:

- total number of Resource Elements per antenna port within a 10 ms radio frame when using the normal cyclic prefix, the 20 MHz channel bandwidth and uplink-downlink subframe configuration 1 is given by 12 × 14 × 100 × 4 = 67 200 Resource Elements
- accounting for 4 antenna ports increases the total number of Resource Elements to 4 × 67 200 = 268 800 Resource Elements
- assuming that the PUCCH and PUCCH Demodulation Reference Signal occupy only a single Resource Block at each edge of the channel bandwidth generates an overhead of 2 / 100 = 2 %
- assuming that there is a single PRACH time and frequency resource allocation within each radio frame, and that the PRACH occupies 6 Resource Blocks in the frequency domain and 1 ms in the time domain, generates an overhead of 6 × 12 × 14 / 67 200 = 1.5 %
- assuming that the PUSCH Demodulation Reference Signal occupies 1 column of Resource Elements within each PUSCH Resource Block generates an overhead of (12 × (98 × 4 × 2 – 6 × 2)) / 67 200 = 13.8 %
- summing the set of overheads generates a total overhead of (2 + 1.5 + 13.8) = 17.3 %

- this overhead reduces the number of Resource Elements available to the PUSCH to 268 800 × (1 – 0.173) = 222 298 Resource Elements. Assuming that 64QAM is applied to each Resource Element generates a physical layer throughput which is given by (222 298 × 6) / 0.010 = 133.379 Mbps
- within a 10 ms radio frame the TDD uplink activity factor is defined by (14 × 4 + 1 × 2) / (14 × 10) = 41.4 % (assuming special subframe configuration 4)
- the spectral efficiency is then obtained by including the activity factor and dividing by the channel bandwidth of 20 MHz, i.e. 133.379 / (20 × 0.414) = 16.1 bps/Hz

★ In practice, the spectral efficiency will be less than 16.1 bps/Hz when using larger PUCCH and PRACH resource allocations

★ Table 241 presents a set of example LTE Advanced application layer throughputs assuming physical layer coding rates of 0.75 and 0.95, a 10 % HARQ retransmission rate and an additional 5 % overhead generated by a combination of RRC signalling and protocol stack headers

			Coding Rate	1.4 MHz	3 MHz	5 MHz	10 MHz	15 MHz	20 MHz
Uplink-Downlink Subframe Configuration 0	2 PUCCH Res. Blocks	2x2 MIMO 64QAM Bit Rate (Mbps)	0.75	2.7	8.6	15.3	31.9	48.5	65.2
			0.95	3.4	10.9	19.4	40.4	61.5	82.5
		4x4 MIMO 64QAM Bit Rate (Mbps)	0.75	5.3	17.3	30.6	63.8	97.1	130.3
			0.95	6.7	21.9	38.7	80.8	123.0	165.1
		4x4 MIMO 64QAM, 2 Component Carriers Bit Rate (Mbps)	0.75	10.6	34.6	61.2	127.7	194.1	260.6
			0.95	13.5	43.8	77.5	161.7	245.9	330.1
		4x4 MIMO 64QAM, 5 Component Carriers Bit Rate (Mbps)	0.75	26.6	86.4	152.9	319.1	485.3	651.6
			0.95	33.7	109.5	193.7	404.2	614.8	825.3
	4 PUCCH Res. Blocks	2x2 MIMO 64QAM Bit Rate (Mbps)	0.75	1.3	7.3	14.0	30.6	47.2	63.8
			0.95	1.7	9.3	17.7	38.7	59.8	80.8
		4x4 MIMO 64QAM Bit Rate (Mbps)	0.75	2.7	14.6	27.9	61.2	94.4	127.7
			0.95	3.4	18.5	35.4	77.5	119.6	161.7
		4x4 MIMO 64QAM, 2 Component Carriers Bit Rate (Mbps)	0.75	5.3	29.3	55.8	122.3	188.8	255.3
			0.95	6.7	37.1	70.7	155.0	239.2	323.4
		4x4 MIMO 64QAM, 5 Component Carriers Bit Rate (Mbps)	0.75	13.3	73.1	139.6	305.8	472.0	638.3
			0.95	16.8	92.6	176.8	387.4	597.9	808.5

Table 241 – TDD application layer throughputs based upon Resource Elements available to the PUSCH (LTE Advanced) (10% retransmissions, 5% additional overheads, PRACH assumed to be within UpPTS)

★ The throughput figures in Table 241 are multiples of the throughputs presented in Table 240 because the use of 2×2 MIMO doubles the peak connection throughput and the use of 4×4 MIMO quadruples the peak connection throughput. It is also assumed that each Component Carrier has the same configuration so the throughput increases in direct proportion to the number of RF carriers

★ 3GPP References: TS 36.211, TR 36.912

21 MEASUREMENTS

21.1 UE MEASUREMENTS

21.1.1 RSRP

★ Reference Signal Received Power (RSRP) is the equivalent of the UMTS CPICH Received Signal Code Power (RSCP)

★ RSRP measurements are used for cell selection, cell reselection and handover. RSRP is also used to estimate the path loss for power control calculations

★ RSRP is the average power received from a single cell specific Reference Signal Resource Element

- o the average is taken in linear units
- o the power measurement is based upon the energy received during the useful part of the OFDMA symbol and excludes the energy of the cyclic prefix
- o the reference point for the RSRP measurement is the antenna connector of the UE

★ RSRP can be based upon the cell specific Reference Signal transmitted by only the first antenna port. Alternatively, RSRP can be based upon the cell specific Reference Signal transmitted by the first and second antenna ports

★ SIB3 informs the UE of the maximum measurement bandwidth for intra-frequency cell reselection. It also informs the UE of whether or not a second antenna port can be used for measurements. SIB5 provides the same information for inter-frequency cell reselection

★ If a UE is equipped with receive diversity then the reported RSRP is never lower than the RSRP measured by any individual receive branch

★ A mapping is applied to RSRP measurements prior to including them within RRC messages. The RSRP measurements are mapped onto an integer value between 0 and 97. This mapping is shown in Table 242

Reported	Actual (dBm)	Reported	Actual (dBm)
0	RSRP < -140	87	-54 ≤ RSRP < -53
1	-140 ≤ RSRP < -139	88	-53 ≤ RSRP < -52
2	-139 ≤ RSRP < -138	89	-52 ≤ RSRP < -51
3	-138 ≤ RSRP < -137	90	-51 ≤ RSRP < -50
4	-137 ≤ RSRP < -136	91	-50 ≤ RSRP < -49
5	-136 ≤ RSRP < -135	92	-49 ≤ RSRP < -48
6	-135 ≤ RSRP < -134	93	-48 ≤ RSRP < -47
7	-134 ≤ RSRP < -133	94	-47 ≤ RSRP < -46
..	...	95	-46 ≤ RSRP < -45
n	n-139 ≤ RSRP < n-140	96	-45 ≤ RSRP < -44
..	...	97	-44 ≤ RSRP

Table 242 – Mapping between reported and measured RSRP values

★ The maximum reportable RSRP is based upon the -25 dBm maximum input power for a UE as specified by 3GPP TS 36.101. The 1.4 MHz channel bandwidth has 72 Resource Elements in the frequency domain. RSRP is based upon the power of a single Resource Element so the maximum RSRP equals -25 - 10×LOG(72) = -44 dBm

★ The minimum reportable value is based upon assumptions of a maximum path loss of 152 dB, a transmit power of 43 dBm and a 5 MHz channel bandwidth (300 Resource Elements). These assumptions lead to a minimum RSRP of 43 - 152 - 10×LOG(300) = -134 dBm. An addition 6 dB has been subtracted to provide some margin

★ The absolute measurement accuracy for intra-frequency RSRP measurements under normal conditions is specified by 3GPP TS 36.133 to be between ± 6 and ± 8 dB. The relative measurement accuracy between two intra-frequency measurements is specified to be between ± 2 and ±3 dB

★ The absolute measurement accuracy for inter-frequency RSRP measurements under normal conditions is specified by 3GPP TS 36.133 to be between ± 6 and ± 8 dB. The relative measurement accuracy between an intra-frequency measurement and an inter-frequency measurement is specified to be ±6 dB

★ 3GPP References: TS 36.133, TS 36.214

21.1.2 RSRQ

★ Reference Signal Received Quality (RSRQ) is the equivalent of UMTS CPICH Ec/Io. RSRQ measurements are used for cell selection, cell reselection and handover

★ RSRQ is defined as

$$RSRQ = \frac{RSRP}{(RSSI / N)}$$

where, N is the number of Resource Blocks over which the Received Signal Strength Indicator (RSSI) is measured

★ The RSSI is calculated as a linear average of the total power measured across OFDMA symbols which contain Reference Symbols transmitted from the first antenna port, e.g. symbols 0 and 4 when MIMO is not used

★ The reference point for the RSRQ measurement is the antenna connector of the UE. If a UE is equipped with receive diversity then the reported RSRQ is never lower than the RSRQ calculated for any individual receive branch

★ A mapping is applied to RSRQ measurements prior to including them within RRC messages. The RSRQ measurements are mapped onto an integer value between 0 and 34. This mapping is shown in Table 243

Reported	Actual (dB)	Reported	Actual (dB)
0	RSRQ < -19.5	18	-11 ≤ RSRQ < -10.5
1	-19.5 ≤ RSRQ < -19	19	-10.5 ≤ RSRQ < -10
2	-19 ≤ RSRQ < -18.5	20	-10 ≤ RSRQ < -9.5
3	-18.5 ≤ RSRQ < -18	21	-9.5 ≤ RSRQ < -9
4	-18 ≤ RSRQ < -17.5	22	-9 ≤ RSRQ < -8.5
5	-17.5 ≤ RSRQ < -17	23	-8.5 ≤ RSRQ < -8
6	-17 ≤ RSRQ < -16.5	24	-8 ≤ RSRQ < -7.5
7	-16.5 ≤ RSRQ < -16	25	-7.5 ≤ RSRQ < -7
8	-16 ≤ RSRQ < -15.5	26	-7 ≤ RSRQ < -6.5
9	-15.5 ≤ RSRQ < -15	27	-6.5 ≤ RSRQ < -6
10	-15 ≤ RSRQ < -14.5	28	-6 ≤ RSRQ < -5.5
11	-14.5 ≤ RSRQ < -14	29	-5.5 ≤ RSRQ < -5
12	-14 ≤ RSRQ < -13.5	30	-5 ≤ RSRQ < -4.5
13	-13.5 ≤ RSRQ < -13	31	-4.5 ≤ RSRQ < -4
14	-13 ≤ RSRQ < -12.5	32	-4 ≤ RSRQ < -3.5
15	-12.5 ≤ RSRQ < -12	33	-3.5 ≤ RSRQ < -3
16	-12 ≤ RSRQ < -11.5	34	-3 ≤ RSRQ
17	-11.5 ≤ RSRQ < -11		

Table 243 – Mapping between reported and measured RSRQ values

★ The maximum reportable RSRQ is based upon the assumption that only the cell specific Reference Signal Resource Elements are occupied, i.e. no traffic is transferred. There are 2 cell specific Reference Signal Resource Elements per OFDMA symbol so the calculation becomes:

$$RSRQ = \frac{RSRP}{(RSSI / N)} = \frac{RSRP}{(2 \times RSRP \times N / N)} = 0.5 = -3dB$$

★ The absolute measurement accuracy for intra-frequency RSRQ measurements under normal conditions is specified by 3GPP TS 36.133 to be between ± 2.5 and ± 3.5 dB. A relative measurement accuracy is not specified for the intra-frequency case

★ The absolute measurement accuracy for inter-frequency RSRQ measurements under normal conditions is specified by 3GPP TS 36.133 to be between ± 2.5 and ± 3.5 dB. The relative measurement accuracy between an intra-frequency measurement and an inter-frequency measurement is specified to be between ±3 and ±4 dB

★ 3GPP References: TS 36.133, TS 36.214

21.1.3 RSTD

- ★ The Reference Signal Time Difference (RSTD) measurement was introduced within the release 9 version of the 3GPP specifications
- ★ RSTD measurements are used for location based services
- ★ The Observed Time Difference of Arrival (OTDOA) positioning method uses Reference Signal Time Difference (RSTD) measurements from the UE. The RSTD quantifies the subframe timing difference between a reference cell and a neighbouring cell. The accuracy of the positioning calculation is improved if the UE can provide RSTD measurements from an increased number of cells
- ★ RSTD measurements can be completed using the positioning Reference Signal which was also introduced within the release 9 version of the 3GPP specifications. Positioning Reference Signals help to improve the 'hearability' of neighbouring cells which can be difficult to measure as a result of them being co-channel with the serving cell, especially at locations where the serving cell signal strength is high
- ★ RSTD is measured in units of Ts (1/30720 ms) and is reported to the Enhanced Serving Mobile Location Centre (E-SMLC) where the location calculation is completed. The E-SMLC is a network element within the Enhanced Packet Core
- ★ The mapping between the reported RSTD and measured RSTD is presented in Table 244. The step size between adjacent values is 5 × Ts for larger time differences, and 1 × Ts for smaller time differences

Reported	Measured (units of Ts)	Step Size (Ts)
0	-15391 > RSTD	5
1	-15391 ≤ RSTD < -15386	5
2	-15386 ≤ RSTD < -15381	5
…	…	…
2258	-4106 ≤ RSTD < -4101	5
2259	-4101 ≤ RSTD < -4096	5
2260	-4096 ≤ RSTD < -4095	1
2261	-4095 ≤ RSTD < -4094	1
…	…	…
6353	-3 ≤ RSTD < -2	1
6354	-2 ≤ RSTD < -1	1
6355	-1 ≤ RSTD ≤ 0	1
6356	0 < RSTD ≤ 1	1
6357	1 < RSTD ≤ 2	1
6358	2 < RSTD ≤ 3	1
…	…	…
10450	4094 < RSTD ≤ 4095	1
10451	4095 < RSTD ≤ 4096	1
10452	4096 < RSTD ≤ 4101	5
10453	4101 < RSTD ≤ 4106	5
…	…	…
12709	15381 < RSTD ≤ 15386	5
12710	15386 < RSTD ≤ 15391	5
12711	15391 < RSTD	5

Table 244 – Mapping between reported and measured RSTD values

- ★ The accuracy of RSTD measurements is greatest for intra-frequency measurements with positioning Reference Signals broadcast across a wide bandwidth. The accuracy of RSTD measurements is reduced for inter-frequency measurements with positioning Reference Signals broadcast across a narrow bandwidth. Accuracy requirements are specified within 3GPP TS 36.133
- ★ 3GPP References: TS 36.133, TS 36.214

21.1.4 RX-TX TIME DIFFERENCE

★ The UE Rx-Tx Time Difference measurement was introduced within the release 9 version of the 3GPP specifications

★ UE Rx-Tx Time Difference measurements are used for location based services

★ The Enhanced Cell Identity method can use UE Rx-Tx Time Difference measurements to help identify the location of a UE within a cell

★ The network based Timing Advance Type 1 measurement is calculated from the UE Rx-Tx Time Difference, i.e. Timing Advance Type 1 = (eNode B Rx-Tx Time Difference) + (UE Rx-Tx Time Difference)

★ UE Rx-Tx Time Difference is measured in units of Ts (1/30720 ms)

★ The mapping between the reported time difference and the measured time difference is presented in Table 245. The step size between adjacent values is 2 × Ts for smaller time differences, and 8 × Ts for larger time differences

Reported	Measured (units of Ts)	Step Size (Ts)
0	UE Rx-Tx < 2	2
1	2 ≤ UE Rx-Tx < 4	2
2	4 ≤ UE Rx-Tx < 6	2
…	…	…
2046	4092 ≤ UE Rx-Tx < 4094	2
2047	4094 ≤ UE Rx-Tx < 4096	2
2048	4096 ≤ UE Rx-Tx < 4104	8
2049	4104 ≤ UE Rx-Tx < 4112	8
…	…	…
4093	20456 ≤ UE Rx-Tx < 20464	8
4094	20464 ≤ UE Rx-Tx < 20472	8
4095	20472 ≤ UE Rx-Tx	8

Table 245 – Mapping between reported and measured UE Rx-Tx Time Difference values

★ The accuracy of UE Rx-Tx Time Difference measurements is greatest for the wider channel bandwidths. Bandwidths less than or equal to 3 MHz are specified with a more relaxed accuracy requirement

★ 3GPP References: TS 36.133, TS 36.214

21.2 NETWORK MEASUREMENTS

21.2.1 REFERENCE SIGNAL TX POWER

- The Downlink Reference Signal Transmit Power measurement was introduced within the release 8 version of the 3GPP specifications
- It is calculated as a linear average of transmit power measurements (in Watts) from Resource Elements occupied by cell specific Reference Signals
- It has been necessary to standardise this measurement because it is signalled to and used by the population of UE
  - Reference Signal Transmit Power is broadcast in SIB2 as part of the PDSCH configuration information
  - it is signalled using a value within the range -60 to 50 dBm so the linear average result must be converted into dBm prior to inclusion within SIB2
- Measurements are based upon antenna port 0, and antenna port 1 when available
- The reference point for the measurement is the transmit antenna connector
- 3GPP References: TS 36.214

21.2.2 RECEIVED INTERFERENCE POWER

- The Received Interference Power measurement was introduced within the release 8 version of the 3GPP specifications
- It is defined as the uplink received interference power (including thermal noise) measured within the bandwidth of a single Resource Block. Measurement results are generated for each Resource Block within the channel bandwidth
- The measurement was introduced to support a standardised approach to detecting and signalling interference overload conditions
  - Inter Cell Interference Coordination (ICIC) can use the X2 Application Protocol: Load Information message to exchange uplink interference power measurements between neighbouring eNode B
  - allows neighbouring eNode B to coordinate their use of the air-interface resources, e.g. an eNode B avoids scheduling a specific set of Resource Blocks to cell edge UE when a neighbouring eNode B reports high uplink interference power within those Resource Blocks
  - the Load Information message reports received interference power as either high, medium or low. The thresholds used to define these levels are provided by the O&M
- The reference point for the measurement is the receive antenna connector. If receive diversity is used then the measurement result is the linear average of the results from each diversity branch
- The mapping between the reported Received Interference Power measurement and the actual measurement is presented in Table 246

Reported	Measured (units of Ts)	Step Size (Ts)
0	RIP < -126.0	2
1	-126.0 ≤ RIP < -125.9	2
2	-125.9 ≤ RIP < -125.8	2
…		…
2046	-75.2 ≤ RIP < -75.1	2
2047	-75.1 ≤ RIP < -75.0	2
2047	-75.0 ≤ RIP	2

Table 246 – Mapping between reported and actual Received Interference Power (RIP) values

- 3GPP References: TS 36.133, TS 36.214

21.2.3 THERMAL NOISE POWER

- The Thermal Noise Power measurement was introduced within the release 8 version of the 3GPP specifications
- It is defined as the uplink thermal noise power measured across the complete channel bandwidth. It is measured at the same time as the Received Interference Power
- The measurement was introduced to support a standardised approach to detecting interference overload conditions
  - an Interference over Thermal (IoT) metric can be generated by subtracting the Thermal Noise Power from the Received Interference Power
  - provides a relative measure of uplink interference rather than the absolute measure provided by the Received Interference Power measurement
- The reference point for the measurement is the receive antenna connector. If receive diversity is used then the measurement result is the linear average of the results from each diversity branch
- 3GPP References: TS 36.214

21.2.4 RX-TX TIME DIFFERENCE

- The eNode B Rx-Tx Time Difference measurement was introduced within the release 9 version of the 3GPP specifications
- eNode B Rx-Tx Time Difference measurements are used for location based services
- The Enhanced Cell Identity method can use eNode B Rx-Tx Time Difference measurements to help identify the location of a UE within a cell
- Both the network based Timing Advance Type 1 and Timing Advance Type 2 make use of the eNode B Rx-Tx Time Difference
  - Timing Advance Type 1 = (eNode B Rx-Tx Time Difference) + (UE Rx-Tx Time Difference)
  - Timing Advance Type 2 = eNode B Rx-Tx Time Difference
- The reference point for the receive timing measurement is the receive antenna connector, whereas the reference point for the transmit timing measurement is the transmit antenna connector
- 3GPP References: TS 36.214

21.2.5 TIMING ADVANCE

- The Timing Advance measurement was introduced within the release 9 version of the 3GPP specifications
- Timing Advance measurements are used for location based services based upon the Enhanced Cell Identity method
- The Timing Advance measurement can be used to identify the distance of a UE from a cell when combined with the Timing Advance instruction used to control the UE transmit timing
  - the Timing Advance instruction (section 24.2) used to control the UE transmit timing represents the complete round trip time from the baseband processing within the eNode B to the baseband processing within the UE
  - when calculating the distance from a cell it is necessary to use the round trip time from the eNode B antenna to the UE antenna, i.e. the air-interface component of the round trip time
- These round trip times are illustrated in Figure 193

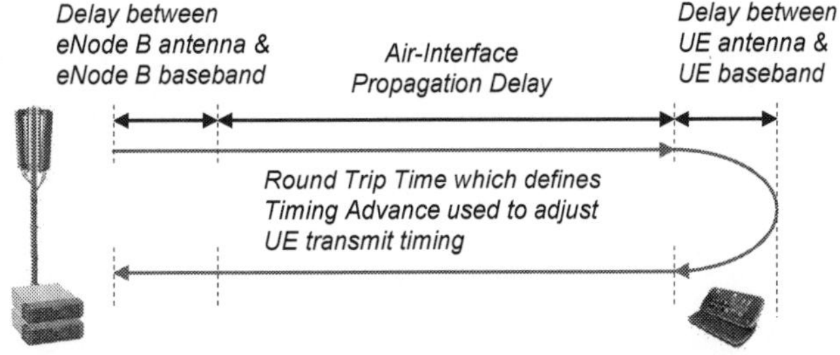

Figure 193 – Components of the total round trip time between the eNode B and UE baseband processing

- 3GPP has defined 2 types of Timing Advance measurements within the release 9 version of the specifications:
  - Timing Advance Type 1 = (eNode B Rx-Tx Time Difference) + (UE Rx-Tx Time Difference)
  - Timing Advance Type 2 = eNode B Rx-Tx Time Difference
- Timing Advance Type 1 is the preferred measurement which provides the greatest accuracy when calculating the UE distance from a cell. This measurement includes both the eNode B and UE components of the round trip time
- Timing Advance Type 2 is applicable when identifying the location of a release 8 UE. The UE Rx-Tx Time Difference measurement was introduced within the release 9 version of the specifications so is not available from release 8 UE
- Subtracting the Timing Advance measurement from the Timing Advance instruction provides the round trip time generated by the air-interface propagation delay
- The calculation should be relatively accurate even for release 8 UE because the eNode B Rx-Tx Time Difference is likely to be greater than the UE Rx-Tx Time Difference. The eNode B Rx-Tx Time Difference will depend upon the feeder length
- The Timing Advance measurement is quantified in units of Ts (1/30720 ms)
- The mapping between the reported Timing Advance measurement and the actual Timing Advance measurement is presented in Table 247. The step size between adjacent values is 2 × Ts for smaller time differences, and 8 × Ts for larger time differences

Reported	Measured (units of Ts)	Step Size (Ts)
0	Tadv < 2	2
1	2 ≤ Tadv < 4	2
2	4 ≤ Tadv < 6	2
...		...
2046	4092 ≤ Tadv < 4094	2
2047	4094 ≤ Tadv < 4096	2
2048	4096 ≤ Tadv < 4104	8
2049	4104 ≤ Tadv < 4112	8
...		...
7688	49216 ≤ Tadv < 49224	8
7689	49224 ≤ Tadv < 49232	8
7690	49232 ≤ Tadv	8

Table 247 – Mapping between reported and actual Timing Advance measurement values

- 3GPP References: TS 36.133, TS 36.214

21.2.6 ANGLE OF ARRIVAL

★ The Angle of Arrival (AoA) measurement was introduced within the release 9 version of the 3GPP specifications

★ AoA measurements are used for location based services using the Enhanced Cell Identity method

★ AoA measurements can also be used for downlink beamforming

★ The AoA measurement is used in combination with the Timing Advance measurement to identify the location of a UE within a cell.

  - the Timing Advance measurement defines a distance from the cell
  - the AoA measurement defines an angle from the cell

  This concept is illustrated in Figure 194

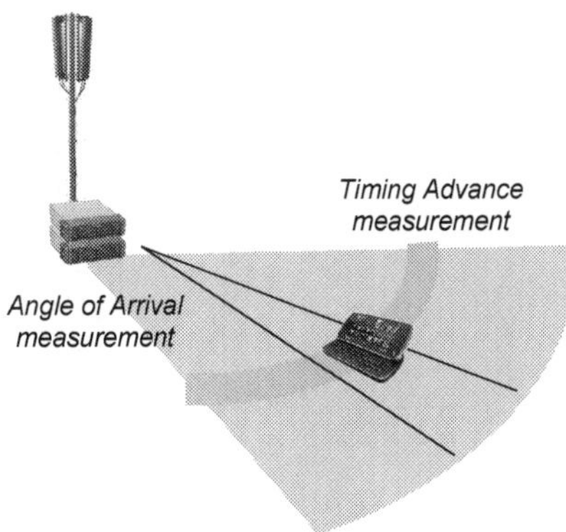

Figure 194 – Combination of Angle of Arrival and Timing Advance measurements to locate UE

★ The AoA measurement is defined as the estimated angle of a UE with respect to the geographical north direction. It is measured such that the result is positive in the counter clockwise direction

★ The mapping between the reported AoA and the measured AoA is presented in Table 248. The step size between adjacent values is 0.5 degrees

Reported	Measured (degrees)
0	$0 \leq AOA < 0.5$
1	$0.5 \leq AOA < 1.0$
2	$1.0 \leq AOA < 1.5$
…	…
717	$358.5 \leq AOA < 359.0$
718	$359.0 \leq AOA < 359.5$
719	$359.5 \leq AOA < 360.0$

Table 248 – Mapping between reported and measured AOA values

★ AoA measurements typically require an antenna array, and the accuracy of measurements increases for a larger number of antenna elements. Measurements are completed by identifying the relative delays (phases) of the signals received by each antenna element

★ The AoA measurement process is effectively the reverse of the beamforming process where each antenna element transmits the downlink signal with a different phase to steer the transmission in a particular direction

★ The eNode B can use the uplink Sounding Reference Signal (SRS) to complete AoA measurements

★ 3GPP References: TS 36.133, TS 36.214

22 MEASUREMENT REPORTING

22.1 INTRODUCTION

★ The RRC Connection Setup, RRC Connection Reconfiguration and RRC Connection Re-establishment messages are used to configure UE measurement reporting, i.e. they replace the Measurement Control message used by UMTS

★ The eNode B can request the UE to complete the following types of measurements:

- Intra-frequency: measurements on the same RF carrier as the serving cell(s)
- Inter-frequency: measurements on an LTE RF carrier which is not used by the serving cell(s)
- Inter-RAT UTRAN FDD and TDD
- Inter-RAT GERAN
- Inter-RAT CDMA2000 HRPD or CDMA2000 1xRTT

When Carrier Aggregation is used for LTE Advanced there are multiple serving cells and each serving cell has its own RF carrier. Measurements on any RF carrier belonging to a serving cell are categorised as intra-frequency

★ The configuration of a measurement is defined by:

- Measurement Objects
  - intra-frequency object defines a single LTE RF carrier
  - inter-frequency object defines a single LTE RF carrier
  - inter-RAT UTRAN object defines a single UTRAN RF carrier
  - inter-RAT GERAN object defines a set of GERAN RF carriers
  - inter-RAT CDMA2000 object defines a single CDMA2000 RF carrier

  Intra-frequency and inter-frequency measurement objects can specify individual cells to measure (whitelist), and individual cells to exclude from measurements (blacklist). Individual cells are referenced by their Physical layer Cell Identities (PCI). These objects can also request the UE to report the Cell Global Identity (CGI) for a specific PCI

  Inter-RAT UTRAN measurement objects can specify individual cells to measure (whitelist) by listing their scrambling codes. This object can also request the UE to report the CGI for a specific scrambling code

  Inter-RAT GERAN measurement objects can specify which Network Colour Codes (NCC) the UE is permitted to measure. This object can also request the UE to report the CGI for a specific BSIC

  Inter-RAT CDMA2000 measurement objects can specify individual cells to measure (whitelist) by listing their PN Offsets. This object can also request the UE to report the CGI for a specific PN Offset

- Reporting Configurations
  - for LTE, the triggering mechanism for sending a report can be either event driven or periodic. Event driven reports are based upon events A1, A2, A3, A4, A5 and A6. Periodic reports are based upon the expiry of a timer. The purpose of periodic reporting is signalled to be either 'reporting the strongest cells' or 'reporting a CGI'. The triggering quantity can be specified to be either RSRP or RSRQ, while the reported quantity can be specified to be either the same as the triggering quantity, or both RSRP and RSRQ. The maximum number of cells to report is also specified
  - for inter-RAT, the triggering mechanism for sending a report can be either event driven or periodic. Event driven reports are based upon events B1 and B2. Periodic reports are based upon the expiry of a timer. The purpose of periodic reporting is signalled to be either 'reporting the strongest cells', 'reporting the strongest cells for SON' or 'reporting a CGI'. The maximum number of cells to report is also specified. Triggering quantities are dependent upon the RAT, e.g. UTRAN events can be based upon either CPICH RSCP or CPICH Ec/Io

- Measurement Identities
  - links a measurement object to a reporting configuration. Configuring multiple measurement identities allows multiple measurement objects to be linked to the same reporting configuration, or multiple reporting configurations to be linked to the same measurement object. The measurement identity is used as a reference when the UE sends Measurement Report messages

- Quantity Configurations
  - defines the layer 3 filtering coefficients for measurements prior to event evaluation and measurement reporting. Filtering can be defined independently for each Radio Access Technology (RAT). Measurement quantities can also be specified for UTRAN, GERAN and CDMA2000

- Measurement Gaps
  - measurement gaps are defined to allow inter-frequency and/or inter-RAT measurements
  - the measurement gap configuration defines the gap pattern identity and the gap offset
  - gaps start during SFN and subframes defined by:

    SFN mod (MGRP / 10) = FLOOR (gapOffset / 10)

    Subframe = gapOffset mod 10

  where,

    the MGRP is the Measurement Gap Repetition Period

    the MGRP = 40 ms for gap pattern identity 0

    the MGRP = 80 ms for gap pattern identity 1

    the gapOffset can be configured with a value between 0 and 39 for gap pattern identity 0

    the gapOffset can be configured with a value between 0 and 79 for gap pattern identity 1

  As an example, when gap pattern identity 0 is configured with a gapOffset of 12 then measurement gaps start during subframe 2 of SFN 1, 5, 9, 13, 17, 21, etc

  - measurement gaps for both gap pattern identities 0 and 1 have durations of 6 ms
  - UE are not permitted to transmit any data during measurement gaps

★ Figure 195 illustrates an example of measurement identities being used to link a set of measurement objects to a set of reporting configurations. This example assumes that the UE is using Carrier Aggregation with 2 RF carriers. There is one measurement object per RF carrier, with the exception of GERAN for which there is one measurement object per group of RF carriers. Figure 195 includes examples of multiple measurement objects linked to a single reporting configuration, and multiple reporting configurations linked to a single measurement object

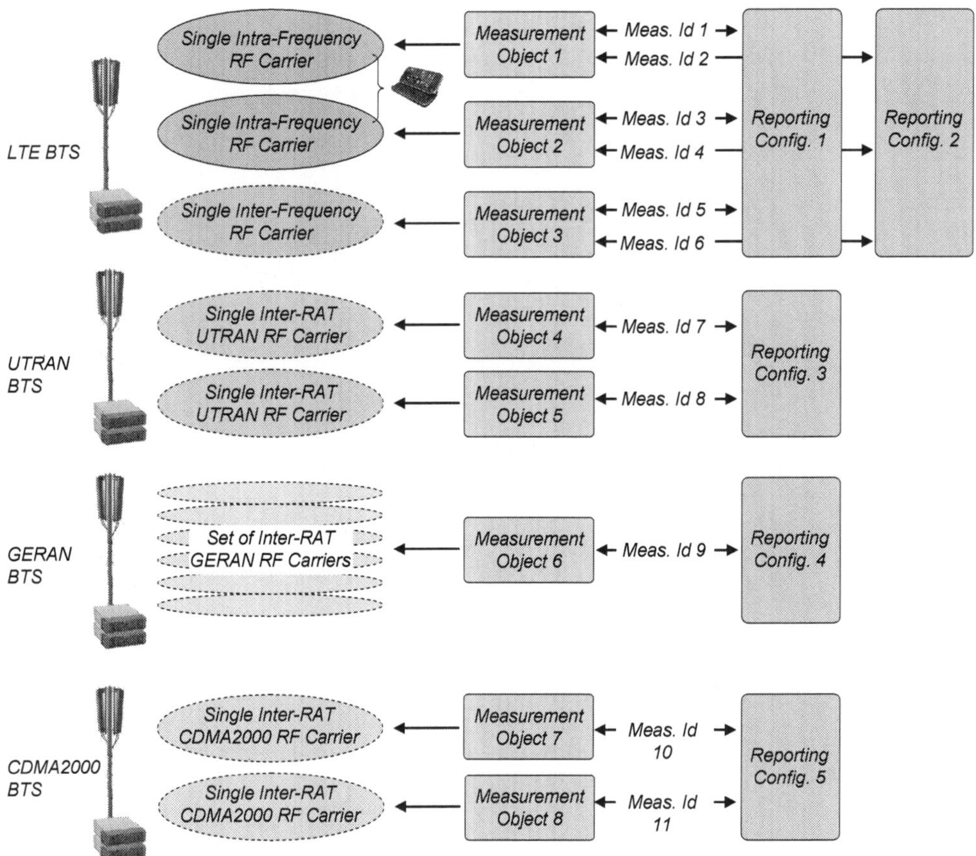

Figure 195 – Example of measurement identities used to link measurement objects to reporting configurations

- ★ It is not possible to configure more than a single measurement object per RF carrier (or set of RF carriers in the case of GERAN). It is possible to configure more than a single instance of the same reporting event by defining multiple reporting configurations with different parameter sets
- ★ Measurement procedures involve 3 types of cells:
    - o serving cell(s) – primary cell, and secondary cell(s) if Carrier Aggregation is used
    - o listed cells – cells which are specified within the measurement object
    - o detected cells – cells which are not specified within the measurement object but are detected by the UE
- ★ For LTE intra-frequency and inter-frequency measurements, the UE measures and reports serving cell(s), listed cells and detected cells
- ★ For UTRAN inter-RAT measurements, the UE measures and reports listed cells, and optionally cells with scrambling codes that are within a specific range (the UTRAN measurement object can specify a range of scrambling codes to measure). The UE can also measure and report detected cells for the purposes of SON
- ★ For GERAN inter-RAT measurements, the UE measures and reports detected cells
- ★ For CDMA2000 inter-RAT measurements, the UE measures and reports listed cells. UE can also measure and report detected cells for the purposes of SON
- ★ The release 8 version of 3GPP TS 36.331 defines five intra-system measurement reporting events:
    - o Event A1: serving cell becomes better than a threshold
    - o Event A2: serving cell becomes worse than a threshold
    - o Event A3: neighbour cell becomes better than the serving cell by an offset
    - o Event A4: neighbour cell becomes better than a threshold
    - o Event A5: serving cell becomes worse than threshold1 while neighbouring cell becomes better than threshold2
- ★ The release 10 version of 3GPP TS 36.331 defines one additional intra-system measurement reporting event which is applicable to LTE Advanced when carrier aggregation is used:
    - o Event A6: neighbour cell becomes better than a secondary cell by an offset
- ★ Each of these events can be based upon either RSRP or RSRQ
- ★ The neighbouring cells in reporting events A3, A4 and A5 can be either intra-frequency or inter-frequency
- ★ The neighbouring cell in reporting event A6 must be on the same frequency as the secondary cell, i.e. this reporting event can be used to trigger a change of the secondary cell on that frequency
- ★ The release 8 version of 3GPP TS 36.331 defines two inter-system measurement reporting events:
    - o Event B1: inter-system neighbour cell becomes better than a threshold
    - o Event B2: serving cell becomes worse than threshold1 while inter-system neighbouring cell becomes better than threshold2
- ★ Inter-system neighbour cell measurements can be based upon CPICH RSCP or CPICH Ec/Io for UMTS; RSSI for GSM and Pilot strength for CDMA2000
- ★ The criteria for triggering and subsequently cancelling each event are evaluated after layer 3 filtering has been applied
- ★ The criteria for each event must be satisfied during at least the time-to-trigger. The time-to-trigger can be configured independently for each reporting event, with values of {0, 40, 64, 80, 100, 128, 160, 256, 320, 480, 512, 640, 1024, 1280, 2560, 5120 ms}
- ★ 3GPP References: TS 36.331

22.2 LAYER 3 FILTERING

★ Layer 3 filtering is applied prior to evaluating whether or not any measurement reporting events have been triggered. It is applied using the equation below:

$$F_n = (1 - a) \times F_{n-1} + a \times Meas_n$$

where, F_n is the updated filtered measurement result
F_{n-1} is the previous filtered measurement result
$a = ½^{(k/4)}$ where k is the appropriate filter coefficient
$Meas_n$ is the latest measurement result received from the physical layer

★ The filter coefficient can be configured with values of {0, 1, 2, 3, 4, 5, 6, 7, 8, 9, 11, 13, 15, 17, 19}. Filtering is not applied if the filter coefficient is set equal to 0, i.e. a = 1

★ Figure 196 illustrates the impulse response for each of the filter coefficient values. The impulse response is generated by feeding a single value of 1 into the filter. This illustrates that the filter has greater memory when using larger filter coefficients

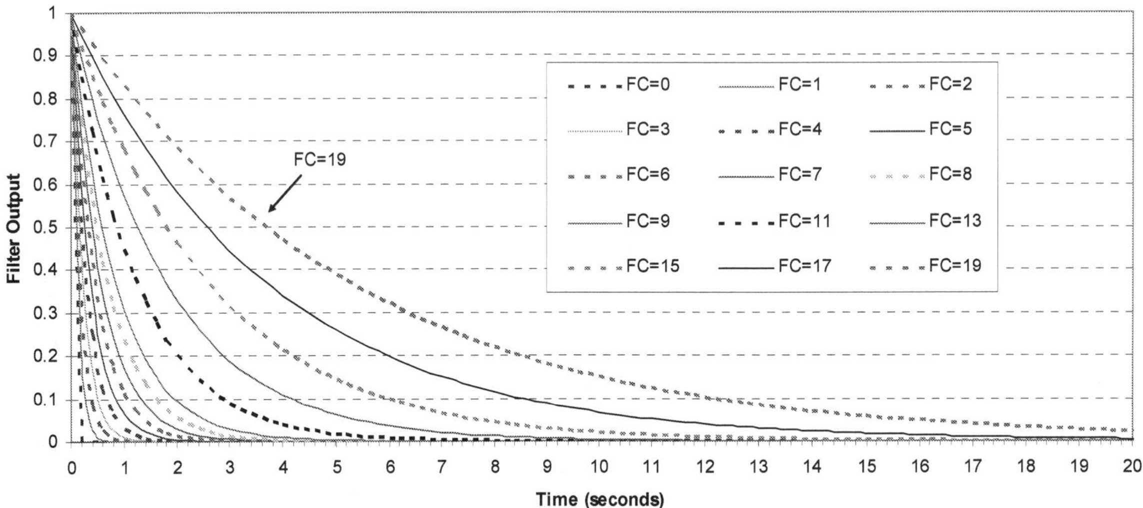

Figure 196 – Impact of filter coefficient

★ The definition of the filter coefficient assumes that measurement results arrive at a rate of one result per 200 ms. The filter is adapted to maintain the same impulse response if measurements arrive at a different rate

★ The filter coefficient can be configured independently for RSRP, RSRQ, UMTS CPICH RSCP, UMTS CPICH Ec/Io and GSM RSSI

★ Filtering is applied in the same domain as the measurements, e.g. filtering is applied in dB for RSRQ measurements, and dBm for RSRP measurements

★ 3GPP References: TS 36.331

22.3 EVENT A1

- Event A1 is triggered when the serving cell becomes better than a threshold. The event is triggered when the following condition is true:

$$Meas_{serv} - Hyst > Threshold$$

- Triggering of the event is subsequently cancelled when the following condition is true:

$$Meas_{serv} + Hyst < Threshold$$

- The hysteresis can be configured with a value between 0 and 30 dB
- When using RSRP, the threshold can be configured with a value between -140 and -44 dBm. The value of the threshold is signalled using the mapping presented in Table 242 (using a signalled value of between 0 and 97)
- When using RSRQ, the threshold can be configured with a value between -3 and -19.5 dB. The value of the threshold is signalled using the mapping presented in Table 243 (using a signalled value of between 0 and 34)

22.4 EVENT A2

- Event A2 is triggered when the serving cell becomes worse than a threshold. The event is triggered when the following condition is true:

$$Meas_{serv} + Hyst < Threshold$$

- Triggering of the event is subsequently cancelled when the following condition is true:

$$Meas_{serv} - Hyst > Threshold$$

- The hysteresis can be configured with a value between 0 and 30 dB
- When using RSRP, the threshold can be configured with a value between -140 and -44 dBm. The value of the threshold is signalled using the mapping presented in Table 242 (using a signalled value of between 0 and 97)
- When using RSRQ, the threshold can be configured with a value between -3 and -19.5 dB. The value of the threshold is signalled using the mapping presented in Table 243 (using a signalled value of between 0 and 34)

22.5 EVENT A3

- Event A3 is triggered when a neighbouring cell becomes better than the serving cell by an offset. The offset can be either positive or negative. The event is triggered when the following condition is true:

$$Meas_{neigh} + O_{neigh,freq} + O_{neigh,cell} - Hyst > Meas_{serv} + O_{serv,freq} + O_{serv,cell} + Offset$$

- Triggering of the event is subsequently cancelled when the following condition is true:

$$Meas_{neigh} + O_{neigh,freq} + O_{neigh,cell} + Hyst < Meas_{serv} + O_{serv,freq} + O_{serv,cell} + Offset$$

- Both the neighbour and serving cell can have frequency specific and cell specific offsets applied to their measurements. Each of these offsets can be configured with values between -24 and +24 dB
- The additional Offset added to the serving cell measurement can be configured with a value between -30 and +30 dB
- The hysteresis can be configured with a value between 0 and 30 dB

22.6 EVENT A4

- Event A4 is triggered when a neighbouring cell becomes better than a threshold. The event is triggered when the following condition is true:

$$Meas_{neigh} + O_{neigh,freq} + O_{neigh,cell} - Hyst > Threshold$$

- Triggering of the event is subsequently cancelled when the following condition is true:

$$Meas_{neigh} + O_{neigh,freq} + O_{neigh,cell} + Hyst < Threshold$$

- The neighbour cell can have frequency specific and cell specific offsets applied to its measurements. Both offsets can be configured with values between -24 and +24 dB
- The hysteresis can be configured with a value between 0 and 30 dB
- When using RSRP, the threshold can be configured with a value between -140 and -44 dBm. The value of the threshold is signalled using the mapping presented in Table 242 (using a signalled value of between 0 and 97)
- When using RSRQ, the threshold can be configured with a value between -3 and -19.5 dB. The value of the threshold is signalled using the mapping presented in Table 243 (using a signalled value of between 0 and 34)
- UE support for event A4 is intended to be mandatory, but event A4 may not have been implemented and tested for some early devices. UE use release 8 Feature Group Indicator (FGI) bit 14 to indicate whether or not event A4 has been implemented and tested. The FGI bit string can be included within a UE Capability Information message

22.7 EVENT A5

- Event A5 is triggered when the serving cell becomes worse than threshold1 while a neighbouring cell becomes better than threshold2. The event is triggered when both of the following conditions are true:

$$Meas_{serv} + Hyst < Threshold1$$
$$Meas_{neigh} + O_{neigh,freq} + O_{neigh,cell} - Hyst > Threshold2$$

- Triggering of the event is subsequently cancelled when either of the following conditions are true:

$$Meas_{serv} - Hyst > Threshold1$$
$$Meas_{neigh} + O_{neigh,freq} + O_{neigh,cell} + Hyst < Threshold2$$

- The neighbour cell can have frequency specific and cell specific offsets applied to its measurements. Both offsets can be configured with a value between -24 and +24 dB
- The hysteresis can be configured with a value between 0 and 30 dB
- When using RSRP, the threshold can be configured with a value between -140 and -44 dBm. The value of the threshold is signalled using the mapping presented in Table 242 (using a signalled value of between 0 and 97)
- When using RSRQ, the threshold can be configured with a value between -3 and -19.5 dB. The value of the threshold is signalled using the mapping presented in Table 243 (using a signalled value of between 0 and 34)
- UE support for event A5 is intended to be mandatory, but event A5 may not have been implemented and tested for some early devices. UE use release 8 Feature Group Indicator (FGI) bit 14 to indicate whether or not event A5 has been implemented and tested. The FGI bit string can be included within a UE Capability Information message

22.8 EVENT A6

- Event A6 is triggered when a neighbouring cell becomes better than a secondary cell by an offset. The offset can be either positive or negative.
- This measurement reporting event is applicable to LTE Advanced connections using Carrier Aggregation, i.e. connections which have secondary serving cells (in addition to a primary serving cell)
- The event is triggered when the following condition is true:

$$Meas_{neigh} + O_{neigh,cell} - Hyst > Meas_{\sec} + O_{\sec,cell} + Offset$$

- Triggering of the event is subsequently cancelled when the following condition is true:

$$Meas_{neigh} + O_{neigh,cell} + Hyst < Meas_{\sec} + O_{\sec,cell} + Offset$$

- The neighbour cell has to be on the same frequency as the secondary serving cell
- Both the neighbour and secondary serving cell can have cell specific offsets applied to their measurements. Each of these offsets can be configured with values between -24 and +24 dB
- The additional Offset added to the secondary serving cell measurement can be configured with a value between -30 and +30 dB
- The hysteresis can be configured with a value between 0 and 30 dB
- UE support for event A6 is intended to be mandatory for all 3GPP release 10 UE which support Carrier Aggregation. Nevertheless, event A6 may not have been implemented and tested for some early release 10 devices. UE use release 10 Feature Group Indicator (FGI) bit 11 to indicate whether or not event A6 has been implemented and tested. UE are only permitted to signal their support for event A6 if they support Carrier Aggregation. The FGI bit string can be included within a UE Capability Information message

22.9 EVENT B1

★ Event B1 is triggered when a neighbouring inter-system cell becomes better than a threshold. The event is triggered when the following condition is true:

$$Meas_{neigh} + O_{neigh,freq} - Hyst > Threshold$$

★ Triggering of the event is subsequently cancelled when the following condition is true:

$$Meas_{neigh} + O_{neigh,freq} + Hyst < Threshold$$

★ The inter-system neighbour cell can have a frequency specific offset applied to its measurements. Cell specific offsets are not applicable to inter-system measurements. The offset can be configured with a value between -15 and +15 dB

★ The hysteresis can be configured with a value between 0 and 30 dB

★ When using CPICH RSCP, the threshold can be configured with a value between -120 and -25 dBm. The value of the threshold is signalled using the mapping specified in 3GPP TS 25.133 (using a signalled value of between -5 and 91)

★ When using CPICH Ec/Io, the threshold can be configured with a value between -24 and 0 dB. The value of the threshold is signalled using the mapping specified in 3GPP TS 25.133 (using a signalled value of between 0 and 49)

★ When using GSM RSSI, the threshold can be configured with a value between -110 and -48 dB. The value of the threshold is signalled using the mapping specified in 3GPP TS 45.008 (using a signalled value of between 0 and 63)

★ UE support for event B1 is intended to be mandatory for UE which support inter-RAT measurements. Nevertheless, event B1 may not have been implemented and tested for some early devices. UE use release 8 Feature Group Indicator (FGI) bit 15 to indicate whether or not event B1 has been implemented and tested. UE are only permitted to signal their support for event B1 if they support inter-RAT measurements for UTRAN, GERAN, CDMA2000 1xRTT or CDMA2000 HRPD (FGI bits 22, 23, 24 and 26 respectively). The FGI bit string can be included within a UE Capability Information message

22.10 EVENT B2

★ Event B2 is triggered when the serving cell becomes worse than threshold1 while a neighbouring inter-system cell becomes better than threshold2. The event is triggered when both of the following conditions are true:

$$Meas_{serv} + Hyst < Threshold1$$
$$Meas_{neigh} + O_{neigh,freq} - Hyst > Threshold2$$

★ Triggering of the event is subsequently cancelled when either of the following conditions are true:

$$Meas_{serv} - Hyst > Threshold1$$
$$Meas_{neigh} + O_{neigh,freq} + Hyst < Threshold2$$

★ The inter-system neighbour cell can have a frequency specific offset applied to its measurements. Cell specific offsets are not applicable to inter-system measurements. The offset can be configured with a value between -15 and +15 dB

★ The hysteresis can be configured with a value between 0 and 30 dB

★ When using CPICH RSCP, threshold2 can be configured with a value between -120 and -25 dBm. The value of the threshold is signalled using the mapping specified in 3GPP TS 25.133 (using a signalled value of between -5 and 91)

★ When using CPICH Ec/Io, threshold2 can be configured with a value between -24 and 0 dB. The value of the threshold is signalled using the mapping specified in 3GPP TS 25.133 (using a signalled value of between 0 and 49)

★ When using GSM RSSI, threshold2 can be configured with a value between -110 and -48 dB. The value of the threshold is signalled using the mapping specified in 3GPP TS 45.008 (using a signalled value of between 0 and 63)

★ UE support for event B2 is intended to be mandatory for UE which support inter-RAT measurements. Nevertheless, event B2 may not have been implemented and tested for some early devices. UE use release 8 Feature Group Indicator (FGI) bits 22, 23, 24 and 26 to indicate whether or not event B2 has been implemented and tested for UTRAN, GERAN, CDMA2000 1xRTT and CDMA2000 HRPD respectively. These FGI bits also indicate UE support for the relevant inter-RAT measurements. The FGI bit string can be included within a UE Capability Information message

23 IDLE MODE PROCEDURES

23.1 PLMN SELECTION

- ★ The UE is responsible for selecting a Public Land Mobile Network (PLMN) for subsequent cell selection. A PLMN is identified by its PLMN identity broadcast within System Information Block 1 (SIB1). A single cell can belong to multiple PLMN so SIB1 may broadcast a list of PLMN identities
- ★ The UE Non-Access Stratum (NAS) layer is responsible for requesting the UE Access Stratum (AS) layer to report available PLMN:
  - o the UE scans all RF channels within its supported frequency bands
  - o the UE searches for the strongest cell on each carrier and reads the system information to identity the PLMN
  - o PLMN are reported to the NAS as high quality if their RSRP ≥ -110 dBm
  - o PLMN which do not satisfy the high quality criteria are reported to the NAS together with their RSRP measurement
  - o UE can optimise the PLMN search procedure using stored information, e.g. carrier frequencies and cell parameters
  - o the NAS layer can stop the search at any time, e.g. after finding the home PLMN
- ★ The NAS layer is responsible for selecting a PLMN from the list of reported PLMN. The NAS layer uses information from the USIM to help with PLMN selection:
  - o International Mobile Subscriber Identity (IMSI): defines the Home PLMN (HPLMN)
  - o HPLMN Selector with Access Technology: defines the priority of each technology associated with the HPLMN
  - o User Controlled PLMN Selector with Access Technology: allows the end-user to prioritise the PLMN and technology
  - o Operator Controlled PLMN Selector with Access Technology: allows the operator to prioritise the PLMN and access technology
  - o Forbidden PLMNs: defines PLMN which the UE does not automatically attempt to access. Forbidden PLMN are added to the list when the network rejects an Attach Request using the cause value 'PLMN not allowed'
  - o Equivalent HPLMN (EHPLMN): defines a set of PLMN which are treated as equivalent to the PLMN with which the UE is registering. This list can be updated or deleted by the network during the attach or tracking area update procedures. EHPLMN are ordered in terms of their priority (high priority EHPLMN appear first)
- ★ The PLMN can be selected either automatically or manually
- ★ In the case of automatic selection, the UE selects the PLMN and access technology using the following order of priority:
  - i) HPLMN (if EHPLMN list is not available) or the highest priority EHPLMN (if EHPLMN list is available)
  - ii) PLMN and access technology combinations defined within the User Controlled PLMN Selector
  - iii) PLMN and access technology combinations defined within the Operator Controlled PLMN Selector
  - iv) other PLMN reported as high quality PLMN, selected in random order
  - v) other PLMN selected in order of decreasing signal quality
- ★ The UE searches all supported access technologies before selecting a PLMN when using steps iv) and v)
- ★ In the case of manual selection, the UE presents the end-user with the available PLMN, listing them in the following order:
  - i) HPLMN or Equivalent HPLMN (EHPLMN)
  - ii) PLMN and access technology combinations defined within the User Controlled PLMN Selector
  - iii) PLMN and access technology combinations defined within the Operator Controlled PLMN Selector
  - iv) other PLMN reported as high quality PLMN, selected in random order
  - v) other PLMN selected in order of decreasing signal quality

  The end-user is then able to select which PLMN the UE should attempt to access
- ★ 3GPP references: TS 36.304, TS 23.122, TS 31.102, TS 24.301

23.2 CELL SELECTION

★ Cell selection allows a UE to camp on a cell

★ There are 3 types of cell selection
  - initial cell selection
  - stored information cell selection
  - cell selection when leaving RRC Connected mode

  These are illustrated in Figure 197

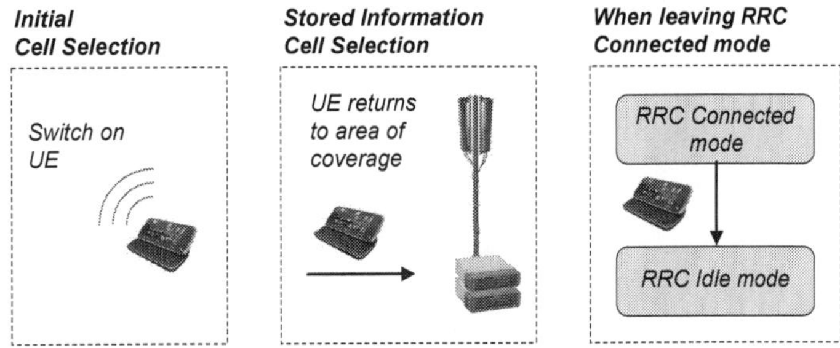

Figure 197 – Types of cell selection

★ Initial cell selection does not require any prior knowledge. The UE scans all RF carriers while searching for a suitable cell

★ Stored information cell selection uses RF carrier and potentially cell parameter information to help identify a candidate suitable cell. Initial cell selection is completed if stored information cell selection fails

★ Cell selection when leaving RRC Connected mode applies to UE making the transition from RRC Connected mode to RRC Idle mode. The eNode B can direct the UE towards a specific RF carrier by including 'redirected carrier information' within the RRC Connection Release message. Otherwise the UE is free to search any RF carrier

★ All 3 types of cell selection involve the UE searching for a 'suitable' cell. A suitable cell is defined as a cell which
  - belongs to the selected PLMN, the registered PLMN or an Equivalent PLMN
  - is not barred
  - belongs to at least one Tracking Area which is not forbidden
  - satisfies the cell selection criteria

★ Within the release 8 version of the 3GPP specifications, the cell selection criteria (also known as the 'S' criteria) is defined as:

Srxlev > 0

where, Srxlev = Qrxlevmeas - (Qrxlevmin + Qrxlevminoffset) - Pcompensation

★ Qrxlevmeas is the RSRP measured by the UE in dBm

★ Qrxlevmin is the minimum required RSRP signalled within System Information Block 1 (SIB1)

★ Qrxlevminoffset is a measurement offset which is only applied when a cell is evaluated for cell selection as a result of a periodic search for a higher priority PLMN while camped on a Visited PLMN. Qrxlevminoffset is optionally included within SIB1. A value of 0 dB is applied if Qrxlevminoffset is excluded from SIB1

★ Pcompensation is defined as MAX($P_{EMAX} - P_{UMAX}$, 0) where P_{EMAX} is the maximum allowed uplink transmit power within the cell, and P_{UMAX} is the maximum transmit power capability of the UE. If the UE transmit power capability is less than the maximum allowed transmit power, Pcompensation makes the cell selection criteria more difficult to achieve. This avoids UE camping on cells for which they have insufficient transmit power to access, i.e. towards cell edge

★ The release 9, and newer versions of the 3GPP specifications add a quality check to the cell selection criteria ('S' criteria):

Srxlev > 0 AND Squal > 0

where, Srxlev = Qrxlevmeas - (Qrxlevmin + Qrxlevminoffset) - Pcompensation

Squal = Qqualmeas - (Qqualmin + Qqualminoffset)

- Qqualmeas is the RSRQ measured by the UE in dB
- Qqualmin is the minimum required RSRQ signalled within System Information Block 1 (SIB1)
- Qqualminoffset is a measurement offset which is only applied when a cell is evaluated for cell selection as a result of a periodic search for a higher priority PLMN while camped on a Visited PLMN. Qqualminoffset is optionally included within SIB1. A value of 0 dB is applied if Qqualminoffset is excluded from SIB1
- Priorities between RF carriers and different radio access technologies are not used during cell selection
- 3GPP reference: TS 36.304

23.3 CELL RESELECTION

23.3.1 PRIORITIES

- Absolute priorities can be allocated to RF carriers belonging to both LTE and other Radio Access Technologies (RAT)
- Absolute priorities can be broadcast within the system information on the BCCH:
  - SIB3 - absolute priority for the current LTE RF carrier
  - SIB5 - absolute priorities for other LTE RF carriers
  - SIB6 - absolute priorities for UMTS RF carriers
  - SIB7 - absolute priorities for sets of GSM RF carriers
  - SIB8 - absolute priorities for CDMA2000 RF carriers
- Absolute priorities can also be signalled directly to individual UE within an RRC Connection Release message, i.e. when UE are returning to RRC Idle mode from RRC Connected mode. All absolute priorities within the system information are ignored if priorities are included within an RRC Connection Release message
- An RRC Connection Release message can also define the T320 timer. This timer is started as soon as it is received. Absolute priorities included within the RRC Connection Release message are deleted when T320 expires. T320 can be allocated values between 5 minutes and 3 hours
- UE also delete priorities included within an RRC Connection Release message when entering RRC Connected mode or when performing PLMN selection
- Absolute priorities can also be inherited from another RAT when completing inter-RAT cell reselection. UE inherit priorities when they have been allocated using dedicated signalling. UE also inherit the remaining validity time when a validity time has been allocated (T320 for LTE, T322 for UMTS and T3230 for GSM)
- Absolute priorities can be allocated values between 0 and 7, where 0 represents the lowest priority
- The allocation of equal priorities to different RAT is not supported
- UE only perform cell reselection evaluation for LTE RF carriers and inter-RAT RF carriers which are defined within the system information and for which the UE has a priority
- Blacklisted cells are excluded from the list of candidate cells for cell reselection
- When a UE is camped on a cell which defines priorities for other network layers but not for the current network layer, the UE assumes the current RF carrier to have the lowest priority (lower than all 8 of the possible signalled values)
- When a UE is camped on a suitable Closed Subscriber Group (CSG) cell, the UE assumes the current RF carrier to have the highest priority (higher than all 8 of the possible signalled values)
- 3GPP references: TS 36.304, TS 36.331

23.3.2 TRIGGERING MEASUREMENTS

★ Cell reselection measurement rules are based upon the parameters broadcast by the serving cell rather than those broadcast by neighbouring cells

★ Cell reselection measurement rules are intended to reduce the quantity of measurements completed by a UE. This is done by triggering measurements only when necessary. Reducing the quantity of measurements increases the UE battery life

★ Sservingcell is defined as Srxlev measured from the serving cell, where Srxlev is defined in section 23.2

★ Table 249 presents the set of rules used to trigger intra-frequency measurements for 3GPP release 8 UE. These are dependent upon whether or not the Sintrasearch parameter is broadcast within System Information Block 3 (SIB3)

	Sservingcell > Sintrasearch	Sservingcell ≤ Sintrasearch
Sintrasearch is broadcast	Measurements not mandatory	Measurements mandatory
Sintrasearch is not broadcast	Measurements mandatory	

Table 249 – Intra-frequency measurement triggering for cell reselection (3GPP release 8)

★ Intra-frequency measurements are triggered when the serving cell signal becomes relatively weak, or if Sintrasearch is not included within SIB3

★ Sintrasearch is signalled using a value within the range 0 to 31, while the actual value of Sintrasearch = signalled value × 2

★ 3GPP updated the rules for triggering intra-frequency measurements within the release 9 version of the specifications. 2 new parameters were introduced for this purpose: SintrasearchP and SintrasearchQ. Both of these parameters can be broadcast in SIB3

★ Table 250 presents the updated rules for triggering intra-frequency measurements based upon SintrasearchP and SintrasearchQ

	Srxlev > SintrasearchP AND Squal > SintrasearchQ	Srxlev ≤ SintrasearchP OR Squal ≤ SintrasearchQ
SintrasearchP and SintrasearchQ are broadcast	Measurements not mandatory	Measurements mandatory
SintrasearchP and SintrasearchQ are not broadcast	Measurements mandatory	

Table 250 – Intra-frequency measurement triggering for cell reselection (3GPP releases 9, and newer)

★ Both SintrasearchP and SintrasearchQ are signalled using values within the range 0 to 31. In the case of SintrasearchP, the actual value equals the signalled value × 2, whereas in the case of SintrasearchQ, the actual value equals the signalled value

★ Table 251 presents the set of rules used to trigger inter-frequency and inter-RAT measurements for 3GPP release 8 UE. These are dependent upon whether or not the Snonintrasearch parameter is broadcast within SIB3

		Candidate Layer Priority relative to Serving Cell Priority				
		Lower		Equal		Higher
		Sservingcell > Snonintrasearch	Sservingcell ≤ Snonintrasearch	Sservingcell > Snonintrasearch	Sservingcell ≤ Snonintrasearch	
Inter-Frequency Candidate Layer	Snonintrasearch is broadcast	Measurements not mandatory	Measurements mandatory	Measurements not mandatory	Measurements mandatory	Measurements mandatory
	Snonintrasearch is not broadcast	Measurements mandatory				
Inter-RAT Candidate Layer	Snonintrasearch is broadcast	Measurements not mandatory	Measurements mandatory	Not Applicable		
	Snonintrasearch is not broadcast	Measurements mandatory				

Table 251 – Inter-frequency and inter-RAT measurement triggering for cell reselection (3GPP release 8)

★ Inter-frequency and inter-RAT measurements are always triggered when the candidate layer has a higher priority than the serving cell

★ Inter-frequency and inter-RAT measurements are also triggered when the serving cell signal becomes relatively weak, or if Snonintrasearch is not included within SIB3

★ The allocation of equal priorities to different RAT is not supported

★ Snonintrasearch is signalled using a value within the range 0 to 31. The signalled value is mapped into the actual value using the equation: actual value of Sintrasearch = signalled value × 2

★ 3GPP updated the rules for triggering inter-frequency and inter-RAT measurements within the release 9 version of the specifications. 2 new parameters were introduced for this purpose: SnonintrasearchP and SnonintrasearchQ. Both of these parameters can be broadcast in SIB3

★ Table 252 presents the set of rules used to trigger inter-frequency and inter-RAT measurements for 3GPP release 9, and newer UE

<table>
<tr><td colspan="2" rowspan="3"></td><th colspan="5">Candidate Layer Priority relative to Serving Cell Priority</th></tr>
<tr><th colspan="2">Lower</th><th colspan="2">Equal</th><th rowspan="2">Higher</th></tr>
<tr><th>Srxlev > SnonintrasearchP AND Squal > SnonintrasearchQ</th><th>Srxlev ≤ SnonintrasearchP OR Squal ≤ SnonintrasearchQ</th><th>Srxlev > SnonintrasearchP AND Squal > SnonintrasearchQ</th><th>Srxlev ≤ SnonintrasearchP OR Squal ≤ SnonintrasearchQ</th></tr>
<tr><td rowspan="2">Inter-Frequency Candidate Layer</td><td>SnonintrasearchP and SnonintrasearchQ are broadcast</td><td>Measurements not mandatory</td><td>Measurements mandatory</td><td>Measurements not mandatory</td><td>Measurements mandatory</td><td rowspan="4">Measurements mandatory</td></tr>
<tr><td>SnonintrasearchP and SnonintrasearchQ are not broadcast</td><td colspan="4">Measurements mandatory</td></tr>
<tr><td rowspan="2">Inter-RAT Candidate Layer</td><td>SnonintrasearchP and SnonintrasearchQ are broadcast</td><td>Measurements not mandatory</td><td>Measurements mandatory</td><td colspan="2" rowspan="2">Not Applicable</td></tr>
<tr><td>SnonintrasearchP and SnonintrasearchQ are not broadcast</td><td colspan="2">Measurements mandatory</td></tr>
</table>

Table 252 – Inter-frequency and inter-RAT measurement triggering for cell reselection (3GPP releases 9, and newer)

★ Both SnonintrasearchP and SnonintrasearchQ are signalled using values within the range 0 to 31. In the case of SnonintrasearchP, the actual value equals the signalled value × 2, whereas in the case of SnonintrasearchQ, the actual value equals the signalled value

★ 3GPP references: TS 36.304, TS 36.331

23.3.3 MOBILITY STATES

- The following mobility states can be used for cell reselection:
    - normal mobility
    - medium mobility
    - high mobility

  The high and medium mobility states are applicable if the optional N_{CR_H}, N_{CR_M}, T_{CRmax} and $T_{CRmaxHyst}$ parameters are broadcast within System Information Block 3 (SIB3). N_{CR_H} and N_{CR_M} can be configured with values between 1 and 16, while T_{CRmax} and $T_{CRmaxHyst}$ can be configured with values of 30, 60, 120, 180 and 240 seconds
- A UE has normal mobility by default
- The criteria for medium mobility is satisfied if the number of cell reselections during time period T_{CRmax} exceeds N_{CR_M} but does not exceed N_{CR_H}
- The criteria for high mobility is satisfied if the number of cell reselections during time period T_{CRmax} exceeds N_{CR_H}
- A UE returns from the medium or high mobility states to the normal mobility state if the criteria for neither medium nor high mobility is detected during time period $T_{CRmaxHyst}$
- The UE excludes ping-pongs between two cells when counting the number of cell reselections
- If the UE is in the high or medium mobility state then speed dependent scaling rules are applied
- If high mobility is detected
    - if 'sf-high' for 'Qhyst' is included within SIB3 then add the value of 'sf-high' to Qhyst
    - if 'sf-high' for 'speed state scale factors' is included within the SIB then multiply Treselection by 'sf-high'
- If medium mobility is detected
    - if 'sf-medium' for 'Qhyst' is included within SIB3 then add the value of 'sf-Medium' to Qhyst
    - if 'sf-medium' for 'speed state scale factors' is included within the SIB then multiply Treselection by 'sf-medium'
- 'sf-high' and 'sf-medium' for 'Qhyst' can be configured with values of -6, -4, -2 and 0 dB, i.e. they tend to decrease the value of Qhyst to make cell reselection easier for UE with increased mobility
- 'sf-high' and 'sf-medium' for 'speed state scale factors' can be configured with values of 0.25, 0.5, 0.75 and 1.0, i.e. they tend to decrease Treselection to make cell reselection easier for UE with increased mobility
- 'sf-high' and 'sf-medium' for 'Qhyst' are optional within SIB3, while Qhyst is mandatory within SIB3
- 'sf-high' and 'sf-medium' for 'speed state scale factors' are optional within the SIB, while Treselection is mandatory. They can be found within the SIB:
    - SIB3 for intra-frequency cell reselection
    - SIB5 for inter-frequency cell reselection
    - SIB6 for inter-system cell reselection to UMTS
    - SIB7 for inter-system cell reselection to GSM
    - SIB8 for inter-system cell reselection to CDMA200
- When Treselection is scaled, the result is rounded up to the nearest second
- 3GPP references: TS 36.304, TS 36.331

23.3.4 RANKING

INTRA-FREQUENCY AND EQUAL PRIORITY INTER-FREQUENCY

★ The UE ranks all cells which satisfy the cell selection 'S' criteria presented in section 23.2

  - Qrxlevmin and Qqualmin for candidate intra-frequency cells are broadcast in SIB3. Qqualmin is only applicable to 3GPP release 9 and newer UE. The inclusion of Qqualmin is optional and if excluded the UE assumes a value of minus infinity
  - Qrxlevmin and Qqualmin for candidate inter-frequency cells are broadcast in SIB5. Qqualmin is only applicable to 3GPP release 9 and newer UE. The inclusion of Qqualmin is optional and if excluded the UE assumes a value of minus infinity

★ The UE can exclude Closed Subscriber Group (CSG) cells which the UE knows to be not allowed

★ Cell reselection is completed if a neighbouring cell is ranked better than the current serving cell during a time period defined by Treselection, and if more than 1 second has passed since the UE camped on the current serving cell. Treselection for intra-frequency cells is broadcast in SIB3, whereas Treselection for inter-frequency cells is broadcast in SIB5

★ The cell reselection ranking criteria (also known as the 'R' criteria) is defined as:

$$Rs = Q_{meas,s} + Q_{Hyst}$$ (calculated for the serving cell)

$$Rn = Q_{meas,n} - Q_{Offset}$$ (calculated for the neighbouring cell)

★ Qmeas is the RSRP measured by the UE in dBm

★ For intra-frequency cells, Qoffset is the optional, cell specific Qoffsets,n value broadcast in SIB4. A value of 0 is assumed if the candidate cell does not have a value broadcast in SIB4

★ For inter-frequency cells, Qoffset is given by:

Qoffset = Qoffsets,n + Qoffsetfrequency

where,

Qoffsets,n is the optional, cell specific offset broadcast in SIB5
Qoffsetfrequency is the optional, RF carrier specific offset broadcast in SIB5

Default values of 0 are assumed for Qoffsets,n and Qoffsetfrequency if they are not broadcast in SIB5

★ UE always camp on the highest ranked cell to help minimise levels of intercell interference, i.e. UE connect to the best server. If a cell is allocated a measurement offset to increase its ranking then UE can camp on that cell without it being the best server. In that case, the radio access network can experience an increase in interference

INTER-SYSTEM AND DIFFERENT PRIORITY INTER-FREQUENCY

★ The rules for cell reselection towards an inter-system cell, or towards an inter-frequency cell with a different priority to the current serving cell, were changed within the release 9 version of the 3GPP specifications. The 2 sets of rules are presented separately below:

3GPP release 8

★ Cell reselection to a higher priority inter-frequency or inter-RAT cell is completed if:

  - SnonServingCell > Thresh,high during a time period defined by Treselection, and
  - more than 1 second has passed since the UE camped on the current serving cell

★ Cell reselection to a lower priority inter-frequency or inter-RAT cell is completed if:

  - no cell on a higher priority inter-frequency or inter-RAT layer can be selected, and
  - no intra-frequency cell or equal priority inter-frequency cell can be selected, and
  - SservingCell < Threshserving,low & SnonServingCell > Thresh,low during the Treselection time interval, and
  - more than 1 second has passed since the UE camped on the current serving cell

★ In the case of LTE, UMTS and GSM, SnonServingCell is the Srxlev of the candidate cell. The value of Qrxlevmin used in these calculations is broadcast in SIB5, SIB6 and SIB7 for LTE, UMTS and GSM cells respectively

- ★ In the case of FDD UMTS, Squal must also be greater than 0. The value of Qqualmin for UMTS is broadcast in SIB6
- ★ In the case of CDMA2000, SnonServingCell is equal to –FLOOR(-20 × LOG(Ec/Io)) in units of 0.5 dB
- ★ Thresh,high, Thresh,low and Threshserving,low are signalled using values between 0 and 31, while the actual value of these parameters = signalled value × 2
- ★ In the case of CDMA2000, Thresh,high and Thresh,low are multiplied by -1 prior to comparing against SnonServingCell
- ★ The value of Treselection is scaled when the UE has high or medium mobility (section 23.3.3)
- ★ Treselection and the associated scaling factor values are broadcast in SIB5, SIB6, SIB7 and SIB8 for inter-frequency LTE, UMTS, GSM and CDMA2000 cells respectively
- ★ If more than one candidate meets the criteria for cell reselection, the cells are ranked using the intra-frequency ranking criteria

3GPP release 9, and newer

- ★ Cell reselection to a higher priority inter-frequency or inter-RAT cell is completed if:

  if ThreshServingLowQ is broadcast by SIB3,

  LTE and UMTS FDD

  - o neighbour cell Squal > Threshx,high,Q during time Treselection

  UMTS TDD, GSM or CDMA2000

  - o neighbour cell Srxlev > Threshx,high,P during time Treselection

  else,

  - o neighbour cell Srxlev > Threshx,high,P during time Treselection

  In all cases, cell reselection is only permitted if more than 1 second has passed since the UE camped on the current serving cell

- ★ Cell reselection to a lower priority inter-frequency or inter-RAT cell is completed if:

  if ThreshServingLowQ is broadcast by SIB3,

  LTE and UMTS FDD

  - o serving cell Squal < ThreshServingLowQ and neighbour cell Squal > Threshx,low,Q during time Treselection

  UMTS TDD, GSM or CDMA2000

  - o serving cell Squal < ThreshServingLowQ and neighbour cell Srxlev > Threshx,low,P during time Treselection

  else,

  - o serving cell Srxlev < ThreshServingLowP and neighbour cell Srxlev > Threshx,low,P during time Treselection

  In all cases, cell reselection is only permitted if more than 1 second has passed since the UE camped on the current serving cell

- ★ In the case of FDD UMTS, Squal must also be greater than 0. The value of Qqualmin for UMTS is broadcast in SIB6
- ★ In the case of CDMA2000, Srxlev is equal to –FLOOR(-20 × LOG(Ec/Io)) in units of 0.5 dB, and Threshx,high,P and Threshx,low,P are multiplied by -1 prior to comparing against Srxlev
- ★ The values of Qqualmin used to calculate Squal for LTE inter-frequency neighbours and UMTS FDD neighbours are broadcast in SIB5 and SIB6 respectively. Similarly, the values of Threshx,high,Q and Threshx,low,Q are broadcast in SIB5 and SIB6
- ★ The values of Qrxlevmin used to calculate Srxlev are broadcast in SIB5, SIB6, SIB7 and SIB8 for inter-frequency LTE, UMTS, GSM and CDMA2000 cells respectively. The values of Threshx,high,P and Threshx,low,P are given by the 3GPP release 8 parameters Threshx,high and Threshx,low within SIB5, SIB6, SIB7 and SIB8
- ★ The values of ThreshServingLowQ and ThreshServingLowP are broadcast in SIB3. ThreshServingLowP is given by the 3GPP release 8 parameter ThreshServingLow
- ★ The value of Treselection is scaled when the UE has high or medium mobility (section 23.3.3)
- ★ If more than one candidate meets the criteria for cell reselection, the target cell is selected according to:
  - o if the highest priority RF carrier is an LTE RF carrier, then cells are ranked using the intra-frequency ranking criteria
  - o if the highest priority RF carrier belongs to a different RAT, then cells are ranked using the ranking criteria for that RAT
- ★ 3GPP references: TS 36.304, TS 36.331

23.3.5 CLOSED SUBSCRIBER GROUP CELLS

★ Closed Subscriber Group (CSG) cells are cells which only specific UE are allowed to access

★ The concept of CSG cells can be applied to Home eNode B (femto cells)

★ USIM can be programmed to include one or more allowed CSG identities, i.e. there can be a whitelist which defines CSG identities

★ SIB1 broadcasts the CSG identity of a cell when a cell belongs to a CSG

★ UE which have at least 1 allowed CSG identity within their USIM, use an autonomous search function to detect, at least previously visited CSG cells on other frequencies and Radio Access Technologies (RAT). This autonomous search function operates in parallel to normal cell reselection. The UE can also use an autonomous search function on the serving frequency

★ If a UE detects a CSG cell on a different frequency or RAT, the UE completes cell reselection onto that CSG cell irrespective of the priority of the frequency as long as the CSG cell is the highest ranked cell on that frequency or RAT

★ UE use the normal intra-frequency cell reselection rules to camp on intra-frequency CSG cells

★ Once a UE is camped on a CSG cell, the normal cell reselection rules are applied

★ 3GPP references: TS 36.304, TS 36.331, TS 31.102

23.4 CELL STATUS AND CELL RESERVATIONS

★ Cell status and cell reservation information allows operators to impose access restrictions

★ Cell status refers to whether or not a cell is barred

★ System Information Block 1 (SIB1) includes the following information:

- Cell Barred (barred/not barred) which is applicable to all PLMN listed in SIB1
- Cell Reserved for Operator Use (reserved/not reserved) which is specified per PLMN

★ Table 253 presents the cell selection and reselection rules depending upon whether or not cells are barred or reserved for operator use

	Not Barred	Barred
Not Reserved	All UE treat the cell as a candidate for cell selection and cell reselection	UE are not allowed to select or reselect the cell (not even for emergency calls) If the cell is a Closed Subscriber Group (CSG) cell then UE are allowed to select another cell on the same frequency If the cell is not a CSG cell then the UE has to check the 'intra-frequency reselection' information within SIB1 to determine whether or not it is allowed to select another cell on the same frequency
Reserved	UE with Access Class 11 or 15 operating in their HPLMN or Equivalent HPLMN treat the cell as a candidate for cell selection and cell reselection UE with Access Class 0 to 9, or 12 to 14 treat the cell as barred	

Table 253 – UE behaviour as a function of cell barring and cell reserved for operator use

★ Table 253 does not specify the behaviour of Access Class 11 or 15 when UE are not camped on their HPLMN or Equivalent HPLMN because these access classes are only valid within the HPLMN or Equivalent HPLMN

★ Cell barring and cell reserved for operator use impact cell selection and reselection but they do not impact incoming handovers, e.g. a UE which is already in RRC Connected mode can complete a handover onto a barred cell

★ 3GPP references: TS 36.304, TS 36.331

23.5 ACCESS CONTROL

★ Access control can be used to prevent UE from sending RRC Connection Request messages. Access control does not prevent a UE from camping on a cell, nor does it prevent a UE from completing an incoming handover

★ System Information Block 2 (SIB2) can be used to broadcast access control information. This information is presented in Table 254

Information Elements	
AC Barring for Emergency Calls (True / False)	
AC Barring for MO Signalling	Barring Factor (0 to 0.95, step 0.05)
	Barring Time (4, 8, 16, 32, 64, 128, 256, 512 seconds)
	Barring for Special AC
AC Barring for MO Data	Barring Factor (0 to 0.95, step 0.05)
	Barring Time (4, 8, 16, 32, 64, 128, 256, 512 seconds)
	Barring for Special AC
SSAC Barring for MMTEL Voice (rel. 9)	Barring Factor (0 to 0.95, step 0.05)
	Barring Time (4, 8, 16, 32, 64, 128, 256, 512 seconds)
	Barring for Special AC
SSAC Barring for MMTEL Video (rel. 9)	Barring Factor (0 to 0.95, step 0.05)
	Barring Time (4, 8, 16, 32, 64, 128, 256, 512 seconds)
	Barring for Special AC
AC Barring for CS Fallback (rel. 10)	Barring Factor (0 to 0.95, step 0.05)
	Barring Time (4, 8, 16, 32, 64, 128, 256, 512 seconds)
	Barring for Special AC

Table 254 – Access control information within System Information Block 2 (SIB2)

★ Access Class (AC) barring for emergency calls determines whether or not UE with AC 0 to 9 should treat the cell as barred for emergency calls. The same rule is applicable to UE without a USIM (AC information is stored on the USIM). UE with AC 11 to 15 are able to make emergency calls unless the cell is both AC class barred for emergency calls and AC class barred for mobile originating data calls

★ Access Class (AC) barring for Mobile Originated (MO) signalling determines whether or not UE should treat the cell as barred when initiating a MO signalling connection. UE with AC 0 to 9 generate a random number between 0 and 1. If it is less than the Barring Factor then the cell is treated as not barred. UE with AC 11 to 15 check the Barring for Special AC to determine whether or not it is necessary to generate a random number with which to compare against the Barring Factor

★ If the cell is to be treated as barred for MO signalling connections then the Barring Time is used to define a penalty time during which the UE is not allowed to re-attempt access. The penalty time is defined as T305 = (0.7 + 0.6 × RAND) × Barring Time, where RAND is a random number between 0 and 1

★ Similar AC barring information is provided for UE establishing MO data connections. In this case, if the cell is to be treated as barred, the penalty time is defined as T303 = (0.7 + 0.6 × RAND) × Barring Time, where RAND is a random number between 0 and 1

★ The release 9 version of the 3GPP specifications introduced Service Specific Access Class (SSAC) barring for Multimedia Telephony (MMTEL) voice and video calls. The parameters in SIB2 are used to determine whether or not UE should treat the cell as barred when initiating MMTEL voice and data calls

  o UE with AC 0 to 9 generate a random number between 0 and 1. If it is less than the Barring Factor then the cell is treated as not barred. UE with AC 11 to 15 check the Barring for Special AC to determine whether or not it is necessary to generate a random number with which to compare against the Barring Factor

  o If the cell is to be treated as barred then the Barring Time is used to define a penalty time during which the UE is not allowed to re-attempt access. The penalty time is defined as (0.7 + 0.6 × RAND) × Barring Time, where RAND is a random number between 0 and 1

★ The release 10 version of the 3GPP specifications introduced a similar set of parameters for UE establishing mobile originating CS Fallback connections

★ 3GPP references: TS 36.304, TS 36.331

23.6 PAGING PROCEDURE

★ The paging procedure can be used to:

- initiate a mobile terminating PS data connection
- initiate a mobile terminating SMS connection
- initiate a mobile terminating CS fallback connection
- trigger a UE to re-acquire system information
- provide an Earthquake and Tsunami Warning System (ETWS) notification
- provide a Commercial Mobile Alert Service (CMAS) notification

★ The paging procedure is applicable to UE in both RRC Idle mode and RRC Connected mode. For example, a UE in RRC Idle mode could be paged to initiate a mobile terminating CS fallback connection, while a UE in RRC Connected mode could be paged to signal a CMAS notification

★ The paging procedure can be initiated by either the MME or the eNode B. For example, the MME can use the paging procedure to initiate a mobile terminating PS data connection, while the eNode B could use the paging procedure to trigger a UE to re-acquire the system information

★ The MME initiated paging procedure (also known as the S1 paging procedure) is illustrated in Figure 198

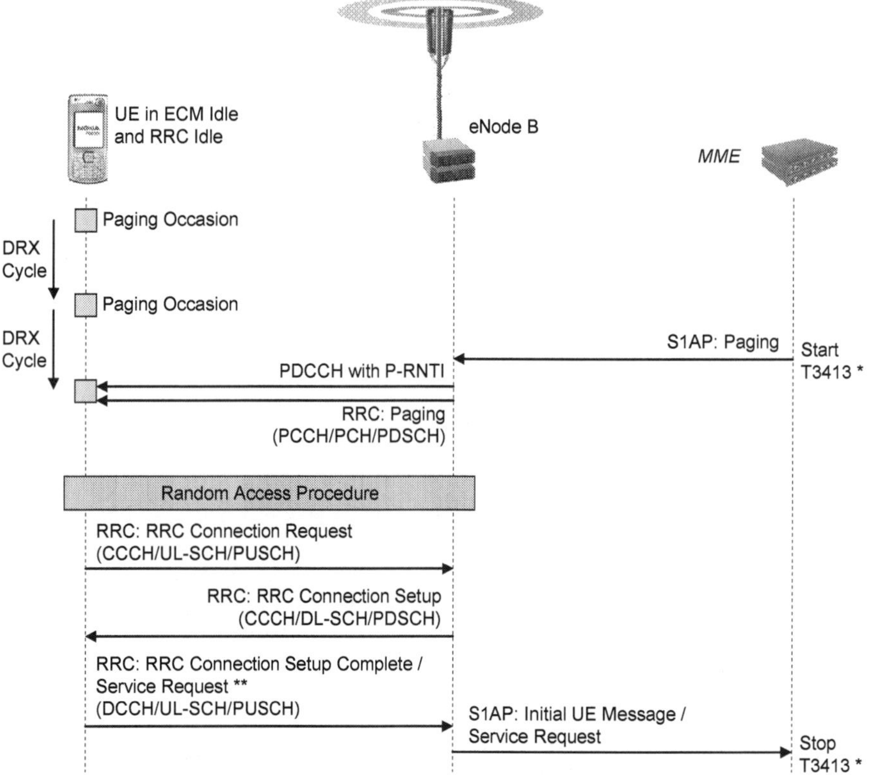

Figure 198 – High level Paging procedure initiated by the MME

★ The MME starts the procedure by sending an S1 Application Protocol: Paging message to each eNode B with cells belonging to the relevant Tracking Area(s). The location of a UE in EMM Registered, ECM Idle state is known by the MME with a resolution of one or more Tracking Areas. The requirement to broadcast the paging message across multiple Tracking Areas is generated by UE which are registered with more than a single Tracking Area. In general, a single Tracking Area will include multiple eNode B so the paging message is broadcast by multiple cells belonging to multiple eNode B

★ If the paging procedure is being used to establish a PS data connection and the UE is being addressed by its S-TMSI then the MME starts timer T3413 when sending the S1AP: Paging message. Timer T3413 is not used when paging a UE for a CS fallback connection nor when paging an IMSI attached UE for an incoming SMS. Timer T3413 is also not used when addressing a UE by its IMSI (it is usual to address a registered UE by its S-TMSI. A UE may be paged by its IMSI during a network error recovery situation)

★ T3413 is a supervision timer for the paging procedure. Its expiry time is implementation dependent and is not specified by 3GPP. The MME can re-attempt the paging procedure if T3413 expires before a response is received

★ The content of the S1AP: Paging message is shown in Table 255

Information Elements	
UE Identity Index (0 to 1023)	
UE Paging Identity (S-TMSI or IMSI)	
Paging DRX (32, 64, 128, 256)	
Core Network Domain (PS or CS)	
List of Tracking Area Identities (TAI) (1 to 256 instances)	TAI
List of Closed Subscriber Group (CSG) Identities (0 to 256 instances)	CSG Id
Paging Priority	

Table 255 – Content of an S1AP: Paging message

★ The UE Identity Index is used by the eNode B to calculate the Paging Frame (PF). It is defined as:

UE Identity Index = UE IMSI mod 1024

This preserves user confidentiality by avoiding transmission of the USIM permanent identity across the S1 interface

★ The UE Paging Identity is included as either an S-TMSI or IMSI. UE are addressed by their S-TMSI under normal circumstances. UE can be addressed by their IMSI if the network is recovering from a failure

★ The Paging DRX is optional. It is used by the eNode B when calculating the Paging Frame if the UE is using a UE specific DRX cycle length. The UE can specify a DRX cycle length within the Attach Request or Tracking Area Update messages

★ The Core Network Domain indicates Packet Switched (PS) if the paging message is for data transfer or an incoming SMS. The Circuit Switched (CS) value is used if the paging message is for CS fallback services, e.g. an incoming CS speech call

★ The list of Tracking Area Identities (TAI) informs the eNode B of which Tracking Areas (TA) the paging message should be broadcast. UE can be registered with multiple TA so it may be necessary to broadcast across more than a single TA

★ If available, the set of Closed Subscriber Group (CSG) Identities to which the UE is subscribed can be included. This avoids the eNode B paging the UE within CSG cells with which the UE is not subscribed

★ The Paging Priority is an optional field within the Paging message which was introduced within the release 10 version of the 3GPP specifications. It was introduced for the purposes of Multimedia Priority Services (MPS) which are intended to prioritise specific connections during periods of congestion, e.g. to provide the emergency services with priority during an emergency situation. MPS is described within 3GPP TS 22.153. Both the MME and eNode B can use the priority information during periods of congestion

★ The eNode B receives the S1AP: Paging message from the MME and constructs an RRC: Paging message. A single RRC: Paging message can include information from multiple S1AP: Paging messages, i.e. a single RRC: Paging message can include multiple paging records to page multiple UE. The structure of the RRC: Paging message is presented in Table 256

★ The RRC: Paging message is transferred using the PCCH logical channel, the PCH transport channel and the PDSCH physical channel. It uses transparent mode at the RLC layer

Information Elements		
Paging Record List	Paging Record (1 to 16 instances)	UE Identity (S-TMSI or IMSI)
		Core Network Domain (CS or PS)
System Information Modification (True, False)		
ETWS Indication (True, False)		
CMAS Indication (True, False)		

Table 256 – Content of an RRC: Paging message

★ All 4 fields within the RRC: Paging message are optional although at least one has to be present to generate the message

★ A maximum of 16 paging records can be included within the paging record list. Each paging record specifies the UE identity and the core network domain, i.e. forwarding the content of the S1AP: Paging message

★ The system information modification flag is generated by the eNode B rather than extracted from the S1AP: Paging message. This flag is used to trigger UE to re-acquire the set of System Information Blocks (SIB)

- ★ The Earthquake and Tsunami Warning System (ETWS) flag is also generated by the eNode B rather than extracted from the S1AP: Paging message. This flag is only applicable to ETWS capable UE and can be used to trigger those UE to re-acquire SIB1. The content of SIB1 directs the UE to the ETWS content within SIB10 and SIB11
- ★ The Commercial Mobile Alert Service (CMAS) flag is also generated by the eNode B rather than extracted from the S1AP: Paging message. This flag is only applicable to CMAS capable UE and can be used to trigger those UE to re-acquire SIB1. The content of SIB1 directs the UE to the CMAS content within SIB12
- ★ UE in RRC Idle mode check for paging messages once every Discontinuous Receive (DRX) cycle. The Paging Occasion within the Paging Frame (PF) defines the specific subframe during which a UE checks for a paging message (described in section 23.7)
- ★ UE in RRC Connected mode do not have specific Paging Occasions during which they need to check for paging messages. However, they are expected to check for paging messages at a similar rate to UE in RRC Idle mode. This allows them to receive paging messages which trigger re-acquisition of the system information, or trigger the reception of ETWS or CMAS information. These paging messages are not directed towards individual UE. They are intended to be received by all UE. The eNode B will broadcast them during multiple subframes to allow sufficient time for their reception by all UE
- ★ UE in RRC Idle mode search for the P-RNTI within the PDCCH of the subframe belonging to the Paging Occasion. The P-RNTI has a single fixed value of FFFE and serves as a flag to indicate that the UE may have a paging message on the PDSCH. UE in RRC Connected mode also search for the P-RNTI within the PDCCH when checking for paging messages
- ★ If a UE finds the P-RNTI within the PDCCH then it proceeds to decode the resource allocation information from within the PDCCH. This information directs the UE to the PDSCH Resource Blocks within which the paging message is sent
- ★ The UE decodes the RRC: Paging message from the PDSCH Resource Blocks and checks the UE identity within each of the paging records. If the UE does not find its identity within a paging record then it returns to checking the PDCCH for the P-RNTI at each Paging Occasion
- ★ If the UE finds its identity within a paging record then it triggers the Random Access procedure to establish an RRC connection. The UE sends an RRC Connection Request message while the eNode B responds with an RRC Connection Setup message
- ★ If the paging procedure is for a PS data connection, the UE includes a 'Service Request' Non-Access Stratum (NAS) message within the RRC Connection Setup Complete message
- ★ If the paging procedure is for a CS fallback connection, the UE includes an 'Extended Service Request' NAS message within the RRC Connection Setup Complete message
- ★ The eNode B forwards the NAS message to the MME which then stops T3413 if it was running, and proceeds to establish a connection with the UE
- ★ A paging retransmission can be triggered if T3413 expires prior to the MME receiving a NAS message from the UE
- ★ The UE also checks the RRC: Paging message for a System Information Modification flag, an Earthquake and Tsunami Warning System (ETWS) flag and a Commercial Mobile Alert Service (CMAS) flag. If any of these are present, the UE reads the relevant sections of the system information. UE support for ETWS and CMAS is optional so only UE which support these services react to these flags within the paging message
- ★ 3GPP references: TS 36.304, TS 36.331, TS 24.301

23.7 PAGING OCCASIONS

- ★ UE in RRC Idle mode use Discontinuous Reception (DRX) to reduce power consumption
- ★ The DRX cycle determines how frequently UE check for paging messages
- ★ The default DRX cycle is broadcast within System Information Block 2 (SIB2). It can have values of 32, 64, 128 or 256 radio frames. These figures correspond to time intervals of 320, 640, 1280 and 2560 ms
- ★ The UE can also propose its own DRX cycle length within the Attach Request and Tracking Area Update Request messages. The set of allowed values are the same as those used within SIB2
- ★ When the UE proposes its own DRX cycle, the smaller of the 2 DRX cycles is used, i.e. the minimum of the default DRX cycle and the UE specific DRX cycle
- ★ Paging Frames (PF) are the specific radio frames during which a UE checks for a paging message. PF occur when the following equation is satisfied:

$$\text{SFN mod T} = (\text{T} / \text{N}) \times (\text{UE\_ID mod N})$$

where,

T = DRX cycle length in radio frames

N = Min(T, nB)

nB is broadcast in SIB2 and can have values of {4T, 2T, T, T/2, T/4, T/8, T/16, T/32}, i.e. N can have values of {T, T/2, T/4, T/8, T/16, T/32}

UE_ID = IMSI mod 1024

- ★ The UE_ID is the same as the UE Identity Index provided to the eNode B by the MME within the S1AP: Paging message
- ★ Assuming a DRX cycle of 32 radio frames, the left hand part of the equation cycles between 0 and 31 as the SFN increases from 0 to 1023. Assuming, an N value of T/4, and a UE_ID of 51, the right hand part of the equation equals 4 × (51 mod 8) = 4 × 3 = 12. Based upon these example figures, the UE checks for paging messages within radio frames with SFN = 12, 44, 76, 108, 140, …
- ★ A UE does not need to check all 1 ms subframes within its Paging Frames. A UE only needs to check the subframe identified by its Paging Occasion (PO)
- ★ The PO is extracted from a look-up table which is indexed using:
  - o Ns = Max(1, nB / T)
  - o i_s = Floor(UE_ID / N) mod Ns
- ★ The PO look-up table for FDD is presented in Table 257

	i_s = 0	i_s = 1	i_s = 2	i_s = 3
Ns = 1	subframe 9	N/A	N/A	N/A
Ns = 2	subframe 4	subframe 9	N/A	N/A
Ns = 4	subframe 0	subframe 4	subframe 5	subframe 9

Table 257 – Paging Occasion in terms of subframe number as a function of Ns and i_s (FDD)

- ★ The value of Ns is cell specific so cells carrying low quantities of traffic can be configured with a value of 1 to provide relatively low paging capacity, while cells carrying higher quantities of traffic can be configured with a value of 4 to provide relatively high paging capacity
- ★ 3GPP selected subframes 4 and 9 for the lower capacity configurations because they are adjacent to subframes 5 and 0 which carry the Synchronisation Signals. This allows the UE to check the Synchronisation Signals at the same time as 'waking up' to check for a paging message. Subframes 5 and 0 were not selected for the lower capacity configurations (likely to have a smaller channel bandwidth) because subframe 0 already includes the Master Information Block (MIB) and subframe 5 already includes System Information Block 1 (SIB1). For the higher capacity configuration (likely to have a larger channel bandwidth) there should be sufficient capacity to accommodate both paging messages and the MIB/SIB1 within the same subframe. Selecting subframes 0 and 5 helps to minimise the impact upon MBSFN which cannot use the same subframes as paging messages nor can it use the same subframes as the MIB/SIB1
- ★ The equivalent PO look-up table for TDD is presented in Table 258
- ★ Similar to FDD, the value of Ns can be configured to match the paging capacity

★ 3GPP selected subframes 0 and 5 for all capacity configurations because these subframes are always downlink subframes irrespective of the uplink-downlink subframe configuration. In this case, it is accepted that the paging subframes will clash with the MIB/SIB1 subframes for both the low and high capacity configurations

★ The high capacity configuration also uses subframes 1 and 6. Subframe 1 is always a special subframe whereas subframe 6 can be either a special subframe or a downlink subframe (depending upon the uplink-downlink subframe configuration). Special subframes tend to be PDCCH capacity limited so these subframes were not used for the lower capacity configurations (Table 48 in section 6.2 shows that a maximum of 2 symbols can be allocated to the PDCCH within subframes 1 and 6)

	i_s = 0	i_s = 1	i_s = 2	i_s = 3
Ns = 1	subframe 0	N/A	N/A	N/A
Ns = 2	subframe 0	subframe 5	N/A	N/A
Ns = 4	subframe 0	subframe 1	subframe 5	subframe 6

Table 258 – Paging Occasion in terms of subframe number as a function of Ns and i_s (TDD)

★ UE check for the P-RNTI within the PDCCH belonging to the subframe indicated by the PO

★ 3GPP references: TS 36.304

23.8 IDLE MODE SIGNALLING REDUCTION

- ★ UE normally have to complete registration procedures after cell reselection between technologies in Idle mode. For example, if a UE moves out of LTE coverage and completes cell reselection onto UMTS then the UE has to register with UMTS to inform the network that the transition has been made
- ★ Idle mode Signalling Reduction (ISR) removes the requirement for these inter-RAT registration procedures by allowing the UE to register with multiple technologies. The UE can be simultaneously registered with the LTE, UTRAN and GERAN core networks. This allows the UE to move freely between technologies without having to perform any signalling
- ★ The general concept of ISR is illustrated in Figure 199. A UE which is at the edge of LTE coverage could ping-pong between camping on LTE and camping on UMTS. Registration procedures are required after each transition when ISR is not used. Using ISR removes the requirement for the registration procedures because the UE is simultaneously registered with both technologies

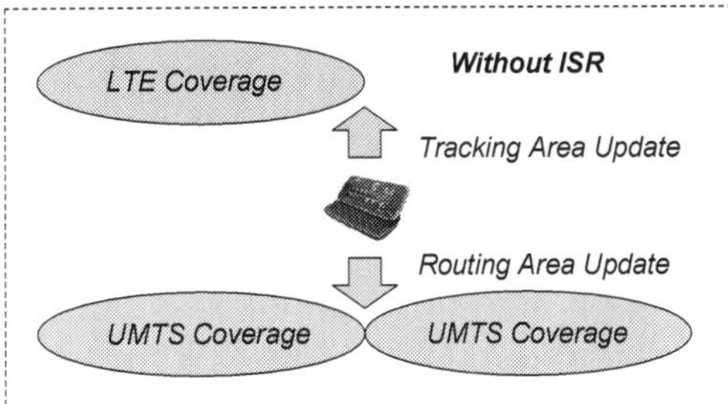

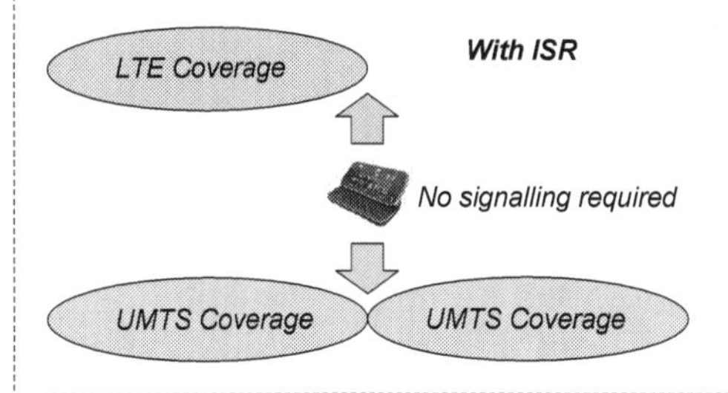

Figure 199 – Comparison of inter-RAT cell reselection with and without ISR

- ★ The general concept of ISR can be compared to co-planning GERAN and UTRAN routing areas. UE are not required to complete routing area update procedures when moving between GERAN and UTRAN when the cells belonging to each technology are allocated to the same routing area
- ★ The drawback of using ISR is an increased paging load. When using ISR, the network does not know which technology the UE is camped upon because the UE can move freely between technologies without informing the network. This means that all technologies have to broadcast paging messages
- ★ ISR is specified by 3GPP within TS 23.401. It is mandatory for LTE UE to support ISR if they have a UTRAN/GERAN dual mode capability. An example signalling flow for ISR activation is shown in Figure 200. This example is based upon a UE initially registering with the LTE network before completing a cell reselection onto UMTS

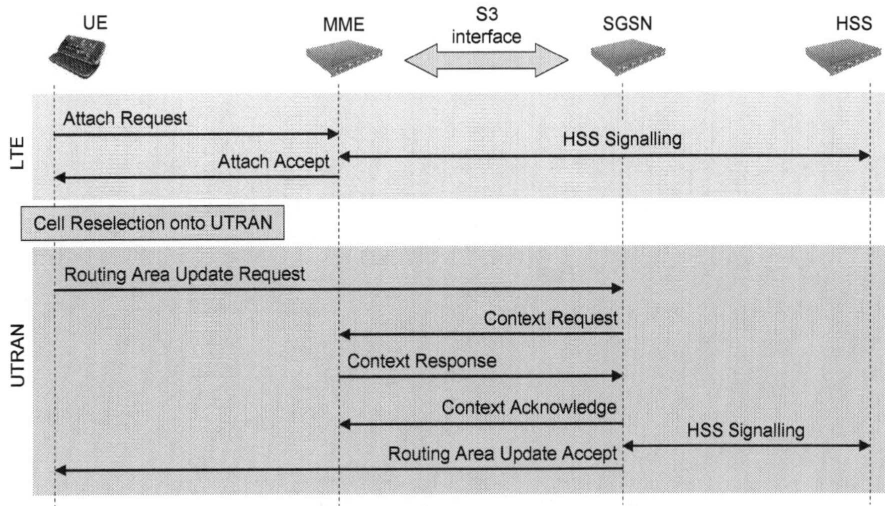

Figure 200 – Example ISR activation signalling

- ★ The initial attach shown in Figure 200 is a normal attach procedure and does not reference ISR within any of the messages. The UE deletes any existing ISR information during this procedure and ISR remains deactivated. The Attach Accept message allocates a Globally Unique Temporary UE Identity (GUTI) to the UE. The structure of the GUTI is presented in section 33.2
- ★ It is assumed that the UE then completes a cell reselection onto UMTS. This could be caused by LTE coverage becoming relatively weak. ISR is not yet activated so the UE has to complete a registration procedure with the UMTS core network. This registration procedure provides an opportunity to activate ISR

★ The UE starts the routing area updating procedure by sending a Routing Area Update Request message to the SGSN. The routing area updating procedure is specified within 3GPP TS 24.008. The UE uses the previously assigned GUTI to populate the RAI, P-TMSI and P-TMSI Signature fields within the Routing Area Update Request message. This is done using the mapping rules specified in 3GPP TS 23.003

★ The SGSN extracts the RAI, P-TMSI and P-TMSI Signature fields and inserts them within a Context Request message. The Context Request message belongs to the GTPv2-C protocol specified in 3GPP TS 29.274. It is forwarded across the S3 interface to the old MME

★ The old MME is then able to apply the reverse mapping of RAI, P-TMSI and P-TMSI Signature to re-generate the GUTI and identify the UE context being requested by the SGSN. The MME responds with a GTPv2-C: Context Response message which provides the SGSN with the IMSI belonging to the UE, and details of the UE context. The MME also includes a flag to indicate that ISR is supported

★ The SGSN responds to the MME using a Context Acknowledge to confirm that the UE context details have been received and that ISR is being used

★ The SGSN then responds to the UE using a Routing Area Update Accept message. This message allocates a P-TMSI to the UE and also includes an 'Update Result' field which can be allocated values of:

  o RA Updated

  o Combined RA/LA Updated

  o RA Updated and ISR Activated

  o Combined RA/LA Updated and ISR Activated

  The third and forth of these values are used to activate ISR at the UE. With ISR activated, the UE stores both the GUTI allocated by the LTE core network, and the P-TMSI allocated by the UMTS core network. The UE is then registered with both the MME and the SGSN

★ The UE can then reselect between LTE and UMTS without having to update the network, as long as the UE remains within the registered Tracking Area(s) on LTE and the registered Routing Area on UMTS

★ The example ISR activation procedure illustrated in Figure 200 is based upon a UE reselecting UMTS after registering with LTE. Alternatively, ISR could be activated when a UE reselects LTE after registering with UMTS. ISR can also be activated between LTE and GERAN

★ When activation is based upon a UE reselecting LTE after registering with UMTS, the Tracking Area Update Accept message is used to instruct the UE to activate ISR. The 'EPS Update Result' field within the Tracking Area Update Accept message is used for this purpose and can be allocated values of:

  o TA Updated

  o Combined TA/LA Updated

  o TA Updated and ISR Activated

  o Combined TA/LA Updated and ISR Activated

★ Figure 201 illustrates the signalling associated with the arrival of downlink data when ISR is active. The Packet Gateway (P-GW) forwards the data towards the Serving Gateway (S-GW)

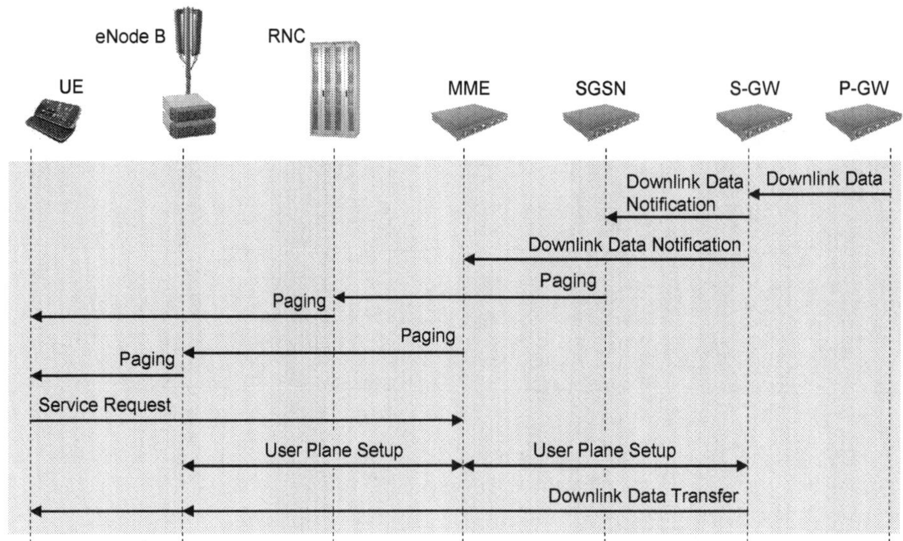

Figure 201 – Downlink data transfer with ISR

- ★ The S11 interface connects the S-GW to the MME, whereas the S4 interface connects the S-GW to the SGSN. These interfaces provide the S-GW with control connections to both the SGSN and MME so Downlink Data Notifications can be sent to both. The Downlink Data Notification message belongs to the GTPv2-C protocol. It specifies the identity of the EPS bearer to which the data belongs
- ★ The SGSN forwards a RANAP: Paging message to all RNC associated with the registered routing area. Each RNC then pages the UE across all cells within that routing area. Meanwhile, the MME forwards an S1: Paging message to all eNode B associated with the registered tracking area. Each eNode B then pages the UE across all cells within that tracking area
- ★ The example illustrated in Figure 201 assumes that the UE is camped on LTE so the UE responds to the MME with a NAS: Service Request message. User plane connections are then established to the S-GW and the transfer of data to the UE begins
- ★ The 'Temporary Identity used in Next update' (TIN) is used to define the type of identity which the UE includes within the next Routing Area Update Request, Tracking Area Update Request or Attach Request. The TIN is also used to identify the status of ISR at the UE. The TIN can be set to P-TMSI, GUTI or RAT-related TMSI. Under normal circumstances, the TIN is set to RAT-related TMSI when ISR is activated. The TIN is set to the temporary identity belonging to the currently used technology when ISR is not active, i.e. either GUTI or P-TMSI
- ★ While ISR is active, the UE and network use a periodic tracking area update timer for LTE (T3412) and a periodic routing area update timer for UTRAN/GERAN (T3312). ISR is deactivated in the UE if a periodic update cannot be completed. The corresponding core network node expecting the periodic update deletes the session context and the control connection to the S-GW if a periodic update is not received
- ★ The network can deactivate ISR whenever the UE moves into a new Routing Area or Tracking Area. Sending a Routing Area Update Accept message or a Tracking Area Update Accept message which does not indicate ISR is activated informs the UE that ISR is to be deactivated
- ★ 3GPP references: TS 23.003, TS 23.401, TS 24.008, TS 29.274

24 PHYSICAL AND MAC LAYER PROCEDURES

24.1 RANDOM ACCESS

★ The random access procedure is required when:

  - making the transition from RRC Idle mode to RRC Connected mode
  - completing an intra-system handover
  - uplink or downlink data arrives while the UE is in the 'non-synchronised' RRC Connected mode state
  - re-establishing an RRC connection

★ The random access procedure can be either contention based or non-contention based. The contention based procedure involves the UE selecting a random access resource, whereas the non-contention based procedure involves the eNode B allocating the random access resource. The contention based procedure can be used for all random access reasons. The non-contention based procedure can be used for intra-system handover and the arrival of downlink data

★ There are two possibilities for the contention based random access procedure. These are illustrated in Figure 202

  - RRC connection establishment and RRC connection re-establishment procedures: the initial layer 3 message is transferred on the CCCH logical channel, contention resolution is based upon the reception of a Contention Resolution Identity (CRI) MAC control element, and a new C-RNTI is allocated
  - intra-system handover and the arrival of uplink or downlink data while non-synchronised: the initial layer 3 message is transferred on the DCCH logical channel and contention resolution is based upon the reception of a PDCCH whose CRC bits have been scrambled by the already allocated C-RNTI

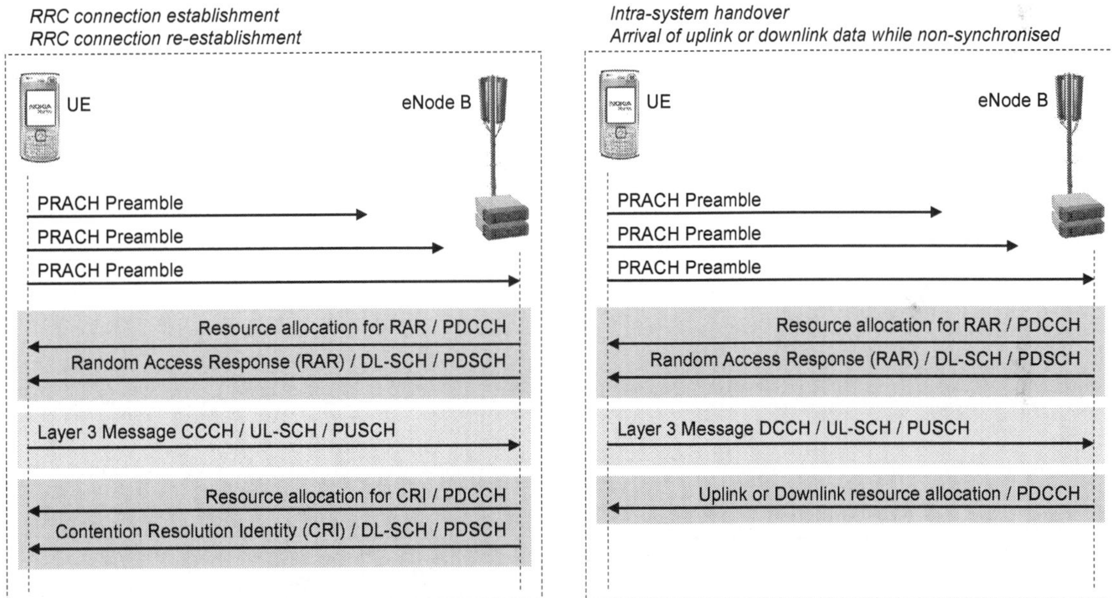

Figure 202 – Signalling for contention based random access procedure

★ The contention based procedure starts with the UE selecting a set of resources for the PRACH in terms of a preamble sequence, and the next available subframe for PRACH transmission. In the case of TDD, the UE also has to select the frequency domain resources because there can be more than a single set of Resource Blocks allocated to the PRACH during each subframe

★ The preamble sequence is used to differentiate between multiple UE using the same set of PRACH Resource Blocks. There is a total of 64 preamble sequences which can be divided into the contention based random access groups A and B, and the non-contention based random access group. Section 14.1 describes the rules for selecting a preamble sequence from the two contention based groups

★ The next available subframe is defined by the PRACH configuration index broadcast within SIB2, or signalled to the UE within an RRC Connection Reconfiguration message. Table 348 within section 31.6.2 presents the relationship between the PRACH configuration index and the set of subframes which can be used for PRACH transmission for FDD. The equivalent relationship for TDD is presented in Table 349

★ In the case of TDD, the PRACH configuration index also determines the number of frequency domain resources available to the PRACH. Smaller PRACH configuration indices define only a single set of frequency resources for the PRACH. Larger PRACH configuration indices define an increased number of frequency domain resources for the PRACH. These frequency domain resources are shown in Table 350 within section 31.6.2. The UE selects one set of frequency domain resources at random. This defines the set of Resource Blocks used for PRACH transmission during the relevant subframe

★ The UE proceeds to transmit the PRACH preamble using a transmit power defined by:

PRACH Preamble Transmit Power = min{P_{CMAX}, PL + PreambleRxTargetPower)}

where,

P_{CMAX} is the UE maximum transmit power according to the UE power class, e.g. 23 dBm for power class 3

PL is the downlink path loss calculated by the UE using a combination of RSRP measurements and knowledge of the Reference Signal transmit power, i.e. PL = Reference Signal transmit power – RSRP measurement

PreambleRxTargetPower = InitialRxTargetPower + DeltaPreamble + (PreambleCounter – 1) × RampingStep

where,

InitialRxTargetPower is broadcast within SIB2, or sent to the UE within an RRC Connection Reconfiguration message. Its value can range from -120 dBm to -90 dBm

DeltaPreamble defines a power offset which is dependent upon the random access preamble format. The preamble format is identified by the PRACH configuration index presented in Table 348 (FDD) and Table 349 (TDD). Preamble formats 0 and 1 use a 0 dB power offset, whereas preamble formats 2 and 3 use a -3 dB power offset

PreambleCounter is a counter maintained by the UE. Its value is initialised to 1 at the start of the random access procedure. It is incremented by 1 if no response is received after transmitting a PRACH preamble. It is used as a multiplying factor to increase the preamble transmit power after receiving no response. It is also used to determine when the maximum allowed number of preamble transmissions has been reached

RampingStep is broadcast within SIB2, or sent to the UE within an RRC Connection Reconfiguration message. It determines the rate at which the preamble transmit power is increased after receiving no response. The step size can be configured with a value of 0, 2, 4 or 6 dB

★ The structure of the PRACH preamble is presented in section 14.1. Preamble formats 0 to 3 occupy 1, 2 or 3 subframes in the time domain (1, 2 or 3 ms) and 839 subcarriers in the frequency domain (subcarrier spacing of 1.25 kHz so occupies 1.05 MHz). Preamble format 4 is only applicable to TDD, and is transmitted during the DwPTS field within special subframes. It occupies 0.16 ms in the time domain and 139 subcarriers in the frequency domain (subcarrier spacing of 7.5 kHz so occupies 1.04 MHz). All preamble formats occupy 6 Resource Blocks in the frequency domain (1.08 MHz) after accounting for the guard band either side of the PRACH subcarriers

★ After transmitting the PRACH preamble, the UE searches for a response during the time domain window defined by the Random Access (RA) response window. The RA response window starts during the third subframe after the preamble, and has a length defined by the response window size which is broadcast in SIB2, or can be signalled to the UE within an RRC Connection Reconfiguration message. The response window size can be configured as 2, 3, 4, 5, 6, 7, 8 or 10 subframes. An example response window is shown in Figure 203

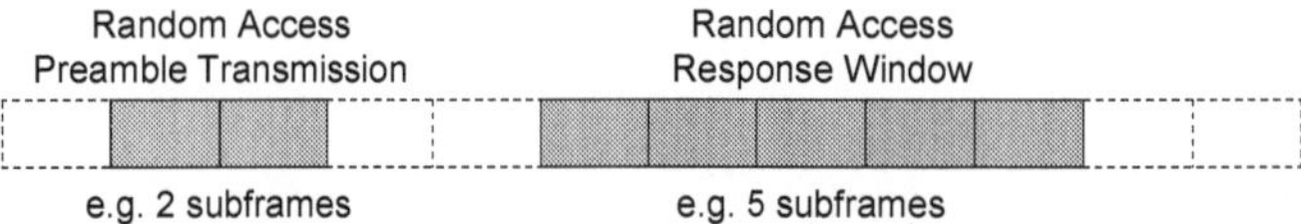

Figure 203 – Random access response window

★ The UE checks each subframe within the search window for a PDCCH whose CRC bits have been scrambled by the relevant RA-RNTI. There is a one-to-one mapping between RA-RNTI and the time/frequency resources used by the PRACH preamble. The RA-RNTI associated with a specific preamble is defined by:

RA-RNTI = 1 + t_id + (10 × f_id)

Where t_id is the index of the subframe within which the start of the preamble was transmitted ($0 \leq t\_id < 10$), and f_id is the frequency domain index of the PRACH within that subframe ($0 \leq f\_id < 6$)

★ In the case of FDD, there is a maximum of one set of Resource Blocks allocated to the PRACH within each subframe so f_id always equals 0 and the equation simplifies to:

RA-RNTI (FDD) = 1 + t_id

i.e. FDD PRACH preambles can be associated with RA-RNTI values of 1 to 10. All UE using the same subframes for PRACH preamble transmission share the same RA-RNTI. These UE are differentiated by their preamble sequence. Contention occurs if multiple UE have selected the same preamble sequence for transmission during the same set of subframes

★ f_id is applicable to TDD because there can be more than a single set of frequency domain resources for the PRACH during each subframe. The PRACH configuration index presented in Table 350 (section 31.6.2) determines the number of frequency domain resources for the PRACH. The number of resources can range from 1 to 6 and these resources are indexed from 0 to 5

★ Figure 204 illustrates 3 FDD UE transmitting preambles starting in subframe 2 (sharing RA-RNTI = 3), and a further 2 UE transmitting preambles starting in subframe 7 (sharing RA-RNTI = 8). Each UE can be differentiated by its RA-RNTI and preamble sequence combination. Preamble sequences are identified by their Random Access Preamble Identity (RAPID). There is no contention because all UE sharing the same subframes use different preamble sequences

★ Figure 204 illustrates that the response windows for the 2 sets of UE overlap but each set of UE attempts to find its own RA-RNTI within a PDCCH. Response windows only overlap when they are associated with different RA-RNTI. The maximum window size of 10 subframes means that search windows associated with the same RA-RNTI never overlap

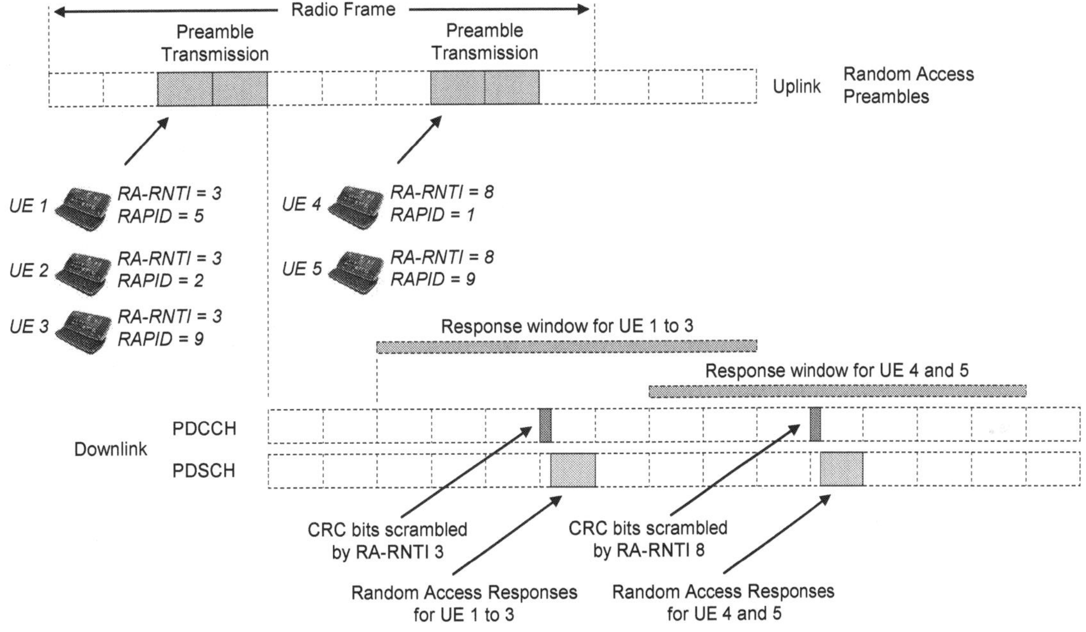

Figure 204 – Random access responses for 5 UE using 2 sets of random access subframes

★ If the UE does not find a PDCCH with its RA-RNTI within the response window, or if the UE finds a PDCCH with its RA-RNTI but the associated Random Access Response (RAR) within the PDSCH does not include any information addressed to the UE, then the UE:

- o increments the value of PreambleCounter by 1
- o checks whether or not the maximum allowed number of preamble transmissions has been reached. The UE exits the random access procedure if the maximum number of transmissions has been reached. The maximum allowed number of transmissions is broadcast within SIB2, or can be signalled to the UE within an RRC Connection Reconfiguration message. Its value can be 3, 4, 5, 6, 7, 8, 10, 20, 50 100 or 200
- o waits until the backoff timer has expired. The backoff timer is set to a value of 0 ms at the start of the random access procedure, so by default a UE does not have to wait before identifying the next subframe for PRACH transmission. The eNode B can increase the value of the back-off timer by attaching an E/T/R/R/BI subheader to a Random Access Response (RAR) message on the PDSCH. This may be done during periods of congestion. The backoff timer is signalled as a Backoff Indicator (BI) which has a length of 4 bits. The relationship between the signalled BI and the actual backoff timer is presented in Table 259
- o selects another PRACH resource (preamble sequence, subframe and frequency domain resource) for the next PRACH transmission

BI	Backoff
0	0 ms
1	10 ms
2	20 ms
3	30 ms

BI	Backoff
4	40 ms
5	60 ms
6	80 ms
7	120 ms

BI	Backoff
8	160 ms
9	240 ms
10	320 ms
11	480 ms

BI	Backoff
12	960 ms
13	reserved
14	reserved
15	reserved

Table 259 – Relationship between signalled Backoff Indicator (BI) and backoff timer

★ If the UE finds a PDCCH whose CRC bits have been scrambled by its RA-RNTI then it proceeds to read the content of the Downlink Control Information (DCI) within the PDCCH. DCI formats 1A and 1C can have their CRC bits scrambled by an RA-RNTI. The downlink resource allocation information is read to identify the position of the Random Access Response (RAR) within the PDSCH of that subframe

★ The structure of the subheaders and payload belonging to a RAR on the PDSCH is illustrated in Figure 205

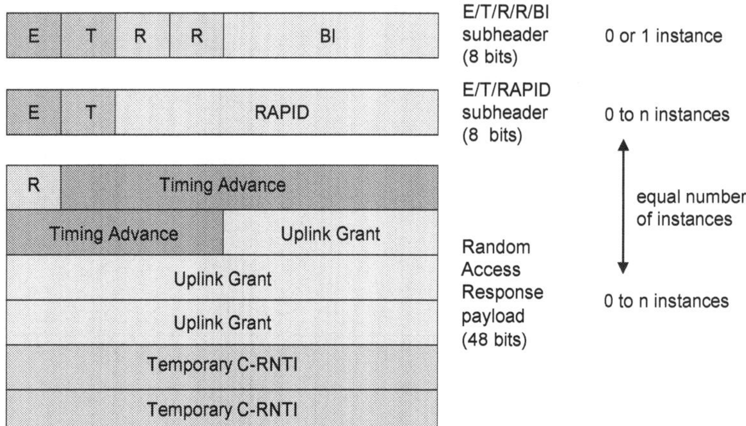

Figure 205 – Format of subheaders and payload for Random Access Response (RAR) message on PDSCH

★ The E/T/R/R/BI subheader is included if the eNode B wishes to define a backoff timer. It is applicable to all UE with the relevant RA-RNTI. There can be a maximum of 1 instance of the E/T/R/R/BI subheader within a RAR message

★ The E/T/RAPID subheader is used to address specific UE. There is 1 instance of this subheader for each UE being addressed. Every instance of the E/T/RAPID subheader has its own instance of the RAR payload

★ The various fields within the RAR subheaders are:

- the Extension (E) field indicates whether or not there are any further subheaders
- the Type (T) field indicates the format of the subheader (E/T/R/R/BI or E/T/RAPID)
- the Reserved (R) field is not used
- the Backoff Indicator (BI) field is used to signal the backoff timer
- the Random Access Preamble Identity (RAPID) field addresses a UE by the index of its preamble sequence

★ There is an instance of the RAR payload for every instance of the E/T/RAPID subheader. The payload includes:

- Timing Advance information to ensure that subsequent UE transmissions are synchronised with other UE when arriving at the eNode B. Timing Advance is described in section 24.2
- Uplink Grant information to allocate uplink Resource Blocks to the UE for transmission on the PUSCH. For example, when establishing an RRC connection, this allocation allows the UE to transmit the RRC Connection Request message. The Resource Block allocation is signalled using the approach described in section 9.4
- a Temporary C-RNTI which becomes the C-RNTI after successful completion of the random access procedure for RRC connection establishment and RRC connection re-establishment

★ The overall structure of an example RAR message is illustrated in Figure 206. This example includes an E/T/R/R/BI subheader to define a backoff timer for all UE with the relevant RA-RNTI. It also includes 3 instances of the E/T/RAPID subheader and a corresponding 3 instances of the RAR payload

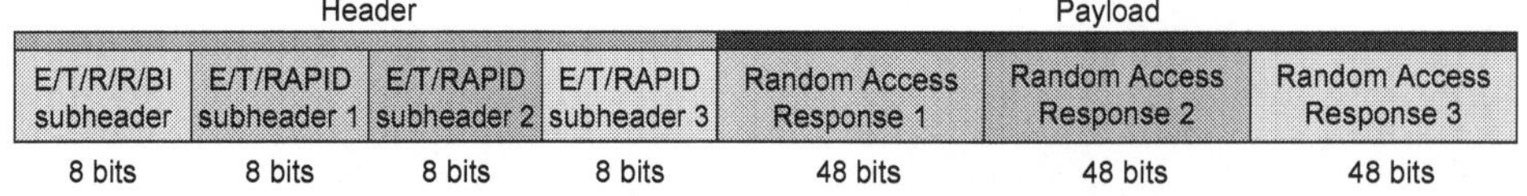

Figure 206 – Example Random Access Response (RAR) message

★ If the UE identifies an E/T/RAPID subheader with the appropriate random access preamble identity, then it reads the corresponding instance of the RAR message payload and identifies the allocated uplink Resource Blocks for transmission on the PUSCH. The UE then proceeds to transmit its initial layer 3 message using a transmit power defined by the equation given in section 24.4.2 with $j = 2$

★ The initial layer 3 message can be transmitted using either the CCCH or DCCH logical channels:

- RRC connection establishment: RRC Connection Request message is sent using the CCCH
- RRC connection re-establishment: RRC Connection Re-establishment Request message is sent using the CCCH
- intra-system handover: RRC Connection Reconfiguration Complete message is sent using the DCCH
- arrival of uplink or downlink data while non-synchronised: Uplink Information Transfer message using the DCCH. The Uplink Information Transfer message contains a Non-Access Stratum (NAS) Service Request message

- When using the CCCH, the Temporary C-RNTI is used as an input when scrambling the PUSCH physical layer bits prior to modulation. When using the DCCH, the existing C-RNTI is used as an input
- The C-RNTI MAC control element is included as part of the MAC header when using the DCCH to transfer the initial layer 3 message. This control element uses 16 bits to specify the existing C-RNTI associated with the UE
- The final stage of the random access procedure is contention resolution. This stage is used to determine whether or not multiple UE used the same combination of RA-RNTI and preamble sequence. If contention occurred then multiple UE would have read the same section of payload within the RAR message and would have transmitted on the same set of uplink Resource Blocks
- The UE starts a contention resolution timer after transmitting the initial layer 3 message. The contention resolution timer is broadcast in SIB2, or can be signalled directly to the UE within an RRC Connection Reconfiguration message. It can be configured with values of 8, 16, 24, 32, 40, 48, 56 and 64 subframes. If the UE does not receive a response from the eNode B before the contention resolution timer expires then the UE returns to transmitting PRACH preambles
- If the UE sent the initial layer 3 message using the CCCH then contention resolution is based upon the eNode B responding with a UE Contention Resolution Identity (CRI) MAC control element. In this case, the UE searches for a PDCCH whose CRC bits have been scrambled by the Temporary C-RNTI. This PDCCH directs the UE to a downlink Resource Block allocation within the same subframe. These PDSCH Resource Blocks include a combination of the R/R/E/LCID MAC subheader and CRI MAC control element. The structure of this PDU is illustrated in Figure 207

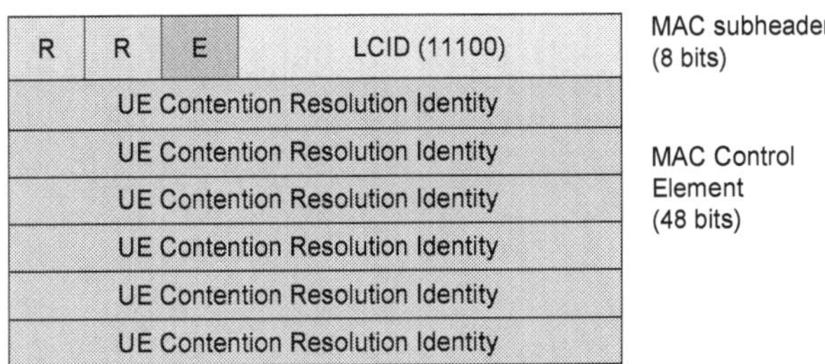

Figure 207 – R/R/E/LCID MAC subheader with a UE Contention Resolution Identity MAC control element

- The Logical Channel Identity (LCID) value of 11100 identifies the subsequent CRI MAC control element. The control element itself includes a reflection of the initial layer 3 message sent by the UE using the CCCH
- If the UE manages to successfully decode the MAC PDU and identify its own initial layer 3 message then contention resolution and the overall random access procedure is successful and the Temporary C-RNTI becomes the C-RNTI. If the UE fails to decode the MAC PDU, or does not find its own initial layer 3 message within the MAC control element then contention resolution fails and the UE returns to transmitting PRACH preambles
- If the UE sent the initial layer 3 message using the DCCH then contention resolution is based upon the eNode B responding with a PDCCH whose CRC bits have been scrambled by the C-RNTI specified by the UE within the preceding C-RNTI MAC control element. If the UE manages to successfully decode a PDCCH associated with its C-RNTI then contention resolution and the overall random access procedure is successful and the Temporary C-RNTI is discarded. If the UE fails to identify a PDCCH associated with its C-RNTI before the contention resolution timer expires, then contention resolution fails and the UE returns to transmitting PRACH preambles
- The non-contention based random access procedure avoids the possibility of multiple UE selecting the same PRACH resource (subframe, preamble sequence and frequency domain resource). This is achieved by the eNode B instructing the UE to use a specific resource which is outside the pool available to UE completing the contention based random access procedure, i.e. some of the 64 preamble sequences can be reserved for allocation by the eNode B
- There are two possibilities for the non-contention based random access procedure. These are illustrated in Figure 208.
    - intra-system handover: the PRACH resource is signalled to the UE within an RRC Connection Reconfiguration message. The initial layer 3 message is an RRC Connection Reconfiguration Complete message on the DCCH logical channel
    - arrival of downlink data while non-synchronised: the PRACH resource is signalled to the UE within a Downlink Control Information (DCI) format 1A PDCCH. The initial layer 3 message is an Uplink Information Transfer message on the DCCH logical channel

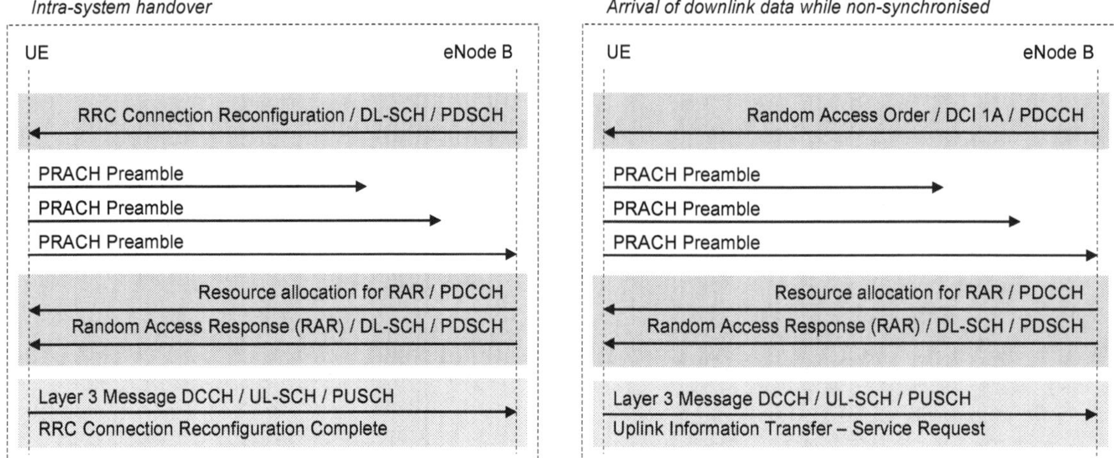

Figure 208 – Signalling for non-contention based random access procedure

★ Once the UE has been instructed to initiate the random access procedure using a specific PRACH resource, PRACH preambles are transmitted in the same way as for the contention based procedure. Also similar to the contention based procedure, the UE searches for a PDCCH whose CRC bits have been scrambled by the relevant RA-RNTI

★ The non-contention based random access procedure is successful as soon as the UE receives a Random Access Response (RAR) message which includes an E/T/RAPID subheader with the appropriate Random Access Preamble Identity (RAPID). The UE can then proceed to transfer its layer 3 message using the uplink Resource Block allocation signalled within the RAR

★ 3GPP References: TS 36.213, TS 36.321

24.2 TIMING ADVANCE

★ Timing advance is used to control the uplink transmission timing of individual UE. It is applicable to the PUSCH, PUCCH and Sounding Reference Signal. The PRACH does not use timing advance because the UE transmits it before the network is able to provide any timing instructions

★ Timing advance helps to ensure that transmissions from all UE are synchronised when received by the eNode B

★ The general concept of timing advance is shown in Figure 209. This figure illustrates 2 UE. The UE furthest from the eNode B requires a larger timing advance to compensate for the larger propagation delay

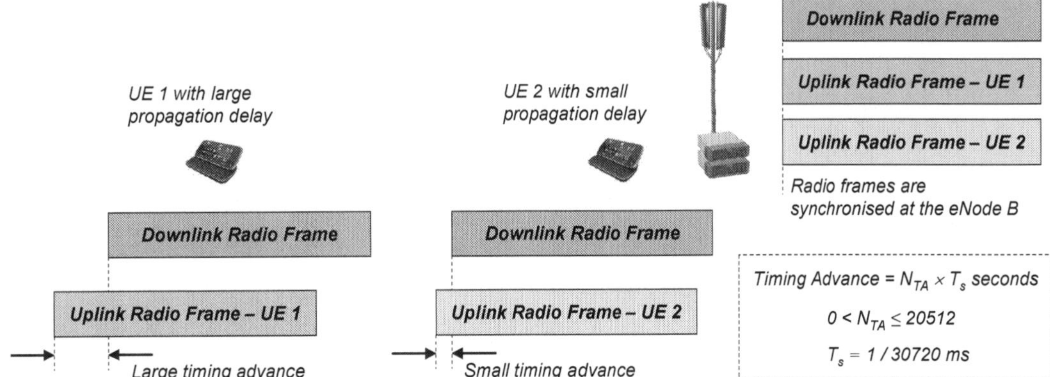

Figure 209 – Concepts of timing advance and uplink / downlink radio frame timing

★ Figure 210 illustrates the radio frame timing for the 2 UE shown in Figure 209. The downlink radio frame arrives at UE 1 relatively late as a result of the larger propagation delay. This figure illustrates that the timing advance equals 2 × propagation delay

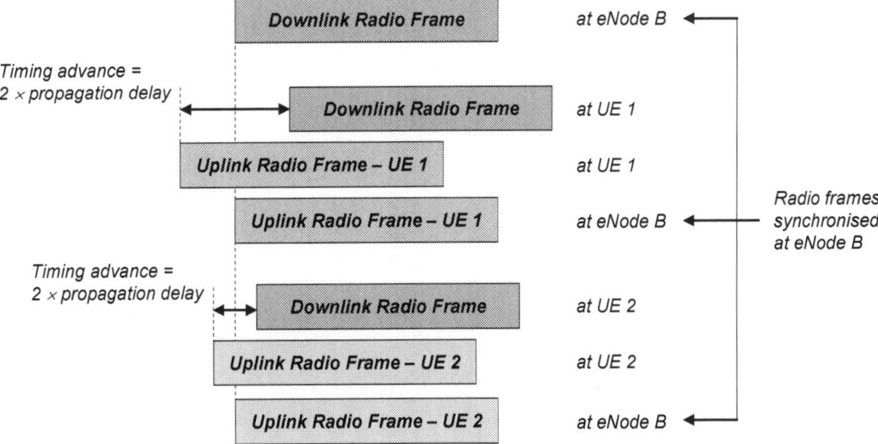

Figure 210 – Uplink / downlink radio frame timing from the UE and eNode B perspectives

★ UE are first provided with timing advance information during the Random Access procedure used to make the transition from RRC Idle mode to RRC Connected mode. An 11 bit timing advance command is included within the Random Access Response

★ These 11 bits are used to signal a value between 0 and 1282. The variable N_{TA} = signalled value × 16, which leads to a maximum value of 1282 × 16 = 20512

★ The actual timing advance applied by the UE is given by:

- FDD: timing advance = $N_{TA} \times Ts$
- TDD: timing advance = $(N_{TA} + 624) \times Ts$

where the value of Ts is given by 1 / 30720 ms

★ The TDD case includes a fixed additional 624 T_S which is equivalent to 20 μs. This additional offset allows time for the eNode B to switch from receiving to transmitting. This additional guard period is illustrated in Figure 17 within section 3.2.2. The timing advance instructed by the eNode B corresponds to N_{TA} so the UE adds this additional offset of 624 T_S when defining its transmission timing

★ The maximum value of N_{TA} corresponds to 0.6677 ms when multiplied by Ts. Based upon the speed of light (3×10^8 ms^{-1}), this corresponds to a round trip distance of 200 km, and a cell range of 100 km

- After the Random Access procedure, timing advance commands are provided using the Timing Advance MAC Control Element which can be included as part of the MAC header. The Timing Advance MAC Control Element includes a 6 bit timing advance command which provides a range from 0 to 63. This control element is illustrated in Figure 211

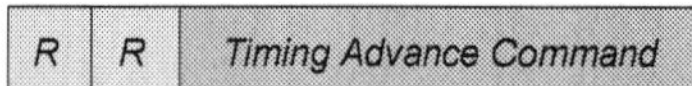

Figure 211 – Timing advance command MAC control element

- The signalled value of the timing advance command within the MAC Control Element corresponds to T_A within the equation:

$$N_{TAnew} = N_{TAold} + (T_A - 31) \times 16$$

Subtracting 31 from the value of T_A allows the eNode B to shift the timing advance in both positive and negative directions, i.e. the timing advance command provided during the Random Access procedure is an absolute timing advance, whereas the subsequent timing advance commands provided within the MAC Control Elements are relative and define changes to the existing timing advance

- The value of N_{TAnew} is applied within the FDD and TDD timing advance equations shown earlier in this section
- Timing advance commands received during downlink subframe 'n' are applied to uplink subframe 'n+6'
- When a timing advance command causes subframe 'm+1' to overlap with subframe 'm' the UE transmits all of subframe 'm' but does not transmit the overlapping part of subframe 'm+1'
- 3GPP References: TS 36.211, TS 36.213, TS 36.321

24.3 HARQ

- Automatic Repeat reQuest (ARQ) refers to a retransmission protocol in which the receiver checks for errors within the received data and if an error is detected than the receiver discards the data and requests a retransmission from the sender. The RLC layer uses an ARQ protocol
- Hybrid ARQ (HARQ) refers to a retransmission protocol in which the receiver checks for errors in the received data and if an error is detected then the receiver buffers the data and requests a retransmission from the sender. A HARQ receiver is then able to combine the buffered data with the re-transmitted data prior to channel decoding and error detection. This improves the performance of the retransmissions. The MAC layer uses a HARQ protocol
- HARQ retransmissions can benefit from either chase combining or incremental redundancy
- Chase combining means the physical layer applies the same puncturing pattern to both the original transmission and each retransmission. This results in retransmissions which include the same set of physical layer bits as the original transmission. The principle of chase combining is shown in Figure 212. The benefits of chase combining are its simplicity and lower UE memory requirement

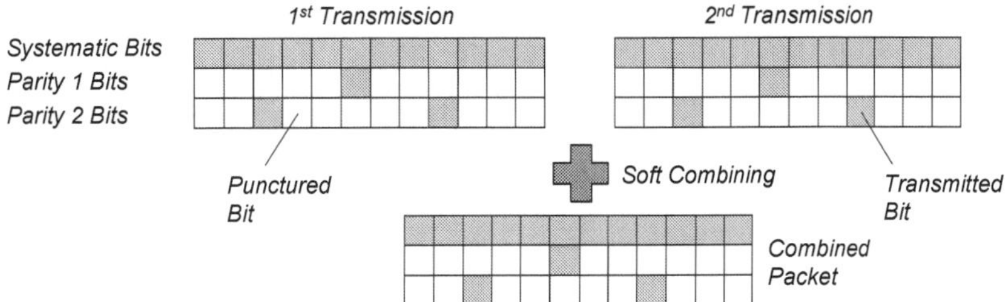

Figure 212 – Principle of HARQ chase combining

- Turbo coding generates systematic, parity 1 and parity 2 bits. The systematic bits are the same as the original bit sequence. These bits are the most important to the receiver and are provided with the greatest priority, i.e. the parity 1 and parity 2 bits are punctured in preference to the systematic bits
- A receiver using chase combining benefits from a soft combining gain because the received symbols are combined prior to demodulation. This improves the received signal to noise ratio if it is assumed that the signal power is correlated while the noise power is uncorrelated
- Incremental redundancy means the physical layer applies different puncturing patterns to the original transmission and retransmissions. This results in retransmissions which include a different set of physical layer bits to the original transmission. The principle of incremental redundancy is illustrated in Figure 213. The drawbacks associated with incremental redundancy are its increased complexity and increased UE memory requirements

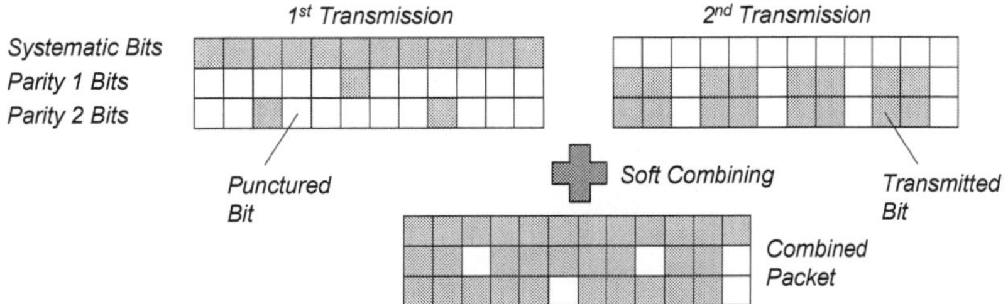

Figure 213 – Principle of HARQ incremental redundancy

- The first transmission provides the systematic bits with the greatest priority while subsequent retransmissions can provide either the systematic or the parity 1 and parity 2 bits with the greatest priority
- The performance of incremental redundancy is similar to the performance of chase combining when the coding rate is low, i.e. there is little puncturing. The performance of incremental redundancy becomes greater when there is an increased quantity of puncturing. Incremental redundancy performs better than chase combining when the coding rate is high because channel coding gain is greater than soft combining gain
- The HARQ protocol relies upon the sender receiving acknowledgements from the receiver. The round trip time, which includes both the sender and receiver processing times as well as the propagation delays, means that these acknowledgements are not received instantaneously. In general, the propagation delays are negligible compared to the processing delays
- The impact of waiting for an acknowledgement is shown in Figure 214. The sender becomes inactive while waiting for the acknowledgement so the average throughput is relatively low. This corresponds to using a single Stop And Wait (SAW) process. A SAW process stops and waits for an acknowledgement before proceeding to transfer any further data

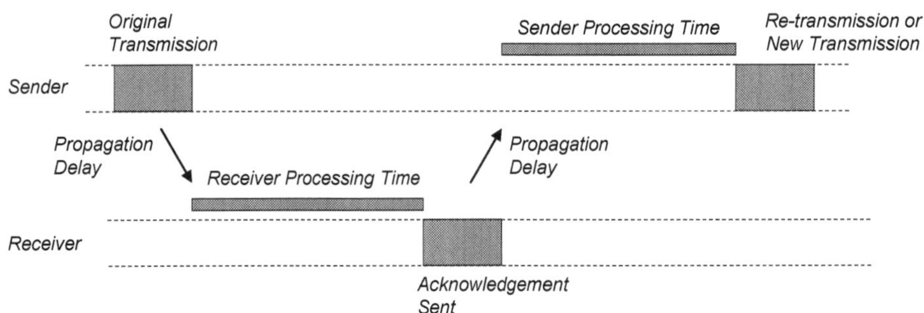

Figure 214 – Data transfer using a single Stop And Wait (SAW) process

★ Multiple parallel SAW processes are used to avoid the round trip time having an impact upon throughput, i.e. additional SAW processes transfer data while the first SAW process is waiting for its acknowledgement. Larger round trip times require an increased number of parallel SAW processes. The concept of multiple parallel SAW processes is illustrated in Figure 215

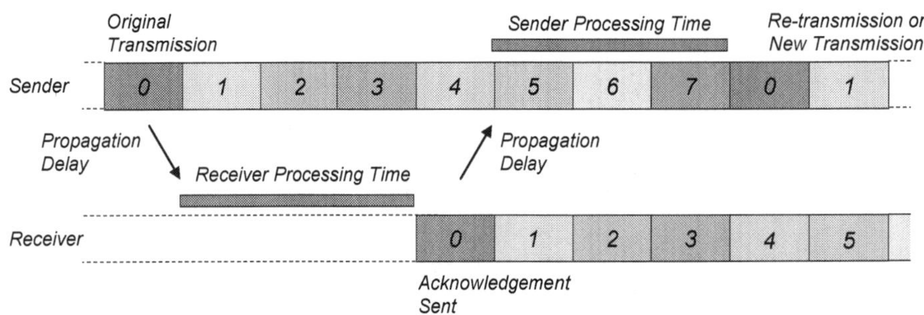

Figure 215 – Data transfer using multiple parallel Stop And Wait (SAW) process

★ The example in Figure 215 requires 8 parallel SAW processes to maintain a constant flow of data. These SAW processes are also referred to as HARQ processes. The HARQ entity within the MAC layer manages these multiple HARQ processes

★ The duration associated with each transmission corresponds to the Transmission Time Interval (TTI) which is represented by the 1 ms subframe in LTE. Figure 215 illustrates 8 parallel HARQ processes so there is an 8 ms cycle between sending data from a specific SAW process and being able to send a retransmission or new data from the same SAW process. This allows for a total round trip time of 8 ms (UMTS HSDPA uses 6 HARQ processes and a 2 ms TTI to allow for a 12 ms round trip time)

★ The sender buffers the transmitted data until a positive acknowledgement has been received in case a retransmission is required. Data is cleared from the transmit buffer once a positive acknowledgement has been received or the maximum number of allowed retransmissions has been reached. New data can be sent by a specific HARQ process once its transmit buffer has been cleared

★ The uplink and downlink of LTE are both specified to use parallel HARQ processes. However, the uplink and downlink differ in some respects, e.g. the uplink uses synchronous HARQ whereas the downlink uses asynchronous HARQ. Asynchronous HARQ provides greater flexibility but requires additional overhead in terms of signalling which HARQ process is being used at any point in time

★ 3GPP References: TS 36.213, TS 36.321

24.3.1 FDD UPLINK

★ HARQ for the uplink refers to transferring uplink data on the PUSCH while receiving downlink acknowledgements on the PHICH

★ HARQ is synchronous in the uplink direction. This means that acknowledgements and retransmissions are scheduled at fixed time intervals relative to the initial transmission. Synchronous HARQ does not require the HARQ process identity to be signalled so generates a lower overhead

★ There is one HARQ entity at the UE for each serving cell. There are multiple serving cells and multiple HARQ entities when LTE Advanced Carrier Aggregation is used

★ Each HARQ entity maintains a number of synchronous parallel HARQ processes. The total number of HARQ processes is summarised in Table 260. TTI bundling and its associated HARQ are described in section 27.6.1. TTI bundling is not used when Transmission Mode 2 is configured, i.e. it is not used with uplink MIMO

★ In the case of Transmission Mode 2, the number of HARQ processes doubles for both 2×2 MIMO and 4×4 MIMO. The number of HARQ processes does not quadruple for 4×4 MIMO because it uses 2 transport blocks which are subsequently segmented into 4 blocks prior to transmission across the air-interface

	Transmission Mode 1	Transmission Mode 2
Normal Operation	8	16
LTE Advanced Carrier Aggregation	8 per serving cell	16 per serving cell
TTI Bundling	4	not applicable

Table 260 – Total number of HARQ processes for the uplink of FDD

★ The use of 8 HARQ processes for FDD Transmission Mode 1 is illustrated in Figure 216. 8 parallel HARQ Stop And Wait (SAW) processes are used to provide continuous data transmission while allowing for a maximum round trip time of 8 ms

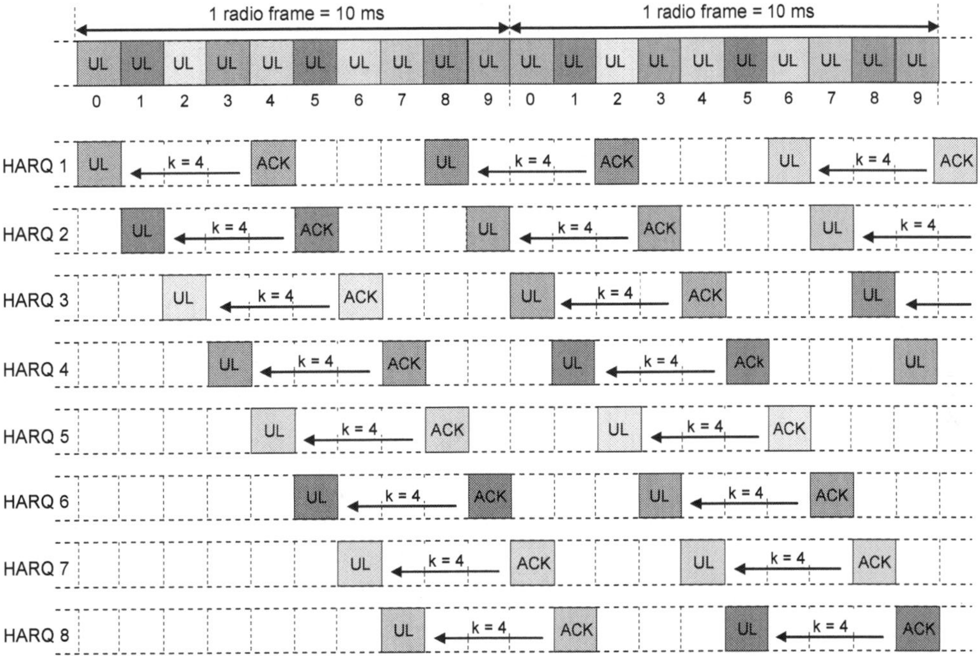

Figure 216 – Timing of uplink HARQ processes for FDD uplink

★ HARQ acknowledgements are provided on the PHICH. They are sent 4 subframes after uplink transmission on the PUSCH. A retransmission or new data transmission using the same HARQ process can then follow 4 subframes after the acknowledgement

★ The requirement for acknowledgements doubles when spatial multiplexing is used and 2 transport blocks are transferred during each subframe. The pair of acknowledgements associated with the pair of transport blocks are transferred during the same subframe but using different PHICH resources (combination of PHICH group and orthogonal sequence)

★ The variable 'k' defines the difference between the subframe number associated with the HARQ acknowledgement and the subframe number associated with the PUSCH transmission. This variable is always equal to 4 for FDD

★ Incremental redundancy is always possible in the uplink direction because the eNode B has a relatively large memory capability. The physical layer puncturing pattern is determined by the Redundancy Version (RV) which uses a value of 0 for the first transmission to provide the systematic bits with priority:

- o the RV cycles through the values 2, 3, 1, 0 when retransmissions are triggered by a negative acknowledgement on the PHICH (non-adaptive retransmission)
- o the RV is signalled explicitly as part of the Modulation and Coding Scheme (MCS) when retransmissions are triggered by the PDCCH (adaptive retransmission)

★ Retransmissions can be either non-adaptive or adaptive:

- o non-adaptive retransmissions are triggered by a negative acknowledgement on the PHICH. Non-adaptive retransmissions use the same set of resources as the previous transmission, i.e. the modulation scheme and the set of allocated Resource Blocks remain unchanged. However, the RV does change between transmissions
- o adaptive retransmissions can be triggered by the PDCCH Downlink Control Information (DCI) formats 0 and 4. The New Data Indicator (NDI) flag triggers a retransmission if its value is not toggled relative to the previous transmission. Adaptive retransmissions allow the set of allocated Resource Blocks as well as the RV to change

★ Table 261 summarises the difference between non-adaptive and adaptive retransmissions

	Non-Adaptive Retransmission	Adaptive Retransmission
Triggered by	PHICH Negative Ack.	PDCCH DCI formats 0 and 4 NDI
Transport Block Size	no change	no change
Modulation Scheme		
Allocated Resource Blocks		changes
Redundancy Version	changes	

Table 261 – Comparison between non-adaptive and adaptive retransmissions

★ PDCCH DCI formats 0 and 4 takes precedence over the PHICH. The PHICH can indicate a positive acknowledgement but a retransmission is still triggered if a PDCCH is received during the relevant subframe with the NDI bit untoggled. Both PHICH acknowledgements and PDCCH instructions are received 4 subframes after transmission on the PUSCH. The rules for interpreting the combination of PHICH and PDCCH acknowledgements are summarised in Table 262

PHICH Feedback Received by UE	PDCCH Received by UE	UE Transmission
ACK or NACK	New Transmission	New transmission
	Retransmission	Adaptive retransmission
ACK	None	No transmission nor retransmission but keep data in HARQ buffer
NACK		Non-adaptive retransmission

Table 262 – Retransmission triggering using the PHICH and PDCCH

24.3.2 TDD UPLINK

★ HARQ for the uplink refers to transferring uplink data on the PUSCH while receiving downlink acknowledgements on the PHICH

★ Similar to FDD, HARQ for TDD is synchronous in the uplink direction. This means that acknowledgements and retransmissions are scheduled at fixed time intervals relative to the initial transmission. In the case of TDD, those fixed time intervals depend upon the uplink-downlink subframe configuration. Synchronous HARQ does not require the HARQ process identity to be signalled so generates a lower overhead

★ There is one HARQ entity at the UE for each serving cell. There are multiple serving cells and multiple HARQ entities when LTE Advanced Carrier Aggregation is used

★ Each HARQ entity maintains a number of synchronous parallel HARQ processes. The total number of HARQ processes as a function of the TDD uplink-downlink subframe configuration is summarised in Table 263. TTI bundling and its associated HARQ are described in section 27.6.1. TTI bundling is not used when Transmission Mode 2 is configured, i.e. it is not used with uplink MIMO

UL/DL Subframe Configuration	Uplink Subframes per Radio Frame	Transmission Mode 1			Transmission Mode 2	
		Normal Operation	LTE Advanced Carrier Aggregation	TTI Bundling	Normal Operation	LTE Advanced Carrier Aggregation
0	6	7	7 per Serving Cell	3	14	14 per Serving Cell
1	4	4	4 per Serving Cell	2	8	8 per Serving Cell
2	2	2	2 per Serving Cell	not applicable	4	4 per Serving Cell
3	3	3	3 per Serving Cell	not applicable	6	6 per Serving Cell
4	2	2	2 per Serving Cell	not applicable	4	4 per Serving Cell
5	1	1	1 per Serving Cell	not applicable	2	2 per Serving Cell
6	5	6	6 per Serving Cell	3	12	12 per Serving Cell

Table 263 – Total number of HARQ processes for the uplink of TDD

★ In the case of uplink-downlink subframe configurations 1 to 5, when using normal operation with Transmission Mode 1, the number of HARQ processes is equal to the number of uplink subframes within each radio frame. In the case of uplink-downlink subframe configurations 0 and 6 an additional HARQ process is required

★ The HARQ process timing for uplink-downlink subframe configuration 1 is illustrated in Figure 217. The delay between the UE sending some uplink data on the PUSCH and receiving an acknowledgement on the PHICH varies depending upon the subframe being used. This delay is shown using the variable 'k' in Figure 217. The value of 'k' never decrease below 4

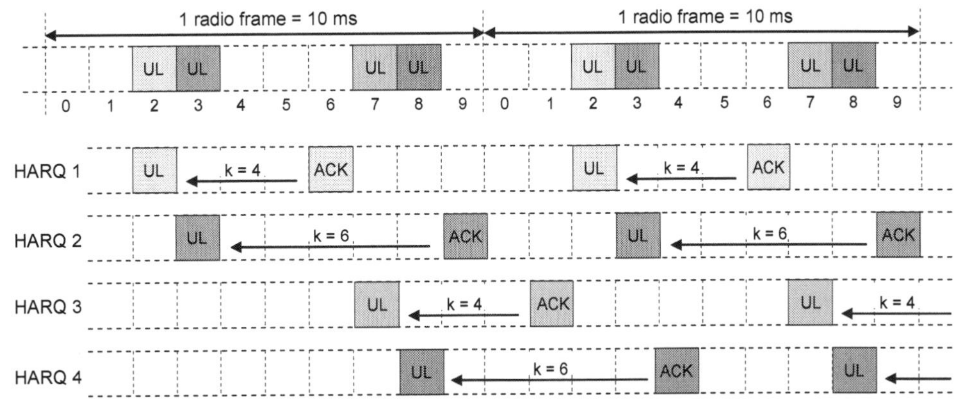

Figure 217 – Timing of HARQ processes for TDD uplink-downlink subframe configuration 1 (normal operation, transmission mode 1)

- ★ The pattern is less regular than FDD because acknowledgements cannot always be sent after 4 subframes, i.e. the 4th subframe can be an uplink subframe so the eNode B has to wait until the next available downlink subframe before being able to send the acknowledgement. HARQ acknowledgements can be sent during the DwPTS of the special subframes introduced in section 3.2.2
- ★ Similarly, the delay between receiving an acknowledgement on the PHICH and being able to send either new data or a retransmission using the same HARQ process is not constant because the 4th subframe after an acknowledgement can be a downlink subframe so the UE has to wait until the next available uplink subframe
- ★ Figure 218 illustrates the HARQ process timing for uplink-downlink subframe configuration 6 when assuming only 5 HARQ processes are used. This is not a real configuration but illustrates the reason for requiring 6 HARQ processes

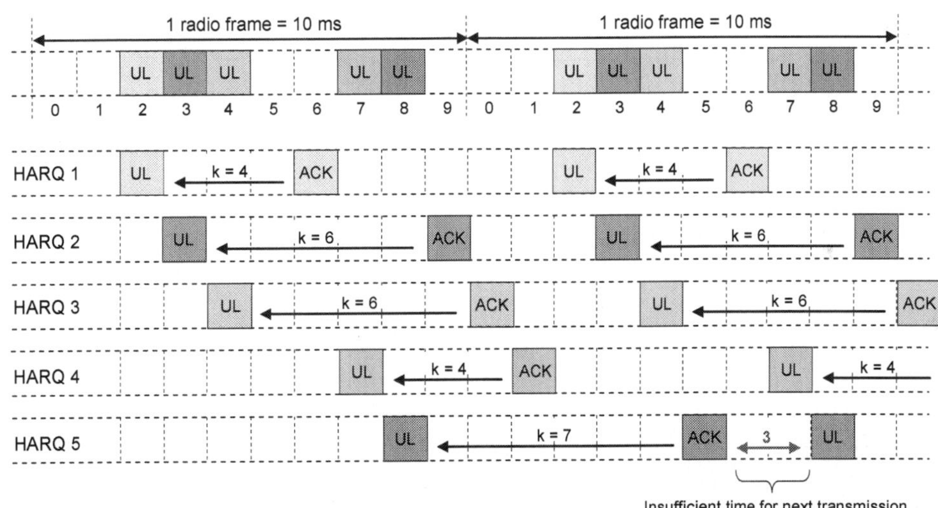

Figure 218 – Timing of HARQ processes for TDD uplink-downlink subframe configuration 6 if assuming only 5 HARQ processes (note this is not a real configuration – real configuration uses 6 HARQ processes)

- ★ Within Figure 218, HARQ processes 1 to 4 appear normal but HARQ process 5 has only 3 subframes difference between the UE receiving its acknowledgement and having to send the next transmission. A minimum of 4 subframes is required to allow sufficient processing time so this triggers the requirement for an additional HARQ process
- ★ Figure 219 illustrates the real HARQ process timing for uplink-downlink subframe configuration 6. This demonstrates that an additional HARQ process allows a minimum of 4 subframes difference between receiving an acknowledgement and sending the next block of data
- ★ The value of the 'k' variable for each uplink-downlink subframe configuration is presented in Table 264. This table shows that when an acknowledgement is received on the PHICH within the nth subframe, then it refers to the PUSCH data sent during subframe n - k
- ★ The uplink-downlink subframe configuration 0 represents a unique case because two sets of acknowledgements are received in subframes 0 and 5, i.e. acknowledgements are received in these subframes which refer to PUSCH transmissions 7 subframes earlier, and 6 subframes earlier. This approach is necessary because there are 6 uplink subframes available for the PUSCH but only 4 downlink subframes available for the PHICH

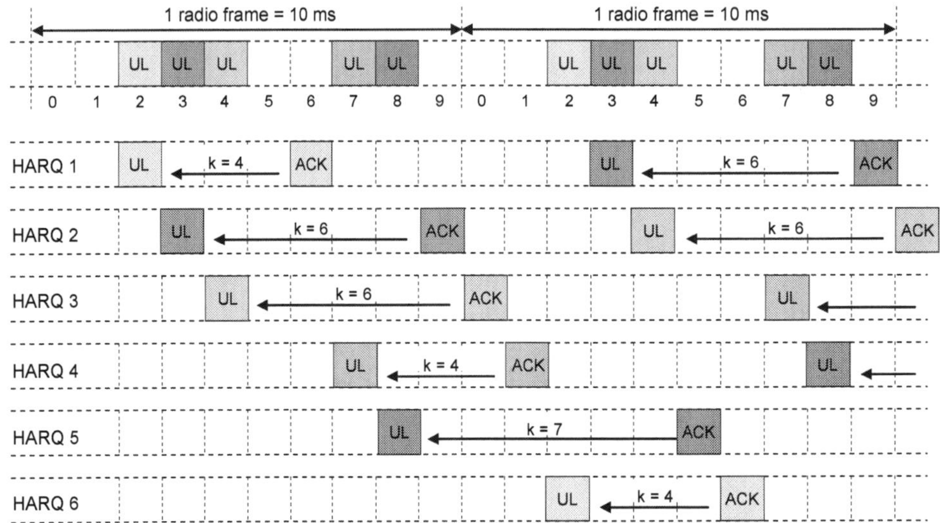

Figure 219 – Timing of HARQ processes for TDD uplink-downlink subframe configuration 6 (real configuration)

UL-DL Subframe Configuration	Subframe Number									
	0	1	2	3	4	5	6	7	8	9
0	7 / 6	4				7 / 6	4			
1		4			6		4			6
2				6					6	
3	6								6	6
4									6	6
5									6	
6	6	4				7	4			6

Table 264 – 'k' variable for each uplink-downlink subframe configuration

★ Figure 220 illustrates the HARQ process timing for uplink-downlink subframe configuration 0. The 2 PHICH acknowledgements sent during subframe 0 (and also subframe 5) are differentiated by their PHICH group. 3GPP TS 36.213 specifies which PHICH group is allocated to each acknowledgement. This allows the UE to link each acknowledgement to the appropriate PUSCH transmission

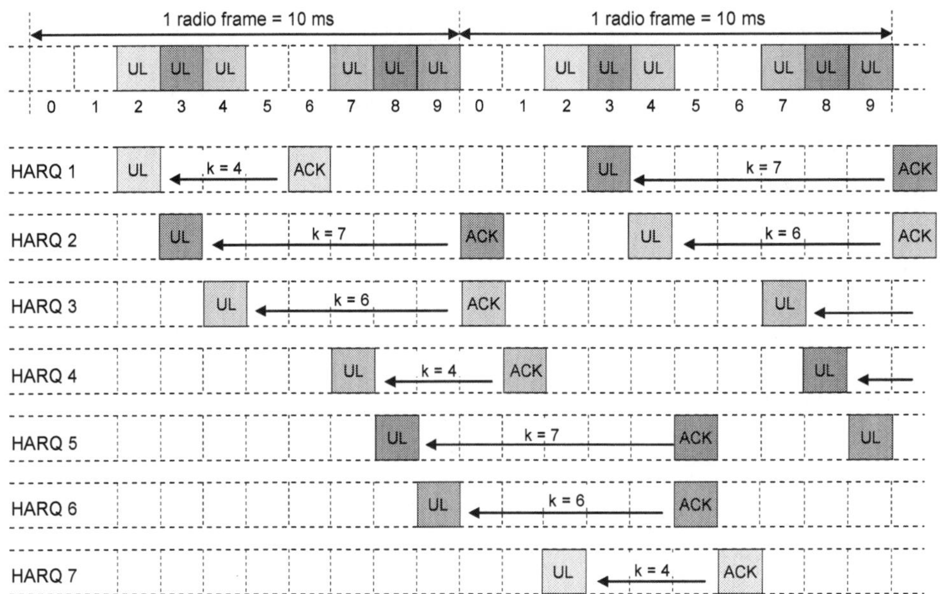

Figure 220 – Timing of HARQ processes for TDD uplink-downlink subframe configuration 0

★ Table 265 presents the number of HARQ acknowledgements per subframe on the PHICH for each uplink-downlink subframe configuration. Within the context of this table, special subframes are counted as downlink subframes. Multiple HARQ acknowledgements can be multiplexed within the same subframe by allocating different PHICH groups and orthogonal sequences

			Without Spatial Multiplexing		With Spatial Multiplexing	
UL-DL Subframe Configuration	Uplink Subframes	Downlink Subframes	HARQ ACK to Send in Downlink	ACK to send per Subframe	HARQ ACK to Send in Downlink	ACK to send per Subframe
0	6	4	6	1 or 2	12	2 or 4
1	4	6	4	1 or 0	8	2 or 0
2	2	8	2	1 or 0	4	2 or 0
3	3	7	3	1 or 0	6	2 or 0
4	2	8	2	1 or 0	4	2 or 0
5	1	9	1	1 or 0	2	2 or 0
6	5	5	5	1	10	2

Table 265 – The number of HARQ acknowledgements per subframe on the PHICH

★ Without spatial multiplexing, the PHICH has to transfer a maximum of 2 HARQ acknowledgements per subframe per connection. The majority of uplink-downlink subframe configurations only require a single HARQ acknowledgement per subframe because there are more downlink subframes than uplink subframes

★ The requirement for acknowledgements doubles when spatial multiplexing is used and 2 transport blocks are transferred during each subframe. The pair of acknowledgements associated with the pair of transport blocks are transferred during the same subframe but using different PHICH resources (combination of PHICH group and orthogonal sequence)

24.3.3 FDD DOWNLINK

★ HARQ for the downlink refers to transferring downlink data on the PDSCH while receiving uplink acknowledgements on the PUCCH or PUSCH

★ There is one HARQ entity at the UE for each serving cell. There are multiple serving cells and multiple HARQ entities when LTE Advanced Carrier Aggregation is used

★ Each HARQ entity maintains a number of asynchronous parallel HARQ processes. The total number of HARQ processes is summarised in Table 266. TTI bundling is not applicable to the downlink

	All Transmission Modes
Normal Operation	up to 8
LTE Advanced Carrier Aggregation	up to 8 per serving cell

Table 266 – Total number of HARQ processes for the downlink of FDD

★ Table 266 illustrates 2 main differences relative to the uplink scenario presented in Table 260:

- o in the case of the downlink, there is no dependency upon whether or not spatial multiplexing is used. In the downlink direction, when 2 transport blocks are transferred during the same subframe using spatial multiplexing, then those 2 transport blocks are associated with the same HARQ process rather than 2 different HARQ processes. This avoids the requirement to double the number of HARQ processes when using spatial multiplexing
- o in the case of the downlink, the number of HARQ processes is 'up to' 8 and 'up to' 4 rather than exactly 8 or exactly 4. This results from HARQ being asynchronous in the downlink direction so fewer HARQ processes can be used when a connection is not scheduled data during every subframe

★ HARQ is asynchronous in the downlink direction. The eNode B provides the UE with instructions regarding which HARQ process to use during each subframe that resources are allocated. The HARQ process identity is included within PDCCH DCI formats 1, 1A, 1B, 1D, 2, 2A, 2B and 2C

★ Asynchronous HARQ increases the signalling overhead as a result of the requirement to include the HARQ process identity within the DCI. However, asynchronous HARQ also increases flexibility because retransmissions do not have to be scheduled during specific subframes. The eNode B may wish to postpone a retransmission because other UE are being scheduled during that subframe, or because the channel conditions have become relatively poor

★ Up to 8 parallel HARQ Stop And Wait (SAW) processes are used to provide continuous data transmission while allowing for a maximum round trip time of 8 ms. The use of asynchronous HARQ means that less than 8 HARQ processes can be used if the UE is not scheduled during every subframe. This is in contrast to synchronous HARQ in the uplink which always cycles through the set of 8 HARQ processes irrespective of whether or not resources are allocated during a specific subframe. Figure 221 compares synchronous and asynchronous HARQ processes

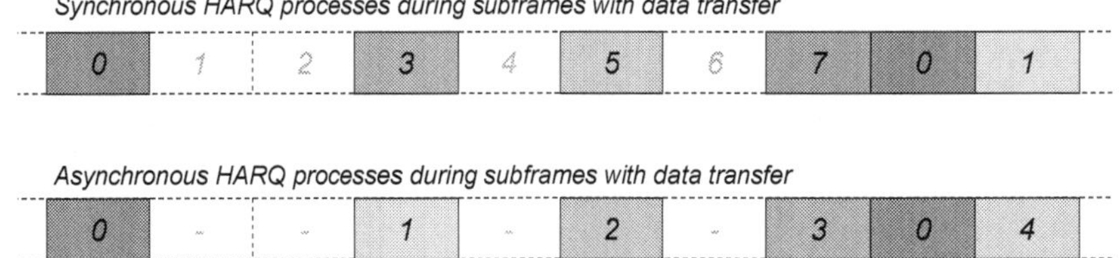

Figure 221 – Synchronous and asynchronous HARQ process identities when data transfer is not continuous

★ When data transfer is continuous and the UE is scheduled during every subframe then 8 HARQ processes are required and the timing is the same as for the uplink, i.e. as illustrated in Figure 216

★ Both chase combining and incremental redundancy are used in the downlink. Incremental redundancy is the preferred method but chase combining is selected when incremental redundancy is not possible. 3GPP TS 36.306 specifies a 'Total Number of Soft Channel Bits' capability for each UE category. This represents a buffer size used to store the data received by the set of HARQ processes. The eNode B keeps track of the UE buffer occupancy and switches to chase combining when there is insufficient capacity to support incremental redundancy

★ Chase combining is most likely to be required when UE are allocated their maximum throughput capability, i.e. the quantity of data filling the UE buffer is large. When using chase combining, the Redundancy Version (RV) is set to 0 to provide the systematic bits with the greatest priority

★ The RV value is signalled explicitly with the downlink resource allocations within PDCCH Downlink Control Information (DCI) formats 1, 1A, 1B, 1D, 2, 2A, 2B and 2C. DCI format 1C does not specify an RV value because this DCI is used to schedule resources for random access responses, paging messages, and system information. Random access responses and paging messages always use an RV value of 0. System information uses RV values which are calculated from the System Frame Number (SFN) and subframe number

★ The selection of RV values is implementation dependent but in general the uplink pattern specified by 3GPP can be used for the downlink, i.e. original transmission uses RV=0 while retransmissions cycle through the values 2, 3, 1, 0

★ HARQ retransmissions in the downlink can always be adaptive because the PDCCH can allocate a different set of resources for the original transmission and each retransmission. The MCS indices 29, 30 and 31 within Table 387 illustrate that the modulation scheme can be changed between the original transmission and each retransmission. This is in contrast to the uplink MCS indices 29, 30 and 31 within Table 395 which only allow the RV value to change

★ The UE returns acknowledgements to the eNode B using either the PUCCH or PUSCH. In the case of FDD without Carrier Aggregation, it is necessary to transfer 1 HARQ acknowledgement per subframe when spatial multiplexing is not used, and 2 HARQ acknowledgements per subframe when spatial multiplexing is used. These requirements can be satisfied by the capacity of the PUCCH without any additional processing

★ The eNode B uses the New Data Indicator (NDI) flag within the PDCCH DCI to inform the UE of whether it is sending a new transport block or a retransmission. Toggling the value relative to the previous value used for the same HARQ process indicates that new data is being sent rather than a retransmission

24.3.4 TDD DOWNLINK

★ HARQ for the downlink refers to transferring downlink data on the PDSCH while receiving uplink acknowledgements on the PUCCH or PUSCH

★ There is one HARQ entity at the UE for each serving cell. There are multiple serving cells and multiple HARQ entities when LTE Advanced Carrier Aggregation is used

★ Each HARQ entity maintains a number of asynchronous parallel HARQ processes. The total number of HARQ processes is summarised in Table 267. TTI bundling is not applicable to the downlink because it is aimed at improving uplink coverage

★ Similar to the downlink of FDD, Table 267 illustrates 2 main differences relative to the uplink

- o in the case of the downlink, there is no dependency upon whether or not spatial multiplexing is used. In the downlink direction, when 2 transport blocks are transferred during the same subframe using spatial multiplexing, then those 2 transport blocks are associated with the same HARQ process rather than 2 different HARQ processes
- o in the case of the downlink, the number of HARQ processes is 'up to' *x* and 'up to' *y*, rather than exactly *x* or exactly *y*. This results from HARQ being asynchronous in the downlink direction so fewer HARQ processes can be used when a connection is not scheduled data during every subframe

UL-DL Subframe Configuration	Downlink Subframes per Radio Frame	All transmission modes	
		Normal Operation	LTE Advanced Carrier Aggregation
0	4	up to 4	up to 4 per Serving Cell
1	6	up to 7	up to 7 per Serving Cell
2	8	up to 10	up to 10 per Serving Cell
3	7	up to 9	up to 9 per Serving Cell
4	8	up to 12	up to 12 per Serving Cell
5	9	up to 15	up to 15 per Serving Cell
6	5	up to 6	up to 6 per Serving Cell

Table 267 – Total number of HARQ processes for the downlink of TDD

★ Similar to the downlink of FDD, HARQ for TDD is asynchronous in the downlink direction. This means that the eNode B provides the UE with instructions regarding which HARQ process to use during each subframe that resources are allocated. The HARQ process identity is included within PDCCH DCI formats 1, 1A, 1B, 1D, 2, 2A, 2B and 2C

★ Asynchronous HARQ increases the signalling overhead as a result of the requirement to include the HARQ process identity within the DCI. However, asynchronous HARQ also increases flexibility because retransmissions do not have to be scheduled during specific subframes. The eNode B may wish to postpone a retransmission because other UE are being scheduled during that subframe, or because the channel conditions have become relatively poor

★ Table 268 presents the number of HARQ acknowledgements per subframe which would ideally be supported by the PUCCH (in practice, the PUCCH does not have this much capacity). Within the context of this table, special subframes are counted as downlink subframes

UL-DL Subframe Configuration	Downlink Subframes	Uplink Subframes	Without Spatial Multiplexing		With Spatial Multiplexing	
			HARQ ACK to Send in Downlink	ACK to send per Subframe	HARQ ACK to Send in Downlink	ACK to send per Subframe
0	4	6	4	1 or 0	8	2 or 0
1	6	4	6	1 or 2	12	2 or 4
2	8	2	8	4	16	8
3	7	3	7	2 or 3	14	4 or 6
4	8	2	8	4	16	8
5	9	1	9	9	18	18
6	5	5	5	1	10	2

Table 268 – Requirement for HARQ acknowledgements

★ Many of these figures exceed the capacity of the PUCCH. The number of acknowledgements supported by the PUCCH is presented in Table 172 within section 14.2. Channel selection can be used with PUCCH format 1b to allow up to 4 HARQ acknowledgements to be transferred during a single subframe. However, uplink-downlink subframe configuration 5 requires 9 acknowledgements to be sent within a single subframe (assuming spatial multiplexing is not used and downlink data is sent during every downlink subframe)

★ 3GPP TS 36.213 specifies 2 HARQ feedback modes for TDD which reduce the number of acknowledgements to be transferred:

- o HARQ-ACK bundling
- o HARQ-ACK multiplexing

★ All uplink-downlink subframe configurations support both feedback modes, with the exception of uplink-downlink subframe configuration 5 which only supports HARQ-ACK bundling. HARQ multiplexing is not supported for configuration 5 because there is only a single uplink subframe to support the transfer of acknowledgements for 9 downlink subframes (including the DwPTS field within the special subframe). This means that HARQ-ACK bundling is required to reduce the number of acknowledgements

★ A UE can be instructed to use either HARQ-ACK bundling or HARQ-ACK multiplexing within an RRC Connection Setup, RRC Connection Reconfiguration or RRC Connection Re-establishment message. The instruction is provided as part of the PUCCH configuration information

★ HARQ-ACK bundling reduces the number of HARQ acknowledgements by performing a logical AND operation between the acknowledgements belonging to multiple downlink subframes, i.e. bundling is completed by applying a logical AND operation in the time domain

★ For example, uplink-downlink subframe configuration 2 includes 6 downlink subframes, 2 special subframes and 2 uplink subframes. Downlink data can be transferred during the DwPTS section of special subframes so there are effectively 8 downlink subframes and 2 uplink subframes. If the downlink subframes are divided into 2 groups and a single acknowledgement generated from each group then the uplink subframes are only required to transfer a single HARQ acknowledgement. Figure 222 illustrates the concept of HARQ-ACK bundling for uplink-downlink subframe configuration 2

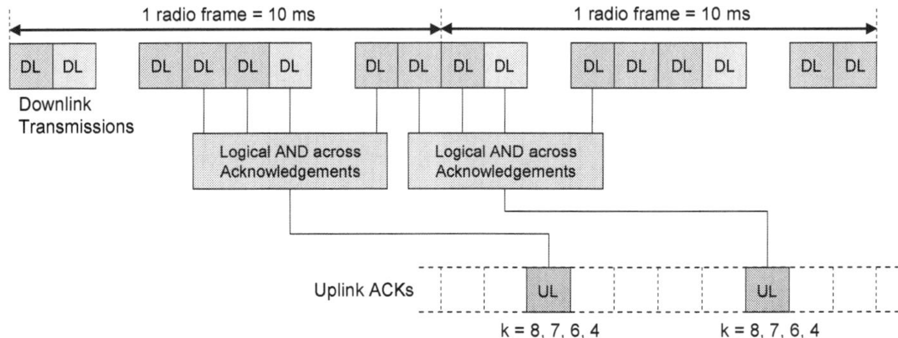

Figure 222 – Concept of HARQ bundling for uplink-downlink subframe configuration 2 (without spatial multiplexing)

★ The example illustrated in Figure 222 assumes that spatial multiplexing is not used. This results in a single HARQ acknowledgement for each uplink subframe. When spatial multiplexing is used, the logical AND operations are completed independently for the two transport blocks, i.e. there is one logical AND operation for the set of transport block 1's, and a second logical AND operation for the set of transport block 2's. This results in 2 HARQ acknowledgements for each uplink subframe

★ When spatial multiplexing is not used, and a single HARQ acknowledgement is generated, that HARQ acknowledgement can be transferred using PUCCH format 1a. When spatial multiplexing is used, and 2 HARQ acknowledgements are generated, those HARQ acknowledgements can be transferred using PUCCH format 1b

★ Receiving a single transport block in error results in the UE returning a negative acknowledgement for the complete 'bundle' of downlink transport blocks. This leads to the eNode B re-transmitting all transport blocks within the bundle

★ The benefit of HARQ-ACK bundling is a reduced capacity requirement for the PUCCH

★ Table 269 presents the number of downlink transmissions within each 'bundle' for each uplink-downlink subframe configuration. Uplink-downlink configurations 0 and 6 have sufficient uplink subframes to avoid the requirement for bundling

	Subframe Number									
UL-DL Subframe Configuration	0	1	2	3	4	5	6	7	8	9
0			1	0	1			1	0	1
1			2	1				2	1	
2			4					4		
3			3	2	2					
4			4	4						
5			9							
6			1	1	1			1	1	

Table 269 – The number of downlink transmissions within each 'bundle' for HARQ-ACK bundling

★ The 1 or 2 'bundled' acknowledgements transferred during subframe 'n' are generated from the downlink data received during subframes 'n - k', where k is specified in Table 270

	Subframe Number									
UL-DL Subframe Configuration	0	1	2	3	4	5	6	7	8	9
0			6	-	4			6	-	4
1			7, 6	4				7, 6	4	
2			8, 7, 4, 6					8, 7, 4, 6		
3			7, 6, 11	6, 5	5, 4					
4			12, 8, 7, 11	6, 5, 4, 7						
5			13, 12, 9, 8, 7, 5, 4, 11, 6							
6			7	7	5			7	7	

Table 270 – 'k' variable for each uplink-downlink subframe configuration

★ HARQ-ACK multiplexing reduces the number of HARQ acknowledgements by performing a logical AND operation between the acknowledgements belonging to the pair of transport blocks associated with spatial multiplexing, i.e. HARQ bundling applies the AND operation in the time domain whereas HARQ multiplexing applies the AND operation in the spatial domain

★ Figure 223 illustrates the concept HARQ-ACK multiplexing for uplink-downlink subframe configuration 2. A total of 16 acknowledgements per radio frame is reduced to a total of 8 acknowledgements. These acknowledgements are transferred using 2 uplink subframes. PUCCH format 1b can be used to transfer 4 acknowledgements when channel selection is used, i.e. the PUCCH payload transfers 2 acknowledgments and the selection of PUCCH resources (1 resource out of a set of 4 resources) transfers a further 2 acknowledgements. The number of acknowledgements supported by each PUCCH format is presented in Table 172 within section 14.2

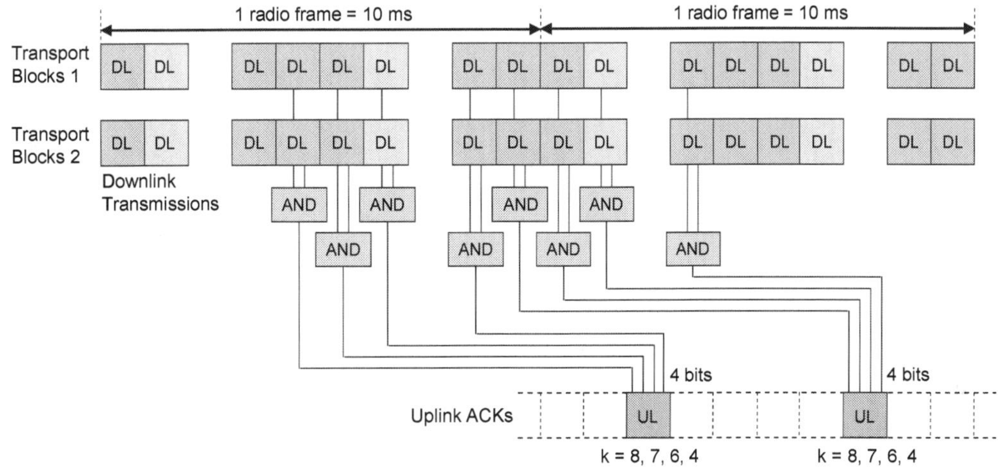

Figure 223 – Concept of HARQ multiplexing for uplink-downlink subframe configuration 2

★ HARQ-ACK multiplexing is only applicable to uplink subframes which provide acknowledgements for more than a single downlink subframe, i.e. entries within Table 269 which have values greater than 1. It is not necessary to apply HARQ multiplexing for uplink subframes which acknowledge only a single downlink subframe because there is only 1 or 2 acknowledgements to transfer and PUCCH formats 1a and 1b can accommodate these acknowledgements without HARQ-ACK multiplexing

★ Receiving a single transport block in error results in the UE returning a negative acknowledgement for the pair of downlink transport blocks. This leads to the eNode B re-transmitting both transport blocks

★ The benefit of HARQ-ACK multiplexing is a reduced capacity requirement for the PUCCH

★ 3GPP References: TS 36.213, TS 36.321, TS 36.331

24.3.5 REPETITION

★ The eNode B can use an RRC Connection Setup, RRC Connection Reconfiguration or RRC Connection Re-establishment message to instruct a UE to repeat its HARQ acknowledgements sent in the uplink direction. This is intended to improve the reliability of signalling for UE located in relatively weak coverage

★ The eNode B can request 1, 3 or 5 repetitions of the HARQ acknowledgements after the original transmission

★ Repetition is applicable to both FDD and TDD, but is only applicable to TDD when using HARQ-ACK bundling

★ 3GPP References: TS 36.213, TS 36.331

24.4 UPLINK POWER CONTROL

- The main objectives of uplink power control are to:
  - limit intracell and intercell interference
  - reduce UE power consumption
- 3GPP TS 36.213 specifies uplink power control separately for the PUSCH, PUCCH and Sounding Reference Signal. A prerequisite to applying uplink power control is defining the UE maximum transmit power

24.4.1 MAXIMUM POWER

- Power control uses the UE configured maximum output power (P_{CMAX}) as an upper limit
- 3GPP TS 36.101 specifies that the UE sets its own configured maximum output power to satisfy:

$$P_{CMAX\_L} \leq P_{CMAX} \leq P_{CMAX\_H}$$

where,

$$P_{CMAX\_L} = \text{MIN}\{P_{EMAX} - \Delta T_C, P_{Powerclass} - \text{MAX}(\text{MPR} + \text{A-MPR}, \text{P-MPR}) - \Delta T_C\}$$

$$P_{CMAX\_H} = \text{MIN}\{P_{EMAX}, P_{Powerclass}\}$$

- P_{EMAX} is the maximum allowed transmit power specified by the eNode B. The eNode B broadcasts a common value for this maximum power within SIB1. The eNode B can also send dedicated values to individual UE using RRC Connection Reconfiguration messages
- $P_{Powerclass}$ is the maximum transmit power associated with the UE power class. The release 8, 9 and 10 versions of the 3GPP specifications define only a single UE power class: power class 3 which has a maximum transmit power of 23 dBm. $P_{Powerclass}$ does not account for the tolerance associated with the UE power class maximum transmit power, e.g. UE power class 3 defines a maximum transmit power of 23 ± 2 dBm when using operating band 1 so the value of $P_{Powerclass}$ is 23 dBm
- ΔT_C is only applicable to UE which are transmitting at either the upper or lower edges of the operating band. $\Delta T_C = 1.5$ dB if the UE transmission bandwidth is confined to either the lowest 4 MHz of the operating band, or the highest 4 MHz of the operating band. Otherwise, $\Delta T_C = 0$ dB. This factor helps to limit emissions outside the operating band by defining a reduced transmit power at the operating band edges
- MPR is the Maximum Power Reduction which is a function of the uplink modulation scheme and the transmission bandwidth relative to the channel bandwidth. 3GPP TS 36.101 specifies the MPR for UE power class 3 using the figures presented in Table 271. Additional requirements are applicable when the UE uses Carrier Aggregation or uplink MIMO

	Channel Bandwidth / Transmission Bandwidth in Resource Blocks						
Modulation	1.4 MHz	3 MHz	5 MHz	10 MHz	15 MHz	20 MHz	MPR
QPSK	> 5 RB	> 4 RB	> 8 RB	> 12 RB	> 16 RB	> 18 RB	≤ 1 dB
16QAM	≤ 5 RB	≤ 4 RB	≤ 8 RB	≤ 12 RB	≤ 16 RB	≤ 18 RB	≤ 1 dB
16QAM	> 5 RB	> 4 RB	> 8 RB	> 12 RB	> 16 RB	> 18 RB	≤ 2 dB

Table 271 – Maximum Power Reduction (MPR) for power class 3

- A-MPR is the Additional Maximum Power Reduction which has been defined to allow a reduction in the UE transmit power when the spectrum emission requirements are relatively stringent, i.e. reducing the UE transmit power helps to achieve the spectrum emission requirements. SIB2 broadcasts the 'Additional Spectrum Emission' parameter. This parameter indexes a row within a table specified by 3GPP TS 36.101. A value of 1 indicates that only the general spectrum emission requirements are applicable and the A-MPR = 0 dB. Other values can indicate more stringent spectrum emission requirements and the A-MPR is then typically ≤ 1, ≤ 2 or ≤ 3 dB
- P-MPR is the Power Management Maximum Power Reduction. This factor was introduced within the release 10 version of the 3GPP specifications. It is intended to allow transmit power reductions in scenarios where the UE is simultaneously transmitting multiple radio access technologies, or in scenarios where the UE reduces its maximum transmit power to comply with electromagnetic energy absorption requirements
- P-MPR is not summed with MPR and A-MPR because MPR and A-MPR help the UE to satisfy the requirements associated with P-MPR, i.e. if the sum of MPR and A-MPR is relatively large then the requirements associated with P-MPR will be satisfied
- As an example, $P_{CMAX\_H} = 23$ dBm and $P_{CMAX\_L} = 21$ dBm (assuming $\Delta T_C = 0$ dB, MPR = 2 dB, A-MPR = 0 dB, P-MPR = 0 dB). In this case, the UE would be permitted to define its maximum output power using a value between 23 and 21 dBm
- 3GPP References: TS 36.101, TS 36.331

24.4.2 PUSCH

★ Conventional power control schemes attempt to maintain a constant Signal to Interference plus Noise Ratio (SINR) at the receiver. UE increase their transmit power to fully compensate any increases in path loss

★ Fractional power control schemes allow the received SINR to decrease as the path loss increases, i.e. the received SINR decreases as the UE moves towards cell edge. The UE transmit power increases at a reduced rate as the path loss increases, when compared to a conventional power control scheme, i.e. increases in path loss are only partially compensated. Fractional power control schemes improve air-interface efficiency and increase average cell throughputs by reducing intercell interference

★ The concepts of conventional and fractional power control schemes are illustrated in Figure 224

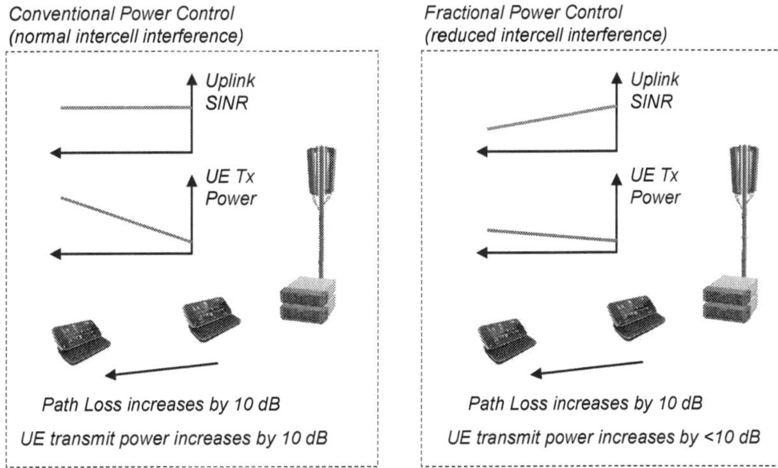

Figure 224 – Conventional and fractional power control schemes

★ 3GPP TS 36.213 specifies a fractional power control scheme for the PUSCH, but with the option to disable it and revert to a conventional power control scheme

★ Both open and closed loop power control components are specified for the PUSCH. The fractional power control scheme forms part of the open loop component. The closed loop component is based upon receiving feedback from the eNode B in the form of Transmit Power Control (TPC) commands

★ The uplink transmit power of the PUSCH during subframe 'i' is defined as:

$$\text{PUSCH Transmit Power (i)} = \min\{P_{UPPER}(i), P_{CALCULATED}(i)\}$$

where,

$P_{UPPER}(i) = 10 \times LOG(P_{CMAX\_LINEAR}(i) - P_{PUCCH\_LINEAR}(i))$

P_{CMAX_LINEAR} is the value of P_{CMAX} described in section 24.4.1 expressed in linear units

P_{PUCCH_LINEAR} is only applicable when the UE transmits both the PUSCH and PUCCH during the same subframe. This capability was introduced within the release 10 version of the 3GPP specifications. When both the PUSCH and PUCCH are transmitted during the same subframe, the control information on the PUCCH is given priority to the UE transmit power so the maximum power available to the PUSCH is reduced by the power used by the PUCCH. P_{PUCCH_LINEAR} is set to 0 when the PUSCH is transmitted without the PUCCH

$P_{CALCULATED}(i) = P_{CALCULATED\_OPEN\_LOOP}(i) + P_{CALCULATED\_CLOSED\_LOOP}(i)$

$P_{CALCULATED\_OPEN\_LOOP}(i) = 10 \times LOG(M_{PUSCH}(i)) + P_{O\_PUSCH}(j) + [PL \times \alpha(j)] + \Delta_{TF}(i)$

$P_{CALCULATED\_CLOSED\_LOOP}(i) = f(i)$

★ The open loop component of $P_{CALCULATED}(i)$ is based upon a transmit power calculation which accounts for the measured Path Loss (PL), or a fraction of the measured Path Loss. Path Loss measurements include the impact of longer term channel variations, e.g. slow fading

★ The closed loop component of $P_{CALCULATED}(i)$ involves the eNode B providing feedback to the UE in the form of Transmit Power Control (TPC) commands. TPC commands for the PUSCH can be signalled to the UE within Downlink Control Information (DCI) formats 0, 3, 3A and 4

★ $M_{PUSCH}(i)$ is the PUSCH bandwidth during subframe 'i' expressed in terms of Resource Blocks. This variable is used to increase the UE transmit power for larger Resource Block allocations. The UE transmit power is increased in direct proportion to the number of allocated Resource Blocks, i.e. the transmit power per Resource Block remains constant if all other factors remain unchanged. This is often referred to as maintaining a constant Power Spectral Density (PSD)

★ A number of variables depend upon 'j' which can have a value of 0, 1 or 2

j = 0 for PUSCH transmissions associated with a semi-persistent grant (allocations applicable to multiple subframes)

j = 1 for PUSCH transmissions associated with a dynamic scheduled grant (allocations applicable to a single subframe)

j = 2 for PUSCH transmissions associated with a random access response grant

★ $P_{O\_PUSCH}(j)$ represents the eNode B received power per Resource Block when assuming a path loss of 0 dB. The received power per Resource Block is maintained as the path loss increases when using conventional power control, i.e. fractional power control is disabled. The received power per Resource Block decreases as the path loss increases when using fractional power control

★ $P_{O\_PUSCH}(j)$ is calculated by the eNode B using an equation of the form:

$$P_{O\_PUSCH} = \alpha \times (SINR + IN) + (1 - \alpha) \times (P_{CMAX} - 10 \times LOG(M_{PUSCH}))$$

where,

α is used to configure the use of fractional power control. Fractional power control is disabled when $\alpha = 1$. Fractional power control is enabled when $\alpha < 1$

SINR represents the target SINR at the eNode B receiver

IN represents the Interference plus Noise power per Resource Block at the eNode B receiver

P_{CMAX} is the UE maximum transmit power according to the UE power class

$M_{PUSCH}(i)$ is the number of allocated Resource Blocks

A set of example P_{O_PUSCH} values for a range of α are presented in Table 272. These values are based upon a target SINR of 10 dB, an IN of -106 dBm, a UE maximum transmit power capability of 23 dBm and a PUSCH allocation of 25 Resource Blocks

	$\alpha = 0$	$\alpha = 0.4$	$\alpha = 0.6$	$\alpha = 0.8$	$\alpha = 1.0$
P_{O_PUSCH}	9 dBm	-33 dBm	-54 dBm	-75 dBm	-96 dBm

Table 272 – Example P_{O_PUSCH} values for a range of α

The value of P_{O_PUSCH} is small when $\alpha = 1$ to allow sufficient UE transmit power headroom for when the path loss increases, i.e. the UE transmit power should fully compensate the path loss when fractional power control is disabled

The value of P_{O_PUSCH} increases as the value of α decreases because less UE transmit power headroom is required for when the path loss increases, i.e. the UE transmit power only partially compensates the path loss when fractional power control is enabled

The value of P_{O_PUSCH} is at its maximum when $\alpha = 0$. This corresponds to the UE transmitting at its maximum capability irrespective of the path loss

★ $P_{O\_PUSCH}(j)$ is signalled to the UE using a combination of cell specific and UE specific components:

$$P_{O\_PUSCH}(j) = P_{O\_NOMINAL\_PUSCH}(j) + P_{O\_UE\_PUSCH}(j)$$

$P_{O\_NOMINAL\_PUSCH}(j)$ represents the cell specific component

$P_{O\_NOMINAL\_PUSCH}(0)$ can be signalled within the RRC Connection Setup, RRC Connection Reconfiguration and RRC Connection Re-establishment messages. Its value can range from -126 to 24 dBm

$P_{O\_NOMINAL\_PUSCH}(1)$ can be broadcast within SIB2 or signalled within an RRC Connection Reconfiguration message. Its value can range from -126 to 24 dBm

$P_{O\_NOMINAL\_PUSCH}(2) = P_{O\_PRE} + \Delta_{PREAMBLE\_MSG3}$

P_{O_PRE} is the PRACH preamble initial received target power broadcast within SIB2 or signalled within an RRC Connection Reconfiguration message. Its value can range from -120 to -90 dBm

$\Delta_{PREAMBLE\_MSG3}$ is the power offset between the acknowledged PRACH preamble and the subsequent PUSCH transmission. It is signalled using a value between -1 and 6 but the actual value in dB = signalled value × 2

$P_{O\_UE\_PUSCH}(j)$ represents the UE specific component

$P_{O\_UE\_PUSCH}(0)$ and $P_{O\_UE\_PUSCH}(1)$ can be signalled within the RRC Connection Setup, RRC Connection Reconfiguration and RRC Connection Re-establishment messages. Their values can range from -8 to 7 dB
$P_{O\_UE\_PUSCH}(2) = 0$

★ PL is the downlink path loss calculated by the UE using a combination of RSRP measurements and knowledge of the Reference Signal transmit power, i.e. PL = Reference Signal transmit power – RSRP measurement. The Reference Signal transmit power is broadcast within SIB2 and can also be signalled within an RRC Connection Reconfiguration message. Its value can range from -60 to 50 dBm. The RSRP measurement is provided as an input after the layer 3 filtering described in section 22.2

★ $\alpha(j)$ is used to configure the use of fractional power control. This is the same variable as that used by the eNode B when calculating $P_{O\_PUSCH}(j)$. A value of 1 disables fractional power control

$\alpha(0)$ and $\alpha(1)$ are broadcast within SIB2 and can also be signalled to the UE within an RRC Connection Reconfiguration message. They can be allocated values of {0, 0.4, 0.5, 0.6, 0.7, 0.8, 0.9, 1}

$\alpha(2) = 1$, i.e. fractional power control is not used for PUSCH transmissions associated with a random access response

★ Figure 225 illustrates the impact of fractional power control upon the UE transmit power as a function of path loss. This example is based upon the P_{O_PUSCH} values presented in Table 272. The UE transmit power is calculated using the expression: $10 \times LOG(M_{PUSCH}) + P_{O\_PUSCH} + [PL \times \alpha]$. Figure 225 also illustrates the resultant received Power Spectral Density (PSD) at the eNode B

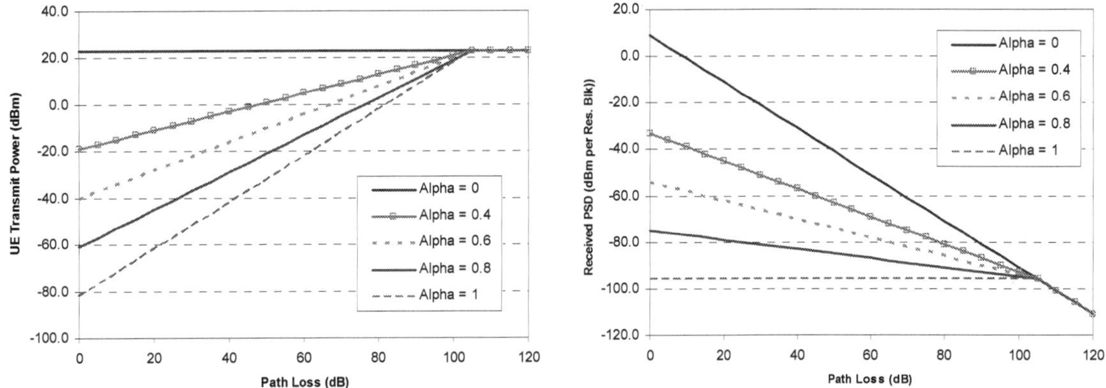

Figure 225 – UE transmit power and received Power Spectral Density (PSD) as a function of path loss (fractional power control)

When $\alpha = 1$, the path loss is fully compensated by the UE transmit power and the PSD remains constant. When $\alpha < 1$, the path loss is only partially compensated by the UE transmit power so the PSD decreases as the path loss increases

★ Figure 226 compares fractional power control with conventional power control when assuming an equal average received PSD. Fractional power control benefits from increased PSD when the path loss is small. It also reduces intercell interference by reducing the UE transmit power when the path loss becomes greater

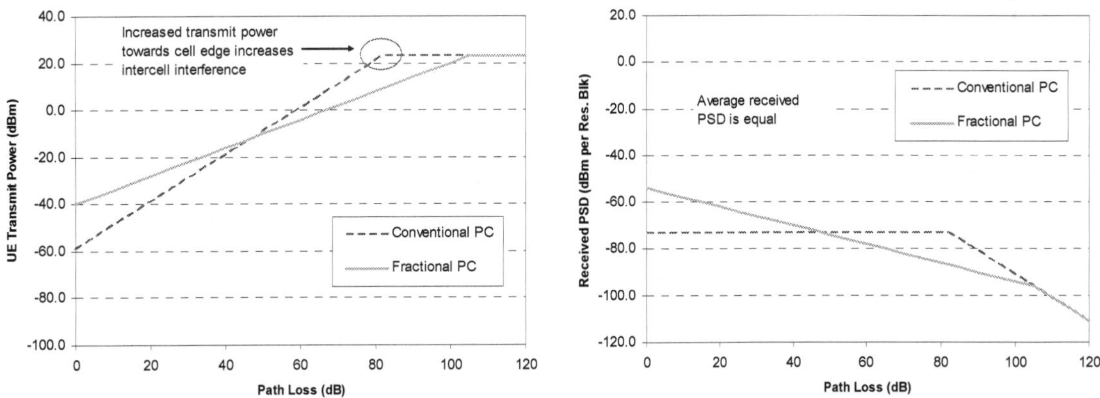

Figure 226 – UE transmit power and received Power Spectral Density (PSD) as a function of path loss (conventional power control)

★ $\Delta_{TF}(i)$ increases the UE transmit power when transferring a large number of bits per Resource Element. This links the UE transmit power to the Modulation and Coding Scheme (MCS). The number of bits per Resource Element is high when using 64QAM and a large transport block size. The number of number of bits per Resource Element is low when using QPSK and a small transport block size. Increasing the UE transmit power helps to achieve the SINR requirements associated with higher order modulation schemes and high coding rates

★ $\Delta_{TF}(i)$ is dependent upon the value of 'Ks' which is configured using the 'delta MCS Enabled' parameter within the RRC Connection Setup, RRC Connection Reconfiguration and RRC Connection Re-establishment messages. 'Ks' can be configured with a value of 0 or 1.25. 'Ks' = 0 when using uplink transmission mode 2 (uplink spatial multiplexing)

$\Delta_{TF}(i) = 0$ when ks = 0 (MCS power boost disabled)

$\Delta_{TF}(i) = 10 \times \text{LOG}[(2^{\text{MPR} \times \text{Ks}} - 1) \times \beta_{offset}^{PUSCH}]$ when ks = 1.25 (MCS power boost enabled)

where,

$$\text{MPR} = \frac{O_{CQI}}{N_{RE}}$$ when the PUSCH is used to transfer only control data

O_{CQI} is the number of CQI bits after the attachment of CRC bits

N_{RE} is the number of allocated Resource Elements

$$\text{MPR} = \sum_{r=0}^{C-1} \frac{Kr}{N_{RE}}$$ when the PUSCH is used to transfer more than just control data

C is the number of code blocks processed by the physical layer

Kr is the size of codeblock 'r' after the attachment of CRC bits

$\beta_{offset}^{PUSCH} = \beta_{offset}^{CQI}$ for control data sent using the PUSCH. Otherwise, $\beta_{offset}^{PUSCH} = 1$

★ The value of Ks = 1.25 reduces the number of allocated Resource Elements to 80% of N_{RE}. This is done to account for Resource Elements used by the uplink Reference Signal and any other control overheads

★ f(i) represents the closed loop power control component which makes use of feedback from the eNode B. Feedback is provided in terms of Transmit Power Control (TPC) commands which instruct the UE to increase or decrease its transmit power. TPC commands for the PUSCH can be signalled to the UE using Downlink Control Information (DCI) formats 0, 3, 3A and 4

★ Interpretation and application of TPC commands depends upon whether or not TPC command accumulation is enabled. The UE is instructed whether or not to use accumulation within the RRC Connection Setup, RRC Connection Reconfiguration and RRC Connection Re-establishment messages

★ If TPC command accumulation is enabled then:

$f(i) = f(i-1) + \delta_{PUSCH}(i - K_{PUSCH})$

$\delta_{PUSCH}(i - K_{PUSCH})$ is the TPC command received during subframe 'i - K_{PUSCH}'. $K_{PUSCH} = 4$ for FDD. In the case of TDD, the value of K_{PUSCH} is dependent upon the uplink-downlink subframe configuration and the subframe number

Table 273 presents the interpretation of TPC commands when accumulation is enabled. The 2 bit TPC commands within DCI formats 0, 3 and 4 allow the UE transmit power to be increased relatively rapidly by including a 3 dB step size. Larger positive step sizes are useful if radio conditions suddenly become worse

		TPC Command			
		0	1	2	3
δ_{PUSCH}	2 bit TPC Command (DCI 0, 3, 4)	-1 dB	0 dB	1 dB	3 dB
	1 bit TPC Command (DCI 3A)	-1 dB	1 dB	-	-

Table 273 – Interpretation of TPC commands when accumulation is enabled

Positive TPC commands are ignored if the UE reaches its maximum power. Similarly, negative TPC commands are ignored if the UE reaches its minimum power

Accumulation is reset if the value of $P_{O\_UE\_PUSCH}$ is changed or if the UE receives a Random Access Response message

$f(0) = \Delta P_{rampup} + \delta_{msg2}$ where ΔP_{rampup} equals the total power ramp-up from the first to the last PRACH preamble, and δ_{msg2} equals the TPC command included within the Random Access Response

★ If TPC command accumulation is disabled then:

$f(i) = \delta_{PUSCH}(i - K_{PUSCH})$

$\delta_{PUSCH}(i - K_{PUSCH})$ is the TPC command received during subframe 'i - K_{PUSCH}'. Table 274 presents the interpretation of TPC commands when accumulation is disabled. This mode of operation only uses TPC commands sent on DCI formats 0 and 4. The dynamic range of the commands is greater than when accumulation is enabled to account for not being able to sum multiple commands over time

		TPC Command			
		0	1	2	3
δ_{PUSCH}	2 bit TPC Command (DCI 0)	-4 dB	-1 dB	1 dB	4 dB

Table 274 – Interpretation of TPC commands when accumulation is disabled

24.4.3 PUCCH

★ 3GPP TS 36.213 specifies a conventional power control scheme for the PUCCH, i.e. the UE transmit power fully compensates the path loss until the maximum UE transmit power is reached. This is in contrast to the PUSCH which allows the use of a fractional power control scheme. Similar to the PUSCH, both open and closed loop power control components are specified for the PUCCH

★ The uplink transmit power of the PUCCH during subframe 'i' is defined as:

$$\text{PUCCH Transmit Power (i)} = \min\{P_{CMAX}, P_{CALCULATED}(i)\}$$

where,

P_{CMAX} is the UE maximum transmit power described in section 24.4.1

$P_{CALCULATED}(i) = P_{CALCULATED\_OPEN\_LOOP}(i) + P_{CALCULATED\_CLOSED\_LOOP}(i)$

$P_{CALCULATED\_OPEN\_LOOP}(i) = P_{O\_PUCCH} + PL + h(n_{CQI}, n_{HARQ}, n_{SR}) + \Delta_{F\_PUCCH}(F) + \Delta_{TxD}(F')$

$P_{CALCULATED\_CLOSED\_LOOP}(i) = g(i)$

★ The open loop component of $P_{CALCULATED}(i)$ is based upon a transmit power calculation which accounts for the measured Path Loss (PL). Path Loss measurements include the impact of longer term channel variations, e.g. slow fading

★ The closed loop component of $P_{CALCULATED}(i)$ involves the eNode B providing feedback to the UE in the form of Transmit Power Control (TPC) commands. TPC commands for the PUCCH can be signalled to the UE within Downlink Control Information (DCI) formats 1, 1A, 1B, 1D, 2, 2A, 2B, 2C, 3 and 3A

★ P_{O_PUCCH} represents the eNode B received power per Resource Block. The received power per Resource Block is maintained as the path loss increases as long as the UE doesn't reach its maximum transmit power capability

P_{O_PUCCH} is signalled to the UE using a combination of cell specific and UE specific components

$P_{O\_PUCCH} = P_{O\_NOMINAL\_PUCCH} + P_{O\_UE\_PUCCH}$

$P_{O\_NOMINAL\_PUCCH}$ represents the cell specific component. It can be broadcast within SIB2 or signalled within an RRC Connection Reconfiguration message. Its value can range from -127 to -96 dBm

$P_{O\_UE\_PUCCH}$ represents the UE specific component. It can be signalled within the RRC Connection Setup, RRC Connection Reconfiguration and RRC Connection Re-establishment messages. Its value can range from -8 to 7 dB

★ PL is the downlink path loss calculated by the UE using a combination of RSRP measurements and knowledge of the Reference Signal transmit power, i.e. PL = Reference Signal transmit power – RSRP measurement. The Reference Signal transmit power is broadcast within SIB2 and can also be signalled within an RRC Connection Reconfiguration message. Its value can range from -60 to 50 dBm. The RSRP measurement is provided as an input after the layer 3 filtering described in section 22.2

★ $h(n_{CQI}, n_{HARQ}, n_{SR})$ increases the UE transmit power when the number of control bits transferred by the PUCCH is relatively large. n_{CQI} corresponds to the number of information bits for the Channel Quality Information, whereas n_{HARQ} corresponds to the number of information bits for the Hybrid Automatic Repeat Request acknowledgements

For PUCCH formats 1, 1a and 1b:

$h(n_{CQI}, n_{HARQ}, n_{SR}) = 0$ dB (because the number of control bits being transferred is small)

For PUCCH formats 2, 2a and 2b using the normal cyclic prefix:

$h(n_{CQI}, n_{HARQ}, n_{SR}) = 0$ if $n_{CQI} < 4$

$h(n_{CQI}, n_{HARQ}, n_{SR}) = 10 \times \text{LOG}[n_{CQI} / 4]$ if $n_{CQI} \geq 4$

For PUCCH formats 2 using the extended cyclic prefix:

$h(n_{CQI}, n_{HARQ}, n_{SR}) = 0$ if $n_{CQI} + n_{HARQ} < 4$

$h(n_{CQI}, n_{HARQ}, n_{SR}) = 10 \times \text{LOG}[(n_{CQI} + n_{HARQ}) / 4]$ if $n_{CQI} + n_{HARQ} \geq 4$

For PUCCH format 3:

$h(n_{CQI}, n_{HARQ}, n_{SR}) = (n_{HARQ} + n_{SR} - 1) / 3$ if $n_{HARQ} + n_{SR} > 11$

$h(n_{CQI}, n_{HARQ}, n_{SR}) = (n_{HARQ} + n_{SR} - 1) / 2$ if $n_{HARQ} + n_{SR} \leq 11$

★ $\Delta_{F\_PUCCH}(F)$ defines a transmit power offset for PUCCH formats 1, 1b, 2, 2a, 2b and 3 relative to format 1a, i.e. P_{O_PUCCH} represents the received power for PUCCH format 1a, while $\Delta_{F\_PUCCH}(F)$ allows offsets to be applied to define the received powers for the other PUCCH formats. The variable 'F' represents the PUCCH format

★ $\Delta_{F\_PUCCH}(F)$ can be signalled to the UE within SIB2 or within an RRC Connection Reconfiguration message. It is signalled as part of the common uplink power control information shown in Table 275. The release 10 version of the 3GPP specifications introduces the offsets for PUCCH format 3, and PUCCH format 1b with channel selection

★ The offsets for PUCCH format 1b are all positive because format 1b transfers more information than format 1a without benefiting from an $h(n_{CQI}, n_{HARQ}, n_{SR})$ transmit power increase

Information Elements	
p0-NominalPUSCH (-126 to 24 dBm)	
alpha (0, 0.4, 0.5, 0.6, 0.7, 0.8, 0.9, 1.0)	
p0-NominalPUCCH (-127 to -96 dBm)	
deltaFList-PUCCH	deltaF-PUCCH-Format1 (-2, 0, 2 dB)
	deltaF-PUCCH-Format1b (1, 3, 5 dB)
	deltaF-PUCCH-Format2 (-2, 0, 1, 2 dB)
	deltaF-PUCCH-Format2a (-2, 0, 2 dB)
	deltaF-PUCCH-Format2b (-2, 0, 2 dB)
	deltaF-PUCCH-Format3 (-1, 0, 1, 2, 3, 4, 5, 6 dB)
	deltaF-PUCCH-Format1b_ChSel (1, 2 dB)
deltaPreambleMsg3 (-2 to 12 dB)	

Table 275 – Common uplink power control information

★ $\Delta_{TxD}(F')$ is set equal to 0 dB unless the PUCCH is configured with transmit diversity. Transmit diversity for the PUCCH was introduced within the release 10 version of the 3GPP specifications. When transmit diversity is used, the eNode B signals the value of $\Delta_{TxD}(F')$ to the UE for PUCCH formats 1,1a, 1b, 2, 2a, 2b and 3. Values of 0 and -2 dB can be specified by the eNode B

★ g(i) represents the closed loop power control component which makes use of feedback from the eNode B. Feedback is provided in terms of Transmit Power Control (TPC) commands which instruct the UE to increase or decrease its transmit power. TPC commands for the PUCCH can be signalled to the UE using Downlink Control Information (DCI) formats 1, 1A, 1B, 1D, 2, 2A, 3 and 3A

$$g(i) = g(i-1) + \sum_{m=0}^{M-1} \delta_{PUCCH}(i - k_m)$$

★ $\delta_{PUSCH}(i - k_m)$ is the TPC command received during subframe 'i - k_m'. In the case of FDD, M = 1 so the summation includes only a single entry, and $k_m = 4$. In the case of TDD, the values of M and k_m are dependent upon the uplink-downlink subframe configuration and the subframe number. The value of M is defined using the values presented in Table 269, while the values of k_m are defined using the values presented in Table 270

★ Table 276 presents the interpretation of TPC commands. The 2 bit TPC commands allow the UE transmit power to be increased relatively rapidly by including a 3 dB step size. Larger positive step sizes are useful if radio conditions suddenly become worse

		TPC Command			
		0	1	2	3
δ_{PUCCH}	2 bit TPC Command (DCI 1, 1A, 1B, 1D, 2, 2A, 2B, 2C, 3)	-1 dB	0 dB	1 dB	3 dB
	1 bit TPC Command (DCI 3A)	-1 dB	1 dB	-	-

Table 276 – Interpretation of TPC commands for the PUCCH within DCI formats 1, 1A, 1B, 1D, 2 and 2A

★ Positive TPC commands are ignored if the UE reaches its maximum power. Similarly, negative TPC commands are ignored if the UE reaches its minimum power

★ g(i) is reset if the UE changes cell, enters or leaves RRC active state, if the value of $P_{O\_UE\_PUSCH}$ is changed, or if the UE receives a Random Access Response message

★ $g(0) = \Delta P_{rampup} + \delta_{msg2}$ where ΔP_{rampup} equals the total power ramp-up from the first to the last PRACH preamble, and δ_{msg2} equals the TPC command included within the Random Access Response

24.4.4 SOUNDING REFERENCE SIGNAL

- 3GPP TS 36.213 specifies a fractional power control scheme for the Sounding Reference Signal (SRS), but with the option to disable it and revert to a conventional power control scheme
- Both open and closed loop power control components are specified for the SRS. The SRS does not have its own Transmit Power Control (TPC) commands for the closed loop component but shares those with the PUSCH
- The uplink transmit power of the SRS during subframe 'i' is defined as:

$$\text{SRS Transmit Power (i)} = \min\{P_{CMAX}, P_{CALCULATED}(i)\}$$

where,

P_{CMAX} is the UE maximum transmit power described in section 24.4.1

$P_{CALCULATED}(i) = P_{CALCULATED\_OPEN\_LOOP}(i) + P_{CALCULATED\_CLOSED\_LOOP}(i)$

$P_{CALCULATED\_OPEN\_LOOP}(i) = P_{SRS\_OFFSET}(m) + 10\times LOG(M_{SRS}) + P_{O\_PUSCH}(j) + [PL \times \alpha(j)]$

$P_{CALCULATED\_CLOSED\_LOOP}(i) = f(i)$

- $P_{SRS\_OFFSET}(m)$ defines a UE transmit power offset which can be signalled within the RRC Connection Setup, RRC Connection Reconfiguration and RRC Connection Re-establishment messages. The dependency upon 'm' was introduced within the release 10 version of the 3GPP specifications
  - 'm = 0' corresponds to triggering type 0 for the SRS: periodic transmission configured by RRC signalling
  - 'm = 1' corresponds to triggering type 1 for the SRS: aperiodic transmission triggered by an instruction within the DCI

  The eNode B can signal separate values for $P_{SRS\_OFFSET}(0)$ and $P_{SRS\_OFFSET}(1)$. In both cases, the signalled value ranges from 0 to 15. The mapping to the actual value depends upon the configuration of the Ks variable used to configure the PUSCH MCS dependent transmit power increase:

  actual value = signalled value – 3   when Ks = 1.25

  actual value = -10.5 + 1.5 × signalled value   when Ks = 0

- $M_{SRS}(i)$ is the bandwidth of the SRS during subframe 'i' expressed in Resource Blocks. This variable is used to increase the UE transmit power in direct proportion to the bandwidth of the SRS transmission
- $P_{O\_PUSCH}(j)$ is the same parameter as that used for power control of the PUSCH. It represents the eNode B received power per Resource Block when assuming a path loss of 0 dB. The received power per Resource Block is maintained as the path loss increases when using conventional power control, i.e. fractional power control is disabled. The received power per Resource Block decreases as the path loss increases when using fractional power control
- PL is the downlink path loss calculated by the UE using a combination of RSRP measurements and knowledge of the Reference Signal transmit power, i.e. PL = Reference Signal transmit power – RSRP measurement. The Reference Signal transmit power is broadcast within SIB2 and can also be signalled within an RRC Connection Reconfiguration message. Its value can range from -60 to 50 dBm. The RSRP measurement is provided as an input after the layer 3 filtering described in section 22.2
- $\alpha(j)$ is the same parameter as that used for power control of the PUSCH. It is used to configure the use of fractional power control. It also impacts the calculation completed by the eNode B when generating $P_{O\_PUSCH}(j)$. A value of 1 disables fractional power control whereas values less than 1 allow the use of fractional power control
- f(i) is the same parameter as that used for power control of the PUSCH. It represents the closed loop power control component based upon feedback from the eNode B. Feedback is provided in terms of Transmit Power Control (TPC) commands which instruct the UE to increase or decrease its transmit power. The same set of power control commands and the same accumulation are used for both the PUSCH and SRS. TPC commands for the PUSCH and SRS can be signalled to the UE using Downlink Control Information (DCI) formats 0, 3, 3A and 4
- 3GPP References: TS 36.213, TS 36.331

24.5 CHANNEL QUALITY INDICATOR

★ Channel Quality Indicator (CQI) reports provide a measure of the downlink channel conditions experienced by the UE. UE generate CQI values and report them to the eNode B using either the PUCCH or PUSCH

★ The eNode B uses CQI reports within its scheduling and link adaptation algorithms. If a proportional fair scheduler is used then UE reporting high CQI values relative to their average CQI are more likely to be scheduled. Link adaptation is more likely to allocate high throughputs to UE reporting high CQI values

★ The set of CQI values is presented in Table 277. The range from 0 to 15 can be signalled using 4 bits. The 'bits per symbol' figure is given by the modulation order multiplied by the coding rate (modulation orders are QPSK: 2, 16QAM: 4, 64QAM: 6)

CQI Index	Modulation	Coding Rate	Bits per Symbol
0	Out of Range		
1	QPSK	0.08	0.15
2	QPSK	0.12	0.23
3	QPSK	0.19	0.38
4	QPSK	0.30	0.60
5	QPSK	0.44	0.88
6	QPSK	0.59	1.18
7	16QAM	0.37	1.48
8	16QAM	0.48	1.91
9	16QAM	0.60	2.41
10	64QAM	0.46	2.73
11	64QAM	0.55	3.32
12	64QAM	0.65	3.90
13	64QAM	0.75	4.52
14	64QAM	0.85	5.12
15	64QAM	0.93	5.55

Table 277 – CQI values

★ UE report the CQI value which corresponds to the largest transport block that can be received with a transport block error rate which does not exceed 10 %. A CQI of 0 is reported if the UE cannot receive the transport block associated with a CQI of 1 with an error rate which does not exceed 10 %

★ When generating CQI values, UE assume they are allocated the modulation scheme and coding rate indicated by Table 277. Transport block sizes are determined using a combination of the coding rate and the CQI Reference Resource

★ The CQI Reference Resource is defined by a specific set of Resource Blocks in the frequency domain and a single subframe in the time domain. The set of Resources Blocks in the frequency domain covers the complete channel bandwidth when reporting wideband CQI, and a subset of the complete channel bandwidth when reporting sub-band CQI. Sub-band sizes are presented in sections 17.3 and 17.4

★ UE make the following assumptions regarding the CQI Reference Resource:

  - the first 3 OFDMA symbols are occupied by the PCFICH, PHICH and PDCCH
  - Resource Elements are not occupied by the Primary or Secondary Synchronisation Signals., nor by the PBCH
  - the cyclic prefix length equals the value configured for non-MBSFN subframes
  - Redundancy Version (RV) 0 is used
  - no Resource Elements are allocated for the:
    - CSI Reference Signal
    - zero power CSI Reference Signal
    - Positioning Reference Signal

★ UE also account for the configured transmission mode when generating the CQI value

★ UE configured to use transmission mode 9 derive CQI values from the CSI Reference Signal when configured to report Precoding Matrix Indicators (PMI) and Rank Indicators (RI). Otherwise they derive CQI values from the cell specific Reference Signal

★ UE configured with transmission modes 1 to 8 derive CQI values from the cell specific Reference Signal

★ If the cell specific Reference Signal is used to derive the reported CQI then the UE accounts for the ratio of the PDSCH Energy Per Resource Element (EPRE) to the cell specific Reference Signal EPRE. Likewise, if the CSI Reference Signal is used to derive the reported CQI then the UE accounts for the ratio of the PDSCH EPRE to the CSI Reference Signal EPRE

★ The throughput associated with each CQI value depends upon the size of the Reference Resource. For a specific CQI value, larger Reference Resources correspond to higher bit rates

★ Table 278 presents a set of example throughputs as a function of CQI and Reference Resource size. These examples assume a normal cyclic prefix and single antenna element transmission

★ The number of Resource Elements available to the PDSCH within each 1ms subframe is given by (2 × 84 - 36 - 6) × Number of Resource Blocks = 126 × Number of Resource Blocks. The throughputs are then given by 126 × Number of Resource Blocks × Bits per Symbol.

CQI Index	Modulation	Wideband CQI Reporting			Sub-band CQI Reporting		
		20 MHz Channel	10 MHz Channel	5 MHz Channel	8 Res.Blk	6 Res.Blk	4 Res.Blk
1	QPSK	1.9	0.9	0.5	0.2	0.1	0.1
2	QPSK	2.9	1.4	0.7	0.2	0.2	0.1
3	QPSK	4.8	2.4	1.2	0.4	0.3	0.2
4	QPSK	7.6	3.8	1.9	0.6	0.5	0.3
5	QPSK	11.1	5.5	2.8	0.9	0.7	0.4
6	QPSK	14.9	7.4	3.7	1.2	0.9	0.6
7	16QAM	18.6	9.3	4.7	1.5	1.1	0.7
8	16QAM	24.1	12.0	6.0	1.9	1.4	1.0
9	16QAM	30.4	15.2	7.6	2.4	1.8	1.2
10	64QAM	34.4	17.2	8.6	2.8	2.1	1.4
11	64QAM	41.8	20.9	10.5	3.3	2.5	1.7
12	64QAM	49.1	24.6	12.3	3.9	2.9	2.0
13	64QAM	57.0	28.5	14.2	4.6	3.4	2.3
14	64QAM	64.5	32.3	16.1	5.2	3.9	2.6
15	64QAM	69.9	35.0	17.5	5.6	4.2	2.8

Table 278 – Example throughputs associated with each CQI value (all figures in Mbps)

★ 3GPP References: TS 36.213

24.6 DISCONTINUOUS RECEPTION (DRX)

★ Both UMTS and LTE networks use Discontinuous Reception (DRX) in RRC Idle mode when monitoring for paging messages. This avoids the UE having to receive all downlink subframes and helps to conserve UE battery power. The LTE paging occasions defined by the idle mode DRX cycle are described in section 23.7

★ UMTS networks use CELL_PCH and URA_PCH to help conserve UE battery power while the UE is in RRC Connected mode. Continuous Packet Connectivity (CPC) can also be used by UMTS networks to help conserve UE battery power while the UE is in CELL_DCH. These solutions can also help to conserve air-interface and BTS processing resources

★ LTE networks can use Discontinuous Reception (DRX) in RRC Connected mode to help conserve UE battery power. The DRX cycle determines the subframes during which the UE monitors for a PDCCH addressed using its C-RNTI, TPC-PUCCH-RNTI, TPC-PUSCH-RNTI or SPS C-RNTI. When DRX is used, the UE only has to monitor the PDCCH during specific subframes. When DRX is not used, the UE has to monitor the PDCCH during every subframe

★ The main drawback of using a DRX cycle is that it causes an increased delay when there is a requirement to transfer data. On average, this delay increases as the duration of the DRX cycle increases

★ The general concept of RRC Connected mode DRX for LTE is illustrated in Figure 227

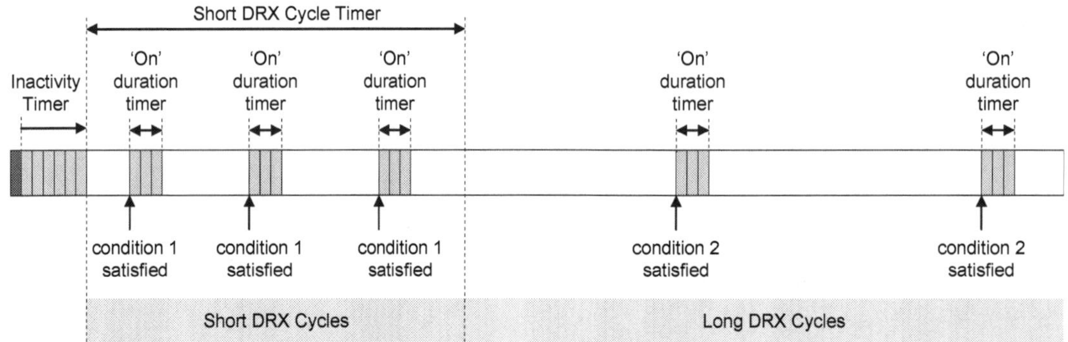

Condition 1

[(SFN × 10) + subframe number] modulo (shortDRX-Cycle) = (drxStartOffset) modulo (shortDRX-Cycle)

Condition 2

[(SFN × 10) + subframe number] modulo (longDRX-Cycle) = drxStartOffset

Figure 227 – DRX pattern (both short and long DRX cycles configured)

★ An inactivity timer is started after each period of activity. This timer defines the number of consecutive inactive subframes which the UE must experience before using DRX mode. The timer is stopped and re-started if there is any activity while it is running. The UE monitors the PDCCH continuously while the inactivity timer is running

★ Assuming the inactivity timer expires, there is an optional period of short DRX cycles. Short DRX cycles can be used initially because in general, the probability of further activity is greater during the time window immediately after any previous activity. The probability of further activity tends to decrease as the period of inactivity increases

★ The eNode B instructs the UE whether or not it should use the period of short DRX cycles. If short DRX cycles are not used then the period of long DRX cycles starts directly

★ During the period of short DRX cycles, the UE triggers its 'on' periods when the following condition is satisfied:

[(SFN × 10) + subframe number] modulo (shortDRX-Cycle) = (drxStartOffset) modulo (shortDRX-Cycle)

For example, shortDRX-Cycle = 20 subframes, and drxStartOffset = 25 leads to:

[(SFN × 10) + subframe number] modulo 20 = 25 modulo 20 = 5

i.e. the 'on' periods start during SFN 0 subframe 5, SFN 2 subframe 5, SFN 4 subframe 5, etc

The eNode B signals the 'on' period duration to the UE. The UE monitors the PDCCH continuously during the 'on' period

★ The period of short DRX cycles continues for a specific number of short cycles before the mode of operation switches to long DRX cycles. During the period of long DRX cycles, the UE triggers its 'on' periods when the following condition is satisfied:

[(SFN × 10) + subframe number] modulo (longDRX-Cycle) = drxStartOffset

For example, longDRX-Cycle = 40 subframes, and drxStartOffset = 25 leads to:

[(SFN × 10) + subframe number] modulo 40 = 25

i.e. the 'on' periods start during SFN 2 subframe 5, SFN 6 subframe 5, SFN 10 subframe 5, etc

★ DRX configuration information can be signalled to the UE within an RRC Connection Setup, RRC Connection Reconfiguration or RRC Connection Re-establishment message. The content of the DRX configuration information is presented in Table 279

<table>
<tr><th colspan="5">Information Elements</th></tr>
<tr><th colspan="5">CHOICE</th></tr>
<tr><td colspan="5">Release</td></tr>
<tr><td rowspan="7">Setup</td><td colspan="4">onDurationTimer</td></tr>
<tr><td colspan="4">drx-InactivityTimer</td></tr>
<tr><td colspan="4">drx-RetransmissionTimer</td></tr>
<tr><td colspan="2" rowspan="2">longDRX-CycleStartOffset</td><td colspan="2">CHOICE</td></tr>
<tr><td>longDRX-Cycle</td><td>drxStartOffset</td></tr>
<tr><td rowspan="2">shortDRX</td><td colspan="3">shortDRX-Cycle</td></tr>
<tr><td colspan="3">drxShortCycleTimer</td></tr>
</table>

Table 279 – Content of DRX configuration information

★ onDurationTimer can be configured with values of 1, 2, 3, 4, 5, 6, 8, 10, 20, 30, 40, 50, 60, 80, 100 and 200 subframes

★ drx-InactivityTimer can be configured with values of 0, 1, 2, 3, 4, 5, 6, 8, 10, 20, 30, 40, 50, 60, 80, 100, 200, 300, 500, 750, 1280, 1920, 2560 subframes

★ drx-RetransmissionTimer can be configured with values of 1, 2, 4, 6, 8, 16, 24 and 33 subframes

★ longDRX-Cycle can be configured with values of 10, 20, 32, 40, 64, 80, 128, 160, 256, 320, 512, 640, 1024, 1280, 2048 and 2560 subframes

★ drxStartOffset can be configured with any integer value between 0 and (longDRX-Cycle - 1)

★ shortDRX-Cycle is optional and can be configured with values of 2, 5, 8, 10, 16, 20, 32, 40, 64, 80, 128, 160, 256, 320, 512, 640 subframes

★ drxShortCycleTimer is optional and can be configured with any integer value between 1 and 16

★ The retransmission timer defines an additional 'on' period when a downlink retransmission is expected. This prevents the UE from missing a retransmission after entering DRX mode. The retransmission timer is started after the HARQ Round Trip Time (RTT) expires if the corresponding downlink transmission was not successfully decoded. The HARQ RTT is fixed at 8 subframes for FDD, and is given by k + 4 subframes for TDD, where 'k' is defined in Table 270 (section 24.3.4)

★ The eNode B can instruct the UE to enter DRX mode immediately using the DRX Command MAC control element. The DRX Command MAC control element does not have any payload. It is signalled simply by sending a DL-SCH MAC subheader with the Logical Channel Identity (LCID) set to 11110

★ DRX patterns should account for the end-user application. DRX can be applied to Voice over IP (VoIP) connections if the DRX pattern is designed to have 'on' periods which coincide with the periodic arrival of speech information

★ UE in RRC Connected mode are required to complete handovers when moving from one cell to another. UE must exit DRX to send an uplink RRC: Measurement Report message when the signal strength from a neighbouring cell becomes sufficiently strong to trigger a handover

★ 3GPP References: TS 36.321

25 PROTOCOL STACKS

25.1 USER PLANE

★ The user plane protocol stack is responsible for transferring application data

★ Figure 228 illustrates the user plane protocol stack between a UE and an application server. The eNode B, Serving Gateway and Packet Data Network (PDN) Gateway relay the application data between the UE and application server. The Mobility Management Entity (MME) within the Evolved Packet Core (EPC) is part of the control plane protocol stack but is not part of the user plane protocol stack

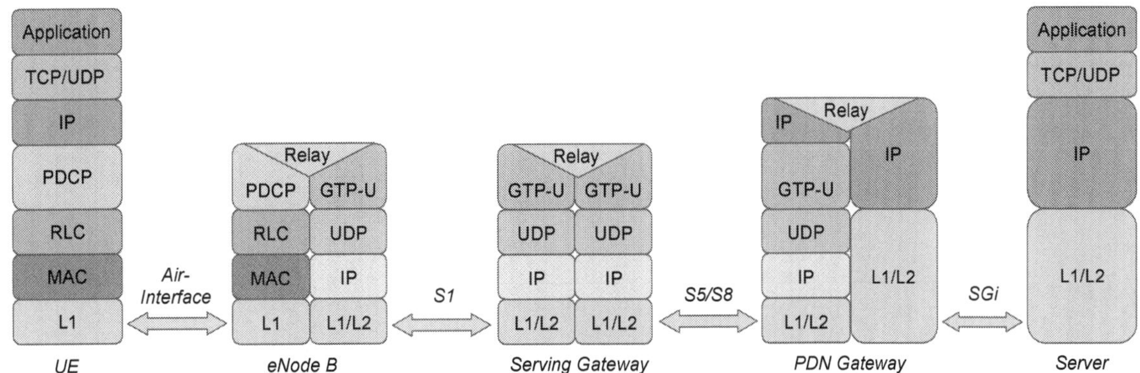

Figure 228 – User plane protocol stack

★ The application layer represents the highest layer within the user plane protocol stack and is present at both the UE and application server. The application layer uses the lower layers to provide a data transfer service. A TCP or UDP layer is immediately below the application layer. Example applications using either TCP or UDP are:

- Hypertext Transfer Protocol (HTTP) over TCP can be used for internet browsing
- File Transfer Protocol (FTP) over TCP can be used to transfer files
- Simple Mail Transfer Protocol (SMTP) over TCP can be used to send and receive email
- Telnet over TCP can be used for remote access
- Real-time Transport Protocol (RTP) over UDP can be used to stream audio and video content
- RTP over UDP can also be used for Voice over IP (VoIP) connections

★ Transmission Control Protocol (TCP) provides reliable data transfer between the UE and application server. Retransmissions can be triggered when packets are not received successfully. Retransmissions at the RLC and MAC layers should help to minimise the requirement for TCP retransmissions. TCP includes a flow control mechanism to ensure that the rate at which data is sent matches the bandwidth of the connection. TCP headers include a sequence number to allow in-order delivery of the packets. TCP headers are typically 20 bytes but can be larger if optional content is included

★ User Datagram Protocol (UDP) provides less reliable data transfer between the UE and application server. UDP does not trigger retransmissions, nor does the header include a sequence number so packets can be delivered out-of-order. UDP is relatively lightweight because it does not require any signalling to establish a connection, i.e. UDP is a connectionless protocol, in contrast to TCP which is a connection oriented protocol. The application layer above UDP may have its own flow control and error checking mechanisms. UDP generates a lower overhead than TCP with a header size of 8 bytes

★ The Internet Protocol (IP) layer at the UE, PDN Gateway and application server is used to route user plane data across the packet switched network. Figure 228 illustrates a direct connection between the PDN Gateway and application server. In practice, there may be a series of IP routers which use the IP layer to forward traffic in the appropriate direction. The UE uses the IP address allocated by the PDN Gateway during connection establishment. The IP layer can be based upon either IPv4 or IPv6. IPv4 headers are typically 20 bytes whereas IPv6 headers are typically 40 bytes. Header sizes can be greater if optional content is included

★ The Packet Data Convergence Protocol (PDCP) layer is specified by 3GPP TS 36.323. It is the highest of the radio access network protocol stack layers, i.e. it operates between the UE and eNode B. The PDCP layer provides header compression and ciphering for user plane data. Header compression is completed using the Robust Header Compression (ROHC) protocol. ROHC has been designed for connections with relatively high packet loss, e.g. wireless connections. It is capable of compressing IP, UDP, RTP and TCP headers. Header compression is important because the overheads generated by the higher layers can become large and without compression they would consume valuable air-interface resources. Header compression is less important across the fixed network because bandwidths are significantly greater. The PDCP header for user plane data is either 1 or 2 bytes depending upon the length of sequence number used

★ The Radio Link Control (RLC) layer is specified by 3GPP TS 36.322. Similar to the PDCP layer, it belongs to the radio access network and operates between the UE and eNode B. The RLC layer can be used to transfer user plane data in one of two modes – unacknowledged or acknowledged (transparent RLC is not applicable to the user plane):

- Unacknowledged Mode (UM) supports segmentation and concatenation of higher layer packets. The RLC header includes a sequence number to ensure in-sequence delivery to the higher layers. The sequence number can be either 5 or 10 bits leading to header sizes of 1 or 2 bytes respectively. Length indicators are included when necessary and further increase the size of the RLC header. Length indicators are used to identify the boundaries between different higher layer packets contained within the same RLC PDU
- Acknowledged Mode (AM) also supports segmentation and concatenation of higher layer packets. The RLC header includes a 10 bit sequence number to ensure in-sequence delivery to the higher layers. The minimum header size is 2 bytes, but this increases when length indicators are included. Acknowledged mode uses an Automatic Repeat Request (ARQ) protocol to provide reliable transmission. Status PDU are used to provide acknowledgement information to the sender. RLC retransmissions are prioritised over new transmissions

★ The Medium Access Control (MAC) layer is specified by 3GPP TS 36.321. It also belongs to the radio access network and operates between the UE and eNode B. The MAC layer is responsible for prioritising and multiplexing logical channel data. The MAC layer within the eNode B is responsible for both uplink and downlink link adaptation, i.e. selecting an appropriate modulation scheme and transport block size for the air-interface. The MAC layer generates transport blocks from the logical channel data and passed down to the physical layer. The MAC layer also supports the Hybrid Automatic Repeat Request (HARQ) protocol described in section 24.3. The MAC header has a variable size dependent upon the number of logical channels being multiplexed and the inclusion of any length indicators

★ Layer 1 (L1) between the UE and eNode B is specified by 3GPP TS 36.211, TS 36,212, TS 36.213 and TS 36.214. It provides physical layer processing as well as transmission and reception across the air-interface. Cyclic Redundancy Check (CRC) bits are added to each transport block at the transmitter to allow error detection at the receiver. Rate 1/3 Turbo coding is applied to generate redundancy and increase resilience to the radio propagation channel. Interleaving is applied to improve the performance of decoding at the receiver by randomising bursts of contiguous bit errors generated by fades in the propagation channel. Modulation is applied prior to mapping onto the air-interface resources, using OFDMA in the downlink and SC-FDMA in the uplink.

★ The connection between the eNode B and the Serving Gateway is based upon the S1 user plane protocol stack, specified by 3GPP TS 36.414. The connection between the Serving Gateway and PDN Gateway is based upon the S5/S8 user plane protocol stack which is specified to use GTP-U. The S5 interface provides connectivity between a home Serving Gateway and a home PDN Gateway. The S8 interface provides roaming connectivity between a visited Serving Gateway and a home PDN Gateway

★ GPRS Tunnelling Protocol User Plane (GTP-U) is specified by 3GPP TS 29.281. GTP-U tunnels are used to transfer user plane data between a pair of GTP-U tunnel endpoints. The eNode B and Serving Gateway represent one pair of endpoints, while the Serving Gateway and PDN Gateway represent another pair of endpoints. The Tunnel Endpoint Identifier (TEID) within the GTP-U header defines the tunnel to which the data belongs. TEID values for the tunnel across the S1 interface are communicated using S1 Application Protocol (S1-AP) signalling. The GTP-U header has a size of at least 8 bytes. GTP-U packets are transferred using UDP over IP

★ Layer 1 (L1) and Layer 2 (L2) of the transport network are both based upon Ethernet. L2 corresponds to the Data Link layer which uses the Carrier Sense Multiple Access (CSMA) with Collision Detection (CD) protocol. L2 is responsible for adding header information and generating the Ethernet packets. L1 corresponds to the Physical layer which is likely to be based upon either an electrical or optical connection. Both electrical and optical connections can be configured as Gigabit Ethernet, i.e. electrical as 1000Base-T, and optical as 1000Base-SX/LX/ZX

★ There is an additional user plane protocol stack between neighbouring eNode B. This protocol stack is used when forwarding data during the handover procedure, i.e. the source eNode B forwards data to the target eNode B during intra-system handover. This protocol stack uses the X2 interface, and is known as the X2 user plane protocol stack. It is shown in Figure 229

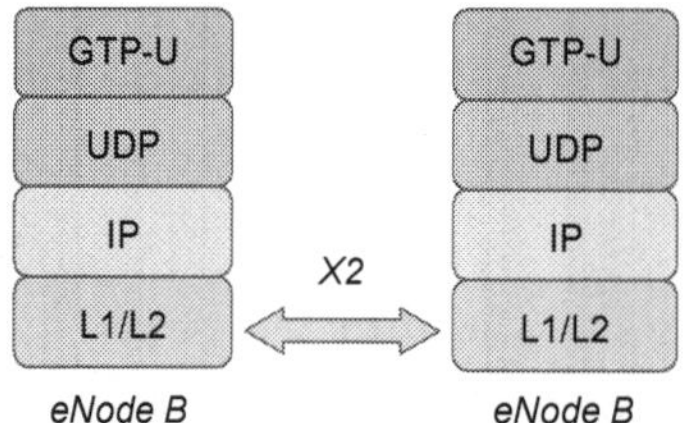

Figure 229 – X2 user plane protocol stack

★ The X2 user plane protocol stack has the same layers as the user plane protocol stacks across the S1 and S5/S8 interfaces. A GTP-U tunnel is used to transfer data from one eNode B to another. The Tunnel Endpoint Identifier (TEID) values for the tunnel are communicated using X2 Application Protocol (S1-AP) signalling. The GTP-U tunnel runs over UDP, IP and Ethernet

25.2 CONTROL PLANE

★ The control plane protocol stack is responsible for transferring signalling messages

★ The Radio Resource Control (RRC) control plane protocol stack is used for signalling between the UE and eNode B. The equivalent UMTS protocol stack allows signalling between the UE and RNC. The RRC control plane protocol stack is fully contained within the radio access network, and is shown in Figure 230

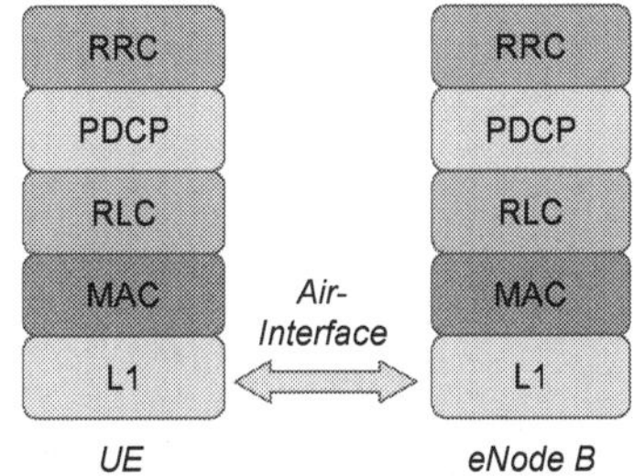

Figure 230 – RRC control plane protocol stack

★ The RRC layer is specified by 3GPP TS 36.331. RRC messages are transferred using the set of Signalling Radio Bearers (SRB) described in section 2.9. The set of SRB includes SRB 0 which transfers messages using the CCCH logical channel, and SRB 1 and SRB 2 which transfer messages using the DCCH logical channel. Example RRC signalling procedures include paging, RRC connection establishment, RRC connection reconfiguration, and RRC connection release

★ The Packet Data Convergence Protocol (PDCP) layer is specified by 3GPP TS 36.323. In the case of UMTS, the PDCP layer is excluded from the control plane protocol stack. In the case of LTE, the PDCP layer is included within the control plane protocol stack for ciphering and integrity protection. The header compression function of the PDCP layer is not applicable to the control plane. The PDCP layer adds both header and tail information to RRC messages. The header size is 1 byte and includes a 5 bit sequence number to allow in-order delivery of messages. The tail size is 4 bytes and includes a Message Authentication Code for Integrity (MAC-I) field used for integrity protection

★ The Radio Link Control (RLC) layer is specified by 3GPP TS 36.322. The RLC layer can be used to transfer control plane signalling in one of two modes – transparent or acknowledged (unacknowledged RLC is not applicable to the control plane):

- o Transparent Mode (TM) is used by SRB 0 and the CCCH logical channel. TM does not segment nor concatenate the higher layer PDU. Nor does it add any header information
- o Acknowledged Mode (AM) supports segmentation and concatenation of higher layer packets. The RLC header includes a 10 bit sequence number to ensure in-sequence delivery to the higher layers. The minimum header size is 2 bytes, but this increases when length indicators are included. Acknowledged mode uses an Automatic Repeat Request (ARQ) protocol to provide reliable transmission. Status PDU are used to provide acknowledgement information to the sender. RLC retransmissions are prioritised over new transmissions

★ The Medium Access Control (MAC) layer is specified by 3GPP TS 36.321. The MAC layer is responsible for prioritising and multiplexing logical channel data, i.e. user plane and control plane data is prioritised and multiplexed at the MAC layer. The MAC layer within the eNode B is responsible for both uplink and downlink link adaptation, i.e. selecting an appropriate modulation scheme and transport block size for the air-interface. The MAC layer generates transport blocks from the logical channel data and passed down to the physical layer. The MAC layer also supports the Hybrid Automatic Repeat Request (HARQ) protocol described in section 24.3. The MAC header has a variable size dependent upon the number of logical channels being multiplexed and the inclusion of any length indicators

★ Layer 1 (L1) between the UE and eNode B is specified by 3GPP TS 36.211, TS 36,212, TS 36.213 and TS 36.214. It provides physical layer processing as well as transmission and reception across the air-interface. Cyclic Redundancy Check (CRC) bits are added to each transport block at the transmitter to allow error detection at the receiver. Rate 1/3 Turbo coding is applied to generate redundancy and increase resilience to the radio propagation channel. Interleaving is applied to improve the performance of decoding at the receiver by randomising bursts of contiguous bit errors generated by fades in the propagation channel. Modulation is applied prior to mapping onto the air-interface resources, using OFDMA in the downlink and SC-FDMA in the uplink

★ The S1 control plane protocol stack is used for signalling between an eNode B and Mobility Management Entity (MME), i.e. it allows signalling between the radio access network and the packet core network. The S1 control plane protocol stack differs from the S1 user plane protocol stack in terms of its termination point within the packet core, i.e. the S1 user plane protocol stack terminates at the Serving Gateway rather than the MME. The S1 control plane protocol stack is shown in Figure 231

Figure 231 – S1 control plane protocol stack

★ The S1 Application Protocol (S1-AP) is specified by 3GPP TS 36.413. The S1-AP defines procedures and messages for signalling between the eNode B and MME. These procedures can be associated with an end-user connection, or a network management function. Example procedures associated with an end-user connection are listed in Table 280

Paging
E-RAB Setup, E-RAB Modify, E-RAB Release Indication, E-RAB Release
Initial Context Setup, UE Context Modification, UE Context Release, UE Context Release Request
Handover Preparation, Handover Resource Allocation, Path Switch Request, Handover Cancellation, Handover Notification
Initial UE Message, Downlink NAS Transport, Uplink NAS Transport

Table 280 – S1-AP procedures associated with an end-user connection

★ The paging procedure simply transfers paging messages from the MME to the eNode B. Enhanced Radio Access Bearer (E-RAB) management procedures allow E-RAB to be setup, modified and released. Likewise, UE Context management procedures allow UE contexts to be setup, modified or released. The set of handover management procedures provide support for both intra and inter-system handovers. The Initial UE message and the NAS Transport messages allow Non-Access Stratum (NAS) messages to be transferred across the S1 interface

★ Example S1-AP procedures associated with a network management function include S1 Setup, E-Node B Configuration Update, MME Configuration Update, E-Node B Status Transfer, MME Status Transfer and Reset.

★ The Stream Control Transmission Protocol (SCTP) is specified by the IETF within RFC 4960. SCTP provides an alternative to TCP and UDP. SCTP provides a reliable data transfer service which supports retransmissions. SCTP allows signalling messages to be transferred using parallel streams (multi-streaming capability). This avoids retransmissions from one signalling procedure delaying the transfer of messages from other signalling procedures. SCTP also includes a flow control mechanism, similar to that used by TCP. An SCTP packet forms the payload of an IP packet. The SCTP layer adds a common header of 12 bytes. Additional header information is included for each 'chunk' of data within the SCTP packet. Chunks form the payload of the SCTP packet

★ The Internet Protocol (IP) layer is used to route SCTP data between the eNode B and MME. Figure 231 illustrates a direct connection between the eNode B and MME. In practice, there may be a series of IP routers which use the IP layer to forward traffic in the appropriate direction. The IP layer can be based upon either IPv4 or IPv6. IPv4 headers are typically 20 bytes whereas IPv6 headers are typically 40 bytes. Header sizes can be greater if optional content is included

★ Layer 1 (L1) and Layer 2 (L2) of the transport network are both based upon Ethernet. L2 corresponds to the Data Link layer which uses the Carrier Sense Multiple Access (CSMA) with Collision Detection (CD) protocol. L2 is responsible for adding header information and generating the Ethernet packets. L1 corresponds to the Physical layer which is likely to be based upon either an electrical or optical connection. Both electrical and optical connections can be configured as Gigabit Ethernet, i.e. electrical as 1000Base-T, and optical as 1000Base-SX/LX/ZX

★ The RRC protocol stack shown in Figure 230, and the S1 protocol stack shown in Figure 231 can be combined to generate the control plane protocol stack used to transfer Non-Access Stratum (NAS) messages between the UE and MME. The NAS control plane protocol stack is shown in Figure 232

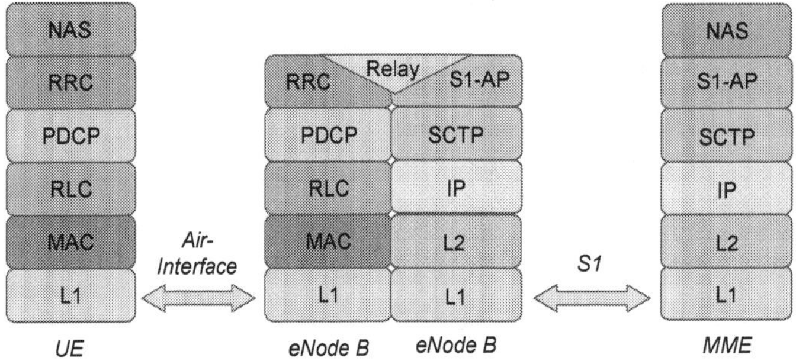

Figure 232 – Non-Access Stratum (NAS) control plane protocol stack

★ NAS messages allow signalling between the UE and MME. NAS messages are transferred via the eNode B but the eNode B only acts as a relay and does not interpret the content of the messages. NAS signalling procedures and messages are specified within 3GPP TS 24.301. NAS messages are categorised as either EPS Mobility Management or EPS Session Management. These two groups of messages are presented in Table 281

EPS Mobility Management messages	Attach Request, Attach Accept, Attach Complete, Attach Reject Authentication Request, Authentication Response, Authentication Reject, Authentication Failure CS Service Notification Detach Request, Detach Accept Downlink NAS Transport, Uplink NAS Transport EMM Information, EMM Status GUTI Reallocation Command, GUTI Reallocation Complete Identity Request, Identity Response Security Mode Command, Security Mode Complete, Security Mode Reject Security Protected NAS Message Service Request, Service Reject, Extended Service Request Tracking Area Update Request, Tracking Area Update Accept, Tracking Area Update Complete, Tracking Area Update Reject
EPS Session Management messages	Activate Dedicated EPS Bearer Context Request, Activate Dedicated EPS Bearer Context Accept, Activate Dedicated EPS Bearer Context Reject Activate Default EPS Bearer Context Request, Activate Default EPS Bearer Context Accept, Activate Default EPS Bearer Context Reject Modify EPS Bearer Context Request, Modify EPS Bearer Context Accept, Modify EPS Bearer Context Reject Deactivate EPS Bearer Context Request, Deactivate EPS Bearer Context Accept Bearer Resource Allocation Request, Bearer Resource Allocation Reject Bearer Resource Modification Request, Bearer Resource Modification Reject ESM Information Request, ESM Information Response, ESM Status PDN Connectivity Request, PDN Connectivity Reject PDN Disconnect Request, PDN Disconnect Reject

Table 281 – Non Access Stratum (NAS) messages for EPS mobility management and session management

★ Example EPS Mobility Management NAS messages include those used to register with the network (Attach Request) and those used to provide location updates (Tracking Area Update Request). The Service Request message is also included to request a NAS signalling connection across both the air-interface and S1 interface. The Uplink and Downlink NAS Transport messages are used to transfer Short Message Service (SMS) content

★ Example EPS Session Management NAS messages include those used to establish, modify and release an EPS Bearer Context. The PDN Connectivity Request can be used to initiate the setup of a Packet Data Network (PDN) connection. The ESM Information Request is used by the MME to request EPS Session Management (ESM) information from the UE. This can include protocol configuration options and Access Point Name (APN) data

★ The RRC messages, Uplink Information Transfer and Downlink Information Transfer are used to encapsulate NAS messages for transfer between the UE and eNode B using the RRC control plane protocol stack. These messages are transferred using acknowledged mode RLC on SRB 2. They can also be transferred on SRB 1 if SRB 2 is not available

★ The S1-AP messages, Initial UE Message, Downlink NAS Transport and Uplink NAS Transport are used to encapsulate NAS messages for transfer between the eNode B and MME using the S1 control plane protocol stack

★ The X2 control plane protocol stack is used for signalling between eNode B. It is similar to the S1 control plane protocol stack (shown in Figure 231) used to signal between an eNode B and MME. The X2 control plane protocol stack is shown in Figure 233

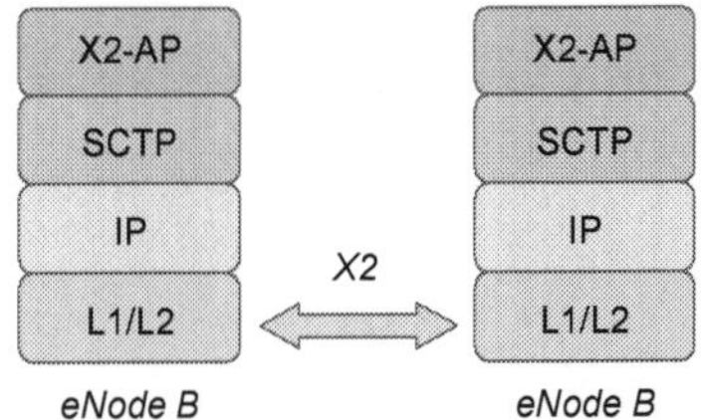

Figure 233 – X2 control plane protocol stack

★ The X2 Application Protocol (X2-AP) is specified by 3GPP TS 36.423. The X2-AP defines procedures for signalling between eNode B. These procedures are categorised as either Mobility or Global. Table 282 presents the X2-AP messages associated with both the Mobility and Global procedures.

Mobility messages	Handover Request, Handover Request Acknowledge, Handover Preparation Failure, Handover Cancel SN Status Transfer UE Context Release
Global messages	X2 Setup Request, X2 Setup Response, X2 Setup Failure Reset Request, Reset Response Load Information Error Indication ENB Configuration Update, ENB Configuration Update Acknowledge, ENB Configuration Update Failure Resource Status Request, Resource Status Response, Resource Status Failure, Resource Status Update

Table 282 – X2 Application Protocol (X2-AP) messages for mobility and global procedures

★ The set of Mobility messages are primarily associated with inter-eNode B handovers, i.e. they provide the capability to request, acknowledge and cancel handovers. The SN Status Transfer message is used to signal PDCP sequence numbers from the source eNode B to the target eNode B. This allows sequence numbers to be preserved during handovers. The UE Context Release message is sent from the target eNode B to the source eNode B at the end of a successful handover to inform the source eNode B that its resources for the connection can be released

★ The set of Global messages are associated with a range of network management functions. The X2 setup messages can be used to establish an X2 connection between eNode B. These messages are used during the automatic definition of neighbours (shown in section 32.4, Figure 317). The Load Information message allows neighbouring eNode B to co-ordinate their resource allocations by exchanging load and interference information

★ The layers below the X2-AP are the same as those below the S1-AP within the S1 control plane protocol stack. SCTP is used as an alternative to TCP and UDP. It offers a multi-streaming capability to avoid retransmissions of one message delaying the transfer of another message. This is in contrast to TCP which treats all messages belonging to a single connection as a single stream of bytes. The IP layer is used to route messages across the packet switched network, while L1 and L2 are based upon Ethernet connections which could be electrical or optical

26 SIGNALLING PROCEDURES

26.1 RRC CONNECTION ESTABLISHMENT

★ RRC connection establishment is used to make the transition from RRC Idle mode to RRC Connected mode. UE must make the transition to RRC Connected mode before transferring any application data, or completing any signalling procedures

★ The RRC connection establishment procedure is always initiated by the UE but can be triggered by either the UE or the network. For example, the UE triggers RRC connection establishment if the end-user starts an application to browse the internet, or to send an email. Similarly, the UE triggers RRC connection establishment if the UE moves into a new Tracking Area and has to complete the Tracking Area Update signalling procedure. The network triggers the RRC connection establishment procedure by sending a Paging message. This could be used to allow the delivery of an incoming SMS or notification of an incoming voice call

★ RRC connection establishment for LTE is relatively simple compared to UMTS. The UMTS procedure requires NBAP and ALCAP signalling across the Iub interface between the Node B and RNC. These signalling protocols are used to setup a radio link and new transport connection. The flat network architecture for LTE removes the requirement for these signalling procedures

★ In the case of LTE, the initial Non-Access Stratum (NAS) message is transferred as part of the RRC connection establishment procedure. In the case of UMTS, the initial NAS message is transferred after the RRC connection establishment procedure. The approach used by LTE helps to reduce connection establishment delay

★ RRC connection establishment configures Signalling Radio Bearer (SRB) 1 and allows subsequent signalling to use the Dedicated Control Channel (DCCH) rather than the Common Control Channel (CCCH) used by SRB 0

★ The signalling for RRC connection establishment is shown in Figure 234. The entire procedure is completed using only RRC signalling. A 3-way handshake is used to move the UE into RRC connected mode

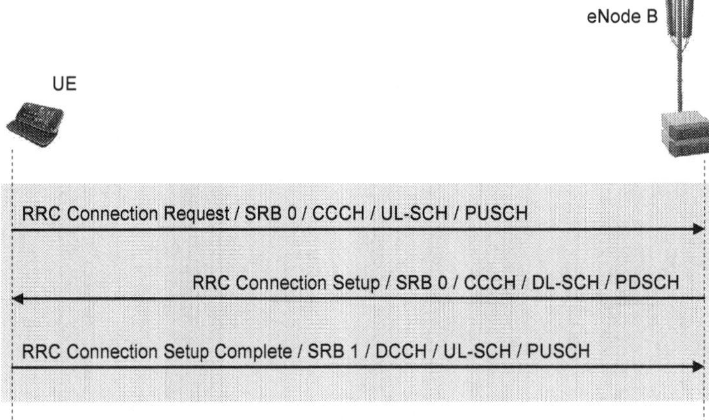

Figure 234 – Signalling for RRC connection establishment

★ The RRC Connection Request message is sent as part of the Random Access procedure. It corresponds to the initial Layer 3 message shown in Figure 202 (section 24.1). It is transferred using SRB 0 on the Common Control Channel (CCCH) because neither SRB 1 nor a Dedicated Control Channel (DCCH) have been setup at this point. The uplink Resource Block allocation for the RRC Connection Request message is signalled within the Random Access Response message

★ The content of the RRC Connection Request message is shown in Table 283. It includes a UE identity and an establishment cause. There is no scope for the UE to report any measurements within the RRC Connection Request message. The UMTS version of the RRC Connection Request message allows the UE to report CPICH measurements which can subsequently be used for open loop power control calculations

Information Elements	
UE Identity	CHOICE
	S-TMSI
	Random Value
Establishment Cause	CHOICE
	Emergency
	High Priority Access
	Mobile Terminating Access
	Mobile Originating Signalling
	Mobile Originating Data
	Delay Tolerant Access (3GPP release 10)

Table 283 – Content of RRC Connection Request message

★ The UE identity is signalled using the SAE Temporary Mobile Subscriber Identity (S-TMSI) if the UE is registered with the Tracking Area to which the current cell belongs. Otherwise, the UE selects a random number in the range from 0 to $2^{40} - 1$ to represent the UE identity. The S-TMSI is described in section 33.2

★ The establishment cause within the RRC Connection Request message is determined by the Non-Access Stratum (NAS) procedure for which the connection is being established. The relationship between establishment cause and NAS procedure is specified by 3GPP TS 24.301. This relationship is presented in Table 284

NAS Procedure		RRC Establishment Cause
Attach		Mobile Originating Signalling Delay Tolerant Access Emergency
Detach		Mobile Originating Signalling
Tracking Area Update		Mobile Originating Signalling Delay Tolerant Access Emergency
Service Request	User plane radio resources request	Mobile Originating Data Delay Tolerant Access Emergency
	Uplink signalling resources request	Mobile Originating Data Delay Tolerant Access
	Paging response for PS core network domain	Mobile Terminating Access
	PDN connectivity request with cause 'emergency'	Emergency
Extended Service Request	Mobile originating CS fallback	Mobile Originating Data Delay Tolerant Access
	Mobile terminating CS fallback	Mobile Terminating Access
	Mobile originating CS fallback emergency call	Emergency
	Packet services via S1	Mobile Terminating Access Delay Tolerant Access Emergency

Table 284 – Relationship between higher layer establishment cause and RRC establishment cause

★ In the case of the Attach procedure:

  - the 'Mobile Originating Signalling' cause value is used by default
  - the 'Delay Tolerant Access' cause value is used if the UE has been configured for 'low priority NAS signalling'. This RRC establishment cause was introduced within the release 10 version of the 3GPP specifications. The concept of 'low priority NAS signalling' is intended to provide a mechanism for congestion control, i.e. low priority signalling is dropped prior to higher priority signalling during periods of congestion. The 'NAS Signalling Priority Tag' within the USIM defines whether or not the device has been configured for low priority NAS signalling. This priority can also be used to impact charging, i.e. devices using low priority signalling could be charged less. Machine to machine type communications could use low priority signalling if their traffic is primarily background and best effort
  - the 'Emergency' cause value is used if the EPS Attach Type within the Attach Request message is set to 'EPS Emergency Attach'. The 'Emergency' cause value can also be used if the higher layers within the UE indicate the requirement to establish emergency bearer services, even when the EPS Attach Type is not set to 'EPS Emergency Attach'

★ In the case of the Detach procedure, the 'Mobile Originating Signalling' cause value is used

★ In the case of the Tracking Area Update procedure:

  - the 'Mobile Originating Signalling' cause value is used by default
  - the 'Delay Tolerant Access' cause value is used if the UE has been configured for 'low priority NAS signalling'
  - the 'Emergency' cause value is used if the UE already has a Packet Data Network (PDN) connection established for emergency bearer services, or if the UE is establishing a PDN connection for emergency bearer services

★ In the case of the Service Request procedure:

  - the 'Mobile Originating Signalling' cause value is used by default when the Service Request is used to request either user plane radio resources or uplink signalling resources
  - the 'Delay Tolerant Access' cause value is used when the Service Request is used to request either user plane radio resources or uplink signalling resources, and the UE has been configured for 'low priority NAS signalling'

  - the 'Mobile Terminating Access' cause value is used when the Service Request is a response to paging where the core network domain indicator is set to Packet Switched (PS)
  - the 'Emergency' cause value is used if the Service Request is being used to request user plane radio resources for emergency bearer services, or if the Service Request is triggered by a PDN connectivity request with an emergency cause value
- In the case of the Extended Service Request procedure:
  - the 'Mobile Originating Data' cause value is used by default if the Extended Service Request is used for mobile originating CS fallback
  - the 'Delay Tolerant Access' cause value is used for mobile originating CS fallback and packet services via the S1 when the UE has been configured for 'low priority NAS signalling'
  - the 'Mobile Terminating Access' cause value is used by default if the Extended Service Request is used for packet services via S1. It is also used when the Extended Service Request is used for mobile terminating CS fallback
  - the 'Emergency' cause value is used if the Extended Service Request is being used for a mobile originating CS fallback emergency call. It is also used if the Extended Service Request is being used for packet services via S1 when requesting radio resources for emergency bearer services
- In all cases, the RRC establishment cause is set to 'High Priority Access' if the UE uses Access Class (AC) 11 to 15 within its home PLMN
- The UE starts the T300 timer after transmitting the RRC Connection Request message. The value of T300 is broadcast within SIB 2. UMTS uses T300 in combination with N300 to manage retransmissions of the RRC Connection Request message. LTE does not have an N300 parameter and the RRC layer sends the RRC Connection Request message only once per establishment procedure. HARQ retransmissions from the MAC layer can be used to improve the reliability of transferring the RRC Connection Request and RRC Connection Setup messages. LTE uses the T300 timer to define how long the UE waits for a response to the RRC Connection Request message. The establishment procedure fails if T300 expires before receiving an RRC Connection Setup message. The procedure also fails if the UE completes a cell re-selection prior to receiving the RRC Connection Setup message
- Random access contention can occur after sending the RRC Connection Request message. Section 24.1 explains that contention occurs when multiple UE select the same subframe and preamble sequence for PRACH transmission. Contention requires the UE to repeat transmission of the PRACH preamble and the subsequent RRC Connection Request message. This increases the delay associated with connection establishment but does not cause the overall procedure to fail unless the maximum number of preamble transmissions has been reached
- Assuming that random access contention does not occur, the UE proceeds to wait for an RRC Connection Setup message from the eNode B. The UE has successfully completed the random access procedure so has been allocated a C-RNTI (signalled within the random access response message). The UE monitors the PDCCH for a downlink allocation addressed to its C-RNTI. The PDCCH specifies the set of PDSCH Resource Blocks used to transfer the RRC Connection Setup message. The RRC Connection Setup message is transferred using SRB 0 on the CCCH
- The RRC Connection Setup message contains configuration information for SRB 1. This allows subsequent signalling to use the DCCH logical channel. SRB 2 is always configured after security activation so the RRC Connection Setup message does not include any information regarding SRB 2. The eNode B can instruct the UE to apply a default configuration for SRB 1, or it can instruct the UE to apply a specific configuration.
- The default configuration for SRB 1 is presented in Table 285. This default configuration has been specified by 3GPP within TS 36.331. Using the default configuration helps to reduce the signalling requirement. The default configuration for SRB 2 is also presented in Table 285 for information. SRB 2 has a lower priority than SRB 1, i.e. a value of 3 represents a lower priority than a value of 1. Both SRB 1 and 2 always use acknowledged mode RLC

			SRB 1	SRB 2
RLC Configuration	Uplink	Poll Retransmission Timer	45	45
		Poll PDU	Infinity	Infinity
		Poll Byte	Infinity	Infinity
		Max Retransmission Threshold	4	4
	Downlink	Re-ordering Timer	35	35
		Status Prohibit Timer	0	0
Logical Channel Configuration		Priority	1	3
		Prioritised Bit Rate	Infinity	Infinity
		Bucket Size Duration	N/A	N/A
		Logical Channel Group	0	0

Table 285 – Default configurations for SRB 1 and SRB 2

- The RRC Connection Setup message can also define configuration information for the PDSCH, PUCCH and PUSCH physical channels. It can also include information regarding uplink power control, CQI reporting, the Sounding Reference Signal, antenna configuration and scheduling requests
- Upon receiving an RRC Connection Setup message, the UE stops the T300 timer and makes the transition to RRC Connected mode. The UE then proceeds to complete the procedure by sending an RRC Connection Setup Complete message. The content of the RRC Connection Setup Complete message is shown in Table 286

Information Elements	
RRC Transaction Identifier (0 to 3)	
Selected PLMN Identity (1 to 6)	
Registered MME	PLMN Identity
	MMEGI
	MMEC
Dedicated NAS Information	
GUMMEI Type (3GPP release 10)	
Radio Link Failure Information Available (3GPP release 10)	
Logged Measurements Available (3GPP release 10)	
Relay Node Subframe Configuration Requested (3GPP release 10)	

Table 286 – Content of RRC Connection Setup Complete message

- The Transaction Identifier, combined with the message type, identifies the RRC procedure with the UE
- The Selected PLMN Identity defines a pointer to a PLMN listed within SIB1, i.e. UE select the PLMN to which they want to connect when a cell belongs to more than a single PLMN
- The Registered MME information is optional, and is included when available. It becomes available after a UE has registered with an MME. The MME is identified by its Globally Unique MME Identity (GUMMEI) which is a concatenation of the PLMN identity, MME Group Identity (MMEGI) and MME Code (MMEC). The MMEC identifies the MME within its group
- The UE also includes its initial Non-Access Stratum (NAS) message within the RRC Connection Setup Complete message. NAS messages are specified within 3GPP TS 24.301. As indicated within Table 284, the NAS message could be an Attach, Detach, Tracking Area Update, Service Request or Extended Service Request message
- The Globally Unique MME Identity (GUMMEI) Type information was added within the release 10 version of the specifications. This can be signalled using values of 'native' or 'mapped'. The 'native' value indicates that the GUMMEI has been assigned by the Evolved Packet Core (EPC), whereas the 'mapped' value indicates that the GUMMEI has been derived from 2G/3G identifiers. This information can impact the selection of an MME for the UE
- The Radio Link Failure Information Available flag was added within the release 10 version of the specifications. This flag can be used for the mobility robustness optimisation component of Self Organising Networks (SON). Mobility robustness optimisation and radio link failure reporting is described in section 32.9
- The Logged Measurements Available flag was added within the release 10 version of the specifications. This flag can be used to indicate that information is available to be reported for the Minimisation of Drive Tests (MDT). MDT is described in section 32.15
- The Relay Node Subframe Configuration Requested information element was also added within the release 10 version of the specifications. It is used to indicate that the RRC connection establishment is for a relay node. It is also used to indicate whether or not the relay node would like a subframe configuration to be allocated, i.e. when included, it can be signalled using values of 'required' or 'not required'. Relay nodes are described in section 30.7
- The eNode B extracts the NAS message from the RRC Connection Setup Complete message and forwards it to an MME using the S1 Application Protocol (S1-AP) Initial UE Message. Forwarding this message does not form part of the RRC establishment procedure but is described within this section for completeness
- The content of the S1-AP Initial UE Message is shown in Table 287. The eNode B sends this message to the appropriate MME based upon its NAS Node Selection Function (NNSF). In the case of a Service Request, the S-TMSI included within the RRC Connection Request is used to identify the appropriate MME (S-TMSI includes the MMEC). In the case of an Attach or Tracking Area Update, the eNode B uses the GUMMEI included within the RRC Connection Setup Complete message. The eNode B is free to select an MME when the UE does not have an S-TMSI nor GUMMEI
- The eNode B allocates the 'eNode B UE S1-AP Identity' to allow the eNode B to identify the UE within S1 signalling procedures. The MME UE S1-AP Identity (not included within the Initial UE Message) allows the MME to identify the UE within S1 signalling procedures
- The release 9 version of the specifications added the Cell Access Mode information to the S1-AP: Initial UE message. This is used to indicate whether or not the source cell is operating in hybrid mode. Hybrid mode allows the definition of a Closed Subscriber Group (CSG) but also allows any UE to gain access. UE which are registered with the CSG are provided with priority over non-registered UE

Information Elements	Presence
eNode B UE S1-AP Identity	Mandatory
NAS PDU	Mandatory
Tracking Area Identity (TAI)	Mandatory
E-UTRAN Cell Global Identity (CGI)	Mandatory
RRC Establishment Cause	Mandatory
S-TMSI	Optional
CSG Identity	Optional
Globally Unique MME Identity (GUMMEI)	Optional
Cell Access Mode (3GPP release 9)	Optional
GW Transport Layer Address (3GPP release 10)	Optional
Relay Node Indicator (3GPP release 10)	Optional

Table 287 – Content of S1 Application Protocol (S1-AP) Initial UE Message

- ★ The release 10 version of the specifications added the Gateway (GW) Transport Layer Address to the S1-AP: Initial UE message. This can be used to define a GW transport layer address when the GW is collocated with the eNode B. This is relevant when the RRC connection is established for a relay node, or when the eNode B has a Local Gateway function for Local IP Access (LIPA)
- ★ The release 10 version of the specifications also added the Relay Node Indicator to the S1-AP: Initial UE message. This flag is used to indicate that the message is originating from a relay node rather than a UE
- ★ Figure 235 illustrates the signalling associated with the RRC connection establishment procedure when the eNode B rejects the RRC Connection Request. The reject message is returned to the UE using SRB 0 on the CCCH logical channel. The eNode B may reject the connection establishment request as a result of congestion

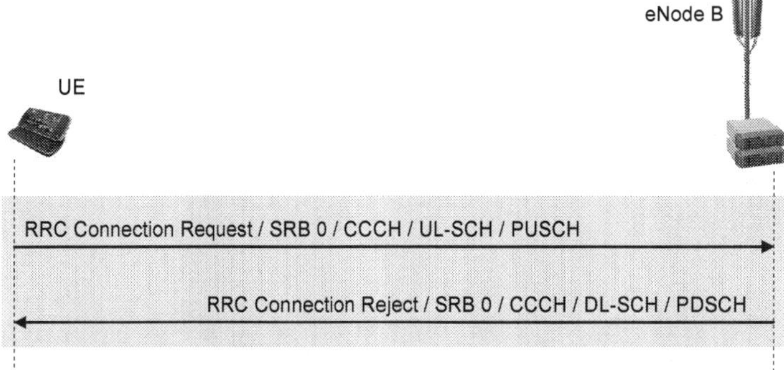

Figure 235 – Signalling for rejected RRC connection establishment

- ★ The content of the RRC Connection Reject message is presented in Table 288. This message only includes a Wait Time. This is in contrast to the equivalent UMTS message which also includes a rejection cause, although the UMTS rejection cause can only be defined as congestion or unspecified. The UMTS message can also include redirection information to direct the UE towards another RF carrier, or Radio Access Technology (RAT)
- ★ The Extended Wait Time was added by the release 10 version of the specifications. This is applicable when the connection is being established by a UE which has been configured for 'low priority NAS signalling', e.g. machine to machine type communications

Information Elements
Wait Time (1 to 16 seconds)
Extended Wait Time (1 to 1800 seconds) (3GPP release 10)

Table 288 – Content of RRC Connection Reject message

- ★ Upon receiving an RRC Connection Reject message, the UE starts the T302 timer with its value set equal to the Wait Time. Access Class barring for mobile originating calls, mobile originating signalling and mobile terminating access is applied until T302 expires, i.e. the UE is not allowed to send another RRC Connection Request for those connection types, and to the same cell, until T302 expires. T302 is stopped if the UE completes cell reselection. In that case, the UE is permitted to send an RRC Connection Request to the new cell. If included, the Extended Wait Time is forwarded to the upper layers
- ★ In contrast to UMTS, LTE requires the higher layers to initiate a new connection establishment procedure after the UE receives an RRC Connection Reject message. UMTS allows the RRC Connection Request message to be repeated from the RRC layer, based upon the value of N300
- ★ 3GPP References: TS 36.331, TS 36.413

26.2 ATTACH AND DEFAULT BEARER ESTABLISHMENT

★ The attach procedure is used to register with the Evolved Packet Core (EPC). This allows the subsequent use of packet services within the Evolved Packet System (EPS). The UE can also use the attach procedure to register for non-EPS services, e.g. the speech service based upon CS fallback

★ The attach procedure also involves establishing a default bearer between the UE and PDN Gateway. This provides always-on connectivity for the UE. In addition, the attach procedure can trigger the establishment of dedicated bearers

★ The signalling associated with the attach and default bearer establishment procedure is illustrated in Figure 236. Some signalling messages are nested within other messages, e.g. the PDN Connectivity Request message is sent within an Attach Request message, which is sent within an RRC Connection Setup Complete message

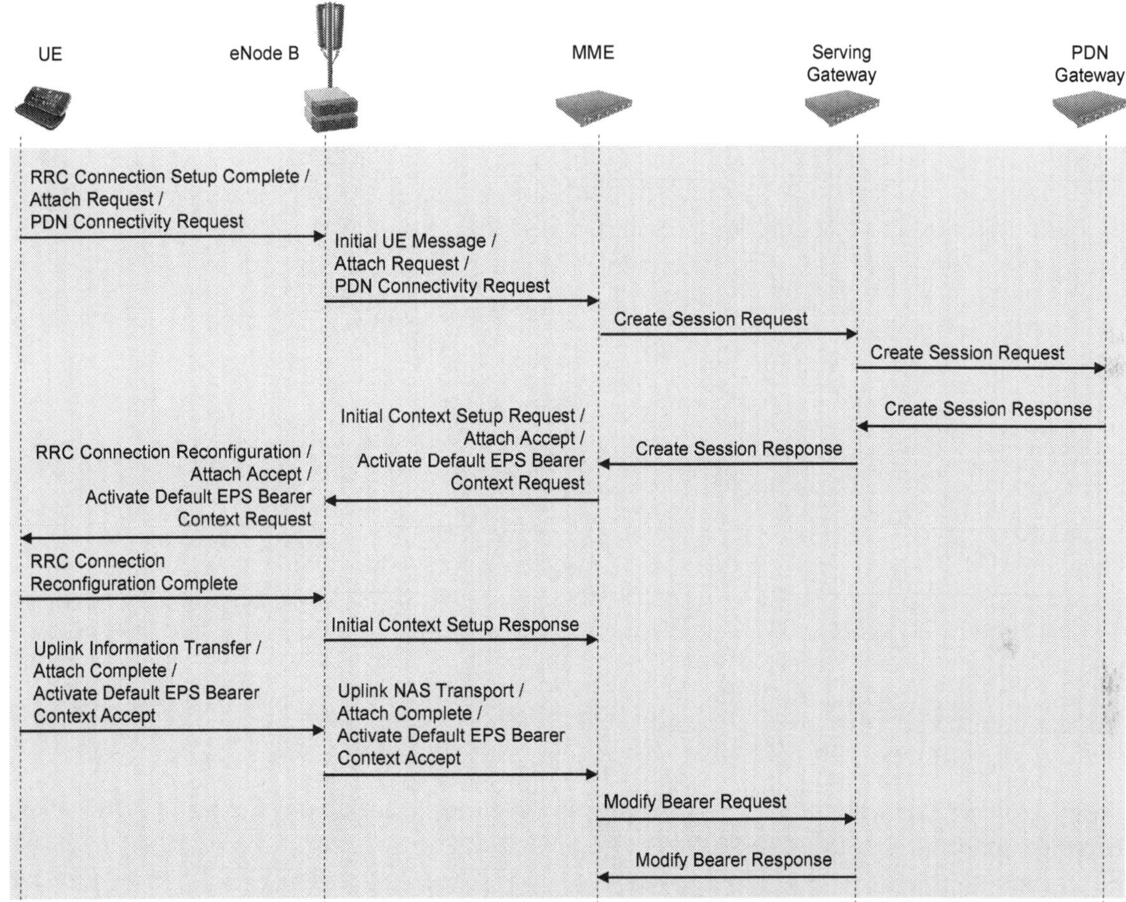

Figure 236 – Signalling for the attach and default bearer establishment procedure

★ The attach procedure starts with the RRC connection establishment procedure which takes the UE from RRC Idle mode to RRC connected mode. The Attach Request message is included as the 'Dedicated NAS Information' within the RRC Setup Complete message. The full content of the RRC Setup Complete message is presented in Table 286 within section 26.1. The Attach Request is a Non-Access Stratum (NAS) message so its content is specified by 3GPP TS 24.301. This content is presented in Table 289

★ The first 3 information elements within the Attach Request message represent the message header. The Protocol Discriminator indicates that the subsequent message is an EPS mobility management message. Its value is specified within 3GPP TS 24.007. The Security Header Type indicates whether or not the NAS message is integrity protected and ciphered. The Attach Request Message Identity indicates that the message itself is an Attach Request

★ The main content of the Attach Request message starts with the EPS Attach Type which can be configured with values of 'EPS Attach', 'Combined EPS/IMSI Attach' or 'EPS Emergency Attach'. The first two values correspond to registering for only EPS packet services, and registering for both EPS packet services and non-EPS services. The most common non-EPS service is likely to be the CS speech service. The 'EPS Emergency Attach' value was added by the release 9 version of the 3GPP specifications

★ The NAS Key Set Identifier (KSI) is used to identify the security parameters for authentication, integrity protection and ciphering. Security is not activated at this point in time when sending the Attach Request message within an RRC Connection Setup Complete message so the KSI information element is not relevant

Information Elements	Presence
Protocol Discriminator	*Mandatory*
Security Header Type	*Mandatory*
Attach Request Message Identity	*Mandatory*
EPS Attach Type	Mandatory
NAS Key Set Identifier	Mandatory
EPS Mobile Identity	Mandatory
UE Network Capability	Mandatory
ESM Message Container	Mandatory
Old P-TMSI Signature	Optional
Additional GUTI	Optional
Last Visited Registered TAI	Optional
DRX Parameter	Optional
MS Network Capability	Optional
Old Location Area Identification	Optional
TMSI Status	Optional
Mobile Station Classmark 2	Optional
Mobile Station Classmark 3	Optional
Supported Codecs	Optional
Additional Update Type	Optional
Voice Domain Preference and UE Usage Setting	Optional
Device Properties	Optional
Old GUTI Type	Optional
MS Network Feature Support	Optional

Table 289 – Content of Attach Request message

- ★ The EPS Mobile Identity field was updated in the release 9 version of the 3GPP specifications to allow the use of the Globally Unique Temporary UE Identity (GUTI), IMSI or IMEI. The release 8 version was limited to using the GUTI or IMSI. In some scenarios, a previous P-TMSI and the associated RAI can be mapped onto the GUTI field. An IMSI is used when the required GUTI or P-TMSI is not available. An IMEI is only used when the UE is attaching for emergency bearer services and neither GUTI, P-TMSI nor IMSI are available
- ★ The UE Network Capability provides information regarding the encryption and integrity protection algorithms supported by the UE. It also signals whether or not the UE has a preference for the default alphabet over the Universal Character Set (UCS) 2. Support for Single Radio Voice Call Continuity (SRVCC), LTE Positioning Protocol (LPP), Access Class Control for CS Fallback, Notification procedure and Location Services Notification are also indicated
- ★ The ESM Message Container is used to include a PDN Connectivity Request message. The PDN Connectivity Request is another NAS message, so one NAS message is transferred within another NAS message. The content of the PDN Connectivity Request message is shown in Table 290
- ★ The first 4 information elements within the PDN Connectivity Request message represent the message header. The Protocol Discriminator indicates that the subsequent message is an EPS session management message
- ★ The main content of the PDN Connectivity Request message starts with the Request Type which can be configured with values of 'Initial Request', 'Handover' or 'Emergency'. The 'Handover' value can be used for an incoming inter-system transition from another technology
- ★ The PDN Type indicates the IP capability of the UE. It can be signalled using values of IPv4, IPv6 or IPv4v6. The IPv4v6 value indicates that the UE supports the dual stack configuration which uses both IPv4 and IPv6 address allocations
- ★ The optional ESM Information Transfer Flag is included if the UE has protocol configuration options that need to be transferred security protected, or if the UE wishes to provide an Access Point Name (APN) for the PDN connection to be established during the attach procedure
- ★ The optional Access Point Name information element is always excluded when the PDN Connectivity Request message is included within an Attach Request message, i.e. the APN is transferred during a subsequent ESM Information Request procedure. Transfer of this information is delayed to allow its transfer after security has been enabled, i.e. to allow ciphering of the content
- ★ The optional Device Properties is included if the UE has been configured for 'low priority NAS signalling'. This flag was introduced within the release 10 version of the 3GPP specifications

Information Elements	Presence
Protocol Discriminator	*Mandatory*
EPS Bearer Identity	*Mandatory*
Procedure Transaction Identity	*Mandatory*
PDN Connectivity Request Message Identity	*Mandatory*
Request Type	Mandatory
PDN Type	Mandatory
ESM Information Transfer Flag	Optional
Access Point Name	Optional
Protocol Configuration Options	Optional
Device Properties	Optional

Table 290 – Content of PDN Connectivity Request message

★ The UE starts timer T3410 after sending the Attach Request message. 3GPP has standardised a fixed value of 15 s for the T3410 timer. If T3410 expires prior to the UE receiving a response from the MME, then the current attach attempt is aborted and the attach attempt count is incremented by 1. The UE starts timer T3411 if the resultant attach attempt count is less than 5. 3GPP has standardised a fixed value of 10 s for the T3411 timer. The UE re-transmits the Attach Request message once T3411 has expired

★ If the attach attempt count reaches 5, the UE starts timer T3402, and deletes its GUTI, TAI and list of equivalent PLMN. 3GPP has standardised a default value of 12 minutes for the T3402 timer. The network can signal a different value for T3402 within the Attach Accept and Tracking Area Update Accept messages. A value signalled by the network is only applicable within the set of assigned Tracking Areas. The default value is re-adopted if the UE receives an Attach Accept or Tracking Area Update Accept message which does not specify a T3402 value. The attach procedure can be re-initiated once T3402 expires

★ Once the eNode B receives the RRC Connection Setup Complete message, the Attach Request message is extracted and placed within an S1 Application Protocol (S1-AP) Initial UE Message. The content of this message is presented in Table 287 within section 26.1. The eNode B includes information regarding the Tracking Area Identity (TAI) and Cell Global Identity (CGI). The S-TMSI can also be included. The eNode B proceeds to forward the Initial UE Message to the MME

★ Once the MME has received the Attach Request message, it determines the IMSI belonging to the UE. The IMSI may have been included explicitly within the Attach Request. However, if the UE included an Old GUTI instead of an IMSI then the MME uses the Old GUTI to identify the address of the old MME (or SGSN) to which the UE was previously registered. The MME then sends an Identity Request message to the old MME (or SGSN) asking for the IMSI. If the MME cannot determine the IMSI using this approach, an Identity Request is sent to the UE asking for its IMSI. The signalling used to determine the IMSI is not shown in Figure 236

★ The MME can then complete authentication and NAS security setup. Authentication is completed using an Equipment Identity Register (EIR) which can be connected to the MME using the S13 interface. Security setup is used to enable integrity protection and ciphering for any subsequent signalling. The signalling used for authentication and NAS security setup is not shown in Figure 236

★ If the UE included the ESM Information Transfer Flag within the PDN Connectivity Request message, the MME initiates the ESM Information Request procedure. This procedure allows the UE to provide an APN as well as any protocol configuration options. The additional signalling for the ESM Information Request procedure is shown in Figure 237.

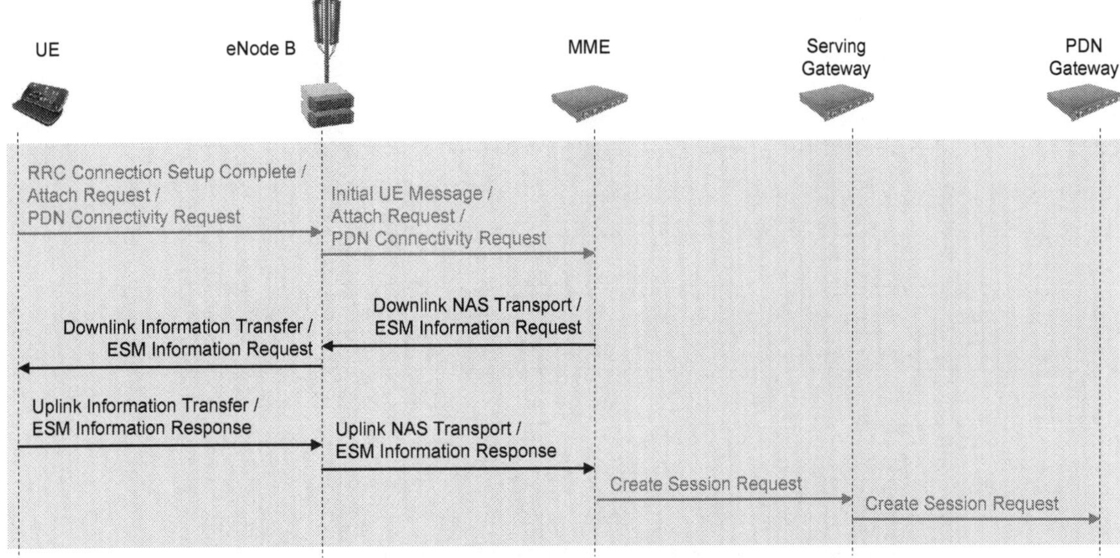

Figure 237 – Additional signalling for the ESM Information Request procedure

★ The content of the ESM Information Request and ESM Information Response messages is presented in Table 291. The request message serves as a simple prompt which triggers the UE to return a response. The response message can include an Access Point Name (APN) and a set of Protocol Configuration Options. The protocol configuration options can define the username and password for the APN

ESM Information Request

Information Elements	Presence
Protocol Discriminator	*Mandatory*
EPS Bearer Identity	*Mandatory*
Procedure Transaction Identity	*Mandatory*
ESM Information Request Message Identity	*Mandatory*

ESM Information Response

Information Elements	Presence
Protocol Discriminator	*Mandatory*
EPS Bearer Identity	*Mandatory*
Procedure Transaction Identity	*Mandatory*
ESM Information Response Message Identity	*Mandatory*
Access Point Name	Optional
Protocol Configuration Options	Optional

Table 291 – Content of ESM Information Request and ESM Information Response messages

★ The default APN can be used if the UE did not include the ESM Information Transfer Flag within the PDN Connectivity Request message. Once the MME has finalised the APN to be used, it sends a Create Session Request message to the Serving Gateway. This message belongs to the GTP control plane protocol (GTPv2-C) and is specified within 3GPP TS 29.274

★ The Serving Gateway creates an entry within its EPS Bearer table and forwards the Create Session Request message to the PDN Gateway. The PDN Gateway creates a new entry within its EPS Bearer Context table and generates a charging identity. The new context allows the PDN Gateway to start routing user plane data between the Packet Data Network (PDN) and the Serving Gateway, and to start charging

★ The PDN Gateway returns a Create Session Response message to the Serving Gateway. This message includes the IP address that has been allocated to the UE by the PDN Gateway, i.e. the PDN Gateway is responsible for managing the allocation of IP addresses

★ The Serving Gateway forwards the Create Session Response message to the MME, which then proceeds to initiate the S1-AP initial context setup procedure. This is done by sending an Initial Context Setup Request message to the eNode B. The content of the Initial Context Setup Request message is presented in Table 292. It allows the eNode B to configure itself for the new E-RAB

<table>
<tr><th colspan="3">Information Elements</th></tr>
<tr><td colspan="3">Message Type</td></tr>
<tr><td colspan="3">MME UE S1AP ID</td></tr>
<tr><td colspan="3">eNB UE S1AP ID</td></tr>
<tr><td colspan="3">UE Aggregate Maximum Bit Rate</td></tr>
<tr><td rowspan="6">E-RAB to be Setup List</td><td rowspan="6">E-RAB to be Setup</td><td>E-RAB ID</td></tr>
<tr><td>E-RAB Level QoS Parameters</td></tr>
<tr><td>Transport Layer Address</td></tr>
<tr><td>GTP TEID</td></tr>
<tr><td>NAS PDU</td></tr>
<tr><td>Correlation Identity</td></tr>
<tr><td colspan="3">UE Security Capabilities</td></tr>
<tr><td colspan="3">Security Key</td></tr>
<tr><td colspan="3">Trace Activation</td></tr>
<tr><td colspan="3">Handover Restriction List</td></tr>
<tr><td colspan="3">UE Radio Capability</td></tr>
<tr><td colspan="3">Subscriber Profile ID for RAT/Frequency Priority</td></tr>
<tr><td colspan="3">CS Fallback Indicator</td></tr>
<tr><td colspan="3">SRVCC Operation Possible</td></tr>
<tr><td colspan="3">CSG Membership Status</td></tr>
<tr><td colspan="3">Registered LAI</td></tr>
<tr><td colspan="3">GUMMEI ID</td></tr>
<tr><td colspan="3">MME UE S1 AP ID 2</td></tr>
<tr><td colspan="3">Management Based MDT Allowed</td></tr>
</table>

Table 292 – Content of Initial Context Setup Request message

- The UE Aggregate Maximum Bit Rate (UE-AMBR) defines the maximum allowed throughput for a UE based upon the sum of all its non-GBR bearers. This parameter forms part of the QoS profile described in section 2.10. It can be configured with values between 0 and 10 000 Mbps, and can be specified independently for the uplink and downlink
- The E-RAB Level QoS Parameters include:
  - QoS Class Identifier (QCI)
  - Allocation and Retention Priority (ARP)
  - Guaranteed Bit Rates (GBR)
  - Maximum Bit Rates (MBR)

  These parameters form part of the QoS profile described in section 2.10. The relationship between QCI and its associated QoS parameters is presented in Table 8. QCI values of 8 and 9 are intended for the default bearer. QCI 8 can be used for premium subscribers, whereas QCI 9 can be used for basic subscribers. The GBR and MBR parameters are only applicable to GBR bearers
- The transport layer address defines an IP address which can subsequently be used for transferring user plane data from the eNode B to the Serving Gateway
- The GTP Tunnel Endpoint Identifier (TEID) defines the GTP-U tunnel for the user plane data. Multiple tunnels will share the same transport layer address
- The Non-Access Stratum (NAS) PDU is used to transfer the Attach Accept message which acknowledges the original Attach Request message
- The Correlation Identity was introduced by the release 10 version of the 3GPP specifications. It is used to define the GTP Tunnel Endpoint Identifier or Generic Routing Encapsulation (GRE) key to be used for the user plane transport between an eNode B and a Local Gateway (L-GW). A L-GW can be used in combination with a Home eNode B (femtocell) on a local network
- The MME starts timer T3450 after sending the Attach Accept message. 3GPP has standardised a fixed value of 6 s for the T3450 timer. If T3450 expires prior to the MME receiving a response from the UE, then the Attach Accept message is re-sent. The Attach Accept message can be re-sent four times. The overall attach procedure is aborted if T3450 expires a fifth time
- The content of the Attach Accept message is shown in Table 293
- The first 3 information elements within the Attach Accept message represent the message header. The Protocol Discriminator indicates that the subsequent message is an EPS mobility management message. Its value is specified within 3GPP 24.007. The Security Header Type indicates whether or not the NAS message is integrity protected and ciphered. The Attach Accept Message Identity indicates that the message itself is an Attach Accept

Information Elements	Presence
Protocol Discriminator	*Mandatory*
Security Header Type	*Mandatory*
Attach Accept Message Identity	*Mandatory*
EPS Attach Result	Mandatory
T3412 Value	Mandatory
TAI List	Mandatory
ESM Message Container	Mandatory
GUTI	Optional
Location Area Identification	Optional
MS Identity	Optional
EMM Cause	Optional
T3402 Value	Optional
T3423 Value	Optional
Equivalent PLMN	Optional
Emergency Number List	Optional
EPS Network Feature Support	Optional
Additional Update Result	Optional
T3412 Extended Value	Optional

Table 293 – Content of Attach Accept message

- The main content of the Attach Accept message starts with the EPS Attach Result which can be configured with values of 'EPS Only' or 'Combined EPS/IMSI Attach'. These values correspond to EPS Attach Type specified within the request message

★ T3412 defines the periodic Tracking Area Update (TAU) timer. UE trigger a periodic TAU procedure when this timer expires. It can be configured with a value up to 186 minutes

★ The Tracking Area Identity (TAI) list defines the set of allocated tracking areas, i.e. the tracking areas with which the UE is registered within. The TAI list can include tracking areas belonging to more than a single PLMN when a shared network is being used. Tracking Area Updates are subsequently triggered if the UE moves into a tracking area which is not included within the TAI list

★ The ESM Message Container is used to transfer the Activate Default EPS Bearer Context Request message which represents a response to the PDN Connectivity Request message. The content of the PDN Connectivity Request message is shown in Table 294

Information Elements	Presence
Protocol Discriminator	*Mandatory*
EPS Bearer Identity	*Mandatory*
Procedure Transaction Identity	*Mandatory*
Activate Default EPS Bearer Context Request Message Identity	*Mandatory*
EPS QoS	Mandatory
Access Point Name	Mandatory
PDN Address	Mandatory
Transaction Identifier	Optional
Negotiated QoS	Optional
Negotiated LLC SAPI	Optional
Radio Priority	Optional
Packet Flow Identifier	Optional
APN-AMBR	Optional
ESM Cause	Optional
Protocol Configuration Options	Optional

Table 294 – Content of Activate Default EPS Bearer Context Request message

★ The first 4 information elements within the Activate Default EPS Bearer Context Request message represent the message header. The Protocol Discriminator indicates that the subsequent message is an EPS session management message

★ The EPS QoS parameter defines:

- o QoS Class Identifier (QCI)
- o Guaranteed Bit Rates (GBR)
- o Maximum Bit Rates (MBR)

These parameters form part of the QoS profile described in section 2.10. The relationship between QCI and its associated QoS parameters is presented in Table 8. The GBR and MBR parameters are only applicable to GBR bearers. The Allocation and Retention Priority (ARP) has been signalled to the eNode B but is not signalled to the UE, i.e. ARP values are used by the network rather than by the UE

★ The Access Point Name (APN) informs the UE of the APN to which the default bearer is connected

★ The PDN Address allocates the UE with an IP address

★ The eNode B extracts the Attach Accept message from within the Initial Context Setup Request, and packages it within an RRC Connection Reconfiguration message. The RRC Connection Reconfiguration is transferred to the UE across the air-interface. This message provides the UE with all of the information necessary to establish the default bearer

★ The UE responds to the eNode B using the RRC Connection Reconfiguration Complete message. This message is a simple acknowledgement. The release 10 version of the specifications allows it to include flags indicating the availability of radio link failure and logged measurement reports

★ The RRC Connection Reconfiguration Complete message does not offer the ability to transfer NAS messages so the UE has to send the Attach Complete within a subsequent Uplink Information Transfer message

★ The eNode B completes the S1-AP Initial Context Setup procedure by forwarding an Initial Context Setup Response to the MME. The content of this message is shown in Table 295. It acknowledges that the E-RAB for the default bearer has been successfully setup. A transport layer address is provided in terms of an IP address for downlink data transfer from the Serving Gateway to the eNode B. A GTP Tunnel Endpoint Identifier (TEID) is also provided to define the GTP-U tunnel for the user plane data. A failure cause is specified if the E-RAB setup failed

<table>
<tr><th colspan="3">Information Elements</th></tr>
<tr><td colspan="3">Message Type</td></tr>
<tr><td colspan="3">MME UE S1AP ID</td></tr>
<tr><td colspan="3">eNB UE S1AP ID</td></tr>
<tr><td rowspan="3">E-RAB Setup List</td><td rowspan="3">E-RAB Setup Item</td><td>E-RAB ID</td></tr>
<tr><td>Transport Layer Address</td></tr>
<tr><td>GTP TEID</td></tr>
<tr><td rowspan="2">E-RAB Failed to Setup List</td><td rowspan="2">E-RAB List</td><td>E-RAB ID</td></tr>
<tr><td>Cause</td></tr>
<tr><td colspan="3">Criticality Diagnostics</td></tr>
</table>

Table 295 – Content of Initial Context Setup Response message

- The content of the Attach Complete message is shown in Table 296. It is a simple acknowledgement but also contains an EPS Session Management (ESM) NAS message. The ESM NAS message is an Activate Default EPS Bearer Context Accept message

Information Elements	Presence
Protocol Discriminator	*Mandatory*
Security Header Type	*Mandatory*
Attach Complete Message Identity	*Mandatory*
ESM Message Container	Mandatory

Table 296 – Content of Attach Complete message

- The content of the Activate Default EPS Bearer Context Accept message is shown in Table 297. It is another simple acknowledgement but may also contain some protocol configuration options

Information Elements	Presence
Protocol Discriminator	*Mandatory*
EPS Bearer Identity	*Mandatory*
Procedure Transaction Identity	*Mandatory*
Activate Default EPS Bearer Context Accept Message Identity	*Mandatory*
Protocol Configuration Options	Optional

Table 297 – Content of Activate Default EPS Bearer Context Accept message

- After receiving both the Initial Context Setup Response and Attach Complete messages, the MME sends a Modify Bearer Request message to the Serving Gateway. This message belongs to the GTP control plane protocol (GTPv2-C). It provides the Serving Gateway with the downlink transport layer address and the associated GTP TEID. Both of these are taken from the S1-AP Initial Context Setup Response message
- The Serving Gateway completes the procedure by sending a Modify Bearer Response message to the MME. This message is a simple acknowledgement
- 3GPP References: TS 24.301, TS 23.401, TS 36.413, TS 29.274

26.3 DEDICATED BEARER ESTABLISHMENT

- A dedicated bearer is required if the end-user requires a different Quality of Service (QoS) to that offered by the default bearer, or if the end-user requires connectivity to a different Packet Data Network (PDN) to that provided by the default bearer. Dedicated bearers are configured to run in parallel to the existing default bearer
- The dedicated bearer establishment procedure is initiated by the network but may be requested by the UE. The UE can request a dedicated bearer by sending a Non-Access Stratum (NAS) Bearer Resource Allocation Request to the MME
- The UE starts the T3480 timer when sending a Bearer Resource Allocation Request message. The message is re-transmitted if T3480 expires prior to the UE receiving a response from the MME. 3GPP have specified a fixed value of 8 s for the T3480 timer. The Bearer Resource Allocation Request message can be re-transmitted a maximum of 4 times before the overall procedure is aborted
- The signalling associated with dedicated bearer establishment is shown in Figure 238

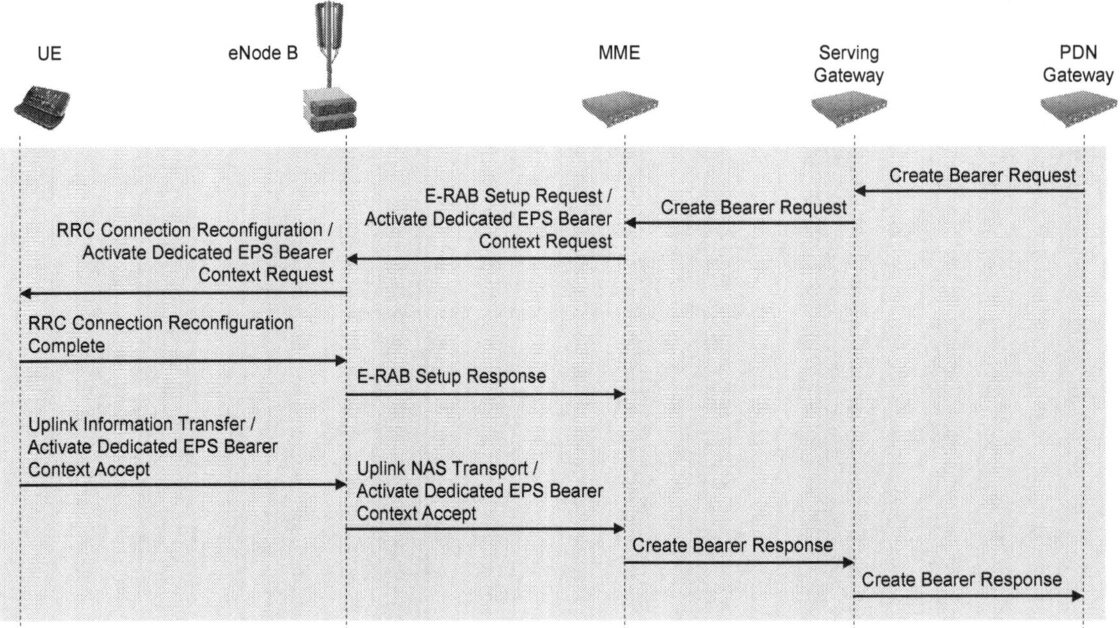

Figure 238 – Signalling for the dedicated bearer establishment procedure

- The PDN Gateway starts by sending a Create Bearer Request message to the Serving Gateway. This message belongs to the GTPv2-C protocol so is specified within 3GPP TS 29.274. It includes a definition of the dedicated bearer QoS requirements, i.e. QoS Class Identifier (QCI), Allocation and Retention Priority (ARP), Guaranteed Bit Rate (GBR) and Maximum Bit Rate (MBR)
- The Create Bearer Request message also specifies the uplink GTP-U tunnel endpoint address at the PDN Gateway. This Tunnel Endpoint Identity (TEID) defines the GTP-U tunnel to be used by the Serving Gateway when forwarding user plane data to the PDN Gateway. The Create Bearer Request also includes the identity of the default bearer so the dedicated bearer and default bearer can be linked. Protocol configuration options can be included to transfer application layer parameters between the PDN and UE. These parameters are sent transparently through the Serving Gateway and MME
- The Serving Gateway generates its own version of the Create Bearer Request message. The content is similar to that received from the PDN Gateway but the Serving Gateway replaces the uplink GTP-U TEID at the PDN Gateway with an uplink GTP-U TEID at the Serving Gateway. The MME is responsible for forwarding this TEID to the eNode B to allow the eNode B to forward uplink user plane data to the appropriate address within the Serving Gateway
- The MME generates an E-RAB Setup Request message which encapsulates an Activate Dedicated EPS Bearer Context Request message. The E-RAB Setup Request belongs to the S1 Application Protocol (S1-AP) so is specified within 3GPP TS 36.413. The Activate Dedicated EPS Bearer Context Request belongs to the session management component of the Non-Access Stratum (NAS) protocol so is specified within 3GPP TS 24.301. The content of the E-RAB Setup Request is shown in Table 298
- The UE Aggregate Maximum Bit Rate is used to update the value currently applied by the eNode B. This parameter defines a maximum total throughput for all non-GBR bearers belonging to the UE. The E-RAB QoS parameters are based upon those originating from the PDN Gateway. The GTP TEID corresponds to the uplink tunnel specified by the Serving Gateway. The transport layer address identifies the Serving Gateway itself

Information Elements		
Message Type		
MME UE S1AP ID		
eNB UE S1AP ID		
UE Aggregate Maximum Bit Rate		
E-RAB to be Setup List	E-RAB to be Setup	E-RAB ID
		E-RAB Level QoS Parameters
		Transport Layer Address
		GTP TEID
		NAS PDU
		Correlation Identity

Table 298 – Content of E-RAB Setup Request message

★ The NAS PDU within the E-RAB Setup Request corresponds to the Activate Dedicated EPS Bearer Context Request. This message is directed to the UE rather than the eNode B so the eNode B is responsible for forwarding it across the air-interface. The content of the Activate Dedicated EPS Bearer Context Request is shown in Table 299

Information Elements	Presence
Protocol Discriminator	*Mandatory*
EPS Bearer Identity	*Mandatory*
Procedure Transaction Identity	*Mandatory*
Activate Dedicated EPS Bearer Context Request Message Identity	*Mandatory*
Linked EPS Bearer Identity	Mandatory
EPS QoS	Mandatory
TFT	Mandatory
Transaction Identifier	Optional
Negotiated QoS	Optional
Negotiated LLC SAPI	Optional
Radio Priority	Optional
Packet Flow Identifier	Optional
Protocol Configuration Options	Optional

Table 299 – Content of Activate Dedicated EPS Bearer Context Request message

★ The Linked EPS Bearer Identity is used to link the dedicated bearer to the default bearer

★ The QoS information within the Activate Dedicated EPS Bearer Context Request is similar to that within the E-RAB Setup Request but it excludes the Allocation and Retention Priority (ARP) information. ARP is used by the network rather than the UE so is not sent to the UE

★ The Traffic Flow Template (TFT) defines a set of packet filters for the UE to apply in the uplink direction. These filters can be based upon various criteria, including source and destination port ranges, type of service, remote address types and protocol identifiers

★ The eNode B uses the E-RAB Setup Request to configure itself for the dedicated bearer. It also extracts the Activate Dedicated EPS Bearer Context Request and packages it within an RRC Connection Reconfiguration message before forwarding to the UE. The RRC Connection Reconfiguration message belongs to the RRC protocol so is specified within 3GPP TS 36.331. This message provides the UE with all of the information necessary for it to configure itself for the dedicated RAB. It is transferred across the air-interface using SRB 1 and acknowledged mode RLC

★ The UE acknowledges reception of the RRC Connection Reconfiguration message using the RRC Connection Reconfiguration Complete message. This message also acknowledges that the UE has applied the configuration for the dedicated bearer. The RRC Connection Reconfiguration Complete message is a simple acknowledgement and is not able to transfer any NAS signalling. This means that the response for the Activate Dedicated EPS Bearer Context Request has to be sent in a separate message

★ Once the eNode B has received the RRC Connection Reconfiguration Complete message, it responds to the E-RAB Setup Request from the MME using an E-RAB Setup Response. The content of this message is shown in Table 300. It provides the MME with a transport layer address and GTP-U Tunnel Endpoint Identity (TEID) for the downlink user plane data

<table>
<tr><th colspan="3">Information Elements</th></tr>
<tr><td colspan="3">Message Type</td></tr>
<tr><td colspan="3">MME UE S1AP ID</td></tr>
<tr><td colspan="3">eNB UE S1AP ID</td></tr>
<tr><td rowspan="3">E-RAB Setup List</td><td rowspan="3">E-RAB Setup</td><td>E-RAB ID</td></tr>
<tr><td>Transport Layer Address</td></tr>
<tr><td>GTP TEID</td></tr>
<tr><td rowspan="2">E-RAB Failed to Setup List</td><td rowspan="2">E-RAB List</td><td>E-RAB ID</td></tr>
<tr><td>Cause</td></tr>
<tr><td colspan="3">Criticality Diagnostics</td></tr>
</table>

Table 300 – Content of E-RAB Setup Response message

★ The UE generates an Activate Dedicated EPS Bearer Context Accept message to respond to the Activate Dedicated EPS Bearer Context Request from the MME. This message belongs to the NAS protocol so is specified within 3GPP TS 24.301. Its content is shown in Table 301. The UE packages the NAS message within an Uplink Information Transfer message before sending it to the eNode B. The eNode B extracts the NAS message and packages it within an S1-AP Uplink NAS Transport message before sending it to the MME

Information Elements	Presence
Protocol Discriminator	*Mandatory*
EPS Bearer Identity	*Mandatory*
Procedure Transaction Identity	*Mandatory*
Activate Dedicated EPS Bearer Context Accept Message Identity	*Mandatory*
Protocol Configuration Options	Optional

Table 301 – Content of Activate Dedicated EPS Bearer Context Accept message

★ The MME then uses the GTPv2-C protocol to respond to the Serving Gateway by sending a Create Bearer Response. This message provides the Serving Gateway with the GTP-U tunnel identity for the downlink user plane data, i.e. it allows the Serving Gateway to forward downlink data to the eNode B. The eNode B provided the MME with this information within the E-RAB Setup Response

★ The Serving Gateway completes the procedure by sending its own version of the Create Bearer Response to the PDN Gateway. This version includes the downlink GTP-U tunnel identity at the Serving Gateway so allows the PDN Gateway to transfer downlink data to the Serving Gateway

★ 3GPP References: TS 24.301, TS 23.401, TS 36.413, TS 29.274, TS 36.331

26.4 TRACKING AREA UPDATE

★ Tracking Area Updates (TAU) become applicable after the UE has completed the Attach procedure and has moved into the EPS Mobility Management (EMM) REGISTERED state. They can be triggered in either RRC Idle mode or RRC Connected mode, but the UE must be in RRC Connected mode to actually complete the procedure

★ The TAU procedure is used for:

  - normal TAU due to mobility, i.e. when the UE enters a tracking area which is not included within the list of tracking areas with which the UE is registered
  - periodic TAU after the T3412 timer expires. The value of T3412 is initialised during the Attach procedure within the Attach Accept message shown in Table 293. Periodic TAU confirm to the network that the UE is still present
  - registering with the CS domain for non-EPS services when the UE is already attached for EPS services. This involves completing an IMSI attach as part of the TAU procedure
  - registering with the EPS after an incoming inter-system change, e.g. reselecting the LTE network after being camped on the UMTS network
  - re-registering with the EPS after a CS fallback connection has been completed
  - MME load balancing. The UE initiates a TAU procedure if the eNode B releases its RRC connection using a cause value of 'Load Balancing TAU Required'
  - updating the UE specific DRX cycle. UE are allowed to select their own DRX cycle and signal it to the MME. The selected DRX cycle can be specified within the Tracking Area Update Request message
  - indicating that a UE has selected a Closed Subscriber Group (CSG) cell whose CSG Identity is not included within the UE's allowed CSG list

★ The TAU procedure may involve a change of Serving Gateway and/or a change of MME

★ The signalling associated with the TAU procedure is illustrated in Figure 239. This example includes a change of MME but maintains the same Serving Gateway. The example assumes that the UE starts in RRC Idle mode so the Tracking Area Update Request is sent within an RRC Connection Setup Complete message. If the UE is already in RRC Connected mode when the TAU procedure is triggered, the Tracking Area Update Request is sent within an Uplink Information Transfer message

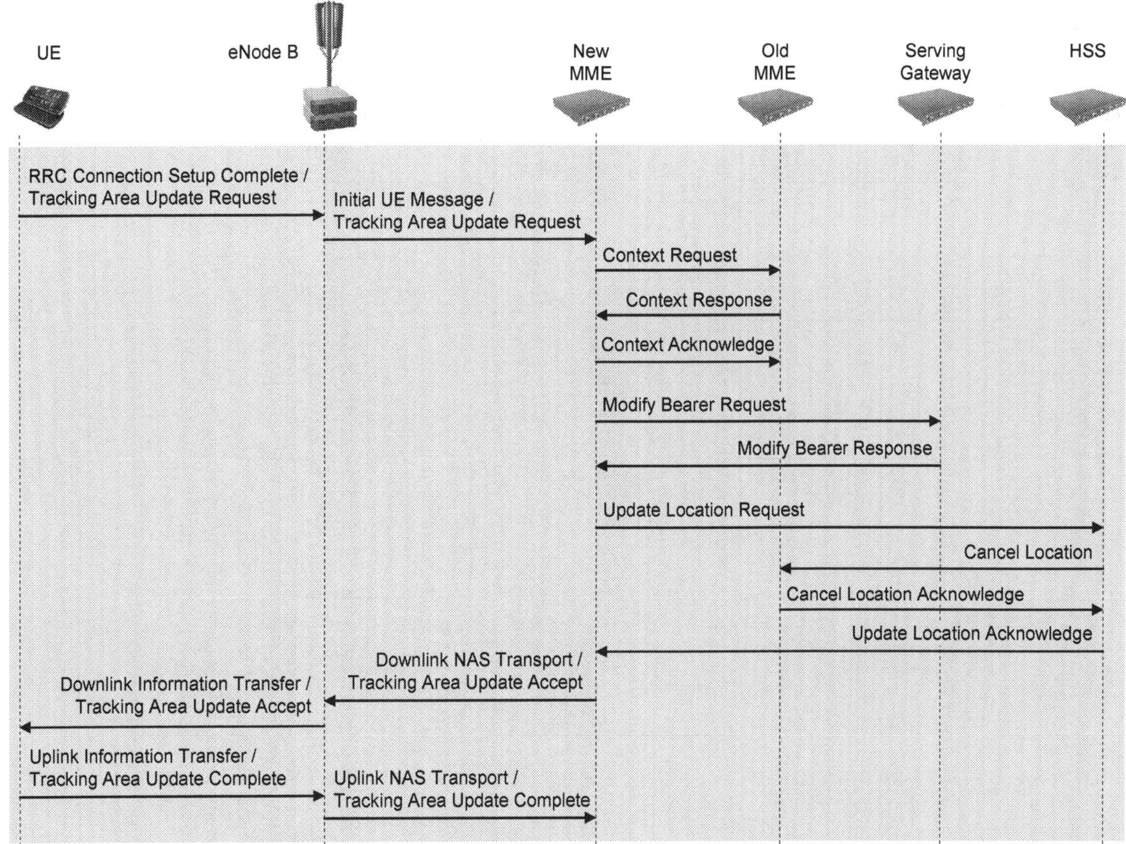

Figure 239 – Signalling for the Tracking Area Update (TAU) procedure (with change of MME)

★ The UE starts timer T3430 after sending the Tracking Area Update Request message. 3GPP has standardised a fixed value of 15 s for the T3430 timer. If T3430 expires prior to the UE receiving a response from the MME, then the current TAU attempt is aborted and the TAU attempt count is incremented by 1. The UE starts timer T3411 if the resultant attempt count is less than 5. 3GPP has standardised a fixed value of 10 s for the T3411 timer. The UE re-transmits the Tracking Area Update Request message once T3411 has expired

★ If the TAU attempt count reaches 5, the UE starts timer T3402, and deletes its list of equivalent PLMN. 3GPP has standardised a default value of 12 minutes for the T3402 timer. The network can signal a different value for T3402 within the Attach Accept and Tracking Area Update Accept messages. A value signalled by the network is only applicable within the set of assigned Tracking Areas. The default value is re-adopted if the UE receives an Attach Accept or Tracking Area Update Accept message which does not specify a T3402 value. The TAU procedure can be re-initiated once T3402 expires

★ The content of the Tracking Area Update Request message is shown in Table 302

Information Elements	Presence
Protocol Discriminator	*Mandatory*
Security Header Type	*Mandatory*
Tracking Area Update Request Message Identity	*Mandatory*
EPS Update Type	Mandatory
NAS Key Set Identifier	Mandatory
Old GUTI	Mandatory
Non-Current Native NAS Key Set Identifier	Optional
GPRS Ciphering Key Sequence Number	Optional
Old P-TMSI Signature	Optional
Additional GUTI	Optional
Nonce UE	Optional
UE Network Capability	Optional
Last Visited Registered TAI	Optional
DRX Parameter	Optional
UE Radio Capability Information Update Needed	Optional
EPS Bearer Context Status	Optional
MS Network Capability	Optional
Old Location Area Identification	Optional
TMSI Status	Optional
Mobile Station Classmark 2	Optional
Mobile Station Classmark 3	Optional
Supported Codecs	Optional
Additional Update Type	Optional
Voice Domain Preference and UE Usage Setting	Optional
Old GUTI Type	Optional
Device Properties	Optional
MS Network Feature Support	Optional

Table 302 – Content of Tracking Area Update Request message

★ The first 3 information elements within the Tracking Area Update Request message represent the message header. The Protocol Discriminator indicates that the subsequent message is an EPS mobility management message. Its value is specified within 3GPP TS 24.007. The Security Header Type indicates whether or not the NAS message is integrity protected and ciphered. The Tracking Area Update Request Message Identity indicates that the message itself is a Tracking Area Update Request

★ The main content of the Tracking Area Update Request message starts with the EPS Update Type which can be configured with values of 'TA Updating', 'Combined TA/LA Updating', 'Combined TA/LA Updating with IMSI Attach' and 'Periodic Updating'. The 'TA Updating' value is used if the UE is only registered with the EPS, i.e. only registered for PS services. The 'Combined TA/LA Updating' value is used if the UE is registered for both EPS and non-EPS services, i.e. both CS and PS services. The 'Combined TA/LA Updating with IMSI Attach' value is used if the UE is registered for EPS services but wishes to also register for non-EPS services. The 'Periodic Updating' value is used when the T3412 timer expires

- ★ The NAS Key Set Identifier (KSI) is used to identify the security parameters for authentication, integrity protection and ciphering. Security is not activated at this point in time when sending the Tracking Area Update Request message within an RRC Connection Setup Complete message so the KSI information element is not relevant
- ★ The Old Globally Unique Temporary UE Identity (GUTI) information element provides the network with an identity for the UE. The GUTI can be derived from a previous P-TMSI and the associated RAI if they are available
- ★ Once the eNode B receives the RRC Connection Setup Complete message, the Tracking Area Update Request message is extracted and placed within an S1 Application Protocol (S1-AP) Initial UE Message. The content of this message is presented in Table 287 within section 26.1. The eNode B includes information regarding the new Tracking Area Identity (TAI) and Cell Global Identity (CGI). The S-TMSI can also be included. The eNode B proceeds to forward the Initial UE Message to the new MME (this example assumes the TAU procedure involves a change of MME)
- ★ The new MME uses the old GUTI information from within the Tracking Area Update Request to derive the address of the old MME. The new MME then sends a Context Request message to the old MME. This message belongs to the GTP control plane protocol (GTPv2-C) and is specified within 3GPP TS 29.274. It is used to request user information from the old MME.
- ★ The new MME responds with a Context Response message which includes the requested UE information. The new MME finalises the 3-way handshake by sending a Context Acknowledge message to the old MME. This triggers the old MME to mark its gateway and Home Subscriber Server (HSS) information for the UE as invalid. The context itself is maintained and is not deleted at this stage
- ★ The new MME sends a Modify Bearer Request message to the Serving Gateway. This message also belongs to the GTP control plane protocol (GTPv2-C). It informs the Serving Gateway that the MME associated with the existing PDN connection has changed. Additional Modify Bearer Request messages are sent if the UE has multiple PDN connections. The Serving Gateway acknowledges the new information using the Modify Bearer Response message
- ★ The new MME then proceeds to send an Update Location Request message to the HSS. This message belongs to the Diameter protocol and is specified within 3GPP TS 29.272. The MME is connected to the HSS using the S6a interface. The Update Location Request identifies the UE by its IMSI and informs the HSS that the UE has moved to the new MME. The HSS contacts the old MME using the Cancel Location message which also uses the Diameter protocol. The old MME then deletes the context for the UE and responds with a Cancel Location Acknowledge. Finally, the HSS uses the Update Location Acknowledge message to confirm to the new MME that is has updated its records for the UE
- ★ The new MME then progresses the TAU procedure by sending a Tracking Area Update Accept message to the UE. This message is transferred across the S1 interface using an S1AP: Downlink NAS Transport message. The content of the Tracking Area Update Accept message is shown in Table 303

Information Elements	Presence
Protocol Discriminator	*Mandatory*
Security Header Type	*Mandatory*
Tracking Area Update Accept Message Identity	*Mandatory*
EPS Update Result	Mandatory
T3412 Value	Optional
GUTI	Optional
TAI List	Optional
EPS Bearer Context Status	Optional
Location Area Identification	Optional
MS Identity	Optional
EMM Cause	Optional
T3402 Value	Optional
T3423 Value	Optional
Equivalent PLMN	Optional
Emergency Number List	Optional
EPS Network Feature Support	Optional
Additional Update Result	Optional
T3412 Extended Value	Optional

Table 303 – Content of Tracking Area Update Accept message

- ★ The first 3 information elements within the Tracking Area Update Accept message represent the message header. The Protocol Discriminator indicates that the subsequent message is an EPS mobility management message. Its value is specified within 3GPP TS 24.007. The Security Header Type indicates whether or not the NAS message is integrity protected and ciphered. The Tracking Area Update Accept Message Identity indicates that the message itself is an Tracking Area Update Accept

- ★ The EPS Update Result can be allocated values of 'TA Updated', 'Combined TA/LA Updated', 'TA Updated and ISR Activated' and 'Combined TA/LA updated and ISR Activated'. Idle mode Signalling Reduction (ISR) is applicable to E-UTRAN UE which support GERAN and/or UTRAN. It allows a reduction of the signalling between the UE and the network, as well as a reduction of the signalling within the network
- ★ The MME starts timer T3450 after sending the Tracking Area Update Accept message. 3GPP has standardised a fixed value of 6 s for the T3450 timer. If T3450 expires prior to the MME receiving a response from the UE, then the Tracking Area Update Accept message is re-sent. The Tracking Area Update Accept message can be re-sent four times. The overall TAU procedure is aborted if T3450 expires a fifth time
- ★ The eNode B extracts the Tracking Area Update Accept from within the S1AP message and packages it within an RRC: Downlink Information Transfer message. This message is then sent to the UE across the air-interface
- ★ The UE completes the overall procedure by sending a Tracking Area Update Complete message to the MME. This message is transferred across the air-interface using an RRC: Uplink Information Transfer message. The content of the Tracking Area Update Complete message is shown in Table 304

Information Elements	Presence
Protocol Discriminator	*Mandatory*
Security Header Type	*Mandatory*
Tracking Area Update Complete Message Identity	*Mandatory*

Table 304 – Content of Tracking Area Update Complete message

- ★ The Tracking Area Update Complete message is a simple acknowledgement which only includes header information, i.e. Protocol Discriminator, Security Header Type and Message Identity
- ★ The eNode B extracts the Tracking Area Update Complete from within the RRC message and packages it within an S1AP: Uplink NAS Transport message. This message is then sent to the new MME across the S1 interface
- ★ 3GPP References: TS 24.301, TS 36.413, TS 36.331

26.5 INTRA-SYSTEM HANDOVER

★ The intra-system handover procedure is applicable to UE in the ECM-CONNECTED state. UE trigger handovers as they move from one cell to another. Handovers are UE assisted but network controlled, i.e. UE provide measurements which help the eNode B to make handover decisions

★ LTE does not support soft handover so all handovers are hard handovers. Data is forwarded from the source eNode B to the target eNode B during the handover procedure to ensure that handovers are lossless

★ The intra-system handover procedure may involve a change of eNode B and/or a change of Serving Gateway

★ The signalling associated with the intra-system handover procedure is illustrated in Figure 240. This example includes a change of eNode B but maintains the same Serving Gateway. It is referred to as an X2-based handover because it uses the X2 interface between neighbouring eNode B for both control plane signalling and user plane data transfer

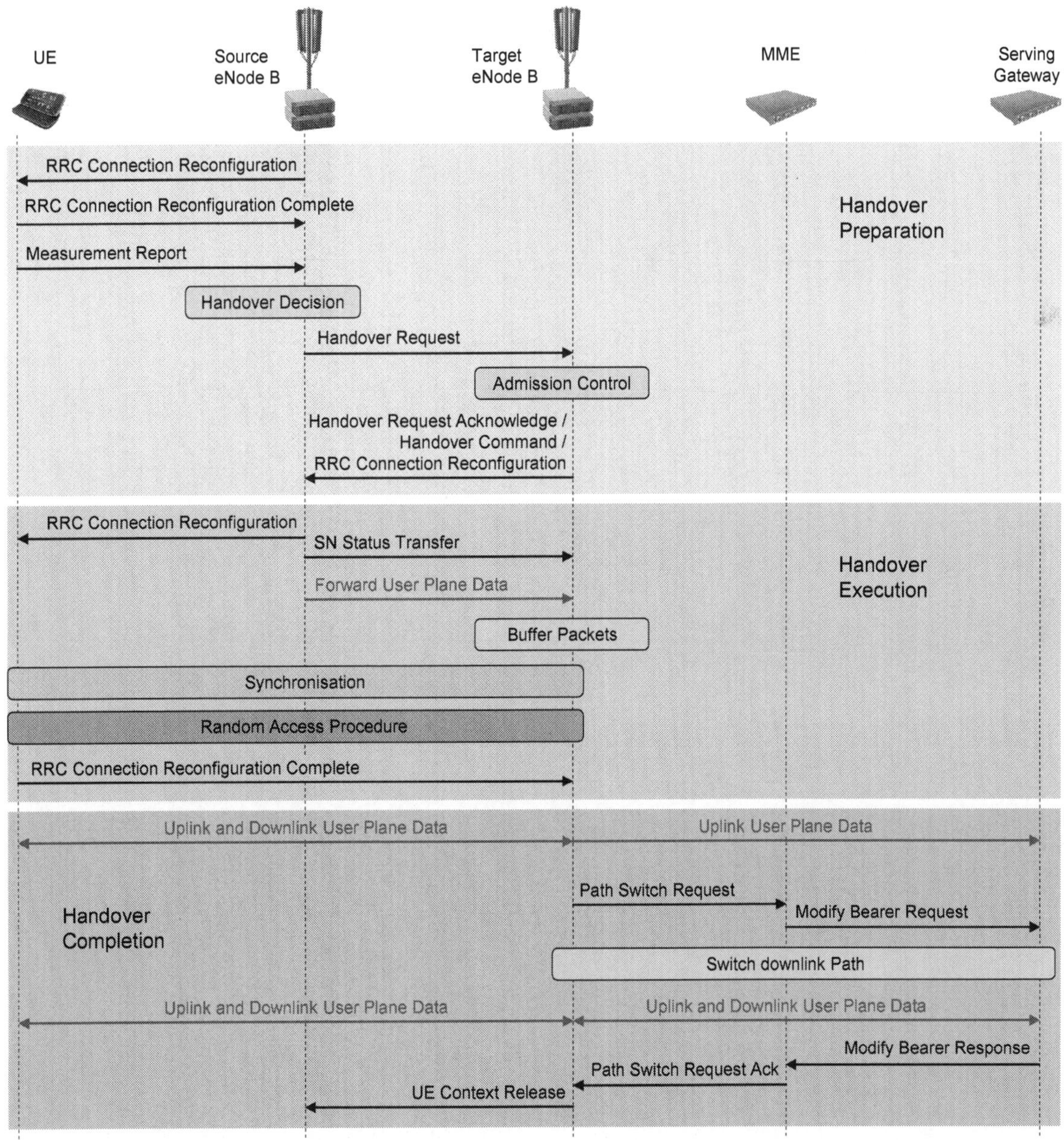

Figure 240 – Signalling for the intra-system handover procedure (with change of eNode B)

★ The overall handover procedure is divided into 3 phases: handover preparation, handover execution and handover completion. Handover preparation and handover execution involve only the radio access network. Handover completion involves both the core network and radio access network

★ Handover preparation starts with configuring the UE to send Measurement Report messages. The UE can be configured to send these periodically, or based upon one or more measurement reporting events. The RRC Connection Reconfiguration message is used to configure the UE to send Measurement Report messages. This message replaces the Measurement Control message used by UMTS. Intra-frequency measurement reporting events are described in section 22. Events A3 and A5 are the most likely to be used for triggering intra-frequency handovers

- event A3 – neighbouring cell becomes x dB better than the serving cell
- event A5 – serving cell becomes worse than threshold 1 and neighbouring cell becomes better than threshold 2

★ The UE sends a Measurement Report to the eNode B when one of its reporting events is triggered, or when the periodic reporting timer expires if periodic reporting has been configured. The complete content of a Measurement Report is shown in Table 305. Many of the items are optional and in practice, only a subset of the content is included

<table>
<tr><th colspan="5">Information Elements</th></tr>
<tr><td colspan="5">Measurement Identity</td></tr>
<tr><td colspan="2" rowspan="2">Primary Serving Cell Measurement Results</td><td colspan="3">RSRP</td></tr>
<tr><td colspan="3">RSRQ</td></tr>
<tr><td rowspan="27">Neighbouring Cell Measurement Results</td><td rowspan="8">E-UTRA Result List</td><td colspan="3">Physical Cell Identity</td></tr>
<tr><td rowspan="3">CGI Information</td><td colspan="2">Cell Global Identity</td></tr>
<tr><td colspan="2">Tracking Area Code</td></tr>
<tr><td colspan="2">PLMN Identity List</td></tr>
<tr><td rowspan="2">Measurement Results</td><td colspan="2">RSRP</td></tr>
<tr><td colspan="2">RSRQ</td></tr>
<tr><td rowspan="2">Additional System Information</td><td colspan="2">CSG Membership Status</td></tr>
<tr><td colspan="2">CSG Identity</td></tr>
<tr><td rowspan="9">UTRA Result List</td><td colspan="3">Physical Cell Identity</td></tr>
<tr><td rowspan="4">CGI Information</td><td colspan="2">Cell Global Identity</td></tr>
<tr><td colspan="2">Location Area Code</td></tr>
<tr><td colspan="2">Routing Area Code</td></tr>
<tr><td colspan="2">PLMN Identity List</td></tr>
<tr><td rowspan="2">Measurement Results</td><td colspan="2">CPICH RSCP</td></tr>
<tr><td colspan="2">CPICH Ec/Io</td></tr>
<tr><td rowspan="2">Additional System Information</td><td colspan="2">CSG Membership Status</td></tr>
<tr><td colspan="2">CSG Identity</td></tr>
<tr><td rowspan="5">GERAN Result List</td><td colspan="3">Carrier Frequency</td></tr>
<tr><td colspan="3">Physical Cell Identity</td></tr>
<tr><td rowspan="2">CGI Information</td><td colspan="2">Cell Global Identity</td></tr>
<tr><td colspan="2">Routing Area Code</td></tr>
<tr><td>Measurement Results</td><td colspan="2">RSSI</td></tr>
<tr><td rowspan="5">CDMA2000 Results</td><td colspan="3">HRPD Pre-Registration Status</td></tr>
<tr><td rowspan="4">CDMA2000 Result List</td><td colspan="2">Physical Cell Identity</td></tr>
<tr><td colspan="2">CGI Information</td></tr>
<tr><td rowspan="2">Measurement Results</td><td>Pilot PN Phase</td></tr>
<tr><td>Pilot Strength</td></tr>
<tr><td rowspan="2">Measurement Result for ECID</td><td colspan="4">UE Rx-Tx Time Difference Result</td></tr>
<tr><td colspan="4">Current SFN</td></tr>
<tr><td colspan="5">Location Information</td></tr>
<tr><td rowspan="6">Serving Frequency List Measurement Results</td><td>Serving Frequency Identity</td><td colspan="3">Serving Cell Index</td></tr>
<tr><td rowspan="2">Secondary Serving Cell Measurement Results</td><td colspan="3">RSRP</td></tr>
<tr><td colspan="3">RSRQ</td></tr>
<tr><td rowspan="3">Best Neighbour Cell Measurement Results</td><td colspan="3">Physical Cell Identity</td></tr>
<tr><td colspan="3">RSRP</td></tr>
<tr><td colspan="3">RSRQ</td></tr>
</table>

Table 305 – Content of Measurement Report message

- ★ The Measurement Report identifies its triggering mechanism using the Measurement Identity. This identity links the message content to the configuration information within the RRC Connection Reconfiguration message. It does not explicitly indicate the triggering mechanism. For example, the Measurement Identity could be signalled as '4' and then measurement identity '4' within the RRC Connection Reconfiguration message references event A3
- ★ Serving cell measurements are always included within the Measurement Report. These measurements allow the eNode B to evaluate the coverage provided by the existing serving cell, and to compare it against the coverage reported for each of the neighbouring cells
- ★ In the case of intra-frequency handovers, the Measurement Report includes the E-UTRA list of measurement results but excludes the inter-system lists of measurement results, i.e. the UTRA, GERAN and CDMA2000 measurement results are excluded
- ★ The E-UTRA list of measurement results identifies each neighbouring cell by its Physical layer Cell Identity (PCI). The Cell Global Identity (CGI) information is typically excluded at this stage. This information is only included when specifically requested by the eNode B. It is not included by default because the UE has to decode the MIB and SIB to extract the CGI. This makes it relatively processor intensive compared to just recording Reference Signal measurements. The eNode B may request the CGI in a separate Measurement Report if the Automatic Neighbour Relation (ANR) procedure described in section 32.4 is triggered. The CGI information is not required if the neighbouring cell is already included within the eNode B database of neighbours, i.e. the eNode B can identify the neighbouring cell using its PCI
- ★ RSRP and/or RSRQ measurements from the neighbouring cells are reported to allow the eNode B to evaluate their coverage. If the eNode B can identify a candidate neighbouring cell from its PCI, and if the coverage conditions indicate that a handover should be completed, the eNode B initiates the handover procedure by sending an X2-AP: Handover Request message to the target eNode B. The content of this message is presented in Table 306

Information Elements			Presence
Message Type			*Mandatory*
Old eNode B UE X2AP ID			*Mandatory*
Cause			Mandatory
Target Cell Identity			Mandatory
GUMMEI			Mandatory
UE Context Information	MME UE S1AP ID		Mandatory
	UE Security Capabilities		Mandatory
	AS Security Information		Mandatory
	UE Aggregate Maximum Bit Rate		Mandatory
	Subscriber Profile Id for RAT/Frequency Priority		Optional
	E-RAB Setup List	E-RAB Identity	Mandatory
		E-RAB Level QoS Parameters	Mandatory
		Downlink Forwarding	Optional
		Uplink GTP Tunnel Endpoint	Mandatory
	RRC Context		Mandatory
	Handover Restriction List		Optional
	Location Reporting Information		Optional
	Management based MDT Allowed		Optional
UE History Information			Mandatory
Trace Activation			Optional
SRVCC Operation Possible			Optional
CSG Membership Status			Optional

Table 306 – Content of the X2-AP: Handover Request message

- ★ The Cause value within the X2-AP: Handover Request is typically configured as 'Handover Desirable for Radio Reasons'. It can also be configured as 'Time Critical Handover', 'Resource Optimisation Handover' or 'Reduce Load in Serving Cell'. The time critical cause indicates that the connection is likely to drop if a handover is not completed. The resource optimisation cause indicates that the serving eNode B is attempting to balance the network load between itself and its neighbours
- ★ The Target Cell Identity is signalled as the E-UTRAN CGI. This information is extracted from the eNode B database of neighbours using the PCI as a reference. The Automatic Neighbour Relation (ANR) procedure is triggered if the PCI and the corresponding CGI cannot be found within the database (assuming the network supports the ANR procedure)
- ★ The Globally Unique MME Identity (GUMMEI) is included to inform the target eNode B of which MME hosts the UE connection. This allows the target eNode B to address the S1-AP 'Path Switch Request' message to the appropriate MME during a later stage of the handover procedure

★ UE Context Information informs the target eNode B of which bearers need to be established to support the incoming handover request. The set of E-RAB belonging to the UE are specified along with their Quality of Service (QoS) requirements. The uplink GTP Tunnel Endpoint is also included to allow the target eNode B to send uplink user plane data across the S1 interface towards the appropriate termination point at the Serving Gateway

★ The target eNode B applies its admission control after receiving the X2-AP: Handover Request. Assuming a positive outcome, the target eNode B returns an X2-AP: Handover Request Acknowledge message. The content of this message is presented in Table 307

Information Elements		Presence
Message Type		*Mandatory*
Old eNode B UE X2AP ID		*Mandatory*
New eNode B UE X2AP ID		*Mandatory*
E-RAB Admitted List	E-RAB Identity	Mandatory
	Uplink GTP Tunnel Endpoint	Optional
	Downlink GTP Tunnel Endpoint	Optional
E-RAB Not Admitted List		Optional
Target eNode B to Source eNode B Transparent Container		Mandatory
Criticality Diagnostics		Optional

Table 307 – Content of the X2-AP: Handover Request Acknowledge message

★ The target eNode B uses the Handover Request Acknowledge message to inform the source eNode B of which E-RAB have been admitted. GTP tunnel endpoints are specified to allow the source eNode B to forward uplink and downlink user plane data across the X2 interface. The target eNode B can also specify any E-RAB which have not been admitted

★ The Target eNode B to Source eNode B Transparent Container includes a Handover Command message. The Handover Command message encapsulates an RRC Connection Reconfiguration message which has been generated by the target eNode B. The RRC Connection Reconfiguration message is sent to the source eNode B to be forwarded across the air-interface to the UE. This message provides the UE with instructions for completing the handover onto the target cell

★ The source eNode B extracts the RRC Connection Reconfiguration message from the X2-AP message and initiates the Handover Execution phase by forwarding it across the air-interface to the UE

★ The RRC Connection Reconfiguration message includes a section which specifies mobility control information. This section of the message is presented in Table 308

Information Elements	
Target Physical Cell Identity	
Carrier Frequency	Uplink Carrier Frequency
	Downlink Carrier Frequency
Carrier Bandwidth	
Additional Spectrum Emission Requirements	
T304	
New UE Identity	C-RNTI
Common Radio Resource Configuration	RACH Configuration
	PRACH Configuration
	PDSCH Configuration
	PUSCH Configuration
	PHICH Configuration
	PUCCH Configuration
	Sounding Reference Signal Configuration
	Uplink Power Control
	Antenna Information
	Pmax
	Uplink Cyclic Prefix Length
Dedicated RACH Configuration	Random Access Preamble Index
	Random Access PRACH Mask Index

Table 308 – Content of Mobility Control Information from within an RRC Connection Reconfiguration message

★ The Physical layer Cell Identity (PCI) is specified to direct the UE onto the appropriate cell. The uplink and downlink RF carriers are specified in terms of their E-UTRAN ARFCN. The channel bandwidth is also specified in combination with any additional spectrum emission requirements

★ The T304 timer is specified using a value of 50, 100, 150, 200, 500, 1000 or 2000 ms. The UE starts this timer upon receiving the mobility control information. If T304 expires prior to the UE successfully completing the random access procedure on the target cell, the UE initiates the RRC Connection Re-establishment procedure

★ The UE is allocated a new C-RNTI by the target eNode B. This C-RNTI is used to address the UE within the target cell

★ The Common Radio Resource Configuration provides information regarding various transport and physical channels at the target cell. The maximum uplink transmit power and uplink cyclic prefix length are also specified

★ The dedicated RACH Configuration provides support for allowing the UE to complete the non-contention based Random Access procedure. The UE is allocated a dedicated preamble sequence rather than allowing it to select a sequence at random from the pool of common sequences. The PRACH mask defines the subframe during which the UE is allowed to use the dedicated preamble sequence This approach helps to improve the reliability of the overall handover procedure

★ The source eNode B proceeds to send an X2-AP: Sequence Number (SN) Status Transfer message to the target eNode B. This message provides the target eNode B with uplink and downlink PDCP sequence number information for E-RAB using acknowledged mode RLC. This helps to ensure that sequence numbering is preserved during the handover and that the handover is lossless. PDCP sequence numbers for unacknowledged mode RLC are reset in the target eNode B. The content of the SN Status Transfer message is shown in Table 309

Information Elements		Presence
Message Type		*Mandatory*
Old eNode B UE X2AP ID		*Mandatory*
New eNode B UE X2AP ID		*Mandatory*
E-RAB Subject to Status Transfer List	E-RAB Identity	Mandatory
	Receive Status of Uplink PDCP SDU	Optional
	Uplink COUNT value	Mandatory
	Downlink COUNT value	Mandatory

Table 309 – Content of the X2-AP: SN Status Transfer message

★ The optional Receive Status of Uplink PDCP SDU information defines a bitmap to indicate which uplink SDU have been successfully received, and which have not been received

★ The Uplink COUNT value defines the PDCP sequence number and Hyper Frame Number (HFN) of the first missing uplink SDU. The Downlink COUNT value defines the PDCP sequence number and HFN that the target eNode B should allocate to the next downlink SDU that has not already been allocated a sequence number

★ The source eNode B starts forwarding user plane data across the X2 interface to the target eNode B after receiving the X2-AP: Handover Request Acknowledge message. Forwarding makes use of the X2 uplink and downlink GTP Tunnel Endpoints specified within the Handover Request Acknowledge message. At this point in time, the target eNode B buffers the packets without forwarding

★ The UE proceeds to achieve air-interface synchronisation with the target cell using the Primary and Secondary Synchronisation Signals. The UE then initiates the random access procedure which is non-contention based if the target eNode B included Dedicated RACH Configuration information within the Mobility Control section of the RRC Connection Reconfiguration message. The random access procedure is contention based if the Dedicated RACH Configuration information was excluded

★ The Random Access Response (RAR) message provides the UE with timing advance information for the target cell. It also provides an uplink resource allocation for the initial layer 3 message. In this case, the initial layer 3 message is the RRC Connection Reconfiguration Complete message. The UE completes the handover procedure from the UE perspective by sending this message

★ Both uplink and downlink data can now be transferred between the UE and target eNode B. This represents the start of the Handover Completion phase. Downlink data from the Serving Gateway continues to be sent to the source eNode B, from where it is forwarded to the target eNode B across the X2 interface. The target eNode B can forward uplink data to the Serving Gateway using the S1 uplink GTP Tunnel Endpoint specified within the X2-AP: Handover Request message

★ After receiving the RRC Connection Reconfiguration Complete message, the target eNode B initiates the procedure for switching the downlink GTP tunnel from the source eNode B to the target eNode B. This is done by sending an S1-AP Path Switch Request to the MME. The content of this message is presented in Table 310

★ The Path Switch Request specifies both the transport layer address and the GTP Tunnel Endpoint Identity (TEID) which are to be used when forwarding downlink data to the target eNode B. The MME is also informed of the Tracking Area Identity (TAI) and Cell Global Identity (CGI) belonging to the cell with which the UE is now connected

Information Elements		Presence
Message Type		*Mandatory*
eNode B UE S1AP ID		*Mandatory*
E-RAB to be Switched in Downlink List	E-RAB Identity	Mandatory
	Transport Layer Address	Mandatory
	GTP-TEID	Mandatory
Source MME UE S1AP ID		Mandatory
E-UTRAN CGI		Mandatory
TAI		Mandatory
UE Security Capabilities		Mandatory
CSG Identity		*Optional*
Cell Access Mode		*Optional*
Source MME GUMMEI		*Optional*

Table 310 – Content of the S1-AP Path Switch Request message

- ★ The MME decides whether or not the existing Serving Gateway can continue providing services to the UE. The example shown in Figure 240 assumes there is no requirement to change Serving Gateway. If the Serving Gateway needs to be changed, the MME forwards a GTPv2-C Create Session Request message to the new Serving Gateway, and the new Serving Gateway contacts the PDN Gateway to request re-direction of the existing GTP tunnel
- ★ Assuming the Serving Gateway remains unchanged, the MME forwards a GTPv2-C Modify Bearer Request message to the existing Serving Gateway. This message instructs the Serving Gateway to start forwarding downlink data towards the target eNode B. The Serving Gateway proceeds to switch the downlink path, and at this point both uplink and downlink data is transferred between the UE and Serving Gateway without any forwarding via the source eNode B
- ★ The Serving Gateway acknowledges the switch to the MME using a GTPv2-C Modify Bearer Response message. The MME then acknowledges the switch to the target eNode B using an S1-AP Path Switch Request Acknowledge message
- ★ The target eNode B completes the handover procedure by sending an X2-AP: UE Context Release message to the source eNode B. This X2-AP message informs the source eNode B that it can release its resources for the UE which has now moved across to the target eNode B
- ★ 3GPP References: TS 36.300, TS 23.401, TS 36.331, TS 36.423, TS 29.274

27 VOICE SERVICES

27.1 INTRODUCTION

- Voice is an important service which generates significant revenues for mobile network operators and service providers
- Voice services have traditionally been provided using the Circuit Switched (CS) core network domain
- LTE does not have a CS core network domain so requires a different approach
- The main options for LTE voice services are summarised in Figure 241

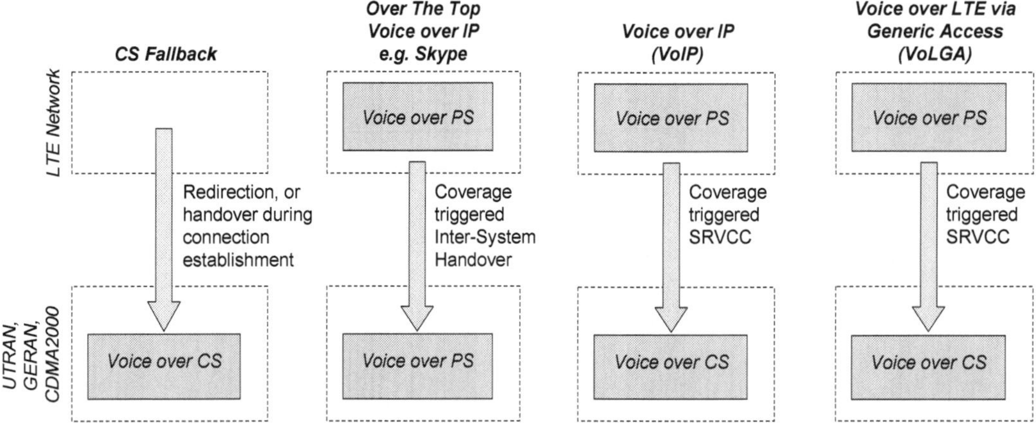

Figure 241 – Voice service options with LTE

- CS Fallback involves moving the UE onto a Radio Access Technology (RAT) with a CS core network domain during connection establishment. CS Fallback is described in greater detail in section 27.2. The main characteristics are:
  - it allows the re-use of a legacy network, which could be either UMTS, GSM or CDMA2000
  - it is simple to implement but requires support from both the network and UE. In the case of UMTS and GSM, it requires the introduction of the SGs interface between the MME and MSC/VLR
  - the transition to a different RAT can have an impact upon the connection establishment delay
  - the UE registers with both the LTE PS core network domain, and the legacy network CS core network domain
  - it helps to retain the revenue stream from CS voice services
  - it is viewed as a short to medium term solution
- Over The Top (OTT) software clients such as 'Skype' and 'Google Voice' use normal data connections to support Voice over IP (VoIP) services. The main characteristics are:
  - they does not require any specific support from the end-user device nor the network
  - software clients are widely available and relatively well known
  - they do not benefit from being prioritised by procedures using Quality of Service (QoS) unless Deep Packet Inspection (DPI) identifies the voice traffic and prioritises it
  - they are free to use from the end-user perspective as long as data quotas are not exceeded
  - they erode the operator's revenue stream from CS voice services. Some operators may use DPI to block traffic originating from OTT speech services
- Voice over IP (VoIP):
  - is viewed as the longer term solution
  - requires support from both the end-user device and the network
  - connections are managed using the Session Initiation Protocol (SIP) with an IP Multimedia Subsystem (IMS)
  - separate data bearers with different QoS are used for the SIP signalling and voice information
  - header compression is required to minimise the impact of the protocol stack overheads
  - services benefit from QoS to help provide carrier grade performance
  - allows the operator to retain greater control and to maintain a voice service revenue stream

★ Voice over LTE via Generic Access (VoLGA) use the LTE network in combination with a VoLGA Access Network Controller (VANC) to generate a tunnel between the UE and CS core network. VoLGA is described in greater detail in section 27.4. The main characteristics are:

  - it allows access to CS services while connected to the LTE network
  - it does not require any changes to the LTE network
  - it requires support from the UE and the introduction of the VANC network element
  - it increases utilisation of the LTE network and retains the revenue stream from CS voice services
  - it is standardised by the VoLGA Forum rather than by 3GPP

★ Single Radio Voice Call Continuity (SRVCC) provides mobility between Voice over IP (VoIP) and a circuit switched voice call. It is applicable to UE which have LTE VoIP connections and are moving towards the edge of LTE coverage. Those UE can be moved across to UTRAN, GERAN or CDMA2000, and their voice connections can be switched across to the CS core network domain. SRVCC is described in greater detail in section 27.5

★ The global Multimedia Telephony (MMTel) standard is applicable to VoIP over LTE. MMTEL defines an IMS service set which provides support for real time multimedia services. Video, chat and file sharing are supported in addition to voice. The MMTel standard is intended to help IP based voice services replace existing fixed and mobile CS domain voice services. The MMTel standard has been jointly developed by 3GPP and ETSI TISPAN. It allows interworking between operators and vendors

★ The GSM Association (GSMA) established the Voice over LTE (VoLTE) initiative to define a single globally adopted approach for delivering voice and messaging services over LTE. The initiative has adopted the work of the One Voice initiative which has significant support from mobile operators and network vendors. All participating organisations support the principle of a single IMS based voice solution. The motivation behind the GSMA VoLTE initiative is that defining a single solution will help to promote scale, reduce complexity and enable roaming. The GSMA have generated Permanent Reference Document IR92 "IMS Profile for Voice and SMS" to define a minimum mandatory set of 3GPP features which a mobile device and the network must support to allow an interoperable, high quality IMS based VoLTE service

27.2 CS FALLBACK

- CS Fallback is specified by 3GPP within TS 23.272
- CS Fallback involves moving the UE away from LTE and onto a mobile network with a circuit switched core network domain at the time of initiating a mobile originated or mobile terminated voice call
- CS Fallback requires the introduction of the SGs interface illustrated in Figure 242. This interface connects the MME (LTE PS core network) to the MSC (2G and 3G CS core network). The MSC can use this interface to forward CS core network domain paging messages to the UE on LTE

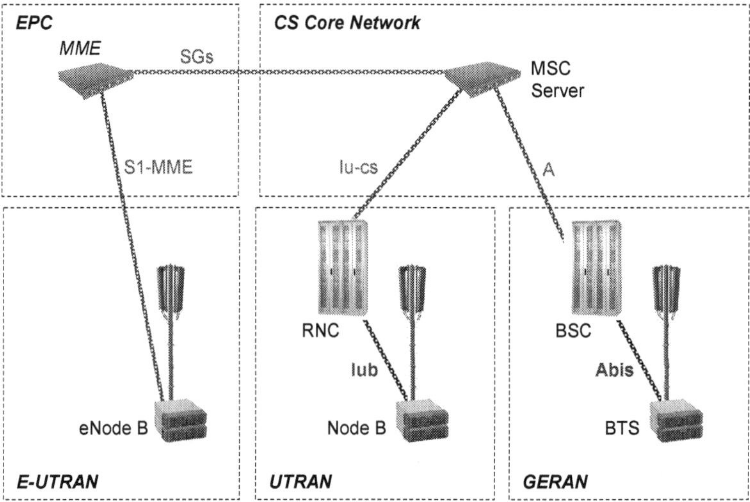

Figure 242 – Network architecture for CS Fallback

- Signalling across the SGs interface is based upon the SGs Application Protocol (SGsAP). The SGs interface and its associated signalling messages are specified within 3GPP TS 29.118
- A UE using CS Fallback must be registered with both the EPC and CS core network domains. This is achieved by completing a combined 'EPS/IMSI' Attach procedure from the LTE network. This signalling procedure is illustrated in Figure 243

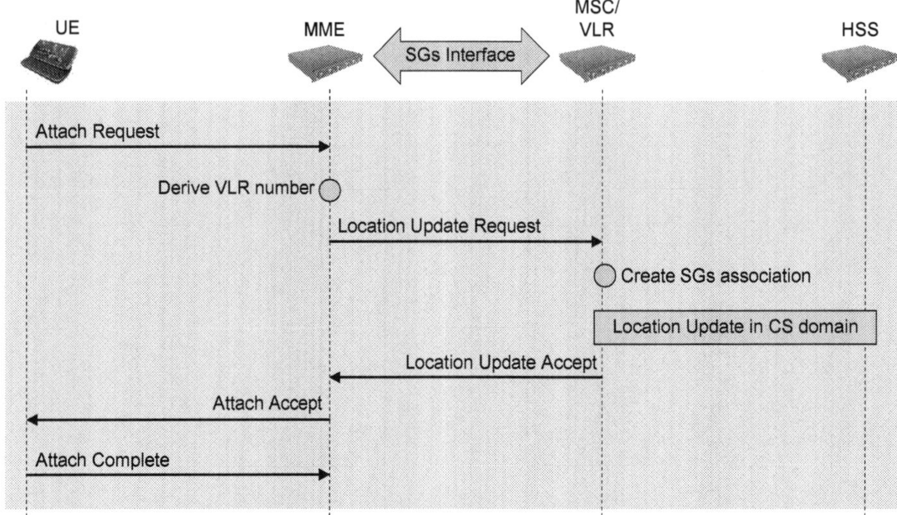

Figure 243 – Combined EPS/IMSI Attach procedure

- The UE sends a Non-Access Stratum (NAS) Attach Request message to the MME as if it was initiating a normal Attach procedure. The content of the Attach Request message is specified within 3GPP TS 24.301. CS Fallback requires that the Attach message is sent with:
  - the EPS Attach Type set to 'Combined EPS/IMSI Attach'
  - the EMM Combined Procedures Capability within the MS Network Capability set to 'Mobile Station supports EMM Combined Procedures'

★ The MME then recognises that it must register the UE with the MSC within the CS core network domain. The MME derives a VLR number and Location Area Identity (LAI) from the Tracking Area Identity (TAI) of the current cell. Multiple MSC/VLR can serve the same LAI so the VLR number is used to uniquely identify the MSC/VLR to be used

★ The MME then forwards an SGsAP: Location Update Request message to the target MSC/VLR. The content of this Location Update Request message is shown in Table 311. The EPS Location Update Type can be allocated values of either 'IMSI Attach' or 'Normal Location Update'

Information Element	Inclusion
Message Type	Mandatory
IMSI	Mandatory
MME Name	Mandatory
EPS Location Update Type	Mandatory
New Location Area Identity	Mandatory
Old Location Area Identity	Optional
TMSI Status	Optional
IMEISV	Optional

Table 311 – Content of SGsAP: Location Update Request message

★ The MSC/VLR creates an association with the MME by storing the MME name. It then proceeds by completing normal location update procedures within the CS core network to register the UE. This includes checking subscription information with the Home Subscriber Server (HSS)

★ The MSC/VLR responds to the MME with an SGsAP: Location Update Accept message. This message can allocate the UE with a TMSI, or instruct the UE to use its IMSI as an identity. If neither are included then the UE continues to use its existing identity. The content of the SGsAP: Location Update Accept message is shown in Table 312

Information Element	Inclusion
Message Type	Mandatory
IMSI	Mandatory
Location Area Identity	Mandatory
New TMSI, or IMSI	Optional

Table 312 – Content of SGsAP: Location Update Accept message

★ The MME responds to the UE with a NAS: Attach Accept message before the UE acknowledges with a NAS: Attach Complete message. The UE is then able to use the CS Fallback procedure

★ There are multiple scenarios for the CS Fallback procedure:

  o mobile originated or mobile terminated

  o starting from Idle mode or Connected mode

  o using an inter-system handover or a redirection instruction

These possibilities result in a total of 8 different scenarios. The inter-system handover possibility is applicable to the idle mode scenario as well as the connected mode scenario because UE in Idle mode are moved into connected mode prior to redirection

★ The signalling for a mobile terminated CS Fallback procedure starting from Idle mode and using a redirection instruction is illustrated in Figure 244

★ The MME receives a SGsAP: Paging Request message from the MSC/VLR. This message includes the IMSI of the UE to be paged. It also includes a Service Indicator to differentiate between a paging request for an SMS and a paging request for a CS call. The Paging Request may also contain the TMSI belonging to the UE and Calling Line Identification (CLI) information, i.e. the phone number of the calling device

★ The MME generates an S1AP: Paging message from the SGsAP: Paging Request. This message informs the eNode B that the paging message has originated from the CS core network domain. The MME uses the IMSI to identify the S-TMSI associated with the UE. Both the IMSI and S-TMSI are included within the S1AP: Paging message. The CLI information is not passed down to the eNode B within the S1AP: Paging message. The MME forwards the S1AP: Paging message to all eNode B with cells belonging to the appropriate Tracking Area

★ The eNode B generates an RRC: Paging message from the S1AP: Paging message to inform the UE about the incoming voice call. This paging message informs the UE that the incoming call is originating from the CS core network domain

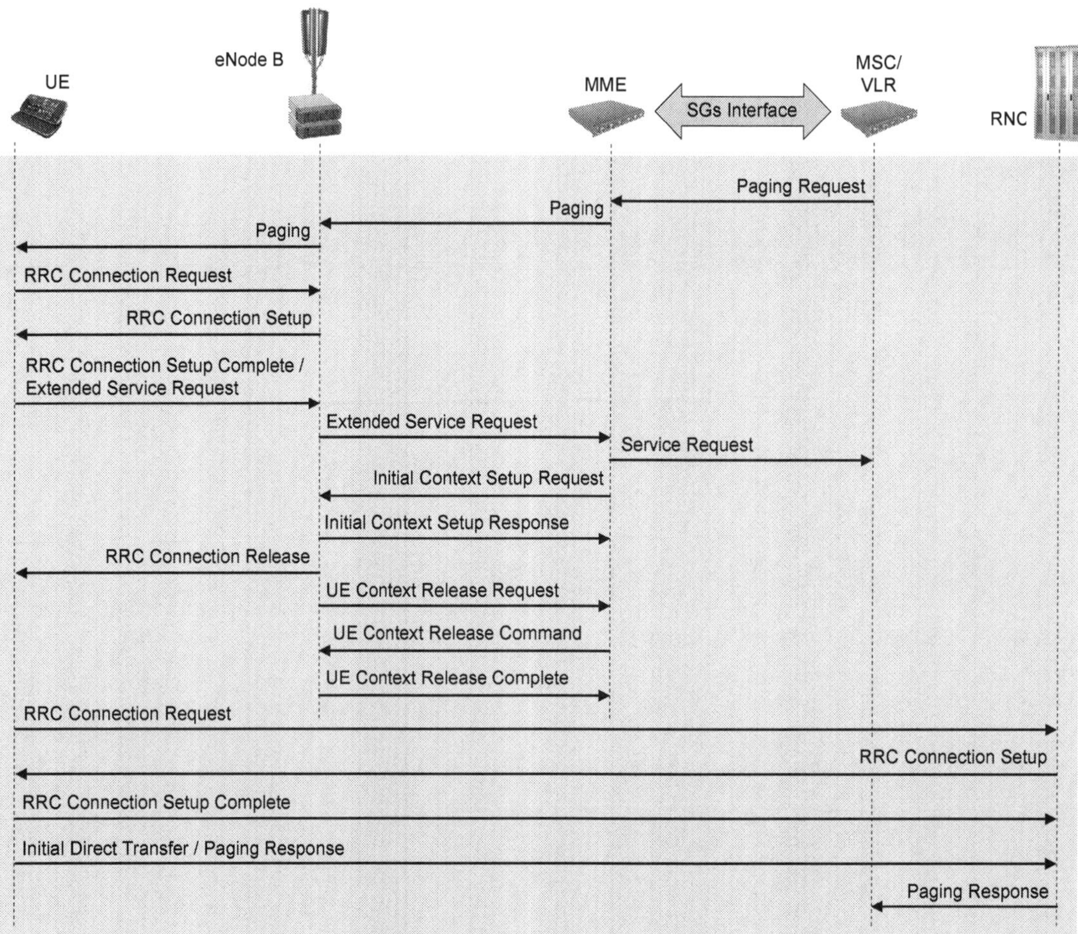

Figure 244 – Signalling for mobile terminated CS Fallback from Idle mode

★ The UE responds to the Paging message by establishing an RRC connection. The UE encapsulates a NAS: Extended Service Request message within the RRC Connection Setup Complete message. This message includes a CS Fallback Response field which informs the MME of whether or not the CS Fallback procedure has been accepted or rejected by the UE. The eNode B forwards the Extended Service Request to the MME within the S1AP: Initial UE Message

★ The MME sends an SGsAP: Service Request message to the MSC/VLR. This message includes a UE EMM Mode field which is used to inform the MSC/VLR of the EMM state of the UE when the SGsAP: Paging Request was received. In this case the UE was in EMM Idle mode, and this allows the MSC/VLR to deduce that the previously sent CLI information was not forwarded to the UE. Receipt of the Service Request message prevents the MSC/VLR from re-transmitting the Paging Request

★ The MME also sends an S1AP: Initial Context Setup Request message to the eNode B. This message includes a CS Fallback Indicator which instructs the eNode B to move the UE across to UTRAN/GERAN. A Registered Location Area Identity (LAI) can also be included to specify the registered PLMN identity for the UE (PLMN identity forms part of the LAI identity). The eNode B acknowledges the Initial Context Setup Request message using an Initial Context Setup Response message

★ The eNode B then forwards an RRC Connection Release message to the UE. This message includes the redirection information instructing the UE to move across to UTRAN/GERAN. The content of the RRC Connection Release message is presented in Table 313

★ In general, the Release Cause can be set to either 'Load Balancing Tracking Area Update Required', 'Other' or 'CS Fallback High Priority'. The eNode B uses the 'CS Fallback High Priority' value if the MME had previously signalled a value of 'CS Fallback High Priority' within the S1AP: Initial Context Setup Request and the target Radio Access Technology (RAT) is UTRA FDD or UTRA TDD. Otherwise, the release cause 'Other' is used

★ The eNode B selects the target Radio Access Technology (RAT) and target RF carrier. In the case of GERAN, a group of RF carriers can be specified. The target RF carrier, or group of RF carriers is included within the RRC Connection Release message using the Redirected Carrier Information

★ Cell reselection priorities can be signalled for various RF carriers using the Idle Mode Mobility Control Information. These priorities are dedicated to the UE rather than common to all UE. The RRC Connection Release message can also define the T320 timer. This timer is started as soon as the release message is received by the UE. The priorities within the RRC Connection Release message are deleted after T320 expires. T320 can be allocated values between 5 minutes and 3 hours

Information Elements		
Release Cause		
Redirected Carrier Information	CHOICE	
	E-UTRA ARFCN	
	GERAN Carrier Frequencies	
	FDD UTRA ARFCN	
	TDD UTRA ARFCN	
	HRPD CDMA2000 Carrier Frequency	
	1xRTT CDMA2000 Carrier Frequency	
Idle Mode Mobility Control Information	E-UTRA Frequency Priority List	
	GERAN Frequency Priority List	
	FDD UTRA Frequency Priority List	
	TDD UTRA Frequency Priority List	
	HRPD CDMA2000 Band Class Priority List	
	1xRTT CDMA2000 Band Class Priority List	
	T320	
Cell Information List	GERAN Cell Information List	LIST
		Physical Cell Identity
		Carrier Frequency
		System Information List
	FDD UTRA Cell Information List	LIST
		Physical Cell Identity
		BCCH Container
	TDD UTRA Cell Information List	LIST
		Physical Cell Identity
		BCCH Container
Extended Wait Time		

Table 313 – Content of RRC Connection Release message

★ The Cell Information list allows the eNode B to provide the UE with system information from the target cell. This allows the UE to access the target cell without having to wait while the system information is read from the target cell BCCH, i.e. the inclusion of this information helps to reduce delay

★ Reception of system information within the RRC Connection Release message is an optional UE capability (see Table 213 in section 19.2). It is known as Enhanced Redirection to UTRA or Enhanced Redirection to GERAN, and was introduced within the release 9 version of the 3GPP specifications

★ An Extended Wait Time may be included within the RRC Connection Release message for delay tolerant connections. It can be configured with values between 1 second and 30 minutes

★ Returning to Figure 244, after sending the RRC Connection Release message, the eNode B releases the UE context using a 3-way handshake with the MME

★ The UE moves across to the target RF carrier (assumed to be UTRAN in Figure 244) and establishes an RRC connection. An RRC Connection Request cause value of 'Terminating High Priority Signalling' is used if the eNode B had previously indicated 'CS Fallback High Priority' within the RRC Connection Release message

★ The UE can include a flag within the RRC Connection Setup Complete message to indicate that it has applied 'Deferred Measurement Control Reading'. This informs the RNC that the UE has not waited to read SIB11, SIB11bis and SIB12 before establishing its RRC connection, i.e. the UE requires the RNC to provide neighbour cell information using dedicated RRC signalling

★ The UE then forwards a NAS: Paging Response message to the RNC using an RRC: Initial Direct Transfer message. The RNC extracts the Paging Response and forwards to the MSC/VLR. This acknowledges the original Paging Request sent to the MME. Normal voice call connection establishment signalling proceeds from this point

★ At the end of the speech call, the UE can either remain on the current RAT, or return to LTE using normal cell reselection procedures

★ The CS Fallback procedure for a UE in Connected mode is similar to that for a UE in Idle mode. The main difference is that the MME generates a NAS: CS Service Notification message instead of a Paging message after receiving the Paging Request from the

MSC/VLR. This message can include the Calling Line Identification (CLI) information provided by the MSC/VLR (in contrast to the Paging message which excludes it). Providing the UE with this information allows the end-user to decide whether or not to accept the call, while the end-user is potentially in the middle of using a data service

★ The CS Fallback procedure for a mobile originated connection is similar to that for a mobile terminated connection. The paging procedure does not exist for a mobile originated connection, i.e. the UE sends the Extended Service Request without having to be paged. Then when the UE has reached the target RAT, it sends a CM Service Request to the MSC instead of a Paging Response

★ The CS Fallback procedure based upon handover is similar to that based upon redirection. The main difference is that the redirection instruction within the RRC Connection Release message is replaced by a PS inter-system handover procedure. The handover procedure is PS rather than CS because the speech connection has not yet been established and the UE is connected to the PS core network domain while on LTE. The handover procedure can include inter-system measurements to identify the best target cell

★ The impact of CS Fallback upon the connection establishment delay depends upon the scenario. Mobile originating call setup time can be increased by 1 to 2 seconds. Mobile terminating call setup time is not impacted when starting from Idle mode because the UE does not start to ring until the UE is on the target RAT, i.e. the initial signalling is transparent to the end-user. The impact upon mobile terminating call setup time when starting from Connected mode depends upon the user-interface. The UE is likely to notify the end-user of the incoming call while the UE is still on LTE in case the end-user is in the middle of an active data session

★ Enhanced Redirection to UTRA or Enhanced Redirection to GERAN can be used to reduce connection establishment delays by using dedicated signalling to provide the UE with system information belonging to the target RAT. In the case of CDMA2000, Enhanced CS Fallback to CDMA2000 1xRTT can be used to avoid the requirement for an RRC connection release with re-direction. The enhanced CS fallback procedure tunnels handover signalling between the UE and the 1xRTT network. This allows the transition between radio access technologies to be handled as a handover rather than a release with re-direction

★ 3GPP References: TS 23.272, TS 24.301, TS 29.118

27.3 VOICE OVER IP (VoIP)

★ Voice over IP (VoIP) is viewed as the long term solution for voice services over fixed and mobile networks

★ VoIP can be provided using:

  o Over The Top (OTT) software clients such as 'Skype' and 'Google Voice' which use normal data connections to transfer voice information. This type of service does not require any specific support from the end-user device nor the network. These services are less likely to benefit from Quality of Service (QoS). They tend to reduce the operator's voice service revenue stream and the operator has relatively little control over them

  o a VoIP specific protocol stack with VoIP specific signalling and speech bearers. This type of service requires support from both the end-user device and the network. Connections are managed using the Session Initiation Protocol (SIP) with an IP Multimedia Subsystem (IMS). These VoIP services benefit from QoS to help provide carrier grade performance. They allow the operator to retain greater control and to maintain a voice service revenue stream

This section focuses upon the second of these two options for VoIP services

★ VoIP requires the implementation of Quality of Service (QoS) to help ensure carrier grade performance. 3GPP has specified the set of QoS Class Identities (QCI) introduced in section 2.10. QCI values 1 and 5 presented in Table 314 are directly applicable to VoIP

QCI	Resource Type	Priority	Packet Delay Budget	Packet Error Loss Rate	Services
1	GBR	2	100 ms	10^{-2}	Conversational Voice
5	Non-GBR	1	100 ms	10^{-6}	IMS Signalling

Table 314 – Quality of Service Class Identities (QCI) applicable to VoIP

★ QCI value 1 is applicable to the speech information itself, transferred using the RTP/UDP/IP protocol stack. It is also applicable to the associated Real time Transport Control Protocol (RTCP) signalling transferred using the UDP/IP protocol stack. RTCP partners with RTP during the transfer of speech

★ QCI value 1 defines a Guaranteed Bit Rate (GBR) bearer with a priority of 2. The GBR characteristic ensures that there is always sufficient bandwidth to transfer the speech information. The priority of 2 represents a high priority to help reduce packet queuing times. The specified packet delay budget of 100 ms is low enough to provide a high quality end-user experience

★ QCI value 5 is applicable to the IMS signalling used to establish and release the speech connection. This signalling is based upon the Session Initiation Protocol (SIP). It does not require the GBR characteristic but requires a high priority to ensure reliability, and a delay budget which is comparable to the speech information itself

★ The Data Radio Bearers (DRB) directly associated with a VoIP connection are summarised in Figure 245

- there is a first DRB for the combination of speech and RTCP signalling. This DRB uses Robust Header Compression (ROHC) within the PDCP layer, and unacknowledged mode RLC.
- there is a second DRB for SIP signalling to the IMS. This DRB uses acknowledged mode RLC but does not use ROHC

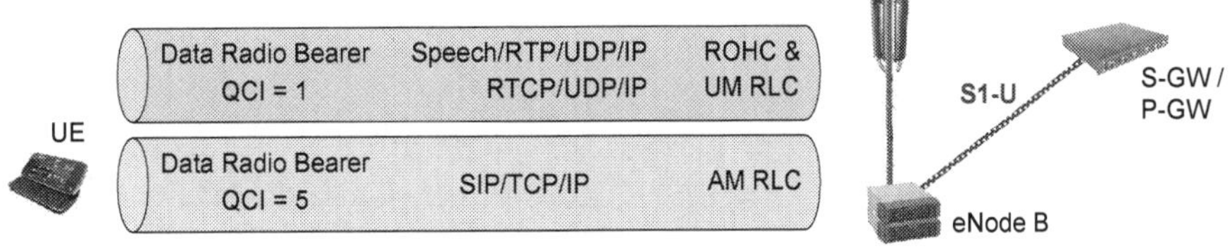

Figure 245 – Data bearers used by a VoIP service

★ In addition to the DRB shown in the Figure 245, the UE is likely to be configured with one or more DRB for data services, i.e. to allow parallel transfer of data during the VoIP connection. The UE will also be configured with a set of Signalling Radio Bearers (SRB) to support signalling to the eNode B and MME

★ VoIP requires protocol stack support from the UE. The UE signals its capability to the eNode B using the RRC: UE Capability Information message. This message includes a UE EUTRA Capability field which specifies:

- PDCP Parameters which indicate which Robust Header Compression (ROHC) profiles are supported, i.e. whether or not header compression is supported for the RTP/UDP/IP and UDP/IP protocol stacks
- a bit string of 3GPP release 8 Feature Group Indicators (FGI) (unless all of the associated features have been implemented and tested, in which case the set of FGI are not required and are excluded from the message):
  - bit 3 should be set to '1' to indicate UE support for 5 bit Unacknowledged Mode RLC sequence numbers, and 7 bit PDCP sequence numbers
  - bit 7 should be set to '1' to indicate UE support for Unacknowledged Mode RLC

★ VoIP also requires protocol stack support (including header compression) from the eNode B. In addition, the eNode B should support QoS to ensure that voice packets receive appropriate prioritisation during scheduling. The eNode B can also provide support for TTI Bundling (described in section 27.6.1) and Semi-Persistent Scheduling (described in section 27.6.2)

★ The longer term network architecture for VoIP should include IMS. This allows the use of globally standardised service sets such as MMTEL, which provide support for voice, video, chat and file sharing. In the shorter term, VoIP services can be provided with appropriate upgrades to the core network

27.3.1 PROTOCOL STACK

★ The radio access network protocol stacks for VoIP are illustrated in Figure 246. There are 3 main protocol stacks visible within the UE - the speech service user plane protocol stack, the RTCP protocol stack and the SIP signalling protocol stack. The MAC, RLC and PDCP layers within the eNode B partner those within the UE. The eNode B does not decode the higher layer information above the PDCP layer. Instead, the IP packets are transferred to and from the core network across the S1 interface. The S1 interface has its own protocol stack (GTP-U/UDP/IP) for transferring these packets

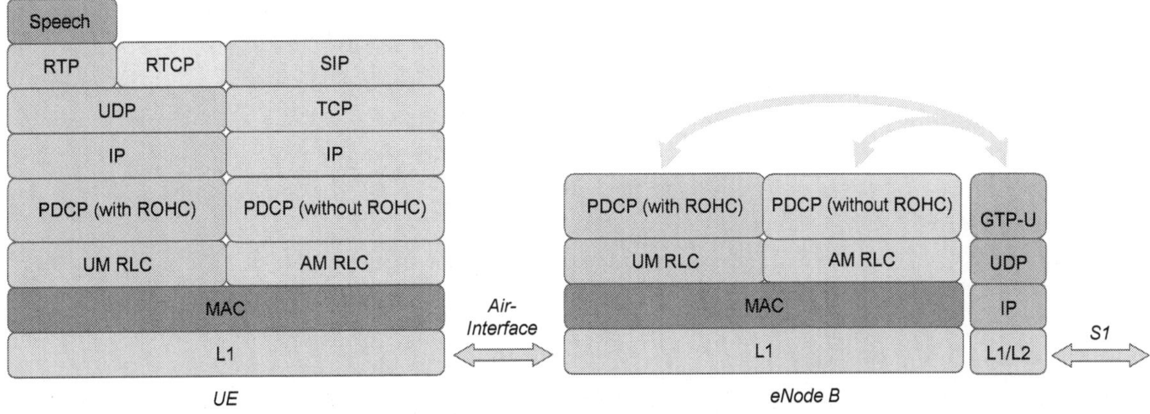

Figure 246 – User plane protocol stack for Voice over IP (VoIP)

★ A speech codec periodically generates blocks of data at the top of the protocol stack. There is a wide range of possible speech codecs but it is assumed that the UE and network will support the Adaptive Multi-Rate (AMR) speech codec with its full set of 8 modes, i.e. 8 different bit rates. There may also be support for the AMR wideband codec with its full set of 9 modes. Both AMR and WB-AMR use a 20 ms frame structure

★ The speech information generated by these codecs is packaged slightly differently when using the RTP layer. This can be done based upon IETF RFC 4867 which specifies 'bandwidth efficient' and 'octet aligned' formats. The 'bandwidth efficient' format adds 4 bits of Codec Mode Request (CMR) and 6 bits of Table of Contents (ToC) as header information. Padding is then added to ensure that the total number of bits is an integer number of bytes

★ This header information and padding creates a relatively small additional overhead. The total number of bits associated with each AMR mode is presented in Table 315. The 12.2 kbps mode generates a total of 32 bytes every 20 ms which corresponds to an aggregate throughput of 12.8 kbps. The header and padding have a greater impact upon the lower bit rate modes, e.g. the 4.75 kbps mode generates a total of 14 bytes every 20 ms which corresponds to an aggregate throughput of 5.6 kbps

Mode	Bit Rate (kbps)	Speech Payload (bits)	Header (bits)	Padding (bits)	Total Size (bits)	Total Size (bytes)	Aggregate Bit Rate (kbps)
0	4.75	95	10	7	112	14	5.6
1	5.15	103	10	7	120	15	6.0
2	5.90	118	10	0	128	16	6.4
3	6.70	134	10	0	144	18	7.2
4	7.40	148	10	2	160	20	8.0
5	7.95	159	10	7	176	22	8.8
6	10.2	204	10	2	216	27	10.8
7	12.2	244	10	2	256	32	12.8

Table 315 – Number of bits associated with each AMR mode (IETF RFC 4867 bandwidth efficient format)

★ The Real Time Protocol (RTP) layer is specified by the IETF within RFC 3550. It has been designed to transfer real time audio and video content across IP networks. The RTP layer provides services which include payload type identification, sequence numbering and time stamping. The RTP layer does not provide support for retransmissions

★ The RTP header has a minimum size of 12 bytes. These 12 bytes are illustrated in Figure 247

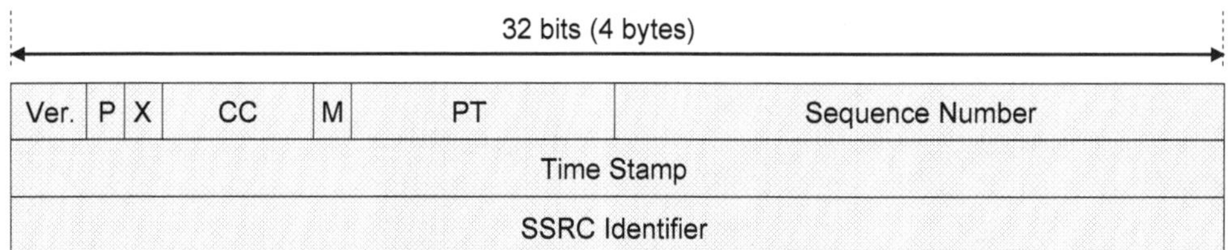

Figure 247 – Content of RTP header (12 bytes)

★ The RTP header fields include:

- Version: signals the version of the RTP protocol being used (2 bits)
- Padding (P): indicates whether or not there is any additional padding at the end of the RTP packet (1 bit)
- Extension (X): indicates whether or not an extension header is present between the fixed header and the payload (only fixed header is shown in Figure 247) (1 bit)
- CSRC Count (CC): defines the number of Contributing Source (CSRC) identifiers which follow the fixed header. These are applicable when the data stream originates from multiple sources (4 bits)
- Marker (M): defines a flag which can be used by the application layer. It is intended to allow events such as frame boundaries to be marked in the packet stream (1 bit)
- Payload Type (PT): signals the content of the payload so the receiving application layer can interpret it appropriately. The payload type defines the codec as well as whether the content is audio or video (7 bits)
- Sequence Number: is incremented by 1 for every RTP packet which is sent. The sequence number is used for re-ordering and packet loss detection at the receiver (16 bits)
- Time Stamp: is used by the receiver to playback the received data frames at appropriate time intervals (32 bits)
- Synchronisation Source (SSRC) Identifier: is used to identify the source of a data stream (32 bits)

- The Real time Transport Control Protocol (RTCP) layer is specified by the IETF within RFC 3550. It operates in conjunction with the RTP layer. The RTCP layer within the receiving device collects statistics regarding the quality of the received speech signal. These statistics are sent back to the transmitting device where they can be used to adapt the transmitted signal, e.g. step down to a lower codec rate when the received signal quality is poor
- The User Datagram Protocol (UDP) layer is specified by the IETF within RFC 768. It provides a simple solution for transferring data without providing services for retransmissions nor sequence numbering. UDP is often used rather than TCP for time sensitive applications because losing packets can be preferable to waiting for delayed packets
- The UDP layer differentiates between the RTP and RTCP information using the Port Numbers within the header of the UDP packet. Packets belonging to the RTP layer are allocated an even Port Number, whereas packets belonging to the RTCP layer are allocated the next higher odd Port Number. The same Port Numbers should be used for both sending and receiving the packets. RTP and RTCP information is sent in separate UDP packets
- The UDP header has a fixed size of 8 bytes. These 8 bytes are illustrated in Figure 248

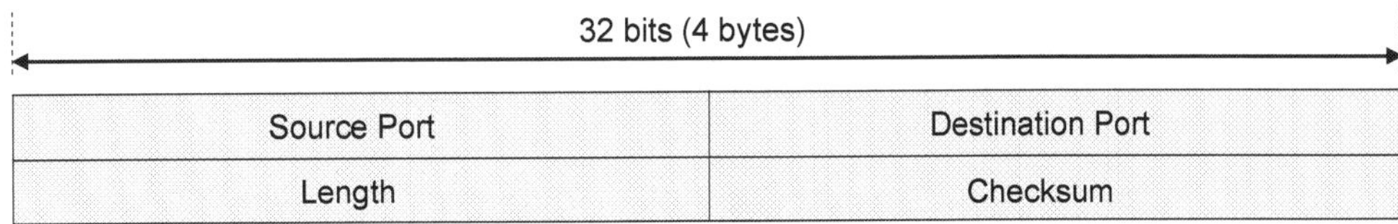

Figure 248 – Content of UDP header (8 bytes)

- The UDP header fields include:
  - Source Port: identifies the port number used for transmission on the sending device (16 bits)
  - Destination Port: identifies the port number used for reception on the receiving device (16 bits)
  - Length: indicates the total number of bytes within both the UDP header and payload (16 bits)
  - Checksum: can be used by the receiver for error checking across both the UDP header and payload (16 bits)
- Internet Protocol version 4 (IPv4) is specified by the IETF within RFC 791, whereas Internet Protocol version 6 (IPv6) is specified by the IETF within RFC 2460. This section focuses upon IPv4. The IP layer provides support for packet delivery from a source IP address to a destination IP address. The RTP/UDP and RTCP/UDP packets are sent in separate IP packets
- The IPv4 header has a minimum size of 20 bytes (the IPv6 header has a minimum size of 40 bytes). The minimum 20 bytes of IPv4 header are illustrated in Figure 249

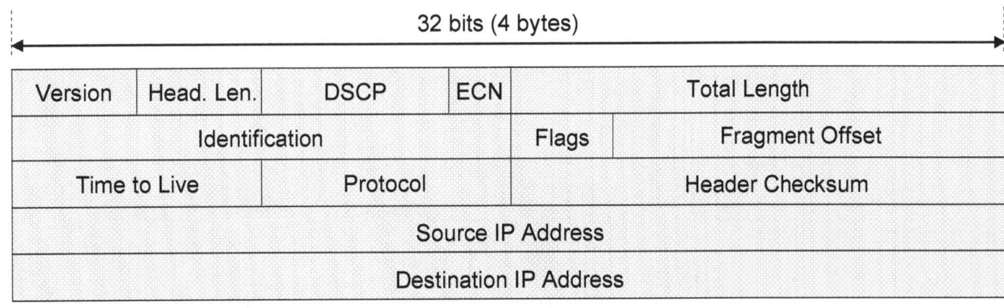

Figure 249 – Content of IPv4 header (20 bytes)

- The IPv4 header fields include:
  - Version: identifies the version of the IP protocol, e.g. version 4 or version 6 (4 bits)
  - Header Length: specifies the length of the IP header in units of 32 bit words, i.e. the minimum value is 5 (4 bits)
  - Differentiated Services Code Point (DSCP): this field was originally used to define the type of service associated with the IP packet, but is now used for Differentiated Services (DiffServ) as specified within IETF RFC 2474. The DSCP determines the Per Hop Behaviour (PHB) that the packet experiences at each node within the IP network, i.e. the DSCP determines the prioritisation of the packet when queuing to be served (6 bits)
  - Explicit Congestion Notification (ECN): allows notification of network congestion according to IETF RFC 3168. Use of the ECN field is optional (2 bits)
  - Total Length: specifies the total length of the IP packet (header plus payload) in units of bytes (16 bits)
  - Identification: is used to help re-assembly after fragmentation of higher layer packets has been applied (16 bits)
  - Flags: are used to help manage fragmentation. There is a 'don't fragment' flag and a 'more fragments' flag (3 bits)

- Fragment Offset: defines the offset of a specific fragment relative to the start of the original unfragmented packet. The offset is specified in units of 8 bytes (13 bits)
- Time To Live (TTL): is used to limit the number of hops made by a packet when being delivered to the destination address. A packet is discarded if the maximum number of hops is reached before reaching the destination (8 bits)
- Protocol: defines the content of the IP packet payload. In the case of VoIP, this field indicates that the payload is a UDP packet (8 bits)
- Header Checksum: allows each IP network node to check for errors within the IP header. The header checksum is re-calculated every time the Time to Live field is updated (16 bits)
- Source IP Address: is the IP address of the sending device. It can be changed during the transit of the IP packet by a Network Address Translation (NAT) procedure (32 bits)
- Destination IP Address: is the IP address of the target device. It can be changed during the transit of the IP packet by a Network Address Translation (NAT) procedure (32 bits)

★ The Packet Data Convergence Protocol (PDCP) layer is specified by 3GPP within TS 36.323. The PDCP layer provides header compression for both the RTP and RTCP packets. Header compression significantly reduces the protocol stack overheads generated by the higher layers. Header compression is described in section 27.3.2. The PDCP layer also provides ciphering to help keep the connection between the eNode B and UE secure

★ After applying header compression and ciphering, the PDCP layer adds its own header. This header can be 1 or 2 bytes depending upon the Sequence Number (SN) length. Figure 250 illustrates the format of the PDCP header for each SN length. The shorter 7 bit SN length is most appropriate for VoIP applications because it results in a lower PDCP overhead. VoIP does not require a large sequence number range because delays are kept to a minimum and the RLC layer does not complete retransmissions

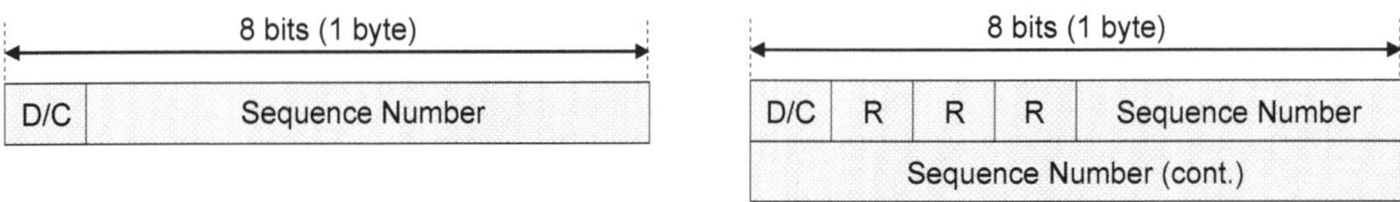

Figure 250 – Content of PDCP header for user plane data PDU (1 or 2 bytes)

★ The PDCP header fields include:

- D/C: indicates whether the PDCP PDU contains data or control information (1 bit)
- R: are reserved bits and are ignored by the receiver (1 bit)
- Sequence Number: used for re-ordering and detection of lost packets at the receiver (7 or 12 bits)

★ The PDCP layer also generates control PDU which do not transfer any user plane data. There are 2 types of control PDU:

- ROHC Feedback Packet: provides feedback for the header compression algorithm
- PDCP Status Report: includes the sequence number of the First Missing SDU (FMS) and a bitmap to indicate which other SDU have been successfully received

The PDCP Status Report control PDU are only applicable when Acknowledged Mode RLC is used, so are not applicable to the RTP and RTCP protocol stacks belonging to VoIP connections

★ The Radio Link Control (RLC) layer is specified by 3GPP within TS 36.322. The RTP and RTCP protocol stacks belonging to VoIP connections use Unacknowledged Mode (UM) RLC to provides in-sequence delivery without retransmissions. The UM RLC header size can be 1 or 2 bytes depending upon the Sequence Number (SN) length (assuming Length Indicators (LI) are not required for VoIP packets)

★ Figure 251 illustrates the format of the UM RLC header for each SN length. The shorter 5 bit SN length is the most appropriate for VoIP applications because it results in a lower RLC overhead. VoIP does not require a large sequence number range because delays are kept to a minimum and the number of retransmissions completed by the physical layer is relatively small

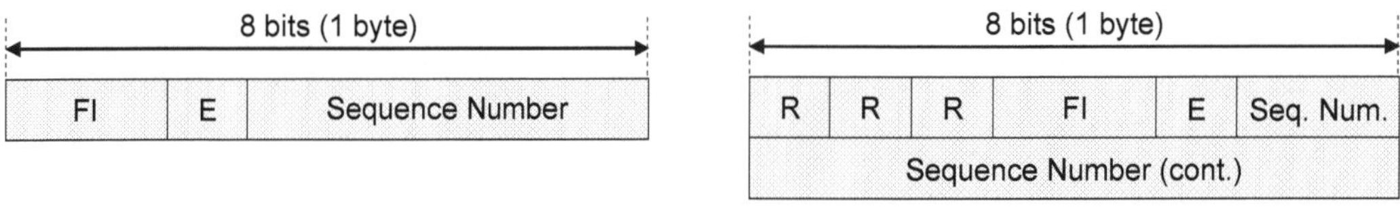

Figure 251 – Content of Unacknowledged Mode RLC header without Length Indicators (1 or 2 bytes)

★ The RLC header fields include:

  - Framing Information (FI): indicates whether the RLC SDU has been segmented at the beginning or end of the RLC PDU payload (2 bits)
  - Extension (E): indicates whether or not Length Indicator (LI) information follows (1 bit)
  - R: are reserved bits and are ignored by the receiver (1 bit)
  - Sequence Number: used for re-ordering and detection of lost packets at the receiver (5 or 10 bits)

★ The Medium Access Control (MAC) layer is specified by 3GPP within TS 36.321. The MAC layer can multiplex packets belonging to different logical channels so the RTP and RTCP packets can be multiplexed with the SIP packets. The MAC header specifies the logical channel identity associated with each SDU so the receiving device can separate them for the higher layers

★ Figure 252 illustrates the format of the MAC subheaders. There is a MAC subheader for each MAC SDU and for each MAC control element. MAC control elements can be multiplexed with MAC SDU for control plane purposes, e.g. to provide a timing advance instruction, or to report the UE transmit power headroom.

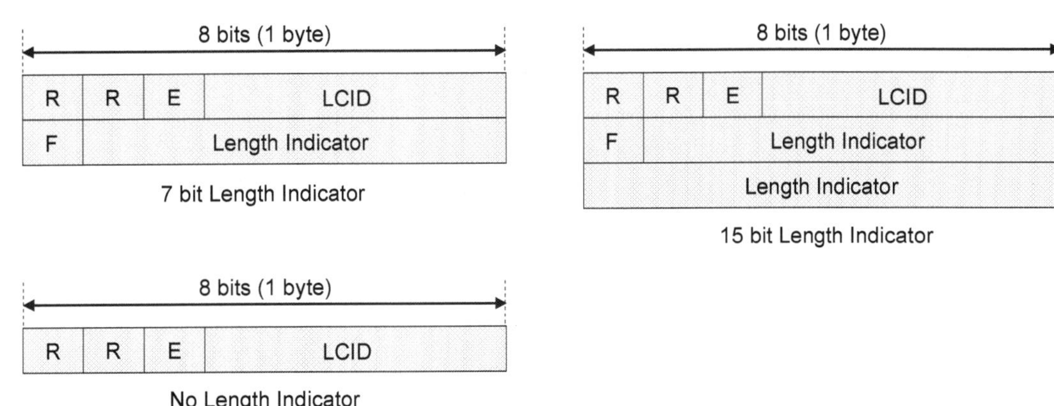

Figure 252 – Content of MAC subheader (1, 2 or 3 bytes)

★ The MAC subheader fields include:

  - R: are reserved bits and are ignored by the receiver (1 bit)
  - Extension (E): indicates whether or not there are further MAC subheaders (1 bit)
  - Logical Channel Identity (LCID): identifies either the logical channel associated with a MAC SDU, or the type of MAC control element (5 bits)
  - Format (F): specifies the number of bits allocated to the Length Indicator. Longer Length Indicators are only required for large MAC SDU (1 bit)
  - Length Indicator: specifies the length of the associated MAC SDU or MAC control element in units of bytes. It is not necessary to include length indicators for fixed size MAC control elements. In addition, the last subheader does not require a length indicator

★ The Session Initiation Protocol (SIP) protocol stack runs in parallel to the RTP and RTCP protocol stacks. SIP uses its own bearer whereas RTP and RTCP share the same bearer. SIP is used for signalling with the IMS when establishing and releasing VoIP connections. Registration and authentication procedures are also supported by SIP. SIP itself is specified by the IETF within RFC 3261, whereas the signalling procedures applicable to LTE are specified by 3GPP within TS 24.229

★ The SIP protocol stack uses Transmission Control Protocol (TCP) rather than UDP to help ensure reliable delivery. TCP provides support for retransmissions and re-ordering. The IP layer provides support for packet delivery from a source IP address to a destination IP address

★ The SIP protocol stack uses the PDCP layer but for ciphering rather than a combination of header compression and ciphering. The RLC layer is used in acknowledged mode to provide further support for retransmissions and re-ordering

★ IETF References: RFC 4867, RFC 3550, RFC 768, RFC 791, RFC 2460, RFC 2474, RFC 3168, RFC 3261

★ 3GPP References: TS 36.323, TS 36.322, TS 36.321, TS 24.229

27.3.2 HEADER COMPRESSION

★ Header compression is a requirement for VoIP due to the significant overhead added by the RTP, UDP and IP protocol stack layers. Speech information has overheads added by all 3 of these layers whereas RTCP signalling has overheads added by the UDP and IP layers

★ Header compression is provided by the Packet Data Convergence Protocol (PDCP) layer. 3GPP has specified the PDCP layer within TS 36.323. Header compression is based upon the Robust Header Compression (ROHC) framework specified within IETF RFC 3095 and RFC 4815

★ Figure 253 illustrates the protocol stack layers targeted by header compression. In the uplink direction, the UE applies header compression while the eNode B applies header decompression. In the downlink direction, the eNode B applies header compression while the UE applies header decompression. This minimises the throughput requirement across the air-interface

★ The eNode B does not terminate the IP, UDP nor RTP layers but receives downlink packets from the core network which already include headers from these layers. Likewise, the eNode B transfers uplink packets towards the core network which include headers from these layers after header decompression

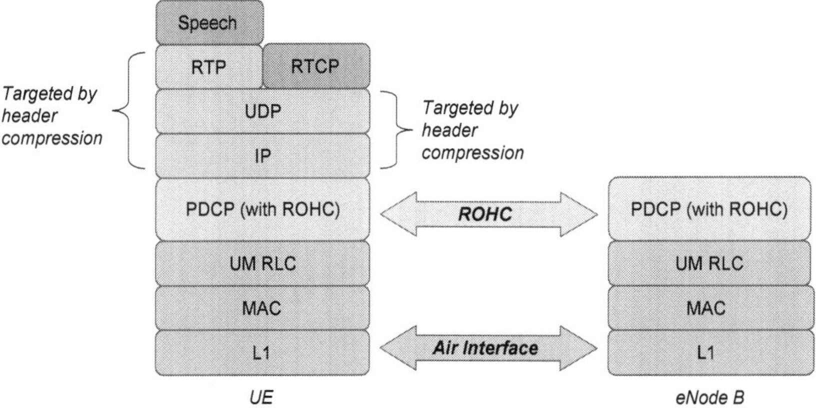

Figure 253 – VoIP protocol stack layers targeted by header compression

★ Figure 254 compares the throughput requirements of CS voice over 3G when using a Dedicated Channel (DCH), with VoIP over LTE, with and without header compression. The comparison assumes an AMR 12.2 kbps codec

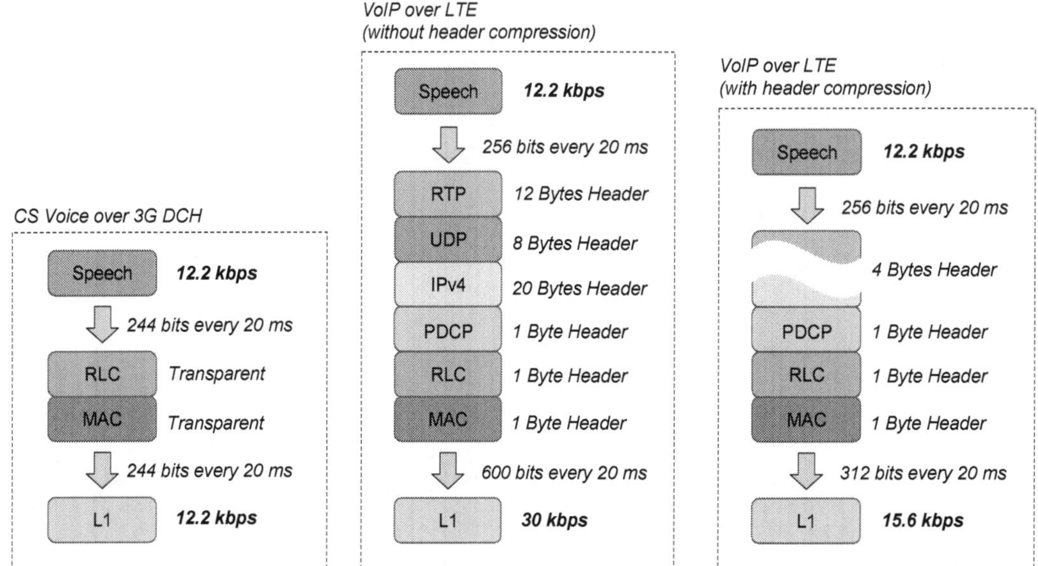

Figure 254 – Comparison of CS Voice over 3G with VoIP over LTE, with and without header compression

★ CS voice over 3G DCH uses transparent RLC and MAC layers so the resultant throughput requirement is equal to the bit rate generated by the AMR codec

★ VoIP over LTE without header compression experiences a significant overhead from the RTP/UDP/IP protocol stack layers. The resultant throughput requirement at the top of the physical layer is 30 kbps, i.e. more than double the equivalent figure for CS voice over 3G DCH. The 244 bits generated by the AMR codec are increased to 256 bits for use by RTP as presented in Table 315. Short sequence numbers are assumed so the PDCP and RLC layer headers have a size of 1 byte. The MAC layer header is also assumed to be 1 byte although in practice it could frequently be greater than 1 byte, e.g. when a MAC control element is being sent

★ VoIP over LTE with header compression experiences a reduced overhead from the RTP/UDP/IP layers. During steady state conditions, ROHC is able to reduce the overhead from 40 bytes to 4 bytes. This results in a throughput requirement of 15.6 kbps, i.e. 28 % greater than the 12.2 kbps for CS voice over 3G DCH

★ The IETF defines multiple ROHC profiles to support different protocol stack combinations. The ROHC profiles specified for use by LTE within 3GPP TS 36.323 are presented within Table 316

★ Profiles 1 and 101 are applicable to the speech protocol stack, whereas profiles 2 and 102 are applicable to the RTCP protocol stack. Profiles 101 and 102 are updated versions of profiles 1 and 2

Profile Identity	Application	Reference
0000	No compression	RFC 4995
0001	RTP/UDP/IP	RFC 3095, RFC 4815
0002	UDP/IP	RFC 3095, RFC 4815
0003	ESP/IP	RFC 3095, RFC 4815
0004	IP	RFC 3843, RFC 4815
0006	TSCP/IP	RFC 4996
0101	RTP/UDP/IP	RFC 5225
0102	UDP/IP	RFC 5225
0103	ESP/IP	RFC 5225
0104	IP	RFC 5225

Table 316 – Header compression profiles supported by the PDCP layer

★ It is not mandatory for UE to support all of these profiles. UE report their PDCP capability within the RRC: UE Capability Information message. UE specify which ROHC profiles are supported and also the maximum number of concurrently active ROHC contexts

★ The main principles of header compression are:

  o avoid sending fields which do not change between consecutive packets

  o allow the receiver to deduce some changing fields without sending them

  o apply efficient coding to other changing fields

For example, within an IPv4 header, the Source Address (4 bytes) and Destination Address (4 bytes) do not change for a specific connection. In addition, this information is not used to transfer the packets between the eNode B and UE. Thus, the sending device can remove both of these fields while the receiving device can re-insert them. This is possible after the receiving device has learnt the Source and Destination Addresses by receiving at least one non-compressed header

★ ROHC categorises header information according to how it can be compressed. IETF RFC 3095 specifies the categories presented in Table 317

Category		Characteristics
Inferred		Values can be deduced from other values
Static		Values which remain constant for the packet stream
Static-Defined		Static values which define a packet stream
Static-Known		Values are well known
Changing	Static	Changing values which are categorised as static due to certain assumptions
	Semi-Static	Values which change occasionally and revert to their original values
	Rarely Changing	Values which change occasionally and then keep their new values
	Alternating	Values which alternate between a small set of different values
	Irregular	Values for which no useful change pattern can be identified

Table 317 – Header field categories used by ROHC (IETF RFC 3095)

★ ROHC uses state machines at both the sender (compressor) and receiver (decompressor). Both state machines start in the lowest compression state before making transitions to higher compression states

- ★ The compressor states are Initialisation and Refresh (IR), First Order (FO) and Second Order (SO). The compressor starts in the IR state where packets are sent without header compression. Partial compression is used in the FO state which allows the decompressor to acquire changing header field patterns. Compression is optimal in the SO state
- ★ The decompressor states are No Context (NC), Static Context (SC) and Full Context (FC). The NC state is used prior to successfully decompressing a header. The SC state is used if the decompressor fails to successfully decompress a number of headers while in the FC state. The decompressor moves from the SC state to the NC state if decompression continues to fail. The FC state is used after the decompressor has successfully decompressed its first header
- ★ ROHC also uses 3 modes of operation at both the compressor and decompressor:
  - o Unidirectional (U): packets are sent in one direction. ROHC starts in this mode prior to the decompressor providing feedback to indicate that one of the bi-directional modes can be used
  - o Bi-directional Optimistic (O): the feedback channel is used to some extent. The decompressor can send error recovery requests and acknowledgements
  - o Bi-directional Reliable (R): the feedback channel is used intensively, and there is a stricter logic for state transitions to help ensure synchronisation between the compressor and decompressor
- ★ IETF References: RFC 3095, RFC 4815
- ★ 3GPP References: TS 36.323

27.4 VOICE OVER LTE VIA GENERIC ACCESS (VoLGA)

★ Voice over LTE via Generic Access (VoLGA) has been specified by the VoLGA Forum rather than by 3GPP

★ It is based upon the concept of a Generic Access Network (GAN) specified by 3GPP within TS 43.318

★ The general concept of a GAN is illustrated in Figure 255. This example illustrates a GAN interfacing with a UMTS network. The Generic Access Network Controller (GANC) provides connectivity between the GAN and the UMTS core network. This allows the UE to benefit from the services offered by UMTS while connected to a GAN. The GAN could be a WiFi network

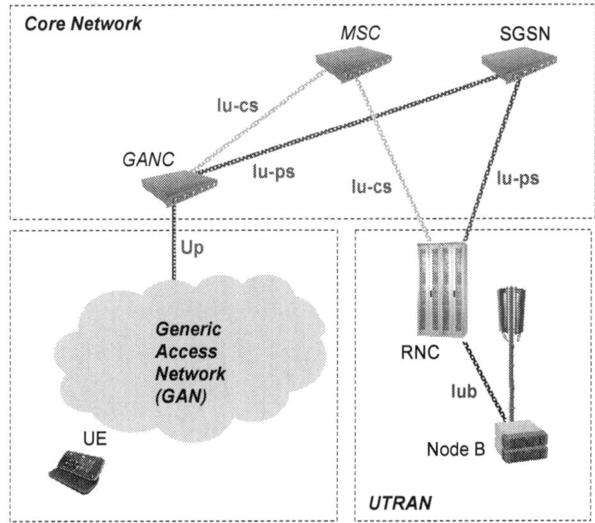

Figure 255 – General concept of a Generic Access Network (GAN)

★ The general architecture for VoLGA with UTRAN is illustrated in Figure 256. The main characteristics are:

- o the GAN is provided by the LTE network
- o the GANC is replaced by a VoLGA Access Network Controller (VANC)
- o VoLGA is focused upon voice services so connectivity is provided to the MSC but not the SGSN

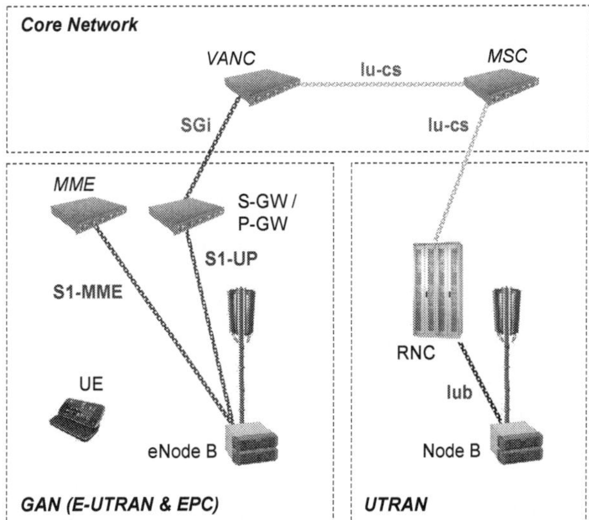

Figure 256 – General architecture for VoLGA with UTRAN

★ The use of VoLGA does not require any changes to the LTE network. The LTE network just has to provide an IP connection to the VANC. The combination of IP connection and VANC defines a tunnel between the UE and CS core network domain. Both the control plane signalling and the user plane data belonging speech calls are passed through this tunnel.

★ The control plane protocol stack for VoLGA is presented in Figure 257. When establishing a speech connection, the UE sends control plane messages to the VANC using the GA-RRC protocol specified by the VoLGA Forum. The VANC converts the GA-RRC messages into 3GPP RANAP messages before forwarding to the MSC/VLR

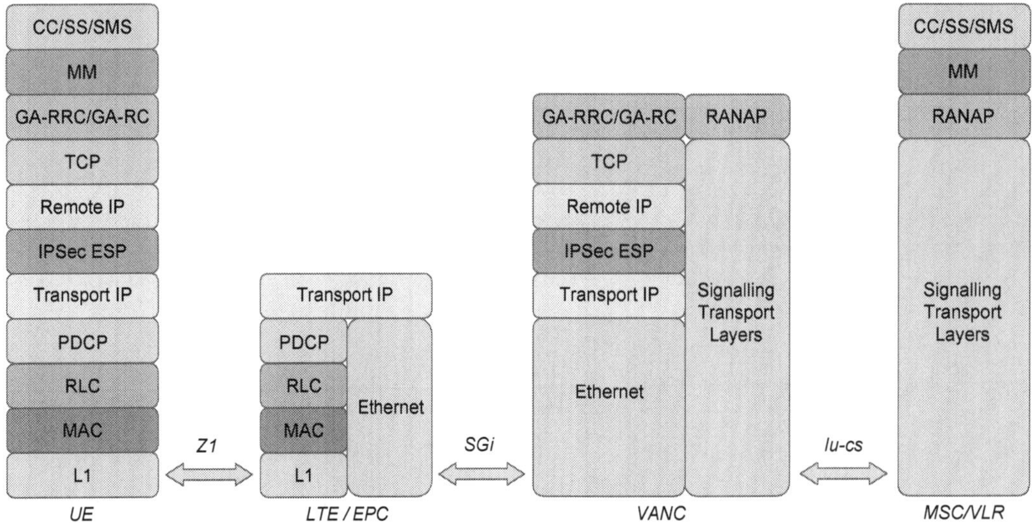

Figure 257 – Control plane protocol stack for VoLGA over UTRAN

- The VANC appears as an RNC connected to the MSC using a standardised Iu-cs interface connection
- The VANC appears an IP application server to the P-GW using a standardised SGi interface connection
- The user plane protocol stack for VoLGA is presented in Figure 258. The protocol stack between UE and VANC corresponds to a VoIP connection using RTP/UDP/IP, whereas the protocol stack between the VANC and MSC corresponds to a CS voice connection using only the Iu-cs user plane protocol stack

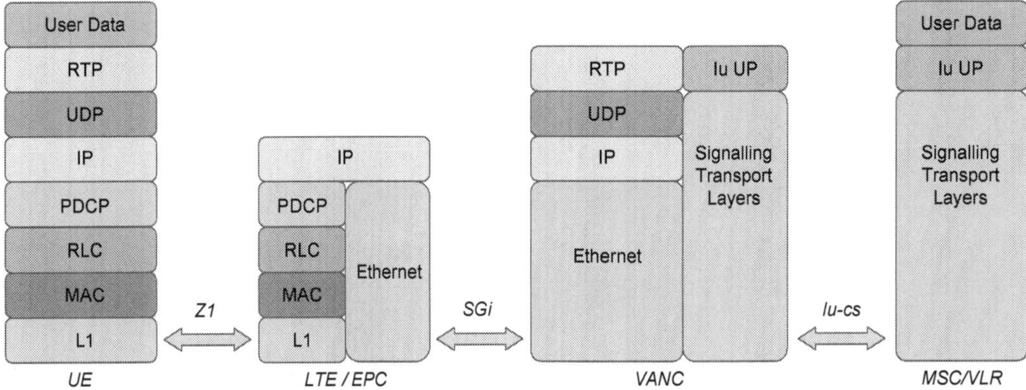

Figure 258 – User plane protocol stack for VoLGA over UTRAN

- The VoLGA specifications also support connectivity to the CS core network belonging to GERAN, i.e. either UTRAN or GERAN CS core networks can be used
- 3GPP References: TS 43.318

27.5 SINGLE RADIO VOICE CALL CONTINUITY (SRVCC)

★ Single Radio Voice Call Continuity (SRVCC) is specified by 3GPP within TS 23.216

★ SRVCC provides mobility between Voice over IP (VoIP) and a circuit switched voice call. An LTE VoIP connection uses the SRVCC procedure when the UE moves into an area which has poor LTE coverage but relatively good UMTS or GERAN coverage, i.e. the connection is continued by completing a handover towards the UMTS or GERAN circuit switched domain

★ The 'Single Radio' part of SRVCC refers to the UE only having to use one access network at a time

★ 3GPP provides support for various scenarios including mobility from UTRAN VoIP to the UTRAN circuit switched domain. SRVCC from LTE VoIP to the GERAN circuit switched domain, and from LTE VoIP to 3GPP2 1xCS are also supported. This section focuses upon mobility from LTE VoIP to the UTRAN circuit switched domain

★ SRVCC requires the IP Multimedia Subsystem (IMS) within the core network to manage the switching between the CS and PS core network domains. The network architecture required for SRVCC between LTE and UTRAN is illustrated in Figure 259

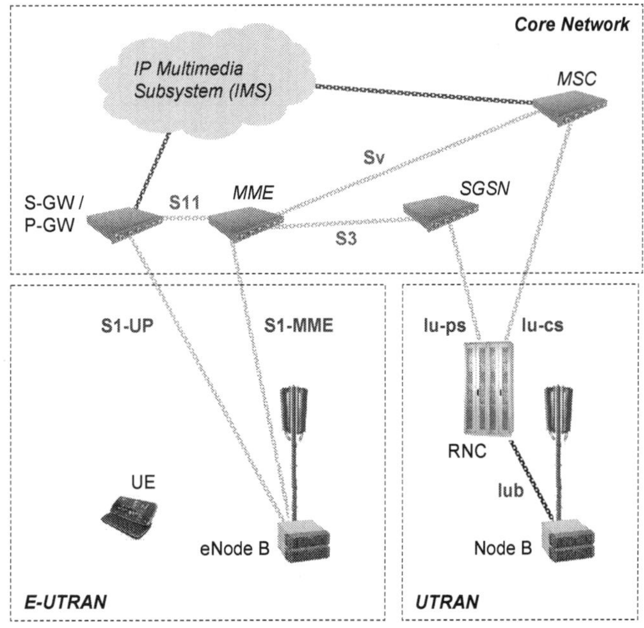

Figure 259 – Architecture for SRVCC between LTE and UTRAN

★ The MME is connected to both the MSC and SGSN so it is able to request that resources are pre-allocated for connections that have triggered SRVCC from LTE to UTRAN. The content of messages sent across the Sv interface between the MME and MSC is specified within 3GPP TS 29.280. Messages sent across the S3 interface between the MME and SGSN are based upon the GTP-C protocol specified in 3GPP TS 29.274

★ The Service Centralisation and Continuity Application Server (SCC AS) within IMS provides the control functions for switching voice calls between the IP and CS domains. This application server is introduced within 3GPP TS 23.237. Voice Call Continuity (VCC) between IP and CS is specified within 3GPP TS 23.206

★ SRVCC requires support from both the network and the UE:

- the UE signals its capability to the MME during the attach procedure. The Attach Request message includes a flag within the MS Network Capability section which indicates whether or not the UE supports SRVCC. 3GPP TS 24.301 specifies the content of the Attach Request message, while 3GPP TS 24.008 specifies the content of the MS Network Capability section
- the UE signals its capability to the eNode B during the UE capability enquiry procedure. The UE Capability Information message can include a UE Capability RAT Container for E-UTRA. This container can include a Feature Group Indicators (FGI) bit string. Bit 27 of this bit string indicates whether or not the UE supports SRVCC. 3GPP TS 36.331 specifies the content of the UE Capability Information message
- the MME signals a flag to the eNode B during the initial context setup procedure. The S1AP: Initial Context Setup Request message includes an SRVCC Operation Possible field to indicate whether or not the eNode B can apply the SRVCC procedure based upon both the UE and MME capability, i.e. this flag reflects a combination of the UE and MME capabilities. The S1 Application Protocol is specified within 3GPP TS 36.413

★ The SRVCC procedure can be triggered with or without a PS inter-system handover for the data bearer, i.e. the procedure can move only the voice connection across to UTRAN/GERAN, or a combination of the voice and data connections

★ The first phase of signalling for the SRVCC procedure with PS inter-system handover is presented in Figure 260. Both an MSC Server and a target MSC are shown. The MSC Server must be upgraded to support SRVCC procedures whereas the MSC can remain without any specific SRVCC support

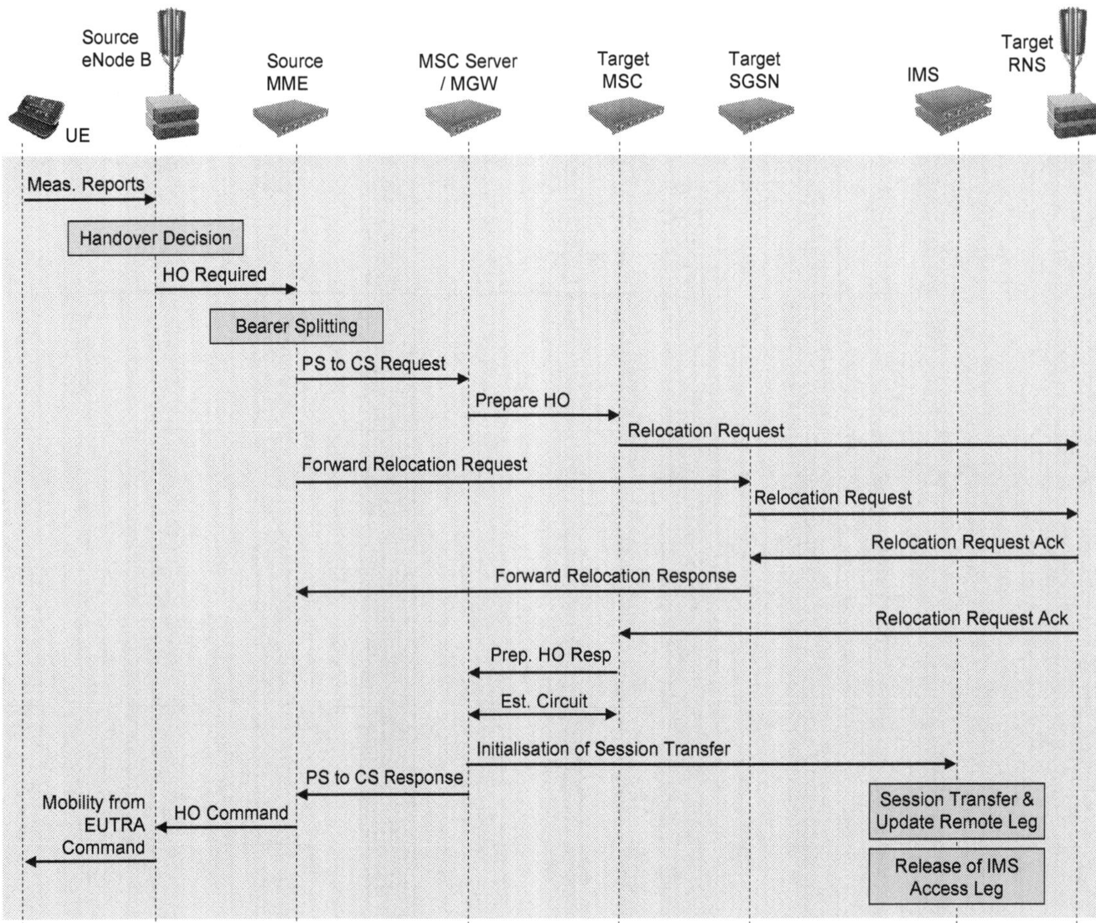

Figure 260 – Signalling for SRVCC with PS handover from LTE towards UTRAN (part 1)

★ The procedure is triggered by the eNode B after the UE sends a measurement report to indicate that LTE coverage is becoming relatively poor. This could be based upon measurement reporting event A2 which is triggered when the serving cell coverage becomes worse than a specific threshold. The UE would then be requested to complete some inter-RAT measurements on the UMTS system to help identify a target cell for the SRVCC procedure

★ The eNode B takes a decision on whether or not the SRVCC procedure should be completed, e.g. depending upon whether or not an appropriate target cell has been identified. Assuming the SRVCC procedure is to be completed, the eNode B sends an S1AP: Handover Required message to the MME. The most relevant fields within this message are:

- 'Handover Type' which is set to 'LTE to UTRAN' (when SRVCC uses UMTS as a target RAT)
- 'Target ID' which is configured with the target RNC identity. This includes the LAI as well as the actual RNC identity. The RAC is also included when completing a PS inter-system handover
- 'SRVCC HO Indication' which specifies whether the handover is for PS and CS, or CS only. This field is set to 'PS and CS' when the SRVCC procedure includes a PS inter-system handover
- 'Source to Target Transparent Container' which includes details of the target cell as well as the connections to be established on the target RAT. This information is transparent to the core network but is forwarded across to the target RNC using the subsequent signalling messages

★ The source MME separates the voice bearer (which is to be directed towards the UTRAN CS domain) from the data bearers (which are to be directed towards the UTRAN PS domain). Separation is based upon the QoS Class Identifiers (QCI) belonging to each bearer. Speech bearers are assumed to have QCI = 1

★ The MME forwards a PS to CS Request message across the Sv interface towards the MSC Server, and a Forward Relocation Request message across the S3 interface towards the SGSN. Both messages include the 'Source to Target Transparent Container' from the eNode B and the target RNC identity. The PS to CS Request message sent to the MSC Server also includes a Session Transfer Number for SRVCC (STN-SR). The STN-SR is a unique number for each UE configured within the Home Subscriber Server (HSS). The

MME previously received this number from the HSS during initial attachment to the EPC. The MSC Server will subsequently use the STN-SR to reference the voice call when contacting the IMS

★ The MSC Server forwards a Prepare Handover message to the target MSC (assuming the MSC Server is not the final MSC which parents the target cell). The target MSC then forwards a RANAP: Relocation Request message to the target RNC, requesting setup of the speech connection. RANAP messages are specified within 3GPP TS 25.413. The Relocation Request message provides the RNC with the 'Source to Target Transparent Container' originally generated by the eNode B. Similarly, the target SGSN forwards a RANAP: Relocation Request message to the target RNC, requesting setup of the data connections. This provides the RNC with the same 'Source to Target Transparent Container'

★ The RNC proceeds to setup each connection with the relevant Node B, and returns Relocation Request Acknowledge messages to the MSC and SGSN. These messages include 'Target RNC to Source RNC Transparent Container' fields which are to be returned to the source eNode B using the subsequent signalling and sent to the UE within the Mobility from EUTRA Command

★ The MSC informs the MSC Server that the RNC has provided a positive acknowledgement using a Prepare Handover Response message. The MSC server and MSC establish a circuit connection between themselves using for example, the ISDN User Part (ISUP) Initial Address Message (IAM) and Address Complete Message (ACM)

★ The MSC Server then initiates the session transfer within the IMS domain. This is done by sending a request for a new voice call towards a special number known as a Session Transfer Number for SRVCC (STN-SR). The request is sent in the form of an ISUP Initial Address Message (IAM). This request is routed to the Service Centralisation and Continuity Application Server (SCC AS) within the IMS. The SCC AS compares the received STN-SR with its active IMS voice sessions. Once the relevant connection has been identified, the SCC AS initiates the re-direction of the voice session towards the MSC Server, i.e. the MSC which has been enhanced for SRVCC. IMS service continuity procedures are specified within 3GPP TS 23.237

★ The MSC Server responds to the MME using the PS to CS Response message which includes the 'Target RNC to Source RNC Transparent Container' field provided by the target RNC. This is cascaded to the source eNode B using the S1AP: Handover Command message. The eNode B then provides the UE with the RRC: Mobility from EUTRA Command, instructing the UE to move across to the target cell on UMTS

★ The second phase of signalling for the SRVCC procedure with PS inter-system handover is presented in Figure 261

★ After receiving the Mobility from EUTRA Command from the eNode B, the UE re-tunes its transmitter and receiver to the target cell. The UE achieves downlink synchronisation before starting to transmit the DPCCH. The DPCCH transmissions allow the Node B to achieve uplink synchronisation and RANAP: Relocation Detect messages are then sent by the RNC to the target MSC and SGSN. These messages are simple acknowledgements and do not have any content. They acknowledge that the UE has been detected on UTRAN

★ The UE sends an RRC: Handover to UTRAN Complete message after starting to transmit the uplink DPCCH. This message is sent using the configuration information provided within the 'Target RNC to Source RNC Transparent Container' field received by the UE within the Mobility from EUTRA Command

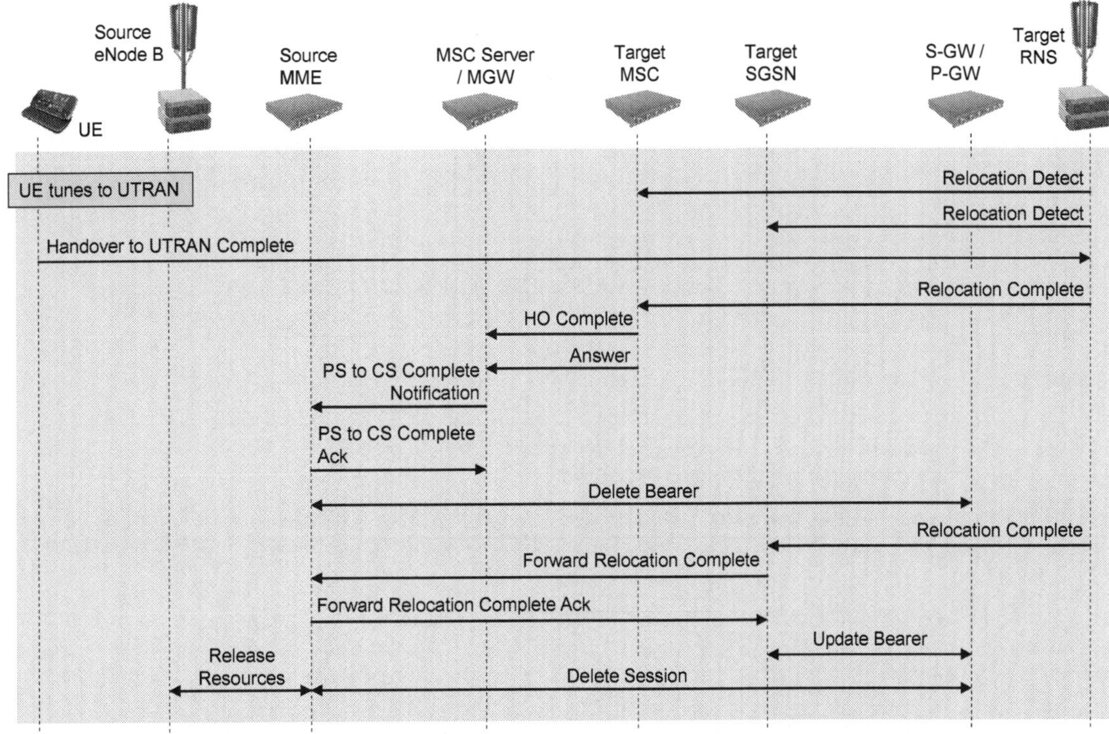

Figure 261 – Signalling for SRVCC with PS handover from LTE towards UTRAN (part 2)

- The RNC sends RANAP: Relocation Complete messages to both the target MSC and SGSN. These messages are simple acknowledgements and do not have any content. They acknowledge that the Node B has received the Handover to UTRAN Complete message
- The Target MSC signals with the MSC Server to complete the procedure for routing the speech call through the CS core network. The MSC Server is then able to send a PS to CS Complete Notification message to the source MME. This is acknowledged by the MME, before the MME proceeds to delete the speech bearer between itself and the S-GW/P-GW
- The Target SGSN signals with the source MME to complete the procedure for routing the data connection through the PS core network. The session within the EPC is deleted and resources at the eNode B are released
- 3GPP References: TS 23.216, TS 23.237, TS 23.206

27.6 RADIO RESOURCE MANAGEMENT

27.6.1 TTI BUNDLING

- TTI bundling is intended to improve the uplink coverage performance of Voice over IP (VoIP), i.e. improve air-interface performance in scenarios where coverage is limited by the UE transmit power capability. TTI bundling is necessary to achieve a similar uplink coverage performance as can be achieved with 3G CS voice and VoIP with HSUPA
- The general concept of TTI bundling is illustrated in Figure 262. Each VoIP transport block is passed to the Physical layer where it has CRC bits added before being channel coded using rate 1/3 Turbo coding. 4 duplicates of the channel coded transport block are generated prior to rate matching. Each duplicate is processed using a different Redundancy Version (RV). This provides the eNode B receiver with an Incremental Redundancy soft combining gain. The set of 4 codewords are modulated and mapped onto 4 consecutive uplink subframes

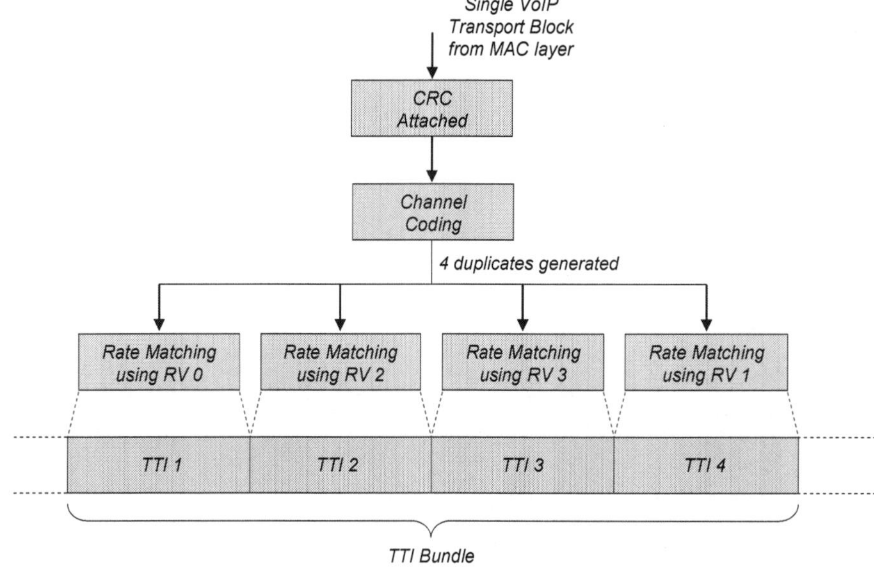

Figure 262 – Concept of TTI bundling

- TTI bundling groups 4 consecutive uplink TTI to generate an effective TTI duration of 4 ms. These 4 consecutive TTI define the bundle size. 3GPP TS 36.321 specifies a fixed bundle size of 4
- Transmissions belonging to each bundle are sent without waiting for any HARQ acknowledgements. This corresponds to using autonomous retransmissions. Each bundle of 4 TTI requires a single resource allocation from the eNode B and a single HARQ acknowledgement
- 3GPP TS 36.213 specifies that a maximum of 3 Resource Blocks can be allocated when TTI bundling is used. In addition, the modulation scheme is limited to QPSK. The combination of up to 3 Resource Blocks and QPSK provides sufficient capacity to transfer the bit rates required by VoIP
- The same Modulation and Coding Scheme (MCS) and frequency bandwidth are used for all 4 TTI belonging to the bundle

★ The eNode B can instruct the UE whether or not to use TTI bundling within the RRC Connection Setup, RRC Connection Reconfiguration or RRC Connection Re-establishment messages

★ TTI bundling is applicable to both FDD and TDD

★ In the case of FDD, the number of HARQ processes is halved from 8 to 4. The eNode B generates a single HARQ acknowledgement for each bundle of 4 TTI. The timing of the HARQ acknowledgement is based upon the timing of the last TTI within the bundle, i.e. the acknowledgement is sent 4 subframes after the last TTI in the bundle

★ The complete bundle of 4 TTI is re-transmitted when a HARQ acknowledgement indicates that a retransmission is required. The retransmission delay is 16 subframes (16 ms) when using TTI bundling, compared to the retransmission delay of 8 subframes (8ms) when TTI bundling is not used

★ Figure 263 illustrates the set of 4 HARQ processes for TTI bundling with FDD. Each transmission belonging to a specific HARQ process is separated by 16 subframes

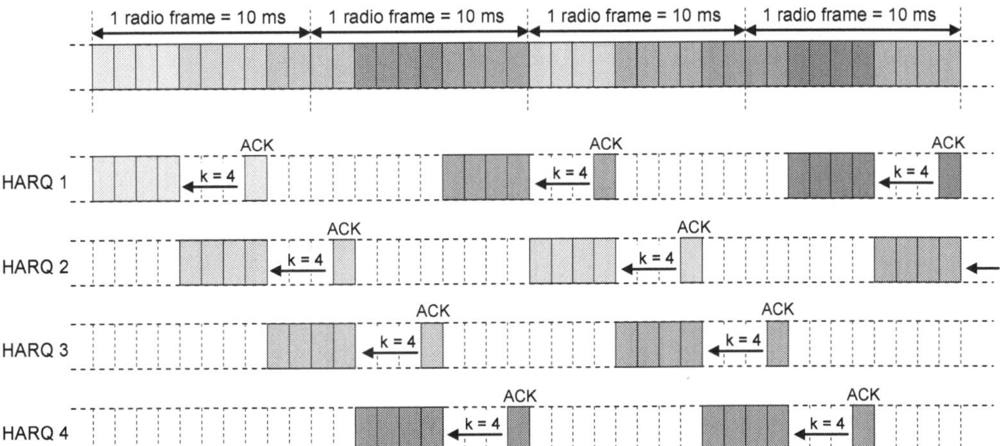

Figure 263 – FDD HARQ processes for TTI bundling

★ In the case of TDD, TTI bundling is only applicable to uplink-downlink subframe configurations 0, 1 and 6. The number of HARQ processes is 3 when using configurations 0 and 6, and 2 when using configuration 1. Configuration 1 requires less HARQ processes because there are fewer uplink subframes within each radio frame. This is illustrated in Table 318

Uplink – Downlink Configuration	Subframe Number										TTI Bundling Supported	Number of HARQ Processes
	0	1	2	3	4	5	6	7	8	9		
0	D	S	U	U	U	D	S	U	U	U	Yes	3
1	D	S	U	U	D	D	S	U	U	D	Yes	2
2	D	S	U	D	D	D	S	U	D	D	No	-
3	D	S	U	U	U	D	D	D	D	D	No	-
4	D	S	U	U	D	D	D	D	D	D	No	-
5	D	S	U	D	D	D	D	D	D	D	No	-
6	D	S	U	U	U	D	S	U	U	D	Yes	3

Table 318 – Support of TTI bundling for each TDD uplink-downlink subframe configuration

★ Figure 264 illustrates the set of 3 HARQ processes for TTI bundling with TDD uplink-downlink subframe configuration 0. In this case, the timing of the HARQ acknowledgement depends upon the timing of the TTI bundle within the radio frame. The 'k' variable defines the difference between the HARQ acknowledgement subframe number and the last subframe number of the TTI bundle

★ Figure 265 illustrates the 2 HARQ processes for TTI bundling with TDD uplink-downlink subframe configuration 1. In this case, there are 4 uplink subframes within each radio frame so a single bundle can be sent every 10 ms

★ Similar to FDD, the complete bundle of 4 TTI is re-transmitted when a HARQ acknowledgement indicates that a retransmission is required

★ 3GPP TS 36.321 specifies that TTI bundling is not supported if the UE has been configured to use LTE Advanced Carrier Aggregation in the uplink

★ TTI bundling provides an alternative to RLC segmentation which can also be used to improve cell edge coverage performance. RLC segmentation generates a number of smaller data blocks from the higher layer VoIP packet. These smaller packets are processed and transmitted independently using separate HARQ processes. The smaller data block size allows the physical layer processing to include greater redundancy and so increases the probability of successful reception

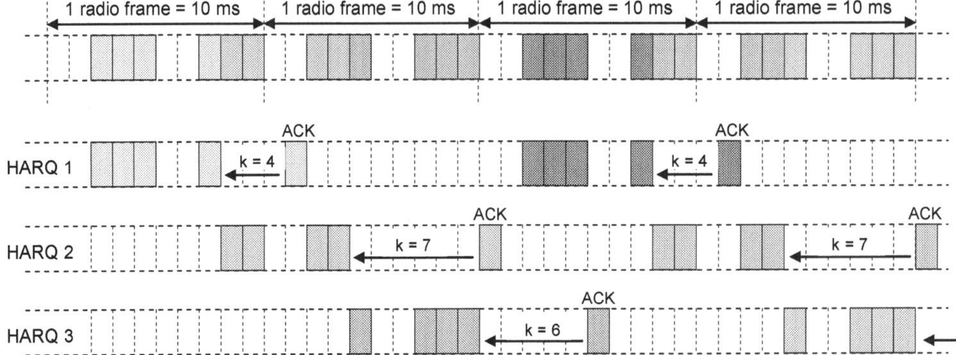

Figure 264 – TDD HARQ processes for TTI bundling when using uplink-downlink subframe configuration 0

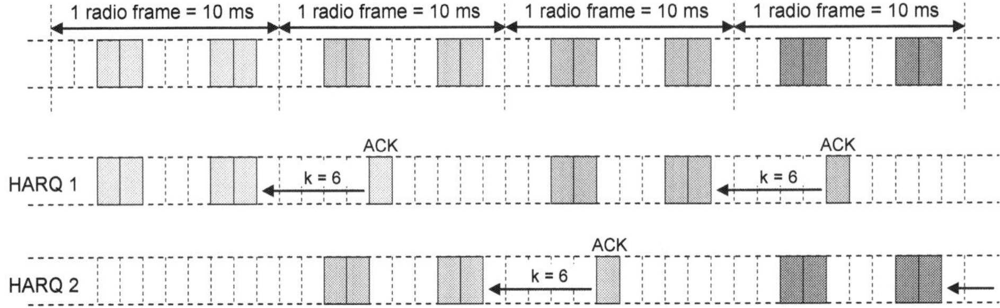

Figure 265 – TDD HARQ processes for TTI bundling when using uplink-downlink subframe configuration 1

- 3GPP References: TS 36.321, TS 36.213

27.6.2 SEMI-PERSISTENT SCHEDULING

- Semi-Persistent Scheduling (SPS) refers to a combination of persistent and dynamic scheduling:
    - persistent scheduling is used to allocate periodic resources which are intended for the first transmission of transport blocks
    - dynamic scheduling is used to allocate resources for retransmissions, as and when required.
- SPS is suitable for applications like Voice over IP (VoIP) where data arrives in periodic bursts, i.e. VoIP packets arrive once every 20 ms. The use of persistent scheduling reduces the overhead generated by Downlink Control Information (DCI) on the PDCCH. The PDCCH is normally required to allocate uplink and downlink resources every time a connection needs to transfer data. The persistent scheduling component of SPS does not require any PDCCH signalling once SPS has been activated
- SPS is applicable to both the uplink and downlink, of both FDD and TDD. SPS is only applicable to the primary cell when LTE Advanced Carrier Aggregation is used
- The eNode B provides the UE with the initial SPS configuration using either an RRC Connection Setup, RRC Connection Reconfiguration or RRC Connection Re-establishment message. These messages do not activate SPS but provide the UE with sufficient information to allow subsequent activation by the PDCCH. The content of the SPS Configuration information is presented in Table 319
- The UE is allocated a Semi-Persistent Scheduling C-RNTI (SPS C-RNTI). This is used to address the UE within any subsequent PDCCH transmissions relevant to the allocation and release of SPS resources
- The SPS Downlink Interval defines the periodicity with which persistent downlink resources will be allocated to the UE. The eNode B can signal values of {10, 20, 32, 40, 64, 80, 128, 160, 320, 640} subframes. For example, the VoIP service could be configured with an interval of 20 subframes so speech frames can be transferred at a rate of one per 20 ms. In the case of TDD, the value is rounded down to the nearest multiple of 10 subframes, e.g. 32 subframes becomes 30 subframes, and 64 subframes becomes 60 subframes. This is necessary for TDD because not all subframes within a 10 ms radio frame are available for downlink transmission
- The number of HARQ processes for SPS in the downlink can be configured with a value between 1 and 8. Configuring a relatively high number of HARQ processes allows greater scope for retransmissions without impacting the flow of data. For example, if only 2 HARQ processes are configured and both of those processes are occupied with retransmissions then no new data can be transferred until the retransmissions have completed

Information Elements				
Semi-Persistent Scheduling C-RNTI				
Downlink Configuration	CHOICE			
	Release			
	Setup	SPS Downlink Interval		
		Number of SPS HARQ Processes		
		PUCCH Resource List		
		2nd Antenna Port	CHOICE	
			Release	
			Setup	PUCCH Resource List for 2nd antenna
Uplink Configuration	CHOICE			
	Release			
	Setup	SPS Uplink Interval		
		Implicit Release After		
		P0 Persistent	P0 Nominal PUSCH Persistent	
			P0 UE PUSCH Persistent	
		Two Intervals Configuration		

Table 319 – Semi-Persistent Scheduling Configuration information element

★ The PUCCH Resource List provides up to 4 values for the variable $n_{PUCCH}^{(1,p)}$. These values are within the range 0 to 2047. The PUCCH is used to return HARQ acknowledgements to the eNode B after the UE has received a downlink transport block. The 2 Antenna Port information is included if the UE has been configured to use the PUCCH on 2 antenna ports. In this case, a second set of up to 4 values for the variable $n_{PUCCH}^{(1,p)}$ are specified

★ The SPS Uplink Interval defines the periodicity with which persistent uplink resources will be allocated to the UE. The eNode B can signal values of {10, 20, 32, 40, 64, 80, 128, 160, 320, 640} subframes. Similar to the downlink, when using TDD the value is rounded down to the nearest multiple of 10 subframes. In the case of TDD, the 'Two Intervals Configuration' flag (also shown in Table 319) can impact the timing of the resource allocations

★ The 'Implicit Release After' information defines the number of consecutive empty transmissions after which the UE releases the uplink SPS resource allocation. This prevents the UE from holding on to an unused resource allocation. In the case of VoIP, this could trigger during periods of uplink inactivity while the other person is talking. The parameter can be configured with values of 2, 3, 4, or 8 empty transmissions

★ The P0 parameters define values of $n_{0\_NOMINAL\_PUSCH}(0)$ and $n_{0\_UE\_PUSCH}(0)$. They are optional within the SPS Configuration information

★ The Two Intervals Configuration flag is optional and only applicable to TDD. This parameter can be used to help avoid retransmissions colliding with new data transmissions (illustrated later in this section)

★ Once the UE has been configured for SPS using the RRC signalling information presented in Table 319, the eNode B uses the PDCCH to activate the SPS transmission pattern during a specific subframe. The PDCCH is coded in a specific way to indicate that its content refers to an SPS resource allocation

★ In the case of the downlink, DCI formats 1, 1A, 2, 2A, 2B, 2C can be used to activate SPS during a specific subframe. DCI Format 1A can subsequently be used to release the SPS resource allocation. The coding of the fields within the DCI to activate and release downlink SPS is presented in Table 320

	Activate SPS		Release SPS
	DCI Formats 1, 1A	DCI Formats 2, 2A,2B, 2C	DCI Format 1A
HARQ Process Number	FDD: 000 TDD: 0000		
Modulation and Coding Scheme	MSB is set to 0	MSB is set to 0 for the enabled transport block	11111
Redundancy Version	00	00 for the enabled transport block	00
Resource Block Assignment	-	-	set to all 1's

Table 320 – Coding of DCI fields used to activate or release SPS in the downlink

★ Once the UE has received an activate downlink SPS command on the PDCCH, it periodically receives data on the PDSCH. The subframes during which the UE expects to receive downlink data are defined by:

$$(10 \times \text{SFN} + \text{subframe}) = [(10 \times \text{SFN}_{start} + \text{subframe}_{start}) + \text{N} \times \textit{SPS Downlink Interval}] \text{ modulo } 10240$$

where,
N can have all integer values > 0
SFN_{start} and subframe_{start} are the SFN and subframe during which the SPS activation instruction was received
SPS Downlink Interval is the value signalled by the RRC layer, shown in Table 319

The equation is modulo 10240 because this corresponds to the number of subframes within a complete SFN cycle, i.e. the SFN range is from 0 to 1023

★ In the case of the uplink, DCI format 0 can be used to activate and release SPS during a specific subframe. The coding of the fields within DCI format 0 to activate and release uplink SPS is presented in Table 321

	Activate SPS	Release SPS
TPC Command	00	00
Cyclic Shift DM RS	000	000
MCS and Redundancy Version	MSB is set to 0	11111
Resource Block Assignment and Hopping Resource Allocation	-	set to all 1's

Table 321 – Coding of DCI format 0 fields used to activate or release SPS in the uplink

★ Once the UE has received an activate uplink SPS command on the PDCCH, it is able to periodically transmit data on the PUSCH. The subframes during which the UE is allowed to transmit are defined by

$$(10 \times \text{SFN} + \text{subframe}) = [(10 \times \text{SFN}_{start} + \text{subframe}_{start}) + \text{N} \times \textit{SPS Uplink Interval} + \textit{Offset} \times (\text{N modulo } 2)\ ] \text{ modulo } 10240$$

where,
N can have all integer values > 0
SFN_{start} and subframe_{start} are the SFN and subframe during which the SPS activation instruction was received
SPS Uplink Interval is the value signlled by the RRC layer, shown in Table 319
Offset = 0 if the 'Two Intervals Configuration' flag was not included within the RRC message used to configure SPS (it can only be included for TDD). Otherwise, it is defined by the values presented in Table 322

Uplink-Downlink Subframe Configuration	Position of initial SPS activation (subframe)	Subframe Offset (ms)
0	offset not applicable	0
1	2, 7	1
	3, 8	-1
2	2	5
	7	-5
3	2, 3	1
	4	-2
4	2	1
	3	-1
5	offset not applicable	0
6	offset not applicable	0

Table 322 – Offset values applied to uplink SPS when using the 'Two Intervals Configuration'

★ Figure 266 illustrates an example TDD scenario when the 'Two Intervals Configuration' is not used. This example is based upon uplink-downlink subframe configuration 1. SPS has been activated with a 20 ms uplink interval using the 3rd subframe within every second radio frame

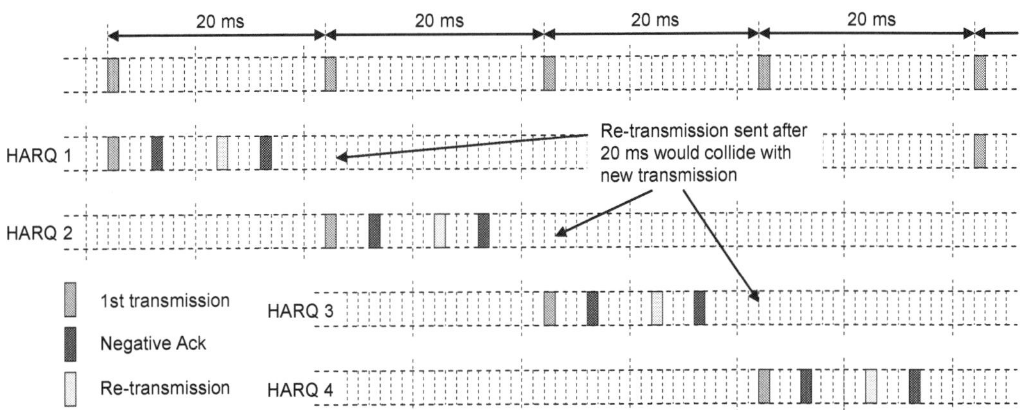

Figure 266 – Uplink SPS for TDD without using the 'Two Intervals Configuration'

★ With this example, only a single HARQ retransmission is possible. A second retransmission would collide with the first transmission from the next HARQ process. This is caused by the synchronous nature of uplink HARQ which means that retransmissions have to occur at specific times. This issue is alleviated by using the 'Two Intervals Configuration'

★ Figure 267 illustrates an example TDD scenario using the 'Two Intervals Configuration'. This example is also based upon uplink-downlink subframe configuration 1. SPS has been activated with a 20 ms uplink interval starting with the 3rd subframe within the first radio frame

★ Table 322 specifies that when using uplink-downlink configuration 1, and initiating SPS during subframe 2 (3rd subframe within the radio frame based upon 0, 1, 2 numbering), then the subframe offset has a value of 1 ms. This means that every second resource allocation is delayed by 1 ms. This prevents the second retransmission from colliding with the next new transmission. 3 retransmissions can be completed before a retransmission would collide with the next new transmission

★ Thus, the 'Two Intervals Configuration' has increased the maximum allowed number of retransmissions for the uplink of TDD when using SPS. Referring back to Table 322 it can be seen that uplink-downlink subframe configurations 0, 5 and 6 have 0 ms offsets associated with them. It is not possible to apply an offset to configuration 5 because there is only a single uplink subframe within each radio frame. 3GPP did not view it necessary to specify offsets for configurations 0 and 6 because in these cases the collision between the retransmission and new data transmission occurs after 70 ms and 60 ms respectively, i.e. after a relatively large number of retransmissions

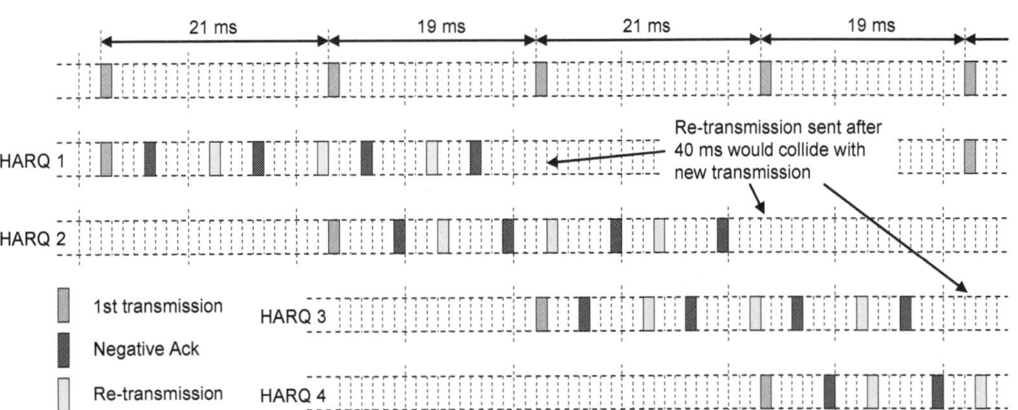

Figure 267 – Uplink SPS for TDD using the 'Two Intervals Configuration'

★ Figure 268 illustrates the uplink scenario for SPS with FDD. Uplink HARQ is also synchronous for FDD but the timing allows 4 retransmissions before an uplink collision would occur. This illustrates why it is not necessary for an offset type solution for FDD

★ In the case of the downlink, resources for retransmissions are dynamically allocated when necessary using the PDCCH. In subframes where the UE has semi-persistent downlink resources, if the UE cannot find its C-RNTI then it assumes an SPS transmission. If the UE finds its C-RNTI on the PDCCH then that PDCCH allocation overrides the semi-persistent allocation and the UE does not decode the SPS resources. When LTE Advanced Carrier Aggregation is configured only PDCCH allocations for the primary cell can override the SPS allocation

★ In the case of the uplink, in subframes where the UE has semi-persistent uplink resources, if the UE cannot find its C-RNTI then it assumes an SPS transmission. The network performs decoding of the pre-defined Resource Blocks according to the pre-defined Modulation and Coding Scheme (MCS). Otherwise, if the UE finds its C-RNTI on the PDCCH during subframes that it has an SPS allocation, the PDCCH allocation overrides the persistent allocation for that subframe. Retransmissions are either implicitly allocated in which case the UE uses the SPS uplink allocation, or explicitly allocated via PDCCH in which case the UE does not follow the SPS allocation. Similarly as for the downlink, when LTE Advanced Carrier Aggregation is configured only PDCCH allocations for the primary cell can override the SPS allocation

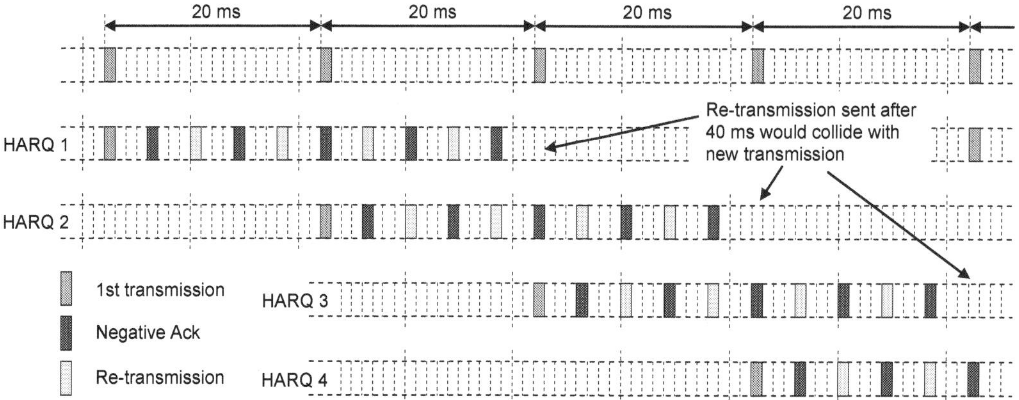

Figure 268 – Uplink SPS for FDD

★ Interpretation of the downlink transmission modes changes when SPS is used. Only single antenna port transmission and transmit diversity are supported when SPS is used. Spatial multiplexing is not required because SPS is not intended to support high throughputs. Table 323 presents the interpretation of each transmission mode when using SPS. This can be compared with Table 20 which presents the equivalent interpretation when the PDCCH is scrambled using the C-RNTI

Mode	PDSCH Transmission Scheme	DCI Format	Search Space
1	Single Antenna Port, port 0	1A	Common and UE Specific
		1	UE Specific
2	Transmit Diversity	1A	Common and UE Specific
		1	UE Specific
3	Transmit Diversity	1A	Common and UE Specific
		2A	UE Specific
4	Transmit Diversity	1A	Common and UE Specific
		2	UE Specific
5	Transmit Diversity	1A	Common and UE Specific
6	Transmit Diversity	1A	Common and UE Specific
7	Single Antenna Port, port 5	1A	Common and UE Specific
		1	UE Specific
8	Single Antenna Port, port 7	1A	Common and UE Specific
	Single Antenna Port, port 7 or 8	2B	UE Specific
9	Single Antenna Port, port 7	1A	Common and UE Specific
	Single Antenna Port, port 7 or 8	2C	UE Specific

Table 323 – PDSCH transmission modes when using SPS C-RNTI

★ Similarly, interpretation of the uplink transmission modes changes when SPS is used. Only single antenna port transmission is supported when SPS is used. Spatial multiplexing is not required because SPS is not intended to support high throughputs. Table 324 presents the interpretation of each transmission mode when using SPS. This can be compared with Table 153 which presents the equivalent interpretation when the PDCCH is scrambled using the C-RNTI

Mode	PDSCH Transmission Scheme	DCI Format	Search Space
1	Single Antenna Port, port 10	0	Common and UE Specific
2	Single Antenna Port, port 10	0	Common and UE Specific

Table 324 – PUSCH transmission modes when using SPS C-RNTI

★ 3GPP References: TS 36.212, TS 36.213, TS 36.331

28 MULTIMEDIA BROADCAST SERVICES

28.1 INTRODUCTION

★ Multimedia Broadcast Multicast Services (MBMS) provide support for the transmission of multimedia content such as text, pictures, audio and video. MBMS can be used for the transmission of mobile television services

★ In general, MBMS provides support for:

  o broadcast services - 'point-to-multipoint' unidirectional downlink transmissions from a cell to a population of UE. Each UE receives downlink data using the same air-interface resources. This helps to improve the efficiency of data transmission by avoiding the requirement for dedicated resource allocations

  o multicast services - are similar to broadcast services from the perspective of providing 'point-to-multipoint' unidirectional downlink transmissions from a cell to a population of UE. Multicast services require additional UE specific procedures prior to the reception of the MBMS data, i.e. the 'subscription' and 'joining' procedures are required. The UE specific 'leaving' procedure is also used when a UE stops the reception of MBMS data

★ Multicast is likely to be applicable to specialised subscription services, e.g. transmission of results from a sporting event or subscription based television content. Broadcast is likely to be applicable to more general services, e.g. transmission of promotional videos and non-subscription based television content

★ LTE includes support for the broadcast component of MBMS but not the multicast component. This is in contrast to UMTS which includes support for both the broadcast and multicast components

★ Some aspects of MBMS for LTE were defined within the release 8 version of the specifications, e.g. the PMCH physical channel was defined. However, complete support for MBMS was not provided until the release 9 version of the specifications. Enhancements were subsequently introduced within the release 10 version of the specifications, e.g. the UE counting procedure

★ MBMS over a Single Frequency Network (MBSFN) corresponds to the simulcast transmission of MBMS services, i.e. simultaneous transmission of identical data streams from multiple time synchronised cells using the same RF carrier. Identical air-interface resources are used by each cell so the set of multiple transmissions appear as a single transmission from the UE perspective. The concept of MBSFN is illustrated in Figure 269

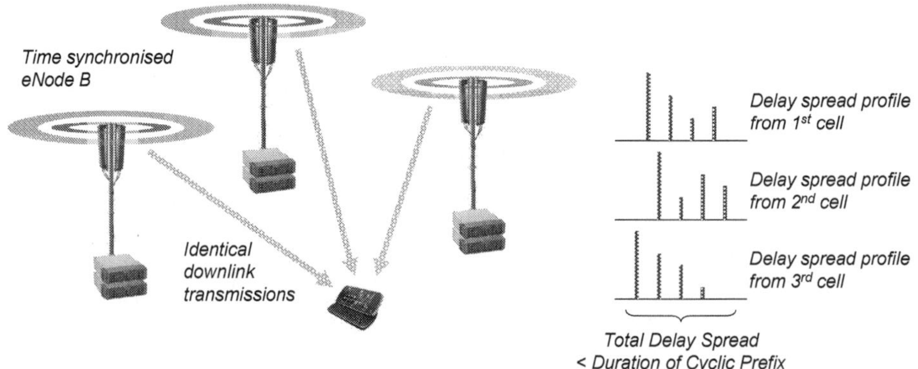

Figure 269 – Concept of MBMS over Single Frequency Network (MBSFN)

★ MBSFN transmissions always use the extended cyclic prefix to allow for inaccuracies in the time synchronisation between neighbouring eNode B. These inaccuracies appear to the UE as additional delay spread and tend to increase the overall delay spread. The overall delay spread must remain less than the duration of the extended cyclic prefix to avoid the delay spread from generating interference

★ MBSFN transmissions help to improve the signal-to-noise ratio towards cell edge. Intercell interference is changed into additional wanted signal power. Cell edge conditions define the upper limit upon the throughput which can be used for broadcast services. Improving the conditions at cell edge allows the use of higher throughput services

★ MBMS uses the hierarchy of area types illustrated in Figure 270

★ An MBMS Service Area is the area within which the data belonging to a specific MBMS service is transmitted. For example, a national news service could be broadcast across the entire country so the MBMS Service Area for that program would be the entire country. A regional news service could be broadcast across only the local area so the MBMS Service Area for that program would be the local area. Figure 270 illustrates a simplistic scenario with only a single MBMS Service Area

★ An MBSFN Synchronisation Area is defined by a collection of eNode B which can be synchronised to perform MBSFN transmissions. Each RF carrier at an eNode B can belong to a maximum of 1 MBSFN Synchronisation Area. The definition of an MBSFN Synchronisation Area is not dependent upon the definition of the MBMS Service Areas, i.e. the MBMS Service Area determines the area within which a service is broadcast, while the MBSFN Synchronisation Area defines a group of eNode B which can be synchronised

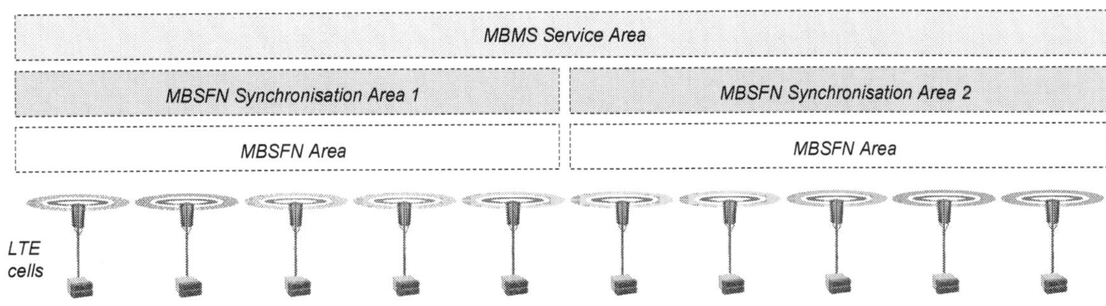

Figure 270 – Hierarchy of areas for MBMS

- ★ A single MBSFN Synchronisation Area can support multiple MBSFN Areas. An MBSFN Area is a group of cells within an MBSFN Synchronisation Area which are co-ordinated to provide an MBSFN transmission for a specific MBMS service. Each cell can belong to multiple MBSFN Areas (up to 8) to provide support for the broadcast of multiple MBMS services. Figure 270 illustrates a simplistic scenario where each cell serves only a single MBSFN Area
- ★ MBSFN Area Reserved Cells (not shown in Figure 270) can be defined within an MBSFN Area. These are cells which do not contribute to the MBSFN transmission. Nevertheless, they coordinate their own transmissions to help minimise any interference towards the MBSFN transmissions, i.e. they may apply Discontinuous Transmission (DTX) or a reduced transmit power for the resources used by the MBSFN transmission
- ★ UE are able to receive MBMS services in both RRC Idle and RRC Connected modes
- ★ MBMS services are supported by both the FDD and TDD variants of LTE. Relay Nodes do not support the transmission of MBMS
- ★ There is no uplink feedback channel for MBMS so retransmissions are not supported. MBMS uses unacknowledged mode RLC so sequence numbers are included by the RLC layer but there is no support for RLC retransmissions. Similarly, MBMS does not use HARQ retransmissions
- ★ The set of System Information Blocks (SIB) include information regarding MBSFN:
  - o SIB 2 (presented in section 10.4)
    - introduced within the release 8 version of the 3GPP specifications
    - the MBSFN subframe configuration list specifies the set of downlink subframes reserved for MBSFN
    - can be used to inform release 8 UE that reception of the cell specific Reference Signal and the PDSCH should not be attempted within the data section of those subframes (the PDCCH and cell specific Reference Signal can still be received within the control section at the start of each subframe)
  - o SIB 13 (presented in section 10.15)
    - introduced within the release 9 version of the 3GPP specifications specifically for the purposes of MBSFN
    - includes information which allows UE to acquire the MCCH and subsequently extract MBMS control information associated with up to 8 MBSFN areas
- ★ Subframes configured for MBSFN can also be used for other purposes. The release 10 version of the 3GPP specifications allows MBSFN subframes to be used as:
  - o Almost Blank Subframes (ABS) for the purposes of Inter-Cell Interference Coordination (ICIC). ABS allow eNode to reduce their transmissions during certain time intervals. This helps to improve the signal to noise ratio conditions for neighbouring eNode B, but reduces the total resources available for transmission. ABS are described in greater detail in section 32.14
  - o 'LTE Advanced' subframes for the purposes of maximising the PDSCH throughput by reducing the overhead generated by the cell specific Reference Signal. UE specific and CSI Reference Signals are sent during 'LTE Advanced' subframes but the cell specific Reference Signal is not transmitted during the data section of the subframe

  Configuring subframes as MBSFN prevents any UE from attempting to measure the cell specific Reference Signal and also prevents 3GPP release 8 and 9 UE from attempting to decode any PDSCH data
- ★ Only a subset of subframes can be configured for MBSFN:
  - o in the case of FDD, subframes 1, 2, 3, 6, 7 and 8 can be used for MBSFN
  - o in the case of TDD, subframes 3, 4, 7, 8 and 9 can be used for MBSFN

  Other subframes are excluded to avoid conflict with the PBCH, synchronisation signals and paging messages. In the case of TDD, subframe 2 is excluded because it is always an uplink subframe
- ★ 3GPP References: TS 36.300, TS 36.331, TS 23.246

28.2 ARCHITECTURE

★ MBMS network elements appear within both the Evolved Packet Core (EPC) and the Evolved UTRAN (E-UTRAN)

★ The network architecture for MBMS is illustrated in Figure 271. There are two general scenarios for the radio access network – with either a standalone or integrated Multi-cell / Multicast Coordination Entity (MCE)

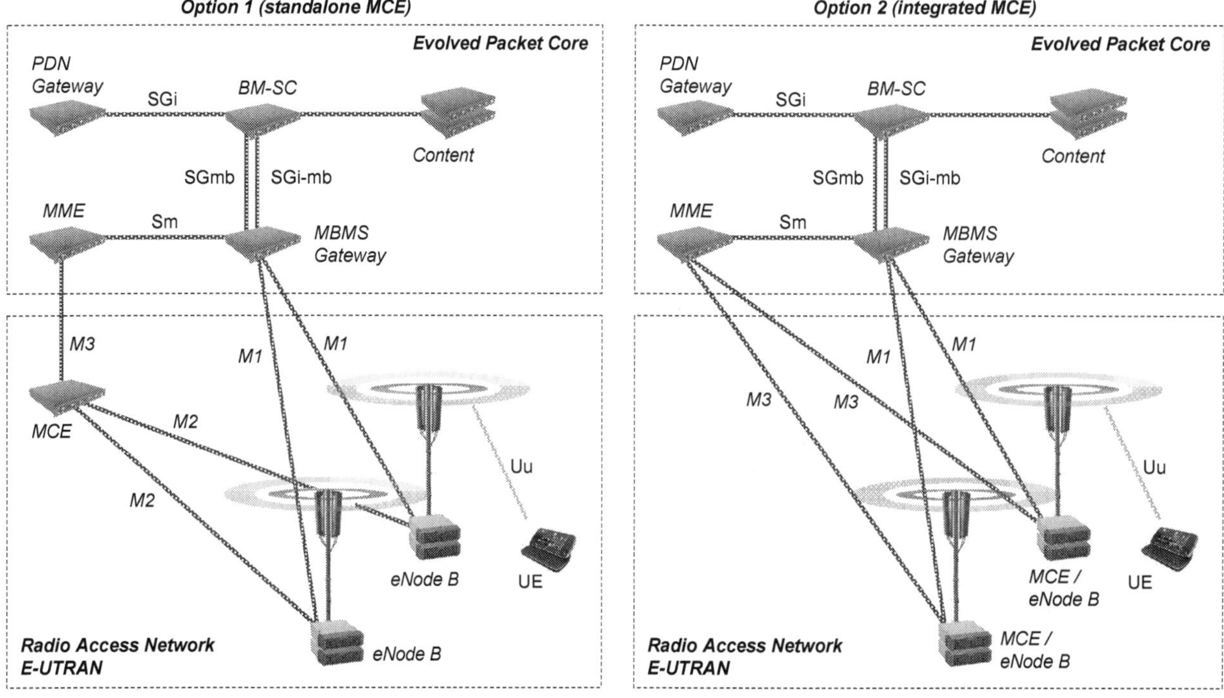

Figure 271 – Network architectures for MBMS

★ The MCE is a logical entity within the radio access network. It can be implemented as a standalone unit which is shared between multiple eNode B, or it can be implemented within each eNode B. The main functions of the MCE are:

- admission control and subsequent allocation of the time and frequency domain radio resources for MBMS
- selection of an appropriate Modulation and Coding Scheme (MCS)
- pre-emption of resources when necessary according to the Allocation and Retention Priority (ARP)
- initiation of the counting procedure and subsequent collection of the results
- use of the counting results to suspend and resume MBMS services when appropriate

★ The MBMS Gateway is a logical entity within the core network. It can be implemented as either a standalone unit or integrated within another network element. For example, it could be integrated within the BM-SC or a Serving Gateway. The main functions of the MBMS Gateway are:

- allocation of an IP multicast address for the transmission of MBMS application data towards the relevant set of eNode B. IP multicast allows IP packets to be sent to a group of destination addresses using a single transmission
- subsequent transmission of MBMS application data using the IP multicast address
- involvement in session control signalling procedures between the BM-SC and the radio access network, i.e. control plane signalling procedures initiated by the BM-SC involve the MBMS Gateway before reaching the MME and radio access network

★ The Broadcast Multicast Service Centre (BM-SC) serves as an entry point for MBMS content. It also provides support for high level management of MBMS services. The main functions of the BM-SC are:

- MBMS session management, e.g. initiating the MBMS session start and stop procedures
- allocation of an identity and Temporary Mobile Group Identity (TMGI) to each MBMS session
- specification of Quality of Service (QoS) parameters associated with each MBMS session
- transmission of MBMS application data using the SYNC protocol to ensure that air-interface transmissions remain synchronised
- service announcements at the application level to provide the end-user with MBMS session scheduling information

★ The M1 interface provides user plane connectivity between the MBMS Gateway and the set of eNode B. There is no control plane signalling across the M1 interface. The M1 interface is specified within 3GPP TS 36.445. GTPv1-U over UDP/IP is used to transfer the application data. GTPv1-U is specified within 3GPP TS 29.281. Diffserv code point marking is used to provide QoS by managing the Per Hop Behaviour (PHB) associated with each IP packet. The SYNC protocol specified within 3GPP TS 25.446 is used when sending application data. This protocol provides support for MBMS content synchronisation and is described in section 28.8

★ The M2 interface provides control plane connectivity between the MCE and eNode B. There is no application data transfer across the M2 interface. 3GPP TS 36.443 defines the M2 Application Protocol (M2-AP) which includes signalling procedures for:

  - MBMS Session Start
  - MBMS Session Stop
  - MBMS Session Update
  - MBMS Scheduling Information
  - eNode B and MCE Configuration Updates
  - MBMS Service Counting

  The M2 interface uses Stream Control Transmission Protocol (SCTP) over IP as a signalling transport mechanism, rather than TCP or UDP. SCTP provides a reliable data transfer service which supports retransmissions. SCTP allows signalling messages to be transferred using parallel streams. This avoids retransmissions from one signalling procedure delaying the transfer of messages from other signalling procedures. SCTP also includes a flow control mechanism, similar to that used by TCP. The use of SCTP is specified within 3GPP TS 36.442

★ The M3 interface provides control plane connectivity between the MCE and MME. There is no application data transfer across the M3 interface. 3GPP TS 36.444 defines the M3 Application Protocol (M3-AP) which includes signalling procedures for:

  - MBMS Session Start
  - MBMS Session Stop
  - MBMS Session Update

  The M3 interface uses SCTP over IP as a signalling transport mechanism. The use of SCTP is specified within 3GPP TS 36.442

★ The Sm interface is used for control plane signalling between the MME and MBMS Gateway. It is based upon the GTPv2-C protocol specified within 3GPP TS 29.274. This protocol includes a relatively small set of messages used to support MBMS, i.e. request and response messages for the MBMS session start, stop and update procedures

★ The SGmb interface is used for control plane signalling between the BM-SC and MBMS Gateway. It is based upon the Diameter protocol specified within IETF RFC 3588. Use of the Diameter protocol for the purposes of MBMS is specified within 3GPP TS 29.061. Signalling messages are defined to support the MBMS session start, stop and update procedures

★ The SGi-mb interface is used for application data transfer between the BM-SC and MBMS Gateway. MBMS data crosses this interface before being sent across the M1 interface towards the set of eNode B

★ 3GPP References: TS 23.246, TS 36.300, TS 36.440, TS 36.442, TS 36.443, TS 36.444, TS 36.445, TS 29.061, TS 29.274, TS 29.281

28.3 CHANNELS AND SIGNALS

★ The logical, transport and physical channels associated with MBMS are illustrated in Figure 272
  - the Multicast Traffic Channel (MTCH) and Multicast Control Channel (MCCH) are mapped onto the Multicast Channel (MCH)
  - the MCH is mapped onto the Physical Multicast Channel (PMCH)

★ The MTCH transfers application data, whereas the MCCH transfers control plane signalling

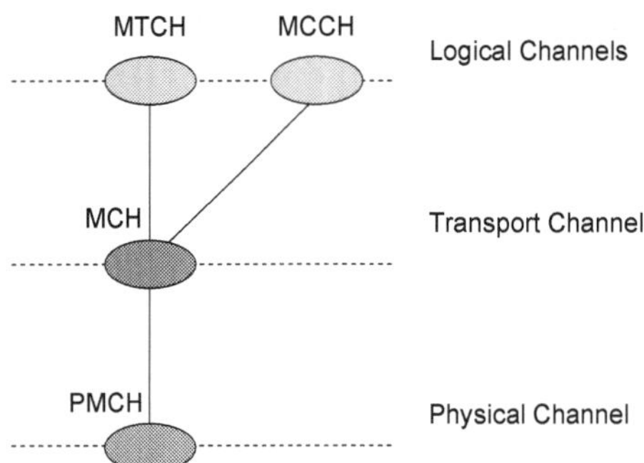

Figure 272 – Logical, transport and physical channels for MBMS

★ The PMCH is described in section 6.6. The main points include:
  - the PMCH always uses the extended cyclic prefix but can use either the 15 kHz or 7.5 kHz subcarrier spacing. The 7.5 kHz subcarrier spacing provides twice as many Resource Elements in the frequency domain but half as many Resource Elements in the time domain, i.e. 24 × 3 Resource Elements within a Resource Block rather than 12 × 6
  - the PMCH can be modulated using either QPSK, 16QAM or 64QAM although the allocated modulation scheme does not change dynamically over time. The modulation scheme applied during MCCH subframes is signalled within SIB13, whereas the modulation scheme applied during MTCH subframes is signalled within the MCCH
  - the PMCH is used to transfer a single MCH transport block during a single 1 ms subframe. All PMCH transmissions use antenna port 4. Neither transmit diversity nor spatial multiplexing are applicable to the PMCH
  - the PMCH can be transmitted:
    - in the case of FDD, during subframes 1, 2, 3, 6, 7 and 8
    - in the case of TDD, during subframes 3, 4, 7, 8 and 9
  - the PMCH uses all Resource Blocks in the frequency domain
  - a non-MBSFN region is reserved at the start of each MBMS subframe. This region can have a duration of either 1 or 2 symbols. The non-MBSFN region can be used for PDCCH signalling, e.g. uplink resource allocations on the PUSCH. The non-MBSFN region uses the same cyclic prefix as subframe 0, i.e. the same cyclic prefix as a non-MBSFN subframe

★ The MCH is described in section 7.4. The main points include:
  - the MCH can transfer 1 transport block per 1 ms subframe
  - processing of the MCH transport channel is the same as that for the DL-SCH transport channel, i.e. it includes CRC attachment, insertion of filler bits, code block segmentation, channel coding (rate 1/3 Turbo coding), rate matching and code block concatenation

★ Each cell broadcasts one MCCH for each MBSFN area. SIB13 provides scheduling information for the MCCH. Scheduling of the MCCH is described in section 28.4

★ The MCCH can be used to transfer the MBSFN Area Configuration and MBMS Counting Request messages
  - the MBSFN Area Configuration message provides scheduling information for the MTCH. Its content is presented in section 28.6
  - The MBMS Counting Request message helps the network to quantify the number of UE interested in a specific MBMS service. It is used as part of the counting procedure described in section 28.9

★ Each cell broadcasts one MTCH for each MBMS session belonging to an MBMS service (an MBMS session can be thought of as a television programme, while an MBMS service can be thought of as a television channel). The set of MTCH can be referenced by their logical channel identities, whereas the set of MBMS sessions can be referenced by the combination of their session identities and Temporary Mobile Group Identities (TMGI). The TMGI identifies the MBMS service and PLMN

★ MTCH scheduling uses a combination of RRC and MAC layer signalling: MBSFN Area Configuration message transferred by the RRC layer and MCH Scheduling Information control element transferred by the MAC layer. These are described in section 28.6

★ The MBSFN Reference Signal is described in section 5.2.2. The main points include:

  - the MBSFN Reference Signal is broadcast with the PMCH on antenna port 4
  - the MBSFN Reference Signal is only broadcast within the data section of subframes allocated to the PMCH
  - the sequence used to generate the MBSFN Reference Signals is a function of the MBSFN area identity
  - the Resource Elements allocated to the MBSFN Reference Signal are fixed, and are not a function of the Physical layer Cell Identity (PCI) nor MBSFN area
  - the overhead generated by the MBSFN Reference Signal is 12.5 % relative to the total number of Resource Elements within a Resource Block (both control and data sections)

★ 3GPP References: TS 36.211, TS 36.212, TS 36.213, TS 36.321, TS 36.322, TS 36.331

28.4 MCCH SCHEDULING

★ SIB13 broadcasts information regarding both MCCH scheduling and MCCH change notification. The content of SIB13 is presented in section 10.15

★ MCCH scheduling information defines the subframes during which the MCCH is broadcast. MCCH scheduling information is provided independently for each MBSFN Area. A single cell can support up to 8 MBSFN areas so SIB13 is able to broadcast scheduling information for up to 8 MCCH

★ SIB13 broadcasts the following parameters for each MCCH:

  - MCCH Repetition Period (32, 64, 128 or 256 radio frames)
  - MCCH Offset (0 to 10)
  - Subframe Allocation Information (bitmap of length 6 bits)

★ The eNode B transmits the MCCH for a specific MBSFN Area during radio frames that satisfy the condition:

System Frame Number (SFN) mod MCCH Repetition Period = MCCH Offset

For example, if the repetition period is 64 radio frames and the offset is 5 then the MCCH is broadcast during radio frames with SFN of 5, 69, 133, 197, etc. The MCCH Repetition Period defines the rate at which the MCCH content is repeated. MCCH transmissions are repeated so that UE can be switched on at any time and be able to acquire the content

★ The Subframe Allocation Information indicates which subframes within the identified radio frames are used for transmission of the MCCH. The bitmap defined by the Subframe Allocation Information maps onto the set of subframes using the mapping shown in Table 325

	Bit 1	Bit 2	Bit 3	Bit 4	Bit 5	Bit 6
FDD	SF 1	SF 2	SF 3	SF 6	SF 7	SF 8
TDD	SF 3	SF 4	SF 7	SF 8	SF 9	Not Used

Table 325 – Mapping between Subframe Allocation Information bits and subframe numbers

★ MCCH scheduling information does not include any frequency domain information because the PMCH uses all Resource Blocks in the frequency domain

★ SIB13 also broadcasts the MCCH Modification Period for each MBSFN Area, i.e. for each MCCH. This can be signalled with values of 512 or 1024 radio frames. Modification boundaries are then defined at radio frames which satisfy:

System Frame Number (SFN) mod MCCH Modification Period = 0

The content of the MCCH is only allowed to change after a modification boundary. The use of a relatively large MCCH Modification Period reduces the rate at which the UE needs to check for changes

★ When checking for changes in the MCCH content, UE do not directly check the content of the MCCH. Change notification indicators are transmitted within Downlink Control Information (DCI) format 1C on the PDCCH

★ DCI format 1C contains a bit string of length 8 bits when used to transmit MCCH change notification indicators. The length of 8 bits allows a single bit to be associated with each of the 8 possible MCCH within a cell. The MBSFN Area Information within SIB13 includes a Notification Indicator which links each MCCH to a specific bit within the bit string, i.e. the Notification Indicator can be signalled with values between 0 and 7

★ UE detect that DCI format 1C is being used for MCCH change notification purposes by detecting that the CRC bits have been scrambled by the M-RNTI. The M-RNTI has a fixed value of FFFD

★ The UE checks for DCI format 1C transmissions which include MCCH change notification indicators during radio frames which satisfy the condition:

System Frame Number (SFN) mod Notification Repetition Period = Notification Offset

The Notification Repetition Period is defined as MIN{MCCH Modification Period, for all MCCH} / Notification Repetition Coefficient, where the Notification Repetition Coefficient is broadcast within SIB3 using a value of either 2 or 4

★ The Notification Offset is also broadcast within SIB13, using a value between 0 and 10

★ The subframe to check within the MCCH change notification radio frame is defined by the Notification Subframe Index. This is signalled within SIB13 using a value between 1 and 6. The value points to a column within Table 325 (only values 1 to 5 are applicable to TDD)

★ An MCCH change notification indicates that the content of the MCCH will change during the next MCCH modification period. After receiving a change notification, the UE decodes the new MCCH information immediately after the start of the new modification period

★ Figure 273 illustrates an example of the timing of the MCCH and their associated change notification indicators for 2 MBSFN areas. This example assumes different modification and repetition periods for each MCCH. It also assumes a notification repetition coefficient of 2

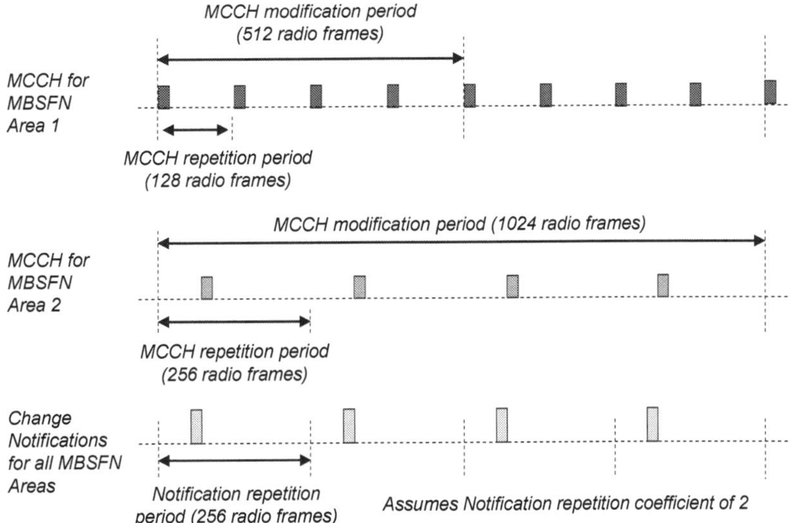

Figure 273 – Example timing of MCCH and Change Notification transmissions for 2 MBSFN areas

★ Figure 274 summarises the general procedure for reading the content of the MCCH based upon an initial reading of SIB13, and subsequent checks for change notifications within DCI format 1C

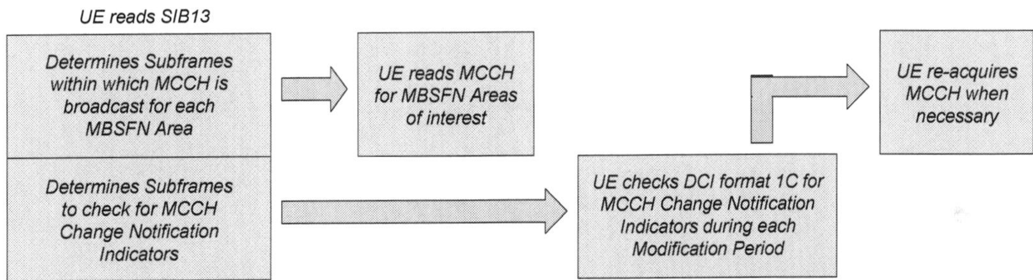

Figure 274 – General procedure for reading the content of the MCCH

★ UE also need to monitor for changes in the content of SIB13. The Paging message is used to indicate that the content of the system information is going to change during the next BCCH modification period, i.e. the paging message can include a System Information Modification flag. UE can also use the System Information Value Tag broadcast within SIB1 to detect whether or not the content of the system information has changed

★ 3GPP References: TS 36.331, TS 36.212

28.5 MCCH MESSAGES

★ There are 2 types of message which can be transferred using the MCCH logical channel. These are presented in Table 326

SRB	Direction	RRC Message	RLC Mode
MCCH	Downlink	MBSFN Area Configuration MBMS Counting Request	Unacknowledged

Table 326 – Messages transferred using the MCCH logical channel

★ The MBSFN Area Configuration message was introduced within the release 9 version of the 3GPP specifications. It provides MTCH scheduling information

★ The MBMS Counting Request message was introduced within the release 10 version of the 3GPP specifications for the purposes of the counting procedure (section 28.9)

★ 3GPP References: TS 36.331

28.6 MTCH SCHEDULING

★ MTCH scheduling uses a combination of RRC and MAC layer signalling:

  o MBSFN Area Configuration message transferred by the RRC layer

  o MCH Scheduling Information control element transferred by the MAC layer

★ The MBSFN Area Configuration message is transferred using the MCCH. There is a separate MCCH for each MBSFN Area so there is also a separate MBSFN Area Configuration message for each MBSFN Area

★ The content of the MBSFN Area Configuration message is presented in Table 327. This message provides the UE with information regarding current MBMS sessions. It also provides the UE with information regarding the radio frames and subframes during which the MTCH is broadcast

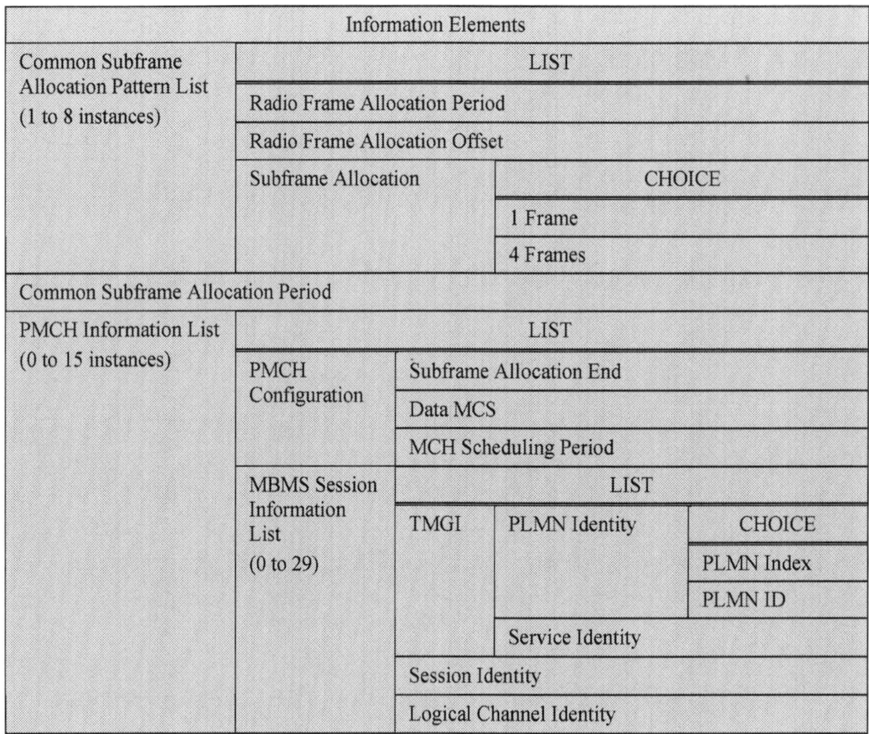

<table>
<tr><td colspan="5">Information Elements</td></tr>
<tr><td rowspan="6">Common Subframe Allocation Pattern List (1 to 8 instances)</td><td colspan="4">LIST</td></tr>
<tr><td colspan="4">Radio Frame Allocation Period</td></tr>
<tr><td colspan="4">Radio Frame Allocation Offset</td></tr>
<tr><td rowspan="3" colspan="2">Subframe Allocation</td><td colspan="2">CHOICE</td></tr>
<tr><td colspan="2">1 Frame</td></tr>
<tr><td colspan="2">4 Frames</td></tr>
<tr><td colspan="5">Common Subframe Allocation Period</td></tr>
<tr><td rowspan="11">PMCH Information List (0 to 15 instances)</td><td colspan="4">LIST</td></tr>
<tr><td rowspan="3">PMCH Configuration</td><td colspan="3">Subframe Allocation End</td></tr>
<tr><td colspan="3">Data MCS</td></tr>
<tr><td colspan="3">MCH Scheduling Period</td></tr>
<tr><td rowspan="7">MBMS Session Information List (0 to 29)</td><td colspan="3">LIST</td></tr>
<tr><td rowspan="4">TMGI</td><td rowspan="3">PLMN Identity</td><td>CHOICE</td></tr>
<tr><td>PLMN Index</td></tr>
<tr><td>PLMN ID</td></tr>
<tr><td colspan="2">Service Identity</td></tr>
<tr><td colspan="3">Session Identity</td></tr>
<tr><td colspan="3">Logical Channel Identity</td></tr>
</table>

Table 327 – Content of MBSFN Area Configuration message

★ The Common Subframe Allocation list defines the aggregate set of radio frames and subframes for the PMCH physical channel transmissions belonging to the relevant MBSFN Area. The union of up to 8 patterns can be used to generate the aggregate set of time domain resources

★ The PMCH physical channel can be transmitted during radio frames which satisfy:

System Frame Number (SFN) mod Radio Frame Allocation Period = Radio Frame Allocation Offset

where, the Radio Frame Allocation Period can be signalled with a value of 1, 2, 4, 8, 16 or 32 radio frames, and the Radio Frame Allocation Offset can be signalled with a value between 0 and 7

★ The Subframe Allocation indicates the subframes during which the PMCH physical channel can be transmitted:

  o in the case of FDD, the PMCH can be transmitted during subframes 1, 2, 3, 6, 7 and 8

  o in the case of TDD, the PMCH can be transmitted during subframes 3, 4, 7, 8 and 9

These are the same subframes as those available to the MCCH allocations signalled within SIB13

★ The MBSFN Area Configuration message allows either a one frame or four frame subframe pattern to be signalled

  o when using the 1 frame pattern, a bitmap of length 6 bits is used to indicate which of the allowed subframes are allocated to the PMCH. In the case of TDD, the last bit is not used because there are only 5 allowed subframes (see Table 325)

  o when using the 4 frame pattern, a bitmap of length 24 bits is used to indicate which of the allowed subframes are allocated to the PMCH over 4 consecutive radio frames. In this case, the radio frame indicated by the period and offset parameters corresponds to the first of 4 consecutive radio frames

★ The Common Subframe Allocation Period defines the number of radio frames within which the allocated subframes are to be divided between the subsequent list of PMCH. The subframe allocations repeat within this period, which can be configured with a value of 4, 8, 16, 32, 64, 128 or 256 radio frames

★ An example common subframe allocation for MBMS using FDD is illustrated in Figure 275. This example assumes 2 entries within the Common Subframe Allocation List – the first with a period of 16 frames and an offset of 0 frames, and the second with a period of 4 frames and an offset of 3 frames

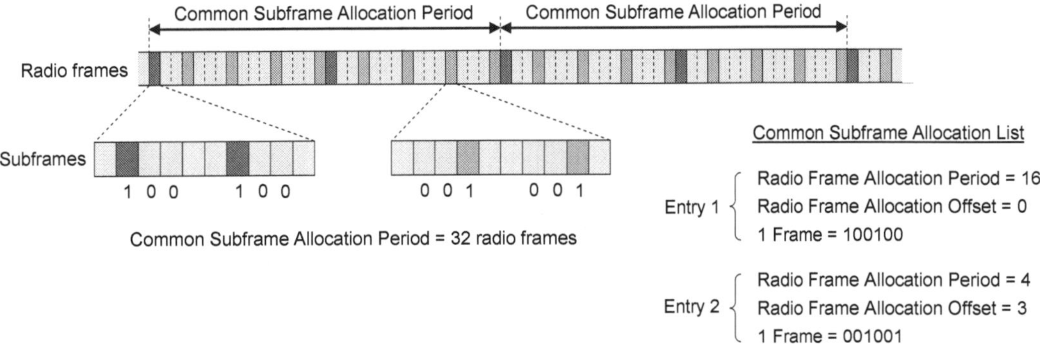

Figure 275 – Example common subframe allocation for MBMS using FDD

★ The PMCH Information list divides the allocated subframes between a set of PMCH. It also associates one or more MTCH logical channels with each PMCH, and each MTCH is given service and session identities

★ The first PMCH uses the first allocated subframe to the mth allocated subframe, where 'm' is defined by the Subframe Allocation End parameter for the first PMCH. The second PMCH uses the m+1th allocated subframe to the nth allocated subframe, where 'n' is defined by the Subframe Allocation End parameter for the second PMCH. This approach is used to allocate sections of successive subframes to each PMCH

★ The Subframe Allocation End parameter can be configured with values between 0 and 1535 to account for the Common Subframe Allocation Period having a maximum value of 256 radio frames and a maximum of 6 subframes being allocated within each radio frame, i.e. 6 × 256 = 1536

★ The example shown in Figure 275 includes 20 subframes within each Common Subframe Allocation Period. Figure 276 illustrates how these subframes could be shared between a set of 3 PMCH. Multiple MTCH logical channels can be associated with each PMCH

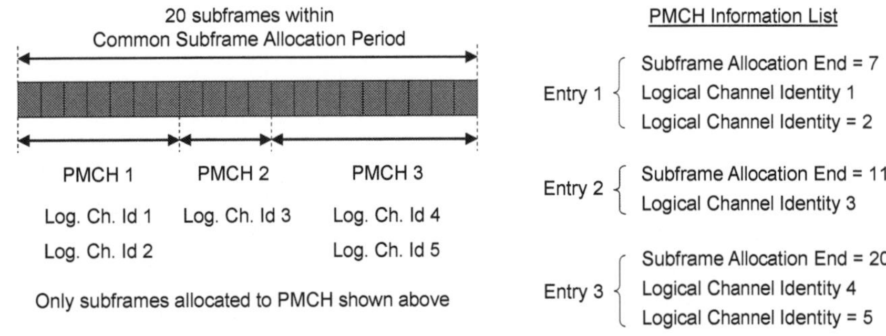

Figure 276 – Example allocation of subframes to a set of 3 PMCH

★ Returning to the MBSFN Area Configuration message shown in Table 327, the Data MCS defines the Modulation and Coding Scheme (MCS) to be used for that PMCH. In the case of the PMCH being used for an MTCH, the MCS can range from 0 to 28. SIB13 broadcasts the equivalent MCS for when the PMCH transfers an MCCH. In this case, only 4 different MCS values are allowed

★ The MCH Scheduling Period (MSP) defines the rate at which the MAC layer sends the 'MCH Scheduling Information' (MSI) MAC control element. This control element specifies the scheduling of each MTCH associated with the PMCH, i.e. it defines how the allocated subframes for the PMCH are distributed between the set of MTCH. It can be signalled using a value of 8, 16, 32, 64, 128, 256, 512 or 1024 radio frames

★ The MBMS Session Information provides details of each MTCH mapped onto the PMCH. This includes the MTCH logical channel identity, as well as the session identity and Temporary Mobile Group Identifier (TMGI). The session identity is a single byte allocated by the BM-SC. It identifies the transmission of a specific MBMS session when combined with the TMGI

★ The Service Identity is used to uniquely identify an MBMS service within a PLMN. It has a length of 3 bytes. A specific MBMS session is associated with an MBMS service, similar to a specific television programme being associated with a specific television channel

★ The structure of the MCH Scheduling Information (MSI) MAC control element is illustrated in Figure 277. This MAC control element is used to specify the scheduling of each MTCH associated with a PMCH. Similar to other MAC control elements, the MSI MAC control element can be included within the payload of an MCH MAC PDU

★ The MSI MAC control element is transmitted during the first subframe allocated to the PMCH within the MCH scheduling period, i.e. its timing will be dependent upon the Subframe Allocation End values

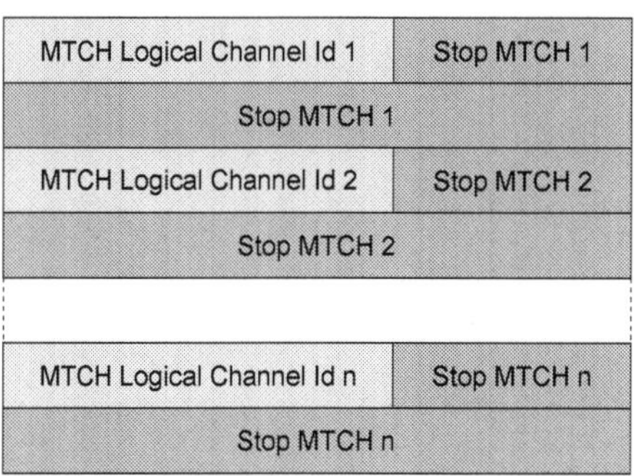

Figure 277 – Structure of MCH Scheduling Information MAC control element

★ The MTCH Logical Channel Identity has a length of 5 bits and is used to identify the MTCH

★ The 'Stop MTCH' field has a length of 11 bits and is used to specify the subframe from within the set of allocated subframes where the MTCH transmission stops, and the next MTCH transmission starts. Thus, the MTCH associated with a specific PMCH are transmitted one after another. A value of 0 corresponds to the first subframe. The special value of 2047 indicates that the MTCH is not scheduled. Values between 2043 and 2046 are not currently used

★ It is possible that there are unused PMCH subframes after the final 'Stop MTCH' value. This depends upon the capacity requirements of the current MTCH

★ 3GPP References: TS 36.331, TS 36.321, TS 29.061, TS 24.008

28.7 SESSION START PROCEDURE

★ The session start procedure is used to initiate the transmission of a new MBMS session. The signalling used for the session start procedure is illustrated in Figure 278

★ The procedure starts from the Broadcast Multicast Service Centre (BM-SC) within the Evolved Packet Core (EPC). The BM-SC forwards a Diameter: Re-Auth-Request (RAR) Command to the MBMS Gateway. The RAR Command belongs to the Diameter protocol and is specified within IETF RFC 3588. This message signals the relevant MBMS Service Areas; Quality of Service information; session duration, session identity and session repetition number; Temporary Mobile Group Identifier (TMGI); and IP address information. The content of this message is presented in 3GPP TS 29.061

★ The MBMS Gateway responds using a Diameter: RE-Auth-Answer (RAA) Command. The content of this message is also presented in 3GPP TS 29.061

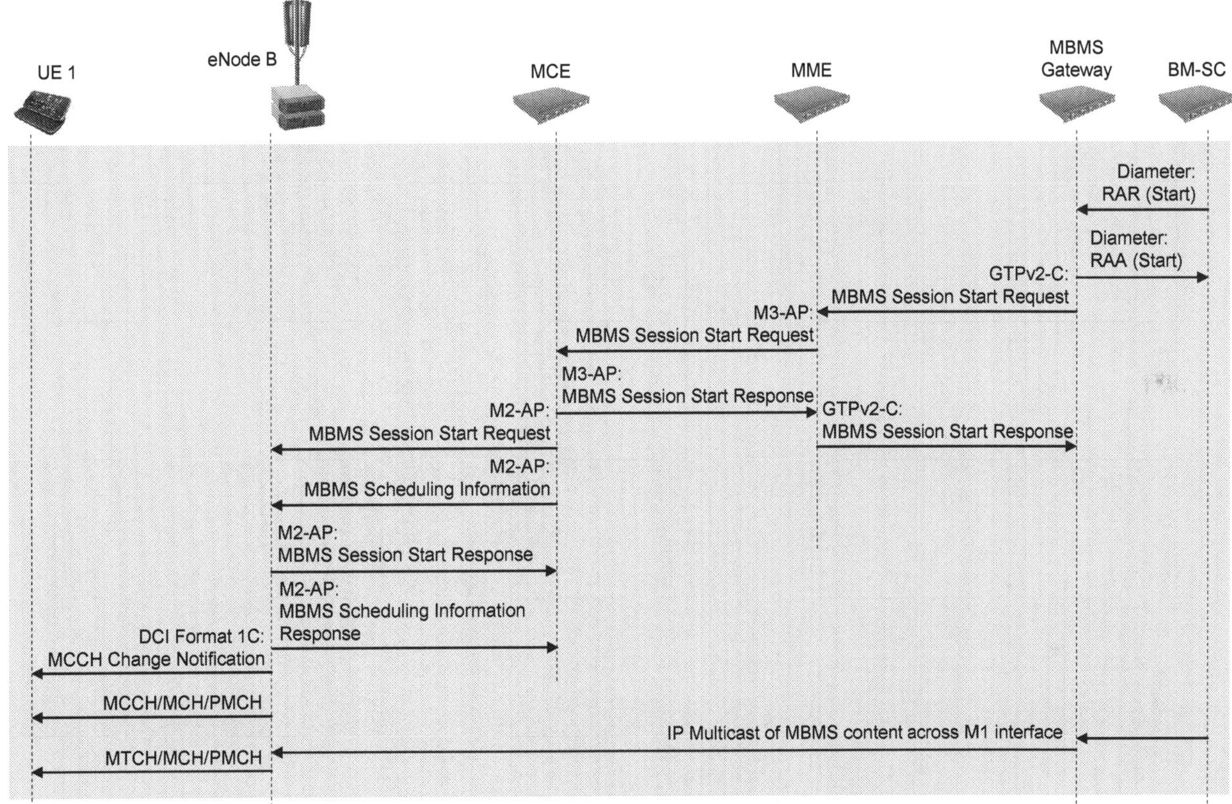

Figure 278 – Signalling associated with the MBMS session start procedure

★ The MBMS Gateway proceeds to send a GTPv2-C: MBMS Session Start Request to the MME. The content of this message is specified within 3GPP TS 29.274. The content is primarily based upon the information provided by the BM-SC. It also includes the IP multicast address allocated by the MBMS Gateway for subsequent transmission of the MBMS IP application data packets

★ The MME then proceeds to send an M3-AP: MBMS Session Start Request to all relevant MCE (only a single MCE is illustrated in Figure 278). The M3 Application Protocol and the content of this message are specified within 3GPP TS 36.444. The content is presented in Table 328

- o The MME MBMS M3AP Identity uniquely identifies the MBMS service within the MME, i.e. it provides a reference which the MCE can subsequently use when signalling to the MME
- o The Temporary Mobile Group Identifier (TMGI) uniquely identifies the MBMS service. As shown in Table 327, it is a combination of the PLMN identity and service identity
- o The MBMS Session Identity provides a reference for the session belonging to the MBMS service
- o Quality of Service (QoS) information is provided in terms of a QoS Class Identifier (QCI), Guaranteed Bit Rate (GBR) information and Allocation and Retention Priority (ARP) information
- o The MBMS Session Duration is signalled using 3 bytes. The coding of these 3 bytes is specified within 3GPP TS 29.061. The duration is defined using a combination of seconds and days. 17 bits are used to signal a duration in seconds between 0 and 86400, while 7 bits are used to signal a duration in days between 0 and 18. 86400 seconds corresponds to 1 day so the maximum duration which can be signalled is 19 days. Setting all bits to 0 indicates an always-on session
- o The MBMS Service Area provides a list of one or more MBMS Service Area identities. These Service Area identities define the area across which the MBMS session is to be broadcast. The coding of this list is specified within 3GPP TS 29.061

- The Minimum Time to MBMS Data Transfer specifies the minimum delay between sending the M3-AP: MBMS Session Start Request to the MCE and the actual start of the application data transfer. A single byte is used to signal this information using units of 1 second. The actual value = signalled value + 1 so the set of 8 bits allow a range from 1 to 256 seconds. The coding of this information is specified within 3GPP TS 48.018
- The Transport Network Layer (TNL) information provides the IP multicast address which will be used when broadcasting the MBMS session application data across the M1 interface towards the eNode B. It also provides the IP address of the multicast source and the GTP Tunnel Endpoint Identifier (TEID)

<table>
<tr><th colspan="3">Information Elements</th></tr>
<tr><td colspan="3">Message Type</td></tr>
<tr><td colspan="3">MME MBMS M3AP Identity</td></tr>
<tr><td colspan="3">TMGI</td></tr>
<tr><td colspan="3">MBMS Session Identity</td></tr>
<tr><td rowspan="6">MBMS E-RAB QoS Parameters</td><td colspan="2">QCI</td></tr>
<tr><td rowspan="2">GBR QoS Information</td><td>MBMS E-RAB Maximum Bit Rate Downlink</td></tr>
<tr><td>MBMS E-RAB Guaranteed Bit Rate Downlink</td></tr>
<tr><td rowspan="3">Allocation and Retention Priority</td><td>Priority Level</td></tr>
<tr><td>Pre-emption Capability</td></tr>
<tr><td>Pre-emption Vulnerability</td></tr>
<tr><td colspan="3">MBMS Session Duration</td></tr>
<tr><td colspan="3">MBMS Service Area</td></tr>
<tr><td colspan="3">Minimum Time to MBMS Data Transfer</td></tr>
<tr><td rowspan="3">Transport Network Layer Information</td><td colspan="2">IP Multicast Address</td></tr>
<tr><td colspan="2">IP Source Address</td></tr>
<tr><td colspan="2">GTP DL TEID</td></tr>
</table>

Table 328 – M3-AP: MBMS Session Start Request message

- Each MCE responds with an M3-AP: MBMS Session Start Response message. The content of this message is specified within 3GPP TS 36.444, and is presented in Table 329. The MME MBMS M3AP Identity is used to address the session within the MME. The MCE provides the MME with an MCE MBMS M3AP Identity which can be used to address the session within the MCE during any subsequent signalling procedures, e.g. when requesting to stop the session

Information Elements
Message Type
MME MBMS M3AP Identity
MCE MBMS M3AP Identity

Table 329 – M3-AP: MBMS Session Start Response message

- After receiving an M3-AP: MBMS Session Start Response message from at least one MCE, the MME can respond to the MBMS Gateway using a GTPv2-C: MBMS Session Start Response message
- The MCE proceeds to send an M2-AP: MBMS Session Start Request to all relevant eNode B (Figure 278 illustrates the standalone MCE network architecture but the MCE can also be integrated within the eNode B). The content of this message is specified within 3GPP TS 36.443, and is presented in Table 330

<table>
<tr><th colspan="2">Information Elements</th></tr>
<tr><td colspan="2">Message Type</td></tr>
<tr><td colspan="2">MCE MBMS M2AP Identity</td></tr>
<tr><td colspan="2">TMGI</td></tr>
<tr><td colspan="2">MBMS Session Identity</td></tr>
<tr><td colspan="2">MBMS Service Area</td></tr>
<tr><td rowspan="3">Transport Network Layer Information</td><td>IP Multicast Address</td></tr>
<tr><td>IP Source Address</td></tr>
<tr><td>GTP DL TEID</td></tr>
</table>

Table 330 – M2-AP: MBMS Session Start Request message

★ The M2 and M3 version of the MBMS Session Start Request message are similar, but the M2 version of the message excludes the QoS parameters, the session duration and the minimum time to MBMS data transfer. These parameters are used by the MCE rather than the eNode B so are not forwarded, i.e. the MCE is responsible for admission control, allocation of resources and pre-emption

★ The MCE also sends an M2-AP: MBMS Scheduling Information message to all relevant eNode B. The content of this message is specified within 3GPP TS 36.443, and is presented in Table 331. This message allows the eNode B to deduce the scheduling for each MTCH and the corresponding content for the MBSFN Area Configuration message on the MCCH

<table>
<tr><th colspan="4">Information Elements</th></tr>
<tr><td colspan="4">Message Type</td></tr>
<tr><td colspan="4">MCCH Update Time</td></tr>
<tr><td rowspan="10">MBSFN Area Configuration List
(1 to 256 instances)</td><td rowspan="5">PMCH Configuration List
(1 to 15 instances)</td><td rowspan="3">PMCH Configuration</td><td>Allocated Subframes End</td></tr>
<tr><td>Modulation and Coding Scheme</td></tr>
<tr><td>MCH Scheduling Period</td></tr>
<tr><td rowspan="2">MBMS Session List per PMCH
(1 to 29 instances)</td><td>MBMS Service Identity</td></tr>
<tr><td>Logical Channel Identity</td></tr>
<tr><td rowspan="3">Subframes Configuration List
(1 to 8 instances)</td><td rowspan="3">MBSFN Subframe Configuration</td><td>Radio Frame Allocation Period</td></tr>
<tr><td>Radio Frame Allocation Offset</td></tr>
<tr><td>Subframe Allocation</td></tr>
<tr><td colspan="3">Common Subframe Allocation Period</td></tr>
<tr><td colspan="3">MBSFN Area Identity</td></tr>
</table>

Table 331 – M2-AP: MBMS Scheduling Information message

★ The MCCH Update Time specifies the Modification Period during which the MCCH should be updated. A value between 0 and 255 is signalled. Each eNode B maintains a common counter for the number of Modification Periods which have passed so each eNode B is able to apply the change at the same time

★ The M2-AP: MBMS Scheduling Information message then includes configuration information for each MBSFN Area. This information is similar to the content of the MBSFN Area Configuration message broadcast to the UE on the MCCH (Table 327)

★ The eNode B responds with an M2-AP: MBMS Session Start Response message. The content of this message is specified within 3GPP TS 36.443, and is presented in Table 332. The MCE MBMS M2AP Identity is used to address the session within the MCE. The eNode B provides the MCE with an eNode B MBMS M2AP Identity which can be used to address the session within the eNode B during any subsequent signalling procedures

Information Elements
Message Type
MCE MBMS M2AP Identity
eNB MBMS M2AP Identity

Table 332 – M2-AP: MBMS Session Start Response message

★ The eNode B also responds with an M2-AP: MBMS Scheduling Information Response message. This message is a simple acknowledgement without any real content

★ The eNode B can then broadcast an MCCH Change Notification Indication within Downlink Control Information (DCI) format 1C, on the PDCCH. The content of the relevant MCCH and MCH Scheduling Information (MSI) MAC control element are updated and the eNode B starts to receive the downlink MBMS session application data using the IP multicast address. This application data is broadcast across the air-interface using the MTCH logical channel mapped onto the MCH transport channel and PMCH physical channel

★ 3GPP References: TS 36.300, TS 36.443, TS 36.444, TS 29.061, TS 29.274, TS 48.018, TS 23.246

28.8 CONTENT SYNCHRONISATION

★ The concept of MBSFN requires that all eNode B within an MBSFN synchronisation area are able to broadcast the same application data simultaneously. This introduces the requirement for content synchronisation between eNode B

★ 3GPP TS 25.446 specifies the SYNC protocol which operates between the Broadcast Multicast Service Centre (BM-SC) and the set of eNode B which are to have their MBMS content synchronised

★ The SYNC protocol relies upon the BM-SC and the set of eNode B sharing a common time reference. For example, this could be derived from a Global Positioning System (GPS). A common synchronisation period is also defined and is known by both the BM-SC and the set of eNode B

★ 3GPP TS 25.446 specifies a range of SYNC PDU formats. Figure 279 illustrates the format of the header belonging to a type 1 SYNC PDU. This header is attached to each MBMS application data packet

PDU Type		Spare
Time Stamp		
Packet Number		
Elapsed Octet Counter		
Header CRC		Payload CRC
Payload CRC		
Payload Fields		

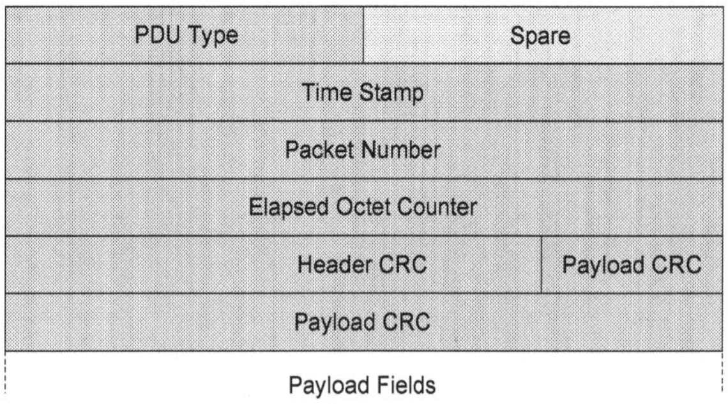

Figure 279 – Format of Type 1 SYNC PDU

★ The PDU Type field has a length of 4 bits and is used to identity the format of the SYNC PDU being sent, e.g. a value of 0001 is sent for a Type 1 SYNC PDU

★ The Time Stamp defines a delay relative to the start of the synchronisation period. It defines the starting time of a synchronisation sequence. All packets belonging to the same synchronisation sequence have the same Time Stamp within their SYNC PDU. There are multiple synchronisation sequences within a synchronisation period

★ The BM-SC is responsible for grouping the MBMS packets into synchronisation sequences. The BM-SC sets the Time Stamp after accounting for the maximum transfer delay from the BM-SC to the group of eNode B, the duration of the synchronisation sequence and any other delays, e.g. processing delay within the eNode B

★ The time stamp has a length of 2 bytes and can be signalled using a value between 0 and 59 999. The actual value equals the signalled value multiplied by 10 ms, i.e. synchronisation periods up to 600 seconds are supported

★ The eNode B detects the start of a new synchronisation sequence when the time stamp changes

★ The duration of the MCH Scheduling Period (MSP) introduced in section 28.6, is always an integer multiple of the synchronisation sequence duration, i.e. there can be 1 or more synchronisation sequences within an MSP

★ The eNode B buffers the received MBMS application data packets until the appropriate transmission time. The eNode B schedules the data packets for transmission during the first MSP following the time corresponding to the Time Stamp

★ The Packet Number shown in Figure 279 specifies the cumulative number of packets within the synchronisation sequence. This allows the eNode B to detect whether or not any packets have been missed. It can also be used to re-order the packets at the eNode B. The packet number is reset at the end of every synchronisation sequence. The Packet Number occupies 2 bytes which provide a range from 0 to $2^{16} - 1$

★ The Elapsed Octet Counter specifies the cumulative number of bytes within the synchronisation sequence. This is used by the eNode B in case of packet loss. The number of missing bytes can be calculated and used to determine the period of time during which the eNode B should mute its transmissions in order to remain synchronised with the other eNode B

★ At the end of a synchronisation sequence, the BM-SC sends a type 0 or type 3 SYNC PDU. A type 0 PDU specifies the total number of packets and bytes within the synchronisation sequence. A type 3 PDU also specifies the total number of packets and bytes within the synchronisation sequence, but in addition specifies the length of each packet, i.e. a list of packet lengths is included

★ Type 2 SYNC PDU are not applicable to the 3GPP release 9 nor 10 versions of LTE. These PDU are applicable to UMTS when the BM-SC applies header compression to the MBMS application data. The release 9 and 10 versions of LTE do not support header compression for MBMS application data

★ 3GPP References: TS 36.300, TS 25.446

28.9 COUNTING PROCEDURE

★ The MBMS counting procedure was introduced within the release 10 version of the 3GPP specifications

★ The counting procedure can be used to help the network select between broadcast and unicast transmission (unicast refers to point-to-point transmission, as used for transmissions on the PDSCH)

  - unicast can be more efficient when the number of UE is small because UE specific link adaptation can be applied, i.e. a reduced number of Resource Blocks can be used to transfer the data when UE are in good coverage. Broadcast transmission uses a relatively large number of Resource Blocks because cell edge coverage is always targeted. Unicast transmissions also allow the eNode B scheduler to have greater flexibility

  - broadcast is more efficient when the number of UE is high because it avoids the eNode B having to transmit the same application data multiple times, i.e. unicast transmission requires dedicated transmissions of the same application data for each UE

★ The counting procedure allows the Multi-cell / Multicast Coordination Entity (MCE) to quantify the number of UE in RRC connected mode which are interested in receiving a specific MBMS service

★ The results from the counting procedure exclude UE in RRC Idle mode. They also exclude 3GPP release 9 UE because these UE do not support the relevant signalling

★ The MBMS counting procedure is illustrated in Figure 280

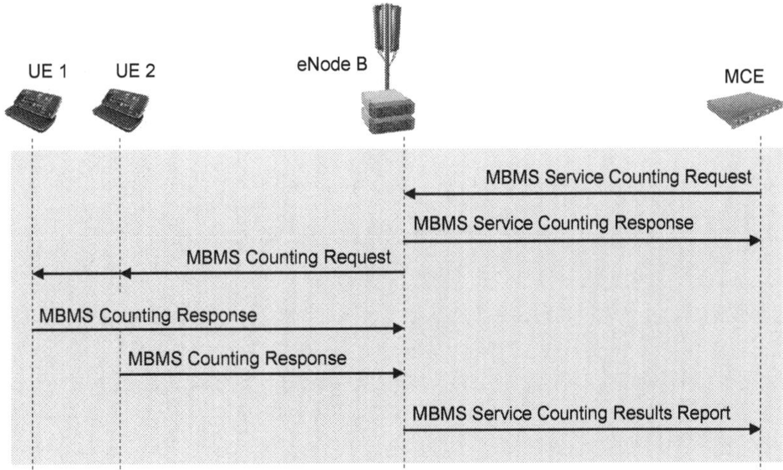

Figure 280 – MBMS counting procedure

★ The MCE initiates the procedure by sending an M2-AP: MBMS Service Counting Request to the relevant eNode B. The content of this message is shown in Table 333.

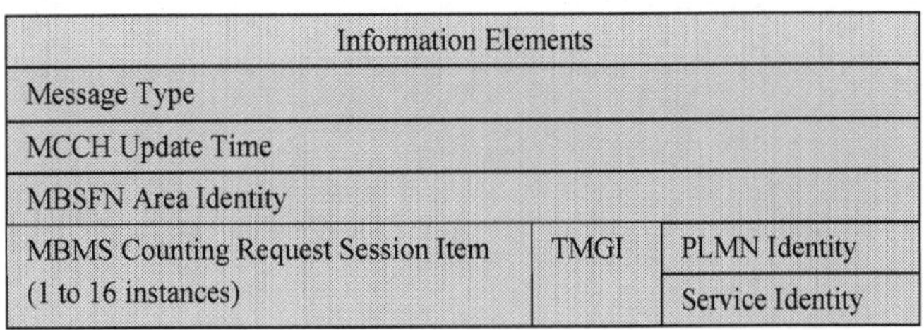

Information Elements		
Message Type		
MCCH Update Time		
MBSFN Area Identity		
MBMS Counting Request Session Item (1 to 16 instances)	TMGI	PLMN Identity
		Service Identity

Table 333 – M2-AP: MBMS Service Counting Request message

★ The subsequent MBMS Counting Request is sent to the UE using the MCCH so the MCE uses the M2-AP: MBMS Service Counting Request message to provide the eNode B with an MCCH Update Time. This specifies the Modification Period during which the MBMS Counting Request should be broadcast to the population of UE. A value between 0 and 255 is signalled. Each eNode B maintains a common counter for the number of Modification Periods which have passed so each eNode B is able to transmit the message at the same time

★ The MBSFN Area Identity allows the eNode B to determine which cells should transmit the MBMS Counting Request message

★ The MCE also uses the M2-AP: MBMS Service Counting Request message to provide the eNode B with a list of Temporary Mobile Group Identifiers (TMGI). These define the services which the counting procedure is targeting. The MCE does not specify individual MBMS sessions so the counting procedure quantifies the number of UE having an interest in a particular MBMS service but does not quantify the interest in sessions belonging to that service

- The eNode B acknowledges receipt of the counting request by returning an M2-AP: MBMS Service Counting Response to the MCE. This message is a simple acknowledgement without any real content
- The eNode B then updates its MCCH transmission to incorporate the RRC: MBMS Counting Request message. An MCCH Change Notification Indication is sent on the PDCCH using Downlink Control Information (DCI) format 1C. The content of the RRC: MBMS Counting Request message is presented in Table 334

Information Elements				
Counting Request List	LIST			
	Counting Request Information	TMGI	PLMN Identity	CHOICE
				PLMN Index
				PLMN ID
			Service Identity	

Table 334 – RRC: MBMS Counting Request message

- This message provides the UE with the list of MBMS services specified by the MCE
- UE which have an interest in one or more of the services respond to the eNode B using an RRC: MBMS Counting Response message. MBMS does not have any uplink channels, so this message is sent using the existing Signalling Radio Bearer (SRB) 1, i.e. the message is sent with acknowledged mode RLC using the Dedicated Control Channel (DCCH) mapped onto the UL-SCH and PUSCH
- The content of the MBMS Counting Response from the UE is shown in Table 335

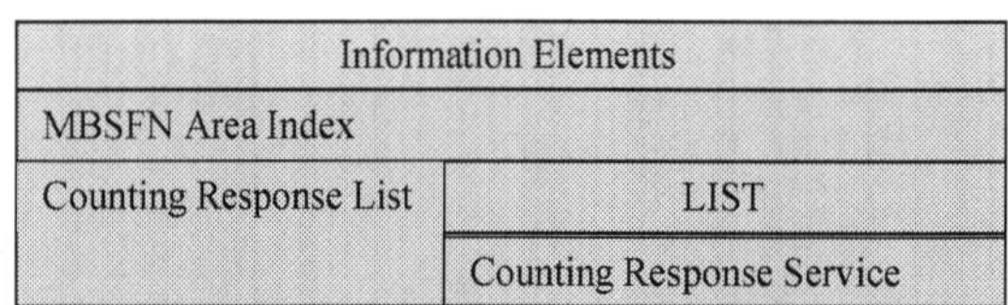

Information Elements	
MBSFN Area Index	
Counting Response List	LIST
	Counting Response Service

Table 335 – RRC: MBMS Counting Response message

- The MBSFN Area Index points to an entry within SIB13. A value of 0 points towards the first MBSFN Area listed in SIB13, a value of 1 points towards the second MBSFN Area, and so on
- The Counting Response Service points to an entry within the MBMS Counting Request message. A value of 0 points towards the first service listed within the MBMS Counting Request, a value of 1 points towards the second service, and so on. This list defines the set of services which the UE is interested in receiving
- The eNode B collects the counting results from the population of UE and generates an M2-AP: MBMS Service Counting Results Report for the MCE. The content of this message is shown in Table 336

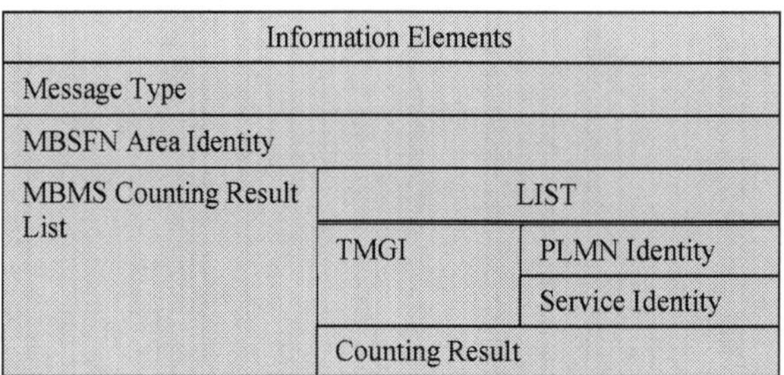

Information Elements		
Message Type		
MBSFN Area Identity		
MBMS Counting Result List	LIST	
	TMGI	PLMN Identity
		Service Identity
	Counting Result	

Table 336 – M2-AP: MBMS Service Counting Results Report message

- This message provides a counting result for each of the MBMS services originally specified by the MCE. Each MBMS service is identified by its TMGI. The counting result is signalled using a value between 0 and 1023. The value of 1023 is interpreted as meaning 1023, or greater
- Having received the counting results from each eNode B, the MCE can take an informed decision on the selection between broadcast and unicast transmission
- 3GPP References: TS 36.300, TS 36.331, TS 36.443

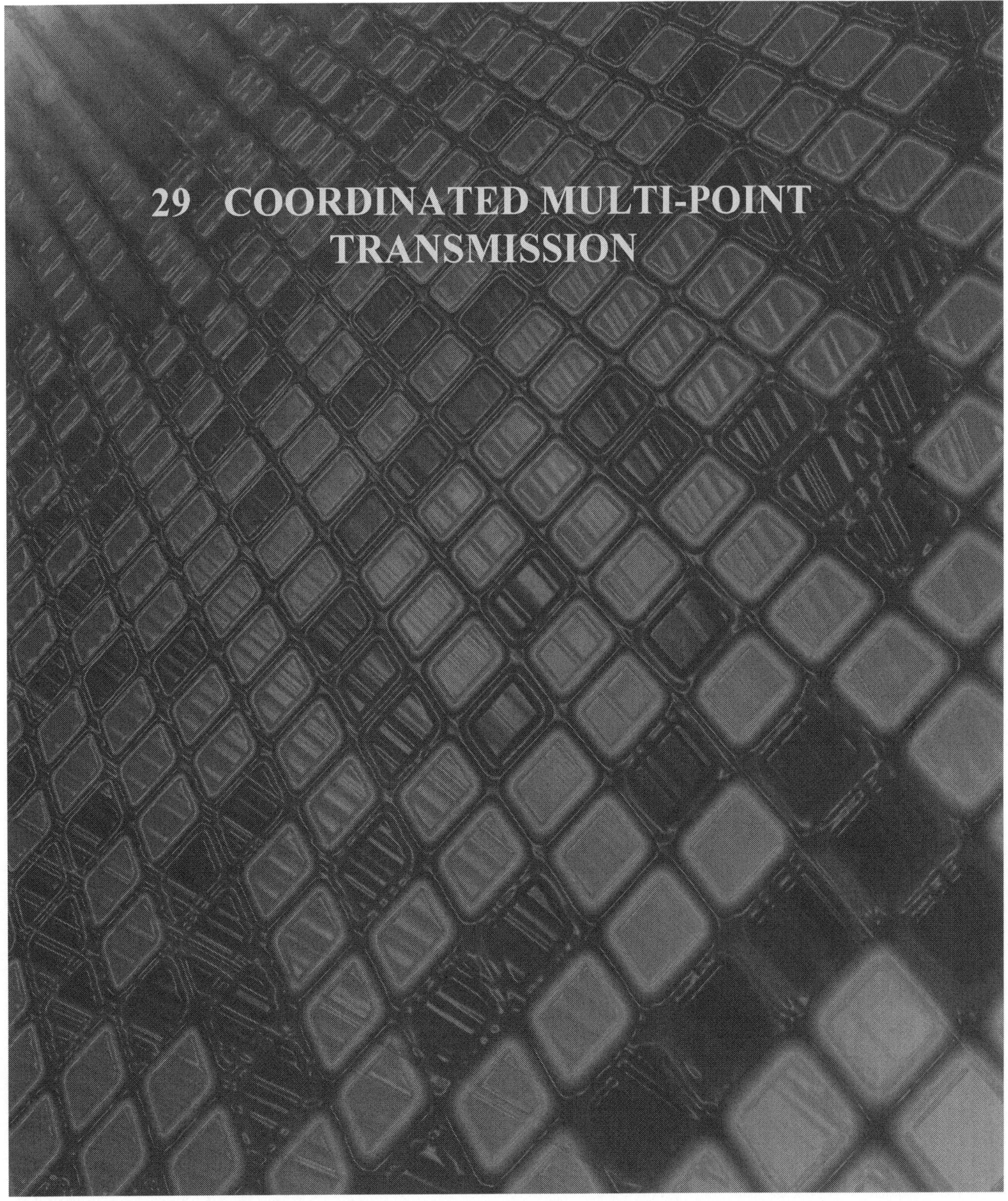

29 COORDINATED MULTI-POINT TRANSMISSION

29.1 INTRODUCTION

★ Coordinated Multi-Point (CoMP) transmission in the downlink and reception in the uplink are LTE-Advanced solutions to help improve the cell edge throughput performance, and consequently improve the overall average cell throughput

★ 3GPP discussed CoMP for inclusion within the release 10 version of the specifications but concluded that:

- o the release 10 version of the specifications will not include any new standardised X2 interface communication for the support of multi-vendor inter-eNode B CoMP
- o CSI Reference Signal design should account for the requirements of CoMP, and should allow accurate inter-cell measurements
- o no additional features will be included within the release 10 version of the specifications to support CoMP

★ 3GPP progressed CoMP as a study item during the timescales of release 10 development. The results from this study item are documented within TR 36.819. A work item has been defined to include CoMP within the release 11 version of the specifications

★ The following sections summarise the study item discussions within the timescales of release 10 development

29.2 DOWNLINK

★ A range of downlink CoMP solutions have been discussed and studied by 3GPP. The main categories include:

- o Joint Processing (JP), which can be further divided into:
  - ▪ Joint Transmission (JT)
  - ▪ Dynamic Point Selection (DPS)
  - ▪ Combination of JT and DPS
- o Coordinated Scheduling (CS) / Coordinated Beamforming (CB), which is characterised by:
  - ▪ Semi-Static Point Selection (SSPS)

★ The concept of Joint Processing (JP) is illustrated in Figure 281. This variant of CoMP is characterised by having user plane data available at more than one location. The locations could be different eNode B, different cells belonging to the same eNode B, or different antenna belonging to the same cell

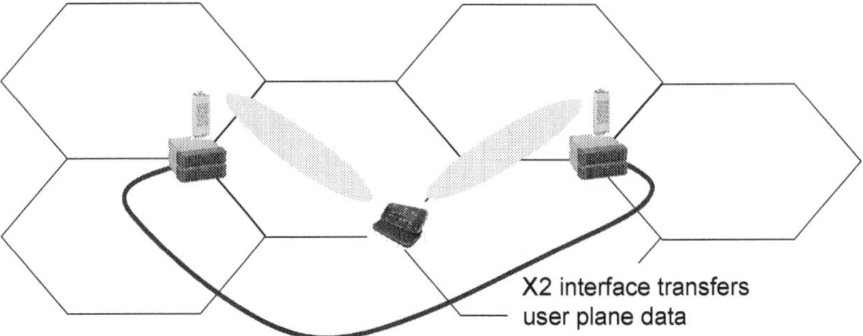

Figure 281 – Joint Processing (JP) CoMP category

★ A challenge associated with JP is the requirement to transfer potentially large quantities of user plane data between locations. This would increase the bandwidth requirements of the X2 interface if each location was a different eNode B

★ The Joint Transmission (JT) variant of JP involves each location simultaneously transmitting to the UE. This provides macro diversity but requires air-interface resources from multiple locations

★ The Dynamic Point Selection (DPS) variant of JP involves transmission from a single location but the location can be changed between 1 ms subframes, or even between 0.5 ms time slots. This solution provides a reduced macro diversity gain but only requires air-interface resources from a single location at any point in time

★ The combination of JT and DPS involves multiple locations simultaneously transmitting downlink data to the UE but those locations can change between subframes or time slots

★ The concept of Coordinated Scheduling (CS) / Coordinated Beamforming (CB) is illustrated in Figure 282. This variant of CoMP is characterised by having user plane data available at only a single location. However scheduling and/or beamforming decisions are coordinated across locations

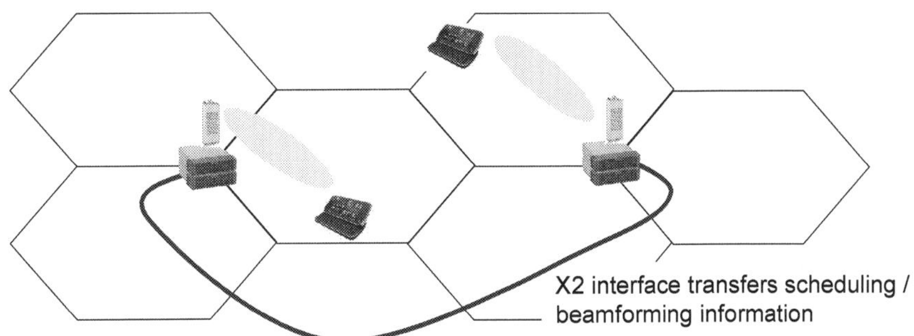

Figure 282 – Coordinated Scheduling (CS) / Coordinated Beamforming (CB) CoMP category

★ A challenge associated with CS/CB is the requirement to exchange scheduling and/or beamforming information between locations with low delay

★ The Semi-Static Point Selection (SSPS) characteristic of CS/CB means that the single location having user plane data can be changed using higher layer reconfiguration procedures, i.e. the single location is not completely fixed but can be changed relatively slowly and infrequently

★ 3GPP has also discussed muting within the context of downlink CoMP. This involves the transmitters at neighbouring locations avoiding the use of co-channel Resource Blocks, i.e. the downlink signal-to-noise ratio is improved for the cell edge UE targeted by the CoMP algorithm

★ The demanding X2 interface requirements can be avoided by focusing upon intra-eNode B scenarios. Figure 283 illustrates intra-site CoMP in which case coordination is applied between the cells belonging to each eNode B

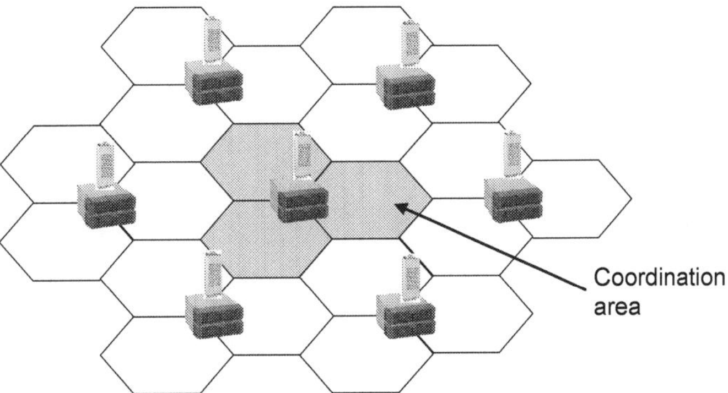

Figure 283 – Intra-site CoMP within a homogeneous network

★ The co-ordination area represents the area over which CoMP is applied. Figure 283 illustrates CoMP being applied across the 3 cells belonging to an eNode B

★ Figure 284 illustrates a second scenario studied by 3GPP. In this case, the coordination area is still based upon a single eNode B, but the eNode B is equipped with high power Remote Radio Heads (RRH). The RRH are distributed geographically to allow a much larger coordination area under the control of a single eNode B

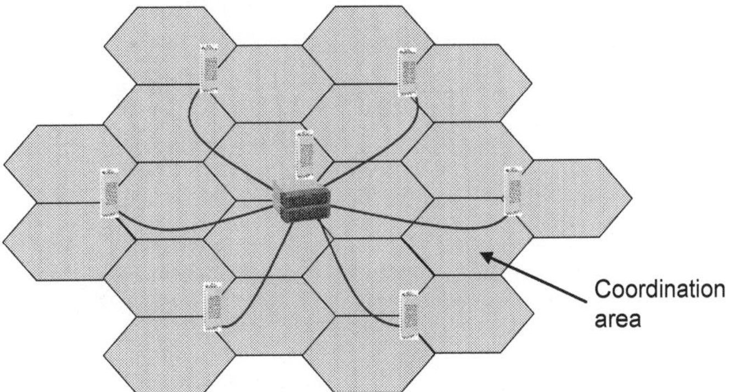

Figure 284 – CoMP within a homogeneous network with high power Remote Radio Heads (RRH)

★ Figure 285 illustrates a third scenario studied by 3GPP. In this case, a single eNode B is equipped with a combination of high power and low power transmitters. The low power transmitters provide relatively small dominance areas within the coverage of the high power transmitters

★ This corresponds to a heterogeneous network design where the low power transmitters can be viewed as microcells, and the high power transmitters as macrocells. The low power transmitters could be used as additional capacity to serve traffic hotspots, or they could be used to improve coverage

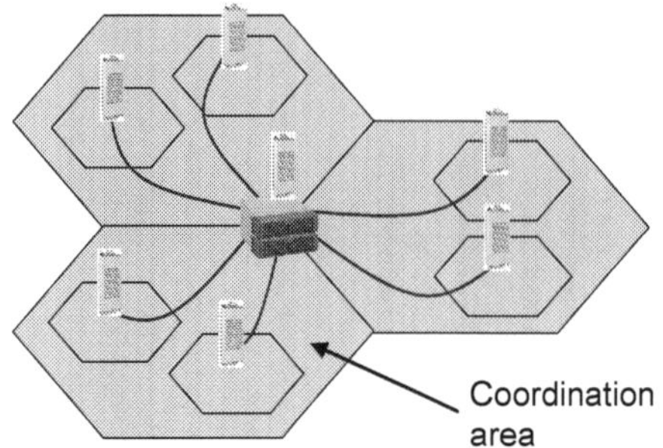

Figure 285 – CoMP within a heterogeneous network with a combination of high and low power Remote Radio Heads (RRH)

★ The smaller cells within Figure 285 could be configured with either the same or different cell identities compared to the overlaying larger cells

★ 3GPP has defined the concepts of:

  - CoMP Cooperating Set
  - CoMP Measurement Set
  - CoMP Resource Management Set / RRM Measurement Set

★ The CoMP Cooperating Set is the set of locations which are either directly or indirectly involved in downlink data transmission. The Cooperating Set corresponds to the coordination area illustrated in the previous figures

  - direct involvement in data transmission refers to locations which are actually transmitting to the UE. Multiple locations transmit when using the Join Transmission (JT) scheme, whereas a single location transmits when using the Coordinated Scheduling / Coordinated Beamforming (CS/CB) scheme
  - indirect involvement refers to locations which are candidates for downlink transmission and which contribute towards making scheduling / beamforming decisions

★ The CoMP Measurement Set is the set of locations which have their channel state measured and potentially reported. The uplink Sounding Reference Signal (SRS) can be used for scenarios where channel reciprocity can be applied. Otherwise downlink channel measurements can be reported by the UE. The CoMP Measurement Set is likely to include a maximum of 2 or 3 CSI Reference Signal resources for the UE to measure

★ 3GPP TR 36.819 defines the RRM Measurement Set but subsequent discussions concluded that this definition should be disregarded and replaced by the CoMP Resource Management Set. The CoMP Resource Management Set defines the set of CSI Reference Signal resources which can be used for measurements and subsequent reporting to the eNode B. These measurements can be used as an input when selecting the CoMP Measurement Set

29.3 UPLINK

★ A range of uplink CoMP solutions have been discussed and studied by 3GPP. The main categories include:

  - Joint Reception (JR): characterised by the network receiving the PUSCH transmission from the UE at more than a single location. This provides a macro diversity gain
  - Coordinated Scheduling (CS) / Coordinated Beamforming (CB): characterised by the network receiving the PUSCH transmission from the UE at only a single location. However, scheduling and beamforming decisions are coordinated between locations

★ In the uplink direction, the CoMP Cooperating Set is defined as a set of locations which can potentially receive the PUSCH transmissions from the UE. The actual locations which receive the PUSCH are known as CoMP reception locations. These are a subset of the Cooperating Set. In the case of JR, there are multiple reception locations, whereas in the case of CS/CB there is only a single reception location

★ 3GPP References: TR 36.819

30 HETEROGENEOUS NETWORKS

30.1 INTRODUCTION

★ Network deployments are generally based upon macrocells which have relatively high transmit powers and above roof-top antennas. Low densities of macrocells provide coverage in low traffic areas, while high densities of macrocells provide capacity in high traffic areas

★ A network composed of only macrocells is known as a homogeneous network. A network composed of multiple site types is known as a heterogeneous network. There are many reasons why a heterogeneous network can be beneficial:

  o site acquisition for macrocells can be difficult. Site acquisition is often easier for microcells and picocells

  o the capacity gain from increasing the density of macrocells tends to saturate as levels of intercell interference increase. Microcells and picocells can be deployed with greater densities to further increase network capacity

  o repeaters and relays can provide coverage and capacity at locations without a transport network connection

  o femto cells can provide coverage at locations outside the reach of the main network. They can also be used to provide coverage at indoor locations where the building penetration loss is too great for outdoor sites

  o indoor solutions can provide dedicated coverage and capacity to serve traffic hotspots or corporate customers

★ Figure 286 illustrates a heterogeneous network which includes macro, micro, pico and femto BTS, as well as relays and repeaters

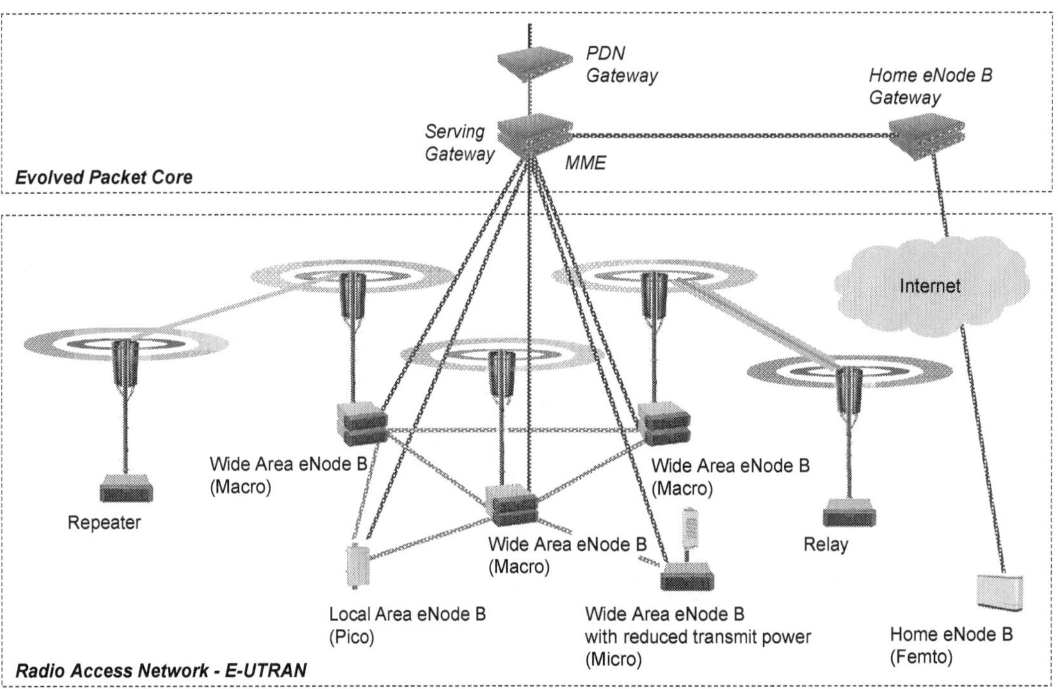

Figure 286 – Heterogeneous network for LTE

★ The various BTS types typically share the same channel bandwidth, so intercell interference can be high because handover boundaries tend to be closer to BTS with lower transmit powers. A picocell could have a transmit power which is 20 dB less than a macrocell. Assuming the handover boundary is defined by equal downlink powers at the UE then the link loss to the picocell would be 20 dB less than the link loss to the macrocell. The picocell would experience uplink interference as a UE approaches the handover boundary while connected to the macrocell

★ Inter Cell Interference Coordination (ICIC) is viewed as being particularly important for heterogeneous networks. ICIC was introduced within the release 8 version of the 3GPP specifications. ICIC allows neighbouring eNode B to coordinate their use of the air-interface resources, e.g. one eNode B could use Resource Blocks towards the upper edge of the channel bandwidth while a second eNode B uses Resource Blocks towards the lower edge of the channel bandwidth

★ ICIC creates a trade-off between improving the signal to noise ratio and reducing the quantity of resources available for transmission. Avoiding the use of Resource Blocks used by neighbouring eNode B improves the signal to noise ratio but reduces the total number of Resource Blocks available for transmission

★ Enhanced ICIC was introduced within the release 10 version of the 3GPP specifications. This includes the use of Almost Blank Subframes (ABS) which allow eNode B to reduce their transmissions during certain time intervals. Fully benefiting from ABS requires neighbouring eNode B to be synchronised in the time domain. Neighbouring eNode B experience improved signal to noise ratios during the ABS

- ★ ICIC and ABS are described in greater detail in section 32.14
- ★ Sites with lower transmit power capabilities may struggle to achieve dominance over sites with higher transmit powers. This can result in reduced cell ranges for the sites with lower transmit power capabilities, and those sites may collect less traffic than originally expected
- ★ Cell range extension can be used to increase the traffic captured by sites with low transmit powers. UE typically complete handovers across boundaries where the downlink received signal from a neighbouring cell becomes stronger than the downlink received signal from the current serving cell. The cell range of a site with a low transmit power can be increased by applying an offset to the downlink received signal measurements, i.e. making the low power site appear more attractive for longer. This concept is illustrated in Figure 287

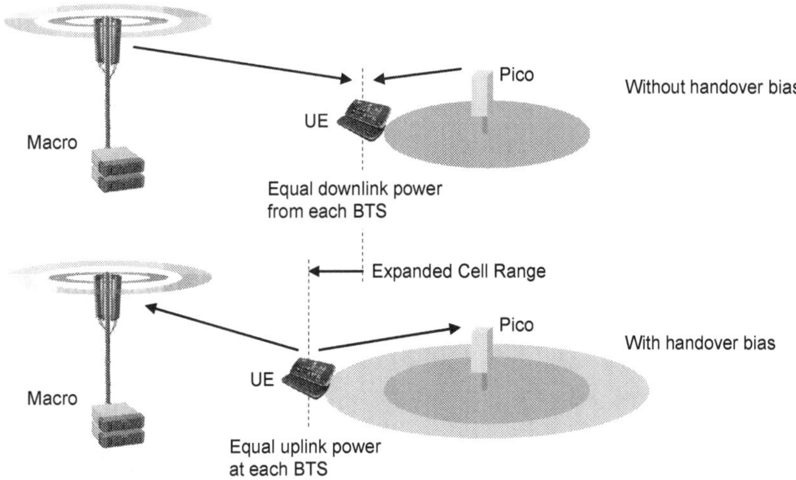

Figure 287 – Expanded cell range with handover bias

- ★ Handover boundaries defined by equal downlink received signal strength are preferable for optimising downlink signal to noise ratio performance, but are not necessarily preferable for optimising uplink signal to noise ratio performance
- ★ The downlink transmit power of a picocell could be 20 dB less than the downlink transmit power of a macrocell. This means that the uplink link loss to the picocell could be 20 dB less than the uplink link loss to the macrocell as a UE completes a handover from a picocell to a macrocell. From an uplink perspective, it would be preferable to remain connected to the picocell until the uplink link loss to the macrocell becomes less than the uplink link loss to the picocell
- ★ This provides some scope for increasing the cell range and balancing the traffic carried by each site, at the cost of downlink signal to noise ratio. The downlink signal to noise ratio performance can be maintained if downlink resource allocations are coordinated between the macrocell and picocell, i.e. by using ICIC. Allocating Resource Blocks which the macrocell does not use to UE within the expanded cell range area of the picocell helps to maintain the downlink signal to noise ratio
- ★ Almost Blank Subframes (ABS) can also be used as a solution to help maintain the downlink signal to noise ratio if UE within the expanded cell range area of the picocell are allocated resource during the ABS of the dominant macrocell
- ★ Interference cancellation at the UE is another solution for improving the downlink signal to noise ratio towards cell edge
- ★ Femto cells (Home eNode B) can be a source of interference when Closed Subscriber Group (CSG) functionality is used to control access. UE which are not authorised to access the femto may not be able to access other cells in the network due to the co-channel interference from the femto. The femto creates a coverage hole for these UE
- ★ A femto could use ABS to allow unauthorised UE to access co-channel macrocells while inside the coverage area of the femto, i.e. interference from the femto would reduce during the ABS so the UE could access a macrocell
- ★ Carrier aggregation can also be used to help CSG femto coexist with macrocells if sufficient spectrum is available. The macrocell network could be allocated a pair of Component Carriers (CC) while the femto could be allocated only a single carrier. The macrocell primary CC would be the carrier which has not been allocated to the femto. UE could then access the macrocell primary CC even when within the coverage of a CSG femto. UE would be able to benefit from both macrocell CC when outside the coverage of the femto

30.2 MACRO BTS

★ The most common BTS type is the macro BTS, which 3GPP categorises as a 'wide area BTS'. The term 'wide area BTS' is used within 3GPP TS 36.104 where radio transmission and reception requirements are specified

★ Macro BTS are characterised by having their antenna above roof-top level so their coverage area is relatively large

★ 3GPP does not define an upper limit for the rated output power of a macro BTS (stated in Table 337). In practice, the national regulator is likely to specify an upper limit. The transmit power of a macrocell is typically 20 to 40 Watts

BTS Class	Rated Output Power
Wide Area BTS	No Upper Limit

Table 337 – Rated output power for Wide Area BTS

★ The 3GPP specifications for wide area BTS have been developed assuming a Minimum Coupling Loss (MCL) of 70 dB between the BTS and UE

★ Macro BTS are specified to have better sensitivity than other BTS types, i.e. high uplink throughputs can be achieved with lower received signal powers. For example, a macro BTS is specified to have a reference sensitivity of -101.5 dBm when using the 20 MHz channel bandwidth, whereas the equivalent figure for a pico BTS is -93.5 dBm

★ Macro BTS could be deployed using an Active Antenna System (AAS) which integrates the RF components of the base station within the antenna housing. Alternatively, macro BTS could be deployed using Remote Radio Heads (RRH) which locate the RF components of the base station adjacent to the antenna. A conventional base station design with the RF components positioned away from the antenna can also be used. In this case, Mast Head Amplifiers (MHA) can be introduced to help compensate the impact of feeder losses in the uplink direction

★ Macro BTS typically use cross polar panel antennas. These antenna are likely to include 2 antenna elements with +/- 45 degrees polarisations. They are able to support 2x2 MIMO in the uplink and downlink directions, as well as 2 branch downlink transmit diversity and 2 branch uplink receive diversity. The size and antenna gain is dependent upon the operating band. Antenna for higher operating bands tend to have a smaller size and a higher gain

★ Macro BTS may also use antenna arrays for the purposes of beamforming and higher order MIMO. An antenna array could include 4 columns of cross polarised antenna elements. This provides support for both beamforming and 8×8 MIMO. An antenna array could also be based upon 8 columns of co-polarised antenna elements. In both cases, the size of the antenna array will depend upon the operating band (columns are typically spaced by half a wavelength)

★ 3GPP References: TS 36.104

30.3 MICRO BTS

★ 3GPP does not explicitly specify a separate power class for micro BTS, but a wide area BTS with reduced transmit power could be designed and used as a micro BTS

★ Micro BTS are characterised by having their antenna below roof-top level so their coverage tends to be limited by the neighbouring buildings. Their transmit power is typically 5 to 10 W

★ Micro BTS can be deployed co-channel to macro BTS

★ Micro BTS often have only a single sector because sectorisation can be made challenging by the high quantity of scattering from neighbouring buildings. This scattering can lead to overlapping coverage areas and poor dominance if sectors do not have sufficient isolation

★ Micro BTS should be placed as close as possible to the source of traffic but should not be placed in areas where co-channel macro BTS have very high signal strengths. The cell range of a micro BTS will depend upon the signal strength of any co-channel macro BTS. A maximum permitted macro BTS RSRP could be defined to avoid micro BTS from being planned in areas where their cell range is likely to be limited

★ UE connected to micro BTS often benefit from line-of-sight propagation with little delay spread

★ A Minimum Coupling Loss (MCL) of 53 dB between the BTS and UE is typically assumed for a micro BTS

★ Micro BTS can use smaller 'bell box' style antenna which typically include a single cross polar pair for 2×2 MIMO, 2 branch transmit diversity and 2 branch receive diversity. Antenna arrays can also be used to support higher order MIMO, although the line-of-sight propagation associated with micro BTS may compromise the performance of higher order MIMO (MIMO requires uncorrelated propagation paths between each transmit antenna and each receive antenna). Scattering may help to reduce the correlation between the propagation paths and so help to improve the performance of MIMO

30.4 PICO BTS

★ 3GPP categorises pico BTS as 'local area BTS'. The term 'local area BTS' is used within 3GPP TS 36.104 where radio transmission and reception requirements are specified

★ Pico BTS are characterised by low transmit powers and relatively small coverage areas. Deploying high densities of Pico BTS increases network capacity. They can also be used as dedicated indoor solutions. Their low transmit power means they need to be deployed close to the source of traffic

★ The maximum rated output powers for pico BTS are presented in Table 338. The rated output power is defined as the average power per RF carrier at the antenna connector. The maximum rated output power decreases as the number of antenna ports increases to maintain the total radiated power

BTS Class	Antenna Configuration	Rated Output Power
Local Area BTS	1 antenna port	≤ 24 dBm
	2 antenna ports	≤ 21 dBm
	4 antenna ports	≤ 18 dBm
	8 antenna ports	< 15 dBm

Table 338 – Rated output power for Local Area BTS

★ Their lower transmit power allows pico BTS to be smaller and lower cost than macro and micro BTS. They are also assumed to be less sensitive than macro and micro BTS so require higher received signal powers

★ Site acquisition tends to be simpler for pico BTS due to their relatively small size. Routing high numbers of transport connections to high densities of pico BTS can be challenging. Wireless transport connections may be considered an appropriate solution

★ Pico BTS can be deployed co-channel to macro and micro BTS, and typically have only a single sector

★ The 3GPP specifications for local area BTS have been developed assuming a Minimum Coupling Loss (MCL) of 45 dB between the BTS and UE

★ Pico BTS are typically deployed with smaller antenna. Pico BTS used for indoor solutions may be equipped with omni directional ceiling mounted antenna, wall mounted panel antenna or wall mounted 'shark fin' antenna. As indicated in Table 338, antenna solutions may be available to support up to 8 antenna ports

★ 3GPP References: TS 36.104

30.5 FEMTO BTS

★ 3GPP categorises femto BTS as 'home BTS'. The term 'home BTS' is used within 3GPP TS 36.104 where radio transmission and reception requirements are specified

★ The maximum rated output powers for home BTS are presented in Table 339. The rated output power is defined as the average power per RF carrier at the antenna connector. The maximum rated output power decreases as the number of antenna ports increases to maintain the total radiated power

BTS Class	Antenna Configuration	Rated Output Power
Home BTS	1 antenna ports	≤ 20 dBm
	2 antenna ports	≤ 17 dBm
	4 antenna ports	≤ 14 dBm
	8 antenna ports	≤ 11 dBm

Table 339 – Rated output power for Home BTS

★ Femto BTS are characterised by their network architecture which differs from other BTS types. Femto BTS are connected to the Evolved Packet Core (EPC) using a Home eNode B Gateway (also known as a Femto Gateway). The connection between the Femto BTS and Femto Gateway typically uses a home broadband connection, e.g. ADSL. The network architecture for a femto BTS is illustrated in Figure 286

★ Femto BTS are intended for use at home, or in small offices. Their low transmit power capability means they should be used in areas where coverage from other BTS types is relatively weak, e.g. indoors. They are able to provide coverage at locations outside the reach of the main network. In contrast to other BTS types, the location of a femto BTS is not usually controlled by the network operator. End-users are free to place their femto BTS wherever they like

★ Femto BTS can be deployed co-channel to macro, micro and pico BTS

- Femto BTS typically select their own Physical layer Cell Identity (PCI) from an allocated pool of PCI. Selection can be completed after scanning for the PCI allocated to neighbouring BTS. This helps to ensure that the femto BTS is able to select a PCI which is unique within the local environment
- Femto BTS can use the concept of Closed Subscriber Groups (CSG) to control access. When operating in CSG mode, only registered subscribers are able to access the femto BTS. These registered subscribers could be members of a family if the femto BTS is providing coverage within a residential home
- UE are able to maintain a 'whitelist' of registered CSG to which access is authorised. UE which have at least 1 allowed CSG identity within their USIM, use an autonomous search function to detect, at least previously visited CSG cells on other frequencies and Radio Access Technologies (RAT). This autonomous search function operates in parallel to normal cell reselection. The UE can also use an autonomous search function on the serving frequency
- If a UE detects a CSG cell on a different frequency or RAT, the UE completes cell reselection onto that CSG cell irrespective of the priority of the frequency as long as the CSG cell is the highest ranked cell on that frequency or RAT
- SIB1 broadcasts the CSG identity, whereas SIB9 can be used to broadcast the name of a femto BTS. The name of the femto BTS can be used by a subscriber when performing a manual selection between BTS
- Femto BTS used in the CSG mode of operation appear as a source of interference to UE which are not authorised to access the femto BTS. An alternative RF carrier can be used in these locations to avoid non-authorised UE experiencing coverage holes
- Femto BTS can also be used in an open mode of operation which allows any subscriber to gain access
- Femto BTS can also be used in a hybrid mode of operation which allows the definition of a CSG but also allows any UE to gain access. UE which are registered with the CSG are provided with priority over non-registered UE
- Proximity detection allows a UE to inform the eNode B when it enters or leaves the coverage of a CSG cell included within its 'whitelist'. The UE uses a Proximity Indication message to specify the RF carrier belonging to the CSG cell and also to specify whether the UE is entering or leaving the coverage area. Proximity detection is an optional UE capability
- Femto BTS are typically equipped with small omni antenna. Multiple antenna may be used to provide a MIMO capability
- 3GPP References: TS 36.104

30.6 REPEATER

- Repeaters can be used to extend the coverage of an existing BTS. They re-transmit the uplink and downlink signals without having to decode any of the content. Repeaters have one antenna directed towards a donor cell, and a second antenna directed towards the target coverage area. The target coverage area could be an indoor location so the second antenna could be indoors. The general network architecture for a repeater is illustrated in Figure 288

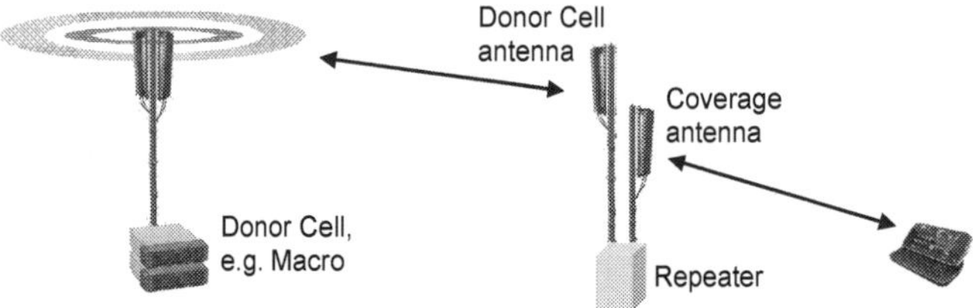

Figure 288 – Repeater network architecture

- Repeaters are relatively transparent to the functionality of the donor cell, i.e. the donor cell may collect an increased quantity of traffic but has no knowledge of the repeater
- The radio transmission and reception requirements for FDD LTE repeaters are specified within 3GPP TS 36.106. This specification defines the same set of operating bands and channel bandwidths as used for FDD BTS. Similar to macro BTS, a maximum output power is not specified
- Repeaters amplify the uplink and downlink signals to provide extended coverage:
  - in the uplink direction, the noise floor of the repeater is amplified and transmitted towards the donor cell. This can desensitise the uplink receiver of the donor cell if the repeater gain is high compared to the link loss between the repeater and donor cell, i.e. the uplink gain of the repeater should account for the link loss between the repeater and donor cell
  - the uplink and downlink gains should be similar otherwise there can be issues with handover reliability. For example, if the downlink gain is set high then UE are likely to handover onto the donor cell/repeater early due to the high cell specific Reference Signal power. If the uplink gain is set low then the donor cell may not be able to receive uplink transmissions from the UE
- Repeater implementations may automatically configure their gains based upon signal strength measurements from the donor cell

★ In general, repeaters require isolation between their donor and coverage antenna. There is a danger of feedback between the two antenna if the isolation is not sufficient, e.g. the donor antenna receiving the downlink signal from the donor cell, could also receive the downlink signal from the coverage antenna. In this case, the repeater's own transmissions are received, amplified and re-transmitted in a loop. Repeater implementations may include interference cancellation technology to help reduce the isolation requirements between the donor and coverage antenna

★ Repeaters typically introduce a small delay which appears as additional air-interface propagation delay to the donor cell. Timing advance will account for this delay so UE transmissions remain synchronised at the donor cell. Delays introduced by repeaters will have some impact upon the maximum cell range

★ In some scenarios, donor cells may receive uplink signals directly from UE, as well as via a repeater, i.e. for UE which are located at the boundary of the donor cell and repeater coverage areas. In this case, the two uplink signals will appear as delay spread components. Similarly, UE may receive downlink signals directly from the donor cell, as well as via a repeater

★ Repeaters can be configured with a high gain directional antenna pointing towards their donor cell. This corresponds to a point-to-point link. The coverage antenna at a repeater is likely to be less directional to provide coverage across a wider target area

★ 3GPP References: TS 36.106

30.7 RELAY NODE

★ Relay Nodes (RN) were introduced within the release 10 version of the 3GPP specifications as part of LTE Advanced

★ Similar to repeaters, relays can be used to extend the coverage of the main network by linking to a donor eNode B. In contrast to repeaters, relays are not transparent to the donor eNode B. The donor eNode B must support relay node protocol stacks to allow signalling and the transfer of user plane data

★ Similar to repeaters, relays do not have a wired backhaul connection into the transport network

★ The network architecture for a relay node is illustrated in Figure 289

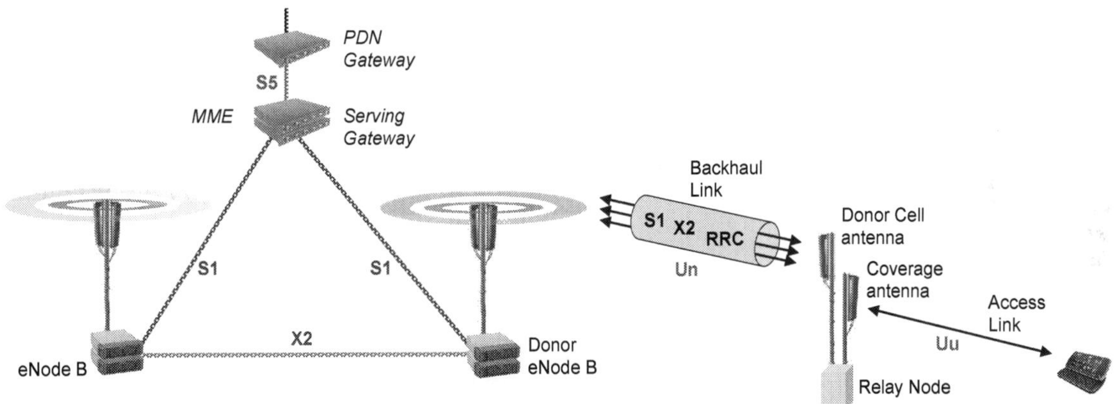

Figure 289 – Relay Node (RN) network architecture

★ Relay nodes support the normal LTE air-interface towards the UE. This allows backwards compatibility for 3GPP release 8 and 9 devices. The 3GPP specifications support both FDD and TDD duplexing methods for relays

★ The relay node communicates with the donor cell using a modified version of the LTE air-interface. This modified version of the air-interface is known as the 'Un' interface (in contrast to the 'Uu' interface). User plane data and control plane signalling are transferred using the 'Un' interface

★ Figure 290 illustrates the user plane protocol stack for the 'Un' radio interface. The relay node appears similar to a UE from the perspective of the donor eNode B

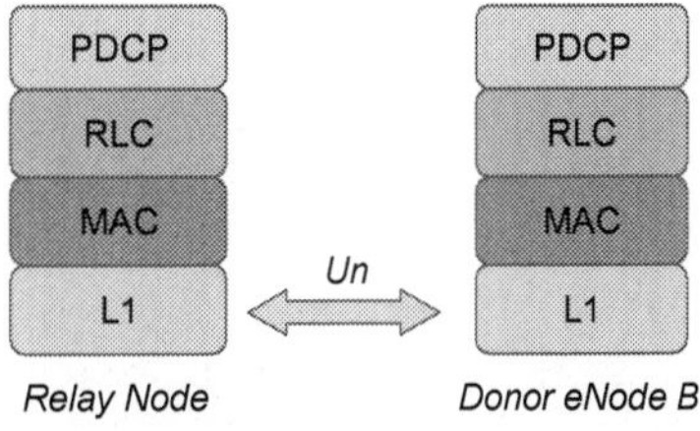

Figure 290 – Radio user plane protocol stack for relay node

★ Figure 291 illustrates the control plane protocol stack for the 'Un' radio interface. This protocol stack allows RRC signalling between the relay node and donor eNode B. It also allows Non-Access Stratum (NAS) signalling between the relay node and MME. The relay node appears similar to a UE from the perspective of the donor eNode B and MME

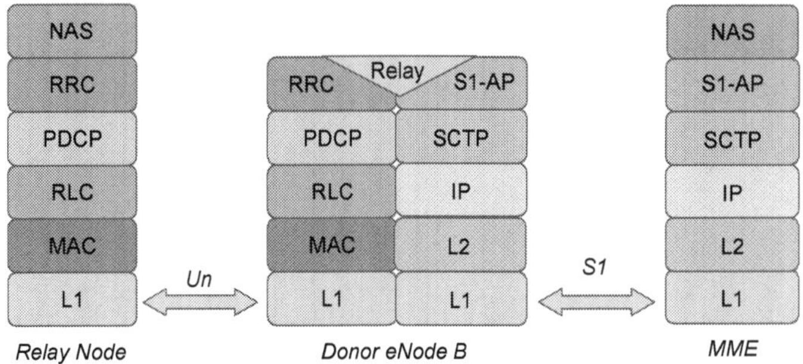

Figure 291 – Radio control plane protocol stack for relay node

★ A relay node establishes an RRC connection with its donor eNode B using the same RRC signalling handshake as a UE establishing an RRC connection with an eNode B, i.e. the relay node sends an RRC Connection Request message, the donor eNode B returns an RRC Connection Setup message and the relay node completes the procedure by sending an RRC Connection Setup Complete message

★ The relay node identifies itself as a relay node within the RRC Connection Setup Complete message. This is done by including the 'relay node subframe configuration required' information element. This information element can be signalled with values of either 'required' or 'not required' to indicate whether or not the relay node would like a subframe configuration to be allocated

★ 3GPP has introduced an additional pair of RRC messages which allow the donor eNode B to configure the relay node. These additional RRC messages are specified within 3GPP TS 36.331 and are illustrated in Figure 292

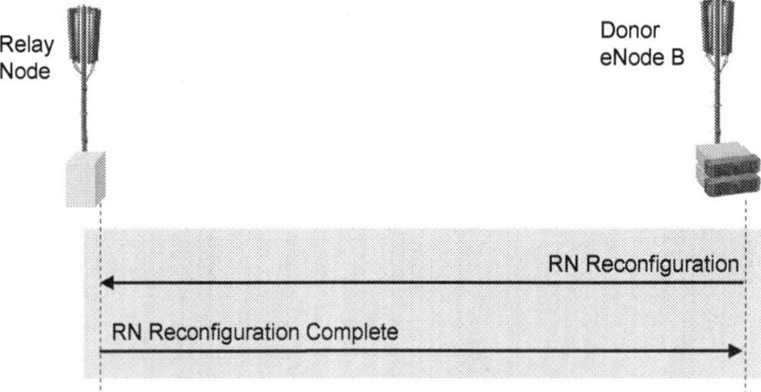

Figure 292 – Relay node reconfiguration procedure

★ The RN Reconfiguration message can be used to provide the relay node with:

- o subframe configuration information which specifies the subframes and Resource Blocks which can be used for the backhaul link between the donor eNode B and relay node
- o the content of SIB1 and SIB2 to be broadcast by the relay node

★ The content of the subframe configuration information is presented in Table 340

- o in the case of FDD, the Subframe Configuration Pattern is a bit string with a length of 8 bits. This bit string indicates the subframes which the donor eNode B can use for downlink transmission towards the relay node. Each bit corresponds to a subframe where the first subframe belonging to the pattern is identified by $\{10 \times \text{SFN} + \text{subframe number}\} \bmod 8 = 0$
- o in the case of FDD, the allocated uplink subframes for the backhaul link follow the allocated downlink subframes by 4 subframes
- o in the case of TDD, the Subframe Configuration Pattern defines a pointer towards a row within a look-up table specified by 3GPP TS 36.216. This look-up table is shown in Table 341
  - ▪ it specifies the set of uplink and downlink subframes for transmission across the backhaul link (shown as U_B and D_B)
  - ▪ it also specifies the uplink-downlink subframe configuration to be used by the relay node for the access link (the access link corresponds to the air-interface between the relay node and the population of UE served by the relay node). Uplink-downlink subframe configurations 1, 2, 3, 4 and 6 can be configured for use. Uplink and downlink subframes for access link transmission are shown as U_A and D_A
  - ▪ all 4 types of subframe (U_B, D_B,U_A and D_A) are time multiplexed to provide isolation between transmissions on the same RF carrier. This limits the number of subframes available for transmission on each link, and will consequently impact the maximum throughputs achieved by the relay

<table>
<tr><td colspan="5">Information Elements</td></tr>
<tr><td rowspan="3">Subframe Configuration Pattern</td><td colspan="4">CHOICE</td></tr>
<tr><td colspan="4">FDD Configuration Pattern</td></tr>
<tr><td colspan="4">TDD Configuration Pattern</td></tr>
<tr><td rowspan="15">R-PDCCH Configuration</td><td colspan="4">Resource Allocation Type</td></tr>
<tr><td rowspan="3">Resource Block Assignment</td><td colspan="3">CHOICE</td></tr>
<tr><td colspan="3">Type 0 or 1</td></tr>
<tr><td colspan="3">Type 2</td></tr>
<tr><td rowspan="3">Demodulation Reference Signal</td><td colspan="3">CHOICE</td></tr>
<tr><td colspan="3">Interleaving</td></tr>
<tr><td colspan="3">No Interleaving</td></tr>
<tr><td colspan="4">PDSCH Start Symbol</td></tr>
<tr><td rowspan="7">PUCCH Configuration</td><td colspan="3">CHOICE</td></tr>
<tr><td rowspan="4">TDD</td><td colspan="2">CHOICE</td></tr>
<tr><td>Channel Selection, Multiplexing, Bundling</td><td>nl PUCCH AN List</td></tr>
<tr><td rowspan="2">Fallback for Format 3</td><td>nl PUCCH AN P0</td></tr>
<tr><td>nl PUCCH AN P1</td></tr>
<tr><td rowspan="2">FDD</td><td colspan="2">nl PUCCH AN P0</td></tr>
<tr><td colspan="2">nl PUCCH AN P1</td></tr>
</table>

Table 340 – Relay node subframe configuration information element

<table>
<tr><td rowspan="2">TDD Subframe Configuration Pattern</td><td rowspan="2">Uplink-Downlink Subframe Configuration for Access Link</td><td colspan="10">Subframe Number</td></tr>
<tr><td>0</td><td>1</td><td>2</td><td>3</td><td>4</td><td>5</td><td>6</td><td>7</td><td>8</td><td>9</td></tr>
<tr><td>0</td><td rowspan="5">1</td><td rowspan="5">D_A</td><td rowspan="5">S_A</td><td rowspan="5">U_A</td><td>U_A</td><td>D_B</td><td rowspan="5">D_A</td><td rowspan="5">S_A</td><td rowspan="5">U_A</td><td>U_B</td><td>D_A</td></tr>
<tr><td>1</td><td>U_B</td><td>D_A</td><td>U_A</td><td rowspan="4">D_B</td></tr>
<tr><td>2</td><td>U_A</td><td rowspan="3">D_B</td><td>U_B</td></tr>
<tr><td>3</td><td rowspan="2">U_B</td><td>U_A</td></tr>
<tr><td>4</td><td>U_B</td></tr>
<tr><td>5</td><td rowspan="6">2</td><td rowspan="6">D_A</td><td rowspan="6">S_A</td><td>U_B</td><td>D_A</td><td rowspan="2">D_A</td><td rowspan="6">D_A</td><td rowspan="6">S_A</td><td>U_A</td><td>D_B</td><td rowspan="3">D_A</td></tr>
<tr><td>6</td><td>U_A</td><td>D_B</td><td>U_B</td><td>D_A</td></tr>
<tr><td>7</td><td>U_B</td><td>D_A</td><td>D_B</td><td>U_A</td><td>D_B</td></tr>
<tr><td>8</td><td>U_A</td><td rowspan="3">D_B</td><td>D_A</td><td>U_B</td><td>D_A</td><td>D_B</td></tr>
<tr><td>9</td><td>U_B</td><td>D_B</td><td>U_A</td><td rowspan="2">D_B</td><td>D_A</td></tr>
<tr><td>10</td><td>U_A</td><td>D_A</td><td>U_B</td><td>D_B</td></tr>
<tr><td>11</td><td rowspan="2">3</td><td rowspan="2">D_A</td><td rowspan="2">S_A</td><td rowspan="2">U_A</td><td rowspan="2">U_B</td><td rowspan="2">U_A</td><td rowspan="2">D_A</td><td rowspan="2">D_A</td><td rowspan="2">D_B</td><td>D_A</td><td rowspan="2">D_B</td></tr>
<tr><td>12</td><td>D_B</td></tr>
<tr><td>13</td><td rowspan="5">4</td><td rowspan="5">D_A</td><td rowspan="5">S_A</td><td rowspan="5">U_A</td><td rowspan="5">U_B</td><td rowspan="4">D_A</td><td rowspan="5">D_A</td><td rowspan="5">D_A</td><td>D_A</td><td rowspan="2">D_A</td><td rowspan="5">D_B</td></tr>
<tr><td>14</td><td>D_B</td></tr>
<tr><td>15</td><td>D_A</td><td rowspan="3">D_B</td></tr>
<tr><td>16</td><td rowspan="2">D_B</td></tr>
<tr><td>17</td><td>D_B</td></tr>
<tr><td>18</td><td>6</td><td>D_A</td><td>S_A</td><td>U_A</td><td>U_A</td><td>U_B</td><td>D_A</td><td>S_A</td><td>U_A</td><td>U_A</td><td>D_B</td></tr>
</table>

Table 341 – TDD uplink and downlink subframes between the relay node and donor eNode B

★ Downlink subframes on the backhaul link are configured as MBSFN subframes on the access link. This is done to avoid the population of UE from expecting any PDSCH or cell specific Reference Signal transmissions from the relay during those subframes

- the relay does not actually transmit any MBSFN information during the MBSFN subframes
- configuring subframes as MBSFN allows the relay to apply Discontinuous Transmission (DTX) during the subframes that it receives downlink transmissions from the donor eNode B
- this corresponds to a half-duplex mode of operation and avoids issues with the relay transmitter interfering with the relay receiver when both are using the same channel bandwidth

- Returning to Table 340, there is a section of information regarding the Relay Physical Downlink Control Channel (R-PDCCH). The R-PDCCH is the backhaul link equivalent of the normal PDCCH used on the access link. The R-PDCCH has some differences compared to the normal PDCCH:
  - the R-PDCCH is not transmitted during the first OFDMA symbols of a downlink subframe (the donor eNode B can use these symbols for normal PDCCH transmissions towards UE which are connected directly to the donor eNode B)
  - R-PDCCH transmissions which include downlink resource allocations for the backhaul link start during the fourth OFDMA symbol of the first time slot and finish during the final OFDMA symbol of the first time slot
  - R-PDCCH transmissions which include uplink resource allocations for the backhaul link start during the first OFDMA symbol of the second time slot and finish during either the final OFDMA symbol, or the sixth OFDMA symbol of the second time slot (transmissions finish earlier when the donor eNode B and relay downlink transmissions are time aligned)
  - these R-PDCCH timing concepts are illustrated in Figure 293. Leaving empty OFDMA symbols at the start of the subframe provides the relay with time to switch between transmitting on the access link and receiving on the backhaul link. Likewise, leaving an empty OFDMA symbol at the end of the subframe provides the relay with time to switch between receiving on the backhaul link and transmitting on the access link

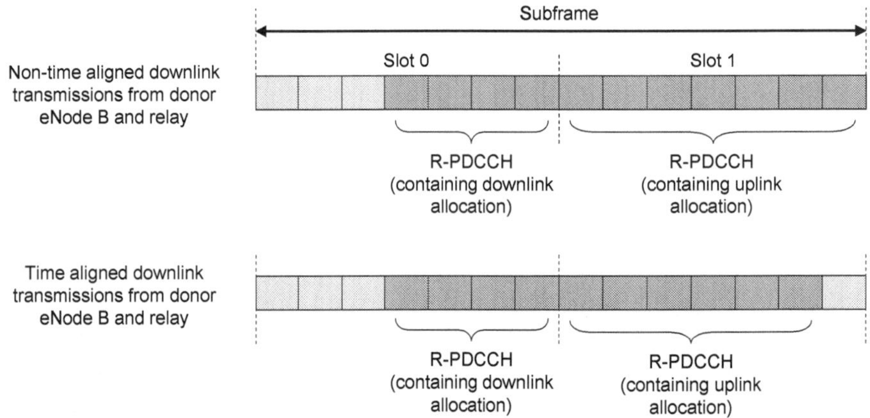

Figure 293 – Timing of R-PDCCH transmissions from donor eNode B

- The R-PDCCH Resource Allocation Type shown in Table 340 can be signalled with values of 0, 1, 2 localised or 2 distributed. These values correspond to the various ways in which downlink Resource Blocks can be allocated (see sections 9.7, 9.8 and 9.9). The subsequent Resource Block Assignment allocates the Resource Blocks themselves when interpreted according to the Resource Allocation Type
- These Resource Block Allocations do not directly indicate the Resource Blocks within which there will be PDSCH transmissions from the donor eNode B. They represent the Resource Blocks within which the relay must check for R-PDCCH transmissions. The R-PDCCH transmissions then make the resource allocations for PDSCH and PUSCH transmissions across the backhaul link, i.e. the relay node is dynamically allocated resources across the backhaul link
- The backhaul link consumes air-interface resources from the donor eNode B so Resource Blocks are only allocated when necessary
- Table 340 shows that the Demodulation Reference Signal can be configured as either Interleaving or Non-Interleaving. The interleaving option allows a pair of R-PDCCH transmissions to use interleaved subcarriers within a common set of Resource Blocks. The non-interleaving option provides each R-PDCCH transmission with its own set of dedicated Resource Blocks
- The PDSCH Start Symbol shown in Table 340 can be signalled with values of 1, 2 or 3. A value of 1 is only permitted if the donor eNode B and relay node downlink transmissions are time aligned. The value of the PDSCH Start Symbol indicates the first OFDMA symbol within the first time slot which can be occupied by a PDSCH transmission across the backhaul link
- The two general scenarios for PDSCH transmission across the backhaul link are illustrated in Figure 294
  - in the first scenario, the PDSCH is allocated the same Resource Block pair as the R-PDCCH. In this case, the PDSCH is only transmitted during the second time slot. This scenario is applicable if there is any overlap between the PDSCH Resource Block allocation and the R-PDCCH Resource Blocks, i.e. the PDSCH does not have to be allocated exactly the same Resource Blocks as the R-PDCCH. The benefit of this approach is that it avoids the second time slot being left unused when there is no R-PDCCH providing an uplink resource allocation
  - in the second scenario, the PDSCH is allocated one or more Resource Block pairs which do not overlap with the R-PDCCH transmission. In this case, the PDSCH transmission is able to use OFDMA symbols in both the first and second time slots. The first symbol used by the PDSCH in the first time slot is defined by the 'PDSCH Start Symbol' parameter (whereas as the first symbol used by the R-PDCCH in the first time slot is always the fourth symbol). This approach could be used if there is an R-PDCCH transmission during the second time slot providing an uplink resource allocation

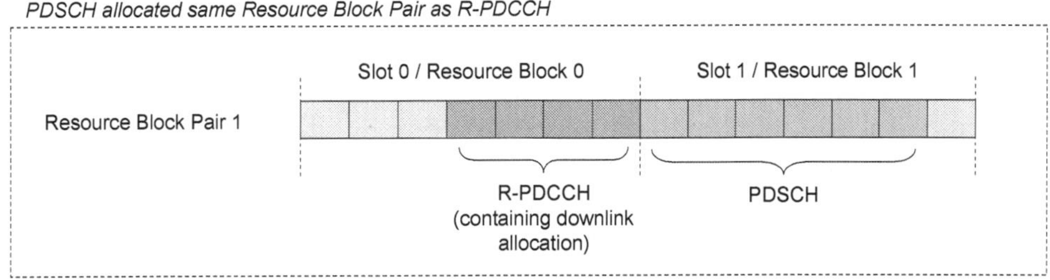

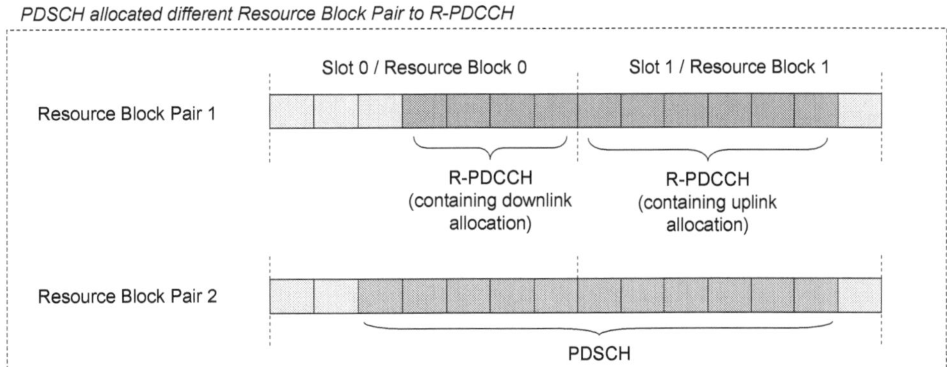

Figure 294 – Resource Block allocations for PDSCH on the backhaul link

- In the case of FDD, the relay node transmits the PUSCH on the backhaul link, 4 subframes after receiving the uplink allocation on the R-PDCCH
- In the case of TDD, the delay between receiving the uplink allocation on the R-PDCCH and transmitting the PUSCH across the backhaul link depends upon the uplink-downlink subframe configuration and the subframe number
- Returning to Table 340, there is a section of information regarding the PUCCH configuration. The PUCCH is used to provide HARQ acknowledgements from the relay node to the donor eNode B, i.e. to acknowledge PDSCH transmissions across the backhaul link
- The PHICH is not used to send HARQ acknowledgments across the backhaul link. Instead, the relay node relies upon receiving the 'New Data Indicator' within a subsequent R-PDCCH transmission to determine whether or not a retransmission is required
- The relay node has its own scheduler for allocating access link resources to the population of UE. These resource allocations are made using the normal PDCCH. The scheduler implementation is likely to be similar to that within an eNode B but must account for the subframes which are reserved for the backhaul link
- The donor eNode B acts as a proxy for the S1 interfaces between the relay node and Serving Gateway (user plane) and between the relay node and MME (control plane). The user plane protocol stack for the S1 interface is shown in Figure 295

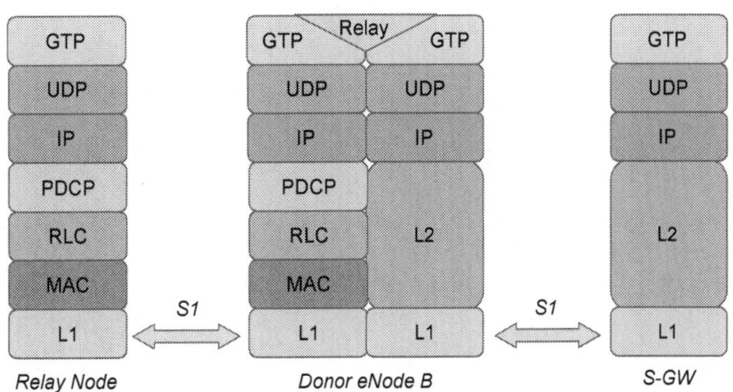

Figure 295 – User plane protocol stack for S1 interface

- This protocol stack makes use of the radio user plane protocol stack shown in Figure 290, i.e. the Un interface is used to transfer S1 related data across the backhaul link. A first GTP tunnel is defined between the relay node and donor eNode B, while a second GTP tunnel is defined between the donor eNode B and Serving Gateway
- A similar protocol stack exists for the S1 control plane connection between relay node and MME. The control plane protocol stack replaces the GTP/UDP/IP layers with S1-Application Protocol and SCTP layers
- Control plane and user plane protocol stacks are also defined for the X2 interface. The donor eNode B acts as a proxy which forwards data to and from neighbouring eNode B

- ★ Relay nodes transmit their own Physical layer Cell Identity (PCI) and appear as separate cells from the UE perspective. Repeaters re-transmit the PCI belonging to the donor cell and appear as an extension of the donor cell from the UE perspective
- ★ Relay nodes introduce additional delay due to the double scheduling required to transfer data, i.e. downlink data has first got to be scheduled for transmission from the donor eNode B to the relay node, and then scheduled for transmission from the relay node to the UE. Similarly, uplink data has to be scheduled twice. Repeaters do not introduce this type of delay because they are simply re-transmitting the uplink and downlink signals
- ★ 3GPP TS 36.300 specifies that a relay node may not use another relay node as its donor cell, i.e. relay nodes cannot be chained
- ★ Relays can be deployed at locations which have a power supply and radio connection back to the donor eNode B. They do not require a fixed transport connection
- ★ 3GPP References: TS 36.300, TS 36.216, TS 36.331

31 NETWORK PLANNING

31.1 RADIO NETWORK PLANNING

★ Radio network planning is used to identify the geographic locations of the eNode B. It is also used to identify the associated antenna configurations in terms of antenna type, height, azimuth and tilt

★ The high level interaction of radio network planning with dimensioning, site acquisition, site design and site build is illustrated in Figure 296

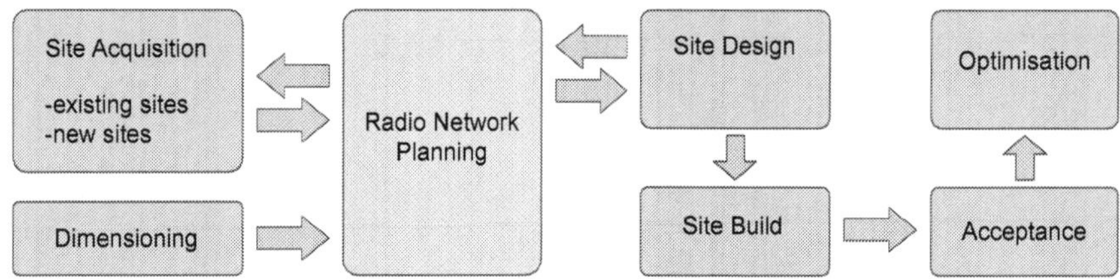

Figure 296 – Interaction of radio network planning with site acquisition, design and build

★ Radio network planning is preceded by system dimensioning. Dimensioning results are used to generate an estimate of the expected site density. This provides a guide to the number of sites required to achieve the target coverage and capacity

★ Site acquisition defines the actual sites which are available to the radio network planners. The site acquisition team may provide radio network planners with a list of candidate sites. Alternatively, radio network planners may identify the requirement for one or more sites at locations where there are currently no candidates

★ Sites represent a relatively expensive investment for the operator. Selecting sites which have a poor location or which limit good site design can lead to reduced system performance irrespective of any subsequent RF and parameter optimisation

★ Evaluating candidate sites should involve a site visit as well as modelling within the radio network planning tool. In practice, the interaction between site acquisition and radio network planning is likely to be a compromise between obtaining ideal site locations and the actual locations which are available

★ If an operator has already deployed a 2G or 3G network then it is likely to be beneficial to re-use as many of the existing site locations as possible. Re-use of existing sites introduces the option of sharing antenna sub-systems between 2G, 3G and LTE, i.e. feeder cables and antennas can be shared

★ A potential drawback associated with sharing antennas is that it may limit the radio network planner's ability to apply independent 2G, 3G and LTE RF optimisation, e.g. if the LTE system requires mechanical antenna downtilt then the impact upon the 2G and 3G systems must also be evaluated

★ Site selection should also account for site design requirements. Site design involves identifying specific locations for the eNode B cabinet and each antenna. It also involves identifying the route for the feeder cable to connect the eNode B cabinet to each antenna. The feeder cable should be routed as directly as possible to help reduce its length and minimise feeder losses

★ Antenna placement represents the part of site design which has a direct impact upon radio network planning:

  - antennas should not be placed behind any obstacles. Roof-top site designs should ensure that antennas are either at the edge of the roof-top or sufficiently high to avoid the edge of the roof-top clipping the antenna gain pattern
  - if antennas are mounted on the side of a building then the azimuth requirements should be considered to ensure that the walls of the building do not shield the horizontal beamwidth
  - if the site already accommodates other antennas then the physical separation to those antennas should be considered to avoid interference

★ There are 2 fundamental approaches to radio network planning – path loss based approach and simulation based approach

  - the path loss based approach is relatively simple and allows the radio network to be planned without modeling any subscriber traffic. This approach uses link budget results to define minimum signal strength thresholds. Results are typically presented in terms of service coverage and best server areas
  - the simulation based approach is more complex and time consuming but generates a greater quantity of information. Simulations are typically based upon static rather than dynamic simulations. Results can be presented in terms of coverage, cell and connection throughputs, transmit powers, interference levels

★ Simulations may be used to supplement the path loss based approach. In this case, the main planning activity is completed using path loss calculations while simulations provide additional and more detailed information for specific sections of the network

31.1.1 PATH LOSS BASED APPROACH

★ The path loss based approach to radio network planning requires a planning tool capable of completing path loss calculations and displaying geographic areas where specific path loss thresholds have been exceeded. The planning tool should also be capable of displaying best server areas

★ The inputs to the planning tool for the path loss based approach are illustrated in Figure 297

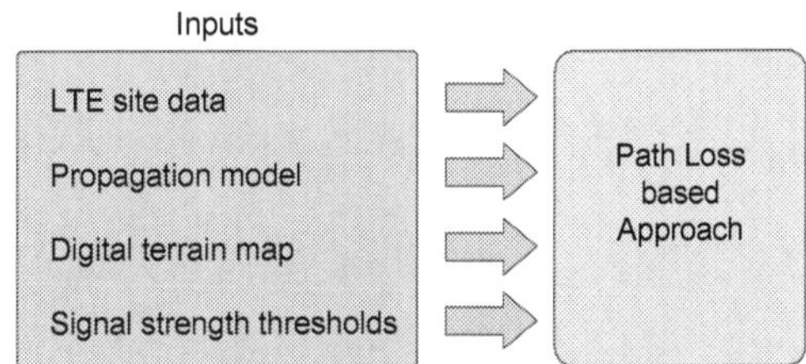

Figure 297 – Inputs for path loss based approach to radio network planning

★ The LTE site data should include site locations, antenna types, antenna heights, antenna tilts, antenna azimuths, feeder types, feeder lengths, RF carrier and eNode B types

★ The propagation model should be tuned from measurements. Accurate propagation modelling is fundamental to radio network planning. Propagation model tuning involves minimising the standard deviation of the error between the predicted and measured propagation loss while maintaining a mean error which is close to 0 dB

★ A different propagation model can be defined for each environment type or environment dependant correction factors can be applied to a common propagation model. Different propagation models can also be defined for different cell ranges, antenna heights and operating bands

★ The digital terrain map should be adequate in terms of resolution, number of clutter categories and accuracy. The resolution should be relatively high for urban and suburban areas but can be reduced for rural areas. The appropriate number of clutter categories depends upon the geographic area. If the number of categories is large then the propagation tuning exercise becomes more difficult

★ If microcells or picocells with below roof-top antennas are to be planned then it may be necessary to purchase a map which includes building vectors. Building vectors can be defined in either two or three dimensions

★ The signal strength thresholds should be based upon a link budget analysis. Link budgets can be used to generate a set of maximum allowed path loss thresholds for a range of target throughputs within each environment type. In general, planning tools display contours of downlink signal strength rather than downlink path loss. This makes it necessary to translate the maximum allowed path loss into a minimum allowed signal strength

★ A relatively arbitrary base station transmit power can be selected and the signal strength thresholds calculated by subtracting the maximum allowed path loss. The transmit power could be set equal to the cell specific Reference Signal transmit power per Resource Element. This allows signal strengths calculated by the planning tool to be interpreted as RSRP

★ An example of the translation from a link budget maximum allowed path loss to a planning tool signal strength threshold is presented in Table 342. In this example, the downlink transmit power is based upon a total downlink transmit power capability of 43 dBm and a channel bandwidth of 20 MHz (1200 Resource Elements), i.e. the cell specific Reference Signal transmit power per Resource Element is 43 - 10×LOG(1200) = 12 dBm (assuming equal transmit powers for all Resource Elements)

★ Table 342 assumes that the maximum allowed path loss from the LTE link budget is 145 dB. This figure could have originated from either an uplink or downlink link budget. In both cases, the planning tool is used to display contours of downlink signal strength, i.e. signal strength is used to quantify path loss

Downlink transmit power configured in the planning tool	12 dBm
Link budget result for maximum allowed path loss	145 dB
eNode B antenna gain assumed in the link budgets	18 dBi
UE antenna gain assumed in the link budgets	0 dBi
Feeder loss assumed in the link budgets	2 dB
Resultant planning tool signal strength threshold	-117 dBm

Table 342 – Translation of maximum allowed path loss to planning tool signal strength threshold

★ Calculating the planning tool signal strength threshold also requires consideration of the antenna gains and feeder loss values assumed in the link budget. The planning tool will apply its own antenna gains and feeder loss values which can be site specific and may not equal the values assumed in the link budget

* The antenna gain values assumed in the link budget can be removed from the maximum allowed path loss result so the planning tool is able to apply its own values on a site by site basis. This is done by subtracting the link budget antenna gain values from the maximum allowed path loss
* Handling feeder loss values can be more complex:
  - if feeder loss values within the planning tool are set to 0 dB then the maximum allowed path loss does not require any feeder loss adjustment
  - if feeder loss values within the planning tool are set to their actual values then the link budget feeder loss should be added to the maximum allowed path loss result. This will generate only an approximate result if the link budget analysis resulted in uplink limited coverage with the use of Mast Head Amplifiers (MHA). The result is approximate because the maximum allowed path loss generated by the link budget is independent of the feeder loss (assuming the MHA exactly compensates the feeder loss) while the signal strength calculated by the planning tool is dependent upon the feeder loss
* The resultant planning tool signal strength threshold in Table 342 is calculated as the downlink transmit power – the adjusted maximum allowed path loss, i.e. 12 - (145 - 18 - 0 + 2) = -117 dBm
* The path loss based approach to radio network planning should include an analysis of the best server areas. This helps to ensure good dominance and a relatively even distribution of network loading. Best server areas should be contiguous and should not be fragmented. Non-contiguous best server areas indicate that there is likely to be relatively poor dominance and increased levels of inter-cell interference. In general, neighbouring best server areas should be of approximately equal size. If there is a known traffic hotspot then an eNode B should be located as close as possible to that hotspot and the dominance area can be smaller

31.1.2 SIMULATION BASED APPROACH

* The simulation based approach to radio network planning requires a more sophisticated radio network planning tool. The majority of radio network planning tools use Monte Carlo simulations. Monte Carlo simulations are static rather than dynamic. This means that system performance is evaluated by considering many independent instants (snap shots) in time. In general, dynamic simulations are more time consuming than static simulations
* Figure 298 compares the principles of static and dynamic simulations:
  - static simulations are able to generate probability distributions and averages. Network behaviour during one simulation snap shot does not impact the behaviour during the next snap shot. Snap shots are random and do not follow any time history. When compared to dynamic simulations, fewer time instants are required to generate results
  - dynamic simulations are able to generate time histories in addition to probability distributions and averages. The time history of network behaviour is modelled with relatively high resolution. Certain aspects of network behaviour can be modelled in a more realistic manner, e.g. a packet scheduler can account for allocations during previous time instants, and hysteresis in handover thresholds can be followed

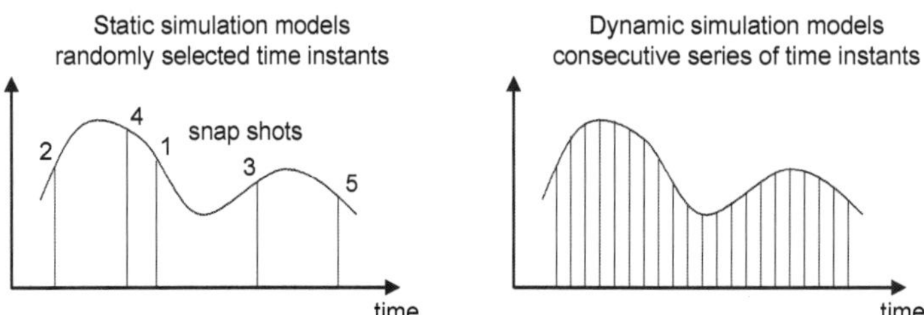

Figure 298 – Comparison of static and dynamic simulation principles

* Both the static and dynamic simulations referenced above are system level simulations rather than link level simulations. The results from link level simulations usually form an input to system level simulations, e.g. the system level simulation look-up table defining the relationship between throughput and signal to noise ratio originates from link level simulations
  - system level simulations model the network at a relatively high level. UE and BTS physical locations are modelled along with their antenna gains, transmit powers and noise floors. System level simulations do not model the transfer of individual modulation symbols. Static simulations calculate the signal to noise ratio conditions and subsequently look-up a corresponding throughput. Dynamic simulations may model the transfer of individual transport blocks and then use look-up tables to determine the probability of successful reception
  - link level simulations model the low level details of the transmitter and receiver. The physical layer algorithms used by the UE and BTS are modelled. These include channel coding, interleaving and error detection. Individual bits are mapped onto modulation symbols before being filtered and transferred across specific propagation channels. Receiver modelling is likely to include synchronisation, channel estimation and equalisation. Bit Error Rates (BER) and Block Error Rates (BLER) are calculated as a function of throughput and signal to noise ratio

★ The general principle for Monte Carlo static simulations is presented in Figure 299

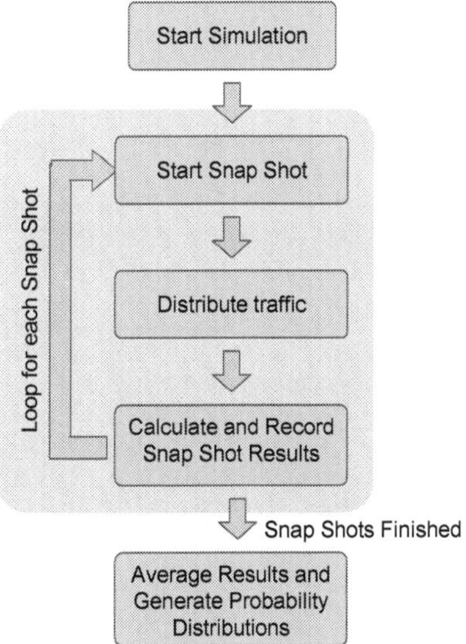

Figure 299 – General principle for Monte Carlo static simulations

★ Each simulation snap shot is started by distributing a population of UE across the simulation area. This distribution could be based upon a uniform random distribution or a weighted random distribution. Weightings can be based upon Erlang maps generated from the traffic belonging to an existing network. Alternatively, weightings could be environment type dependant so for example, urban areas could be specified to have a higher traffic density than suburban areas

★ Once the population of UE have been distributed, the simulation results are generated for that snap shot. The planning tool determines uplink and downlink signal to noise ratios for each physical channel and Reference Signal. Throughputs are derived using look-up tables which define the relationship between throughput and signal to noise ratio. These look-up tables typically originate from separate link level simulations

★ The results are recorded at the end of a simulation snap shot and the process is repeated. This allows the simulation to generate probability distributions and to quantify the probability of certain events occurring, e.g. the probability that a UE will be able to establish a connection at a specific location. The number of snap shots necessary to generate statistically stable simulation results tends to depend upon the quantity of traffic distributed during each snap shot. Distributing relatively little traffic tends to increase the number of snap shots required

★ The inputs to the planning tool for the simulation based approach to radio network planning are illustrated in Figure 300

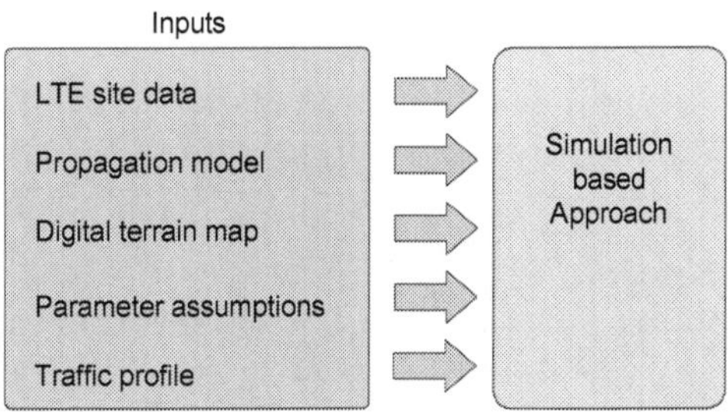

Figure 300 – Inputs for simulation based approach to radio network planning

★ The first three inputs are similar to those required by the path loss based approach. The LTE site data may be more complex in terms of requiring greater information to describe the BTS capability, e.g. baseband processing capability

★ Propagation modelling is also more complex for a simulation if slow fading is modelled. The path loss based approach uses a link budget threshold which includes a slow fade margin. Simulations typically model slow fading explicitly making it necessary to specify a standard deviation and correlation factor. The correlation factor is used to specify the coherence of the fading experienced by the signals between a UE and the set of surrounding BTS. A high correlation factor means that when one signal is experiencing a fade there is a high probability that the other signals will also be experiencing a fade

★ Typical LTE parameter assumptions include channel bandwidth, maximum uplink and downlink transmit powers, Reference Signal transmit powers, uplink and downlink noise figures, look-up tables between signal to noise ratio and throughput for each propagation channel and throughput requirements for a range of services

★ Traffic profiles are specified in terms of the services used by the population of UE. Defining accurate traffic profiles can be difficult and it is reasonable to start a simulation exercise by modelling one service at a time, e.g. generate coverage and capacity results for the data service, and then generate similar results for the VoIP service. This approach provides an indication of the variance which can be expected when changing the traffic profile. The traffic profile also requires the geographic distribution of the UE to be defined. This includes specifying the percentage of UE which are indoors and experience a building penetration loss

★ Identifying the network capacity can be an iterative process requiring the network planner to increase the quantity of traffic loading the network until the probability of blocking reaches a realistic maximum. The absolute maximum network capacity can be quantified by distributing very large quantities of traffic but the blocking probability will become unrealistically high

★ The simulation based approach to radio network planning is more time consuming than the path loss based approach. Results take longer to generate and longer to interpret. Simulation time depends upon the size and resolution of the geographic area, the site density and the quantity of traffic loading the network. It is important to ensure that sufficient simulation snap shots have been completed to generate statistically stable results

★ The main benefit of completing simulations is the relatively large quantity of information which is generated. This information can help guide planning decisions as well as provide more extensive expectations of network performance

31.2 LINK BUDGETS

★ Link budgets are used during both system dimensioning and radio network planning. The dimensioning process provides an estimate of the number of network elements required to achieve a specific coverage and capacity performance. Link budgets are used during dimensioning to estimate the maximum allowed path loss and the corresponding cell range

★ The results from a dimensioning exercise can be used as an input for a business case analysis. The number of network elements and their associated configuration allow the cost of the network to be quantified. The path loss based approach to radio network planning (section 31.1.1) makes use of link budget results to define maximum allowed path loss thresholds

31.2.1 UPLINK

★ A set of example link budgets for the PUSCH is presented in Table 343

	PS Data 500 kbps	PS Data 1 Mbps	PS Data 2 Mbps	
Channel Bandwidth	20	20	20	MHz
Total Resource Blocks	100	100	100	Res. Blks
Allocated Resource Blocks	8	27	25	Res. Blks
Allocated Subcarriers	96	324	300	Subcarriers
Maximum Transmit Power	23	22	22	dBm
UE Antenna Gain	0	0	0	dBi
Body Loss	0	0	0	dB
Transmit EIRP	23	22	22	dBm
Thermal Noise per Subcarrier	-132.2	-132.2	-132.2	dBm
Aggregate Thermal Noise	-112.4	-107.1	-107.4	dBm
Receiver Noise Figure	2.5	2.5	2.5	dB
Interference Margin	2	2	2	dB
Interference Floor	-107.9	-102.6	-102.9	dBm
SINR Requirement	0.5	-2.0	0.9	dB
Received Signal Strength Requirement	-107.4	-104.6	-102.1	dBm
eNode B Antenna Gain		18		dBi
Cable Loss		0		dB
Slow Fading Margin		9		dB
Shadowing Handover Gain		2.5		dB
Building Penetration Loss		18		dB
Isotropic Power Requirement	-100.9	-98.1	-95.6	dBm
Maximum Allowed Path Loss	123.9	120.1	117.6	dB

Table 343 – Example link budgets for the PUSCH

★ The channel Bandwidth is not used explicitly within the link budget but it has an impact upon the Signal to Interference plus Noise Ratio (SINR) requirement. SINR requirements are reduced for larger channel bandwidths when frequency domain scheduling is channel aware. Larger channel bandwidths provide the packet scheduler with greater flexibility to allocate Resource Blocks experiencing good channel conditions. The channel bandwidth also places an upper limit upon the number of Resource Blocks which can be allocated to the UE

★ The total number of Resource Blocks is included for information and is not used explicitly within the link budget

★ The number of allocated Resource Blocks defines a trade-off between the cell capacity and the cell range. Allocating a small number of Resource Blocks to the connection leads to less air-interface redundancy (higher coding rate and less benefit from channel coding). This means the cell range must be limited otherwise the eNode B will not be able to successfully decode the transport blocks. The benefit of allocating a small number of Resource Blocks is that more Resource Blocks are available for other connections and the cell capacity is typically increased

★ The example presented in Table 343 allocates 8 Resource Blocks for a 500 kbps connection. Based upon Table 388, a connection with 8 Resource Blocks can use a transport block size of 552 bits to achieve 500 kbps. Table 395 indicates that QPSK is used as a modulation scheme, which is appropriate for cell edge connections

★ The set of 8 Resource Blocks is a relatively large allocation for a transport block size of 552 bits. A single Resource Block allocation with the normal cyclic prefix can accommodate 2 × 72 = 144 modulation symbols after accounting for the uplink Demodulation Reference Signal and the pairing of Resource Blocks in the time domain. This corresponds to a total allocated capacity of 2304 bits after accounting for all 8 Resource Blocks and the QPSK modulation scheme. The coding rate is then defined by the ratio of 552 / 2304 = 0.24. This represents a low coding rate so the corresponding SINR requirement should be relatively small and the maximum allowed path loss should be relatively large

★ The example presented in Table 343 allocates 27 Resource Blocks for a 1 Mbps connection. Based upon Table 389, a connection with 27 Resource Blocks can use a transport block size of 1192 bits to achieve 1 Mbps. Table 395 indicates that QPSK is used as a modulation scheme. The corresponding coding rate can be calculated as 1192 / 7776 = 0.15

★ The example presented in Table 343 allocates 25 Resource Blocks for a 2 Mbps connection. Based upon Table 389, a connection with 25 Resource Blocks can use a transport block size of 2216 bits to achieve 2 Mbps. Table 395 indicates that QPSK is used as a modulation scheme. The corresponding coding rate can be calculated as 2216 / 7200 = 0.31

★ The number of allocated subcarriers is given by 12 × the number of allocated Resource Blocks. The number of allocated subcarriers is used when aggregating the total thermal noise received by the BTS, i.e. it defines the noise bandwidth

★ The maximum transmit power of 23 dBm corresponds to the capability of UE power class 3 specified by 3GPP within TS 36.101. This capability has a tolerance of ± 2 dB so 23 dBm could be optimistic for some devices. The maximum transmit power has been reduced to 22 dBm for the 1 Mbps and 2 Mbps examples because 3GPP TS 36.101 specifies a 1 dB Maximum Power Reduction (MPR) when more than 18 Resource Blocks are allocated from the 20 MHz channel bandwidth while using QPSK

★ The terminal antenna gain can vary from one UE model to another. Datacards may have a higher antenna gain than handheld devices. UE typically have an antenna gain in the order of 0 dBi

★ Body loss is dependent upon the relative positions of the UE, the end-user and the serving cell. A figure of 3 dB is typically assumed when the UE is held to one side of the end-user's head. A figure of 0 dB is typically assumed when the UE is positioned away from the body

★ The first main result from the uplink link budget is the transmit Effective Isotropic Radiated Power (EIRP). This is defined using the expression below:

$$\textit{Transmit EIRP} = \textit{Maximum Transmit Power} + \textit{UE Antenna Gain} - \textit{Body Loss}$$

★ The thermal noise per subcarrier quantifies the noise power within the bandwidth of a single 15 kHz subcarrier. This is calculated using 10 × LOG(kTB), where k is the Boltzmann constant (1.38×10^{-23}), T is the temperature (290 Kelvin) and B is the bandwidth (15000 Hz)

★ The aggregate thermal noise quantifies the noise power within the total allocated bandwidth. It is given by the thermal noise per subcarrier + 10 × LOG(number of allocated subcarriers)

★ The receiver noise figure assumption reflects the performance of the eNode B receiver subsystem. The noise figure belonging to the eNode B cabinet should be used if the receiver subsystem does not include a Mast Head Amplifier (MHA). If an MHA is included, the noise figure should be the composite noise figure of the MHA, cable/connectors and eNode B cabinet. The composite noise figure can be calculated using Friis' equation:

$$\textit{Composite}\ \text{Noise Figure} = 10 \times LOG\left(NF_{MHA} + \frac{NF_{cable} - 1}{Gain_{MHA}} + \frac{NF_{eNodeB} - 1}{Gain_{MHA} \times Gain_{Cable}} \right)$$

★ With the exception of the composite noise figure result, all of the variables within the preceding equation have linear units. The noise figure of the cable and connectors is equal to their loss. For example, the noise figure is 2 dB in log units and 1.6 in linear units if the cable and connector loss is 2 dB. The gain of the cable and connectors is equal to -1 × their loss, i.e. -2 dB in log units and 0.6 in linear units. Friis' equation illustrates that when the MHA has a high gain, the noise figure of the receiver sub-system is dominated by the noise figure of the MHA. This emphasises the importance of having a low noise, high gain amplifier for the MHA

★ The interference margin is generated by co-channel interference from UE served by neighbouring cells. The interference margin is likely to be greater in urban areas where the site density is relatively high. Heterogeneous network architectures can also lead to increased co-channel interference. Inter-Cell Interference Coordination (ICIC) is intended to help manage levels of co-channel interference

★ The interference floor is defined using the expression below:

$$\textit{Interference Floor} = \textit{Aggregate Thermal Noise} + \textit{Receiver Noise Figure} + \textit{Interference Margin}$$

★ The Signal to Interference plus Noise Ratio (SINR) requirement is defined by link level simulations which model the BTS receiver performance when the allocated transport block size is transferred using the allocated number of Resource Blocks with a specific

Block Error Rate (BLER). Link level simulations model a specific propagation channel so the SINR requirement is specific for that channel. Propagation channel modelling includes fast fading (unless it is a static channel) so the resultant SINR includes the impact of fast fading

★ The SINR examples in Table 343 do not demonstrate an obvious trend between SINR requirement and bit rate requirement, i.e. the 500 kbps example has a higher SINR requirement than the 1 Mbps example, but a lower SINR requirement than the 2 Mbps example. This results from the different Resource Block allocations and different coding rates. The SINR requirement increases as the coding rate increases

★ The second main result from the uplink link budget is the Received Signal Strength Requirement. This is defined using the expression below:

Received Signal Strength Requirement = Interference Floor + SINR Requirement

★ As expected, the Received Signal Strength Requirement increases as the bit rate requirement increases. The higher interference floors associated with the larger Resource Block allocations increase the resultant Received Signal Strength Requirement

★ The eNode B antenna gain should be representative of the antenna type planned for network deployment. In practice, networks may include a range of antenna types. Antenna gains tend to decrease as the horizontal and vertical beamwidths increase and the antenna becomes less directional

★ The antenna gain figure can incorporate a polarisation loss of approximately 0.5 dB. Antenna gains are typically quoted from measurements which have been recorded using a receiving antenna element which has exactly the same polarisation as the transmitting antenna element. In practice, the two antenna elements have different polarisations and a polarisation loss is experienced. Reflections change the polarision of a radio signal and this helps to reduce the loss because many signals with different polarisations can reach the receiving antenna. Cross polar antennas also reduce the potential for polarisation loss because the maximum angular offset is 45° compared to 90° for a vertically polarised antenna

★ The cable loss variable within the uplink link budget is only applicable if it has not already been included as part of the receiver noise figure. The uplink cable loss is included within the composite noise figure if an MHA has been assumed. Otherwise, the cable loss should equal all cable and connector losses between the eNode B cabinet and antenna. The example presented in Table 343 assumes an MHA is used so the cable loss is set to 0 dB

★ The slow fading margin is calculated from an indoor location probability and an indoor standard deviation. The indoor location probability is often specified as an average probability of experiencing indoor coverage across the cell area. This figure is translated to an equivalent cell edge indoor coverage probability before combining with the standard deviation to generate the slow fading margin. The indoor location probability at cell edge will be less than the average as a result of the higher UE transmit power requirement. The indoor standard deviation represents a combination of the outdoor standard deviation associated with slow fading, and the standard deviation generated by the variance of the building penetration loss

★ The shadowing handover gain is generated by allowing the UE to handover onto the best server. When the UE is at cell edge and there are multiple potential serving cells, the UE is able select the best cell to help avoid experiencing fades. The shadowing handover gain is reduced when handovers have hysteresis to avoid ping-pongs between cells. The shadowing handover gain would reduce to 0 dB at the edge of network coverage where there are no neighbouring cells to act as handover candidates. This may also be the case at some indoor locations

★ Including the building penetration loss as part of the link budget generates an outdoor maximum allowed path loss result which includes sufficient margin to allow UE at the cell edge to establish and maintain connections from within buildings. The building penetration loss may be replaced by a vehicle penetration loss if link budgets are generated for a section of motorway or a rural area

★ The assumptions for building penetration loss typically depend upon the environment type, e.g. building penetration could be greater within an urban environment than within a suburban environment. The building penetration loss usually represents the link budget assumption with the greatest uncertainty. SINR figures and shadowing handover gains could have an uncertainty of ± 1 dB while building penetration loss could have an uncertainty of ± 5 dB. This uncertainty is included within the link budget result by calculating the slow fading margin from an indoor standard deviation which incorporates the variance of the building penetration loss. Building penetration loss assumptions are relatively difficult to validate in the field due to the large variance between different buildings and the geometry of those buildings with respect to the radio network plan

★ The third main result from the uplink link budget is the Isotropic Power Requirement. This is defined using the expression below:

Isotropic Power Requirement = Received Signal Requirement – eNode B antenna Gain + Cable Loss + Slow Fading Margin – Shadowing Handover Gain + Building Penetration Loss

★ The Maximum Allowed Path Loss is then calculated as the difference between the transmit EIRP and the isotropic power requirement:

Maximum Allowed Path Loss = Transmit EIRP – Isotropic Power Requirement

★ This maximum allowed path loss result can be compared with the equivalent downlink result to determine whether coverage is uplink or downlink limited. This comparison requires an offset to account for the frequency difference between the uplink and downlink frequency bands. Higher frequencies tend to experience greater path loss so coverage will tend to be downlink limited if both the uplink and downlink link budgets generate equal maximum allowed path loss results. The frequency dependant term within a typical Okumura-Hata path loss model is given by the equation below:

Frequency Dependent Loss = 33.9 × LOG (Frequency (MHz))

This equation indicates that for a fixed cell range, the uplink propagation loss at 2510 MHz is 0.7 dB less than the downlink propagation loss at 2630 MHz, i.e. the uplink maximum allowed path loss can be 0.7 dB less than the downlink maximum allowed path loss before the uplink starts to determine the cell range

31.2.2 DOWNLINK

★ A set of example link budgets for the PDSCH is presented in Table 344

	PS Data 500 kbps	PS Data 1 Mbps	PS Data 2 Mbps	
Channel Bandwidth	20	20	20	MHz
Total Resource Blocks	100	100	100	Res. Blks
Allocated Resource Blocks	8	26	25	Res. Blks
Allocated Subcarriers	96	312	300	Subcarriers
Maximum Transmit Power	43	43	43	dBm
Transmit Power for Allocated Res. Blks.	32.0	37.1	37.0	dBm
Multiple Antenna Increase	3	3	3	dB
eNode B Antenna Gain	18	18	18	dBi
Cable Loss	2	2	2	dB
Transmit EIRP	51.0	56.1	56.0	dBm
Thermal Noise per Subcarrier	-132.2	-132.2	-132.2	dBm
Aggregate Thermal Noise	-112.4	-107.3	-107.4	dBm
Receiver Noise Figure	7.0	7.0	7.0	dB
Interference Margin	2.0	2.0	2.0	dB
Interference Floor	-103.4	-98.3	-98.4	dBm
SINR Requirement	-4.6	-7.1	-4.2	dB
Received Signal Strength Requirement	-108.0	-105.4	-102.6	dBm
UE Antenna Gain		0		dBi
Body Loss		0		dB
Slow Fading Margin		9		dB
Shadowing Handover Gain		2.5		dB
Building Penetration Loss		18		dB
Isotropic Power Requirement	-83.5	-80.9	-78.1	dBm
Maximum Allowed Path Loss	134.5	137.0	134.1	dB

Table 344 – Example link budgets for the PDSCH

★ The channel Bandwidth is not used explicitly within the link budget but it has an impact upon the Signal to Interference plus Noise Ratio (SINR) requirement. SINR requirements are reduced for larger channel bandwidths when frequency domain scheduling is channel aware. The channel bandwidth also places an upper limit upon the number of Resource Blocks which can be allocated to the UE

★ The total number of Resource Blocks is used when calculating the transmit power for the allocated Resource Blocks, i.e. the maximum transmit power is distributed across the total number of Resource Blocks so only a proportion of that power is available for the allocated Resource Blocks

★ The number of allocated Resource Blocks defines a trade-off between the cell capacity and the cell range. Allocating a small number of Resource Blocks to the connection leads to less air-interface redundancy (higher coding rate and less benefit from channel coding). This means the cell range must be limited otherwise the UE will not be able to successfully decode the transport blocks. The benefit of allocating a small number of Resource Blocks is that more Resource Blocks are available for other connections and the cell capacity is typically increased

★ The number of allocated Resource Blocks in Table 344 have been kept similar to those in Table 343 to allow a comparison between uplink and downlink

★ An allocation of 8 Resource Blocks has been assumed for the 500 kbps connection. Similar to the uplink, a transport block size of 552 bits can be used with QPSK as a modulation scheme. 2×2 MIMO is assumed for the link budgets in Table 344 but single layer transmission with a single codeword (single transport block) is assumed at cell edge

★ The set of 8 Resource Blocks is a relatively large allocation for a transport block size of 552 bits. Assuming that the PCFICH, PHICH and PDCCH occupy the first 2 OFDMA symbols within the subframe and that 2×2 MIMO is used (impacts the number of Resource Elements allocated to the cell specific Reference Signal) then the number of Resource Elements remaining for the PDSCH is given by $8 \times [(12 \times 7 \times 2) - (12 \times 2 + 12)] = 1056$. This corresponds to a capacity of 2112 bits when using QPSK. The coding rate is then defined by the ratio of 552 / 2112 bits = 0.26

★ The example presented in Table 344 allocates 26 Resource Blocks for a 1 Mbps connection. Based upon Table 389, a connection with 26 Resource Blocks can use a transport block size of 1160 bits to achieve 1 Mbps. Table 395 indicates that QPSK is used as a modulation scheme. The corresponding coding rate can be calculated as 1160 / 6864 = 0.17

★ The example presented in Table 344 allocates 25 Resource Blocks for a 2 Mbps connection. Based upon Table 389, a connection with 25 Resource Blocks can use a transport block size of 2216 bits to achieve 2 Mbps. Table 395 indicates that QPSK is used as a modulation scheme. The corresponding coding rate can be calculated as 2216 / 6600 = 0.34

★ The number of allocated subcarriers is given by 12 × the number of allocated Resource Blocks. The number of allocated subcarriers is used when aggregating the total thermal noise received by the UE, i.e. it defines the noise bandwidth

★ The maximum transmit power of 43 dBm corresponds to a typical macrocell downlink transmit power capability

★ The downlink transmit power is distributed across the entire channel bandwidth so only a percentage of the total power is available for the Resource Blocks allocated to the connection being considered in the link budget. In the case of the 500 kbps connection, 8 out of 100 Resource Blocks have been allocated so the downlink transmit power for the connection is 43 – 10 × LOG(100 / 8) = 32.0 dBm. Likewise, the transmit powers available for the 1 Mbps and 2 Mbps connections are 37.1 dBm and 37.0 dBm respectively

★ The link budgets presented in Table 344 assume the use of 2×2 MIMO so there are 2 transmitting antenna ports and the downlink transmit powers across the air-interface are increased by 3 dB

★ The eNode B antenna gain should be representative of the antenna type planned for network deployment. In practice, networks may include a range of antenna types. Antenna gains tend to decrease as the horizontal and vertical beamwidths increase and the antenna becomes less directional

★ The cable loss depends upon the site design and the use of remote radio heads or active antenna systems. Table 344 assumes a conventional site design with 2 dB of cable loss in the downlink direction. This cable loss includes the MHA insertion loss when an MHA is present

★ The first main result from the downlink link budget is the transmit EIRP which is defined as:

Transmit EIRP = Transmit Power for Allocated Res Blks + Multiple Antenna Increase + Antenna Gain – Cable Loss

★ The thermal noise per subcarrier quantifies the noise power within the bandwidth of a single 15 kHz subcarrier. This is calculated using 10 × LOG(kTB), where k is the Boltzmann constant (1.38×10^{-23}), T is the temperature (290 Kelvin) and B is the bandwidth (15000 Hz)

★ The aggregate thermal noise quantifies the noise power within the total allocated bandwidth. It is given by the thermal noise per subcarrier + 10 × LOG(number of allocated subcarriers)

★ The receiver noise figure assumption reflects the performance of the UE receiver and is likely to vary between UE models. A noise figure of 7 dB represents a typical assumption

★ The interference margin is generated by co-channel interference from neighbouring eNode B. The interference margin is likely to be greater in urban areas where the site density is relatively high. Heterogeneous network architectures can also lead to increased co-channel interference. Inter-Cell Interference Coordination (ICIC) is intended to help manage levels of co-channel interference

★ The interference floor is defined using the expression below:

Interference Floor = Aggregate Thermal Noise + Receiver Noise Figure + Interference Margin

★ The Signal to Interference plus Noise Ratio (SINR) requirement is defined by link level simulations which model the UE receiver performance when the allocated transport block size is transferred using the allocated number of Resource Blocks with a specific Block Error Rate (BLER). Link level simulations model a specific propagation channel so the SINR requirement is specific for that channel. Propagation channel modelling includes fast fading (unless it is a static channel) so the resultant SINR includes the impact of fast fading

★ The SINR examples in Table 344 show the same trend as for the uplink, i.e. the 500 kbps example has a higher SINR requirement than the 1 Mbps example, but a lower SINR requirement than the 2 Mbps example. This results from the different Resource Block allocations and different coding rates. The SINR requirement increases as the coding rate increases

★ The second main result from the downlink link budget is the Received Signal Strength Requirement. This is defined as:

Received Signal Strength Requirement = Interference Floor + SINR Requirement

★ As expected, the Received Signal Strength Requirement increases as the bit rate requirement increases. The higher interference floors associated with the larger Resource Block allocations increase the resultant Received Signal Strength Requirement

★ The UE antenna gain, body loss, slow fading margin, shadowing handover gain and building penetration loss assumptions are the same as those for the uplink link budget

★ The third main result from the uplink link budget is the Isotropic Power Requirement. This is defined as:

Isotropic Power Requirement = Received Signal Requirement – UE antenna Gain + Body Loss + Slow Fading Margin – Shadowing Handover Gain + Building Penetration Loss

★ The Maximum Allowed Path Loss is then calculated as the difference between the transmit EIRP and the isotropic power requirement:

Maximum Allowed Path Loss = Transmit EIRP – Isotropic Power Requirement

31.3 FREQUENCY PLANNING

★ Frequency planning determines the channel bandwidth available to each cell

★ Figure 301 illustrates the example of an operator with 15 MHz of spectrum:

- the operator can allocate 5 MHz per cell with a frequency reuse of 3, or 15 MHz per cell with a frequency reuse of 1
- the 5 MHz channel bandwidth limits the peak throughput for UE in good coverage but reduces the impact of intercell interference for UE towards cell edge

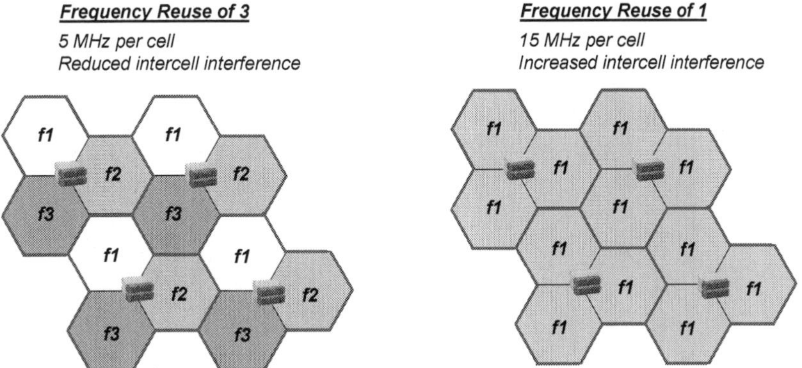

Figure 301 – Trade-off between channel bandwidth and intercell interference

★ Spectrum efficiency is maximised when using a frequency reuse of 1, despite the increased levels of intercell interference

★ Using a frequency reuse of 1 increases the importance of RF planning for good dominance to help minimise the impact of intercell interference

★ Intercell interference can be reduced if the schedulers within neighbouring eNode B co-ordinate with one another to avoid scheduling the same set of subcarriers during the same subframe. This concept of Inter-Cell Interference Coordination (ICIC) is described in section 32.14. ICIC creates a trade-off between improving the signal to noise ratio and reducing the quantity of resources available for transmission

★ A combination of frequency planning and carrier aggregation can be used to help manage co-channel interference within heterogeneous networks. Figure 302 illustrates a Closed Subscriber Group (CSG) femto within the coverage area of a macrocell. If the femto and macro are allocated the same single RF carrier then interference from the femto would generate a coverage hole for any UE which are within femto coverage but are not authorised to use the femto

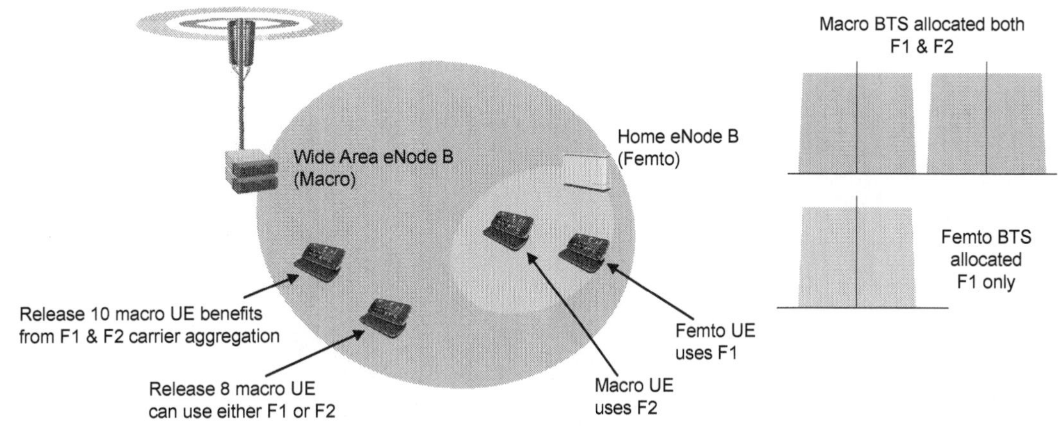

Figure 302 – RF carrier allocation to cope with co-channel interference from heterogeneous network

★ Figure 302 illustrates the case where the macro BTS has been allocated 2 × 20 MHz RF carriers while the femto BTS has been allocated 1 × 20 MHz RF carrier. The RF carrier allocated to the femto is co-channel to one of the macro RF carriers

- 3GPP release 10 UE which support LTE Advanced and carrier aggregation can benefit from using an effective channel bandwidth of 40 MHz while outside the femto coverage area
- 3GPP release 8 or 9 UE can use either of the two 20 MHz RF carriers allocated to the macro while outside the femto coverage area
- UE which are authorised to use the CSG femto can use its 20 MHz RF carrier while inside the femto coverage area

- UE which are not authorised to use the CSG femto can use the second macro 20 MHz RF carrier while inside the femto coverage area

★ Fractional frequency reuse has been proposed as a solution which allows UE in good coverage to benefit from a relatively wide channel bandwidth, while UE towards cell edge have a narrower channel bandwidth but one which benefits from a frequency reuse pattern

★ The concept of fractional frequency reuse is illustrated in Figure 303. If an operator has been allocated 20 MHz of spectrum then f1, f2 and f3 could correspond to 3 MHz channels while f4 could correspond to a 10 MHz channel

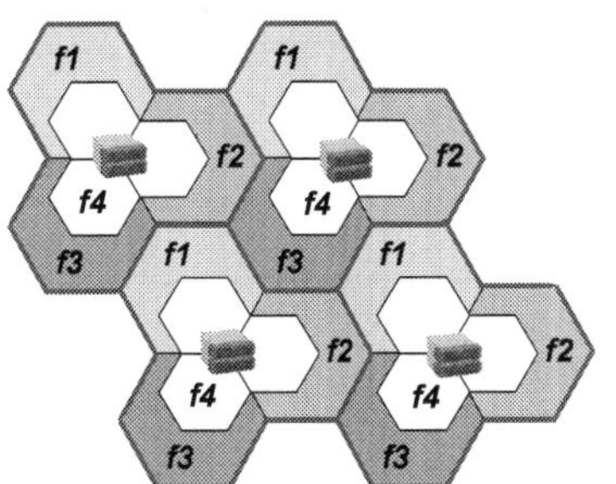

Figure 303 – Concept of fractional frequency reuse

★ In practice, RF planning for a frequency reuse pattern of 3 is difficult and levels of intercell interference would remain relatively high unless a higher frequency reuse pattern is used. This would then reduce the bandwidth available to each channel and consequently limit the peak connection throughput

★ A frequency reuse of 1 represents the generally accepted approach for the deployment of LTE networks

31.4 CYCLIC PREFIX

★ The Cyclic Prefix represents a guard band at the start of each symbol which provides protection against multi-path delay spread. The cyclic prefix also represents an overhead which should be minimised

★ 3GPP TS 36.211 specifies normal and extended cyclic prefix lengths. These are shown in Table 345. The normal cyclic prefix has a longer duration within the first symbol of each time slot. The extended cyclic prefix has an equal duration in all symbols

	Normal		Extended
	First symbol	Remaining symbols	All symbols
Cyclic Prefix Duration	5.2 μs	4.7 μs	16.7 μs
Equivalent Distance	1.6 km	1.4 km	5 km
Overhead	160 / 2048 = 7.8 %	144 / 2048 = 7.0 %	512 / 2048 = 25 %

Table 345 – Normal and extended cyclic prefix lengths

★ The duration of the cyclic prefix should be greater than the multi-path delay spread. The delay spread depends upon the cell range as well as the propagation environment and the presence of any distant reflectors

★ The normal cyclic prefix is intended to be sufficient for the majority of scenarios while the extended cyclic prefix is intended for scenarios with particularly high delay spread

★ Both microcells and indoor solutions typically have delay spreads in the order of 10 to 100 ns, so can use the normal cyclic prefix

★ Macrocells within urban and suburban areas typically have delay spreads in the order of 1 or 2 μs, so can use the normal cyclic prefix. There may be exceptions if signals are reflected by relatively distant buildings or hills

★ Rural areas have a low delay spread when the environment is open and there is line-of-sight propagation. However, the delay spread can also become high as a result of the larger cell range and potentially the presence of any distant reflectors

31.5 PHYSICAL LAYER CELL IDENTITY

★ Physical layer Cell Identity (PCI) planning for LTE is analogous to scrambling code planning for UMTS

- o LTE has 504 PCI which are organised into 168 groups of 3
- o UMTS has 512 scrambling codes organised into 64 groups of 8

★ The most important requirements for planning PCI are to remain 'Collision Free' and 'Confusion Free':

- o Collision Free
  - ▪ the physical separation between cells using the same PCI should be maximised
  - ▪ the physical separation between cells using the same PCI should be sufficiently great to ensure that a UE never simultaneously receives the same PCI from more than a single cell
- o Confusion Free
  - ▪ the physical separation should be sufficiently great to avoid neighbour ambiguity at an eNode B when a UE reports measurements for a specific PCI
  - ▪ a cell should not have neighbouring cells using the same PCI

★ In addition, there are arguments (described below) for further PCI planning rules. UE should not be able to simultaneously receive multiple PCI with equal:

- o 'PCI mod 3' values
- o 'PCI mod 6' values
- o 'PCI mod 30' values

For example, the 'PCI mod 3' rule means that neighbouring cells should not be allocated PCI of 22 and 28 because they both have 'PCI mod 3' values of 1

★ If the 'PCI mod 3' rule is satisfied then the 'PCI mod 6' and 'PCI mod 30' rules will also be satisfied. If it is not practical to satisfy the 'PCI mod 3' rule then the 'PCI mod 6' rule should be targeted. If the 'PCI mod 6' rule is satisfied then the 'PCI mod 30' rule will also be satisfied. If it is not practical to satisfy the 'PCI mod 6' rule then the 'PCI mod 30' rule should be targeted

★ There are 2 arguments for introducing the 'PCI mod 3' rule. Both arguments are applicable when adjacent cells are frame synchronised. Cells belonging to the same eNode B are frame synchronised. eNode B belonging to TDD networks should also be frame synchronised. eNode B belonging to FDD networks are less likely to be frame synchronised although synchronisation may be introduced for Inter-Cell Interference Coordination (ICIC) or MBMS Single Frequency Network

- o simulations have shown that cell acquisition times increase when UE receive the same Primary Synchronisation Signal (PSS) from multiple cells. Receiving the PSS from multiple cells can cause the UE to make misleading channel estimates which are subsequently applied when attempting to detect the Secondary Synchronisation Signal (SSS). The PSS defines a pointer (0, 1, 2) to the PCI within the PCI Group (0, .., 167) so equal 'PCI mod 3' values are analogous to using the same PSS
- o CQI reporting can be affected when UE receive the cell specific Reference Signal on the same subcarriers from multiple cells. CQI values can be based upon measurements from the cell specific Reference Signal. When multiple cells broadcast the cell specific Reference Signal using the same subcarriers they can interfere with one another. This reduces both the signal to noise ratio in general, and the resultant CQI value. The subcarriers used by the cell specific Reference Signal cycle every 3 PCI when using 2×2 or 4×4 MIMO so equal 'PCI mod 3' values are analogous to using the same subcarriers

★ The argument for introducing the 'PCI mod 6' rule is applicable when adjacent cells are frame synchronised and single antenna transmission is used. The argument is based upon the interference levels between cell specific Reference Signals and the resultant reported CQI values. The subcarriers used by the cell specific Reference Signal cycle every 6 PCI when using single antenna transmission so equal 'PCI mod 6' values are analogous to using the same subcarriers

★ The argument for introducing the 'PCI mod 30' rule is based upon the allocation of sequences to the uplink Reference Signals. The full set of sequences is divided into 30 groups. Intercell interference is reduced if different groups are allocated to neighbouring cells. The allocated group is based upon a 'PCI mod 30' calculation so ensuing that neighbouring cells do not have equal 'PCI mod 30' values is analogous to ensuring that neighbouring cells use different uplink Reference Signal sequences. If the 'PCI mod 30' rule cannot be satisfied for practical reasons, group hopping can be enabled to help reduce the impact of any group collisions. Uplink Reference Signal sequence planning is described in greater detail in section 31.7

★ PCI values should be planned in clusters where each cell within a cluster has an equal value of ROUNDDOWN(PCI / 30), i.e. PCI values 0 to 29 form one cluster, while PCI values 30 to 59 form another cluster, etc. This approach helps to reduce interfaces between cells using different uplink Reference Signal base sequence group hopping patterns (if base sequence group hopping is enabled). This requirement is illustrated in Figure 304 and is described further in section 31.7

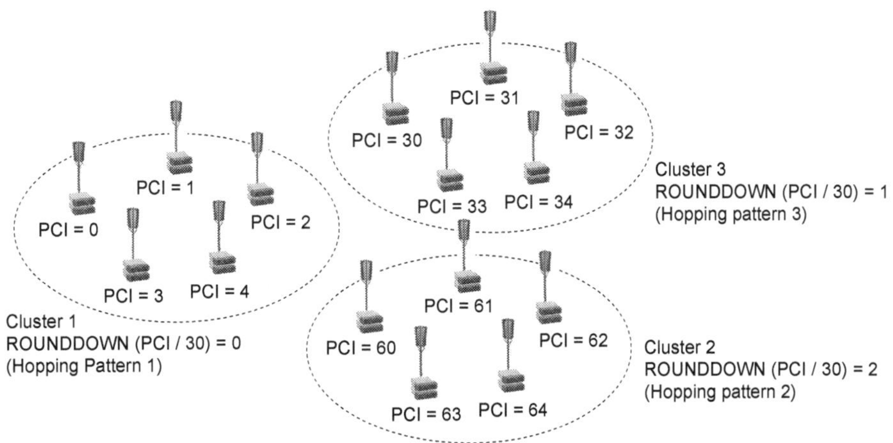

Figure 304 – PCI cluster planning for uplink Reference Signal base sequence group hopping

★ If BTS have 3 sectors and an ideal hexagonal cell layout, the PCI planning rules above can be achieved by allocating one PCI group to each BTS. This is illustrated in Figure 305

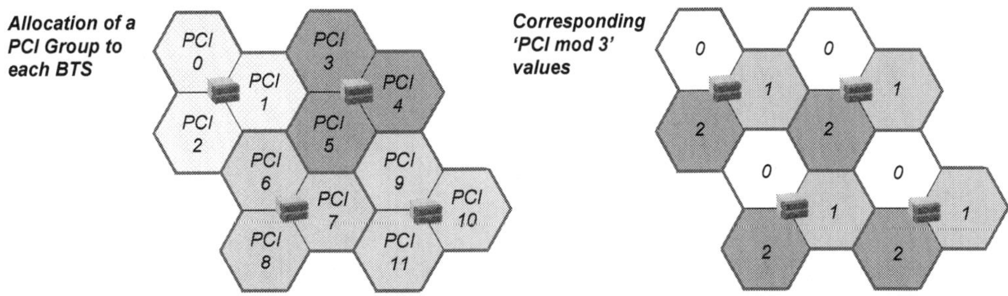

Figure 305 – PCI group planning per eNode B

★ In practice, BTS may have 3 sectors but it is unlikely that an ideal hexagonal cell layout exists. The situation is further complicated by heterogeneous network architectures where a single macrocell could be neighboured with many micro, pico or femto cells. In those cases, the 'PCI mod 3' and 'PCI mod 6' rules may be satisfied intra-BTS but not inter-BTS

★ If BTS have 6 sectors then PCI planning can be based upon allocating 2 PCI groups to each BTS. If all sites were 6-sector then the maximum re-use pattern would reduce from 168 to 84

★ In addition to the PCI planning rules described above, a range of PCI can be excluded from the network plan to allow for future network expansion, or the introduction of Home eNode B (femto). Femto cells could be allocated a subset of the total PCI. The size of the subset would depend upon the expected density of femto cell deployment

★ Additional rules for PCI planning are required at locations close to international borders where there may be another LTE operator using the same RF carrier. These rules are often specified by regulatory organisations. For example, in Europe the Electronic Communications Committee (ECC) within the European Conference of Postal and Telecommunications Administrations (CEPT) has specified ECC Recommendation (08)02. This document states that coordination can be avoided if signal strengths across the international border are below specific thresholds. Otherwise, it recommends that co-channel PCI should be coordinated between neighbouring LTE systems in border areas

★ PCI can be allocated automatically by the self-configuring function of Self Organising Networks (SON). Either centralised or distributed approaches can be used. These are described in section 32.7

31.6 PRACH PARAMETER PLANNING

★ The PRACH parameters shown in Table 346 require configuration when planning an LTE network. These parameters are broadcast to idle mode UE on the BCCH using System Information Block 2 (SIB2). They can also be signalled to individual UE in connected mode using an RRC Connection Reconfiguration message on the DCCH

Parameter	Range	Configuration Criteria
PRACH Configuration Index	0 to 63	Cell range PRACH preamble capacity eNode B processing load Air-interface performance
Zero Correlation Zone Configuration	0 to 15	Cell range Air-interface performance Size of Root Sequence Index re-use pattern
High Speed Flag	True / False	UE mobility Complexity of Root Sequence Index planning
Root Sequence Index	0 to 837	Should be unique in neighbouring cells
PRACH Frequency Offset	0 to 94	PUCCH Resource Block allocation Avoid PUSCH Resource Block fragmentation

Table 346 – PRACH configuration parameters for planning

★ The following sections discuss these parameters in greater detail. An overview is provided here

★ The PRACH Configuration Index determines the preamble format, the preamble density and the preamble version

- o the cell range is the most important criteria when selecting the preamble format. Preamble formats supporting larger cell ranges should not be selected unless required because they generate larger PRACH overheads, i.e. each preamble occupies an increased number of subframes
- o the quantity of traffic is the most important criteria when selecting a preamble density (number of preamble time-frequency resources per 10 ms radio frame). Large preamble densities should not be selected unless required because they generate larger PRACH overheads
- o there is a trade-off between eNode B processing load and air-interface performance when selecting the preamble version (timing of the preamble subframes to achieve a specific preamble density). Allocating different versions to each cell belonging to an eNode B helps to distribute the processing load in time. Allocating the same version to cells which are synchronised on the air-interface avoids PUSCH to PRACH interference

★ The Zero Correlation Zone Configuration has an impact upon cell range and air-interface performance, as well as the size of the re-use pattern used to allocate the Root Sequence Index values

- o larger zero correlation zones are required to support larger cell ranges. Larger zero correlation zones require a greater quantity of root sequences to generate the set of 64 preamble sequences for each cell. Preamble sequences generated from different root sequences are not orthogonal. Using a greater quantity of root sequences for each cell reduces the size of the re-use pattern which can be used when planning the Root Sequence Index values

★ The High Speed Flag can be set to 'false' for the majority of scenarios. It is intended to help tackle the impact of doppler at very high speeds, i.e. typically > 200 km/hr. Setting this flag to 'true' complicates the planning of root sequence index values because the number of preamble sequences generated from each root sequence becomes variable rather than fixed

★ The allocation of Root Sequence Index values to each cell should be based upon a re-use pattern which ensures that neighbouring cells have mutually exclusive sets of root sequences. The size of the zero correlation zone determines the number of root sequences required by each cell. Once this is known, the complete set of root sequences can be divided into groups, and one group allocated to each cell with a re-use pattern determined by the number of groups

★ PRACH Frequency Offset determines the position of the PRACH preambles in the frequency domain. PRACH preambles should be positioned adjacent to the PUCCH allocations at the outer edges of the channel bandwidth. This approach helps to ensure that Resource Blocks available for the PUSCH do not become fragmented

31.6.1 PREAMBLE FORMAT

- Selecting a PRACH preamble format is a prerequisite to selecting the PRACH configuration index
- 3GPP TS 36.211 specifies 5 PRACH preamble formats:
  - formats 0 to 3 can be used by either FDD or TDD, and are based upon a 'long' preamble sequence
  - format 4 can only be used by TDD when using special subframe configurations with UpPTS lengths of 4384 Ts (2 symbols with normal cyclic prefix) and 5120 Ts (2 symbols with extended cyclic prefix). This format is based upon a 'short' preamble sequence
- All preamble formats include a cyclic prefix, a sequence and a guard time
- The duration of the cyclic prefix accounts for the sum of the propagation channel delay spread and the propagation channel round trip time
- The duration of the guard time accounts for the propagation channel round trip time
- Larger cells have longer propagation channel round trip times and tend to experience larger delays spreads so require a longer cyclic prefix and a longer guard time
- The set of PRACH preamble formats is described in greater detail in section 14.1. Their main characteristics from an RF planning perspective are presented in Table 347
- The cell range is calculated from both the duration of the cyclic prefix minus the assumed delay spread, and from the duration of the guard time. The final maximum cell range is the smaller of the two results

Preamble Format	Total Duration (ms)	Cyclic Prefix			Sequence	Guard Time		Maximum Cell Range (km)
		Duration (μs)	Assumed Delay Spread (μs)	Cell Range (km)	Duration (μs)	Duration (μs)	Cell Range (km)	
0	1	103.13	5.20	14.7	800	96.88	14.5	14.5
1	2	684.38	16.67	100.2	800	515.63	77.3	77.3
2	2	203.13	5.20	29.7	1600	196.88	29.5	29.5
3	3	684.38	16.67	100.2	1600	715.63	107.3	100.2
4	0.16	14.58	0	2.2	133	9.38	1.4	1.4

Table 347 – Maximum cell range associated with each Random Access Preamble format

- Format 0 is intended to be the default for small to medium cells (up to ~14 km cell range). It generates a relatively small overhead by occupying only a single subframe
- Format 1 is intended for large cells (up to ~77 km cell range) which have relatively good air-interface conditions. It is necessary to have good air-interface conditions because the cell range is large and the preamble includes a single sequence transmission rather than a double sequence transmission. This format generates an increased overhead by occupying 2 subframes
- Format 2 is intended for medium cells (up to ~29 km cell range). It provides increased robustness by transmitting the preamble sequence twice. It generates the same overhead as format 1 by occupying 2 subframes
- Format 3 is intended for very large cells (up to ~100 km cell range). It transmits the preamble sequence twice but requires a longer cyclic prefix and a longer guard time so generates an increased overhead by occupying 3 subframes
- Format 4 is intended for small cells (less than ~1.4 km cell range) and is only applicable to TDD when using the special subframe configurations specified above. It is based upon a 'short' preamble sequence so requires the better air-interface conditions associated with smaller cells. It benefits from using the UpPTS section of special subframes so PUSCH to PRACH interference does not exist, and the preambles do not generate an overhead from the perspective of the PUSCH
- Different sections of the network can be planned with different preamble formats if the maximum cell range varies from one area to another

31.6.2 CONFIGURATION INDEX

★ The PRACH configuration index can be selected after identifying an appropriate preamble format

★ The set of PRACH configuration indices for FDD is presented in Table 348. The PRACH configuration index determines the preamble format as well as the radio frames and subframes during which UE are permitted to send PRACH preambles

★ The combination of the permitted radio frames and subframes determines the preamble density. Higher preamble densities should be selected for cells carrying higher levels of traffic, i.e. the density reflects the capacity of the PRACH

Config. Index	Preamble Format	Density per 10 ms	SFN	Subframe Number	Config. Index	Preamble Format	Density per 10 ms	SFN	Subframe Number
0	0	0.5	Even	1	32	2	0.5	Even	1
1	0	0.5	Even	4	33	2	0.5	Even	4
2	0	0.5	Even	7	34	2	0.5	Even	7
3	0	1	Any	1	35	2	1	Any	1
4	0	1	Any	4	36	2	1	Any	4
5	0	1	Any	7	37	2	1	Any	7
6	0	2	Any	1, 6	38	2	2	Any	1, 6
7	0	2	Any	2, 7	39	2	2	Any	2, 7
8	0	2	Any	3, 8	40	2	2	Any	3, 8
9	0	3	Any	1, 4, 7	41	2	3	Any	1, 4, 7
10	0	3	Any	2, 5, 8	42	2	3	Any	2, 5, 8
11	0	3	Any	3, 6, 9	43	2	3	Any	3, 6, 9
12	0	5	Any	0, 2, 4, 6, 8	44	2	5	Any	0, 2, 4, 6, 8
13	0	5	Any	1, 3, 5, 7, 9	45	2	5	Any	1, 3, 5, 7, 9
14	0	10	Any	0 to 9	46				
15	0	0.5	Even	9	47	2	0.5	Even	9
16	1	0.5	Even	1	48	3	0.5	Even	1
17	1	0.5	Even	4	49	3	0.5	Even	4
18	1	0.5	Even	7	50	3	0.5	Even	7
19	1	1	Any	1	51	3	1	Any	1
20	1	1	Any	4	52	3	1	Any	4
21	1	1	Any	7	53	3	1	Any	7
22	1	2	Any	1, 6	54	3	2	Any	1, 6
23	1	2	Any	2, 7	55	3	2	Any	2, 7
24	1	2	Any	3, 8	56	3	2	Any	3, 8
25	1	3	Any	1, 4, 7	57	3	3	Any	1, 4, 7
26	1	3	Any	2, 5, 8	58	3	3	Any	2, 5, 8
27	1	3	Any	3, 6, 9	59	3	3	Any	3, 6, 9
28	1	5	Any	0, 2, 4, 6, 8	60				
29	1	5	Any	1, 3, 5, 7, 9	61				
30					62				
31	1	0.5	Even	9	63	3	0.5	Even	9

Table 348 – PRACH configuration indices for FDD

★ There are 16 configurations associated with each preamble format although configurations 30, 46, 60, 61 and 62 are not supported. These configurations are not supported because the associated density of preambles would cause the preambles to overlap in the time domain. For example, configuration 30 is associated with a density of 10 preambles per radio frame (deduced from configuration 14). It is also associated with preamble format 1 which has a duration of 2 ms. Transmitting 10 preambles of 2 ms within a single radio frame would require those preambles to overlap in the time domain

★ In the case of FDD, PRACH preambles are only multiplexed in the time domain. They are not multiplexed in the frequency domain, i.e. there is a maximum of 1 time-frequency resource for PRACH preambles during each subframe

★ Losing some of the higher density configurations for preamble formats 1, 2 and 3 is viewed as acceptable because these formats are intended for cells with larger cell ranges and lower expected quantities of traffic. Preamble format 3 has a maximum density of 3 but supports a cell range of up to 100 km so it not expected to be used in high traffic scenarios

★ There are multiple versions of each density where each version allows different subframe timing, e.g. configurations 0, 1 and 2 represent 3 versions of the same density. These allow cells belonging to the same eNode B to have equal PRACH preamble capacity while distributing the processing requirement over time. Defining a maximum of 3 versions for each density is based upon the assumption that most sites will have up to 3 sectors. Sites with greater sectorisation can re-use versions. The benefit of distributing the processing requirement over time will be eNode B implementation dependent

- ★ In contrast, there is an argument that PRACH to PRACH interference on the air-interface is preferred to PUSCH to PRACH interference. This implies that cells with synchronised air-interface timing should be allocated the same version of the appropriate density. In the case of FDD, cells belonging to the same BTS are synchronised while cells belonging to different BTS are less likely to be synchronised. Inter-site synchronisation may be introduced for the purposes of Inter-Cell Interference Cancellation (ICIC) or MBMS Single Frequency Network (MBSFN)

- ★ Thus, the process for selecting an appropriate PRACH configuration index should involve selection of a preamble format, followed by selection of an appropriate preamble density, followed by selection of one or more versions of that density

- ★ The set of PRACH configuration indices for TDD is presented in Table 349. The definition of indices is more complex for TDD because there is a dependency upon the uplink-downlink subframe configuration

- ★ Table 349 specifies the preamble format associated with each configuration index. The density of preambles and a version number are also specified but without defining any timing information, i.e. the permitted radio frames and subframes are not provided. Similar to FDD, there are multiple versions associated with each density

Config. Index	Preamble Format	Density per 10 ms	Version
0	0	0.5	0
1	0	0.5	1
2	0	0.5	2
3	0	1	0
4	0	1	1
5	0	1	2
6	0	2	0
7	0	2	1
8	0	2	2
9	0	3	0
10	0	3	1
11	0	3	2
12	0	4	0
13	0	4	1
14	0	4	2
15	0	5	0
16	0	5	1
17	0	5	2
18	0	6	0
19	0	6	1
20	1	0.5	0
21	1	0.5	1
22	1	0.5	2
23	1	1	0
24	1	1	1
25	1	2	0
26	1	3	0
27	1	4	0
28	1	5	0
29	1	6	0
30	2	0.5	0
31	2	0.5	1
32	2	0.5	2
33	2	1	0
34	2	1	1
35	2	2	0
36	2	3	0
37	2	4	0
38	2	5	0
39	2	6	0
40	3	0.5	0
41	3	0.5	1
42	3	0.5	2
43	3	1	0
44	3	1	1
45	3	2	0
46	3	3	0
47	3	4	0
48	4	0.5	0
49	4	0.5	1
50	4	0.5	2
51	4	1	0
52	4	1	1
53	4	2	0
54	4	3	0
55	4	4	0
56	4	5	0
57	4	6	0
58			
59			
60			
61			
62			
63			

Table 349 – PRACH configuration indices for TDD

- ★ Densities range from 0.5 to a maximum of 6. It is more challenging to support the higher densities for TDD because there are fewer uplink subframes. TDD allows preambles to be multiplexed in the frequency domain if there are no options for multiplexing in the time domain, i.e. there can be more than 1 time-frequency resource for PRACH preambles during each subframe

- ★ Similar to FDD, the maximum density supported for preamble format 3 is less than the other preamble formats because format 3 is intended for large cell ranges with lower expected quantities of traffic

- ★ 3GPP TS 36.211 provides a second table to define the time-frequency resources available for each configuration index as a function of the TDD uplink-downlink subframe configuration. A section of this 3GPP table is shown in Table 350

			Uplink - Downlink Subframe Configuration						
Config. Index	Preamble Format	Density per 10 ms	0	1	2	3	4	5	6
0	0	0.5	(0, 1, 0, 2)	(0, 1, 0, 1)	(0, 1, 0, 0)	(0, 1, 0, 2)	(0, 1, 0, 1)	(0, 1, 0, 0)	(0, 1, 0, 2)
1	0	0.5	(0, 2, 0, 2)	(0, 2, 0, 1)	(0, 2, 0, 0)	(0, 2, 0, 2)	(0, 2, 0, 1)	(0, 2, 0, 0)	(0, 2, 0, 2)
2	0	0.5	(0, 1, 1, 2)	(0, 1, 1, 1)	(0, 1, 1, 0)	(0, 1, 0, 1)	(0, 1, 0, 0)	N/A	(0, 1, 1, 1)
3	0	1	(0, 0, 0, 2)	(0, 0, 0, 1)	(0, 0, 0, 0)	(0, 0, 0, 2)	(0, 0, 0, 1)	(0, 0, 0, 0)	(0, 0, 0, 2)
4	0	1	(0, 0, 1, 2)	(0, 0, 1, 1)	(0,0,1,0)	(0, 0, 0, 1)	(0, 0, 0, 0)	N/A	(0, 0, 1, 1)
5	0	1	(0, 0, 0, 1)	(0, 0, 0, 0)	N/A	(0, 0, 0, 0)	N/A	N/A	(0, 0, 0, 1)
6	0	2	(0, 0, 0, 2) (0, 0, 1, 2)	(0, 0, 0, 1) (0, 0, 1, 1)	(0, 0, 0, 0) (0, 0, 1, 0)	(0, 0, 0, 1) (0, 0, 0, 2)	(0, 0, 0, 0) (0, 0, 0, 1)	(0, 0, 0, 0) (1, 0, 0, 0)	(0, 0, 0, 2) (0, 0, 1, 1)
7	0	2	(0, 0, 0, 1) (0, 0, 1, 1)	(0, 0, 0, 0) (0, 0, 1, 0)	N/A	(0, 0, 0, 0) (0, 0, 0, 2)	N/A	N/A	(0, 0, 0, 1) (0, 0, 1, 0)
8	0	2	(0, 0, 0, 0) (0, 0, 1, 0)	N/A	N/A	(0, 0, 0, 0) (0, 0, 0, 1)	N/A	N/A	(0, 0, 0, 0) (0, 0, 1, 1)
9	0	3	(0, 0, 0, 1) (0, 0, 0, 2) (0, 0, 1, 2)	(0, 0, 0, 0) (0, 0, 0, 1) (0, 0, 1, 1)	(0, 0, 0, 0) (0, 0, 1, 0) (1, 0, 0, 0)	(0, 0, 0, 0) (0, 0, 0, 1) (0, 0, 0, 2)	(0, 0, 0, 0) (0, 0, 0, 1) (1, 0, 0, 1)	(0, 0, 0, 0) (1, 0, 0, 0) (2, 0, 0, 0)	(0, 0, 0, 1) (0, 0, 0, 2) (0, 0, 1, 1)
10	0	3	(0, 0, 0, 0) (0, 0, 1, 0) (0, 0, 1, 1)	(0, 0, 0, 1) (0, 0, 1, 0) (0, 0, 1, 1)	(0, 0, 0, 0) (0, 0, 1, 0) (1, 0, 1, 0)	N/A	(0, 0, 0, 0) (0, 0, 0, 1) (1, 0, 0, 0)	N/A	(0, 0, 0, 0) (0, 0, 0, 2) (0, 0, 1, 0)
11	0	3	N/A	(0, 0, 0, 0) (0, 0, 0, 1) (0, 0, 1, 0)	N/A	N/A	N/A	N/A	(0, 0, 0, 1) (0, 0, 1, 0) (0, 0, 1, 1)
55	4	4	(0, 0, 0, *) (0, 0, 1, *) (1, 0, 0, *) (1, 0, 1, *)	(0, 0, 0, *) (0, 0, 1, *) (1, 0, 0, *) (1, 0, 1, *)	(0, 0, 0, *) (0, 0, 1, *) (1, 0, 0, *) (1, 0, 1, *)	(0, 0, 0, *) (1, 0, 0, *) (2, 0, 0, *) (3, 0, 0, *)	(0, 0, 0, *) (1, 0, 0, *) (2, 0, 0, *) (3, 0, 0, *)	(0, 0, 0, *) (1, 0, 0, *) (2, 0, 0, *) (3, 0, 0, *)	(0, 0, 0, *) (0, 0, 1, *) (1, 0, 0, *) (1, 0, 1, *)
56	4	5	(0, 0, 0, *) (0, 0, 1, *) (1, 0, 0, *) (1, 0, 1, *) (2, 0, 0, *)	(0, 0, 0, *) (0, 0, 1, *) (1, 0, 0, *) (1, 0, 1, *) (2, 0, 0, *)	(0, 0, 0, *) (0, 0, 1, *) (1, 0, 0, *) (1, 0, 1, *) (2, 0, 0, *)	(0, 0, 0, *) (1, 0, 0, *) (2, 0, 0, *) (3, 0, 0, *) (4, 0, 0, *)	(0, 0, 0, *) (1, 0, 0, *) (2, 0, 0, *) (3, 0, 0, *) (4, 0, 0, *)	(0, 0, 0, *) (1, 0, 0, *) (2, 0, 0, *) (3, 0, 0, *) (4, 0, 0, *)	(0, 0, 0, *) (0, 0, 1, *) (1, 0, 0, *) (1, 0, 1, *) (2, 0, 0, *)
57	4	6	(0, 0, 0, *) (0, 0, 1, *) (1, 0, 0, *) (1, 0, 1, *) (2, 0, 0, *) (2, 0, 1, *)	(0, 0, 0, *) (0, 0, 1, *) (1, 0, 0, *) (1, 0, 1, *) (2, 0, 0, *) (2, 0, 1, *)	(0, 0, 0, *) (0, 0, 1, *) (1, 0, 0, *) (1, 0, 1, *) (2, 0, 0, *) (2, 0, 1, *)	(0, 0, 0, *) (1, 0, 0, *) (2, 0, 0, *) (3, 0, 0, *) (4, 0, 0, *) (5, 0, 0, *)	(0, 0, 0, *) (1, 0, 0, *) (2, 0, 0, *) (3, 0, 0, *) (4, 0, 0, *) (5, 0, 0, *)	(0, 0, 0, *) (1, 0, 0, *) (2, 0, 0, *) (3, 0, 0, *) (4, 0, 0, *) (5, 0, 0, *)	(0, 0, 0, *) (0, 0, 1, *) (1, 0, 0, *) (1, 0, 1, *) (2, 0, 0, *) (2, 0, 1, *)

Table 350 – PRACH configuration parameters for TDD

★ Each entry in Table 350 includes one or more sets of 4 digits (f_{RA}, $t_{RA}(0)$, $t_{RA}(1)$, $t_{RA}(2)$). Each set of digits defines a time-frequency resource for the transmission of PRACH preambles. Higher densities require more resources and so more sets of these 4 digits:

- f_{RA} defines the frequency resource associated with the preamble. This is set to 0 by default but is incremented when time multiplexing of preambles is no longer possible
- $t_{RA}(0)$ defines if preambles can be sent in all radio frames (value 0), even radio frames (value 1) or odd radio frames (value 2)
- $t_{RA}(1)$ defines if the preamble is located in the first half of the radio frame (value 0) or second half of the radio frame (value 1)
- $t_{RA}(2)$ defines the uplink subframe number where the preamble starts. This digit is set to '*' for preamble format 4 because preambles are sent during the UpPTS rather than during a normal uplink subframe

★ A single set of digits corresponds to a density of 0.5 when preambles are only permitted in odd or even radio frames. A single set of digits corresponds to a density of 1 when preambles are allowed in all radio frames. Six sets of digits with preambles allowed in all radio frames provides a density of 6

★ Some entries are shown as Not Applicable (N/A). For example, uplink-downlink subframe configuration 5 includes only a single uplink subframe so there are only two ways of generating a density of 0.5 (transmitting preambles during even radio frames, or transmitting preambles during odd radio frames). This leads to the third version being categorised as N/A

★ Similarly, uplink-downlink subframe configuration 5 has only a single version when generating a density of 1 because the only option is to send preambles during the single uplink subframe in every radio frame. This leads to the second and third versions being categorised as N/A

★ Uplink-downlink subframe configuration 5 is the first uplink-downlink subframe configuration to require frequency multiplexing as the density increases. Generating a density of 2 requires frequency multiplexing because there are not 2 uplink subframes within a radio frame. This results in an entry with f_{RA} incremented to 1

★ Preamble format 4 also uses frequency multiplexing to a relatively large extent because preambles are only sent during the UpPTS sections of special subframes. Some uplink-downlink configurations have 2 UpPTS sections per radio frame whereas other configurations have only 1 UpPTS section per radio frame

★ Similar to FDD, the process for selecting an appropriate PRACH configuration index for TDD should involve selection of a preamble format, followed by selection of an appropriate preamble density, followed by selection of one or more versions of that density

★ Most configurations involve a trade-off between PRACH preamble capacity and PUSCH capacity. The benefit of using preamble format 4 is that the PRACH preambles are sent during the UpPTS and do not impact the PUSCH capacity (PUSCH is not transmitted during the UpPTS)

★ Preamble resources are multiplexed in time rather than in frequency wherever possible to help distribute the eNode B processing load. Distributing preambles in the frequency domain does not help to distribute the eNode B processing load

★ There is a trade-off between distributing the eNode B processing load and the air-interface performance. Distributing the processing load by allocating different preamble subframes to different cells leads to PUSCH to PRACH interference which is generally understood to have greater impact than PRACH to PRACH interference. From the perspective of the air-interface performance it is preferable to have common preamble subframes allocated to all cells (noting that all cells should be synchronised for TDD)

31.6.3 ZERO CORRELATION ZONE

★ Selecting a preamble zero correlation zone is a prerequisite to planning the root sequence index values

★ Each cell has 64 preamble sequences. These preamble sequences allow multiple UE to share the same set of time and frequency resources when transmitting their PRACH preambles

★ 3GPP TS 36.211 specifies that preamble sequences are generated from a set of 838 root sequences when using preamble formats 0 to 3. Each preamble sequence is generated from its root sequence by applying a cyclic shift. The zero correlation zone parameter determines the size of the cyclic shift and the number of preamble sequences which can be generated from each root sequence

- small zero correlation zones:
  - small cyclic shifts used to generate preamble sequences from each root sequence
  - more preamble sequences can be generated from each root sequence
  - smaller cell range
- large zero correlation zones:
  - large cyclic shifts used to generate preamble sequences from each root sequence
  - less preamble sequences can be generated from each root sequence
  - larger cell range

★ It is beneficial to generate as many preamble sequences as possible from the same root sequence because these sequences are orthogonal to one another. Preamble sequences generated from different root sequences are not orthogonal. This has an impact upon air-interface performance

★ Generating a large number of preamble sequences from each root sequence also means that each cell requires fewer root sequences to obtain its set of 64 preamble sequences. This allows larger re-use patterns for the set of 838 root sequences and makes it easier to ensure that the sets of root sequences used by neighbouring cells are mutually exclusive

★ The drawback associated with generating large numbers of preamble sequences from each root sequence is that the maximum supported cell range becomes smaller

★ The set of zero correlation zone values for preamble formats 0 to 3 is presented in Table 351. These values assume that the 'high speed flag' is set to false which allows the use of the 'unrestricted set' of cyclic shifts

Zero Correlation Zone (Cyclic Shift, Ncs)		Preamble Sequences per Root Sequence	Root Sequences Required per Cell	Root Sequence Re-use Pattern	Cell Range
Signalled Value	Actual Value				
1	13	64	1	838	0.76 km
2	15	55	2	419	1.04 km
3	18	46	2	419	1.47 km
4	22	38	2	419	2.04 km
5	26	32	2	419	2.62 km
6	32	26	3	279	3.47 km
7	38	22	3	279	4.33 km
8	46	18	4	209	5.48 km
9	59	14	5	167	7.34 km
10	76	11	6	139	9.77 km
11	93	9	8	104	12.20 km
12	119	7	10	83	15.92 km
13	167	5	13	64	22.78 km
14	279	3	22	38	38.80 km
15	419	2	32	26	58.83 km
0	0	1	64	13	118.90 km

Table 351 – Cell range as a function of the Zero Correlation Zone (preamble formats 0 to 3; high speed flag = false)

★ The zero correlation zone is signalled using a value between 0 and 15 but is mapped onto an actual value defined within 3GPP TS 36.211. The actual value defines the cyclic shift used to generate each of the preamble sequences from each root sequence

★ For example, a signalled value of 7 maps onto an actual value of 38 so preamble sequences are generated by cyclic shifts of each root sequence by 38 positions:

- root sequences and preamble sequences have lengths of 839 when using PRACH preamble formats 0 to 3. A cyclic shift of 38 allows ROUNDDOWN(839 / 38) = 22 preamble sequences to be generated from each root sequence
- obtaining 64 preamble sequences for a single cell requires the use of 3 root sequences, i.e. ROUNDUP(64 / 22) = 3
- dividing the complete set of 838 root sequences into groups of 3 generates 279 groups. These groups can be allocated to 279 cells before any re-use is required

★ The cell range figures in Table 351 are based upon the requirement that the radio propagation round trip time, plus any delay spread is not allowed to exceed the duration of the zero correlation zone:

- a preamble sequence of length 839 is transmitted over a duration of 800 μsecs
- the duration of the zero correlation zone is (Ncs - 1) × 800 / 839 μsecs
- the delay spread is assumed to be 6.4 μsecs when calculating the cell ranges in Table 351

★ For example, a zero correlation zone value of 38 corresponds to a duration of (38 - 1) × 800 / 839 = 35.28 μsecs. The round trip time is then allowed to be 35.28 – 6.4 = 28.88 μsecs. This corresponds to a distance of 8.66 km and a maximum cell range of 4.33 km

★ Based upon the preceding discussion, the zero correlation zone should be configured with the smallest value that the cell range allows, e.g. a cell with a 5 km range could be configured with a zero correlation zone value of 8, supporting a cell range of 5.48 km and requiring 4 root sequences per cell. Some margin should be allowed when evaluating call range values to allow for potentially higher delay spreads. Delay spread assumptions could be reduced for the smaller cell ranges

★ It is unlikely to be practical to configure the zero correlation zone value on a per cell basis so the maximum cell range in an area can be used to define the zero correlation zone value for the whole area

★ 3GPP TS 36.211 specifies that preamble sequences are generated from a set of 138 root sequences when using preamble format 4. In addition, it specifies that each root sequence and each preamble sequence has a length of 139

★ Only 7 zero correlation zone values are supported for preamble format 4. These are presented in Table 352

★ The cell range figures presented in Table 352 assume 0 μsecs delay spread. Assuming a non-zero delay spread will decrease the cell range figures

★ The minimum root sequence re-use pattern size is comparable to the equivalent figure for preamble formats 0 to 3

Zero Correlation Zone (Cyclic Shift, Ncs)		Preamble Sequences per Root Sequence	Root Sequences Required per Cell	Root Sequence Re-use Pattern	Cell Range
Signalled Value	Actual Value				
0	2	69	1	138	0.14 km
1	4	34	2	69	0.43 km
2	6	23	3	46	0.72 km
3	8	17	4	34	1.01 km
4	10	13	5	27	1.29 km
5	12	11	6	23	1.58 km
6	15	9	8	17	2.01 km

Table 352 – Cell range as a function of the Zero Correlation Zone (preamble format 4)

★ The high speed flag is intended to improve performance when UE have high mobility. Frequency offsets generated by doppler can impact the preamble detection performance and false alarm rate

★ The UE speed at which it becomes necessary to use the high speed flag depends upon the operating band. Pedestrian and normal vehicular mobility do not require the high speed flag. High speed trains are more likely to require the flag if their speed exceeds 250 km/hr

★ Setting the high speed flag to 'true' restricts the number of cyclic shifts allowed when generating the set of preamble sequences from the set of root sequences. It also introduces a new set of cyclic shift (Ncs) values

★ When the high speed flag is set to 'false', the number of preamble sequences generated from each root sequence is constant. This makes it relatively easy to group the root sequences and allocate to individual cells. When the high speed flag is set to 'true', the number of preamble sequences generated from each root sequence is no longer constant and this complicates the planning of root sequences

★ Table 353 presents the set of cyclic shift values (Ncs) for when the high mobility flag is set to 'true'. The cell range figures are calculated in the same way as in Table 351. The table also shows the number of cyclic shifts applicable to the first 6 root sequences when generating their preamble sequences

Zero Correlation Zone (Cyclic Shift, Ncs)		Number of Cyclic Shifts for Root Sequence Number 'u' (only 1 to 6 shown out of a possible 838)						Cell Range
Signalled Value	Actual Value	1	2	3	4	5	6	
0	15	0	0	18	14	11	18	1.04 km
1	18	0	0	15	11	9	14	1.47 km
2	22	0	0	12	9	7	12	2.04 km
3	26	0	0	10	8	6	10	2.62 km
4	32	0	0	8	6	5	8	3.47 km
5	38	0	0	7	5	4	6	4.33 km
6	46	0	0	6	4	3	6	5.48 km
7	55	0	0	5	3	3	4	6.76 km
8	68	0	0	4	3	2	4	8.62 km
9	82	0	0	3	2	2	2	10.63 km
10	100	0	0	2	2	1	2	13.20 km
11	128	0	0	2	1	1	2	17.20 km
12	158	0	0	1	1	1	0	21.50 km
13	202	0	0	1	1	0	0	27.79 km
14	237	0	0	1	0	0	0	32.79 km

Table 353 – Cyclic shifts for preamble formats 0 to 3 when the high speed flag is set to 'true'

★ For a specific zero correlation zone value, the number of cyclic shifts is dependent upon the root sequence number. In addition, using the restricted set of cyclic shifts reduces the total number of preamble sequences which can be generated from the set of root sequences. This leads to smaller root sequence re-use patterns

31.6.4 ROOT SEQUENCE INDEX

★ The root sequence index values can be planned after fixing the:
  - PRACH configuration index
  - zero correlation zone value
  - high mobility flag

★ The PRACH configuration index determines the preamble format, which determines the total number of root sequences available for planning. The total number of root sequences available is presented in Table 354

Preamble Formats 0 to 3	Preamble Format 4
838	138

Table 354 – Number of root sequence index values for each preamble format

★ The size of the zero correlation zone and the high mobility flag determine the number of root sequences required by each cell. Once this is known, the complete set of root sequences can be divided into groups, and one group allocated to each cell with a re-use pattern determined by the number of groups

★ For example, PRACH configuration index 6 indicates that preamble format 0 is used so there is a total of 838 root sequences. A zero correlation zone value of 8 with the high mobility flag set to 'false' indicates that each cell requires 4 root sequences. Allocating 4 root sequences to each cell leads to a re-use pattern of 838 / 4 = 209

★ The radio network planner allocates a single logical root sequence number to each cell. The UE and eNode B are then responsible for calculating the number of root sequences required by a cell to obtain the full set of 64 preamble sequences

★ The radio network planner needs to account for the number of root sequences required by each cell when allocating the logical root sequence numbers, e.g. when each cell requires 4 root sequences, the radio network planner could allocate root sequence numbers of 0, 4, 8, 12, 16 and 20 to a set of 6 cells

★ Logical root sequence numbers are mapped onto physical root sequence numbers according to a look-up table specified within 3GPP TS 36.211

★ Figure 306 illustrates an example allocation of logical root sequence numbers where each cell requires 4 root sequences. The subsequent mapping of logical root sequence numbers onto physical root sequence numbers is also shown

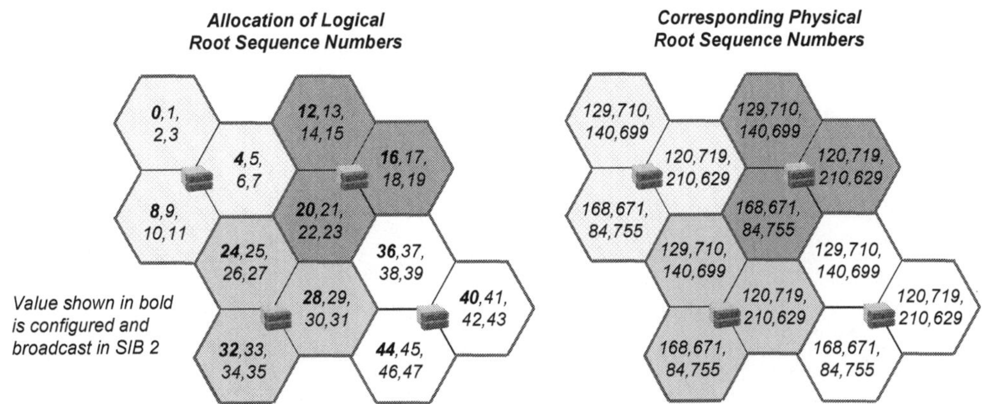

Figure 306 – Example allocation of logical root sequence numbers and their mapping onto physical root sequence numbers

★ The root sequence index values allocated to each cell should ensure that neighbouring cells have mutually exclusive sets of root sequences. Otherwise, one cell could receive and act upon a PRACH preamble being sent to another cell

31.6.5 FREQUENCY OFFSET

- The PRACH frequency offset determines the position of the preambles in the frequency domain
- The main objective when configuring the frequency offset is to avoid fragmentation of the Resource Blocks available to the PUSCH. This requires that the preambles are positioned adjacent to the PUCCH allocations at the edge of the channel bandwidth
- In the case of FDD,
  - there is a maximum of one preamble position within each subframe
  - PRACH preambles occupy 6 Resource Blocks in the frequency domain
  - the frequency offset explicitly signals the index of the first Resource Block to be used for the preambles
  - the frequency offset is applicable to all preamble allocations irrespective of the subframe and radio frame
- Figure 307 illustrates 2 example preamble allocations for FDD. Both allocations avoid fragmentation of the Resource Blocks available to the PUSCH

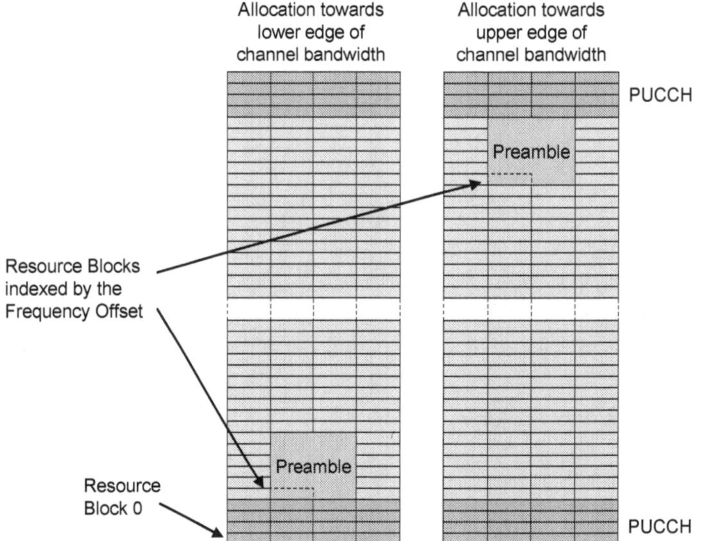

Figure 307 – Example PRACH preamble allocations for FDD

- In the case of TDD preamble formats 0 to 3,
  - there can be multiple preamble positions within each uplink subframe
  - PRACH preambles occupy 6 Resource Blocks in the frequency domain
  - the frequency offset value needs to be combined with the value of f_{RA} to generate the index of the first Resource Block to be used for the preambles
    - when f_{RA} is even, the position of the first preamble Resource Block is defined by:

      Resource Block Index = Frequency Offset + 6 × ROUNDDOWN(f_{RA} / 2)

      this allocates resources towards the lower edge of the channel bandwidth
    - when f_{RA} is odd, the position of the first preamble Resource Block is defined by:

      Resource Block Index = Total UL Res. Blks - 6 - Frequency Offset - 6 × ROUNDDOWN(f_{RA} / 2)

      this allocates resources towards the upper edge of the channel bandwidth
  - the value of f_{RA} is determined by the PRACH configuration index and the TDD uplink-downlink subframe configuration. Examples of f_{RA} are presented in Table 350
- Figure 308 illustrates 3 example preamble allocations for TDD preamble formats 0 to 3. These examples illustrate that PRACH preambles are kept towards the upper and lower edges of the channel bandwidth when multiple time-frequency resources are allocated within a single subframe (assuming an appropriate frequency offset is configured)

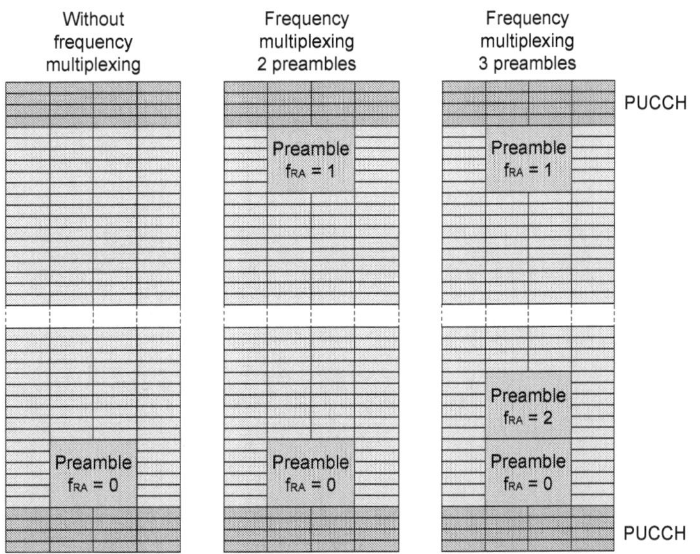

Figure 308 – Example PRACH preamble allocations for TDD preamble formats 0 to 3

★ In the case of TDD preamble format 4,

- there can be multiple preamble positions within each uplink subframe
- PRACH preambles occupy 6 Resource Blocks in the frequency domain
- preambles are sent during the UpPTS section of the special subframe so there are no PUCCH allocations to take into account and the preambles can use the edges of the channel bandwidth
- the frequency offset value needs to be combined with the value of f_{RA} to generate the index of the first Resource Block to be used for the preambles
  - when, $[(n_f \bmod 2) \times (2 - N_{sp}) + t_{RA}(1)] \bmod 2 = 0$ the position of the first preamble Resource Block is defined by:

    Resource Block Index = $6 \times f_{RA}$

    this allocates resources at the lower edge of the channel bandwidth. n_f is the System Frame Number (SFN), N_{sp} is the number of downlink to uplink switch points within the radio frame, i.e. the number of UpPTS fields, and $t_{RA}(1)$ defines whether the preamble is located in the first or second half of the radio frame
  - otherwise, the position of the first preamble Resource Block is defined by:

    Resource Block Index = Total UL Res. Blks - $6 \times (f_{RA} + 1)$

    this allocates resources at the upper edge of the channel bandwidth
- the equations above cause the PRACH preambles to alternate between the upper and lower edges of the channel bandwidth

31.7 UPLINK REFERENCE SIGNAL SEQUENCES

★ There are two types of uplink Reference Signal

- Demodulation Reference Signal (DM-RS), for either the PUSCH or PUCCH
- Sounding Reference Signal (SRS)

These Reference Signals are described in section 13. Both are generated from the same set of base sequences

★ There are 30 groups of base sequences, where each group contains:

- 1 base sequence of length = m × 12, for m = 1 to 5
    - generates 5 base sequences per group
    - PUSCH DM-RS uses these base sequences when the number of allocated Resource Blocks is 5, or less. The length of the base sequence used by the PUSCH DM-RS equals the number of allocated subcarriers, i.e. 12 × the number of allocated Resource Blocks
    - PUCCH DM-RS always uses these base sequences because the number of Resource Blocks used by a PUCCH transmission is always 1, i.e. the length of the base sequence used by the PUCCH DM-RS is always 12
    - SRS uses these base sequences when the SRS transmission bandwidth is 10, or less Resource Blocks (SRS uses every second subcarrier). The length of the base sequence used by the SRS equals half of the number of subcarriers within the transmission bandwidth, i.e. 6 × the number of Resource Blocks
- 2 base sequences of length = m × 12, for m = 6 to maximum number of uplink Resource Blocks
    - generates a number of base sequences dependent upon the channel bandwidth:
        - 2 × 1 = 2 base sequences for 1.4 MHz (total of 6 Resource Blocks)
        - 2 × 10 = 20 base sequences for 3 MHz (total of 15 Resource Blocks)
        - 2 × 20 = 40 base sequences for 5 MHz (total of 25 Resource Blocks)
        - 2 × 45 = 90 base sequences for 10 MHz (total of 50 Resource Blocks)
        - 2 × 70 = 140 base sequences for 15 MHz (total of 75 Resource Blocks)
        - 2 × 95 = 190 base sequences for 20 MHz (total of 100 Resource Blocks)
    - PUSCH DM-RS uses these base sequences when the number of allocated Resource Blocks is 6, or greater.
    - SRS uses these base sequences when the SRS transmission bandwidth is 11, or more Resource Blocks

★ Figure 309 illustrates the set of base sequences associated with the 3 MHz channel bandwidth

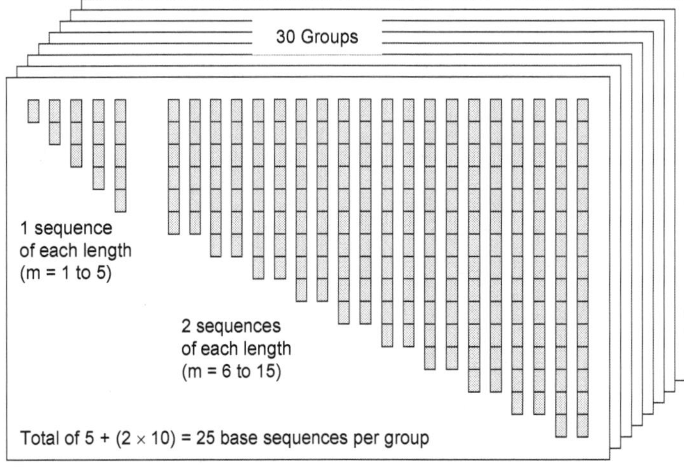

Figure 309 – Base sequences for 3 MHz channel bandwidth

★ The primary requirement for uplink Reference Signal planning is to ensure that neighbouring cells use different base sequence groups

★ The base sequence group used by a cell is defined by:

- PCI mod 30, for the PUCCH DM-RS and SRS
- [(PCI mod 30) + Δss] mod 30, for the PUSCH DM-RS, where 'Δss' is a cell specific parameter (Group Assignment PUSCH) broadcast in SIB2 with a range from 0 to 29

★ These equations indicate that neighbouring cells will automatically use different base sequence groups if the PCI planning 'PCI mod 30' rule is satisfied

★ In practice, PCI planning may not always satisfy the 'PCI mod 30' rule. When the 'PCI mod 30' rule is not satisfied, the 'Δss' parameter can be used to define an offset for the PUSCH Reference Signal base sequence group. This concept is illustrated in Figure 310, where two cells have different PCI values but equal 'PCI mod 30' values

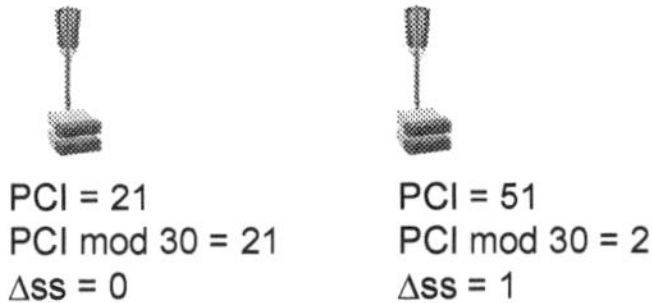

Figure 310 – Use of the Δss parameter to define a base group offset for the PUSCH Reference Signal

★ The 'Δss' parameter provides some flexibility in terms of defining the base sequence group so it does not totally depend upon the PCI value. However, it does not generate additional base sequence groups so the raw limitation remains the re-use pattern of 30

★ Group hopping can be used to randomise clashes of the base sequence groups. Clashes are distributed between cells and over time when group hopping is enabled

★ Group hopping is applicable to the PUSCH DM-RS, PUCCH DM-RS and SRS. It can be enabled on a per cell basis using the 'Group Hopping Enabled' parameter broadcast in SIB2. It can also be disabled for the PUSCH belonging to a specific UE using the 'Disable Sequence Group Hopping' parameter within an RRC Connection Reconfiguration message

★ When group hopping is enabled, the base sequence used by a cell is defined by:

- [(PCI mod 30) + (pseudo random result mod 30)] mod 30, for the PUCCH DM-RS and SRS
- [[(PCI mod 30) + Δss] mod 30 + (pseudo random result mod 30)] mod 30, for the PUSCH

In both cases, the pseudo random result is generated from a sequence initialised by ROUNDDOWN(PCI / 30) at the beginning of every radio frame, i.e. the hopping sequence has a length of 20 corresponding to one hop per time slot

★ Initialising the pseudo random sequence with ROUNDDOWN(PCI / 30) means that PCI 0 to 29 will follow the same group hopping pattern; PCI 30 to 59 will follow the same group hopping pattern, PCI 60 to 89 will follow the same group hopping pattern, etc. This leads to the requirement to plan PCI values in clusters with equal ROUNDDOWN(PCI / 30) values. Satisfying this requirement limits the potential for base sequence group clashes when hopping is enabled

★ When base sequences with a length greater than 5 × 12 = 60 are used, sequence hopping can be enabled between the 2 base sequences of the appropriate length. It can be enabled on a per cell basis using the 'Sequence Hopping Enabled' parameter broadcast in SIB2. It can also be disabled for the PUSCH belonging to a specific UE using the 'Disable Sequence Group Hopping' parameter within an RRC Connection Reconfiguration message

★ Sequence hopping is applicable to the PUSCH DM-RS, PUCCH DM-RS and SRS. It takes advantage of the second base sequence defined for lengths greater than 60 to help reduce the probability of clashes. Otherwise, only the first base sequence from the pair is used

★ Reference Signal sequences are generated from their base sequence using a cyclic shift. This cyclic shift provides a further dimension for differentiating between cells and connections. Cells using the same base sequence group can be allocated different cyclic shifts

★ In the case of the PUSCH DM-RS, the cell specific 'cyclic shift' parameter broadcast in SIB2 is used in combination with the UE specific cyclic shift signalled within the Downlink Control Information (DCI). Cyclic shift hopping is always enabled for the PUSCH DM-RS and is initialised at the start of every radio frame using the PCI

31.8 CELL AND BTS IDENTITY PLANNING

★ Cells are identified at a global level using their E-UTRAN Cell Global Identifier (ECGI)
  - o Mobile Country Code (MCC) and Mobile Network Code (MNC) uniquely identify the operator
  - o E-UTRAN Cell Identifier (ECI) differentiates between cells belonging to a specific operator

★ The MCC, MNC and ECI are broadcast within SIB1

★ The structures of an ECGI and an ECI are illustrated in Figure 311

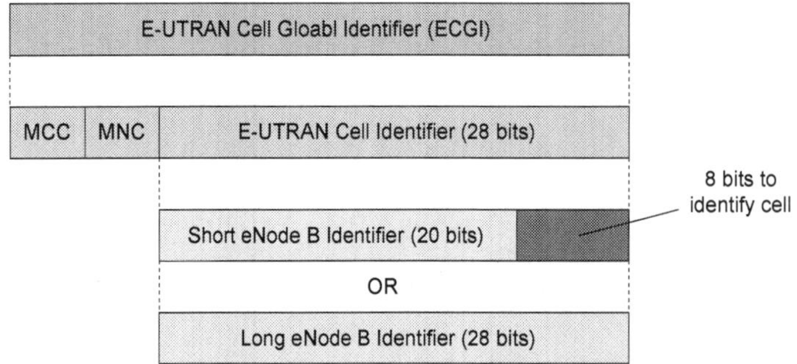

Figure 311 – Structure of E-UTRAN Cell Global Identifier (ECGI)

★ The E-UTRAN Cell Identity has a length of 28 bits and encapsulates the eNode B identifier

★ Either a long or short eNode B identifier can be defined
  - o long eNode B identifier occupies the complete set of 28 bits
    - ▪ provides support for 268 435 456 eNode B within the PLMN
    - ▪ allows only a single cell per eNode B
  - o short eNode B identifier occupies the left most 20 bits of the cell identifier
    - ▪ provides support for 1 048 576 eNode B within the PLMN
    - ▪ provides support for 256 cells per eNode B

★ The short eNode B identifier should be used for any site which can be configured with more than a single cell, i.e. macro and micro BTS. The short eNode B identifier requires an identity (0 to 255) to be allocated to each cell
  - o cells can be allocated sequential identities, e.g. 0, 1 and 2 for a 3-sector site, counting each sector clockwise from the direction of north
  - o different ranges can be used for different RF carriers and operating bands, e.g. values 0 to 9 could be used for operating band 7, while values 10 to 19 could be used for operating band 1
  - o alternatively, different ranges could be used for different site types, e.g. values 0 to 9 could be used for macrocells, while values 10 to 19 could be used for microcells

★ The long eNode B identifier can be used for sites which have only a single cell, i.e. pico and femto BTS. The increased range of values associated with the long identifier allows for having very large numbers of these sites within a network

★ The E-UTRAN Cell Identifier (ECI) and eNode B Identifier must both be unique within a PLMN

31.9 TRACKING AREAS

★ Tracking Areas (TA) represent the LTE equivalent of Routing Areas. LTE does not have a circuit switched domain so does not require Location Areas

★ System Information Block 1 (SIB1) broadcasts the TA to which a cell belongs using the Tracking Area Code (TAC)

★ The Tracking Area Identity (TAI) is constructed from a concatenation of the Mobile Country Code (MCC), Mobile Network Code (MNC) and TAC

★ A cell can broadcast a maximum of 1 TAC but up to 6 PLMN identities (MNC + MCC), so can have a maximum of 6 TAI

★ The TAC has a range from 0 to 65535 (16 bits). Two values are reserved to mark the TAC as deleted. These two TAC values are 0000 and FFFE

★ Paging messages are broadcast across TA when UE are in RRC Idle mode. UE can be registered with more than one TA so paging messages may need to be broadcast across multiple TA

★ The main drawback associated with large TA is an increased paging load

- o each paging message has to be broadcast by a relatively large number of eNode B

★ The main benefit associated with large TA is a reduced requirement for TA updates resulting from mobility

- o TA updates generate signalling load across the network
- o TA updates increase the probability of missing a paging message because there is a delay between the UE crossing a TA boundary and registering with the new TA

★ In general, TA should be planned to be relatively large rather than relatively small. Their size can subsequently be reduced if the paging load becomes high

★ Operators with existing 2G or 3G networks can plan their TA boundaries to coincide with their routing area or location area boundaries. TA boundaries should not traverse dense subscriber areas. Nor should they run close and parallel to major roads or railways

★ An MME pool consists of one or more MME. An MME pool area represents an area within which a UE can be served without the requirement to change MME. MME pool areas can overlap. An MME pool is identified by its MME Group Identity (MMEGI)

★ An MME pool area includes one or more complete TA. TA cannot span multiple MME pools. The concept of TA and MME pools is shown in Figure 312. In this example, TA3 belongs to pool 2 and TA4 belongs to pool 3, while there is a single MME which belongs to both pool 2 and pool 3

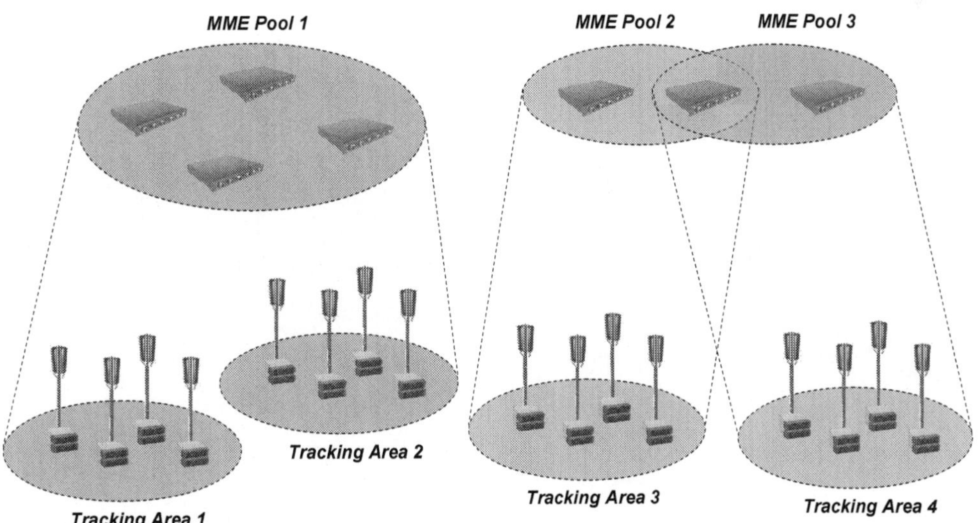

Figure 312 – Tracking areas within MME pools

★ Similar concepts exist for the user plane and Serving Gateway Service Areas. A Serving Gateway Service Area represents an area within which a UE can be served without the requirement to change Serving Gateway

31.10 NEIGHBOUR PLANNING

★ Historically, neighbour lists have been used to support UE mobility:
  - cell reselection for UE in idle mode
  - handovers for UE in connected mode

★ Neighbour lists define the potential routes for UE mobility and missing neighbours can cause UE to either lose coverage or drop connections

★ Neighbour list planning involves identifying adjacent cells to ensure that UE are able to experience seamless mobility

★ Each technology can have its own set of neighbour lists, and those lists can include intra-frequency, inter-frequency and inter-system adjacent cells

31.10.1 WITHIN LTE

★ LTE does not require the definition of individual neighbouring cells for the following types of cell reselection:
  - intra-frequency
  - inter-frequency
  - inter-RAT to UTRAN
  - inter-RAT to GERAN

★ System information broadcast by LTE provides the UE with minimal information to support the detection of neighbouring cells. This reduces both the requirement for network planning and the overhead generated by the system information

★ UE are responsible for identifying neighbouring cells after reading the content of the system information

★ The most relevant information broadcast by the System Information Blocks (SIB) is summarised in Table 355. Section 10 presents the content of the SIB in greater detail

SIB	Neighbour Type	Neighbour Information
4	Intra-Frequency	No individual cells are listed unless measurement offsets are defined for specific neighbours
5	Inter-Frequency	Downlink carrier frequencies are specified using the ARFCN No individual cells are listed unless measurement offsets are defined for specific neighbours
6	Inter-RAT (UTRAN)	Downlink carrier frequencies for FDD and TDD are specified using the ARFCN No individual cells are listed (scrambling code is not specified)
7	Inter-RAT (GERAN)	Downlink carrier frequencies are specified using the ARFCN No individual cells are listed (BSIC is not specified)
8	Inter-RAT (CDMA2000)	Individual cells are listed using a combination of their Band Class, ARFCN and PN Offset

Table 355 – Neighbour information broadcast within the set of System Information Blocks (SIB)

★ RF carriers are specified for inter-frequency and inter-RAT neighbours using their Absolute Radio Frequency Carrier Number (ARFCN)

★ Individual Physical layer Cell Identities (PCI) are only specified for intra-frequency and inter-frequency neighbours if cell specific measurement offsets are defined

★ Neither scrambling codes for UTRAN neighbouring cells, nor Base Station Identity Codes (BSIC) for GERAN neighbouring cells are specified

★ Inter-RAT cell reselection to CDMA2000 is the exception, and individual neighbouring cells are specified within SIB8. This applies to both the 1×RTT (1 times Radio Transmission Technology), and HRPD (High Rate Packet Data) variants. Individual cells are listed using a combination of their Band Class, ARFCN and PN Sequence Offset

★ In contrast to cell reselection, LTE requires the definition of individual neighbouring cells for handovers. Individual neighbours must be defined for handovers so the LTE network knows where to request resources prior to providing the UE with the handover command

★ LTE introduces the concept of Self-Organising Networks (SON) to help remove the requirement for neighbour list planning. Neighbours can be defined automatically using the Automatic Neighbour Relation (ANR) component of SON. Neighbours must be defined by the planner, or generated by a planning tool if ANR has not been implemented

★ ANR is described in sections 32.4 and 32.5. It generates neighbour lists based upon measurement reports from the UE

★ It is not necessary for the LTE network to provide the UE with the handover neighbour lists. Similar to cell re-selection, the UE is only provided with the RF carrier upon which to search for neighbouring cells. This information is signalled within the RRC Connection Reconfiguration message

★ Once provided with the RF carrier, UE are capable of detecting neighbouring cells and reporting them to the eNode B. When reporting neighbouring cells, UE reference them using their physical cell identities:

  - Physical layer Cell Identity (PCI) for LTE neighbours
  - Scrambling code for UMTS neighbours
  - RF carrier and BSIC for GSM neighbours
  - PN Offset for CDMA2000 neighbours

  The eNode uses its neighbour list to link the physical cell identity to a globally unique cell identity. The globally unique identity allows the eNode B to address the target cell for handover

★ This process breaks-down if the planning of physical cell identities has not been completed properly, i.e. if multiple cells within the same geographic area have the same physical cell identity

★ Similar to the SIB used for cell reselection, the RRC Connection Reconfiguration message can be used to define measurement offsets for handovers. These measurement offsets can be defined either for specific RF carriers or for specific neighbouring cells

★ It is possible to define handover measurement offsets for specific RF carriers for all technologies. Cell specific measurement offsets can only be defined for LTE neighbours (both intra and inter-frequency)

31.10.2 WITHIN UMTS

★ UMTS System Information Block 19 (SIB19) provides support for cell reselection from UMTS to LTE. The content of UMTS SIB19 is shown in Table 356

Information Elements		
E-UTRAN frequency and priority (up to 8 instances)	EARFCN	
	Measurement Bandwidth	
	Priority	
	QrxlevminEUTRA	
	ThreshX-High	
	ThreshX-Low	
	Blacklisted Cells (up to 16 instances)	Physical Cell Identity
	E-UTRA detection	

Table 356 – LTE content from UMTS System Information Block 19 (SIB19)

★ UMTS SIB19 does not specify individual LTE cells but specifies the RF carrier upon which the UE should search

★ Inter-system handovers from UMTS to LTE require the definition of individual neighbours within the UMTS network. Neighbours must be defined to provide the UMTS network with sufficient information to request resources prior to instructing the handover

★ The UMTS network does not provide the UE with the neighbour list. The UE is only provided with the RF carrier upon which to search for neighbouring cells. Once provided with the RF carrier, UE are capable of detecting neighbouring LTE cells and reporting them to the RNC using their Physical layer Cell Identities (PCI)

★ The RNC uses its neighbour list to link the PCI to a globally unique cell identity. The globally unique identity allows the RNC to address the target cell for handover

31.10.3 WITHIN GSM

★ GSM System Information Type 2quater (SI 2quater) provides support for cell reselection from GSM to LTE. A section from GSM SI 2quater is shown in Table 357

★ GSM SI 2quater does not specify individual LTE cells but specifies the RF carrier upon which the UE should search

★ Inter-system handovers from GSM to LTE require the definition of individual neighbours within the GSM network. Neighbours must be defined to provide the GSM network with sufficient information to request resources prior to instructing the handover

Information Elements	
E-UTRAN Neighbour Cells	EARFCN
	Measurement Bandwidth
	Priority
	ThreshX-High
	ThreshX-Low
	QrxlevminEUTRA

Table 357 – LTE content from GSM System Information Type 2quater

★ The GSM network does not provide the UE with the neighbour list. The UE is only provided with the RF carrier upon which to search for neighbouring cells. Once provided with the RF carrier, UE are capable of detecting neighbouring LTE cells and reporting them to the BSC using their Physical layer Cell Identities (PCI)

★ The BSC uses its neighbour list to link the PCI to a globally unique cell identity. The globally unique identity allows the BSC to address the target cell for handover

31.11 CO-SITING

★ Co-siting refers to the installation of multiple BTS at the same site. It requires consideration of the isolation requirements between those BTS because the BTS may interfere with one another if the isolation is not sufficiently large

★ An example scenario is illustrated in Figure 313. This example is based upon FDD LTE and GSM BTS operating within the same frequency band. In this case, the duplex spacing between the BTS transmit channels and the BTS receive channels provides isolation in the frequency domain. TDD technologies do not benefit from this frequency domain isolation

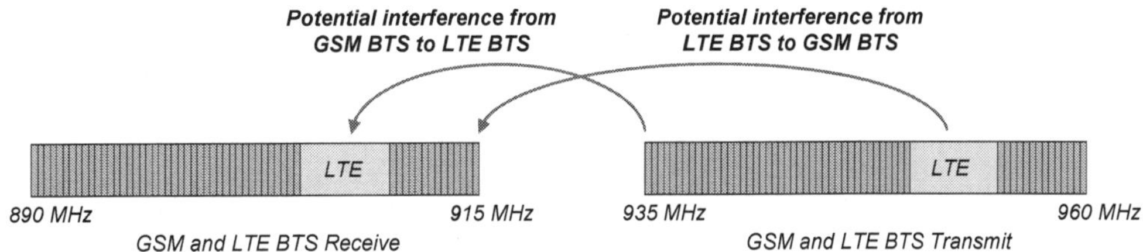

Figure 313 – Example interference scenarios between LTE and GSM BTS operating within the same band

★ The most likely causes of interference between co-sited BTS are spurious emissions (non-ideal transmitter filtering), receiver blocking (non-ideal receiver filtering) and intermodulation (non-ideal mixing)

★ Figure 314 illustrates the spurious emissions and receiver blocking interference mechanisms for a GSM BTS interfering with an LTE BTS. The GSM spurious emissions generate in-band interference to the LTE BTS, whereas the GSM channel itself generates out-of-band interference

★ Figure 314 illustrates a single GSM channel whereas in practice there is likely to be multiple GSM channels and the levels of interference will accumulate

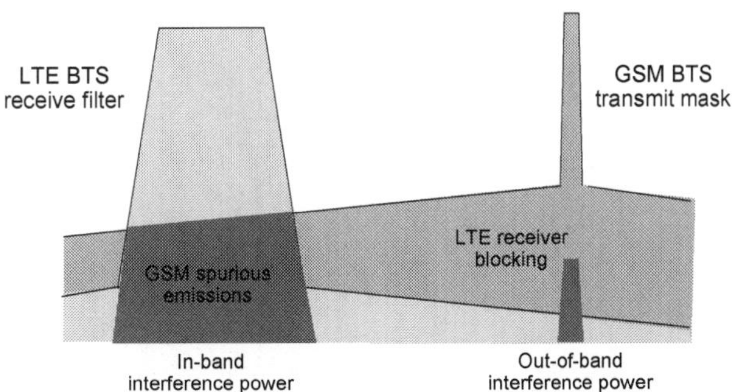

Figure 314 – Example of LTE BTS receiver experiencing interference from a GSM BTS transmitter

★ Isolation requirements between BTS should be evaluated on a case by case basis. The remainder of this section provides an example based upon an FDD LTE BTS co-sited with an FDD UMTS BTS within the 2100 MHz operating band

★ 3GPP TS 36.104 specifies the LTE spurious emissions requirements. Performance requirements have been defined specifically for co-sited scenarios. The relevant requirement is presented in Table 358

Type of Co-Located BTS	Band for Co-location Requirement	Maximum Level	Measurement Bandwidth
UTRA FDD Band I	1920 – 1980 MHz	-96 dBm	100 kHz

Table 358 – LTE spurious emissions requirement extracted from 3GPP TS 36.104

★ The spurious emissions requirement defines the maximum allowed power within a specific bandwidth. The victim UMTS BTS receiver bandwidth is 3.84 MHz so emissions of -96 dBm require increasing by 10×LOG(3840/100) = 16 dB, which results in an interference power of -80 dBm

★ The thermal noise floor of the victim UMTS BTS receiver is given by kTB + Noise Figure = -105 dBm, when assuming a 3 dB noise figure

★ Figure 315 illustrates the impact of the LTE spurious emissions upon the UMTS BTS thermal noise floor, when assuming a range of different isolations. The thermal noise floor is increased by less than 1 dB when the isolation between the two BTS is 40 dB

★ A second curve is included within Figure 315 to illustrate the impact of 2×2 MIMO. The spurious emissions performance requirements are specified per transmit antenna so the use of multiple transmit antennas increases the level of interference power

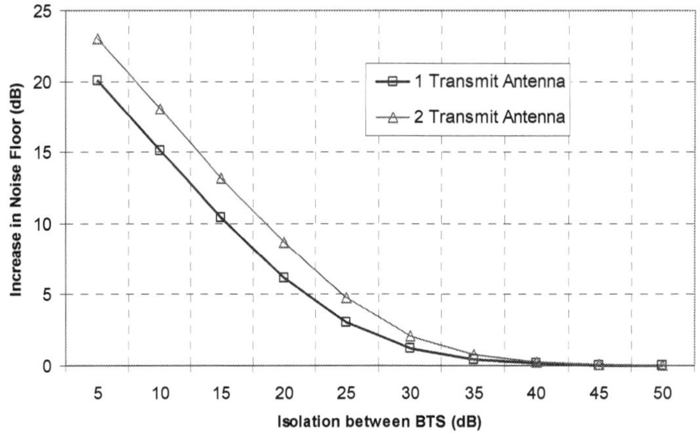

Figure 315 – Increase in UMTS BTS receiver noise floor as a result of LTE BTS spurious emissions

- 3GPP TS 25.104 specifies the UMTS BTS receiver blocking requirements. Performance requirements have been defined specifically for co-sited scenarios. The relevant requirement is presented in Table 359

Type of Co-Located BTS	Center Frequency of Interfering Signal	Interfering Signal Mean Power	Wanted Signal Mean Power	Type of Interfering Signal
EUTRA FDD Band 1	2110 – 2170 MHz	+16 dBm	-115 dBm	CW Carrier

Table 359 – UMTS receiver blocking requirement extracted from 3GPP TS 25.104

- The receiver blocking requirement specifies that the UMTS BTS must be capable of receiving 16 dBm of interference while maintaining its quality requirements with a -115 dBm signal
- Assuming an LTE BTS transmit power of 46 dBm means that 30 dB of isolation is required to attenuate the LTE signal to 16 dBm prior to reaching the UMTS BTS
- Considering the isolation requirement associated with spurious emissions and the isolation requirement associated with receiver blocking indicates that 40 dB of isolation is required between the two BTS
- The same analysis should be completed for the UMTS BTS interfering with the LTE BTS, i.e. considering the spurious emissions performance of the UMTS BTS and the receiver blocking performance of the LTE BTS. The larger of the two isolation requirements should then be incorporated into the site design
- In general, 40 to 45 dB of isolation is sufficient to avoid interference between LTE, UMTS and GSM BTS
- The way in which the isolation is achieved depends upon the site design:
  - shared feeders and shared antenna: RF combiners which allow the BTS to share the same set of feeders typically provide at least 50 dB of isolation
  - dedicated feeders and shared antenna: antenna typically provide at least 45 dB of isolation
  - dedicated feeders and dedicated antenna: physical separation between the antenna can be used to provide sufficient isolation. Separation requirements depend upon the antenna gain patterns but typically 1 to 2 meters of horizontal separation is sufficient when using 90° horizontal beamwidth antenna

32 SELF ORGANISING NETWORK

32.1 INTRODUCTION

★ The concept of Self Organising Networks (SON) aims to reduce operational expenditure and improve end-user experience

★ SON includes 3 main components

- self-configuring network
- self-optimising network
- self-healing network

★ Self-configuring capabilities allow eNode B to be added to the network in a 'plug-and-play' fashion. Self-configuring networks should be relatively simple to deploy and should have less scope for the introduction of errors. Self-configuring capabilities include:

- dynamic configuration of the S1-MME interface (3GPP release 8 capability)
- dynamic configuration of the X2 interface (3GPP release 8 capability)
- Automatic Neighbour Relation (ANR) function (3GPP release 8 capability)
- Transport Network Layer (TNL) address discovery (3GPP release 8 capability)
- Physical layer Cell Identity (PCI) selection (3GPP release 8 capability)

★ Self-optimising refers to automating the procedures associated with network performance improvement. Both measurements and signalling messages have been standardised to support self-optimisation. Self-optimisation capabilities include:

- mobility load balancing (3GPP release 9 capability)
- mobility robustness optimisation (3GPP release 9 capability)
- RACH optimisation (3GPP release 9 capability)
- energy saving (3GPP release 9 capability)
- radio link failure reporting (3GPP release 9 capability)

★ Self-healing refers to automated fault handling. Faults are automatically detected and corrected. This leads to reduced outage times, improved network performance and improved end-user experience. If a complete eNode B experiences outage then self-healing allows neighbouring sites to compensate by uptilting and increasing their transmit powers. Self-healing also takes advantage of redundancy within the network, e.g. switching to a spare baseband processing card if a failure occurs

★ The self-configuring component is of most value during initial network deployment whereas the self-optimising and self-healing components are of value throughout the life time of the network

★ Inter-Cell Interference Coordination (ICIC) is categorised as a Radio Resource Management (RRM) function rather than a SON function. ICIC is described in section 32.14

★ Minimisation of Drive Tests (MDT) has been designed to operate independently from SON although functionality is re-used wherever possible. MDT is described within section 32.15

★ 3GPP References: TS 36.300

32.2 CONFIGURATION OF S1-MME INTERFACE

★ Dynamic configuration of the S1-MME interface allows the eNode B to automatically establish its S1 interface connections towards one or more MME

★ eNode B are connected to multiple MME when the S1 Flex feature (also known as MME Pooling or Multipoint S1) is used. Connecting to multiple MME allows load balancing across MME and also provides redundancy in case of MME failure

★ The eNode B is provided with the IP address of each target MME prior to interface establishment. The one or more IP addresses could be provided as part of an auto-configuration procedure during which the eNode B connects to the Operations Support System (OSS) and subsequently receives configuration data

★ S1-MME interface establishment is then completed in 2 phases. These 2 phases correspond to establishing connections at the upper 2 layers of the S1 control plane protocol stack illustrated in Figure 231

- the first phase involves Stream Control Transmission Protocol (SCTP) association initialisation. This process is described within IETF RFC 4960 and results in SCTP connectivity between the eNode B and MME
- the second phase uses the S1 Setup procedure to establish the connection at the application layer
  - the eNode B forwards an S1 Setup Request message to the MME. This message includes a list of the supported Tracking Area Codes (TAC) and associated PLMN identities. The default paging DRX cycle duration is also included
  - the MME responds with an S1 Setup Response message. This message specifies one or more Globally Unique MME Identities (GUMMEI). The GUMMEI is a concatenation of a PLMN identity, MME Group Identity (MMEGI) and MME Code (MMEC). The MMEC identifies the MME within its group. The relative MME capacity is also included to define the processing capacity of the MME relative to other MME within the group. This information can be used for load balancing

★ Once the S1 Setup procedure has been completed, the S1-MME interface is established and is ready to transfer subsequent S1 Application Protocol (S1-AP) messages

★ 3GPP References: TS 36.300, TS 36.413

32.3 CONFIGURATION OF X2 INTERFACE

★ Dynamic configuration of the X2 interface allows the eNode B to automatically establish its X2 interface connections towards neighbouring eNode B

★ The eNode B is provided with the IP address of each neighbouring eNode B prior to interface establishment. These IP addresses could be provided as part of an auto-configuration procedure during which the eNode B connects to the Operations Support System (OSS) and subsequently receives configuration data

★ X2 interface establishment is then completed in 2 phases. These 2 phases correspond to establishing connections at the upper 2 layers of the X2 control plane protocol stack illustrated in Figure 233

- the first phase involves Stream Control Transmission Protocol (SCTP) association initialisation. This process is described within IETF RFC 4960 and results in SCTP connectivity between the eNode B and its neighbouring eNode B
- the second phase uses the X2 Setup procedure to establish the connection at the application layer
  - the eNode B forwards an X2 Setup Request message to the neighbouring eNode B. This message includes the global identity of the source eNode B as well as information regarding each cell configured at that eNode B. This information includes: Physical layer Cell Identity (PCI), E-UTRAN Cell Global Identity (ECGI), Tracking Area Code (TAC), PLMN identities, RF carriers and channel bandwidths. In the case of TDD, the uplink-downlink subframe configuration and special subframe configuration are also included. Neighbour information is included for each cell configured at the eNode B. This defines neighbours using their ECGI, PCI and RF carrier. The source eNode B also specifies the MME pools to which it belongs
  - the neighbouring eNode B responds with an X2 Setup Response message. The content of this message is similar to the X2 Setup Request message but the information describes the cells at the neighbouring eNode B rather than the source eNode B

★ Once the X2 Setup procedure has been completed, the X2 interface is established and is ready to transfer subsequent X2 Application Protocol (X2-AP) messages

★ 3GPP References: TS 36.300, TS 36.423

32.4 INTRA-FREQUENCY AUTOMATIC NEIGHBOUR RELATIONS

★ Intra-frequency neighbours can be defined automatically as part of the intra-frequency handover procedure. The addition of a neighbour increases the handover signalling requirement but once added, the neighbour can be used by all other UE

★ The intra-frequency handover procedure starts when the serving eNode B sends an RRC Connection Reconfiguration message which instructs the UE to start intra-frequency measurements. This is illustrated in Figure 316

★ Intra-frequency measurements for handover could be based upon reporting event A3, i.e. triggering the UE to send a Measurement Report when a neighbouring cell becomes better than the serving cell by a specific offset

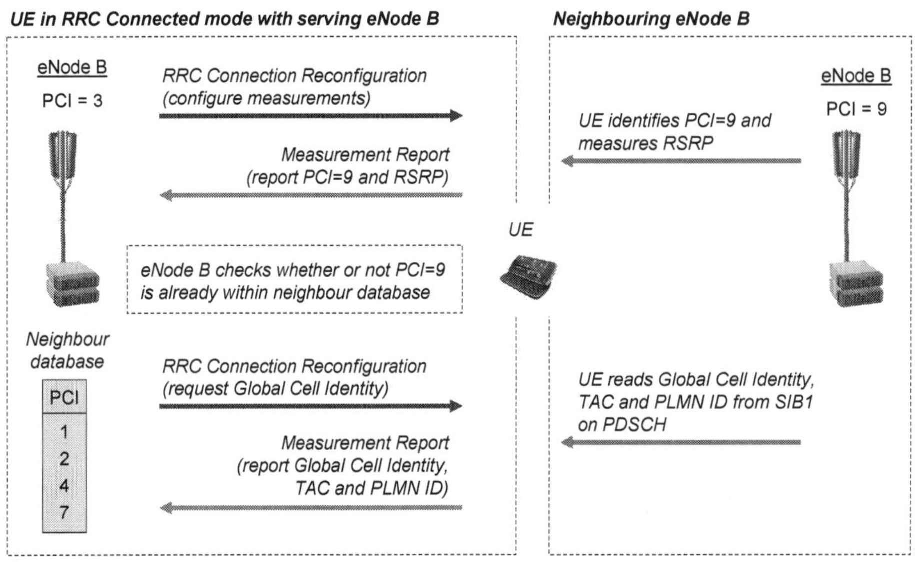

Figure 316 – Intra-frequency automatic neighbour relation definition (part 1)

★ The UE searches for neighbouring cells, identifies their Physical layer Cell Identities (PCI) and measures their RSRP (and/or RSRQ). The UE then follows the instructions of the eNode B in terms of reporting, e.g. when a reporting event is triggered

★ The eNode B receives the neighbour cell measurements and decides whether or not the UE should complete a handover. If a handover is to be completed, the eNode B checks whether or not the reported PCI is included within its existing neighbour database

- o if the PCI is included then the eNode B proceeds with the handover procedure in the normal way (as described in section 26.5)
- o if the PCI is not included then the eNode B proceeds with the automatic neighbour relation definition procedure

The remainder of this section assumes that the automatic neighbour relation definition procedure is initiated

★ The eNode B uses a further RRC Connection Reconfiguration message to instruct the UE to read the Global Cell Identity from the target cell. The UE is also instructed to read the Tracking Area Code and list of PLMN Identities

★ The UE reads the requested information from SIB1 on the PDSCH. This requires the UE to first read the MIB on the PBCH and then the Downlink Control Information (DCI) on the PDCCH. The information is reported back to the eNode B using a Measurement Report message

★ The serving eNode B uses the Global Cell Identity to request the X2 Tunnel Configuration from the MME (assuming an X2 connection to the neighbouring eNode B does not already exist). The information is requested using an S1 Application Protocol (S1-AP) eNode B Configuration Transfer message (shown in Figure 317)

★ The MME uses the Global Cell Identity to interrogate the target neighbouring eNode B for its X2 Tunnel Configuration. This is done using an S1AP: MME Configuration Transfer message. The neighbouring eNode B responds to the MME, which then allows the MME to forward the requested X2 configuration information to the serving eNode B

★ The serving eNode B proceeds to establish an IPSec tunnel and an SCTP connection towards the neighbouring eNode B. Once this has been completed, the two eNode B are able to signal to one another using the X2 Application Protocol (X2-AP)

★ The serving eNode B forwards an X2-AP: X2 Setup Request to the neighbouring eNode B. This message specifies neighbour cell information in terms of Global Cell Identity, PCI and RF carrier. Upon reception, the neighbouring eNode B adds the specified neighbour information to its database

★ The neighbouring eNode B responds using an X2AP: X2 Setup Response message which allows the serving eNode B to update its own neighbour database

★ An X2 connection then exists between the two eNode B and the intra-frequency handover can proceed as normal

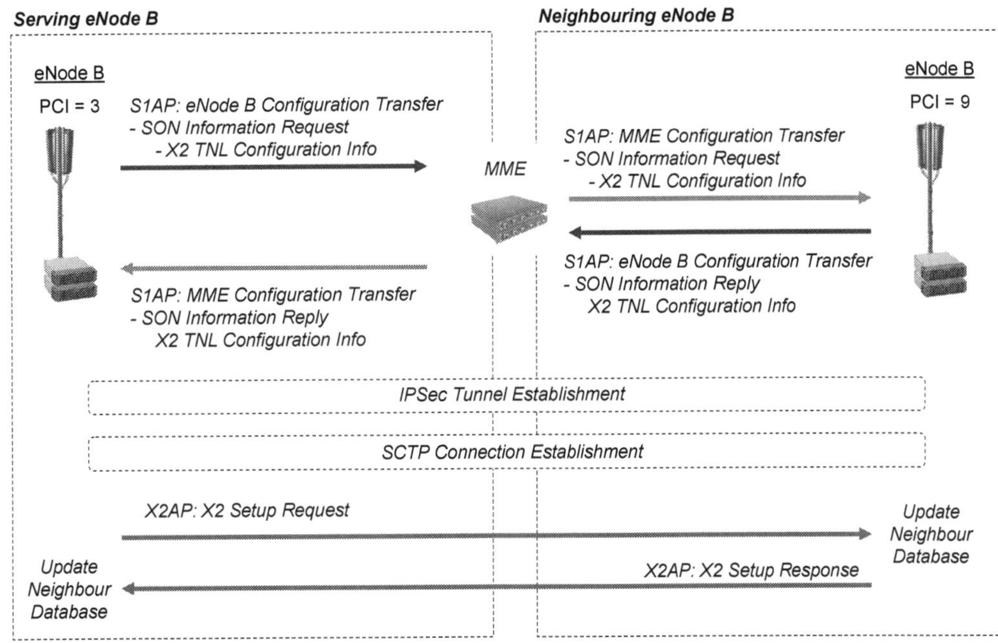

Figure 317 – Intra-frequency automatic neighbour relation definition (part 2)

★ 3GPP References: TS 36.300, TS 36.413, TS 36.423, TS 36.331

32.5 INTER-FREQ/SYSTEM AUTOMATIC NEIGHBOUR RELATIONS

★ Inter-frequency and inter-system neighbours can be defined automatically if the eNode B requests UE to complete the appropriate measurements. This can be done as part of the inter-frequency and inter-system handover procedures

★ Inter-frequency and inter-system measurements require the eNode B to schedule idle periods unless the UE has multiple receivers, i.e. similar to the concept of UMTS compressed mode transmission gaps

★ The eNode B can instruct a UE to start inter-frequency or inter-system measurements using an RRC Connection Reconfiguration message (shown in Figure 318). This could be triggered when the UE moves into an area of relatively weak coverage.

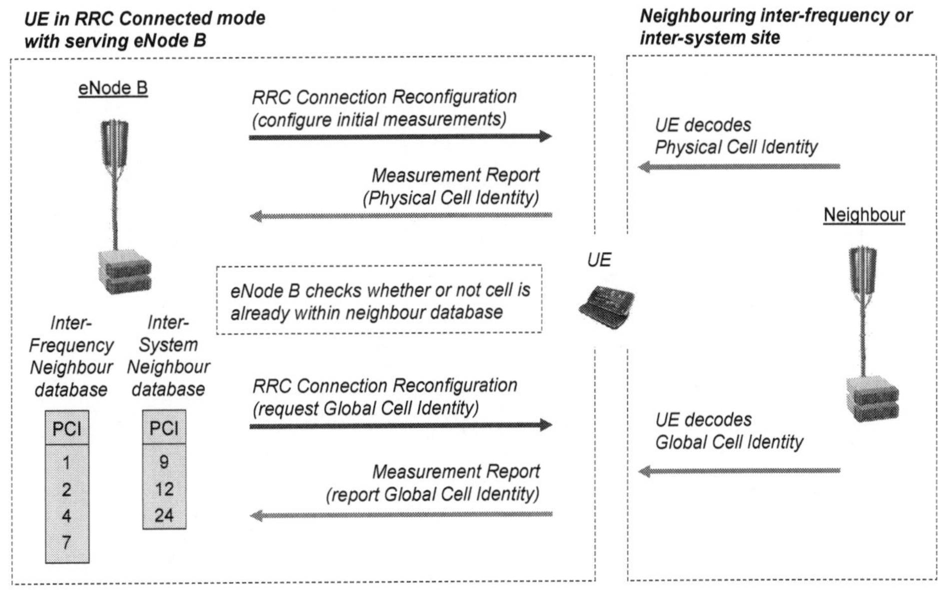

Figure 318 – Inter-frequency and inter-system automatic neighbour relation definition

- The UE searches for neighbouring cells, identifies their Physical Cell Identities and reports them to the eNode B. The format of the Physical Cell Identity depends upon the type of neighbour being measured. Table 360 presents the format of the Physical Cell Identity for each type of neighbour

System	Physical Cell Identity	Global Cell Identity & Additional Info.
LTE	Physical layer Cell Identity (PCI)	Global Cell Identity & TAC & PLMN ID
GSM	Band Indicator, BSIC & RF Carrier	Global Cell Identity & RAC
UMTS	RF Carrier & Scrambling Code	Global Cell Identity, LAC & RAC
CDMA2000	PN Offset	Global Cell Identity

Table 360 – Measurement information reported by UE for neighbour addition

- The eNode B receives the Physical Cell Identity and checks whether or not it is already included within its existing neighbour database. If it is not included then the eNode B uses a further RRC Connection Reconfiguration message to instruct the UE to decode the Global Cell Identity from the system information
- The UE may also be required to decode some additional information, depending upon the system being measured. The requirement for additional information is presented in Table 360, e.g. the Tracking Area Code (TAC) and list of PLMN identities is required for inter-frequency measurements
- The UE reports the Global Cell Identity and any additional information to the eNode B.
  - in the case of inter-system neighbours, the eNode B can update its neighbour database and inter-system handovers can start to benefit from the new neighbour
  - in the case of inter-frequency neighbours, the eNode B checks whether or not there is an existing X2 interface to the neighbouring eNode B. If an X2 interface does not exist then the serving eNode B interrogates the MME for the X2 Tunnel Configuration, similar to the intra-frequency neighbour definition scenario
- If the addition of an inter-system or inter-frequency neighbour is included as part of the normal handover procedure then it may be necessary to trigger the handover relatively early to allow extra time for the neighbour addition process
- 3GPP References: TS 36.300

32.6 TRANSPORT NETWORK LAYER ADDRESS DISCOVERY

- Transport Network Layer (TNL) address discovery refers to an eNode B requesting information regarding the TNL address of another eNode B via the MME. This procedure is illustrated in Figure 317 within the context of automatic neighbour relation definition
- The first eNode sends an S1 Application Protocol (S1-AP) eNode B Configuration Transfer message to the MME. This message includes the global identity of the target eNode B
- The MME interrogates the target eNode B using an S1AP: MME Configuration Transfer message. The target eNode B responds to the MME using an S1-AP: eNode B Configuration Transfer message. The content of this message is relayed to the first eNode B using an S1AP: MME Configuration Transfer message
- The first eNode B can then use the TNL address to establish SCTP and X2 connectivity with the target eNode B
- 3GPP References: TS 36.300, TS 36.413

32.7 PHYSICAL LAYER CELL IDENTITY SELECTION

- Radio network planning rules for the allocation of Physical layer Cell Identities (PCI) are presented in section 31.5
- PCI can be allocated automatically by a Self Organising Network (SON) function rather than planned manually
- The solution will be implementation dependent but in general it can be based upon:
  - a centralised approach
  - a distributed approach
- A centralised approach relies upon a single computer running a PCI allocation algorithm for the complete network. The algorithm has visibility of the complete network so is able to be maximise the re-use distance between cells allocated the same PCI. The algorithm is likely to be similar to that used for planning UMTS scrambling codes. The network management system downloads the PCI plan to the individual eNode B once the allocation algorithm is complete
- A distributed approach relies upon each eNode B running a PCI allocation algorithm. The eNode B may have a downlink receiver to scan the PCI used by neighbouring cells. Using this approach in isolation may not identify the full set of PCI visible to a UE at cell

edge. The distributed allocation algorithm can also make use of PCI reported by UE and by neighbouring eNode B (a neighbouring eNode B reports its PCI and its neighbouring PCI when establishing an X2 interface). The network management system may also signal a specific range of PCI values to be used for allocation

- In practice, the allocation of PCI can be based upon a combination of both the centralised and distributed approaches. The macrocell and microcell layers could be allocated their PCI using the centralised approach while home eNode B (femto) could be allocated their PCI using the distributed approach
- The distributed approach is usually more appropriate for femto because the exact geographic location of the cell is not known to the operator so is not available to provide as an input to the centralised approach
- 3GPP References: TS 36.300

32.8 MOBILITY LOAD BALANCING

- Mobility load balancing was first introduced within the release 9 version of the 3GPP specifications. The inter-system component was enhanced by the release 10 version of the specifications
- Mobility load balancing helps to achieve a balanced load across the network. This increases the overall capacity of the network by helping to avoid scenarios where one cell is congested while its neighbouring cells are relatively unloaded
- Load balancing is achieved by exchanging load information and subsequently triggering handovers or adjusting the parameters used to define the handover triggering thresholds
- Intra-LTE decisions are based upon exchanging load information between neighbouring eNode B. This information exchange procedure is illustrated in Figure 319

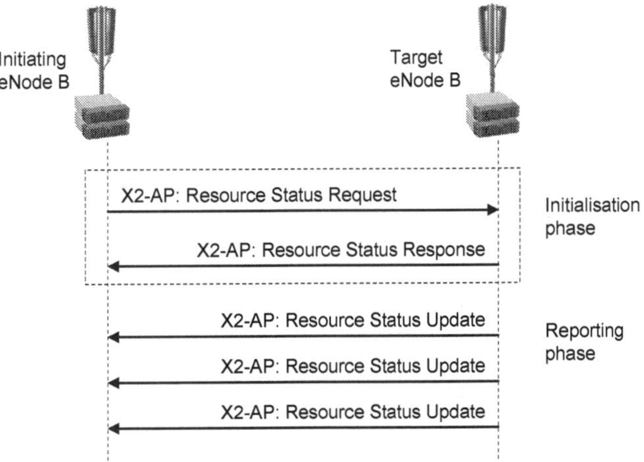

Figure 319 – Procedure for exchanging load information between neighbouring eNode B

- The procedure starts with an initialisation phase where a first eNode B requests reporting from a second eNode B. The X2-AP: Resource Status Request message is used to request reporting. The content of this message is presented in Table 361

Information Elements	
eNode B 1 Measurement Identity	
eNode B 2 Measurement Identity	
Registration Request	
Report Characteristics	PRB Load Indication Periodic
	TNL Load Indication Periodic
	Hardware Load Indication Periodic
	Composite Available Capacity Periodic
	ABS Status Periodic
Cells to Report	LIST
	Cell Identity
Reporting Periodicity	
Partial Success Indicator	

Table 361 – Content of X2-AP: Resource Status Request message

- ★ The eNode B 1 Measurement Identity is allocated by the initiating eNode B. This measurement identity references the reporting from within the initiating eNode B
- ★ The eNode B 2 Measurement Identity is not included within the Resource Status Request message when initiating a measurement reporting procedure. This identity is only included when subsequently requesting the target eNode B to stop reporting, i.e. the Resource Status Request message can be used to request the target eNode B to both start and stop measurements
- ★ The eNode B 2 Measurement Identity is allocated by the target eNode B and is signalled to the initiating eNode B within the X2-AP: Resource Status Response message, i.e. the initiating eNode B does not know the eNode B 2 Measurement Identity when first starting the measurements
- ★ The Registration Request is used to instruct the target eNode B to either start or stop measurements. This will be set to 'start' when initialising a new set of measurements
- ★ The Report Characteristics define the content of the measurement reports to be provided by the target eNode B. This is signalled using a bit string of length 32 bits. Currently, only the first 5 bits are used to request measurement report content:
    - o Physical Resource Block (PRB) Load Indication
    - o Transport Network Layer (TNL) Load Indication
    - o Hardware Load Indication
    - o Composite Available Capacity
    - o Almost Blank Subframes (ABS) Status
- ★ The Cells to Report section of the Resource Status Request message provides a list of the cells at the target eNode B from which load information is requested
- ★ The Reporting Periodicity defines the rate at which the target eNode B should provide reports. The periodicity can be signalled with a value of 1, 2, 5 or 10 seconds
- ★ The Partial Success Indicator is included as a flag when partial success of the measurement reporting configuration is allowed, i.e. the target eNode B can support some of the requested measurements but not all of the requested measurements
- ★ As illustrated in Figure 319, the target eNode B responds with an X2-AP: Resource Status Response message. This message allocates the eNode B 2 Measurement Identity when establishing a new reporting configuration. It also specifies any of the requested measurements which cannot be reported
- ★ The target eNode B then starts periodic reporting of load information using the X2-AP: Resource Status Update message. The content of this message is presented in Table 362

<table>
<tr><th colspan="5">Information Elements</th></tr>
<tr><td colspan="5">eNode B 1 Measurement Identity</td></tr>
<tr><td colspan="5">eNode B 2 Measurement Identity</td></tr>
<tr><td rowspan="16">Cells to Report</td><td colspan="4">LIST</td></tr>
<tr><td colspan="4">Cell Identity</td></tr>
<tr><td rowspan="2">Hardware Load Indication</td><td colspan="3">Downlink Load</td></tr>
<tr><td colspan="3">Uplink Load</td></tr>
<tr><td rowspan="2">S1 TNL Load Indication</td><td colspan="3">Downlink Load</td></tr>
<tr><td colspan="3">Uplink Load</td></tr>
<tr><td rowspan="6">Radio Resource Status</td><td colspan="3">Downlink GBR PRB Usage</td></tr>
<tr><td colspan="3">Uplink GBR PRB Usage</td></tr>
<tr><td colspan="3">Downlink non-GBR PRB Usage</td></tr>
<tr><td colspan="3">Uplink non-GBR PRB Usage</td></tr>
<tr><td colspan="3">Downlink Total PRB Usage</td></tr>
<tr><td colspan="3">Uplink Total PRB Usage</td></tr>
<tr><td rowspan="4">Composite Available Capacity Group</td><td rowspan="2">Downlink</td><td colspan="2">Cell Capacity Class</td></tr>
<tr><td colspan="2">Capacity Value</td></tr>
<tr><td rowspan="2">Uplink</td><td colspan="2">Cell Capacity Class</td></tr>
<tr><td colspan="2">Capacity Value</td></tr>
<tr><td></td><td rowspan="2">ABS Status</td><td colspan="3">Downlink ABS Status</td></tr>
<tr><td></td><td colspan="3">Usable ABS Pattern Information</td></tr>
</table>

Table 362 – Content of X2-AP: Resource Status Update message

★ Both the initiating and target eNode B measurement identities are included within the Resource Status Update message

★ The requested measurements are then provided for each of the cells configured for reporting

★ The Hardware Load Indication is signalled separately for the uplink and downlink directions using a value of 'LowLoad', 'MediumLoad', HighLoad' or 'Overload'. The precise definition of these load states is implementation dependent

★ The S1 Transport Network Layer (TNL) Load Indication is also signalled separately for the uplink and downlink using values of 'LowLoad', 'MediumLoad', HighLoad' or 'Overload'. The precise definition of these load states is implementation dependent

★ The Radio Resource Status provides uplink and downlink information regarding:

  o Guaranteed Bit Rate (GBR) Physical Resource Block (PRB) Usage: corresponds to the percentage of all available PRB used by DTCH transmissions with QoS Class Identifiers (QCI) 1, 2, 3 or 4

  o Non-Guaranteed Bit Rate (non-GBR) Physical Resource Block (PRB) Usage: corresponds to the percentage of all available PRB used by DTCH transmissions with QoS Class Identifiers (QCI) 5, 6, 7, 8 or 9

  o Total Physical Resource Block (PRB) Usage: corresponds to the percentage of all available PRB used for transmission in the downlink, and the percentage of all available PRB allocated for transmission in the uplink

  Each of these measurements is signalled as a percentage between 0 and 100 %. Results are rounded down to the nearest integer. The precise definition of these measurements is specified within 3GPP TS 36.314

★ The Composite Available Capacity Group represents the overall available resources within the cell:

  o the inclusion of a Cell Capacity Class is optional. This information provides a measure of the cell capacity using a value between 1 and 100. A value of 1 means that the cell has a low capacity, whereas a value of 100 indicates that the cell has a high capacity

  o the Capacity Value provides a percentage measure of the available resources within a cell. A value of 0 means that no resources are available, whereas a value of 100 indicates that all resources are available. The reported Capacity Value can be weighted by the ratio of the Cell Capacity Class values to account for the relative capacity of each cell

★ The Almost Blank Subframe (ABS) Status can be used by the source eNode B to evaluate whether or not the existing ABS pattern should be modified. This is used for the optimisation of ABS rather than mobility load balancing

  o the Downlink ABS Status defines the percentage of ABS Resource Blocks which are allocated to UE, i.e. it provides a measure of the utilisation of the ABS by the target eNode B

  o the usable ABS Pattern Information provides a bit string (length 40 for FDD and length 70 for TDD). Each bit within the bit string corresponds to a subframe. A value of 1 indicates an ABS which can be used to allocate resources to a UE and protect that UE from intercell interference. A value of 0 indicates a subframe which cannot be used to protect a UE from intercell interference

★ The initiating eNode B can use the information within the X2-AP: Resource Status Update message to trigger a load based handover. In this case, the X2-AP: Handover Request message includes a cause value which indicates that a load based handover is being triggered, i.e. a cause value of 'Reduce Load in Serving Cell' can be used. This cause value can be used to impact the result of the admission control decision at the target cell

★ The initiating eNode B can also use the information within the X2-AP: Resource Status Update message to optimise both its own and the target eNode B's handover triggering thresholds. The initiating eNode B can adjust its own handover triggering threshold internally without any signalling. The initiating eNode B uses the X2-AP: Mobility Change Request message to request the target eNode B to adjust its handover threshold. The content of the Mobility Change Request message is presented in Table 363

Information Elements	
eNode B 1 Cell Identity	
eNode B 2 Cell Identity	
eNode B 1 Mobility Parameter	Handover Trigger Change
eNode B 2 Proposed Mobility Parameter	Handover Trigger Change
Cause	

Table 363 – Content of X2-AP: Mobility Change Request message

★ Cell identities are specified for both the initiating and target eNode B

★ The eNode B 1 Mobility Parameter is optional. It specifies a change to the handover triggering threshold which the initiating eNode B has already applied for handovers between the initiating and target eNode B. The Handover Triggering Change is signalled using a value between -20 and +20, while the actual value = the signalled value × 0.5 dB. The actual value defines a change to the triggering threshold relative to its previous value, e.g. an actual value of -1 dB could indicate that a threshold has been decreased by 1 dB

★ The eNode B 2 Proposed Mobility Parameter specifies a suggested change to the triggering threshold being applied for handovers from the target eNode B to the source eNode B. The Handover Triggering Change is signalled using a value between -20 and 20, while the actual value = the signalled value × 0.5 dB. The actual value defines a suggested change to the triggering threshold relative to its existing value

★ The Cause value within the Mobility Change Request message specifies the reason for the requested change, e.g. Load Balancing

★ If the proposed handover triggering threshold change is accepted, the target eNode B responds using an X2-AP: Mobility Change Acknowledge message. This message is a simple acknowledgement including the pair of cell identities specified within the an X2-AP: Mobility Change Request message

★ If the proposed handover triggering threshold change is refused, the target eNode B responds using an X2-AP: Mobility Change Failure message. This message can specify upper and lower limits for changes to the handover triggering threshold. These can be used if the source eNode B was requesting a change which was too large

★ Inter-RAT load balancing decisions can be based upon load information exchanged between radio access technologies, e.g. UMTS can report load information to LTE. This exchange of information between radio access technologies uses RAN Information Management (RIM) signalling. The core network is able to relay messages between nodes within the relevant radio access networks

★ An eNode B can request load information reporting from another radio access technology by sending an S1-AP: eNB Direct Information Transfer message to the MME. The content of this message when requesting load information reporting for SON is presented in Table 364

<table>
<tr><th colspan="5">Information Elements</th></tr>
<tr><td rowspan="9">Inter-System Information Transfer Type</td><td rowspan="9">RIM Transfer</td><td rowspan="6">RIM Information</td><td colspan="2">Reporting Cell Identifier</td></tr>
<tr><td rowspan="5">SON Transfer Request Container</td><td>CHOICE</td></tr>
<tr><td>Cell Load Reporting</td></tr>
<tr><td>Multi Cell Load Reporting</td></tr>
<tr><td>Event Triggered Cell Load Reporting</td></tr>
<tr><td>Handover Reporting</td></tr>
<tr><td rowspan="3">RIM Routing Address</td><td colspan="2">CHOICE</td></tr>
<tr><td colspan="2">GERAN Cell Identity</td></tr>
<tr><td colspan="2">UTRAN RNC Identity</td></tr>
</table>

Table 364 – Content of S1-AP: eNB Direct Information Transfer message when used to request SON information transfer

★ The content of the eNB Direct Information Transfer message is specified by 3GPP TS 36.413. Currently, this message is only used to transfer the 'Inter-System Information Transfer Type', which encapsulates the 'RIM Transfer' information element

★ The content of the 'RIM Information' is specified within 3GPP TS 48.018. This specification defines RIM Information for a range of applications including the exchange of System Information between radio access technologies. In the case of requesting load information reporting for the purposes of SON, the 'RIM Information' contains the 'Reporting Cell Identifier' and the 'SON Transfer Request Container'

★ The content of the SON Transfer Request Container is specified within 3GPP TS 36.413. The initiating eNode B can request single cell load reporting, multi-cell load reporting, event triggered load reporting or handover reporting. The multi-cell, event triggered and handover reporting capabilities were added by the release 10 version of the 3GPP specifications. Handover reporting is used for mobility robustness optimisation rather than for mobility load balancing

  o a list of cell identities is provided when requesting multi-cell load reporting

  o a number of measurement reporting levels is specified when requesting event triggered reporting. The cell load scale belonging to the target cell is divided into the specified number of levels. The target cell then provides a measurement report whenever the load changes from one level to another, i.e. reports are sent more frequently when the number of load levels is high

★ The RIM Routing Address can be either a cell identity for GERAN or an RNC identity for UTRAN

★ The MME uses an S1-AP: MME Direct Information Transfer message to forward the reported load information to the initiating eNode B. This message also contains RIM Information but excludes the RIM Routing Address

★ The initiating eNode B can use the inter-RAT load information to trigger a load based inter-system handover (assuming the LTE network is congested while another radio access technology has available capacity)

★ Similarly, other radio access technologies can use RIM to request load information reporting from LTE. This can be used to trigger inter-system handovers towards the LTE network

★ 3GPP References: TS 36.300, TS 36.314, TS 36.413, TS 36.423, TS 48.018

32.9 MOBILITY ROBUSTNESS OPTIMISATION

- Mobility robustness optimisation was first introduced within the release 9 version of the 3GPP specifications. The inter-system component was added by the release 10 version of the specifications
- Mobility robustness optimisation provides support for detecting and helping to correct:
  - connection failures caused by intra-LTE mobility
  - unnecessary inter-system handovers to other radio access technologies
- Figure 320 illustrates the set of 3 intra-LTE mobility scenarios identified for mobility robustness optimisation:
  - handover occurs too late: radio link failure occurs in the source cell before a handover was initiated, and the UE attempts to re-establish its radio link in another cell. Alternatively, radio link failure occurs in the source cell during the handover procedure, and the UE attempts to re-establish its radio link in the target cell
  - handover occurs too early: radio link failure occurs in the target cell after a handover has been completed, and the UE attempts to re-establish its radio link in the source cell. Alternatively, radio link failure occurs in the target cell during the handover procedure, and the UE attempts to re-establish its radio link in the source cell
  - handover to the wrong cell: radio link failure occurs in the target cell after a handover has been completed, and the UE attempts to re-establish its radio link in a cell which is not the source cell nor the target cell. Alternatively, radio link failure occurs in the target cell during the handover procedure, and the UE attempts to re-establish its radio link in a cell which is not the source cell nor the target cell

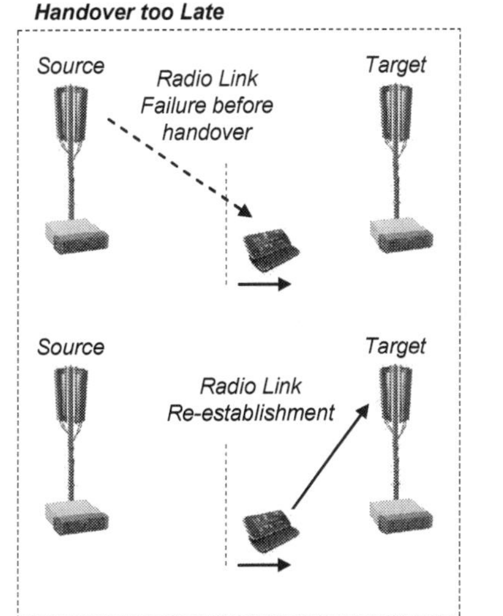

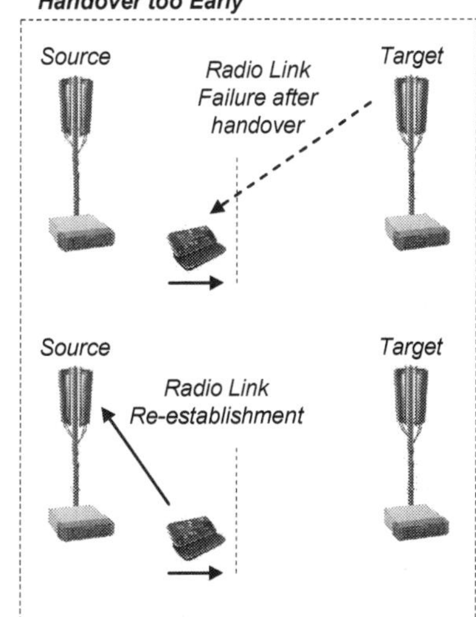

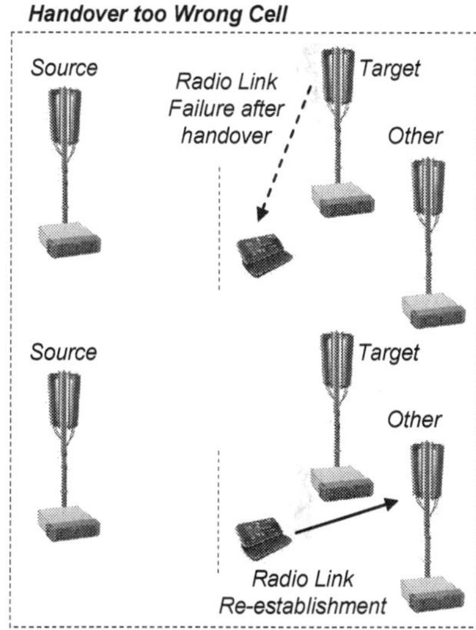

Figure 320 – Handovers too late, too early and to the wrong cell

- Each of the above failure scenarios can be detected after an RRC connection re-establishment, or after an RRC connection setup. UE are able to include a flag within either the RRC Connection Re-establishment Complete message, or the RRC Connection Setup Complete message to indicate that radio link failure information is available for reporting (UE can also include the same flag within an RRC Connection Reconfiguration Complete message)
- After detecting this flag, the eNode B can request the UE to report the radio link failure information using a UE Information Request message. The UE provides its report using a UE Information Response message. This signalling is illustrated in Figure 321
- If RRC connection re-establishment has been completed at a cell which does not belong to the source eNode B then the eNode B which has received the radio link failure information can signal the information to the source eNode B. This is done using the X2-AP: Radio Link Failure Indication message
- The content of the X2-AP: Radio Link Failure Indication message is presented in Table 365
- The Failure Cell Physical layer Cell Identity (PCI) indicates the PCI of the cell to which the UE was connected when the radio link failure occurred. This information is extracted from the RRC Connection Re-establishment Request message from the UE. The Failure Cell PCI information element is not used if the UE used the RRC connection setup procedure rather than the RRC connection re-establishment procedure
- The Re-establishment ECGI provides the identity of the cell where the RRC connection re-establishment procedure was completed, i.e. the identity of one of the cells belonging to the target eNode B shown in Figure 321. This information element is not used if the UE used the RRC connection setup procedure rather than the RRC connection re-establishment procedure

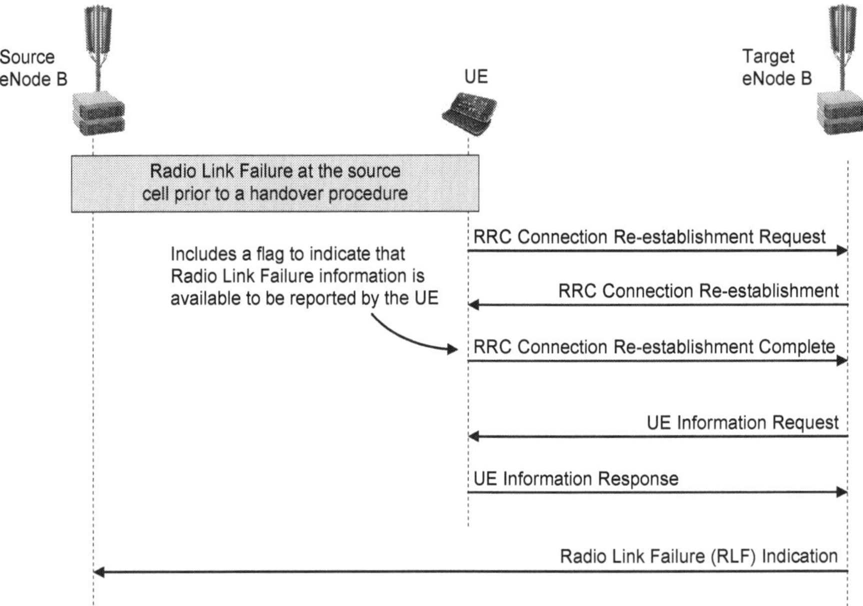

Figure 321 – Reporting of radio link failure information (too late handover)

<table>
<tr><th colspan="3">Information Elements</th></tr>
<tr><td colspan="3">Failure Cell PCI</td></tr>
<tr><td colspan="3">Re-establishment EUTRAN Cell Global Identity (ECGI)</td></tr>
<tr><td colspan="3">C-RNTI</td></tr>
<tr><td colspan="3">ShortMAC-I</td></tr>
<tr><td rowspan="14">UE RLF Report Container</td><td rowspan="2">Measurement Result from Last Serving Cell</td><td>RSRP</td></tr>
<tr><td>RSRQ</td></tr>
<tr><td rowspan="4">Measurement Result from Neighbouring Cells</td><td>Measurement Result List from EUTRAN</td></tr>
<tr><td>Measurement Result List from UTRAN</td></tr>
<tr><td>Measurement Result List from GERAN</td></tr>
<tr><td>Measurement Result List from CDMA2000</td></tr>
<tr><td colspan="2">Location Information</td></tr>
<tr><td rowspan="3">Failed PCell Id</td><td>CHOICE</td></tr>
<tr><td>CGI</td></tr>
<tr><td>PCI & ARFCN</td></tr>
<tr><td>Re-establishment Cell Id</td><td>CGI</td></tr>
<tr><td colspan="2">Time Connection Failure</td></tr>
<tr><td colspan="2">Connection Failure Type</td></tr>
<tr><td>Previous PCell Id</td><td>CGI</td></tr>
<tr><td colspan="3">RRC Connection Setup Indicator</td></tr>
</table>

Table 365 – Content of X2-AP: Radio Link Failure (RLF) Indication message

★ The C-RNTI defines the identity of the UE allocated by the source cell prior to the radio link failure. This information is extracted from the RRC Connection Re-establishment Request message from the UE. The C-RNTI information element is not used if the UE used the RRC connection setup procedure rather than the RRC connection re-establishment procedure

★ The ShortMAC-I provides the 16 least significant bits of the MAC-I calculated using the security configuration belonging to the source cell. This ShortMAC-I information is extracted from the RRC Connection Re-establishment Request message from the UE. The MAC-I information element is not used if the UE used the RRC connection setup procedure rather than the RRC connection re-establishment procedure

★ The UE Radio Link Failure (RLF) Container is extracted from the UE Information Response message. It includes RSRP and optionally, RSRQ measurements from the source cell, i.e. measured from the cell where radio link failure was experienced. It also includes optional measurement results from neighbouring cells. These neighbouring cells can belong to E-UTRAN, UTRAN, GERAN or CDMA2000

- The remaining sections of the UE RLF Report Container were introduced by the release 10 version of the 3GPP specifications and are all optional
- The Location Information allows the UE to report the geographic location where the radio link failure was experienced. It relies upon the UE having knowledge of its location
- The Failed Primary Cell Identity defines the primary cell in which radio link failure was detected or the target primary cell of the failed handover. The cell identity can be reported using either the Cell Global Identity (CGI) or a combination of PCI and ARFCN
- The Re-establishment Cell Identity defines the CGI of the cell where connection re-establishment was attempted after the radio link failure
- The Time Connection Failure defines the time interval between the handover initialisation and the radio link failure. This variable can be signalled using values within the range 0 to 1023. The actual value = signalled value × 100 ms. The maximum value of 102.3 seconds is interpreted as meaning 102.3, or more seconds
- The Connection Failure Type indicates whether the failure was a radio link failure or a handover failure
- The Previous Primary Cell Identity defines the source primary cell of the last handover
- The RRC Connection Setup Indicator is a flag which is included if the UE provided the radio link failure report after an RRC connection setup procedure rather than an RRC connection re-establishment procedure. Inclusion of this flag means that the eNode B receiving the X2-AP: Radio Link Failure Indication message should ignore the Failure Cell PCI, Re-establishment ECGI, C-RNTI and ShortMAC-I information elements
- Having received the X2-AP: Radio Link Failure Indication message, the source eNode B can use the content to adjust its handover triggering configuration
- In the case of a 'too early handover', the UE attempts to re-establish its connection with the source cell so radio link failure information is reported directly to the source eNode B. In this case, the target eNode B can provide the source eNode B with an X2-AP: Handover Report message. The content of this message is presented in Table 366

Information Elements
Handover Report Type
Handover Cause
Source Cell ECGI
Failure Cell ECGI
Re-establishment Cell ECGI

Table 366 – Content of X2-AP: Handover Report message

- The Handover Report Type can be signalled with values of either 'Handover Too Early' or 'Handover to Wrong Cell', i.e. this message can also be used for the 'handover to wrong cell' scenario
- The Handover Cause specifies the reason for the handover, e.g. 'Handover Desirable for Radio Reasons'. This reflects the cause value within the X2-AP: Handover Request message sent from the source cell during the handover procedure
- The source and failure cells are identified by their E-UTRAN Cell Global Identities (ECGI)
- The Re-establishment Cell ECGI is included if the Handover Report Type is set to'Handover to Wrong Cell'
- Figure 322 illustrates example signalling for the 'Handover to Wrong Cell' scenario. In this case, the third eNode B provides the target eNode B with an X2-AP: Radio Link Failure Indication message, while the target eNode B provides the source eNode B with an X2-AP: Handover Report message
- The 'Handover to Wrong Cell' scenario does not always have to involve 3 eNode B. For example, after radio link failure the UE could attempt RRC connection re-establishment at a second cell belonging to the target eNode B
- Having received the X2-AP: Handover Report message, the source eNode B can use the content to adjust its handover triggering configuration
- The second objective of mobility robustness optimisation is to detect and help to correct unnecessary inter-system handovers. In general, UE should be kept on the LTE network unless congestion is detected and load balancing is triggered, or the UE is moving out of LTE coverage. Inter-system handovers generate network signalling load and could result in reduced end-user experience, e.g. lower throughputs on a GERAN network
- Non-optimal handover triggering parameters can lead to coverage based inter-system handovers while LTE coverage remains relatively good. This represents a too early inter-system handover without radio link failure on the target radio access network. The release 10 version of the 3GPP specifications allows this scenario to be detected and reported back to the LTE network
- When completing an inter-system handover towards UTRAN, the eNode B can use the S1-AP: Handover Required message to include 'IRAT Measurement Configuration' information within the 'Source RNC to Target RNC Transparent Container'. Similar information can also be included when completing an inter-system handover towards GERAN

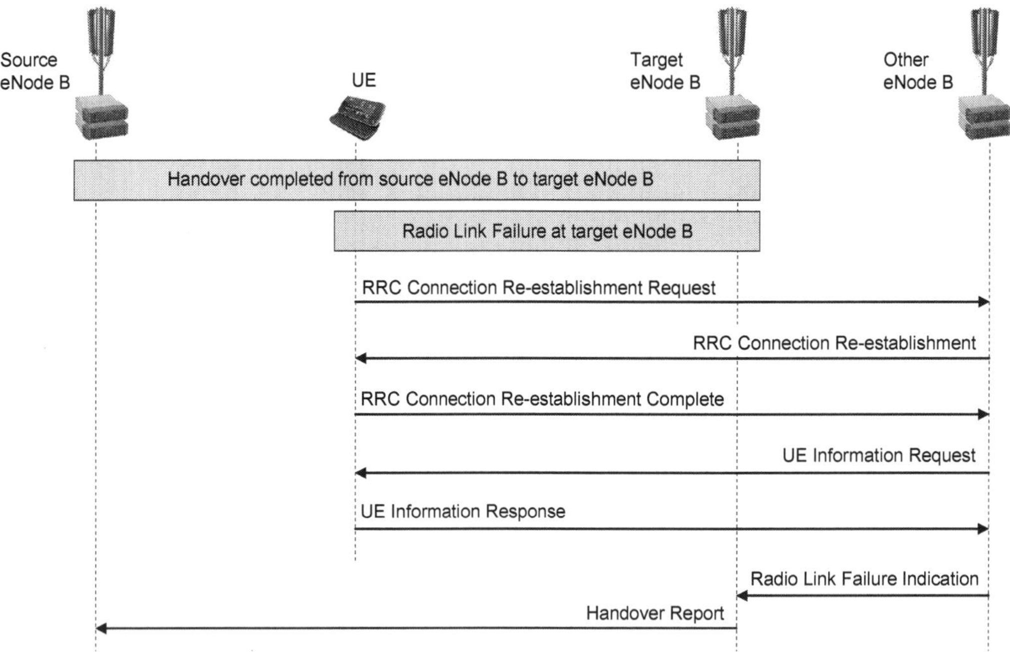

Figure 322 – Reporting of radio link failure information (handover to wrong cell)

★ The 'IRAT Measurement Configuration' information requests the target radio access network to instruct the UE to complete LTE measurements after completing the inter-system handover. The target radio access network can then evaluate whether or not an unnecessary handover has been completed. The source eNode B is subsequently informed if an unnecessary handover has been completed

★ The content of the 'IRAT Measurement Configuration' information, as specified for UTRAN within 3GPP TS 25.413, is presented in Table 367

Information Elements		
RSRP Threshold		
RSRQ Threshold		
IRAT Measurement Parameters	Measurement Duration	
	E-UTRA Frequencies	E-ARFCN
		Measurement Bandwidth

Table 367 – Content of the IRAT Measurement Configuration section of a Source RNC to Target RNC Transparent Container

★ The 'IRAT Measurement Configuration' information can include an RSRP Threshold, or an RSRQ Threshold. Alternatively, both thresholds can be included

- o single threshold included – the target radio access network categorises the inter-system handover as unnecessary if either one LTE cell or a group of LTE cells provide coverage that exceeds the threshold during the complete measurement duration
- o both thresholds included – the target radio access network categorises the inter-system handover as unnecessary if either one LTE cell or a group of LTE cells provide coverage that exceeds both thresholds during the complete measurement duration

★ The Measurement Duration can be signalled with values between 1 and 100 seconds

★ It is optional to specify one or more LTE RF carriers for the UE to measure. RF carriers are specified in terms of their E-ARFCN and measurement bandwidth. Specifying the measurement bandwidth helps to improve measurement accuracy. It is specified in terms of Resource Blocks

★ If the target radio access network categorises the inter-system handover as unnecessary then a Handover Report is sent to the source eNode B. This is sent using RAN Information Management (RIM) signalling. Table 364 illustrates the Handover Report within a SON Transfer Request Container. The content of the SON Transfer Request Container and the Handover Report are specified within 3GPP TS 36.413

★ The content of the Handover Report is presented in Table 368

★ The Handover Type is specified as either 'LTE to UTRAN' or 'LTE to GERAN'

★ The Handover Report Type is specified as 'Unnecessary Handover to another RAT'

Information Elements	
Handover Type	
Handover Report Type	
Handover Source Identity	
Handover Target Identity	
Candidate Cell List	E-UTRAN CGI

Table 368 – Content of Handover Report within SON Transfer Request Container

★ The source and target cell identities for the previously completed inter-system handover are included within the Handover Report

★ The Candidate Cell List specifies the cells which exceed the coverage thresholds within the first measurement report from the UE

★ Having received the Handover Report, the source eNode B can evaluate the requirement for adjusting its inter-system handover triggering thresholds

★ 3GPP References: TS 36.300, TS 36.413, TS 36.423, TS 48.018, TS 25.413

32.10 RACH OPTIMISATION

★ RACH optimisation was introduced as a SON capability within the release 9 version of the 3GPP specifications

★ A UE signals its support for reporting RACH information using a flag within the UE EUTRA Capability section of a UE Capability Information message

★ An eNode B can request a UE to provide a RACH report using a flag within the UE Information Request message (the same message as used to request a radio link failure report for mobility robustness optimisation). The RACH report is then provided by the UE within a UE Information Response message. The content of this report is presented in Table 369

Information Elements
Number of Preambles Sent
Contention Detected

Table 369 – Content of RACH Report within a UE Information Response message

★ The Number of Preambles Sent indicates the number of preambles transmitted during the most recent successfully completed random access procedure. A value between 1 and 200 can be signalled

★ The Contention Detected flag is signalled with a value of 'true' if contention resolution was unsuccessful for at least one of the transmitted preambles from the most recent successfully completed random access procedure

★ Subsequent use of the information contained within the RACH report is implementation dependent. It could be provided to an optimisation team as an input for their network performance improvement activities. Alternatively, it could be used as an input for an automatic optimisation algorithm running in the network

★ Neighbouring eNode B are able to exchange information regarding their RACH configurations during the X2 Setup procedure, i.e. RACH configuration information can be included within both the X2-AP: X2 Setup Request and X2 Setup Response messages. eNode B can also signal they RACH configuration using the X2-AP: eNode B Configuration Update message

★ It is intended that the combination of RACH reporting by the UE, and exchanging information between neighbouring eNode B allows optimisation of the:

- o PRACH configuration index (preamble format, timing and frequency resource allocation)
- o division of PRACH preambles between group A, group B and the group used for non-contention based random access
- o PRACH back-off timer
- o PRACH transmission power control parameters

★ 3GPP References: TS 36.300, TS 36.331

32.11 ENERGY SAVING

★ Energy saving was introduced as a SON capability within the release 9 version of the 3GPP specifications

★ Energy saving allows an eNode B to deactivate one or more cells during periods of low traffic. This helps to reduce power consumption and consequently reduces operating expenditure

★ 2 example scenarios for the use of energy saving are illustrated in Figure 323

- o heterogeneous networks may include microcells and picocells which have been added for capacity below an umbrella macrocell layer. The microcells and picocells can be deactivated if the macrocell layer provides sufficient coverage and capacity during periods of low traffic
- o eNode B may be configured with multiple RF carriers for capacity reasons. One or more RF carriers can be deactivated if a reduced number of RF carriers provides sufficient capacity during periods of low traffic

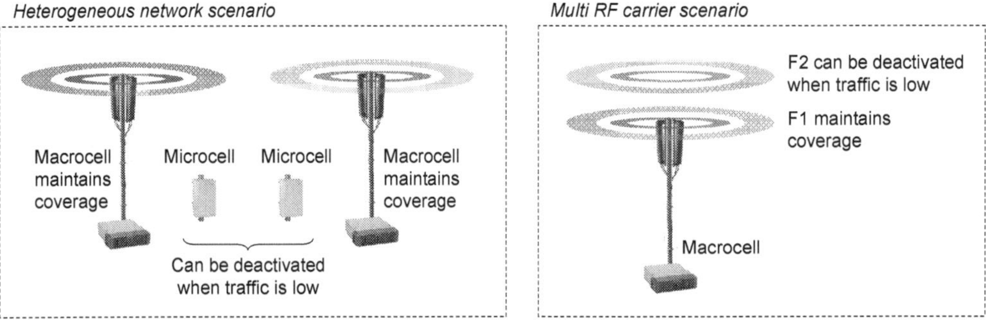

Figure 323 – Example scenarios for the energy saving component of SON

★ Handovers can be triggered prior to deactivating a cell to help provide service continuity for ongoing connections. These handovers could be triggered by ramping down the Reference Signal power until handover measurement reporting events are triggered. Alternatively, the eNode B could autonomously initiate handover procedures for the UE

★ When completing handovers a cause value can be included within the X2-AP: Handover Request message to inform the target cell that the handover is being completed due to deactivation of the source cell, i.e. a cause of value of 'Switch Off Ongoing' can be used. This avoids the target cell from allowing a handover back onto the source cell

★ Once a cell has been shutdown, the eNode B informs its neighbours using an X2-AP: eNode B Configuration Update message. This message allows the inclusion of a cell specific 'Deactivation Indication'

★ Neighbouring eNode B keep a record of the deactivated cells and are able to request reactivation if levels of traffic increase. This is done using an X2-AP: Cell Activation Request message

★ Reactivation could also be triggered by the traffic levels increasing on other cells belonging to the eNode B that has completed the deactivation. Alternatively, reactivation could be triggered when a certain time window expires, e.g. deactivation could be limited to the hours between 1 am and 5 am, so all deactivated cells would be reactivated at 5 o'clock in the morning

★ An eNode B informs its neighbours after reactivation using another eNode B configuration update procedure

★ An example signalling sequence for the energy saving component is SON is illustrated in Figure 324. This example assumes that reactivation is requested by the neighbouring cell which has been the target of handovers prior to deactivation

★ 3GPP References: TS 36.300, TS 36.423

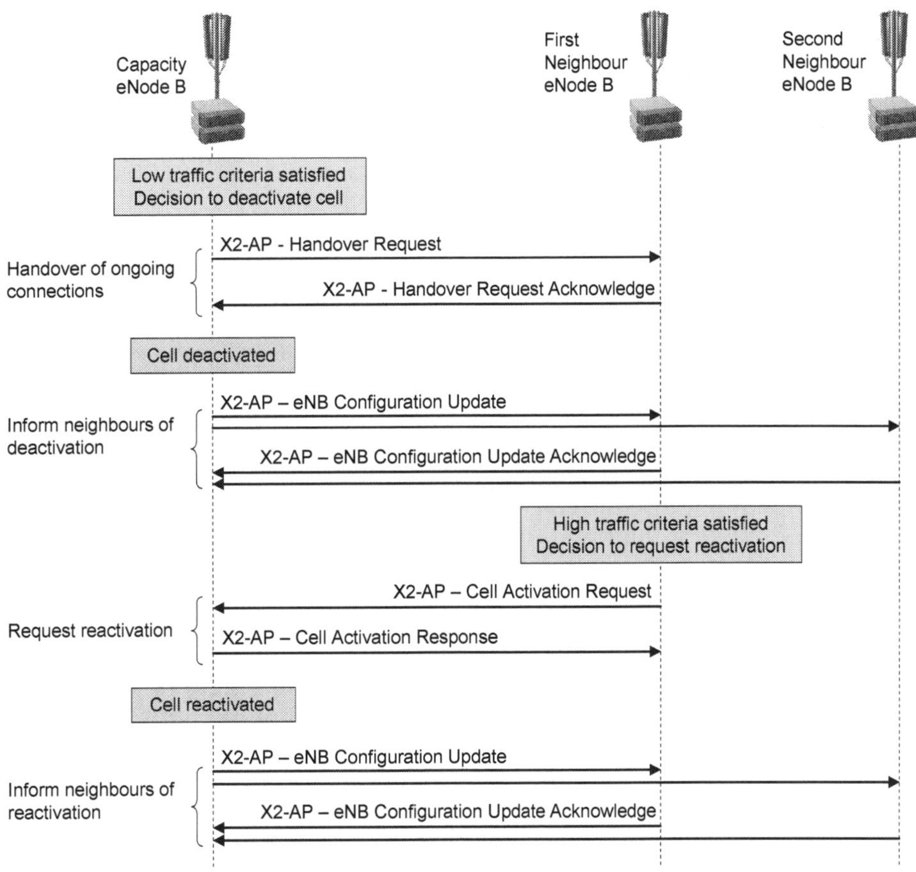

Figure 324 – Example signalling sequence for the energy saving component of SON

32.12 RADIO LINK FAILURE REPORTING

★ Radio link failure reporting was introduced within the release 9 version of the 3GPP specifications to provide support for mobility robustness optimisation. It was subsequently enhanced within the release 10 version of the specifications

★ The release 9 version of the specifications allows the UE to include a flag within the RRC Connection Re-establishment Complete message to indicate that radio link failure information is available for reporting. The release 10 version of the specifications also allows the flag to be included within RRC Connection Setup Complete and RRC Connection Reconfiguration Complete messages

★ The release 10 version of the specifications also expands the content of the radio link failure report to include location information, additional cell identities, timing information and a failure type indicator. The content of a radio link failure report is presented in Table 365 (the section 'UE RLF Report Container' is provided by the UE within a UE Information Response message)

★ The radio link failure report can be used for both mobility robustness optimisation and general coverage optimisation. The inclusion of location information helps to identify locations which would benefit from coverage improvements

★ UE keep a record of their latest radio link failure information for up to 48 hours

★ 3GPP References: TS 36.300, TS 36.331

32.13 UE HISTORY INFORMATION

★ UE History Information was introduced within the release 8 version of the 3GPP specifications

★ UE History Information allows eNode B to maintain a record of the cells visited by a UE. This record includes the time that the UE stayed on each cell. eNode B can use this information to adjust the handover parameters for UE which are experiencing frequent ping-pong transitions between cells, e.g. hysteresis can be increased to reduce the number of handovers

★ The inclusion of UE History Information is mandatory within the X2-AP: Handover Required message, i.e. the source eNode B transfers the UE History Information to the target eNode B during the handover procedure. This message is specified within 3GPP TS 36.423. The content of the UE History Information is presented in Table 370

Information Elements			
Last Visited Cell List (1 to 16 instances)	Last Visited Cell Information	CHOICE	
		E-UTRAN Cell	Global Cell Identity
			Cell Type
			Time UE Stayed in Cell
		UTRAN Cell	UTRAN Cell Identity
			Cell Type
			Time UE Stayed in Cell
		GERAN Cell	Undefined

Table 370 – Content of the UE History Information

★ The source eNode B can signal a list of up to 16 cells visited by the UE. These cells can belong to either LTE or UMTS. The high level message structure provides support for GERAN but the release 8, 9 and 10 versions of the specifications do not define the lower level details required to signal GERAN cell information

★ E-UTRAN (LTE) cells are identified using their Global Cell Identity (GCI) which is a combination of the PLMN identity and the E-UTRAN cell identifier

★ The E-UTRAN Cell Type is signalled using a value of 'very small', 'small', 'medium' or 'large'

★ The E-UTRAN 'Time UE Stayed in Cell' is signalled using a value between 0 and 4095 seconds. The value of 4095 seconds is interpreted as meaning 4095, or more seconds

★ The UTRAN Cell section of the message is specified within 3GPP TS 25.413. This specification defines the RANAP protocol for signalling across the Iu interface

★ The UTRAN Cell Identity is signalled using a combination of the PLMN identity, RNC identity and cell identity

★ The UTRAN Cell Type is signalled using a value of 'femto', 'pico', 'micro' or 'macro'

★ The UTRAN 'Time UE Stayed in Cell' is signalled with a value between 0 and 4095 seconds. The value of 4095 seconds is interpreted as meaning 4095, or more seconds

★ UE History Information can also be included during S1 based handover procedures:

  o in the case of intra-system handovers, the UE History Information is included within the 'Source eNode B to Target eNode B Transparent Container' within an S1-AP: Handover Required message. This transparent container is specified within 3GPP TS 36.413

  o in the case of inter-system handovers towards UMTS, the UE History Information is included within the 'Source RNC to Target RNC Transparent Container' within an S1-AP: Handover Required message. This transparent container is specified within 3GPP TS 25.413

★ Similarly, UE History Information is included within the 'Source eNode B to Target eNode B Transparent Container' belonging to a RANAP: Relocation Required message. The RANAP: Relocation Required message is specified within 3GPP TS 25.413

★ 3GPP References: TS 36.300, TS 36.423, TS 36.413, TS 25.413

32.14 INTER CELL INTERFERENCE COORDINATION

★ Inter-Cell Interference Coordination (ICIC) is categorised as a Radio Resource Management (RRM) function rather than a SON function. Initial support for ICIC was introduced within the release 8 version of the 3GPP specifications. It was subsequently enhanced within the release 10 version of the specifications

★ ICIC allows neighbouring eNode B to coordinate their use of the air-interface resources to help reduce intercell interference, e.g. one eNode B could use Resource Blocks towards the upper edge of the channel bandwidth, while a second eNode B uses Resource Blocks towards the lower edge of the channel bandwidth

★ ICIC creates a trade-off between improving the signal to noise ratio and reducing the quantity of resources available for transmission. Avoiding the use of Resource Blocks used by neighbouring eNode B improves the signal to noise ratio but reduces the total number of Resource Blocks available for transmission

★ ICIC can be applied in both the time and frequency domains. Time domain ICIC involves prioritising the use of individual subframes within specific cells. Frequency domain ICIC involves prioritising the use of individual Resource Blocks within specific cells. Both solutions require a coordination of scheduling between neighbouring eNode B. This necessitates signalling across the X2 interface

★ ICIC is viewed as being particularly important for heterogeneous networks where macrocells are co-channel with micro, pico and femto cells. In this type of scenario, the relatively high uplink and downlink transmit powers associated with a macrocell can cause interference towards the lower power micro, pico and femto cells. This tends to limit the coverage of the lower power cells

★ Figure 325 illustrates an example scenario where a macrocell is interfering with a picocell. The picocell UE in good coverage manages to achieve its uplink and downlink signal to noise ratio requirements without the use of ICIC, but the picocell UE in poor coverage requires ICIC to achieve its uplink and downlink signal to noise ratio requirements. Thus, the picocell should schedule the coordinated air-interface resources (not used by the macrocell) to cell edge UE, and schedule the remaining resources (used by the macrocell) to UE in good coverage

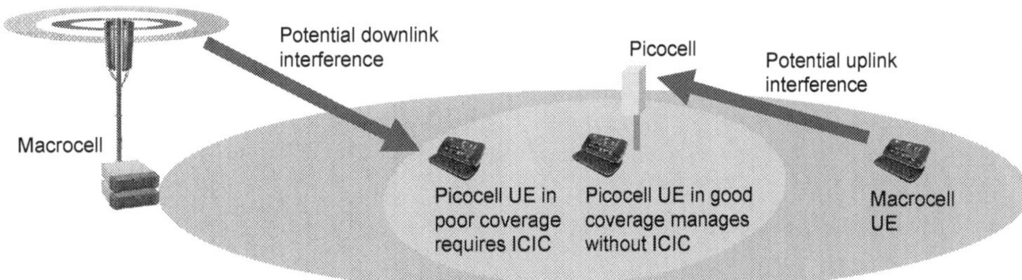

Figure 325 – Intercell interference from a macrocell towards a picocell

★ 3GPP has specified the Load Information message as part of the X2 Application Protocol to allow neighbouring eNode B to exchange uplink and downlink air-interface resource information. This message is presented in Table 371

<table>
<tr><th colspan="3">Information Elements</th></tr>
<tr><td rowspan="14">Cell Information Item
(1 to Number of Cells at source eNode B)</td><td colspan="2">Cell Identity</td></tr>
<tr><td>Uplink Interference Overload Indication
(1 to Number of Resource Blocks)
(3GPP release 8)</td><td>Uplink Interference Overload Indication Result</td></tr>
<tr><td rowspan="2">Uplink High Interference Indication
(1 to Number of Cells at eNode B)
(3GPP release 8)</td><td>Target Cell Identity</td></tr>
<tr><td>Uplink High Interference Indication Bitmap</td></tr>
<tr><td rowspan="5">Relative Narrowband Transmit Power (RNTP)
(3GPP release 8)</td><td>RNTP per PRB Bitmap</td></tr>
<tr><td>RNTP Threshold</td></tr>
<tr><td>Number of Cell Specific Antenna Ports</td></tr>
<tr><td>P_B</td></tr>
<tr><td>PDCCH Interference Impact</td></tr>
<tr><td rowspan="4">ABS Information
(3GPP release 10)</td><td>ABS Pattern Information Bitmap</td></tr>
<tr><td>Number of Cell Specific Antenna Ports</td></tr>
<tr><td>Measurement Subset Bitmap</td></tr>
<tr><td>ABS Inactive</td></tr>
<tr><td colspan="2">Invoke Indication (3GPP release 10)</td></tr>
</table>

Table 371 – Content of X2 Application Protocol Load Information message

- The X2 Application Protocol and the Load Information message are both specified within 3GPP TS 36.423. The Load Information message was introduced within the release 8 version of the specifications but its content was increased within the release 10 version of the specifications as part of the Enhanced ICIC work item
- Figure 326 illustrates an example of the Load Information message being used for uplink ICIC. The content of the Load Information message allows the receiving eNode B to take both reactive and proactive actions

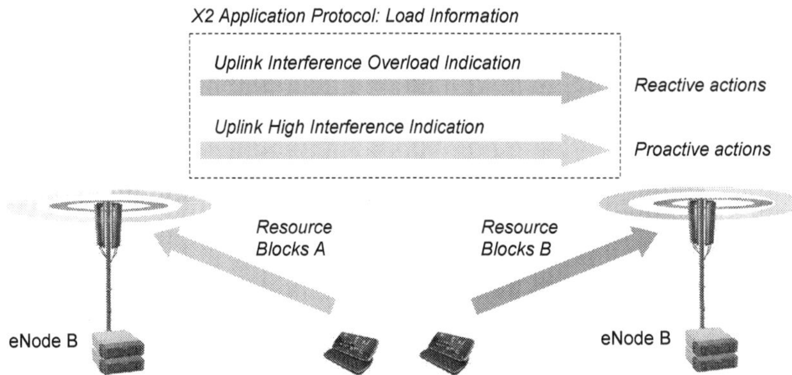

Figure 326 – Concepts of uplink Inter Cell Interference Coordination (ICIC)

- The 'Uplink Interference Overload Indication' reports the level of uplink interference experienced by each Resource Block. The level of uplink interference is categorised as 'high', 'medium' or 'low'. A value can be signalled for each physical Resource Block within the channel bandwidth. This allows the receiving eNode B to be reactive to interference experienced by the sending eNode B
- The 'Uplink High Interference Indication' is used by the sending eNode B to inform the receiving eNode B of which uplink Resource Blocks it plans to use. This allows the receiving eNode B to be proactive about minimising interference on those Resource Blocks. The indication is sent as a bitmap where each bit represents a Resource Block. The information can be addressed towards specific target cells at the receiving eNode B
- Figure 327 illustrates the Load Information message being used for downlink ICIC. The content of the Load Information message allows the receiving eNode B to take proactive actions

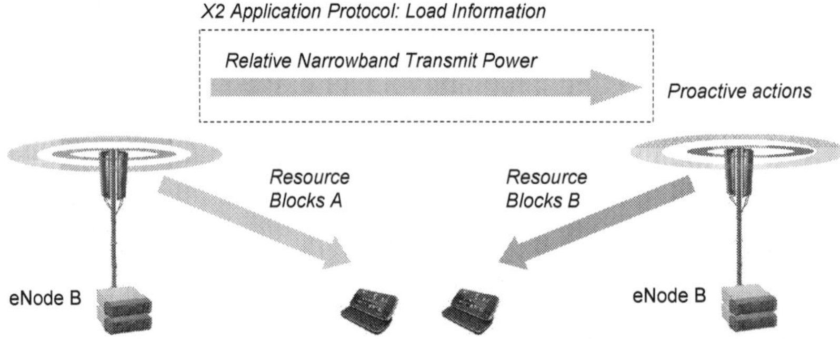

Figure 327 – Concept of downlink Inter Cell Interference Coordination (ICIC)

- The Relative Narrowband Transmit Power (RNTP) informs neighbouring eNode B of which downlink Resource Blocks will have relatively high transmit power. Within the Load Information message, the RNTP is signalled as bitmap, where each bit represents a Resource Block. A bit is set to 0 if the eNode B promises not to exceed a specific relative transmit power with the associated Resource Block. The equation used to define the RNTP is:

$$\text{RNTP} = \begin{cases} 0 & \text{if } \dfrac{\text{EPRE\_MAX}}{\text{EPRE\_MAX\_NOM}} \leq \text{RNTP Threshold} \\ 1 & \text{if no promise about the upper limit of } \dfrac{\text{EPRE\_MAX}}{\text{EPRE\_MAX\_NOM}} \text{ is made} \end{cases}$$

where,

EPRE_MAX is the maximum planned Energy per Resource Element of UE specific PDSCH Resource Elements in OFDMA symbols not containing Reference Signals

EPRE_MAX_NOM is the maximum Energy per Resource Element assuming that the eNode B maximum output is evenly distributed across all Resource Elements within the channel bandwidth

RNTP Threshold is signalled within the Load Information message with a value from the set {-∞, -11, -10, -9, -8, -7, -6, -5, -4, -3, -2, -1, 0, 1, 2, 3 dB}

★ The RNTP section of the Load Information message also specifies the number of antenna ports used for the cell specific Reference Signal (values of 1, 2 or 4 can be signalled). The number of antenna ports is used in combination with the P_B variable to identify a ρB / ρA value using the look-up table presented in Table 372

P_B	ρB / ρA	
	1 Antenna Port	2 and 4 Antenna Ports
0	1	5 / 4
1	4 / 5	1
2	3 / 5	3 / 4
3	2 / 5	1 / 2

Table 372 – Look-up table for the ratio of PDSCH EPRE to Reference Signal EPRE

★ ρA represents the ratio of PDSCH EPRE to cell specific Reference Signal EPRE for OFDMA symbols which do not include the cell specific Reference Signal. ρB represents the same ratio but for OFDMA symbols which include the cell specific Reference Signal

★ The RNTP section of the Load Information message also signals the 'PDCCH Interference Impact'. This informs the neighbouring eNode B of the number of OFDMA symbols that the eNode B plans to allocate for PDCCH transmission. A value from 0 to 4 can be signalled. Values 2 to 4 are used for the 1.4 MHz channel bandwidth, while values 1 to 3 are used for other channel bandwidths. The value 0 indicates that a planned number of PDCCH symbols is not available

★ 'Uplink Interference Overload Indication', 'Uplink High Interference Indication' and RNTP represent forms of frequency domain ICIC. They involve signalling between neighbouring eNode B using the X2 interface. They do not involve signalling to the UE, i.e. these methods of ICIC are transparent to the UE

★ The release 10 version of the 3GPP specifications introduces the concept of Almost Blank Subframes (ABS). This represents a form of time domain ICIC

★ The use of ABS involves eNode B reducing their transmissions during certain subframes. Ideally, eNode B would have zero transmissions during ABS but some transmissions are required for backwards compatibility with 3GPP release 8 and 9 devices

★ ABS can be generated from normal subframes or from MBSFN subframes. Transmissions are reduced by not scheduling PDSCH nor PMCH transmissions during these subframes. These two scenarios are illustrated in Figure 328

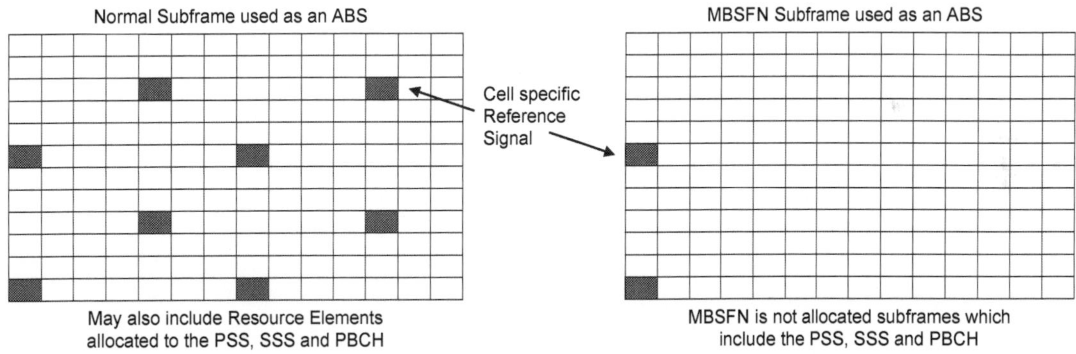

Figure 328 – ABS generated from normal and MBSFN subframes

★ Normal subframes require transmission of the cell specific Reference Signal within both the control and data sections, whereas MBSFN subframes only require transmission of the cell specific Reference Signal within the control section. This makes MBSFN subframes more attractive for ABS

★ Normal subframes may also include transmission of the primary / secondary Synchronisation Signals and the Physical Broadcast Channel (PBCH). In addition, normal subframes may include some limited PDCCH and PDSCH transmissions for the paging channel and the BCCH. MBSFN is not allocated subframes which include these transmissions:

- o in the case of FDD, MBSFN is limited to subframes 1, 2, 3, 6, 7 and 8
- o in the case of TDD, MBSFN is limited to subframes 3, 4, 7, 8 and 9

★ Thus, normal subframes offer greater flexibility in terms of the choice of subframe but MBSFN subframes allow reduced transmissions. In practice, a combination of both normal and MBSFN subframes can be configured as ABS

★ Neighbouring eNode B experience improved signal to noise ratios while transmissions are reduced. If the high power macrocell shown within Figure 325 generates ABS then signal to noise ratios will improve for the lower power picocell. ABS reduces the capacity of the transmitting cell but provides the potential to increase the overall network capacity if levels are intercell interference are carefully managed

★ Table 371 illustrates that ABS information has been added to the X2-AP: Load Information message

★ The ABS Pattern Information Bitmap defines a sequence of bits, where each bit represents a subframe. A value of '1' within the bitmap indicates an ABS subframe during which signal to noise ratios should improve for neighbouring eNode B

- in the case of FDD, the bitmap has a length of 40 bits so spans 4 radio frames. The subframe pattern starts at SFN = 0 and repeats every 4 radio frames
- in the case of TDD, the bitmap has a length which is dependent upon the uplink-downlink subframe configuration.
  - uplink-downlink subframe configurations 1 to 5 use a bitmap length of 20 bits to span 2 radio frames
  - uplink-downlink subframe configuration 6 uses a bitmap length of 60 bits to span 6 radio frames
  - uplink-downlink subframe configuration 0 uses a bitmap length of 70 bits to span 7 radio frames

  The subframe pattern starts at SFN = 0 and is continuously repeated until the next SFN = 0

★ The lengths of these bitmaps have been selected to help satisfy the requirements of the HARQ retransmission protocol, e.g. FDD uses an 8 ms HARQ round trip time, while the timing of the Synchronisation Signals and PBCH is based upon a 10 ms period. The lowest common multiple of 8 ms and 10 ms is 40 ms (4 radio frames)

★ The HARQ retransmission protocol also needs to be taken into account when allocating the set of ABS. Uplink HARQ is synchronous so there is a fixed timing relationship between transmissions, acknowledgements and retransmissions. Figure 329 illustrates an example for FDD where ABS is used to protect the initial resource allocation on the PDCCH as well as the subsequent acknowledgements on the PHICH

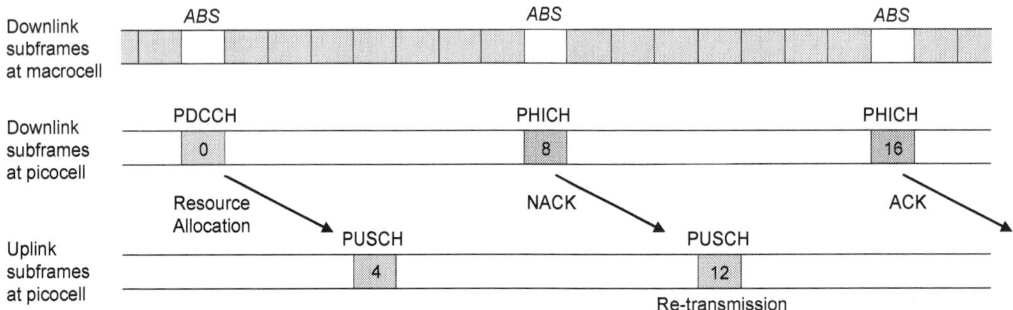

Figure 329 – ABS allowing for synchronous transmission of the PDCCH and PHICH

★ The ABS Information within Table 371 also specifies the number of antenna ports used for the cell specific Reference Signal. This allows the receiving eNode B to estimate the impact of interference from the cell specific Reference Signal during ABS

★ The Measurement Subset Bitmap within the ABS Information is used to signal a subset of the ABS Pattern Information Bitmap. This bitmap has the same length as the ABS Pattern Bitmap and is used to indicate which ABS subframes are recommended for UE measurements, e.g. the set of subframes during which the UE should measure the serving cell RSRP and RSRQ

★ The ABS Inactive flag within the X2-AP: Load Information message indicates that the use of ABS is not active within that cell of the sending eNode B

★ The Invoke Indication within the X2-AP: Load Information message can be used to request ABS information from a neighbouring eNode B. This can also trigger the use of ABS, e.g. a picocell could send an Invoke Indication to a neighbouring macrocell effectively requesting that the macrocell makes use of ABS

★ Figure 330 illustrates an example exchange of X2 Application Protocol messages between a macrocell and picocell. The picocell identifies that its cell edge users would benefit from ABS at a neighbouring co-channel macrocell. The picocell forwards an Invoke Indication to the neighbouring macrocell using an X2-AP: Load Information message

★ The macrocell configures itself to use ABS before proceeding to provide the picocell with its ABS configuration using another X2-AP: Load Information message

★ The picocell then takes advantage of the macrocell ABS to schedule resources for its cell edge users, i.e. the users which are most vulnerable to interference from the macrocell

★ The macrocell can start a resource status reporting initiation procedure to request the picocell to provide Resource Status Update messages. This procedure involves a 2-way handshake using the Resource Status Request and Resource Status Response messages. The Resource Status Request message includes a flag to indicate that ABS Status information is requested

★ The content of the ABS Status information from within an X2-AP: Resource Status Update message is presented in Table 373

★ The Downlink ABS Status defines the percentage of used ABS resources from within the set of usable ABS. The numerator of this calculation is the number of Resource Blocks allocated by the picocell from within the set of usable ABS. The denominator is the total number of Resource Blocks within the set of usable ABS

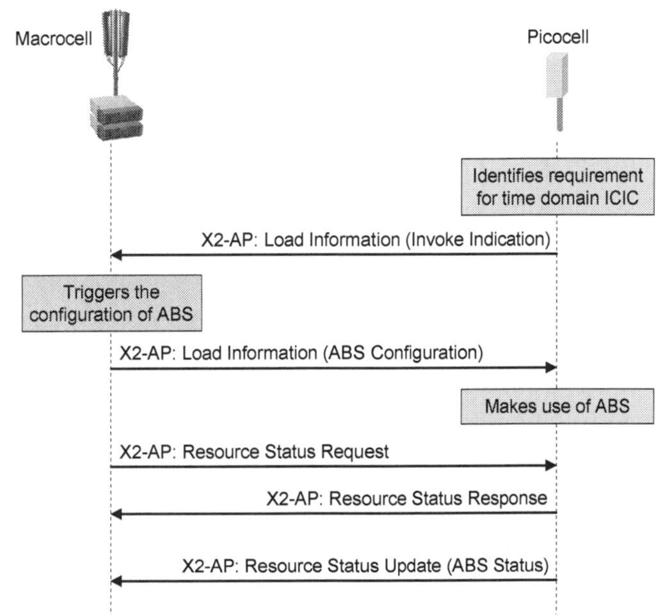

Figure 330 – Example X2 Application Protocol signalling to trigger the use of ABS and subsequently monitor its utilisation

Information Elements
Downlink ABS Status
Usable ABS Pattern Bitmap

Table 373 – Content of the ABS Status information from within an X2-AP: Resource Status Update message

★ The Usable ABS Pattern Bitmap defines the set of ABS which the picocell has been able to use. The length of this bitmap is the same as the length of the bitmap used to define the ABS Pattern. It is either a subset of the ABS Pattern, or equal to the ABS Pattern. A value of 1 indicates an ABS which the picocell has been able to use

★ The ABS Status allows the macrocell to evaluate whether or not it can reduce the number of configured ABS, and consequently increase its own capacity

★ When using ABS, the eNode B can use RRC signalling to provide 3GPP release 10 UE with instructions regarding the subframes to make measurements. Measurement results can vary significantly for picocell UE located towards the edge of coverage. Figure 331 illustrates the increase in signal to noise ratio which can be measured during macrocell ABS

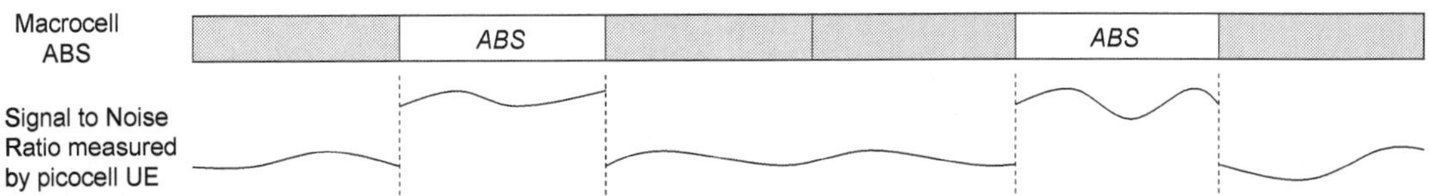

Figure 331 – Variation in serving cell measurement results for a picocell UE towards the edge of coverage when a macrocell uses ABS

★ The picocell UE can be provided with instructions to complete measurements during only the macrocell ABS. This provides the picocell with knowledge of the radio conditions during those ABS. Picocell UE in good coverage may not be provided with these instructions because there will be less dependence upon the macrocell ABS

★ The Measurement Subframe Pattern Primary Cell information presented in Table 374 can be sent to a 3GPP release 10 UE using an RRC Connection Setup, RRC Connection Reconfiguration or RRC Connection Re-establishment message. This information element provides the UE with instructions regarding the set of subframes to use for Radio Link Monitoring (RLM) and Radio Resource Management (RRM) measurements of the serving cell

<table>
<tr><th colspan="3">Information Elements</th></tr>
<tr><td rowspan="3">Measurement Subframe Pattern Primary Cell</td><td colspan="2">CHOICE</td></tr>
<tr><td colspan="2">Release</td></tr>
<tr><td>Setup</td><td>Measurement Subframe Pattern</td></tr>
</table>

Table 374 – Content of the Measurement Subframe Pattern Primary Cell information element

★ Instructing the UE to complete RLM during ABS helps to avoid radio link failure being detected during non-ABS when the levels of downlink interference are relatively high. Completing RRM measurements during ABS quantifies the radio conditions which can be experienced if the UE is scheduled during those subframes

★ The content of the Measurement Subframe Pattern shown in Table 374 is presented in Table 375

Information Elements			
Measurement Subframe Pattern	CHOICE		
	FDD	FDD Subframe Pattern	
	TDD	CHOICE	
		Uplink-Downlink Subframe Configurations 1 to 5	
		Uplink-Downlink Subframe Configuration 0	
		Uplink-Downlink Subframe Configuration 6	

Table 375 – Content of the Measurement Subframe Pattern information element

★ The Measurement Subframe Pattern is a bitmap with a length equal to the ABS Pattern bitmap used within the X2 Application Protocol signalling. The FDD bit map has a length of 40 bits, while the TDD bitmaps have lengths of 20, 70 and 60 bits for uplink-downlink subframe configurations 1 to 5, 0 and 6 respectively. Setting a bit to '1' indicates that the corresponding subframe should be used for measurements

★ 3GPP release 10 UE can also be provided with instructions regarding the set of subframes to use for intra-frequency neighbour cell RRM measurements. The information presented in Table 376 can be sent to a UE within an RRC Connection Reconfiguration message. In this case, a list of one or more Physical layer Cell Identities (PCI) is linked to the Measurement Subframe Pattern. UE are free to complete measurements during any subframe by default. If the PCI belonging to a neighbour is listed then the UE completes RSRP and RSRQ measurements for that neighbour during the specified subframes

Information Elements			
Measurement Subframe Pattern Configuration Neighbour	CHOICE		
	Release		
	Setup	Measurement Subframe Pattern Neighbour	Measurement Subframe Pattern
		Measurement Subframe Cell List	LIST
			PCI

Table 376 – Content of the Measurement Subframe Pattern Configuration Neighbour information element

★ 3GPP release 10 UE can also be provided with instructions regarding the set of subframes to use when generating Channel State Information (CSI). CSI includes CQI, PMI and RI feedback to the serving eNode B. In this case, the information presented in Table 377 can be signalled to the UE within an RRC Connection Setup, RRC Connection Reconfiguration or RRC Connection Re-establishment message

Information Elements			
CSI Subframe Pattern Configuration	CHOICE		
	Release		
	Setup	CSI Measurement Subframe Set 1	Measurement Subframe Pattern
		CSI Measurement Subframe Set 2	Measurement Subframe Pattern

Table 377 – Content of the CSI Subframe Pattern Configuration information element

★ The UE is provided with 2 Measurement Subframe Patterns for CSI measurements. This allows the eNode B to receive CSI feedback based upon both ABS and non-ABS subframes, i.e. one subframe pattern can specify a set of ABS, while the other subframe pattern can specify a set of non-ABS. The eNode B then has sufficient information to schedule resources during both ABS and non-ABS

★ 3GPP References: TS 36.300, TS 36.331, TS 36.423

32.15 MINIMISATION OF DRIVE TESTS

- Operators use drive testing during network optimisation to record measurements from the UE perspective. It represents a relatively expensive and time consuming task
- Minimisation of Drive Tests (MDT) functionality helps to reduce the requirement for drive testing by collecting measurements from the population of subscribers
- MDT has been designed to operate independently from SON although functionality is re-used wherever possible
- There are two general types of MDT:
  - Area based MDT: data is collected from multiple UE within a specific geographic area. A list of cells or tracking areas is used to define the target area within which measurements are to be recorded
  - Signalling based MDT: data is collected from a specific UE addressed by its IMSI or IMEI
- Area based MDT represents an enhancement of management based trace functionality, whereas signalling based MDT represents an enhancement to signalling based subscriber and equipment trace
- Individual subscribers must provide their consent before they can be targeted by MDT functionality. Information regarding consent is stored in the core network but can be transferred to the eNode B
  - the MME can signal subscriber consent information to the eNode B within the S1 Application Protocol messages: Initial Context Setup Request or Handover Request
  - the 'Management based MDT Allowed' information is used for this purpose
  - consent information sent to the eNode B is focused upon area based MDT (management based MDT) because the core network would not request signalling based MDT for a subscriber which has not provided consent
- The general signalling flow for Area based MDT is illustrated in Figure 332

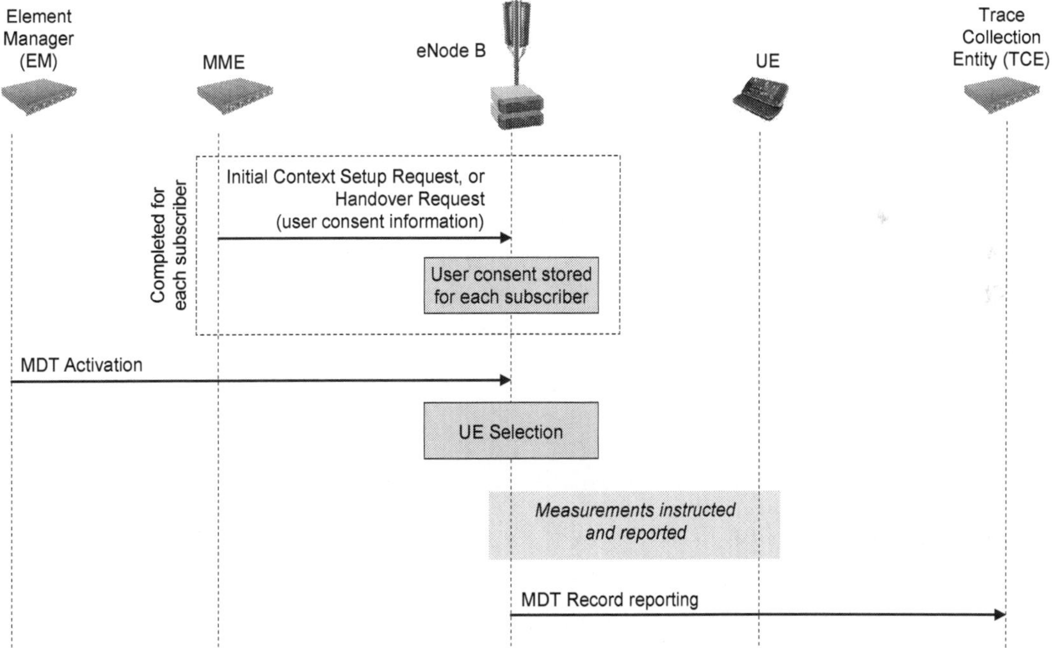

Figure 332 – Example signalling flow for Area based MDT

- The eNode B stores consent information as UE either establish their initial connections or complete incoming handovers
- The Element Manager (EM) for the eNode B is used to trigger MDT activation. The EM provides the eNode B with instructions regarding the type and details of the logging to be completed
- The eNode B selects the set of UE for data collection based upon the consent information, UE capability and whether or not they are located within the target measurement area
- RRC signalling is used to instruct the UE to start measurements and provide reports
- The eNode B then provides the Trace Collection Entity with a copy of the reports

★ The general signalling flow for Signalling based MDT is illustrated in Figure 333

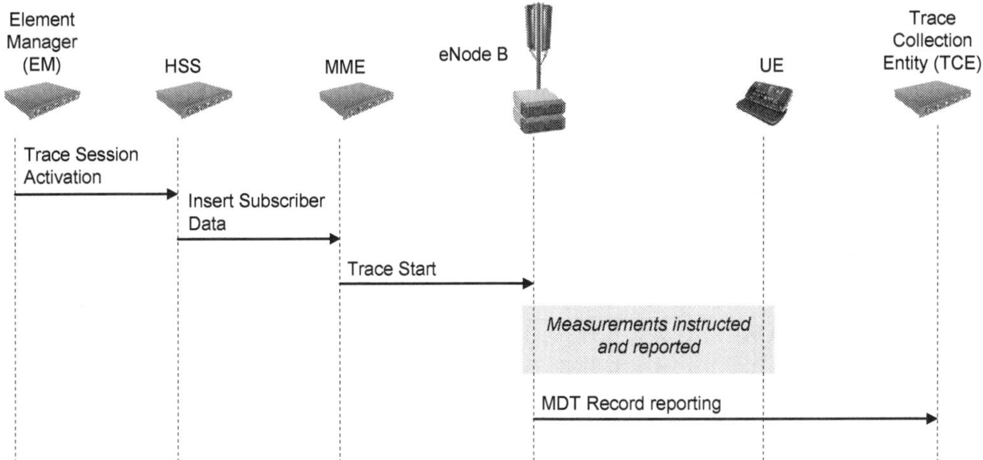

Figure 333 – Example signalling flow for Signalling based MDT

★ The Element Manager (EM) forwards the MDT configuration to the HSS using a Trace Session Activation message. The MDT configuration includes the identity of the target UE (IMSI or IMEI)

★ The MDT configuration is then forwarded to the MME where the user consent information is checked

★ An S1 Application Protocol: Trace Start message is used to provide the eNode B within the MDT configuration

★ RRC signalling is used to instruct the UE to start measurements and provide reports

★ The eNode B then provides the Trace Collection Entity with a copy of the reports

★ Measurements for both Area and Signalling based MDT can be completed using one of two modes of operation:

- Logged MDT
- Immediate MDT

★ Logged MDT measurements are completed by UE in RRC Idle mode, whereas Immediate MDT measurements are completed by UE in RRC connected mode

★ The combinations of Area and Signalling based MDT with Logged and Immediate MDT are presented in Table 378

	Area based MDT	Signalling based MDT
Logged MDT	Group of UE within an area targeted for RRC Idle mode measurements	Individual UE targeted for RRC Idle mode measurements
Immediate MDT	Group of UE within an area targeted for RRC Connected mode measurements	Individual UE targeted for RRC Connected mode measurements

Table 378 – Combinations of Area and Signalling based MDT with Logged and Immediate MDT

★ The following sections describe the Logged and Immediate modes of operation

32.15.1 LOGGED MDT

★ Measurements for 'Logged MDT' are completed while the UE is in RRC Idle mode

★ Prior to entering RRC Idle mode, the eNode B provides the UE with a set of instructions using the 'Logged Measurement Configuration' message. The content of this message is shown in Table 379

<table>
<tr><th colspan="3">Information Elements</th></tr>
<tr><td rowspan="2">Trace Reference</td><td colspan="2">PLMN Identity</td></tr>
<tr><td colspan="2">Trace Identity</td></tr>
<tr><td colspan="3">Trace Recording Session Reference</td></tr>
<tr><td colspan="3">Trace Collection Entity Identity</td></tr>
<tr><td colspan="3">Absolute Time Information</td></tr>
<tr><td rowspan="5">Area Configuration</td><td colspan="2">CHOICE</td></tr>
<tr><td rowspan="2">Cell Global Identity List (1 to 32)</td><td>LIST</td></tr>
<tr><td>EUTRA Cell Global Identity</td></tr>
<tr><td rowspan="2">Tracking Area Code List (1 to 8)</td><td>LIST</td></tr>
<tr><td>Tracking Area Code</td></tr>
<tr><td colspan="3">Logging Duration</td></tr>
<tr><td colspan="3">Logging Interval</td></tr>
</table>

Table 379 – Content of the Logged Measurement Configuration message

★ The Trace Reference (TR) is a globally unique identifier generated from the concatenation of the PLMN Identity and Trace Identity. The Trace Identity has a length of 3 bytes, providing support for more than 16.7 million values

★ The Trace Recording Session Reference (TRSR) defines another identity for the logged information. It has a length of 2 bytes so offers 65 536 values

★ The Trace Collection Entity (TCE) Identity is used as a pointer towards the IP address of the TCE to which the measurements are to be sent. The TCE Identity itself is a string of 8 bits but the network has a look-up table which maps the TCE Identity onto the actual IP address of the target TCE. This mapping is unique within a PLMN

★ The Absolute Time Information provides the UE with a time stamp in the format of YY-MM-DD HH:MM:SS. Binary Coded Decimal (BCD) is used to encode each digit belonging to the time stamp so a total of $12 \times 4 = 48$ bits are required. This is used as a time reference for the UE

★ The Area Configuration defines the area within which the UE is requested to perform measurement logging. The area is defined in terms of either a list of up to 32 Cell Global Identities (CGI), or a list of up to 8 Tracking Area Codes (TAC). The area configuration is an optional information element and if excluded, the UE assumes that measurements are to be completed across the entire PLMN

★ The Logging Duration defines the time period during which the UE is requested to record measurements. Values of 10, 20, 40, 60, 90 and 120 minutes are supported. The timer T330 is allocated the value of the logging duration and is started as soon as the Logged Measurement Configuration message is received. T330 continues to run irrespective of state and radio access technology changes. The UE is permitted to delete any measurements it has recorded 48 hours after T330 expires

★ The Logging Interval defines the rate at which measurement results should be recorded. Values of 1.28, 2.56, 5.12, 10.24, 20.48, 30.72, 40.96 and 61.44 seconds are supported. These values have been chosen to be multiples of the Idle mode DRX cycle, i.e. UE can complete their measurements at the same time as checking for paging messages, and do not need to leave their 'sleep' mode specifically for the purpose of measurements

★ The types of measurements recorded by the UE are listed in Table 380

Type of Cell	Measurement Quantities	Number of Cells to Measure
Serving cell	RSRP, RSRQ	1
Intra-Frequency LTE cells	RSRP, RSRQ	6
Inter-Frequency LTE cells	RSRP, RSRQ	3
UMTS FDD cells	CPICH Ec/Io, CPICH RSCP	3
UMTS TDD cells	P-CCPCH RSCP	3
GSM cells	RSSI	3
CDMA2000 cells	Pilot Phase, Pilot Strength	3

Table 380 – Idle mode measurements for Logged MDT

- These measurements are not configurable but depend upon the UE capability, i.e. a UE will provide measurements from the radio access technologies that it supports
- If the UE memory reserved for MDT becomes full while recording measurements then the UE stops T330 (logging duration) and starts the 48 hour timer which defines the time at which the measurements can be deleted
- UE configured to collect Logged MDT measurements can inform the eNode B that measurements are ready for collection by including a 'LogMeasAvailable' flag within the RRC Connection Setup Complete message. Similarly, this flag can be included within RRC Connection Reconfiguration Complete and RRC Connection Re-establishment Complete messages
- The eNode B subsequently requests the measurements to be uploaded from the UE using the UE Information Request procedure. This procedure involves the 2-way handshake illustrated in Figure 334

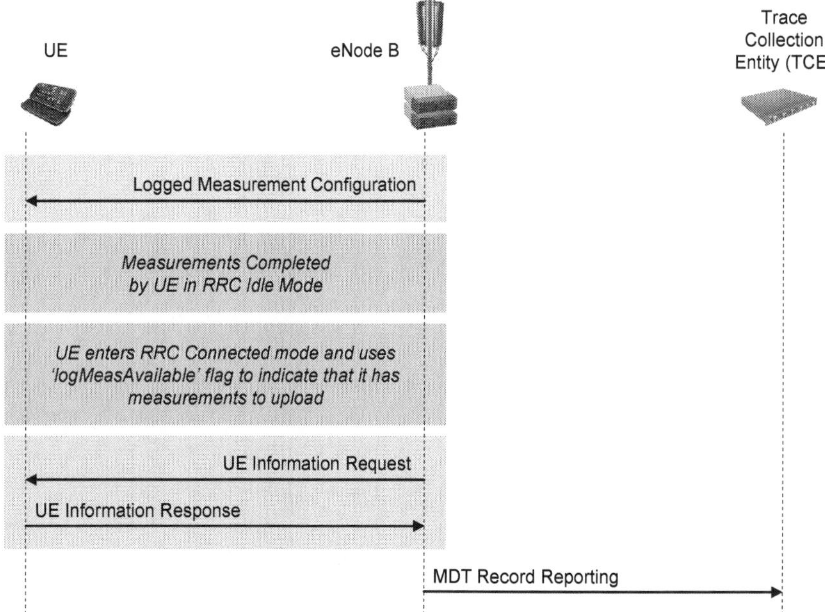

Figure 334 – Measurement configuration and reporting for Logged MDT

- The content of the UE Information Request message is shown in Table 381

Information Elements
RRC Transaction Identifier
RACH Report Request
Radio Link Failure Report Request
Logged Measurement Report Request

Table 381 – Content of the UE Information Request message

- The UE Information Request message includes flags which instruct the UE to provide various types of report. Within the context of MDT, the 'Logged Measurement Report Request' flag is relevant and instructs the UE to send a Logged MDT report
- The content of the UE Information Response message is shown in Table 382. This message is able to provide each of the reports requested by the UE Information Request message. Within the context of MDT, the 'Logged Measurement Report' section of the message is relevant
- The values of the Absolute Time Information, Trace Reference, Trace Recording Session Reference and Trace Collection Entity Identity reflect the values provided within the corresponding Logged Measurement Configuration message
- The message then includes up to 520 logged measurement entries. Each entry can include location information, time stamp information and measurements from each radio access technology
- Location information is optional and its inclusion depends upon the capability of the UE
- The Relative Time Stamp is defined in seconds using a range from 0 to 7200. The value of 7200 corresponds to 120 minutes which is the maximum allowed value for the Logging Duration within the Logged Measurement Configuration message. Time stamp information is relative to the Absolute Time Information
- The Serving Cell Identity is specified in terms of its Cell Global Identity. RSRP and RSRQ measurements are then provided for the serving cell

Information Elements			
RRC Transaction Identifier			
RACH Report	Number of Preambles Sent		
	Contention Detected		
Radio Link Failure Report	Last Serving Cell Measurements	RSRP	
		RSRQ	
	Neighbour Cell Measurements	EUTRA	
		UTRA	
		GERAN	
		CDMA2000	
	Location Information		
	Failed Primary Cell Identity	CHOICE	
		Cell Global Identity	
		PCI ARFCN	PCI
			ARFCN
	Re-establishment Cell Identity	Cell Global Identity	
	Connection Failure Time		
	Connection Failure Type		
	Previous Primary Cell Identity	Cell Global Identity	
Logged Measurement Report	Absolute Time Information		
	Trace Reference		
	Trace Recording Session Reference		
	Trace Collection Entity Identity		
	Log Measurement Information List (1 to 520)	Location Information	
		Relative Time Stamp	
		Serving Cell Identity	Cell Global Identity
		Serving Cell Measurement Results	RSRP
			RSRQ
		Neighbour Cell Measurement Results	EUTRA
			UTRA
			GERAN
			CDMA2000
	Log Measurement Available		

Table 382 – Content of the UE Information Response message

- ★ Neighbour cell measurements are reported when they are available
- ★ E-UTRA neighbour cell measurements can be provided for up to 8 RF carriers. Each RF carrier is identified by its EARFCN and can include measurements from up to 8 cells. Each cell is identified by its Physical layer Cell Identity (PCI). Information can also be included to identify the cell by its Cell Global Identity, Tracking Area Code and PLMN Identity. RSRP and RSRQ measurements are included for each cell
- ★ UTRA neighbour cell measurements can be provided for up to 8 RF carriers. Each RF carrier is identified by its UARFCN and can include measurements from up to 8 cells. Each cell is identified by its scrambling code. Information can also be included to identify the cell by its Cell Global Identity, Location Area Code, Routing Area Code and PLMN Identity. RSCP and Ec/Io measurements can be included for each cell
- ★ GERAN neighbour cell measurements can be provided for up to 3 sets of RF carriers, where each set can include up to 8 RF carriers. Each cell is identified by its RF carrier number as well as its BSIC. Information can also be included to identify the cell by its Cell Global Identity and Routing Area Code. RSSI measurements can be included for each cell
- ★ CDMA2000 neighbour cell measurements can be provided for up to 8 RF carriers. Each RF carrier is identified by its carrier number and can include measurements from up to 8 cells. Each cell is identified by its PN Offset. Information can also be included to identify the cell by its Cell Global Identity. Pilot phase and pilot strength measurements can be included for each cell
- ★ The UE Information Response message can also include a Log Measurement Available flag to indicate that the UE still has some measurements available for upload, i.e. not all measurements could be sent within a single UE Information Response message

32.15.2 IMMEDIATE MDT

★ Measurements for 'Immediate MDT' are completed while the UE is in RRC Connected mode

★ Immediate MDT does not have its own signalling procedures for configuration and reporting. Configuration is completed using an RRC Connection Reconfiguration message, while reporting is completed using Measurement Report messages. An extension is added to the Measurement Report message to allow the UE to report its location

★ The use of RRC Connection Reconfiguration and Measurement Report messages for Immediate MDT is illustrated in Figure 335

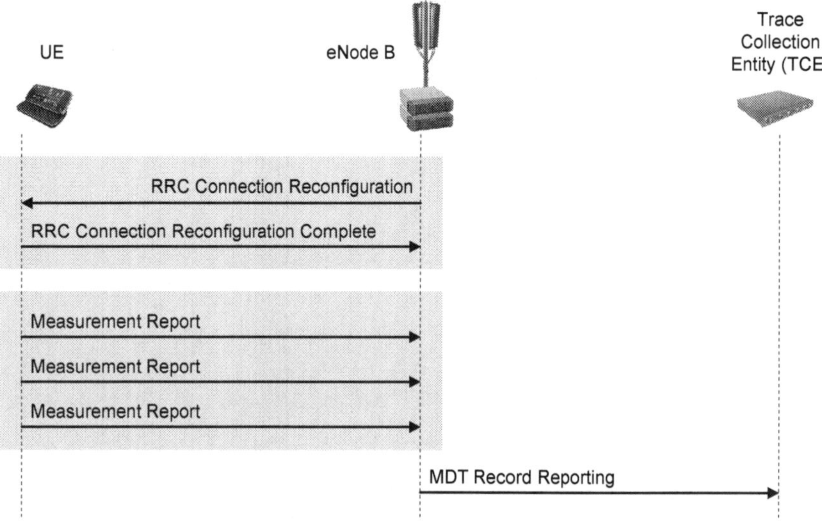

Figure 335 – Measurement configuration and reporting for Immediate MDT

★ Measurement configuration information within the RRC Connection Reconfiguration message can request the UE to provide periodic measurements of RSRP and RSRQ. Alternatively, the UE could be requested to send event triggered measurements, e.g. based upon event A2 when the serving cell becomes worse than an absolute threshold

★ The UE can also be requested to provide Power Headroom Reports (PHR) which provide information regarding the difference between the maximum UE transmit power and the actual UE transmit power. These reports are transferred using MAC control elements rather than RRC messages

★ Time stamps are not reported by the UE for Immediate MDT. The eNode B is responsible for adding time information after receiving a Measurement Report from the UE

★ 3GPP References: TS 32.422, TS 37.320

33 APPENDICES

33.1 RADIO FREQUENCY CHANNEL NUMBERS

★ LTE uses a 100 kHz channel raster, i.e. the center frequency of the channel bandwidth must be a multiple of 100 kHz

★ LTE channel numbers are defined in terms of an E-UTRA Absolute Radio Frequency Channel Number (EARFCN)

★ Table 383 presents the set of parameters used to calculate the uplink and downlink EARFCN for FDD

	Downlink			Uplink		
Operating Band	F_{DL_low} (MHz)	$N_{Offs\text{-}DL}$	Range of EARFCN	F_{UL_low} (MHz)	$N_{Offs\text{-}UL}$	Range of EARFCN
1	2110	0	0 – 599	1920	18000	18000 - 18599
2	1930	600	600 – 1199	1850	18600	18600 – 19199
3	1805	1200	1200 – 1949	1710	19200	19200 – 19949
4	2110	1950	1950 – 2399	1710	19950	19950 – 20399
5	869	2400	2400 – 2649	824	20400	20400 – 20649
6	875	2650	2650 – 2749	830	20650	20650 – 20749
7	2620	2750	2750 – 3449	2500	20750	20750 – 21449
8	925	3450	3450 – 3799	880	21450	21450 – 21799
9	1844.9	3800	3800 – 4149	1749.9	21800	21800 – 22149
10	2110	4150	4150 – 4749	1710	22150	22150 – 22749
11	1475.9	4750	4750 – 4999	1427.9	22750	22750 – 22999
12	728	5000	5000 – 5179	698	23000	23000 – 23179
13	746	5180	5180 – 5279	777	23180	23180 – 23279
14	758	5280	5280 – 5379	788	23280	23280 – 23379
17	734	5730	5730 - 5849	704	23730	23730 – 23849
18	860	5850	5850 – 5999	815	23850	23850 – 23999
19	875	6000	6000 – 6149	830	24000	24000 – 24149
20	791	6150	6150 – 6449	832	24150	24150 – 24449
21	1495.9	6450	6450 – 6599	1447.9	24450	24450 – 24599
22	3510	6600	6600 – 7399	3410	24600	24600 – 25399
23	2180	7500	7500 – 7699	2000	25500	25500 – 25699
24	1525	7700	7700 – 8039	1626.5	25700	25700 – 26039
25	1930	8040	8040 - 8689	1850	26040	26040 – 26689

Table 383 – Parameters used to calculate the uplink and downlink EARFCN for FDD

★ The equivalent set of figures for TDD are presented in Table 384

	Downlink / Uplink		
Operating Band	F_{DL_low}, F_{UL_low} (MHz)	$N_{Offs\text{-}DL}$, $N_{Offs\text{-}UL}$	Range of EARFCN
33	1900	36000	36000 – 36199
34	2010	36200	36200 – 36349
35	1850	36350	36350 – 36949
36	1930	36950	36950 – 37549
37	1910	37550	37550 – 37749
38	2570	37750	37750 – 38249
39	1880	38250	38250 – 38649
40	2300	38650	38650 – 39649
41	2496	39650	39650 – 41589
42	3400	41590	41590 – 43589
43	3600	43590	43590 – 45589

Table 384 – Parameters used to calculate the uplink and downlink EARFCN for TDD

★ The downlink EARFCN is defined by the equation:

$$DL - EARFCN = 10 \times (F_{DL} - F_{DL\_low}) + N_{Offs-DL}$$

where,

F_{DL} is the downlink center frequency of the channel in MHz

F_{DL_low} and $N_{Offs\text{-}DL}$ are dependent upon the operating band and are presented in Table 383 and Table 384

★ As an example of the downlink EARFCN calculation consider an FDD channel centered at 2630 MHz:

$$DL - EARFCN = 10 \times (2630 - 2620) + 2750 = 2850$$

★ The uplink EARFCN is defined by the equation:

$$UL - EARFCN = 10 \times (F_{UL} - F_{UL\_low}) + N_{Offs-UL}$$

where,

F_{UL} is the uplink center frequency of the channel in MHz

F_{UL_low} and $N_{Offs\text{-}UL}$ are dependent upon the operating band and are presented in Table 383 and Table 384

★ As an example of the uplink EARFCN calculation consider a channel centered at 2510 MHz:

$$UL - EARFCN = 10 \times (2510 - 2500) + 20750 = 20850$$

★ Channel numbers towards the edge of each operating band are restricted to avoid any channels from occupying spectrum outside their operating band. Table 385 presents the number of channel numbers not used at the lower and upper edges of the operating band as a function of the channel bandwidth. For example, when using the 5 MHz channel bandwidth, the first 25 channel numbers within the operating band are not used, and likewise the last 24 channel numbers within the operating band are not used

	1.4 MHz	3 MHz	5 MHz	10 MHz	15 MHz	20 MHz
First 'X' channel numbers not used at lower operating band edge	7	15	25	50	75	100
Last 'Y' channel numbers not used at upper operating band edge	6	14	24	49	74	99

Table 385 – The number of channel numbers not used at the lower and upper edges of the operating band

★ 3GPP References: TS 36.101, TS 36.104

33.2 UE IDENTITIES

RNTI

★ Radio Network Temporary Identifiers (RNTI) are applicable within the radio access network. The set of RNTI is presented in Table 386

Identifier	Purpose		Range	Logical Channel	Transport Channel
P-RNTI	Paging		FFFE	PCCH	PCH
SI-RNTI	Broadcast of system information		FFFF	BCCH	DL-SCH
M-RNTI	MCCH information change notification		FFFD	-	-
RA-RNTI	Random access response		0001 to 003C	-	DL-SCH
Temporary C-RNTI	Contention resolution when no C-RNTI is available		0001 to FFF3 (with the exception of values allocated to the RA-RNTI)	CCCH	DL-SCH
	Initial layer 3 message transmission			CCCH DCCH DTCH	UL-SCH
C-RNTI	Dynamically scheduled unicast transmission			DCCH DTCH	UL-SCH
				CCCH DCCH DTCH	DL-SCH
	Triggering PDCCH ordered random access			-	-
SPS C-RNTI	Semi-persistent scheduled unicast transmission	activation, reactivation and retransmission		DCCH DTCH	UL-SCH DL-SCH
		deactivation		-	-
TPC-PUSCH-RNTI	Uplink power control			-	-
TPC-PUCCH-RNTI	Uplink power control			-	-

Table 386 – Radio Network Temporary Identifier (RNTI) types

★ All RNTI have a length of 16 bits

★ The Paging RNTI (P-RNTI) is used during the paging procedure. UE search for Downlink Control Information (DCI) formats 1A and 1C which have had their CRC bits scrambled by the P-RNTI. These DCI allocate PDSCH resources to the PCCH and PCH, i.e. UE can proceed to decode a paging message after finding a DCI whose CRC bits have been scrambled by the P-RNTI. The paging message can include paging records, as well as a system information change notification, Earthquake and Tsunami Warning System (ETWS) notification and Commercial Mobile Alert System (CMAS) indication

★ The System Information RNTI (SI-RNTI) is used during the acquisition of System Information Blocks (SIB). UE search for DCI formats 1A and 1C which have had their CRC bits scrambled by the SI-RNTI. These DCI allocate PDSCH resources to the BCCH and DL-SCH, i.e. UE can proceed to decode system information after finding a DCI whose CRC bits have been scrambled by the SI-RNTI

★ The MBMS RNTI (M-RNTI) is used when receiving an MCCH change notification. UE search for DCI format 1C which have had their CRC bits scrambled by the M-RNTI. These DCI include a bit string of length 8 bits, where each bit corresponds to one of the MBSFN areas listed in SIB13. A value of 1 indicates that the corresponding MCCH will change

★ The Random Access RNTI (RA-RNTI) is used during the random access procedure. DCI formats 1A and 1C which have had their CRC bits scrambled by the RA-RNTI allocate PDSCH resources to Random Access Response messages. There is a one-to-one mapping between the RA-RNTI and the time-frequency resource used by the UE when transmitting the random access preamble. This one-to-one mapping allows the eNode B to address Random Access Response messages to specific UE. Random Access Response messages are transferred using the DL-SCH but are not associated with a logical channel because they are generated by the MAC layer (logical channels transfer information between the RLC and MAC layers)

★ The Temporary Cell RNTI (C-RNTI) is allocated to the UE within the Random Access Response message

- o the Temporary C-RNTI is used for contention resolution when a C-RNTI is not available, e.g. when sending an RRC Connection Request on the CCCH for initial access. After sending the initial layer 3 message, the UE searches for DCI formats 1A or 1C with their CRC bits scrambled by a Temporary C-RNTI which matches the one allocated to the UE within the Random Access Response message. If the Temporary C-RNTI is found, and if contention resolution is successful then the overall random access procedure is successful and the Temporary C-RNTI becomes the C-RNTI

- o the Temporary C-RNTI is also used as an input during PUSCH physical layer scrambling prior to modulation. This is applicable when sending an initial layer 3 message on either the CCCH, DCCH or DTCH

- ★ The Cell RNTI (C-RNTI) is inherited from the Temporary C-RNTI after successful completion of the random access procedure. The C-RNTI is re-allocated during the handover procedure by specifying a new value within the Mobility Control Information section of the RRC Connection Reconfiguration message
  - o the C-RNTI is used during steady state signalling and data transfer within RRC Connected mode. It is used to address a UE by scrambling the CRC bits belonging to DCI formats 0, 1, 1A, 1B, 1D, 2, 2A, 2B, 2C or 4. It is also used as an input during PUSCH and PDSCH physical layer scrambling prior to modulation
  - o the C-RNTI can also be used in combination with DCI format 1A to initiate a random access procedure. The random access procedure is triggered if the CRC bits belonging to DCI format 1A have been scrambled by the C-RNTI and the content of DCI format 1A follows a specific pattern. This pattern is described in section 9.12
- ★ When LTE Advanced Carrier Aggregation is used, a UE uses the same C-RNTI in all serving cells
- ★ The Semi Persistent Scheduling RNTI (SPS-RNTI) is allocated using an RRC Connection Setup or RRC Connection Reconfiguration message. It is applicable when resources are allocated for more than a single subframe, i.e. SPS reduces the control overhead by allowing a single resource allocation to be re-used during multiple subframes. The SPS-RNTI can be used to address a UE by scrambling the CRC bits belonging to DCI formats 0, 1, 1A, 2, 2A, 2B or 2C
- ★ The Transmit Power Control PUSCH RNTI (TPC-PUSCH-RNTI) is used for power control of the PUSCH. DCI formats 3 and 3A can have their CRC bits scrambled by the TPC-PUSCH-RNTI. There are no transport nor logical channels associated with the TPC-PUSCH-RNTI because the power control commands are included within the DCI itself. The TPC-PUSCH-RNTI can be allocated using either an RRC Connection Setup or RRC Connection Reconfiguration message
- ★ The Transmit Power Control PUCCH RNTI (TPC-PUCCH-RNTI) is used for power control of the PUCCH. DCI formats 3 and 3A can have their CRC bits scrambled by the TPC-PUCCH-RNTI. There are no transport nor logical channels associated with the TPC-PUCCH-RNTI because the power control commands are included within the DCI itself. The TPC-PUCCH-RNTI can be allocated using either an RRC Connection Setup or RRC Connection Reconfiguration message

M-TMSI

- ★ An MME Temporary Mobile Subscriber Identity (M-TMSI) is used to identify a UE within a single MME
- ★ An M-TMSI provides the UE with an identity which does not disclose the UE nor the subscriber's permanent identity
- ★ M-TMSI are allocated by the MME and are signalled to the UE within Attach Accept or Tracking Area Update Accept messages
- ★ M-TMSI are typically re-allocated when a UE moves from one MME to another
- ★ M-TMSI have a length of 32 bits

S-TMSI

- ★ An SAE Temporary Mobile Subscriber Identity (S-TMSI) is used to identify a UE within a group of MME
- ★ S-TMSI are generated by concatenating the M-TMSI with the MME Code (MMEC)
  - o MMEC identifies an MME within a group of MME
- ★ The S-TMSI is a shortened form of the GUTI so it provides a more efficient identity for radio interface signalling
- ★ The S-TMSI is used for paging unless it is unavailable in which case the IMSI is used
- ★ S-TMSI have a length of 40 bits

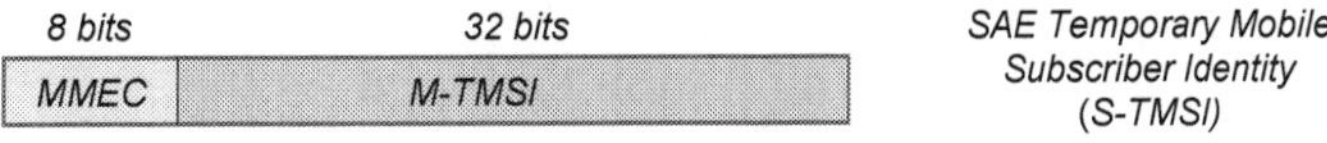

Figure 336 – Structure of S-TMSI

GUTI

- ★ The Globally Unique Temporary UE Identity (GUTI) is used to provide a temporary identity which is globally unique
- ★ The GUTI has three components
  - o PLMN identity – uniquely identifies the network globally
  - o MME Identity (MMEI) - uniquely identifies the MME within the PLMN
  - o MME-TMSI (M-TMSI) – uniquely identifies the UE within the MME
- ★ The PLMN identity is generated from the Mobile Country Code (MCC) and Mobile Network Code (MNC). The MCC is 3 decimal digits, whereas the MNC is either 2 or 3 decimal digits. Each digit is coded using 4 bits, i.e. Binary Coded Decimal (BCD)
- ★ The MMEI is generated from the MME Group ID (MMEGI) and the MME Code (MMEC). The MMEGI is unique within the PLMN while the MMEC is unique within the MME group

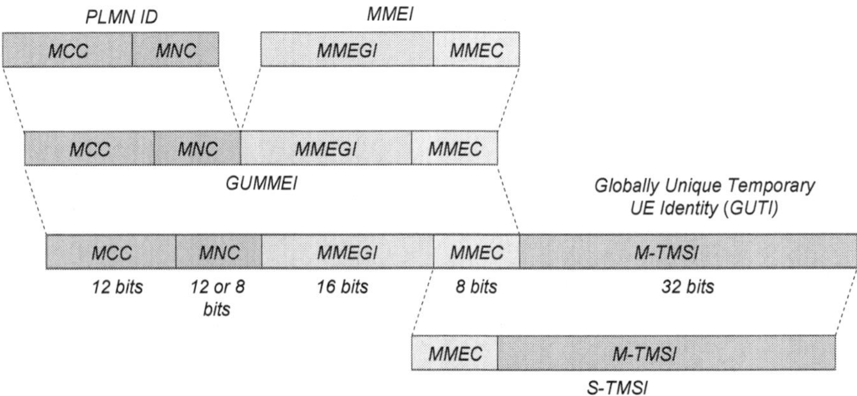

Figure 337 – Structure of GUTI

- ★ The Globally Unique MME Identifier (GUMMEI) is generated from the MCC, MNC and MMEI

IMSI

- ★ The International Mobile Subscriber Identity (IMSI) is a permanent subscriber identity associated with the USIM. An IMSI can be moved between UE by moving the USIM
- ★ The IMSI is globally unique so can be used to identity a subscriber both on the home network and on a visited network when roaming
- ★ The ITU-T specify the structure of an IMSI within recommendation E.212
- ★ The first 3 digits of an IMSI are used to represent the Mobile Country Code (MCC). The ITU-T Telecommunication Standardisation Bureau (TSB) is responsible for maintaining a list of MCC for each country
- ★ The 2 or 3 digits following the MCC are used to represent the Mobile Network Code (MNC). MNC are allocated by national regulators and are used to identify a PLMN associated with a specific MCC
- ★ The MCC-MNC combination defines the PLMN identity and is used by the UE when searching for the home PLMN, i.e. the UE searches for a PLMN which is broadcasting an identity which matches its MCC/MNC combination
- ★ The digits following the PLMN identity define the Mobile Subscriber Identification Number (MSIN). The MSIN is allocated by the operator of the relevant network and has a maximum length of 10 digits. The maximum length of the MSIN is reduced to 9 digits if the PLMN identity occupies 6 rather than 5 digits

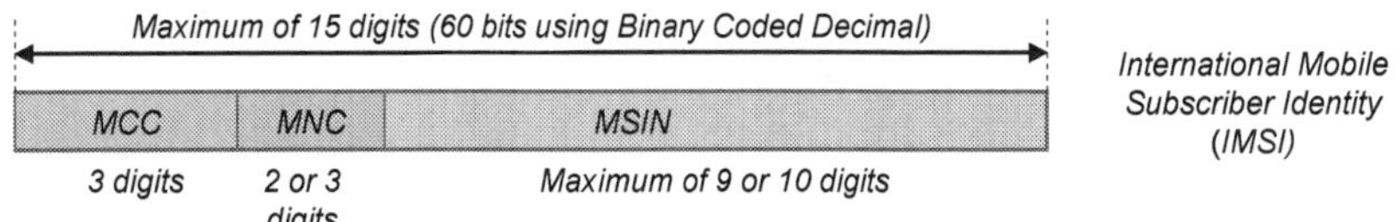

Figure 338 – Structure of IMSI

- ★ 3GPP References: TS 36.321, TS 36.213, TS 23.003

33.3 DL-SCH MODULATION AND TRANSPORT BLOCK SIZES

★ Information regarding the DL-SCH modulation scheme and Transport Block Size (TBS) is signalled within the PDCCH Downlink Control Information (DCI), i.e. the DCI specifies a Modulation and Coding Scheme (MCS) index

★ Table 387 presents the modulation scheme and TBS index associated with each MCS index

MCS Index	Modulation	TBS Index
0	QPSK	0
1		1
2		2
3		3
4		4
5		5
6		6
7		7
8		8
9		9
10	16QAM	
11		10
12		11
13		12
14		13
15		14
16		15

MCS Index	Modulation	TBS Index
17	64QAM	15
18		16
19		17
20		18
21		19
22		20
23		21
24		22
25		23
26		24
27		25
28		26
29	QPSK	Reserved
30	16QAM	
31	64QAM	

Table 387 – Modulation and Coding Scheme (MCS) table for the DL-SCH

★ If the DCI CRC has been scrambled using the P-RNTI, SI-RNTI or RA-RNTI then Table 387 is not applicable

- o the modulation scheme is fixed as QPSK, i.e. QPSK is always used when transferring paging messages, system information and random access responses
- o the TBS Index is set equal to the value of the MCS bits within the DCI. Transport block size selection is then dependent upon the DCI used to allocate resources. Formats 1A and 1C can have their CRC bits scrambled using the P-RNTI, SI-RNTI or RA-RNTI
  - ▪ if DCI format 1C is used then a transport block size is selected from the reduced set of transport block sizes presented in Table 101 (section 9.14)
  - ▪ if DCI format 1A is used then the a transport block size is selected from the set of transport block sizes presented in Table 388. If the least significant bit of the TPC command within DCI format 1A is 0 then the transport block size is selected from column 2, otherwise the transport block size is selected from column 3

★ Otherwise, the MCS Index is set equal to the MCS bits within the DCI and Table 387 is used to identify the corresponding modulation scheme and TBS Index. TBS index 9 is applicable to both QPSK and 16QAM, while TBS index 15 is applicable to both 16QAM and 64QAM

★ MCS indices 29, 30 and 31 are used specifically for HARQ retransmissions. These MCS indices can be used to signal that the modulation scheme has been changed for the retransmission while the TBS has been kept the same as the original transmission

★ Each TBS index points towards a row within the TBS table presented as Table 388 to Table 391. Each column within the TBS table corresponds to the number of allocated Resource Blocks (with the exception of allocations for TDD transmissions within the DwPTS field of a special subframe)

- o in the case of TDD transmissions within the DwPTS field of a special subframe, the column within the TBS table is selected according to the equation:

$$\text{column} = \text{MAX}\ \{\text{Number of Allocated Resource Blocks} \times 0.75\ ,1\}$$

★ TBS index 6 for a single allocated Resource Block specifies a transport block size of 328 bits. Table 387 indicates that TBS index 6 uses QPSK as a modulation scheme so the capacity of a single Resource Block allocation over a subframe is $(12 \times 13 - 6) \times 2 = 300$ bits, when assuming a single OFDMA symbol is allocated to the PDCCH and the cell specific Reference Signal is transmitted for a single antenna port. Thus, the transport block size of 328 bits cannot be fully accommodated by the Resource Block allocation. The transport block size of 328 bits has been specified for the purposes of Voice over IP (VoIP) users allocated QPSK towards cell edge. The use of a single Resource Block helps to maximise system capacity. TTI bundling (presented in section 27.6.1) can be used to transfer the complete set of 328 bits by applying a different puncturing pattern for each transmission within the TTI bundle

★ The maximum TBS for the 20 MHz channel bandwidth is 75 376 bits (assuming all 100 Resource Blocks are allocated to a single connection). This TBS corresponds to a physical layer throughput of 75.376 Mbps

★ The maximum physical layer throughput of 75.376 Mbps is applicable when a single transport block is transferred using a single transmit antenna port. If 2×2 MIMO is used then this maximum physical layer throughput can be doubled to 150.752 Mbps

★ The TBS is determined in a different way when a single transport block is mapped onto 2 layers, e.g. when 4×4 MIMO maps 2 transport blocks onto 4 layers

  o if the number of allocated Resource Blocks is between 1 and 55, then the ‘Number of Allocated Resource Blocks’ is doubled prior to looking-up the TBS within the TBS table

  o if the number of allocated Resource Blocks is greater than 55, then the TBS is obtained from the TBS table in the normal manner but is then mapped onto a new TBS using the mapping presented in Table 392

★ The maximum TBS within Table 392 is 149 776 bits. This TBS can be applicable to the 20 MHz channel bandwidth when all 100 Resource Blocks are allocated to a single connection. The TBS of 149 776 bits corresponds to a physical layer throughput of 149.776 Mbps. This can be doubled to 299.552 Mbps when using 4×4 MIMO

★ Similarly, when a single transport block is mapped onto 3 layers: if the number of allocated Resource Blocks is between 1 and 36, then the ‘Number of Allocated Resource Blocks’ is tripled prior to looking-up the TBS within the TBS table; if the number of allocated Resource Blocks is greater than 36, then the TBS is obtained from the TBS table in the normal manner but is then mapped onto a new TBS using the mapping presented in Table 393

★ Similarly, when a single transport block is mapped onto 4 layers: if the number of allocated Resource Blocks is between 1 and 27, then the ‘Number of Allocated Resource Blocks’ is quadrupled prior to looking-up the TBS within the TBS table; if the number of allocated Resource Blocks is greater than 27, then the TBS is obtained from the TBS table in the normal manner but is then mapped onto a new TBS using the mapping presented in Table 394

★ The maximum TBS in Table 394 is 299 856 bits. This TBS can be applicable to the 20 MHz channel bandwidth when all 100 Resource Blocks are allocated to a single connection. The TBS of 299 856 bits corresponds to a physical layer throughput of 299.856 Mbps. This can be doubled to 599.712 Mbps when using 8×8 MIMO

★ When using LTE Advanced Carrier Aggregation, a pair of transport blocks can be transferred by each Component Carrier so the use of 5 Component Carriers increases the potential throughput to 2998.56 Mbps

★ 3GPP References: TS 36.212, TS 36.213

TBS Index	Number of Allocated Resource Blocks															
	1	2	3	4	5	6	7	8	9	10	11	12	13	14	15	16
0	16	32	56	88	120	152	176	208	224	256	288	328	344	376	392	424
1	24	56	88	144	176	208	224	256	328	344	376	424	456	488	520	568
2	32	72	144	176	208	256	296	328	376	424	472	520	568	616	648	696
3	40	104	176	208	256	328	392	440	504	568	616	680	744	808	872	904
4	56	120	208	256	328	408	488	552	632	696	776	840	904	1000	1064	1128
5	72	144	224	328	424	504	600	680	776	872	968	1032	1128	1224	1320	1384
6	328	176	256	392	504	600	712	808	936	1032	1128	1224	1352	1480	1544	1672
7	104	224	328	472	584	712	840	968	1096	1224	1320	1480	1608	1672	1800	1928
8	120	256	392	536	680	808	968	1096	1256	1384	1544	1672	1800	1928	2088	2216
9	136	296	456	616	776	936	1096	1256	1416	1544	1736	1864	2024	2216	2344	2536
10	144	328	504	680	872	1032	1224	1384	1544	1736	1928	2088	2280	2472	2664	2792
11	176	376	584	776	1000	1192	1384	1608	1800	2024	2216	2408	2600	2792	2984	3240
12	208	440	680	904	1128	1352	1608	1800	2024	2280	2472	2728	2984	3240	3368	3624
13	224	488	744	1000	1256	1544	1800	2024	2280	2536	2856	3112	3368	3624	3880	4136
14	256	552	840	1128	1416	1736	1992	2280	2600	2856	3112	3496	3752	4008	4264	4584
15	280	600	904	1224	1544	1800	2152	2472	2728	3112	3368	3624	4008	4264	4584	4968
16	328	632	968	1288	1608	1928	2280	2600	2984	3240	3624	3880	4264	4584	4968	5160
17	336	696	1064	1416	1800	2152	2536	2856	3240	3624	4008	4392	4776	5160	5352	5736
18	376	776	1160	1544	1992	2344	2792	3112	3624	4008	4392	4776	5160	5544	5992	6200
19	408	840	1288	1736	2152	2600	2984	3496	3880	4264	4776	5160	5544	5992	6456	6968
20	440	904	1384	1864	2344	2792	3240	3752	4136	4584	5160	5544	5992	6456	6968	7480
21	488	1000	1480	1992	2472	2984	3496	4008	4584	4968	5544	5992	6456	6968	7480	7992
22	520	1064	1608	2152	2664	3240	3752	4264	4776	5352	5992	6456	6968	7480	7992	8504
23	552	1128	1736	2280	2856	3496	4008	4584	5160	5736	6200	6968	7480	7992	8504	9144
24	584	1192	1800	2408	2984	3624	4264	4968	5544	5992	6712	7224	7992	8504	9144	9912
25	616	1256	1864	2536	3112	3752	4392	5160	5736	6200	6968	7480	8248	8760	9528	10296
26	712	1480	2216	2984	3752	4392	5160	5992	6712	7480	8248	8760	9528	10296	11064	11832

Table 388 – Transport Block Size (TBS) table for the DL-SCH and UL-SCH (part 1)

TBS Index	Number of Allocated Resource Blocks															
	17	18	19	20	21	22	23	24	25	26	27	28	29	30	31	32
0	456	488	504	536	568	600	616	648	680	712	744	776	776	808	840	872
1	600	632	680	712	744	776	808	872	904	936	968	1000	1032	1064	1128	1160
2	744	776	840	872	936	968	1000	1064	1096	1160	1192	1256	1288	1320	1384	1416
3	968	1032	1096	1160	1224	1256	1320	1384	1416	1480	1544	1608	1672	1736	1800	1864
4	1192	1288	1352	1416	1480	1544	1608	1736	1800	1864	1928	1992	2088	2152	2216	2280
5	1480	1544	1672	1736	1864	1928	2024	2088	2216	2280	2344	2472	2536	2664	2728	2792
6	1736	1864	1992	2088	2216	2280	2408	2472	2600	2728	2792	2984	2984	3112	3240	3368
7	2088	2216	2344	2472	2536	2664	2792	2984	3112	3240	3368	3368	3496	3624	3752	3880
8	2344	2536	2664	2792	2984	3112	3240	3368	3496	3624	3752	3880	4008	4264	4392	4584
9	2664	2856	2984	3112	3368	3496	3624	3752	4008	4136	4264	4392	4584	4776	4968	5160
10	2984	3112	3368	3496	3752	3880	4008	4264	4392	4584	4776	4968	5160	5352	5544	5736
11	3496	3624	3880	4008	4264	4392	4584	4776	4968	5352	5544	5736	5992	5992	6200	6456
12	3880	4136	4392	4584	4776	4968	5352	5544	5736	5992	6200	6456	6712	6712	6968	7224
13	4392	4584	4968	5160	5352	5736	5992	6200	6456	6712	6968	7224	7480	7736	7992	8248
14	4968	5160	5544	5736	5992	6200	6456	6968	7224	7480	7736	7992	8248	8504	8760	9144
15	5160	5544	5736	6200	6456	6712	6968	7224	7736	7992	8248	8504	8760	9144	9528	9912
16	5544	5992	6200	6456	6712	7224	7480	7736	7992	8504	8760	9144	9528	9912	9912	10296
17	6200	6456	6712	7224	7480	7992	8248	8760	9144	9528	9912	10296	10296	10680	11064	11448
18	6712	7224	7480	7992	8248	8760	9144	9528	9912	10296	10680	11064	11448	11832	12216	12576
19	7224	7736	8248	8504	9144	9528	9912	10296	10680	11064	11448	12216	12576	12960	13536	13536
20	7992	8248	8760	9144	9912	10296	10680	11064	11448	12216	12576	12960	13536	14112	14688	14688
21	8504	9144	9528	9912	10680	11064	11448	12216	12576	12960	13536	14112	14688	15264	15840	15840
22	9144	9528	10296	10680	11448	11832	12576	12960	13536	14112	14688	15264	15840	16416	16992	16992
23	9912	10296	11064	11448	12216	12576	12960	13536	14112	14688	15264	15840	16416	16992	17568	18336
24	10296	11064	11448	12216	12960	13536	14112	14688	15264	15840	16416	16992	17568	18336	19080	19848
25	10680	11448	12216	12576	13536	14112	14688	15264	15840	16416	16992	17568	18336	19080	19848	20616
26	12576	13536	14112	14688	15264	16416	16992	17568	18336	19080	19848	20616	21384	22152	22920	23688

TBS Index	Number of Allocated Resource Blocks															
	33	34	35	36	37	38	39	40	41	42	43	44	45	46	47	48
0	904	936	968	1000	1032	1032	1064	1096	1128	1160	1192	1224	1256	1256	1288	1320
1	1192	1224	1256	1288	1352	1384	1416	1416	1480	1544	1544	1608	1608	1672	1736	1736
2	1480	1544	1544	1608	1672	1672	1736	1800	1800	1864	1928	1992	2024	2088	2088	2152
3	1928	1992	2024	2088	2152	2216	2280	2344	2408	2472	2536	2536	2600	2664	2728	2792
4	2344	2408	2472	2600	2664	2728	2792	2856	2984	2984	3112	3112	3240	3240	3368	3496
5	2856	2984	3112	3112	3240	3368	3496	3496	3624	3752	3752	3880	4008	4008	4136	4264
6	3496	3496	3624	3752	3880	4008	4136	4136	4264	4392	4584	4584	4776	4776	4968	4968
7	4008	4136	4264	4392	4584	4584	4776	4968	4968	5160	5352	5352	5544	5736	5736	5992
8	4584	4776	4968	4968	5160	5352	5544	5544	5736	5992	5992	6200	6200	6456	6456	6712
9	5160	5352	5544	5736	5736	5992	6200	6200	6456	6712	6712	6968	6968	7224	7480	7480
10	5736	5992	6200	6200	6456	6712	6712	6968	7224	7480	7480	7736	7992	7992	8248	8504
11	6712	6968	6968	7224	7480	7736	7736	7992	8248	8504	8760	8760	9144	9144	9528	9528
12	7480	7736	7992	8248	8504	8760	8760	9144	9528	9528	9912	9912	10296	10680	10680	11064
13	8504	8760	9144	9144	9528	9912	9912	10296	10680	10680	11064	11448	11448	11832	12216	12216
14	9528	9912	9912	10296	10680	11064	11064	11448	11832	12216	12216	12576	12960	12960	13536	13536
15	10296	10296	10680	11064	11448	11832	11832	12216	12576	12960	12960	13536	13536	14112	14688	14688
16	10680	11064	11448	11832	12216	12216	12576	12960	13536	13536	14112	14112	14688	14688	15264	15840
17	11832	12216	12576	12960	13536	13536	14112	14688	14688	15264	15264	15840	16416	16416	16992	17568
18	12960	13536	14112	14112	14688	15264	15264	15840	16416	16416	16992	17568	17568	18336	18336	19080
19	14112	14688	15264	15264	15840	16416	16992	16992	17568	18336	18336	19080	19080	19848	20616	20616
20	15264	15840	16416	16992	16992	17568	18336	18336	19080	19848	19848	20616	20616	21384	22152	22152
21	16416	16992	17568	18336	18336	19080	19848	19848	20616	21384	21384	22152	22920	22920	23688	24496
22	17568	18336	19080	19080	19848	20616	21384	21384	22152	22920	22920	23688	24496	24496	25456	25456
23	19080	19848	19848	20616	21384	22152	22152	22920	23688	24496	24496	25456	25456	26416	27376	27376
24	19848	20616	21384	22152	22920	22920	23688	24496	25456	25456	26416	26416	27376	28336	28336	29296
25	20616	21384	22152	22920	23688	24496	24496	25456	26416	26416	27376	28336	28336	29296	29296	30576
26	24496	25456	25456	26416	27376	28336	29296	29296	30576	30576	31704	32856	32856	34008	35160	35160

Table 389 – Transport Block Size (TBS) table for the DL-SCH and UL-SCH (parts 2 and 3)

TBS Index	Number of Allocated Resource Blocks															
	49	50	51	52	53	54	55	56	57	58	59	60	61	62	63	64
0	1352	1384	1416	1416	1480	1480	1544	1544	1608	1608	1608	1672	1672	1736	1736	1800
1	1800	1800	1864	1864	1928	1992	1992	2024	2088	2088	2152	2152	2216	2280	2280	2344
2	2216	2216	2280	2344	2344	2408	2472	2536	2536	2600	2664	2664	2728	2792	2856	2856
3	2856	2856	2984	2984	3112	3112	3240	3240	3368	3368	3496	3496	3624	3624	3624	3752
4	3496	3624	3624	3752	3752	3880	4008	4008	4136	4136	4264	4264	4392	4392	4584	4584
5	4392	4392	4584	4584	4776	4776	4776	4968	4968	5160	5160	5352	5352	5544	5544	5736
6	5160	5160	5352	5352	5544	5736	5736	5992	5992	5992	6200	6200	6456	6456	6456	6712
7	5992	6200	6200	6456	6456	6712	6712	6712	6968	6968	7224	7224	7480	7480	7736	7736
8	6968	6968	7224	7224	7480	7480	7736	7736	7992	7992	8248	8504	8504	8760	8760	9144
9	7736	7992	7992	8248	8248	8504	8760	8760	9144	9144	9144	9528	9528	9912	9912	10296
10	8504	8760	9144	9144	9144	9528	9528	9912	9912	10296	10296	10680	10680	11064	11064	11448
11	9912	9912	10296	10680	10680	11064	11064	11448	11448	11832	11832	12216	12216	12576	12576	12960
12	11064	11448	11832	11832	12216	12216	12576	12576	12960	12960	13536	13536	14112	14112	14112	14688
13	12576	12960	12960	13536	13536	14112	14112	14688	14688	14688	15264	15264	15840	15840	16416	16416
14	14112	14112	14688	14688	15264	15264	15840	15840	16416	16416	16992	16992	17568	17568	18336	18336
15	15264	15264	15840	15840	16416	16416	16992	16992	17568	17568	18336	18336	18336	19080	19080	19848
16	15840	16416	16416	16992	16992	17568	17568	18336	18336	19080	19080	19848	19848	19848	20616	20616
17	17568	18336	18336	19080	19080	19848	19848	20616	20616	20616	21384	21384	22152	22152	22920	22920
18	19080	19848	19848	20616	21384	21384	22152	22152	22920	22920	23688	23688	24496	24496	24496	25456
19	21384	21384	22152	22152	22920	22920	23688	24496	24496	25456	25456	25456	26416	26416	27376	27376
20	22920	22920	23688	24496	24496	25456	25456	26416	26416	27376	27376	28336	28336	29296	29296	29296
21	24496	25456	25456	26416	26416	27376	27376	28336	28336	29296	29296	30576	30576	31704	31704	31704
22	26416	27376	27376	28336	28336	29296	29296	30576	30576	31704	31704	32856	32856	34008	34008	34008
23	28336	28336	29296	29296	30576	30576	31704	31704	32856	32856	34008	34008	35160	35160	36696	36696
24	29296	30576	31704	31704	32856	32856	34008	34008	35160	35160	36696	36696	36696	37888	37888	39232
25	31704	31704	32856	32856	34008	34008	35160	35160	36696	36696	37888	37888	39232	39232	40576	40576
26	36696	36696	37888	37888	39232	40576	40576	40576	42368	42368	43816	43816	45352	45352	46888	46888

TBS Index	Number of Allocated Resource Blocks															
	65	66	67	68	69	70	71	72	73	74	75	76	77	78	79	80
0	1800	1800	1864	1864	1928	1928	1992	1992	2024	2088	2088	2088	2152	2152	2216	2216
1	2344	2408	2472	2472	2536	2536	2600	2600	2664	2728	2728	2792	2792	2856	2856	2856
2	2856	2984	2984	3112	3112	3112	3240	3240	3240	3368	3368	3368	3496	3496	3496	3624
3	3752	3880	3880	4008	4008	4136	4136	4264	4264	4392	4392	4392	4584	4584	4584	4776
4	4584	4776	4776	4968	4968	4968	5160	5160	5160	5352	5352	5544	5544	5544	5736	5736
5	5736	5736	5992	5992	5992	6200	6200	6200	6456	6456	6712	6712	6712	6968	6968	6968
6	6712	6968	6968	6968	7224	7224	7480	7480	7736	7736	7736	7992	7992	8248	8248	8248
7	7992	7992	8248	8248	8504	8504	8760	8760	8760	9144	9144	9144	9528	9528	9528	9912
8	9144	9144	9528	9528	9528	9912	9912	9912	10296	10296	10680	10680	10680	11064	11064	11064
9	10296	10296	10680	10680	11064	11064	11064	11448	11448	11832	11832	11832	12216	12216	12576	12576
10	11448	11448	11832	11832	12216	12216	12576	12576	12960	12960	12960	13536	13536	13536	14112	14112
11	12960	13536	13536	13536	14112	14112	14112	14688	14688	14688	15264	15264	15840	15840	15840	16416
12	14688	15264	15264	15264	15840	15840	16416	16416	16416	16992	16992	17568	17568	17568	18336	18336
13	16992	16992	16992	17568	17568	18336	18336	18336	19080	19080	19080	19848	19848	19848	20616	20616
14	18336	19080	19080	19848	19848	19848	20616	20616	20616	21384	21384	22152	22152	22152	22920	22920
15	19848	20616	20616	20616	21384	21384	22152	22152	22152	22920	22920	23688	23688	23688	24496	24496
16	21384	21384	22152	22152	22152	22920	22920	23688	23688	24496	24496	24496	25456	25456	25456	26416
17	23688	23688	24496	24496	24496	25456	25456	26416	26416	26416	27376	27376	27376	28336	28336	29296
18	25456	26416	26416	27376	27376	27376	28336	28336	29296	29296	29296	30576	30576	30576	31704	31704
19	28336	28336	29296	29296	29296	30576	30576	30576	31704	31704	32856	32856	32856	34008	34008	34008
20	30576	30576	31704	31704	31704	32856	32856	34008	34008	34008	35160	35160	35160	36696	36696	36696
21	32856	32856	34008	34008	35160	35160	35160	36696	36696	36696	37888	37888	39232	39232	39232	40576
22	35160	35160	36696	36696	36696	37888	37888	39232	39232	40576	40576	40576	42368	42368	42368	43816
23	37888	37888	37888	39232	39232	40576	40576	40576	42368	42368	43816	43816	43816	45352	45352	45352
24	39232	40576	40576	42368	42368	42368	43816	43816	45352	45352	45352	46888	46888	46888	48936	48936
25	40576	42368	42368	43816	43816	43816	45352	45352	46888	46888	46888	48936	48936	48936	51024	51024
26	48936	48936	48936	51024	51024	52752	52752	52752	55056	55056	55056	55056	57336	57336	57336	59256

Table 390 – Transport Block Size (TBS) table for the DL-SCH and UL-SCH (parts 4 and 5)

TBS Index	Number of Allocated Resource Blocks															
	81	82	83	84	85	86	87	88	89	90	91	92	93	94	95	96
0	2280	2280	2280	2344	2344	2408	2408	2472	2472	2536	2536	2536	2600	2600	2664	2664
1	2984	2984	2984	3112	3112	3112	3240	3240	3240	3240	3368	3368	3368	3496	3496	3496
2	3624	3624	3752	3752	3880	3880	3880	4008	4008	4008	4136	4136	4136	4264	4264	4264
3	4776	4776	4776	4968	4968	4968	5160	5160	5160	5352	5352	5352	5352	5544	5544	5544
4	5736	5992	5992	5992	5992	6200	6200	6200	6456	6456	6456	6456	6712	6712	6712	6968
5	7224	7224	7224	7480	7480	7480	7736	7736	7736	7992	7992	7992	8248	8248	8248	8504
6	8504	8504	8760	8760	8760	9144	9144	9144	9144	9528	9528	9528	9528	9912	9912	9912
7	9912	9912	10296	10296	10296	10680	10680	10680	11064	11064	11064	11448	11448	11448	11448	11832
8	11448	11448	11448	11832	11832	12216	12216	12216	12576	12576	12576	12960	12960	12960	13536	13536
9	12960	12960	12960	13536	13536	13536	13536	14112	14112	14112	14112	14688	14688	14688	15264	15264
10	14112	14688	14688	14688	14688	15264	15264	15264	15840	15840	15840	16416	16416	16416	16992	16992
11	16416	16416	16992	16992	16992	17568	17568	17568	18336	18336	18336	18336	19080	19080	19080	19080
12	18336	19080	19080	19080	19080	19848	19848	19848	20616	20616	20616	21384	21384	21384	21384	22152
13	20616	21384	21384	21384	22152	22152	22152	22920	22920	22920	23688	23688	23688	24496	24496	24496
14	22920	23688	23688	24496	24496	24496	25456	25456	25456	25456	26416	26416	26416	27376	27376	27376
15	24496	25456	25456	25456	26416	26416	26416	27376	27376	27376	28336	28336	28336	29296	29296	29296
16	26416	26416	27376	27376	27376	28336	28336	28336	29296	29296	29296	30576	30576	30576	30576	31704
17	29296	29296	30576	30576	30576	30576	31704	31704	31704	32856	32856	32856	34008	34008	34008	35160
18	31704	32856	32856	32856	34008	34008	34008	35160	35160	35160	36696	36696	36696	37888	37888	37888
19	35160	35160	35160	36696	36696	36696	37888	37888	37888	39232	39232	39232	40576	40576	40576	40576
20	37888	37888	39232	39232	39232	40576	40576	40576	42368	42368	42368	42368	43816	43816	43816	45352
21	40576	40576	42368	42368	42368	43816	43816	43816	45352	45352	45352	46888	46888	46888	46888	48936
22	43816	43816	45352	45352	45352	46888	46888	46888	48936	48936	48936	48936	51024	51024	51024	51024
23	46888	46888	46888	48936	48936	48936	51024	51024	51024	51024	52752	52752	52752	55056	55056	55056
24	48936	51024	51024	51024	52752	52752	52752	52752	55056	55056	55056	57336	57336	57336	57336	59256
25	51024	52752	52752	52752	55056	55056	55056	55056	57336	57336	57336	59256	59256	59256	61664	61664
26	59256	59256	61664	61664	61664	63776	63776	63776	66592	66592	66592	68808	68808	68808	71112	71112

TBS Index	Number of Allocated Resource Blocks													
	97	98	99	100	101	102	103	104	105	106	107	108	109	110
0	2728	2728	2728	2792	2792	2856	2856	2856	2984	2984	2984	2984	2984	3112
1	3496	3624	3624	3624	3752	3752	3752	3752	3880	3880	3880	4008	4008	4008
2	4392	4392	4392	4584	4584	4584	4584	4584	4776	4776	4776	4776	4968	4968
3	5736	5736	5736	5736	5992	5992	5992	5992	6200	6200	6200	6200	6456	6456
4	6968	6968	6968	7224	7224	7224	7480	7480	7480	7480	7736	7736	7736	7992
5	8504	8760	8760	8760	8760	9144	9144	9144	9144	9528	9528	9528	9528	9528
6	10296	10296	10296	10296	10680	10680	10680	10680	11064	11064	11064	11448	11448	11448
7	11832	11832	12216	12216	12216	12576	12576	12576	12960	12960	12960	12960	13536	13536
8	13536	13536	14112	14112	14112	14112	14688	14688	14688	14688	15264	15264	15264	15264
9	15264	15264	15840	15840	15840	16416	16416	16416	16416	16992	16992	16992	16992	17568
10	16992	16992	17568	17568	17568	18336	18336	18336	18336	18336	19080	19080	19080	19080
11	19848	19848	19848	19848	20616	20616	20616	21384	21384	21384	21384	22152	22152	22152
12	22152	22152	22920	22920	22920	23688	23688	23688	23688	24496	24496	24496	24496	25456
13	25456	25456	25456	25456	26416	26416	26416	26416	27376	27376	27376	27376	28336	28336
14	28336	28336	28336	28336	29296	29296	29296	29296	30576	30576	30576	30576	31704	31704
15	29296	30576	30576	30576	30576	31704	31704	31704	31704	32856	32856	32856	34008	34008
16	31704	31704	31704	32856	32856	32856	34008	34008	34008	34008	35160	35160	35160	35160
17	35160	35160	35160	36696	36696	36696	36696	37888	37888	37888	39232	39232	39232	39232
18	37888	39232	39232	39232	40576	40576	40576	40576	42368	42368	42368	42368	43816	43816
19	42368	42368	42368	43816	43816	43816	43816	45352	45352	45352	46888	46888	46888	46888
20	45352	45352	46888	46888	46888	46888	48936	48936	48936	48936	48936	51024	51024	51024
21	48936	48936	48936	51024	51024	51024	51024	52752	52752	52752	52752	55056	55056	55056
22	52752	52752	52752	55056	55056	55056	55056	57336	57336	57336	57336	59256	59256	59256
23	55056	57336	57336	57336	57336	59256	59256	59256	59256	61664	61664	61664	61664	63776
24	59256	59256	61664	61664	61664	61664	63776	63776	63776	63776	66592	66592	66592	66592
25	61664	61664	63776	63776	63776	63776	66592	66592	66592	66592	68808	68808	68808	71112
26	71112	73712	73712	75376	75376	75376	75376	75376	75376	75376	75376	75376	75376	75376

Table 391 – Transport Block Size (TBS) table for the DL-SCH and UL-SCH (parts 6 and 7)

Original TBS	Resultant TBS	Original TBS	Resultant TBS	Original TBS	Resultant TBS	Original TBS	Resultant TBS	Resultant TBS	Resultant TBS
1544	3112	3112	6200	7224	14688	16416	32856	37888	76208
1608	3240	3240	6456	7480	14688	16992	34008	39232	78704
1672	3368	3368	6712	7736	15264	17568	35160	40576	81176
1736	3496	3496	6968	7992	15840	18336	36696	42368	84760
1800	3624	3624	7224	8248	16416	19080	37888	43816	87936
1864	3752	3752	7480	8504	16992	19848	39232	45352	90816
1928	3880	3880	7736	8760	17568	20616	40576	46888	93800
1992	4008	4008	7992	9144	18336	21384	42368	48936	97896
2024	4008	4136	8248	9528	19080	22152	43816	51024	101840
2088	4136	4264	8504	9912	19848	22920	45352	52752	105528
2152	4264	4392	8760	10296	20616	23688	46888	55056	110136
2216	4392	4584	9144	10680	21384	24496	48936	57336	115040
2280	4584	4776	9528	11064	22152	25456	51024	59256	119816
2344	4776	4968	9912	11448	22920	26416	52752	61664	124464
2408	4776	5160	10296	11832	23688	27376	55056	63776	128496
2472	4968	5352	10680	12216	24496	28336	57336	66592	133208
2536	5160	5544	11064	12576	25456	29296	59256	68808	137792
2600	5160	5736	11448	12960	25456	30576	61664	71112	142248
2664	5352	5992	11832	13536	27376	31704	63776	73712	146856
2728	5544	6200	12576	14112	28336	32856	66592	75376	149776
2792	5544	6456	12960	14688	29296	34008	68808		
2856	5736	6712	13536	15264	30576	35160	71112		
2984	5992	6968	14112	15840	31704	36696	73712		

Table 392 – TBS table applicable when single transport block is mapped onto 2 layers and number of allocated Resource Blocks is greater than 55

Original TBS	Resultant TBS	Original TBS	Resultant TBS	Original TBS	Resultant TBS	Original TBS	Resultant TBS	Resultant TBS	Resultant TBS
1032	3112	2280	6712	5352	15840	13536	40576	35160	105528
1064	3240	2344	6968	5544	16416	14112	42368	36696	110136
1096	3240	2408	7224	5736	16992	14688	43816	37888	115040
1128	3368	2472	7480	5992	18336	15264	45352	39232	119816
1160	3496	2536	7480	6200	18336	15840	46888	40576	119816
1192	3624	2600	7736	6456	19080	16416	48936	42368	128496
1224	3624	2664	7992	6712	19848	16992	51024	43816	133208
1256	3752	2728	8248	6968	20616	17568	52752	45352	137792
1288	3880	2792	8248	7224	21384	18336	55056	46888	142248
1320	4008	2856	8504	7480	22152	19080	57336	48936	146856
1352	4008	2984	8760	7736	22920	19848	59256	51024	152976
1384	4136	3112	9144	7992	23688	20616	61664	52752	157432
1416	4264	3240	9528	8248	24496	21384	63776	55056	165216
1480	4392	3368	9912	8504	25456	22152	66592	57336	171888
1544	4584	3496	10296	8760	26416	22920	68808	59256	177816
1608	4776	3624	10680	9144	27376	23688	71112	61664	185728
1672	4968	3752	11064	9528	28336	24496	73712	63776	191720
1736	5160	3880	11448	9912	29296	25456	76208	66592	199824
1800	5352	4008	11832	10296	30576	26416	78704	68808	205880
1864	5544	4136	12576	10680	31704	27376	81176	71112	214176
1928	5736	4264	12960	11064	32856	28336	84760	73712	221680
1992	5992	4392	12960	11448	34008	29296	87936	75376	226416
2024	5992	4584	13536	11832	35160	30576	90816		
2088	6200	4776	14112	12216	36696	31704	93800		
2152	6456	4968	14688	12576	37888	32856	97896		
2216	6712	5160	15264	12960	39232	34008	101840		

Table 393 – TBS table applicable when single transport block is mapped onto 3 layers and number of allocated Resource Blocks is greater than 36

Original TBS	Resultant TBS	Original TBS	Resultant TBS	Original TBS	Resultant TBS	Original TBS	Resultant TBS	Resultant TBS	Resultant TBS
776	3112	1864	7480	4264	16992	11448	45352	30576	124464
808	3240	1928	7736	4392	17568	11832	46888	31704	128496
840	3368	1992	7992	4584	18336	12216	48936	32856	133208
872	3496	2024	7992	4776	19080	12576	51024	34008	137792
904	3624	2088	8248	4968	19848	12960	51024	35160	142248
936	3752	2152	8504	5160	20616	13536	55056	36696	146856
968	3880	2216	8760	5352	21384	14112	57336	37888	151376
1000	4008	2280	9144	5544	22152	14688	59256	39232	157432
1032	4136	2344	9528	5736	22920	15264	61664	40576	161760
1064	4264	2408	9528	5992	23688	15840	63776	42368	169544
1096	4392	2472	9912	6200	24496	16416	66592	43816	175600
1128	4584	2536	10296	6456	25456	16992	68808	45352	181656
1160	4584	2600	10296	6712	26416	17568	71112	46888	187712
1192	4776	2664	10680	6968	28336	18336	73712	48936	195816
1224	4968	2728	11064	7224	29296	19080	76208	51024	203704
1256	4968	2792	11064	7480	29296	19848	78704	52752	211936
1288	5160	2856	11448	7736	30576	20616	81176	55056	220296
1320	5352	2984	11832	7992	31704	21384	84760	57336	230104
1352	5352	3112	12576	8248	32856	22152	87936	59256	236160
1384	5544	3240	12960	8504	34008	22920	90816	61664	245648
1416	5736	3368	13536	8760	35160	23688	93800	63776	254328
1480	5992	3496	14112	9144	36696	24496	97896	66592	266440
1544	6200	3624	14688	9528	37888	25456	101840	68808	275376
1608	6456	3752	15264	9912	39232	26416	105528	71112	284608
1672	6712	3880	15264	10296	40576	27376	110136	73712	293736
1736	6968	4008	15840	10680	42368	28336	115040	75376	299856
1800	7224	4136	16416	11064	43816	29296	115040		

Table 394 – TBS table applicable when single transport block is mapped onto 4 layers and number of allocated Resource Blocks is greater than 27

33.4 UL-SCH MODULATION AND TRANSPORT BLOCK SIZES

★ Information regarding the UL-SCH modulation scheme and Transport Block Size (TBS) is signalled within the PDCCH Downlink Control Information (DCI) formats 0 and 4, i.e. the DCI specifies a Modulation and Coding Scheme (MCS) index

★ Table 395 presents the modulation scheme, TBS index and Redundancy Version associated with each MCS index

MCS Index	Modulation	TBS Index	Redundancy Version
0	QPSK	0	0
1		1	0
2		2	0
3		3	0
4		4	0
5		5	0
6		6	0
7		7	0
8		8	0
9		9	0
10		10	0
11	16QAM		0
12		11	0
13		12	0
14		13	0
15		14	0
16		15	0

MCS Index	Modulation	TBS Index	Redundancy Version
17	16QAM	16	0
18		17	0
19		18	0
20		19	0
21	64QAM		0
22		20	0
23		21	0
24		22	0
25		23	0
26		24	0
27		25	0
28		26	0
29	Reserved		1
30			2
31			3

Table 395 – Modulation and Coding Scheme (MCS) table for the UL-SCH

★ If TTI bundling has been enabled then the modulation is always QPSK and a maximum of 3 Resource Blocks can be allocated

★ If the UE does not support 64QAM, or if the UE has been configured to not use 64QAM, then 16QAM is used rather than 64QAM for MCS indices 21 to 28

★ MCS indices 29, 30 and 31 are used specifically for HARQ retransmissions. These MCS indices can be used to signal that the redundancy version is to be changed for the retransmission while the modulation scheme and Transport Block Size are kept the same as the original transmission

★ When a single transport block is mapped onto a single layer, each TBS index points towards a row within the TBS table presented as Table 388 to Table 391 (the same tables as used for the downlink). Each column within the TBS table corresponds to the number of allocated Resource Blocks

★ The maximum TBS for the 20 MHz channel bandwidth is 75 376 bits (assuming all 100 Resource Blocks are allocated to a single connection and ignoring the PUCCH allocation). This TBS corresponds to a physical layer throughput of 75.376 Mbps

★ When a single transport block is mapped onto a 2 layers (applicable when using 4×4 MIMO in the uplink), the TBS selection is done in the same way as for the downlink:

  - o if the number of allocated Resource Blocks is between 1 and 55, then the 'Number of Allocated Resource Blocks' is doubled prior to looking-up the TBS within the TBS table
  - o if the number of allocated Resource Blocks is greater than 55, then the TBS is obtained from the TBS table in the normal manner but is then mapped onto a new TBS using the mapping presented in Table 392

★ If the MCS index = 29, the CSI request bit within DCI format 0 = 1, and the number of allocated Resource Blocks is less than 5 then the modulation is set to QPSK and there is no transport block for the UL-SCH, i.e. only the control information is transmitted by the UE. 3GPP TS 36.213 also specifies similar conditions for DCI format 4

★ 3GPP References: TS 36.212, TS 36.213

33.5 COMPARISON BETWEEN UMTS AND LTE

★ Table 396 provides a high level comparison between UMTS and LTE

	UMTS	LTE
Core Network Domains	Circuit Switched and Packet Switched	Packet Switched
Flat Architecture	No (includes RNC)	Yes
Channel Bandwidth	5 MHz 10 MHz with 2 carrier HSDPA capability (3GPP release 8) 10 MHz with 2 carrier HSUPA capability (3GPP release 9) 20 MHz with 4 carrier HSDPA capability (3GPP release 10) 40 MHz with 8 carrier HSDPA capability (3GPP release 11)	1.4, 3, 5, 10, 15, 20 MHz 100 MHz with LTE Advanced Carrier Aggregation (5 Component Carriers) Initial support for 40 MHz (2 Component Carriers) with 3GPP release 10
Multiple Access	WCDMA	OFDMA in the downlink SC-FDMA in the uplink
Frequency Re-Use	Re-use of 1	Re-use of 1
Soft Handover Support	Yes, for DCH and HSUPA No, for HSDPA HSDPA Multiflow is a 3GPP release 11 work item	No Coordinated Multi-Point (CoMP) transmission is a 3GPP release 11 work item
Fast Power Control Support	Yes, for DCH and HSUPA No, for HSDPA	No, slower power control used for the uplink
Uplink Modulation	QPSK 16QAM for HSUPA in 3GPP release 7	QPSK, 16QAM, 64QAM
Downlink Modulation	QPSK 16QAM for HSDPA in 3GPP release 5 64QAM for HSDPA in 3GPP release 7	QPSK, 16QAM, 64QAM
Adaptive Modulation	Yes, for HSDPA and HSUPA	Yes
Uplink MIMO	No HSUPA MIMO in 3GPP release 11	No LTE Advanced introduces uplink 4×4 MIMO with 3GPP release 10
Downlink MIMO	2×2 for HSDPA in 3GPP release 7 4×4 for HSDPA in 3GPP release 11	2×2 and 4×4 LTE Advanced introduces downlink 8×8 MIMO with 3GPP release 10
Peak Uplink Throughput (without LTE Advanced)	23 Mbps (10 MHz channel, 16QAM, Coding Rate 1)	85 Mbps (20 MHz channel, 64QAM, Coding Rate 1, normal cyclic prefix, 2 PUCCH Resource Blocks per slot)
Peak Downlink Throughput (without LTE Advanced)	86 Mbps (10 MHz channel, 64QAM, 2×2 MIMO, Coding Rate 1, 15 HS-PDSCH codes per carrier)	325 Mbps (20 MHz channel, 64QAM, 4×4 MIMO, Coding Rate 1, normal cyclic prefix, 1 PDCCH symbol per subframe)
Peak Uplink Throughput in 10 MHz, 16QAM, Coding Rate 1	23 Mbps	28 Mbps (normal cyclic prefix, 2 PUCCH Resource Blocks per slot)
Peak Downlink Throughput in 10 MHz, 64QAM, 2×2 MIMO, Coding Rate 1	86 Mbps	86 Mbps (normal cyclic prefix, 1 PDCCH symbol per subframe)
Hybrid ARQ Support	No, for DCH Yes, for HSDPA and HSUPA	Yes
BTS Scheduling	No, for DCH Yes, for HSDPA and HSUPA	Yes
Neighbour Planning	Yes	No, if Automatic Neighbour Relations (ANR) capability is supported
Scrambling Code Planning	Yes	No
Physical layer Cell Identity Planning	No	Yes

Table 396 – Comparison between UMTS and LTE

33.6 BEAMFORMING PRINCIPLES

- Beamforming uses multiple antenna elements to generate a beam of radiated power in a specific direction
- Beamforming results from constructive and destructive interference between the transmissions from each antenna element
- The distances between each antenna element and a UE directly in front of the antenna are equal. This means that each transmitted signal arrives at the UE with equal phase, and the signals sum constructively
- The distances between each antenna element and a UE to one side of the antenna are unequal. This means that each transmitted signal arrives at the UE with a different phase, and the signals sum less constructively
- Figure 339 illustrates how the difference in distance to each antenna element can be calculated. It is assumed that the distance between the antenna and UE is large when compared to the transmitted signal wavelength so the transmitted signals can be assumed to have parallel paths (simplifies the trigonometry)
- Figure 339 assumes 4 antenna elements but more or less could be used in practice. It is also assumed that the antenna elements are spaced by half a wavelength (λ / 2)

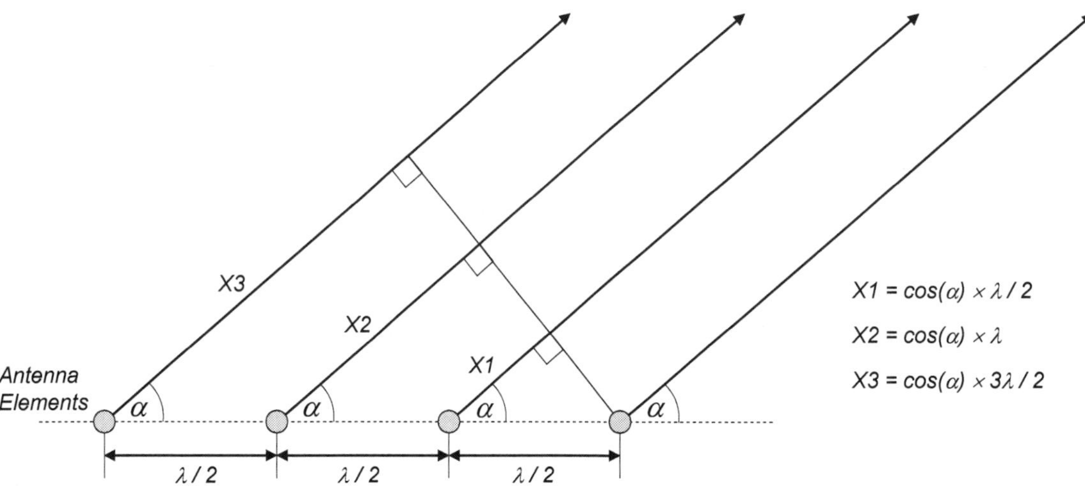

Figure 339 – Calculation of the difference in distance to each antenna element

- The calculations for X1, X2 and X3 can be used to calculate the beam gain as a function of the angle, α
- The beam gain at a specific angle, α is given by:

$$1 + \cos[(X1 / \lambda) \times 2\pi] + \cos[(X2 / \lambda) \times 2\pi] + \cos[(X3 / \lambda) \times 2\pi]$$

This generates the result illustrated in Figure 340

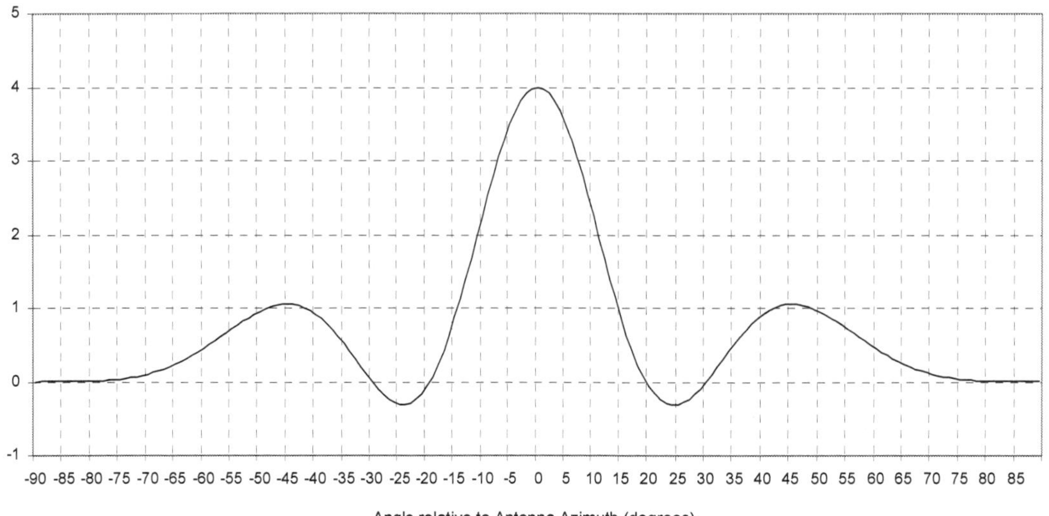

Figure 340 – Antenna beam pattern when using 4 antenna elements with λ / 2 spacing

★ The x-axis of Figure 340 is given by α - 90 degrees. This has been adjusted to make the angle relative to the direction of the antenna

★ Figure 340 clearly indicates that the beam gain is maximised in the direction of the antenna. This corresponds to the scenario in which all 4 transmitted signals arrive at the UE in-phase and combine constructively. Side lobes are generated at angles where the received signals combine constructively to some extent

★ Figure 341 illustrates the equivalent result when using 8 antenna elements spaced by λ / 2. This demonstrates the increase in directivity which can be achieved by increasing the number of antenna elements. Beamforming can be achieved to some extent when using only 2 antenna elements but the directivity is decreased

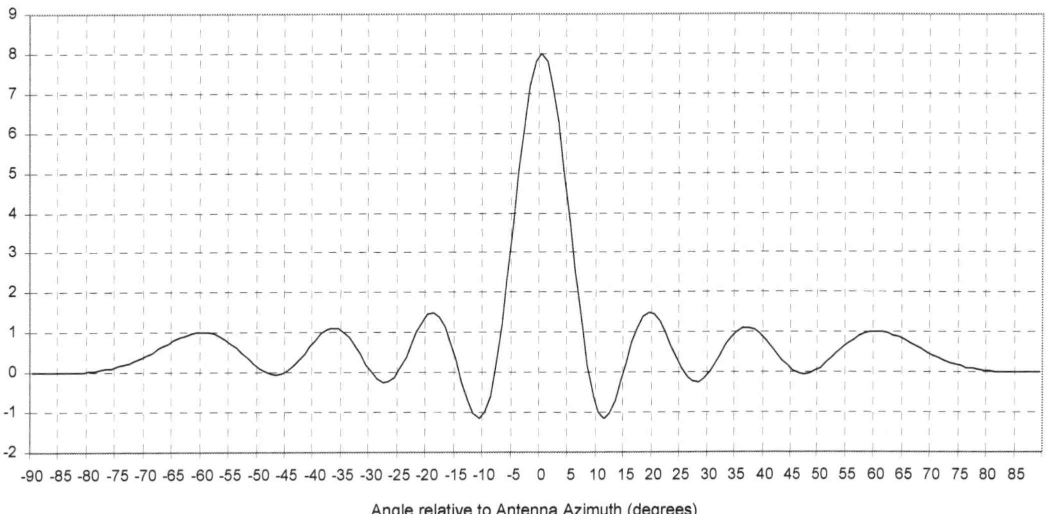

Figure 341 – Antenna beam pattern when using 8 antenna elements with λ / 2 spacing

★ Figure 342 illustrates the equivalent result when using 4 antenna elements spaced by λ / 3. This demonstrates a broadening of both the main beam and the side lobes when the antenna elements are not spaced by λ / 2. The gain of the main beam does not change

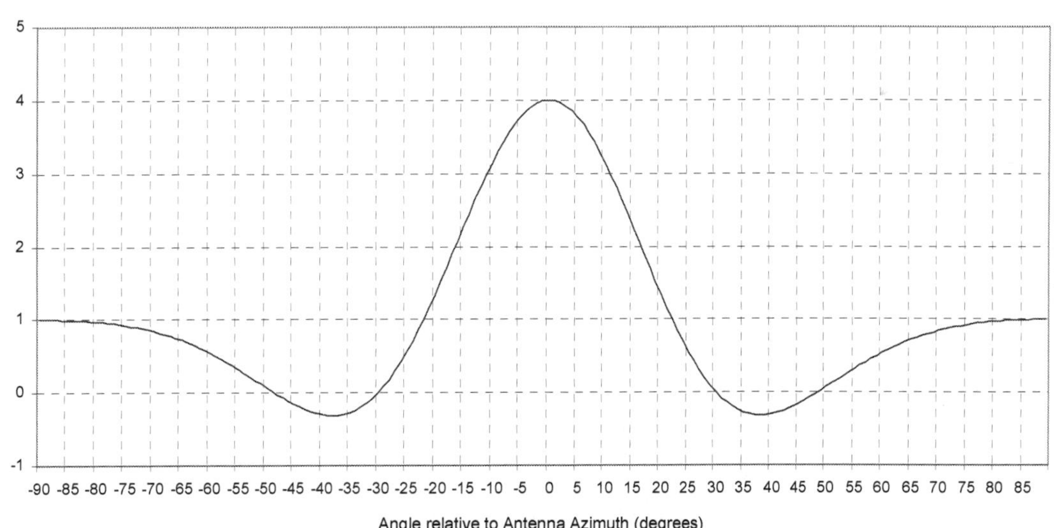

Figure 342 – Antenna beam pattern when using 4 antenna elements with λ / 3 spacing

★ Beamforming typically involves steering the antenna beam pattern towards a specific UE. This can be achieved by applying delays (phase shifts) to each of the transmitted signals. This means that each physical antenna element transmits the same signal but with a different delay. The concept of applying these delays is illustrated in Figure 343

★ The delays associated with each antenna element typically increase in a linear manner. For example, antenna element 1 has zero delay, antenna element 2 has a delay of 'x', antenna element 3 has a delay of '2x' and antenna element 4 has a delay of '3x'

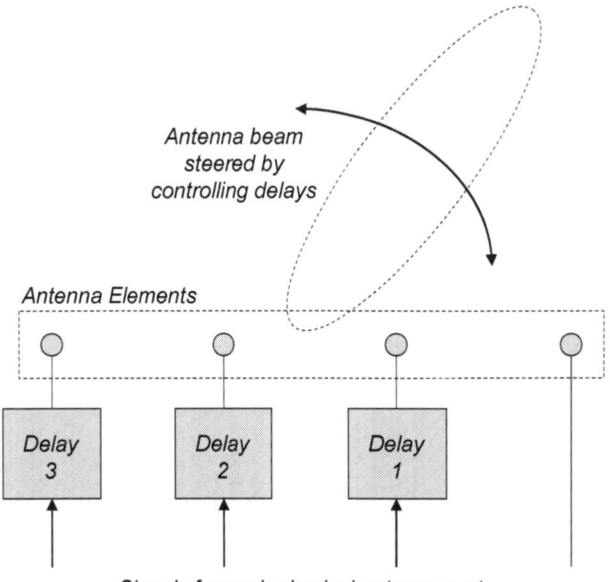

Figure 343 – Steering the beam pattern by applying a set of delays

★ Figure 344 illustrates an example where the antenna beam has been steered to an angle of +15 degrees. This has been achieved by applying phase shifts of 45, 90 and 135 degrees. Assuming a downlink carrier frequency of 2630 MHz, these phase shifts correspond to delays of 0.048, 0,095 and 0.143 nano seconds

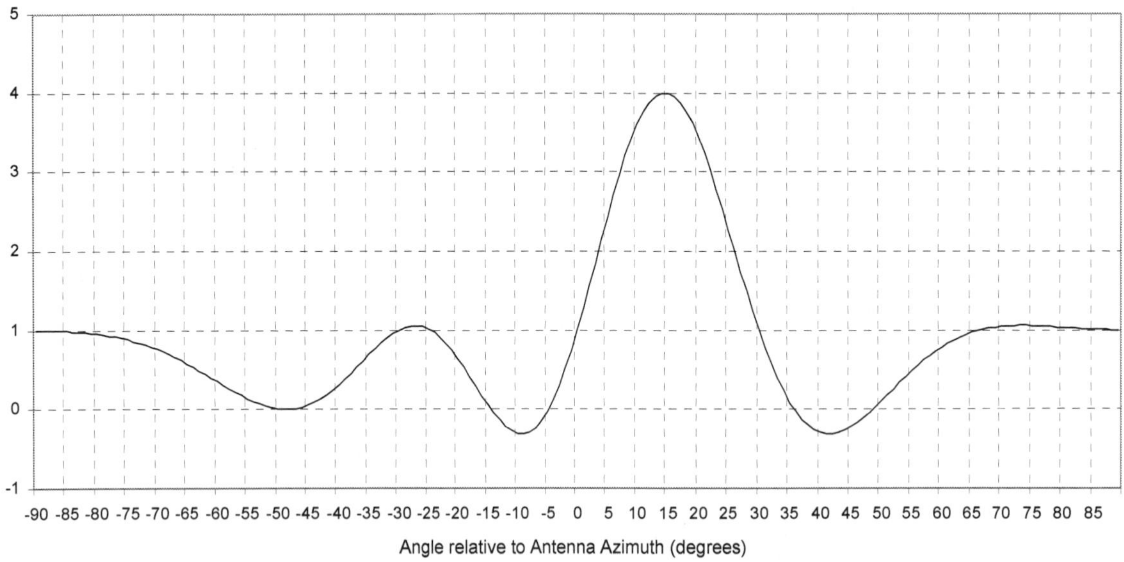

Figure 344 – Antenna beam steered to +15 degrees by applying delays to antenna ports

★ The antenna beam can be steered to -15 degrees by applying phase shifts of -45, -90 and -135 degrees. Similarly the beam can be steered to other angles by applying different phase shifts

34 INDEX

Printed in Great Britain
by Amazon.co.uk, Ltd.,
Marston Gate.